Sustainable Materials for Sensing and Remediation of Noxious Pollutants

Sustainable Materials for Sensing and Remediation of Noxious Pollutants

Edited by

Inderjeet Tyagi
Centre for DNA Taxonomy, Molecular Systematics Division, Zoological Survey of India, Kolkata, West Bengal, India

Joanna Goscianska
Faculty of Chemistry, Adam Mickiewicz University in Poznań, Poznań, Poland

Mohammad Hadi Dehghani
Department of Environmental Health Engineering, School of Public Health, Tehran University of Medical Sciences, Tehran, Iran

Rama Rao Karri
Petroleum and Chemical Engineering, Faculty of Engineering, Universiti Teknologi Brunei, Bandar Seri Begawan, Brunei Darussalam

ELSEVIER

Elsevier
Radarweg 29, PO Box 211, 1000 AE Amsterdam, Netherlands
The Boulevard, Langford Lane, Kidlington, Oxford OX5 1GB, United Kingdom
50 Hampshire Street, 5th Floor, Cambridge, MA 02139, United States

ISBN: 978-0-323-99425-5

For information on all Elsevier publications
visit our website at https://www.elsevier.com/books-and-journals

Publisher: Candice Janco
Acquisitions Editor: Jessica Mack
Editorial Project Manager: Franchezca A. Cabural
Production Project Manager: Paul Prasad Chandramohan
Cover Designer: Matthew Limbert

Typeset by STRAIVE, India

Working together
to grow libraries in
developing countries

www.elsevier.com • www.bookaid.org

Dedication

Education is the ***Most Powerful Weapon*** *which you can use to change the World.*

Nelson Mandela

The accomplishment of the present work is a blessing of many. It has received precious guidance, incessant encouragement, unconditional support, and painstaking efforts of many people; I owe my thanks to all of them. I dedicate this book to my mentor and guru **Late Prof. Vinod Kumar Gupta** and **Smt. Prerna Gupta** for their continuous motivation and support.

I also dedicate this book to my grandparents **Late Shri. Nand Kishore Tyagi** and **Late Smt. Padma Tyagi** and my parents **Shri. Umesh Tyagi** and **Smt. Mamta Tyagi** who have guided and supported me throughout all phases of life.

I am thankful to my lovely wife **Dr. Pratibha Singh**, my siblings, and my in-laws without whose support and understanding this book and my research achievements would not have been possible.

Moreover, I dedicate this book to the frontline warriors and team of scientists for their selfless services day and night in India as well as across the globe to fight against COVID-19 and for developing COVID-19 vaccines.

Dr. Inderjeet Tyagi

I dedicate this book to my beloved parents.

Mom, Dad—thank you for each of your gazes, a mirror of honesty, for your hands stretched out to give me a little sun every day, and for hugs that are a sincere hospitality of the heart.

Thank you for your warm words and for the hopes placed on me.
Thank you for being with me all the time.

Prof. Dr. Joanna Goscianska

In the name of God, the most gracious, the most merciful!

I am thankful to God Almighty that I succeeded in writing this book with the help of my colleagues.

I dedicate this book to my parents, my brothers, and my sister who are always praying for me. I especially appreciate my lovely wife and children (Amir Parsa and Yasamin) who have contributed to my progress and success with their patience and forbearance.

Prof. Dr. Mohammad Hadi Dehghani

I dedicate this book to the memory of my beloved father Karri Sri Ramulu.

I also dedicate this to my mother Karri Kannathalli, who has protected, guided, and supported me all these years. She is a superwoman and my inspiration, driving me to achieve the best.

I also thank my lovely wife Soni, my lovely children Yajna and Jay, and my in-laws L.V. Rao and Prameela. Without their support and understanding, this book and my research achievements would not have been possible.

Dr. Rama Rao Karri

In loving memory of
Prof. Vinod Kumar Gupta

(1953–2021)

Contents

12. Luminescent metal-organic frameworks for sensing of toxic organic pollutants in water and real samples

Luis D. Rosales-Vázquez,
Alejandro Dorazco-González, and
Víctor Sánchez-Mendieta

13. Metal-organic frameworks for remediation of noxious pollutants

Jafar Abdi, Seyyed Hamid Esmaeili-Faraj,
Golshan Mazloom, and
Tahereh Pirhoushyaran

14. Adsorptive removal and concentration of rare-earth elements from aquatic media using various materials: A review

Alexandr Burakov, Inderjeet Tyagi,
Rama Rao Karri, Irina Burakova,
Anastasia Memetova, Vladimir Bogoslovskiy,
Gulnara Shigabaeva, and Evgeny Galunin

15. Ion-selective membranes as potentiometric sensors for noxious ions

Bhavana Sethi and Saurabh Ahalawat

Contributors

Numbers in parentheses indicate the pages on which the authors' contributions begin.

Rafael Abargues (273), UMDO, Institute of Materials Science (ICMUV), University of Valencia, Valencia, Spain

Jafar Abdi (209), Faculty of Chemical and Materials Engineering, Shahrood University of Technology, Shahrood, Iran

Saurabh Ahalawat (247), Central Research and Development, Ultratech Cement LTD, Khor Neemuch, MP, India

Faheem Ahamad (261), Department of Environmental Science, Keral Verma Subharti College of Science (KVSCOS), Swami Vivekanand Subharti University, Meerut, UP, India

Md. Ahmaruzzaman (129), Department of Chemistry, National Institute of Technology, Silchar, Assam, India

Charu Arora (329), Department of Chemistry, Guru Ghasidas University, Bilaspur, Chhattisgarh, India

Anu Bharti (87), Department of Environmental Sciences, Central University of Jammu, Rahya Suchani (Bagla) Samba, Jammu, Jammu and Kashmir, India

Brij Bhushan (65), Department of Chemistry, Graphic Era University, Dehradun, India

Rakesh Bhutiani (261), Limnology and Ecological Modelling Lab, Department of Zoology and Environmental Science, Gurukula Kangri Vishwavidyalaya, Haridwar, UK, India

Manish Biyani (161), Department of Bioscience and Biotechnology, Japan Advanced Institute of Science and Technology, Nomi, Ishikawa, Japan

Vladimir Bogoslovskiy (229), Research School of Chemistry & Applied Biomedical Sciences, Tomsk Polytechnic University, Tomsk, Russian Federation

Alexandr Burakov (229), Department of Technology and Methods of Nanoproducts Manufacturing, Tambov State Technical University, Tambov, Russian Federation

Irina Burakova (229), Department of Technology and Methods of Nanoproducts Manufacturing, Tambov State Technical University, Tambov, Russian Federation

Monika Chaudhary (113, 149), Department of Chemistry, Gurukula Kangri (Deemed to be University), Haridwar, India

Shubham Chaudhary (113, 149), Department of Chemistry, Gurukula Kangri (Deemed to be University), Haridwar, India

Parul Chauhan (261), Department of Agadtantra, Uttarakhand Ayurvedic University, Haridwar, India

Jorge Alberto Vieira Costa (285), Laboratory of Biochemical Engineering, College of Chemistry and Food Engineering, Federal University of Rio Grande, Rio Grande, RS, Brazil

Mohammad Hadi Dehghani (1), Department of Environmental Health Engineering, School of Public Health; Institute for Environmental Research, Center for Solid Waste Research, Tehran University of Medical Sciences, Tehran, Iran

Alejandro Dorazco-González (195), Institute of Chemistry, National Autonomous University of Mexico, Mexico City, Mexico

Partha Dutta (329), Department of Chemistry, Guru Ghasidas University, Bilaspur, Chhattisgarh, India

Aleksander Ejsmont (15), Faculty of Chemistry, Department of Chemical Technology, Adam Mickiewicz University in Poznań, Poznań, Poland

Seyyed Hamid Esmaeili-Faraj (209), Faculty of Chemical and Materials Engineering, Shahrood University of Technology, Shahrood, Iran

Ali Fakhri (315), Department of Chemistry, Nano Smart Science Institute, Tehran, Iran

Aleksandra Galarda (15), Faculty of Chemistry, Department of Chemical Technology, Adam Mickiewicz University in Poznań, Poznań, Poland

Evgeny Galunin (229), Department of Organic and Ecological Chemistry, University of Tyumen, Tyumen, Russian Federation

Kajol Goria (87), Department of Environmental Sciences, Central University of Jammu, Rahya Suchani (Bagla) Samba, Jammu, Jammu and Kashmir, India

Joanna Goscianska (1, 15), Faculty of Chemistry, Department of Chemical Technology, Adam Mickiewicz University in Poznań, Poznań, Poland

Agata Jankowska (15), Faculty of Chemistry, Department of Chemical Technology, Adam Mickiewicz University in Poznań, Poznań, Poland

Rama Rao Karri (1, 229, 315), Petroleum and Chemical Engineering, Faculty of Engineering, Universiti Teknologi Brunei, Bandar Seri Begawan, Brunei Darussalam

Richa Kothari (87), Department of Environmental Sciences, Central University of Jammu, Rahya Suchani (Bagla) Samba, Jammu, Jammu and Kashmir, India

Shreya Kotnala (65), Department of Chemistry, Graphic Era University, Dehradun, India

Gagandeep Kour (87), Department of Environmental Sciences, Central University of Jammu, Rahya Suchani (Bagla) Samba, Jammu, Jammu and Kashmir, India

Nupur Kukretee (65), Department of Chemistry, Graphic Era University, Dehradun, India

Ankur Kumar (149), Central Instrumentation Laboratory, NIFTEM, Sonipat, India

Ravinder Kumar (113), Department of Chemistry, Gurukula Kangri (Deemed to be University), Haridwar, India

Vikas Kumar (1), Centre for DNA Taxonomy, Molecular Systematics Division, Zoological Survey of India, Ministry of Environment, Forest and Climate Change, Government of India, Kolkata, West Bengal, India

Vinod Kumar (113), Special Centre for Nano Sciences, Jawaharlal Nehru University, Delhi, India

Yogendra Kumar (47), Department of Chemical Engineering, Indian Institute of Technology, Madras, Tamil Nadu, India

Tarun Kumar Kumawat (161), Department of Biotechnology, Biyani Girls College, University of Rajasthan, Jaipur, Rajasthan, India

Varsha Kumawat (161), Biyani Institute of Pharmaceutical Sciences, Rajasthan University for Health Sciences, Jaipur, Rajasthan, India

Suelen Goettems Kuntzler (285), Laboratory of Microbiology and Biochemistry, College of Chemistry and Food Engineering, Federal University of Rio Grande, Rio Grande, RS, Brazil

Sarita Kushwaha (113, 149), Department of Chemistry, Gurukula Kangri (Deemed to be University), Haridwar, India

Juan P. Martínez-Pastor (273), UMDO, Institute of Materials Science (ICMUV), University of Valencia, Valencia, Spain

Golshan Mazloom (209), Department of Chemical Engineering, Faculty of Engineering, University of Mazandaran, Babolsar, Iran

Anastasia Memetova (229), Department of Technology and Methods of Nanoproducts Manufacturing, Tambov State Technical University, Tambov, Russian Federation

Jyoti Mittal (329), Department of Chemistry, Maulana Azad National Institute of Technology, Bhopal, Madhya Pradesh, India

Michele Greque de Morais (285), Laboratory of Microbiology and Biochemistry, College of Chemistry and Food Engineering, Federal University of Rio Grande, Rio Grande, RS, Brazil

Juliana Botelho Moreira (285), Laboratory of Microbiology and Biochemistry, College of Chemistry and Food Engineering, Federal University of Rio Grande, Rio Grande, RS, Brazil

Arunima Nayak (65), Department of Chemistry, Graphic Era University, Dehradun, India

Hadi Omidinasab (39), Department of Mechanical Engineering, South Tehran Branch, Islamic Azad University, Tehran, Iran

Hitesh Panchal (39), Mechanical Engineering Department, Government Engineering College Patan, Patan, Gujarat, India

Anjali Pandit (161), Department of Biotechnology, Biyani Girls College, University of Rajasthan, Jaipur, Rajasthan, India

Diksha Praveen Pathak (47), Department of Chemical and Biochemical Engineering, Rajiv Gandhi Institute of Petroleum Technology, Jais, Uttar Pradesh, India

Tahereh Pirhoushyaran (209), Department of Chemical Engineering, Dezful Branch, Islamic Azad University, Dezful, Iran

Nidhi Rai (329), Department of Chemistry, Guru Ghasidas University, Bilaspur, Chhattisgarh, India

Shubham Raina (87), Department of Environmental Sciences, Central University of Jammu, Rahya Suchani (Bagla) Samba, Jammu, Jammu and Kashmir, India

Saravanan Rajendran (177), Department of Mechanical Engineering, Faculty of Engineering, University of Tarapaca, Arica, Chile

Varun Rawat (297), School of Chemistry, Faculty of Exact Science, Tel Aviv University, Tel Aviv, Israel

Sandra Ricart (39), Water and Territory Research Group, Interuniversity Institute of Geography, University of Alicante, Alicante, Spain

Pedro J. Rodríguez-Cantó (273), UMDO, Institute of Materials Science (ICMUV), University of Valencia, Valencia, Spain

Luis D. Rosales-Vázquez (195), Institute of Chemistry, National Autonomous University of Mexico, Mexico City, Mexico

Prerona Roy (129), Department of Chemistry, National Institute of Technology, Silchar, Assam, India

Ala Sadooghi (39), Optimization of Energy Systems' Installations Lab., Faculty of Mechanical Engineering-Energy Division, K.N. Toosi University of Technology, Tehran, Iran

Víctor Sánchez-Mendieta (195), CCIQS—Joint Center for Research in Sustainable Chemistry UAEM-UNAM, Toluca, Estado de México, Mexico

Bhavana Sethi (247), Academy of Business and Engineering Sciences, Ghaziabad, UP, India

Daksha Sharma (297), Department of Chemistry, Vidhya Bhawan Rural Institute, Udaipur, Rajasthan, India

Ved Bhushan Sharma (261), Department of Agadtantra, Uttarakhand Ayurvedic University, Haridwar, India

Vishnu Sharma (161), Department of Biotechnology, Biyani Girls College, University of Rajasthan, Jaipur, Rajasthan, India

Gulnara Shigabaeva (229), Department of Organic and Ecological Chemistry, University of Tyumen, Tyumen, Russian Federation

Har Mohan Singh (87), School of Energy Management, Shri Mata Vaishno Devi University, Katra, Jammu and Kashmir, India

Pooja Singh (261), Department of Agadtantra, Uttarakhand Ayurvedic University, Haridwar, India

Pratibha Singh (1), Department of Chemistry, University of Delhi, New Delhi, India

Ali Sohani (39), Optimization of Energy Systems' Installations Lab., Faculty of Mechanical Engineering-Energy Division, K.N. Toosi University of Technology, Tehran, Iran

Sanju Soni (329), Department of Chemistry, Guru Ghasidas University, Bilaspur, Chhattisgarh, India

Ananthakumar Soosaimanickam (273), UMDO, Institute of Materials Science (ICMUV), University of Valencia, Valencia, Spain

Suhas (113, 149), Department of Chemistry, Gurukula Kangri (Deemed to be University), Haridwar, India

R. Suresh (177), Department of Mechanical Engineering, Faculty of Engineering, University of Tarapaca, Arica, Chile

Ana Luiza Machado Terra (285), Laboratory of Microbiology and Biochemistry, College of Chemistry and Food Engineering, Federal University of Rio Grande, Rio Grande, RS, Brazil

R.C. Tiwari (261), Department of Agadtantra, Uttarakhand Ayurvedic University, Haridwar, India

Inderjeet Tyagi (1, 113, 149, 229, 261, 315), Centre for DNA Taxonomy, Molecular Systematics Division, Zoological Survey of India, Ministry of Environment, Forest and Climate Change, Government of India, Kolkata, West Bengal, India

Kaomud Tyagi (1), Centre for DNA Taxonomy, Molecular Systematics Division, Zoological Survey of India, Ministry of Environment, Forest and Climate Change, Government of India, Kolkata, West Bengal, India

V.V. Tyagi (87), School of Energy Management, Shri Mata Vaishno Devi University, Katra, Jammu and Kashmir, India

Dipti Vaya (297), Department of Chemistry, Amity School of Applied Sciences, Amity University, Gurugram, Haryana, India

Monu Verma (297), Water-Energy Nexus Laboratory, Department of Environmental Engineering, University of Seoul, Seoul, Republic of Korea

Shalu Yadav (47), Department of Basic Science and Humanity, Rajiv Gandhi Institute of Petroleum Technology, Jais, Uttar Pradesh, India

Hüseyin Yagli (39), Gaziantep University, Gaziantep, Turkey

About the Editors

Dr. Inderjeet Tyagi is working as a Scientist at the Zoological Survey of India (ZSI), Ministry of Environment Forest and Climate Change, Kolkata, India. His expertise lies in the fields of wastewater treatment, water quality, environmental metagenomics, and environmental management; he has been working in these fields for the past several years. To his credit, Dr. Tyagi has published 100+ SCI papers with ~7400 citations and has an h-index of 47 with a cumulative impact factor >450. Moreover, he is leading five national and international projects related to wastewater and water quality assessment in heavily polluted areas in India. He is also on the review panel of more than 40 international journals belonging to well-known publishers like Elsevier, Nature, Springer Nature, Taylor & Francis, and ACS. Recently, he was awarded the "India Prime Education Quality Award 2021" in recognition of his outstanding contributions to the field of wastewater treatment and environmental management. He is a lifetime member of the Indian Science Congress Association (ISCA).

Prof. Dr. Joanna Goscianska is an Associate Professor in the Faculty of Chemistry of Adam Mickiewicz University (AMU) in Poznań, Poland. She received her master's degree in 2005, her doctoral degree in 2009, and habilitation in 2019 in the field of chemistry. During her PhD studies, she participated in three research internships at Laboratoire Catalyse et Spectrochimie in Caen (France) and later at the University of Alicante (Spain). She has received a number of prestigious awards, including the Maxima Cum Laude (2005), a scholarship of the city of Poznań for young researchers (2010), a scholarship for outstanding young scientists awarded by the Minister of Science and Higher Education (2016–19), and team awards of the Rector of AMU for achievements in scientific (2014–18), didactic (2014), and organizational work (2014). In 2018, she became a laureate of the habilitation fellowship "L'Oréal-UNESCO for Women in Science." Her principal research interests are focused on the synthesis, modification, and characterization of porous materials (e.g., metal oxides, ordered mesoporous silica and carbons, metal–organic frameworks) and their application in adsorption processes, catalysis, and drug delivery systems. She is the coauthor of 73 papers in journals from the Journal Citation Reports database, 1 book, 9 book chapters, 22 post-conference proceedings, and 170 presentations at international and national conferences.

Prof. Dr. Mohammad Hadi Dehghani is a Full Professor in the Department of Environmental Health Engineering, School of Public Health, Tehran University of Medical Sciences (TUMS), Tehran, Islamic Republic of Iran. His scientific research interests are focused on environmental science. He is the author of various research studies published in national and international journals and conference proceedings and the head of several research projects at TUMS. He has authored 12 books and more than 200 full papers published in peer-reviewed journals. He is an editorial board member, a guest editor, and a reviewer for many national and international journals and is a member of several international science committees around the world. He is supervisor and advisor for many PhD and MSc theses at TUMS. He is currently also a member of the Iranian Association of Environmental Health (IAEH) and the Institute for Environmental Research (IER) at TUMS. He is the editor of six edited books published by Elsevier: (1) *Soft Computing Techniques in Solid Waste and Wastewater Management*, (2) *Environmental and Health Management of Novel Coronavirus Disease (COVID-19)*, (3) *Green Technologies for the Defluoridation of Water*, (4) *Pesticides Remediation Technologies from Water and Wastewater*, (5) *COVID-19 and Sustainable Development Goals*, and (6) *Industrial Wastewater Treatment using Emerging Technologies for Sustainability*.

Dr. Rama Rao Karri is a Senior Assistant Professor in the Faculty of Engineering, Universiti Teknologi Brunei, Brunei Darussalam. He has a PhD from the Indian Institute of Technology (IIT), Delhi, and a master's degree in chemical engineering from IIT Kanpur. He has worked as a postdoctoral research fellow at NUS, Singapore, for about 6 years and has over 18 years of working experience in academics, industry, and research. He has experience of working in multidisciplinary fields with expertise in various evolutionary optimization techniques and process modeling. He has published 150+ research articles in reputed journals, book chapters and conference proceedings with a combined Impact factor of 479.2 and has an h-index of 24 (Scopus - citations: 2100+) and 26 (Google Scholar -citations: 2500+). Among 75 journal publications, 60 articles published are Q1 and high IF journals. He is an editorial board member in 10 renowned journals and peer-review member for more than 93 reputed journals and peer reviewed more than 410 articles. Also, he handled 112 articles as an editor. He also has the distinction of being listed in the top 2% of the world's most influential scientists in the area of environmental science and chemistry for the year 2021. The List of the Top 2% Scientists in the World compiled and published by Stanford University is based on their international scientific publications, number of scientific citations for research, and participation in the review and editing of scientific research. He held the position of editor-in-chief (2019–21) for the *International Journal of Chemoinformatics and Chemical Engineering* (IGI Global, United States). He is also associate editor for *Scientific Reports* (Springer Nature) and *International Journal of Energy and Water Resources* (IJEWR; Springer Inc.). He is also a managing guest editor for the following special issues: (1) "Magnetic nano composites and emerging applications" in the *Journal of Environmental Chemical Engineering* (IF: 5.909); (2) "Novel CoronaVirus (COVID-19) in Environmental Engineering Perspective" in the *Journal of Environmental Science and Pollution Research* (IF: 4.223, Springer); and (3) "Nanocomposites for the Sustainable Environment" in the *Applied Sciences Journal* (IF: 2.679, MDPI). Along with his mentor, Prof. Venkateswarlu, he is authoring the book *Optimal State Estimation for Process Monitoring, Diagnosis and Control* to be published by Elsevier. He is also co-editor and managing editor for seven edited books published by Elsevier- and one edited book each published by Springer and CRC. Elsevier: (1) *Sustainable Nanotechnology for Environmental Remediation*, (2) *Soft Computing Techniques in Solid Waste and Wastewater Management*, (3) *Green Technologies for the Defluoridation of Water*, (4) *Environmental and Health Management of Novel Coronavirus Disease (COVID-19)*, (5) *Pesticides Remediation Technologies from Water and Wastewater: Health Effects and Environmental Remediation*, (6) *Hybrid Nanomaterials for Sustainable Applications*, (7) *Sustainable Materials for Sensing and Remediation of Noxious Pollutants*. Springer: (1) *Industrial Wastewater Treatment using Emerging Technologies for Sustainability*. CRC: (1) *Recent Trends in Advanced Oxidation Processes (AOPs) for Micro-Pollutant Removal*.

Foreword

Water pollution is a global issue and researchers as well as policymakers are coming up with new approaches and mitigation techniques for the remediation of water pollutants. For remediation of noxious impurities, different techniques such as oxidation, membrane processes, coagulation/flocculation, biological treatment methods, and adsorption are applied. However, these techniques have a few limitations and drawbacks.

To present an overview of the technologies and characteristics of materials for application as adsorbents and sensors in the detection and remediation of noxious impurities, the editors reached out to authors from all over the world and compiled the latest research in the sensing and remediation of noxious pollutants.

The editors have structured the contents of this book such that they have covered the basic approaches/concepts/case studies, issues in the existing techniques, synthesis of different adsorbents, and applications in various fields, including environmental, agricultural, energy, and sensor applications.

This book covers the most widely used aspects in the field of wastewater, i.e., sensing and rapid remediation. The compiled case studies provide a better understanding to researchers in this area, in addition to highlighting the benefits and economical methods that have a major societal impact.

In my opinion, this is an excellent reference book for researchers in both academia and industry working in the field of green applied sciences, i.e., sensor and adsorbent development for sensing and remediation activities.

I wish the editors and the authors all the best.

Mika Sillanpaa
University of Johannesburg, South Africa

Acknowledgments

I am thankful to Dr. Dhriti Banerjee, Director, Zoological Survey of India (ZSI), Ministry of Environment, Forest and Climate Change, Kolkata, and the higher management for their constant support and motivation. I am also thankful to my colleagues at ZSI for their valuable suggestions. I thank my co-editors without whose support and cooperation this book would not have been possible. I also thank all the authors who contributed chapters presenting their valuable research. I thank all the reviewers who provided valuable suggestions from time to time to enhance the novelty of the book chapters.

Dr. Inderjeet Tyagi

I thank employees, friends, and colleagues in the Department of Chemical Technology of Adam Mickiewicz University in Poznań for their support at key moments during my scientific endeavors, for surrounding me with a wonderful aura, and for their kindness. Moreover, I thank all the authors of the book chapters whose contributions will allow young scientists to broaden their knowledge of environmental pollution and its negative impact on health and the methods for the detection and removal of environmental pollutants.

Prof. Dr. Joanna Goscianska

I thank my coeditors without whose support and cooperation this book would not have been possible. I also thank my colleagues in the Department of Environmental Health Engineering for their valuable support. I thank all the authors who contributed chapters presenting their valuable research.

Prof. Dr. Mohammad Hadi Dehghani

I thank Prof. Zohrah, Vice Chancellor, Universiti Teknologi Brunei, Prof. Ramesh, and the higher management for their support. I also thank my colleagues Dr. Mubarak and Dr. Malai Zeiti for their constant motivation. Special thanks to my coeditors without whose support and cooperation this book would not have been possible. Finally, I thank all the authors who contributed chapters of high research value.

Dr. Rama Rao Karri

Introduction

Water scarcity is one of the major problems in many parts of the world, and, as a result, water pollution is receiving increasing attention. Water plays a crucial role in accomplishing sustainable livelihood, and achieving clean water is one of the main sustainable development goals (SDGs). Surging human population and resource consumption are the two major issues the human race is trying to address to lead a sustainable lifestyle. Water is an essential component for the sustenance of life. However, the available water resources on the earth are being depleted owing to pollution. It is very evident that urbanization and industrialization have resulted in the exploitation of natural resources all across the globe. Pollution of natural water resources with organic (carbohydrates, dyes, fertilizers, oil and grease, pesticides, pharmaceuticals, plasticizers, polyaromatic hydrocarbons, proteins, etc.) and inorganic (heavy metals) pollutants has become a challenging issue in many countries.

Urban and industrial wastewaters constitute the most important contamination sources of rivers, lakes, reservoirs, and oceans, and increasing industrialization has put pressure on the demand for water resulting in a huge increase in wastewater production as well. Since there is water shortage globally, wastewater treatment and water reuse have become highly significant. Therefore, removing pollutants to prevent pollution of water sources seems to be very necessary. Recent advances have made it possible to partially solve many of the problems associated with water quality and pollution, as well as the protection of water resources using various water treatment techniques. Even though several approaches have been developed for efficient water purification, optimal water purification methods available at a low cost have been undertaken and they are affordable to the developing nations. The adsorption technique using a solid adsorbent meets the above requirements because it offers low installation cost and easy operation with high efficiency and is affordable and environmentally friendly, thus making it one of the preferred methods for water purification.

Green technology appears to be one of the effective strategies to alleviate toxicity by utilizing natural resources to produce nanomaterials and, simultaneously, offers extensive benefits such as reduced operating costs, good biocompatibility, thermal and chemical stability, and reduced environmental impacts. With the salient features of nanotechnology and nanomaterials, in the last decade, there has been a rising interest in the growth of eco-friendly, solid material-based nanoparticles and nanoadsorbents. Nanoadsorbents are widely used to treat contaminated water to remove organic and inorganic contaminants. Nanoparticles have important properties that have enabled them to be widely considered as a suitable adsorbent. They act as a sorbent selective for the adsorption of metal ions and anions. In addition, nanomaterials can be functionalized with different chemical groups to increase their affinity to certain compounds. Due to their high surface area, size, and optical, electronic, and catalytic properties, nanomaterials make it possible to create better and more cost-effective water treatment approaches.

This book provides an overview of the various methods of detecting and removing organic and inorganic contaminants from aqueous solutions. The compiled chapters describe smart and advanced porous nanomaterials, which are applied in the relatively low-cost and effective processes of adsorption and sensing of dyes, pesticides, pharmaceuticals, and heavy metal ions. An important aspect is a thorough analysis of the impact of these pollutants on the environment and human health and highlighting the benefits of removing them from the environment using the most promising technologies, taking into account the possibility of their transfer from the laboratory scale to the industrial scale. In addition, the book presents the most effective methods for removing toxic contaminants from water solutions and air using sustainable nanoporous adsorbents.

This edited book will be very useful for MSc and PhD students who are working in the environmental chemistry field; students who are working in the environmental science field; students who are working on water and wastewater purification technologies; researchers working on environmental nanotechnology; researchers and academics who are working on environmental remediation technologies; researchers and managers who are working in water and wastewater plants and industries; researchers who are working in the environmental toxicology field; researchers and engineers who are working on water treatment and green chemistry; managers and industries that wish to implement professionally advanced technologies; and policymakers working on environmental pollutants and remediation.

Dr. Inderjeet Tyagi
Centre for DNA Taxonomy, Molecular Systematics Division, Zoological Survey of India,
Ministry of Environment, Forest and Climate Change (MoEFCC),
New Alipore, Kolkata, West Bengal, India

Prof. Dr. Joanna Goscianska
Department of Chemical Technology, Faculty of Chemistry, Adam Mickiewicz University in Poznań, Poland

Prof. Dr. Mohammad Hadi Dehghani
Department of Environmental Health Engineering, School of Public Health, Tehran University of
Medical Sciences, Tehran, Iran; Institute for Environmental Research, Center for Solid
Waste Research, Tehran University of Medical Sciences, Tehran, Iran

Dr. Rama Rao Karri
Petroleum and Chemical Engineering, Faculty of Engineering, Universiti Teknologi Brunei (UTB),
Gadong, Brunei Darussalam

Chapter 1

Sustainable materials for sensing and remediation of toxic pollutants: An overview

Inderjeet Tyagi[a,*], Pratibha Singh[b], Rama Rao Karri[c], Mohammad Hadi Dehghani[d,e], Joanna Goscianska[f], Kaomud Tyagi[a], and Vikas Kumar[a]

[a]Centre for DNA Taxonomy, Molecular Systematics Division, Zoological Survey of India, Ministry of Environment, Forest and Climate Change, Government of India, Kolkata, West Bengal, India, [b]Department of Chemistry, University of Delhi, New Delhi, India, [c]Petroleum and Chemical Engineering, Faculty of Engineering, Universiti Teknologi Brunei, Bandar Seri Begawan, Brunei Darussalam, [d]Department of Environmental Health Engineering, School of Public Health, Tehran University of Medical Sciences, Tehran, Iran, [e]Institute for Environmental Research, Center for Solid Waste Research, Tehran University of Medical Sciences, Tehran, Iran, [f]Faculty of Chemistry, Department of Chemical Technology, Adam Mickiewicz University in Poznań, Poznań, Poland

[*]Corresponding author.

Abbreviations

AB1	Acid black 1
AB117	Acid blue 117
AB25	Acid blue 25
AO19	Acid orange 17
AO3	Acid orange 3
AY36	Acid yellow 36
BBG	Basic brown G
BO2	Basic orange 2
BV14	Basic violet 14
CR	Congo red
CV	Crystal violet
DO13	Disperse orange 13
DB3	Disperse blue 3
DB86	Disperse blue 86
DB9	Disperse blue 9
DNB106	Direct navy blue 106
DO26	Direct orange 26
DR11	Direct red 11
DR81	Direct red 81
DV28	Direct violet 28
DY211	Direct yellow 211
DY3	Direct yellow 3
DY50	Direct yellow 50
EBT	Eriochrome Black T
MB	Methylene blue
MG	Malachite green
MO	Methyl orange
MV	Methyl violet
RB 19	Reactive blue 19
RB2	Reactive blue 2
RB74	Reactive blue 74

Sustainable Materials for Sensing and Remediation of Noxious Pollutants. https://doi.org/10.1016/B978-0-323-99425-5.00022-0

RhB	Rhodamine B
RR15, RB2, RB19,RB74	Reactive red 15, Reactive blue 2, 19,74
VBE 1, 4, 5,6,14	Vat blue 1, 4, 5, 6, 14
VBR 1, 68, 72,	Vat brown R 1, 68, 72.
VR13	Vat red 13
VV1	Vat violet 1
VY1	Vat yellow 1

1 Introduction

Water being the most essential element and basic requirement of the living creatures on planet earth, water bodies cover >70% of the entire earth's surface area, out of which only 0.002% is considered as appropriate for human consumption.[1] Due to rapid urbanization and industrialization, it is a tough challenge for us to prevent this important resource from getting polluted from inorganic and organic pollutants.[2]

Inorganic pollutants majorly include trace elements (heavy metals) having a density greater than $4 \pm 1 \, g/cm^{3,}$ such as Nickel (Ni), Lead (Pb), Arsenic (As), Chromium (Cr), Copper (Cu), Cadmium (Cd), Zinc (Zn), Mercury (Hg), Iron (Fe), etc.[3–11] They are introduced into the water or wastewater through natural and anthropogenic activities from sectors such as mining, agricultural activities, industrial outlets (textile, tanneries, nuclear, etc.), domestic sewage, and others. These metal impurities have a detrimental impact on human health and prevailing flora and fauna.[12] Heavy metals such as Copper (Cu) in an excess amount above the permissible limit may lead to liver damage, kidney disorders, muscle impairment, insomnia, and inhibits enzymatic activities.[13] Besides, Chromium (Cr) intake (even in trace amounts) may lead to nausea, headache, and diarrhea.[14] Further, the intoxication of Mercury (Hg) leads to disorders such as circulatory, nervous and rheumatoid arthritis, etc.[13] On the other hand, the characteristics of heavy metals such as high solubility, non-biodegradable nature, and high stability tends them to migrate throughout the aqueous system and gets accumulated in the food chain which in turn hampers the growth of food chain elements. These heavy metals not only deteriorate the faunal and human health but also possesses negative impact on the plants through inhibiting photosynthesis, stunting growth, altering chlorophyll synthesis, altered enzymatic activities, etc.[15]

Organic impurities majorly constitute dyes not just limited to toxic cationic (methylene blue, safranin-O, malachite green, crystal violet, etc.) and anionic dyes (Eriochrome Black T, methyl orange, Congo red, Alizarin red S, etc.).[16–23] Polyaromatic hydrocarbons (PAHs),[24] Chlorophenols (CPs)[25] etc. Thus, it is essential to detect and remediate these impurities from water and wastewater to make it fit for day-to-day activities. These dye molecules and organic impurities pose a serious threat to the ecosystem and lead to several disorders such as cancer, mutation, and other irregularities in vital organs such as reproductive, nephrological, hepatic, and neurological disorders etc.[26–28] In addition to this, these impurities pose a serious threat to the floral ecosystem, it hampers photosynthesis by limiting the transmission of sunlight which in turn affect the food chain of the aquatic community.[29–31] The different sources responsible for the release of dyeing impurities are textile, paint, and pigment industries, etc.[32–36] Approximately more than 0.1 million different types of dyes are commercially available, and these contribute directly or indirectly to the detrimental on the ecosystem due to complex aromatic structure.[37–40] The classification of dyes can be better understood from Table 1.

Keeping in view these noxious pollutants, to date, several efforts have been made to detect and remediate these noxious impurities from water and wastewater. Different materials such as coumarin,[42] rhodamine,[43] Schiff bases,[44] and phthalocyanine tetrafonic acid[45] were used as a sensor for the detection of metal as well as organic impurities. Moreover, the sensor used in environmental applications can be classified into two on-site for real applications and others that require a laboratory environment. Although the second one has great technological advantages, but its non-portable nature and bulkiness are major drawbacks. As a result, it is limited to only labs and not considered for on-site sensing and environmental monitoring. Attempts have been made to subside this major drawback to make them eligible candidates for the on-site environmental monitoring through tuning them under real environmental conditions. As a result, electrochemical and potentiometric sensors have emerged as potential candidates for sensors.[46–50]

For remediation of noxious impurities, different techniques such as oxidation, membrane process, coagulation/flocculation, biological treatment methods, and adsorbents were applied. The advantages and disadvantages associated with different removal techniques explored to date can be better understood from Fig. 1. Some of the novel adsorbents derived from waste products, bio-materials, metal oxides, naturally occurring materials, cellulose lignin derived, nanoparticles, etc., were used as sustainable material across the globe for the remediation of noxious pollutants.[3–13,15–20]

The present chapter presents an overview of the characteristics of sustainable material for an application as adsorbent and sensor for the remediation and detection of noxious impurities, characterization methods, detailed overview of different materials used as adsorbents, and sensors for inorganic and organic impurities.

TABLE 1 Classification of dyes majorly used in textile industries.

Dyes	Examples	Solubility	Substrate	Chemical types
(Ionic)				
Anionic				
Acid	AB1, AY36, AB117, AO19, AB25, AO3, MO, EBT, etc.	Water soluble	Wool, nylon, paper, leather and ink	Anthroquinone; azo; triarylmethane
Reactive	RR15, RB2, RB19, RB74, etc.	Water soluble	Nylon, silk, cotton, wool	Azo; basic; anthraquinone; formazan; oxazine; phthalocyanine
Direct	DR81, DB86, CR, DO26, DY50, DNB106, etc.	Water soluble	Cotton, leather, rayon, paper and nylon	Oxazine; stilbene; phthalocyanine; azo
Cationic				
	MB, MG, RhB, CV, MV, BG, Safranin, BBG, BV14, BO2, etc.	Water soluble	Polyester, treated nylon, paper, inks and polyacrylonitrile	Oxazine; azo; cyanine; diphenylmethane; azine; xanthenes; acridine; diazahemicyanine; triarylmethane; anthraquinone; hemicyaninem
Non-ionic				
Vat	VBL9,8,16; VBE 1, 4, 5,6,14; VBR 1, 68, 72; VR13, VV1; VY1, etc.	Water insoluble	Cotton and wool	Indigoids; anthraquinone
Disperse	DR11, DB3, DY3, DO13, DV28, DY211, DB9, disperse brown, etc.	Water insoluble	Acetate, polyamide, acrylic polyester and plastic	Benzodifuranone; azo; anthraquinone; nitro(e) styryl

Reproduced with permission from ref. 41.

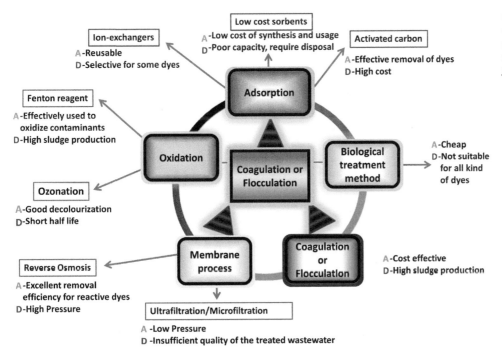

FIG. 1 Advantages (A) and disadvantages (D) associated with removal techniques in general. *(Reproduced with permission from ref. 41.)*

2 Characteristics of sustainable material for adsorbent and sensor

To remove noxious impurities, the sustainable adsorbent must possess characteristic properties such as cost-effectiveness, facile eco-friendly routes of synthesis, high selectivity, exponential adsorption capacity, and long shelf life. Further, other properties which play a significant role in the adsorptive removal of noxious impurities were active sites, specific surface area, pore-volume, pore structure, and surface functional moieties.[51] Based on the pore size, IUPAC classified the porous materials as macroporous (> 50 nm diameter), mesoporous (2 to 50 nm), and microporous (less than 2 nm diameter).[52]

On the other hand, the characteristics properties of sensors have no boundaries as it is a devices used to transform the different information related to chemical or physical properties in the form of a signal. The designing of sensors is always a challenge. It essentially requires a basic understanding of the nature of inter and intramolecular interactions, its ability to create certain nanosized architectures, and computing of the signal/response produced due to the interaction of these architectures. Thus, it can be concluded that the synthesis or designing of modern era sensors has no boundaries, and it works on the interfaces of different science, i.e., chemical, physical, material science, and engineering.[53]

3 Characterization techniques

Characterization is essentially required to understand the properties of the material. It is one of the significant steps before applying the material in different fields, whether it be wastewater treatment or sensing. Furthermore, it has emerged as an independent research field and is widely recommended by researchers across the globe.[54–56] The general overview of different techniques and their role in material characterization can be better understood from Fig. 2. Thus, the commonly adopted characterization techniques are as follows:

- *Proximate and ultimate analyses*: It includes moisture, ash, fixed carbon, volatile matter, elemental analyses such as C, H, O, N, and S, and chemical compounds.
- *Brunauer-Emmett-Teller (BET), Barret-Joyner-Halender (BJH)*: It includes analyzing the specific surface area, pore volume, and pore size distribution.
- *Scanning electron microscopy (SEM), High-resolution transmission electron microscopy (HR-TEM), Scanning transmission electron microscopy (STEM)*: It includes morphological analyses such as surface morphologies and microscopic features.
- *Energy-dispersive X-ray spectroscopy (EDX)*: It is widely used for mineral elements and their dispersion.

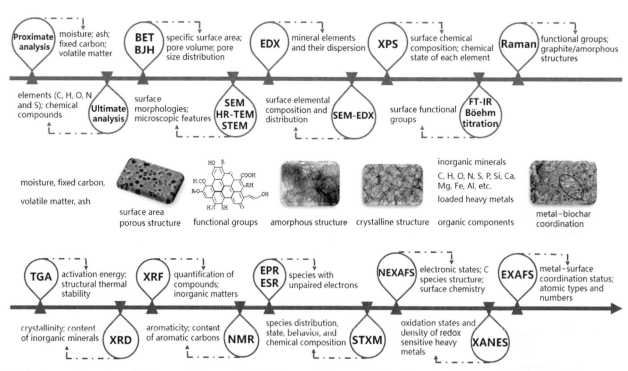

FIG. 2 A general overview of different techniques and their role in material characterization. *(Reproduced with permission from ref. 56.)*

- *SEM-EDX*: It is widely used for surface elemental composition and distribution analyses.
- *X-ray photoelectron spectroscopy (XPS)*: This advanced technology is used to analyze the surface chemical composition and chemical state of each element.
- *Fourier transform-infrared spectroscopy (FT-IR), Raman spectroscopy, Böehm titration*: It is used to analyze the surface functionalization's, i.e., functional group present on the surface of the material and structures such as graphite/amorphous.
- *Thermogravimetric analyzer (TGA)*: It elucidates the activation energy and structural thermal stability.
- *X-ray diffraction (XRD)*: It is used for the mineralogical analyses and nature of the material, i.e., crystalline or amorphous.
- *X-ray fluorescence spectroscopy (XRF)*: It is used to quantify compounds and inorganic matters.
- *Solid-state nuclear magnetic resonance (NMR)*: It is used to elucidate the aromaticity and the content of aromatic carbons.
- *Electron spin resonance (ESR) and Electron paramagnetic resonance (EPR)*: It is used to elucidate the spin of unpaired electrons.
- *Scanning transmission soft X-ray microscopy (STXM)*: It is widely used for species distribution, state, behavior, and chemical composition analyses.
- *Synchrotron-based near-edge X-ray absorption fine structure spectroscopy (NEXAFS)*: It is used to decipher the electronic states, C species structure, and surface chemistry.
- *X-ray absorption near-edge structure spectroscopy (XANES)*: It is used to study the oxidation states and density of redox-sensitive heavy metals.
- *Extended X-ray absorption fine structure spectroscopy (EXAFS)*: It is used to elucidate the metal-surface coordination status, atomic types, and number.

4 Detection techniques for noxious organic and inorganic impurities

Detection of noxious impurities is essentially required before remediation from water or wastewater. Keeping in view the cost-effectiveness of the whole process, the selected technique must possess properties such as eco-friendly, high-selectivity, and precise low limit detection etc.[57] The detailed overview of some of the techniques used to detect noxious pollutants are as follows.

4.1 Advanced instrumental techniques

Noxious organic and inorganic impurities were generally detected with great ease and precision using different spectroscopic methods. For detecting heavy metal impurities, techniques such as AAS, ICP-OES, ICP-MS, XRF/TXRF, ICP-AES, etc.,[57–59] were preferred. The major advantage of these techniques is their ability to detect heavy metals, even trace quantity (Femtomolar range), with great precision.[60] Shirkhanloo and Mousavi used AAS to detect metals such as Cu^{2+}, Pb^{2+}, and Cd^{2+} in aqueous solution even in trace quantities 2, 3, and 0.2 μg/L, respectively.[61] Other advanced techniques such as ICP-MS can detect the metal impurities in the range of parts per billion (ppb) to parts per trillion (ppt).

Further, with the synchronization of these techniques with chromatography and laser ablation, the detection limit can be optimized.[62] Although the spectroscopic technique is most precise, and it involves expensive instruments and a complex mode of operation, which requires trained personnel to run and calibrate from time to time.[63] The Limit of Detection (LOD) of different techniques for different metal ions can be better understood from Table 2.

The detection of organic impurities was carried out using techniques such as HPLC, fluorescence detection (FD), mass spectroscopy, tandem mass spectroscopy, real time-time of flight mass spectrometry, UV-Vis, Raman spectroscopy, and surface-enhanced Raman spectroscopic methods (SERS).[68–74]

4.2 Sensors

4.2.1 Electrochemical

It is a class of chemical sensors in which electrodes act as transducer elements in the presence of analytes. As discussed in Section 2, sensors used diverse characteristics properties to detect environmental pollutants in water and wastewater, and these parameters vary from case to case, whether it be physical, chemical, or biological parameters.[75] It works on the principle of generating electrical signal/response in the presence of analytes.[75] It is widely used as the techniques involving the electrochemistry principles that have great potential with respect to other techniques due to properties such

TABLE 2 Limit of detection (LOD) of different advanced instrumental techniques with respect to particular metal ions.

Heavy metal	Technique	LOD	Refs.
Lead (Pb)	AAS	1 µg/L	57,65–67
	ICP-MS	0.02 µg/L	
	ICP-AES	0.0091 µg/L	
	Potentiometry	1×10^{-6} mol/L	
	Amperometric	1.9×10^{-8} mol/L	
	SWASV	1.8×10^{-9} mol/L	
	CV	9×10^{-8} mol/L	
Cadmium (Cd)	ICP-MS	0.01 µg/L	57,65,66
	AAS	1 µg/L	
	ICP-AES	0.0010 µg/L	
	Potentiometry	1×10^{-7} mol/L	
	Amperometric	1.78×10^{-7} mol/L	
Mercury (Hg)	AAS	0.05 µg/L	57,65
	ICP	0.6 µg/L	
	AAS	5 µg/L	
	Potentiometry	7×10^{-8} mol/L	
	Amperometric	1.8×10^{-8} mol/L	
	CV	1×10^{-9} mol/L	
Nickel (Ni)	ICP-MS	0.1 µg/L	57,65
	AAS	0.5 µg/L	
	ICP-AES	10 µg/L	
Zinc (Zn)	GC-MS	0.1 µg/L	57,65
	GC-FID	1 µg/L	
Chromium (Cr)	AAS	0.05–0.2 µg/L	57,65,66
	ICP-AES	0.0024 µg/L	
Copper (Cu)	ICP-MS	0.02–0.1 µg/L	57,65,66
	ICP-OES	0.3 µg/L	
	AAS	0.5 µg/L	
	ICP-AES	0.0047 µg/L	
	Amperometric	7.4×10^{-9} mol/L	
Arsenic (As)	ICP-MS	0.1 µg/L	57,65,66
	AAS	2 µg/L	

Reproduced with permission from ref. 64.

as cost-effectiveness and facile eco-friendly routes to detect noxious environmental pollutants such as heavy metals, dyeing impurities, complex organic compounds such as pesticides, PAHs, CPs, and herbicides etc.[76] Further, electrochemical sensors are different origins such as amperometric, voltammetric, and potentiometric etc.[77]

Amperometric sensors were based on the redox mechanism, and they measured current as a result of oxidation and reduction of electroactive substance in a particular system undergoing electrochemical reaction.[78,79] Some of the efforts that have been made using amperometric-based electrochemical sensors for the detection of environmental pollutants are as follows.

Tucci et al. developed a microbial amperometric-based sensor for sensing herbicides such as atrazine and diuron in the aqueous samples.[80] On the other hand, amperometric sensors based on competitive reactions at nitroreductase@Layered double hydroxide (NLDH) for the detection of mesotrione were developed by researchers.[81] Ayenimo and Adeloju developed an amperometric sensor from Ultrathin Polypyrrole-based glucose biosensor to detect noxious heavy metal impurities such as lead, mercury, cadmium, and copper in their ionic form (2+).[82] Moreover, an electrochemical for detecting phenols and catechol was developed by Quynh et al. It was a non-enzymatic sensor.[83]

Voltammetric sensors were based on a powerful electroanalytical technique called "Voltammetry." They are widely applied for detecting several environmental pollutants such as metal ions[84–86] and complex organic molecules such as pharmaceutical compounds, pesticides, dyes, and herbicides in the aqueous samples.[87–93]

4.2.2 Optical

The most widely used optical sensors were mainly based on fluorescence sensors and colorimetric sensors, and the details related to each of them are mentioned below.

The former involves the use of specific fluorescence probes, and it was analyzed through changes in the intensity of fluorescence after its interaction with analytes and based on this the concentration of analytes was measured. Apart from the conventional single fluorophore sensor, the advanced technique, i.e., fluorescence resonance energy transfer (FRET), was actively used to detect the impurities with great sensitivity. As a part of the mechanism, the acceptor (Quencher species), when located at a certain distance, can quench the donor (fluorescent species) fluorescence during the energy transfer process, and the change in fluorescence intensity of acceptor or donor species will result in the analysis of the concentration of analytes.[94] On the other hand, colorimetric sensing is a cost-effective, simple and facile technique that generally relies on the change in color of a mixture solution having gold nanoparticles (AuNPs) and the analytes.[95,96] This is because optical absorption of AuNPs is strongly affected by the refractive index of the surrounding media and the interparticle surface plasmon coupling, and the analytes present triggered in the aggregation and redispersion of the AuNPs and results in the significant color change. Further, the NPs were functionalized with specific molecules that only bind with the analytes understudy for high sensitivity and specificity.[95,96] Several optical sensors for different environmental pollutants all[42–45,97–103] have been developed to date, and it is not practically possible to discuss them all.

5   Remediation methods for the noxious organic and inorganic impurities

Water and wastewater treatment technologies for the remediation of noxious impurities in terms of materials and techniques were classified into two, i.e., conventional treatment and non-conventional treatment. The conventional treatment technologies include chemical precipitation, coagulation/flocculation, membrane technologies, ion-exchangers, electrochemical technologies, etc. On the other hand, non-conventional technologies involved adsorption, Fenton-like reactions, microbial fuel cell, and nanotechnology. A general overview of the conventional and non-conventional treatment technologies can be obtained from Figs. 3 and 4.

Further, adsorption is the most widely used wastewater treatment technology, as indicated from the Scopus database. Different sustainable adsorbents such as activated carbon, naturally occurring materials, biochar (agricultural wastes),

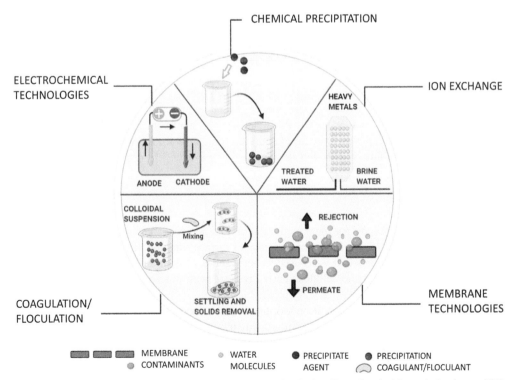

FIG. 3   A general overview of different conventional wastewater treatment technologies. *(Reproduced with permission from ref.64.)*

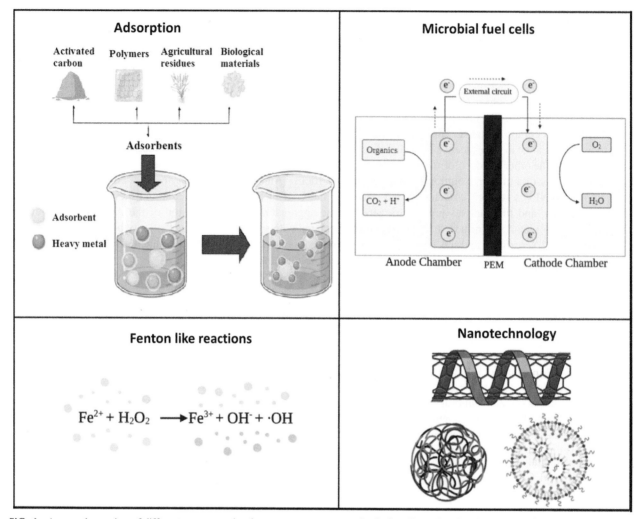

FIG. 4 A general overview of different non-conventional wastewater treatment technologies. *(Reproduced with permission from ref. 64.)*

industrial wastes, nanoparticles, nanomaterials, functionalized adsorbents, etc.,[3–13,15–40] were majorly used for remediation of noxious impurities.

As the present chapter is majorly focused on sustainable materials. So, our major focus will be on the non-conventional wastewater treatments methods that involve various materials. Further, it is not feasible to provide details about all the developed/synthesized adsorbents in this chapter, but details of some most widely used adsorbents are presented below.

5.1 Activated carbon

It is one of the most widely used adsorbents for the remediation of noxious inorganic and organic impurities. Due to properties such as porous nature, high specific surface area, and ease to surface functionalization, it shows excellent adsorbent properties. Some of the common materials that were used as activated carbon upon physical and chemical activation are coal, naturally occurring materials, lignin and cellulose derivatives, agricultural waste, rubber tire, industrial wastes, fruit stones, etc.[104–115]

Ghaedi et al. developed the activated carbon from oak tree wood to remove noxious sunset yellow; results obtained revealed that maximum adsorption of 5.8377–30.1205 mg/g was observed from the adsorbent dosage of 0.05–0.25 g.[104] An adsorbent based on functionalized activated carbon with Ag nanoparticles to remove noxious methylene blue dye. Results revealed that the developed adsorbent possesses excellent adsorption capacity, and ~95% removal of MB dye was observed within 4 min.[105] Nekouei et al. used Ni(OH)$_2$ functionalized activated carbon (Ni(OH)$_2$-NP-AC) for the removal of malachite green (MG). The synthesized adsorbent possesses a specific surface area ~960 m^2/g with excellent removal properties.[106] Dehghanian et al. synthesized SnS NPs modified activated carbon (SnS-NP-AC) to remove toxic

Congo red (CR) dye from the aqueous solution. The synthesized adsorbent possesses excellent adsorption capacity, and ~99% dye removal was observed.[107] In one other study, Al-Aoh used coconut husk fiber-based activated carbon for the adsorptive removal of nitrophenol, methylene blue, and acid red-27 dye.[109] Ghasemi et al. synthesized tetraethylene-pentamine functionalized activated carbon from *Rosa canina* L. The synthesized adsorbent shows a maximum adsorption capacity of 333.3 mg/g.[110] Heibati et al. used activated carbon prepared from walnut and poplar woods to remove noxious AR18 dye. The developed adsorbent took around 90 min for the removal of AR18 dyes.[111] Asfaram et al. used ZnS: Cu NPs modified activated carbon for the removal of Auramine-O; the findings obtained revealed that the optimized parameter for the removal of 99.76% of this noxious dye was 0.02 g (adsorbent dosage), 20 mg/L (initial dye concentration), 7 (pH) and 3 min (sonication time) with the maximum adsorption capacity of 183.15 mg/g.[16]

Gupta et al. used scrap tire as activated carbon to remove noxious Ni (II) ion from the aqueous solution, and the synthesized adsorbent shows maximum adsorption capacity of 25 mg/g with approximately 95% removal within 50 min of contact time.[5] Karmacharya et al. used activated carbon derived from tire and functionalized its alumina composite, and the results obtained revealed that maximum adsorption capacities for both the ionic form of As (V) and As (III) were 23.8 and 14.28 mg/g, respectively.[8] Karri et al. used palm kernel shell-based activated carbon to remove Zn (II) from the aqueous solution. The findings revealed that the optimized values of effective parameters such as pH, residence time, adsorbent dosage, and temperature for the 90% removal were 5, 53.2 min, 44.8 mg/L, 15.5 mg/L, and 40°C, respectively.[10] On the other hand, Sankaran et al. derived activated carbon from eggshell wastes to remove noxious metal impurities such as copper, zinc, nickel, and cobalt along with microbial products.[12] Wong et al. reported the 95% removal of noxious Cr^{4+} onto the activated carbon obtained from rice husk.[116] Wang et al. modified activated carbon with tartaric acid and applied it to remove copper and lead ions from the aqueous solution.[117] Bernard et al. modified activated carbon derived from coconut shell using $ZnCl_2$ for the removal of heavy metals such as lead, iron, copper, and zinc.[118]

5.2 Carbon nanotubes

Carbon nanotubes (CNTs) have developed the great interest of researchers across the globe to be used as adsorbents for the removal of noxious inorganic or organic impurities due to their unique physical and chemical properties. Two well-known forms of CNTs are Single-walled carbon nanotubes (SWCNTs) and Multiwalled carbon nanotubes (MWCNTs).[119] An in-depth analysis revealed that the enhanced adsorption capacities of CNTs were due to the morphology, number of active sites, and fundamental structural features along with π-conjugative structures.[119]

Chen et al. used CNTs for the adsorptive removal of Ni^{2+} and Sr^{2+} and observed that it depends on the effective parameter such as pH and ionic strength. Further, the MWCNTs shows great removal potential for Sr^{2+}.[120] In one another study, Yang et al. carried out the removal of Ni^{2+} onto the MWCNTs, finding obtained revealed that adsorption increases as the pH increases, but the optimized pH for this process was 8.[121] Lu et al. used SWCNTs and MWCNTs as adsorbents for the removal of Ni^{2+} and observed that both the adsorbent possesses excellent adsorption capacity even after several regenerative cycles.[122] Li et al. used MWCNTs to remove Pb^{2+} ions; results obtained revealed that adsorption capacity increases with an increase in temperature with the endothermic nature.[123] Moreover, Atieh functionalized MWCNTs and observed that upon acidification, MWCNTs shows enhanced adsorption capacity up to 20 times and 5 times with respect to un-modified MWCNTs and other surfaces, respectively.[124] Xu et al. applied oxidized MWCNTs for the removal of noxious Pb^{2+} ions and observed that the adsorption process depends on the pH and followed pseudo-second-order model.[125] Likewise in metal impurities, CNTs were used for the remediation of several dyes such as Procion red MX-5B, Sufranine O, acid red 18, bromothymol blue, methyl blue, methyl violet, methylene blue etc.[126–130]

Moreover, the detailed overview of multiple adsorbents is available in a different chapter of the current book, and it will be of no use to furnish similar information in this chapter as it provides only a brief overview of technologies used in wastewater treatment.

6 Conclusion and future scopes

Although water is the most essential element for human life, it is one of the most exploited natural resources. Rapid urbanization and industrialization are majorly responsible for the deterioration of this natural resource through their toxic effluents in the form of inorganic and organic impurities to the nearby aquatic sources. These noxious impurities deteriorate the water and make them unfit for drinking and other day-to-day activities. Approximately 800 million people across the globe do not have access to safe drinking water. Keeping in view the importance of this natural resource, a sustainable material focussed on wastewater treatment and strict environmental policies implementation is essentially required across the globe so that zero-discharge limit of these noxious effluents from their sources may be achieved for the upliftment of the society.

Acknowledgment

Dr. Inderjeet Tyagi is thankful to the Director, Zoological Survey of India, Ministry of Environment, Forest and Climate Change, Kolkata, for continuous motivation and moral support. Further, Dr. Tyagi extends his sincere thanks to the RAMC for considering the project entitled *"DNA Metasystematics studies for the assessment of Macrobiome and Microbiome in fresh water ecosystem with relation to the Noxious Pollutants"* under ZSI core funding.

References

1. Carolin CF, Kumar PS, Saravanan A, Joshiba GJ, Naushad M. Efficient techniques for the removal of toxic heavy metals from aquatic environment: a review. *J Environ Chem Eng*. 2017;5:2782–2799.

2. Vardhan KH, Kumar PS, Panda RC. A review on heavy metal pollution, toxicity and remedial measures: current trends and future perspectives. *J Mol Liq*. 2019;290:111197.

3. Dehghani MH, Taher MM, Bajpai AK, et al. Removal of noxious Cr (VI) ions using single-walled carbon nanotubes and multi-walled carbon nanotubes. *Chem Eng J*. 2015;279:344–352.

4. Al-Khaldi FA, Abusharkh B, Khaled M, et al. Adsorptive removal of cadmium (II) ions from liquid phase using acid modified carbon-based adsorbents. *J Mol Liq*. 2015;204:255–263.

5. Gupta VK, Nayak A, Agarwal S, Chaudhary M, Tyagi I. Removal of Ni (II) ions from water using scrap tire. *J Mol Liq*. 2014;190:215–222.

6. Gupta VK, Chandra R, Tyagi I, Verma M. Removal of hexavalent chromium ions using CuO nanoparticles for water purification applications. *J Colloid Interface Sci*. 2016;478:54–62.

7. Naushad M, Mittal A, Rathore M, Gupta V. Ion-exchange kinetic studies for Cd (II), Co (II), Cu (II), and Pb (II) metal ions over a composite cation exchanger. *Desalin Water Treat*. 2015;54(10):2883–2890.

8. Karmacharya MS, Gupta VK, Tyagi I, Agarwal S, Jha VK. Removal of As (III) and As (V) using rubber tire derived activated carbon modified with alumina composite. *J Mol Liq*. 2016;216:836–844.

9. Verma M, Tyagi I, Chandra R, Gupta VK. Adsorptive removal of Pb (II) ions from aqueous solution using CuO nanoparticles synthesized by sputtering method. *J Mol Liq*. 2017;225:936–944.

10. Karri RR, Sahu JN. Modeling and optimization by particle swarm embedded neural network for adsorption of zinc (II) by palm kernel shell based activated carbon from aqueous environment. *J Environ Manag*. 2018;206:178–191.

11. Lingamdinne LP, Koduru JR, Chang YY, Karri RR. Process optimization and adsorption modeling of Pb (II) on nickel ferrite-reduced graphene oxide nano-composite. *J Mol Liq*. 2018;250:202–211.

12. Sankaran R, Show PL, Ooi CW, et al. Feasibility assessment of removal of heavy metals and soluble microbial products from aqueous solutions using eggshell wastes. *Clean Techn Environ Policy*. 2020;22:773–786.

13. Cheng SY, Show PL, Lau BF, Chang JS, Ling TC. New prospects for modified algae in heavy metal adsorption. *Trends Biotechnol*. 2019;37:1255–1268.

14. Karri RR, Ravindran G, Dehghani MH. Wastewater—sources, toxicity, and their consequences to human health. In: *Soft Computing Techniques in Solid Waste and Wastewater Management*. Elsevier; 2021:3–33.

15. Burakov AE, Galunin EV, Burakova IV, et al. Adsorption of heavy metals on conventional and nanostructured materials for wastewater treatment purposes: a review. *Ecotoxicol Environ Saf*. 2018;148:702–712.

16. Asfaram A, Ghaedi M, Agarwal S, Tyagi I, Gupta VK. Removal of basic dye Auramine-O by ZnS: Cu nanoparticles loaded on activated carbon: optimization of parameters using response surface methodology with central composite design. *RSC Adv*. 2015;5(24):18438–18450.

17. Ghaedi M, Hajjati S, Mahmudi Z, et al. Modeling of competitive ultrasonic assisted removal of the dyes—methylene blue and safranin-O using Fe3O4 nanoparticles. *Chem Eng J*. 2015;268:28–37.

18. Robati D, Mirza B, Rajabi M, et al. Removal of hazardous dyes-BR 12 and methyl orange using graphene oxide as an adsorbent from aqueous phase. *Chem Eng J*. 2016;284:687–697.

19. Nekouei F, Nekouei S, Tyagi I, Gupta VK. Kinetic, thermodynamic and isotherm studies for acid blue 129 removal from liquids using copper oxide nanoparticle-modified activated carbon as a novel adsorbent. *J Mol Liq*. 2015;201:124–133.

20. Karri RR, Tanzifi M, Yaraki MT, Sahu JN. Optimization and modeling of methyl orange adsorption onto polyaniline nano-adsorbent through response surface methodology and differential evolution embedded neural network. *J Environ Manag*. 2018;223:517–529.

21. Khan FSA, Mubarak NM, Tan YH, et al. A comprehensive review on magnetic carbon nanotubes and carbon nanotube-based buckypaper-heavy metal and dyes removal. *J Hazard Mater*. 2021;413:125375.

22. Balarak D, Jaafari J, Hassani G, et al. The use of low-cost adsorbent (canola residues) for the adsorption of methylene blue from aqueous solution: isotherm, kinetic and thermodynamic studies. *Colloids Interface Sci Commun*. 2015;7:16–19.

23. Chaudhary M, Singh R, Tyagi I, Ahmed J, Chaudhary S, Kushwaha S. Microporous activated carbon as adsorbent for the removal of noxious anthraquinone acid dyes: role of adsorbate functionalization. *J Environ Chem Eng*. 2021;9(5):106308.

24. Huang Y, Fulton AN, Keller AA. Simultaneous removal of PAHs and metal contaminants from water using magnetic nanoparticle adsorbents. *Sci Total Environ*. 2016;571:1029–1036.

25. Zheng S, Yang Z, Park YH. Removal of chlorophenols from groundwater by chitosan sorption. *Water Res*. 2004;38(9):2315–2322.

26. Kadirvelu K, Kavipriya M, Karthika C, Radhika M, Vennilamani N, Pattabhi S. Utilization of various agricultural wastes for activated carbon preparation and application for the removal of dyes and metal ions from aqueous solutions. *Bioresour Technol.* 2003;87(1):129–132.

27. Dinçer AR, Güneş Y, Karakaya N, Güneş E. Comparison of activated carbon and bottom ash for removal of reactive dye from aqueous solution. *Bioresour Technol.* 2007;98(4):834–839.

28. Shen D, Fan J, Zhou W, Gao B, Yue Q, Kang Q. Adsorption kinetics and isotherm of anionic dyes onto organo-bentonite from single and multisolute systems. *J Hazard Mater.* 2009;172(1):99–107.

29. Ferreira AM, Coutinho JA, Fernandes AM, Freire MG. Complete removal of textile dyes from aqueous media using ionic-liquid-based aqueous two-phase systems. *Sep Purif Technol.* 2014;128:58–66.

30. Vinu R, Madras G. Kinetics of sonophotocatalytic degradation of anionic dyes with nano-TiO2. *Environ Sci Technol.* 2009;43(2):473–479.

31. Lee JW, Choi SP, Thiruvenkatachari R, Shim WG, Moon H. Evaluation of the performance of adsorption and coagulation processes for the maximum removal of reactive dyes. *Dyes Pigments.* 2006;69(3):196–203.

32. Karnjkar YS, Dinde RM, Dinde NM, et al. Degradation of magenta dye using different approaches based on ultrasonic and ultraviolet irradiations: comparison of effectiveness and effect of additives for intensification. *Ultrason Sonochem.* 2015;27:117–124.

33. Tripathi A, Ranjan MR. Heavy metal removal from wastewater using low cost adsorbents. *J Bioremediat Biodegrad.* 2015;6(6):315.

34. Ahmad T, Belwal T, Li L, et al. Utilization of wastewater from edible oil industry, turning waste into valuable products: a review. *Trends Food Sci Technol.* 2020;99:21–33.

35. Peng Q, Liu M, Zheng J, Zhou C. Adsorption of dyes in aqueous solutions by chitosan–halloysite nanotubes composite hydrogel beads. *Microporous Mesoporous Mater.* 2015;201:190–201.

36. Tan KB, Vakili M, Horri BA, Poh PE, Abdullah AZ, Salamatinia B. Adsorption of dyes by nanomaterials: recent developments and adsorption mechanisms. *Sep Purif Technol.* 2015;150:229–242.

37. Liu L, Gao ZY, Su XP, Chen X, Jiang L, Yao JM. Adsorption removal of dyes from single and binary solutions using a cellulose-based bioadsorbent. *ACS Sustain Chem Eng.* 2015;3(3):432–442.

38. Khan S, Sayed M, Sohail M, Shah LA, Raja MA. Advanced oxidation and reduction processes. In: *Advances in Water Purification Techniques.* Elsevier; 2019:135–164.

39. Sharifpour E, Khafri HZ, Ghaedi M, Asfaram A, Jannesar R. Isotherms and kinetic study of ultrasound-assisted adsorption of malachite green and Pb2+ ions from aqueous samples by copper sulfide nanorods loaded on activated carbon: experimental design optimization. *Ultrason Sonochem.* 2018;40:373–382.

40. May-Lozano M, Mendoza-Escamilla V, Rojas-García E, López-Medina R, Rivadeneyra-Romero G, Martinez-Delgadillo SA. Sonophotocatalytic degradation of Orange II dye using low cost photocatalyst. *J Clean Prod.* 2017;148:836–844.

41. Bushra R, Mohamad S, Alias Y, Jin Y, Ahmad M. Current approaches and methodologies to explore the perceptive adsorption mechanism of dyes on low-cost agricultural waste: a review. *Micropor Mesopor Mat.* 2021;111040. In this issue.

42. Gupta VK, Mergu N, Kumawat LK, Singh AK. Selective naked-eye detection of magnesium (II) ions using a coumarin-derived fluorescent probe. *Sensors Actuators B Chem.* 2015;207:216–223.

43. Gupta VK, Mergu N, Kumawat LK. A new multifunctional rhodamine-derived probe for colorimetric sensing of Cu (II) and Al (III) and fluorometric sensing of Fe (III) in aqueous media. *Sensors Actuators B Chem.* 2016;223:101–113.

44. Gupta VK, Singh AK, Ganjali MR, Norouzi P, Faridbod F, Mergu N. Comparative study of colorimetric sensors based on newly synthesized Schiff bases. *Sensors Actuators B Chem.* 2013;182:642–651.

45. Kumawat LK, Mergu N, Singh AK, Gupta VK. A novel optical sensor for copper ions based on phthalocyanine tetrasulfonic acid. *Sensors Actuators B Chem.* 2015;212:389–394.

46. Crespo GA. Recent advances in ion-selective membrane electrodes for in situ, environmental water analysis. *Electrochim Acta.* 2017;245:1023–1034.

47. Cuartero M, Bakker E. Environmental water analysis with membrane electrodes. *Curr Opin Electrochem.* 2017;3:97–105.

48. Cuartero M, Crespo GA. All-solid-state potentiometric sensors: a new wave for in situ aquatic research. *Curr Opin Electrochem.* 2018;10:98–106.

49. Zhao K, Veksha A, Ge L, Lisak G. Near real-time analysis of para-cresol in wastewater with a laccase-carbon nanotube-based biosensor. *Chemosphere.* 2021;269:128699.

50. Zhao K, Ge L, Wong TI, Zhou X, Lisak G. Gold-silver nanoparticles modified electrochemical sensor array for simultaneous determination of chromium (III) and chromium (VI) in wastewater samples. *Chemosphere.* 2021;281:130880.

51. Pourhakkak P, Taghizadeh M, Taghizadeh A, Ghaedi M. Adsorbent. In: *Interface Science and Technology.* vol. 33. Elsevier; 2021:71–210.

52. Sing KS. Reporting physisorption data for gas/solid systems with special reference to the determination of surface area and porosity (recommendations 1984). *Pure Appl Chem.* 1985;57(4):603–619.

53. Zenkina OV. *Nanomaterials Design for Sensing Applications.* Elsevier; 2019.

54. Igalavithana AD, Mandal S, Niazi NK, et al. Advances and future directions of biochar characterization methods and applications. *Crit Rev Environ Sci Technol.* 2018;47:2275–2330.

55. Khiari B, Ghouma I, Ferjani AI, et al. Kenaf stems: thermal characterization and conversion for biofuel and biochar production. *Fuel.* 2020;262:116654.

56. Liu M, Almatrafi E, Zhang Y, et al. A critical review of biochar-based materials for the remediation of heavy metal contaminated environment: applications and practical evaluations. *Sci Total Environ.* 2022;806:150531.

57. Malik LA, Bashir A, Qureashi A, Pandith AH. Detection and removal of heavy metal ions: a review. *Environ Chem Lett.* 2019;17:1495–1521.

58. Buledi JA, Amin S, Haider SI, Bhanger MI, Solangi AR. A review on detection of heavy metals from aqueous media using nanomaterial-based sensors. *Environ Sci Pollut Res.* 2021;28(42):58994–59002.

59. Bansod B, Kumar T, Thakur R, Rana S, Singh I. A review on various electrochemical techniques for heavy metal ions detection with different sensing platforms. *Biosens Bioelectron.* 2017;94:443–455.

60. Zhong WS, Ren T, Zhao LJ. Determination of Pb (lead), Cd (cadmium), Cr (chromium), Cu (copper), and Ni (nickel) in Chinese tea with high-resolution continuum source graphite furnace atomic absorption spectrometry. *J Food Drug Anal.* 2016;24:46–55.

61. Shirkhanloo H, Mousavi HZ. Preconcentration and determination of heavy metals in water, sediment and biological samples. *J Serb Chem Soc.* 2011;76:1583–1595.

62. Voica C, Kovacs MH, Dehelean A, Ristoiu D, Iordache A. ICP-MS determinations of heavy metals in surface waters from transylvania. *Rom Rep Phys.* 2012;57:1184–1193.

63. Kaur R, Bansod BK, Thakur R. Spectroscopic techniques and electrochemical sensors technologies for heavy metal ions detection: a review. *Int J Adv Eng Manag Sci.* 2016;2:1622–1626.

64. Zamora-Ledezma C, Negrete-Bolagay D, Figueroa F, et al. Heavy metal water pollution: A fresh look about hazards, novel and conventional remediation methods. *Environmental Technology & Innovation.* 2021;22:101504.

65. WHO. *Chemical Fact Sheets. Guidelines for Drinking-Water Quality: Fourth Edition Incorporating the First Addendum.* World Health Organization; 2017:307–442.

66. Tan M, Sudjad, Astuti, Rohman. Validation and quantitative analysis of cadmium, chromium, copper, nickel, and lead in snake fruit by inductively coupled plasma-atomic emission spectroscopy. *J Appl Pharm Sci.* 2018;8:44–48.

67. Bobaker AM, Alakili I, Sarmani SB, Al-Ansari N, Yaseen ZM. Determination and assessment of the toxic heavy metal elements abstracted from the traditional plant cosmetics and medical remedies: case study of Libya. *Int J Environ Res Public Health.* 2019;16:1957.

68. Zughaibi TA, Steiner RR. Differentiating nylons using direct analysis in real time coupled to an AccuTOF time-of-flight mass spectrometer. *J Am Soc Mass Spectrom.* 2020;31(4):982–985.

69. Alvarez-Martin A, Cleland TP, Kavich GM, Janssens K, Newsome GA. Rapid evaluation of the debromination mechanism of eosin in oil paint by direct analysis in real time and direct infusion-electrospray ionization mass spectrometry. *Anal Chem.* 2019;91(16):10856–10863.

70. Duvivier WF, van Putten MR, van Beek TA, Nielen MW. (Un) targeted scanning of locks of hair for drugs of abuse by direct analysis in real time–high-resolution mass spectrometry. *Anal Chem.* 2016;88(4):2489–2496.

71. Cochran KH, Barry JA, Muddiman DC, Hinks D. Direct analysis of textile fabrics and dyes using infrared matrix-assisted laser desorption electrospray ionization mass spectrometry. *Anal Chem.* 2013;85(2):831–836.

72. Day CJ, DeRoo CS, Armitage RA. Developing direct analysis in real time time-of-flight mass spectrometric methods for identification of organic dyes in historic wool textiles. In: *Archaeological Chemistry VIII.* American Chemical Society; 2013:69–85.

73. Geiger J, Armitage RA, DeRoo CS. Identification of organic dyes by direct analysis in real time-time of flight mass spectrometry. In: *Collaborative Endeavors in the Chemical Analysis of Art and Cultural Heritage Materials.* American Chemical Society; 2012:123–129.

74. Selvius DeRoo C, Armitage RA. Direct identification of dyes in textiles by direct analysis in real time-time of flight mass spectrometry. *Anal Chem.* 2011;83(18):6924–6928.

75. Simões FR, Xavier MG. Electrochemical sensors. In: *Nanoscience and Its Applications.* O'Reilly; 2017:155–178.

76. Hussain CM, Keçili R. Electrochemical techniques for environmental analysis. In: *Modern Environmental Analysis Techniques for Pollutants.* Elsevier; 2020:199–222.

77. Pujol L, Evrard D, Groenen-Serrano K, Freyssinier M, Ruffien-Cizsak A, Gros P. Electrochemical sensors and devices for heavy metals assay in water: the French groups' contribution. *Front Chem.* 2014;2:1–24.

78. Thevenot DR, Toth K, Durst RA, Wilson GS. Electrochemical biosensors: recommended definitions and classification. *Biosens Bioelectron.* 2001;16:121–131.

79. Luppa PB, Sokoll LJ, Chan DW. Immunosensors—principles and applications to clinical chemistry. *Clin Chim Acta.* 2005;314:1–26.

80. Tucci M, Grattieri M, Schievano A, Cristiani P, Minteer SD. Microbial amperometric biosensor for online herbicide detection: photocurrent inhibition of *Anabaena variabilis*. *Electrochim Acta.* 2019;302:102–108.

81. Hdiouech S, Bruna F, Batisson I, Besse-Hoggan P, Prevot V, Mousty C. Amperometric detection of the herbicide mesotrione based on competitive reactions at nitroreductase@layered double hydroxide bioelectrode. *J Electroanal Chem.* 2019;835:324–328.

82. Ayenimo JG, Adeloju SB. Rapid amperometric detection of trace metals by inhibition of an ultrathin polypyrrole-based glucose biosensor. *Talanta.* 2016;148:502–510.

83. Quynh BTP, Byun JY, Hoon Kim S. Non-enzymatic amperometric detection of phenol and catecholusing nanoporous gold. *Sensors Actuators B Chem.* 2015;221:191–200.

84. Hatamie A, Jalilian P, Rezvani E, Kakavand A, Simchi A. Fast and ultra-sensitive voltammetric detection of lead ions by twodimensional graphitic carbon nitride (g-C3N4) nanolayers as glassy carbon electrode modifier. *Measurement.* 2019;134:679–687.

85. Liu T, Luo Y, Kong L, Zhu J, Wang W, Tan L. Voltammetric detection of Cu21 using poly(azure a) modified glassycarbon electrode based on mimic peroxidase behavior of copper. *Sensors Actuators B Chem.* 2016;235:568–574.

86. Romero-Cano LA, Zarate-Guzmána AI, Carrasco-Marín F, Gonzalez-Gutiérrez LV. Electrochemical detection of copper in water using carbon paste electrodes prepared from bio-template (grapefruit peels) functionalized with carboxyl groups. *J Electroanal Chem.* 2019;837:22–29.

87. Khosropour H, Rezae B, Ensafi AA. A selective and sensitive detection of residual hazardous textile dyes in wastewaters using voltammetric sensor. *Microchem J.* 2019;146:548–556.

88. Gangadharappa J, Manjunatha G. A novel poly (glycine) biosensor towards the detection of indigo carmine: a voltammetric study. *J Food Drug Anal.* 2018;26:292–299.

89. Vishenkova DA, Korotkova EI, Sokolova VA, Kratochvil B. Electrochemical determination of some triphenylmethane dyes by means of voltammetry. *Procedia Chemistry.* 2015;15:109–114.

90. Shetti NP, Malode SJ, Malladi RS, Nargund SL, Shukla SS, Aminabhavi TM. Electrochemical detection and degradation of textile dye Congo red at graphene oxide modified electrode. *Microchem J.* 2019;146:387–392.

91. de Lima CA, da Silva PS, Spinelli A. Chitosan-stabilized silver nanoparticles for voltammetric detection of nitrocompounds. *Sensors Actuators B Chem.* 2014;196:39–45.

92. Pandey A, Sharma S, Jain R. Voltammetric sensor for the monitoring of hazardous herbicide triclopyr (TCP). *J Hazard Mater.* 2019;367:246–255.

93. Castro SV, Silva MN, Tormin TF, et al. Highly-sensitive voltammetric detection of trinitrotoluene on reduced graphene oxide/carbon nanotube nanocomposite sensor. *Anal Chim Acta.* 2018;1035:14–21.

94. Bi H, Han X. Chemical sensors for environmental pollutant determination. In: *Chemical, Gas, and Biosensors for Internet of Things and Related Applications.* Elsevier; 2019:147–160.

95. Ni J, Lipert RJ, Dawson GB, Porter MD. Immunoassay readout method using extrinsic Raman labels adsorbed on immunogold colloids. *Anal Chem.* 1999;71:4903.

96. Jiang K, Pinchuk AO. Noble metal nanomaterials: synthetic routes, fundamental properties, and promising applications. *Solid State Phys.* 2015;66:131–211.

97. Kim Y, Johnson RC, Hupp JT. Gold nanoparticle-based sensing of "spectroscopically silent" heavy metal ions. *Nano Lett.* 2001;1(4):165–167.

98. Huang CC, Chang HT. Parameters for selective colorimetric sensing of mercury (II) in aqueous solutions using mercaptopropionic acid-modified gold nanoparticles. *Chem Commun.* 2007;12:1215–1217.

99. Lin SY, Wu SH, Chen CH. A simple strategy for prompt visual sensing by gold nanoparticles: general applications of interparticle hydrogen bonds. *Angew Chem.* 2006;118(30):5070–5073.

100. Liu J, Lu Y. A colorimetric lead biosensor using DNAzyme-directed assembly of gold nanoparticles. *J Am Chem Soc.* 2003;125(22):6642–6643.

101. Liu J, Lu Y. Accelerated color change of gold nanoparticles assembled by DNAzymes for simple and fast colorimetric Pb2+ detection. *J Am Chem Soc.* 2004;126(39):12298–12305.

102. Liu J, Lu Y. Optimization of a Pb2+-directed gold nanoparticle/DNAzyme assembly and its application as a colorimetric biosensor for Pb2+. *Chem Mater.* 2004;16(17):3231–3238.

103. Liu J, Lu Y. Stimuli-responsive disassembly of nanoparticle aggregates for light-up colorimetric sensing. *J Am Chem Soc.* 2005;127 (36):12677–12683.

104. Ghaedi AM, Baneshi MM, Vafaei A, et al. Comparison of multiple linear regression and group method of data handling models for predicting sunset yellow dye removal onto activated carbon from oak tree wood. *Environ Technol Innov.* 2018;11:262–275.

105. Ghaedi M, Roosta M, Ghaedi AM, et al. Removal of methylene blue by silver nanoparticles loaded on activated carbon by an ultrasound-assisted device: optimization by experimental design methodology. *Res Chem Intermed.* 2018;44(5):2929–2950.

106. Nekouei F, Kargarzadeh H, Nekouei S, Tyagi I, Agarwal S, Gupta VK. Preparation of nickel hydroxide nanoplates modified activated carbon for malachite green removal from solutions: kinetic, thermodynamic, isotherm and antibacterial studies. *Process Saf Environ Prot.* 2016;102:85–97.

107. Dehghanian N, Ghaedi M, Ansari A, et al. A random forest approach for predicting the removal of Congo red from aqueous solutions by adsorption onto tin sulfide nanoparticles loaded on activated carbon. *Desalin Water Treat.* 2016;57(20):9272–9285.

108. Mashhadi S, Sohrabi R, Javadian H, et al. Rapid removal of Hg (II) from aqueous solution by rice straw activated carbon prepared by microwave-assisted H2SO4 activation: kinetic, isotherm and thermodynamic studies. *J Mol Liq.* 2016;215:144–153.

109. Al-Aoh HA, Mihaina IA, Alzuaibr FM, et al. Optimization of conditions for preparation of activated carbon from coconut husk fiber using responses from measurements of surface area and adsorption. *Asian J Chem.* 2016;28(4):714–724.

110. Ghasemi M, Mashhadi S, Asif M, Tyagi I, Agarwal S, Gupta VK. Microwave-assisted synthesis of tetraethylenepentamine functionalized activated carbon with high adsorption capacity for Malachite green dye. *J Mol Liq.* 2016;213:317–325.

111. Heibati B, Rodriguez-Couto S, Al-Ghouti MA, et al. Kinetics and thermodynamics of enhanced adsorption of the dye AR 18 using activated carbons prepared from walnut and poplar woods. *J Mol Liq.* 2015;208:99–105.

112. Ahmad M, Rajapaksha AU, Lim JE, et al. Biochar as a sorbent for contaminant management in soil and water: a review. *Chemosphere.* 2014;99:19–33.

113. Mohan D, Sarswat A, Ok YS, Pittman Jr CU. Organic and inorganic contaminants removal from water with biochar, a renewable, low cost and sustainable adsorbent—a critical review. *Bioresour Technol.* 2014;160:191–202.

114. Peiris C, Gunatilake SR, Mlsna TE, Mohan D, Vithanage M. Biochar based removal of antibiotic sulfonamides and tetracyclines in aquatic environments: a critical review. *Bioresour Technol.* 2017;246:150–159.

115. Essandoh M, Kunwar B, Pittman Jr CU, Mohan D, Mlsna T. Sorptive removal of salicylic acid and ibuprofen from aqueous solutions using pine wood fast pyrolysis biochar. *Chem Eng J.* 2015;265:219–227.

116. Wong KK, Lee CK, Low KS, Haron MJ. Removal of Cu and Pb by tartaric acid modified rice husk from aqueous solutions. *Chemosphere.* 2003;50:23–28.

117. Wang ZH, Yue BY, Teng J, et al. Tartaric acid modified graphene oxide as a novel adsorbent for high-efficiently removal of Cu (II) and Pb (II) from aqueous solutions. *Journal of the Taiwan Institute of Chemical Engineers.* 2016;66:181–190.

118. Bernard E, Jimoh A, Odigure JO. Heavy metals removal from industrial wastewater by activated carbon prepared from coconut shell. *Res J Chem Sci.* 2013;3:3–9.

119. Gupta VK, Moradi O, Tyagi I, et al. Study on the removal of heavy metal ions from industry waste by carbon nanotubes: effect of the surface modification: a review. *Crit Rev Environ Sci Technol.* 2016;46(2):93–118.

120. Chen C, Hu J, Shao D, Li J, Wang X. Adsorption behavior of multiwall carbon nanotube/iron oxide magnetic composites for Ni (II) and Sr (II). *J Hazard Mater.* 2009;164(2):923–928.

121. Yang S, Li J, Shao D, Hu J, Wang X. Adsorption of Ni (II) on oxidized multiwalled carbon nanotubes: effect of contact time, pH, foreign ions and PAA. *J Hazard Mater.* 2009;166(1):109–116.

122. Lu C, Liu C. Removal of nickel (II) from aqueous solution by carbon nanotubes. *J Chem Technol Biotechnol.* 2006;81(12):1932–1940.

123. Li YH, Di Z, Ding J, Wu D, Luan Z, Zhu Y. Adsorption thermodynamic, kinetic and desorption studies of Pb2+ on carbon nanotubes. *Water Res.* 2005;39(4):605–609.

124. Atieh MA. Removal of chromium (VI) from polluted water using carbon nanotubes supported with activated carbon. *Procedia Environ Sci.* 2011;4:281–293.

125. Xu D, Tan X, Chen C, Wang X. Removal of Pb (II) from aqueous solution by oxidized multiwalled carbon nanotubes. *J Hazard Mater.* 2008;154 (1):407–416.

126. Wu CH. Adsorption of reactive dye onto carbon nanotubes: equilibrium, kinetics and thermodynamics. *J Hazard Mater.* 2007;144(1–2):93–100.

127. Shirmardi M, Mahvi AH, Hashemzadeh B, Naeimabadi A, Hassani G, Niri MV. The adsorption of malachite green (MG) as a cationic dye onto functionalized multi walled carbon nanotubes. *Korean J Chem Eng.* 2013;30(8):1603–1608.

128. Ghaedi M, Khajehsharifi H, Yadkuri AH, Roosta M, Asghari A. Oxidized multiwalled carbon nanotubes as efficient adsorbent for bromothymol blue. *Toxicol Environ Chem.* 2012;94(5):873–883.

129. Yao Y, Xu F, Chen M, Xu Z, Zhu Z. Adsorption behavior of methylene blue on carbon nanotubes. *Bioresour Technol.* 2010;101(9):3040–3046.

130. Shahryari Z, Goharrizi AS, Azadi M. Experimental study of methylene blue adsorption from aqueous solutions onto carbon nano tubes. *Int J Water Resour Environ Eng.* 2010;2(2):016–028.

Chapter 2

The outcome of human exposure to environmental contaminants. Importance of water and air purification processes

Agata Jankowska, Aleksander Ejsmont, Aleksandra Galarda, and Joanna Goscianska[*]
Faculty of Chemistry, Department of Chemical Technology, Adam Mickiewicz University in Poznań, Poznań, Poland
[*]Corresponding author.

1 Introduction

The rapid development of modern civilization and the ever-increasing population on earth intensify industrial and manufacturing production and with it the creation of waste (Fig. 1). In developed countries, strong consumerism contributes to the exploitation of the planet through increased production of food and everyday goods. On the other hand, developing countries, where a lot of investment is directed into the industry, cannot keep up in neutralizing the pollution produced. Similarly, in poorer countries where poverty, hunger, and corruption are fought, care for the environment is not a priority. In these countries, there are no legal regulations for environmental protection, which is why so much pollution ends up in soil and water, then reaching flora, fauna, and humans without control.[1–4]

Population growth has resulted in a huge expansion of agriculture, which requires the use of highly poisonous pest control agents in order to produce effective crops. Pesticides are characterized by high toxicity in addition to high stability or highly hazardous decomposition products. Their number is impressive due to the many diverse purposes of their use, hence the negative effects of these pollutants are broad and often unexpected.[5] Another group of chemicals that is widely used in food, medical, textile, and many other industries are dyes. Although safe dyes are being developed, the amount in which they are used and their tendency to permeate to the environment contribute to the disruption of biochemical processes not only in plants, but also in animals, and consequently in humans.[6] Other pollutants are heavy metals, which occur naturally in the environment. However, due to a lack of awareness during the industrial revolution, they have been released into the natural cycle and, to a large extent, into the spheres of human life. Moreover, they still get into the environment through geogenic, pharmaceutical, domestic effluents, and agricultural sources.[7]

Despite increased human life span and better health care, high consumption of drugs and easy access to them is another source of environmental contamination and potential harm to humans. Not all drugs are fully metabolized; furthermore, they are often disposed of with wastewater or utilized without proper control.[8] Their high bioactivity causes many diseases characteristic of their side effects rather than their specific therapeutic action. Moreover, their metabolites, derivatives, mixtures, as well as their ability to accumulate can induce uncontrolled changes in the body.[9]

So many sources and types of pollution result in many new diseases appearing in different age groups, communities, and genders. Pollutants have a devastating effect on the human body, which has no protective systems against their high concentrations or long-term exposure[10]. Therefore, it is necessary to reduce the sources of production of pollutants, which is extremely difficult. For this purpose, it is obligatory to constantly develop methods that regulate, remove, isolate, degrade, or reduce the toxicity of chemicals.

In this chapter, the effects of pesticides, dyes, heavy metals, and drugs on human health according to recent reports have been presented. In addition, currently used as well as developing methods of removing contaminants are proposed for specific pollutants.

FIG. 1 The major pollutants with a high impact on human health.

2 Pesticides as pollutants

In an era of widespread cleanliness and tightly controlled agriculture, an overused tactic is the use of highly toxic chemicals—pesticides (-*cides*, lat. *to kill*). According to the United States Environmental Protection Agency (EPA), there are nine general types of pesticides such as insecticides, herbicides, rodenticides, fungicides, disinfectants, attractants, plant defoliants, swimming pool treatments, and plant growth regulators. Moreover, they can be further divided into 24 groups, which specify their target, e.g., desiccants for drying unwanted living tissue; antifoulants, which kill barnacles attaching to boats, or ovicides focusing on the eggs of insects and mites.[11] Since nearly every area of life has some kind of agent that contributes to pest prevention, around 2 million tonnes per year of pesticides are produced.[12] Some of them resist degradation, especially the persistent organic pollutants (POPs) such as organochlorines or dioxins,[13–15] others (e.g., organophosphorus compounds) exhibit very harmful products of degradation.[16] Therefore, it is an inevitable large part of the chemicals, which are not fully selective and often completely uncontrolled get into aquatic systems, groundwater, sewage, consequently reaching the ecosystems and humans themselves (Fig. 2).[17,18]

2.1 The impact of pesticides on human health

The formulations of organic and inorganic pesticides are found to vary from liquids, powders, granules, baits, dust to ultra-low volume liquids.[19] Their form is dictated by the mode of action, i.e., the pathway that is most effective in controlling the pest. It can be either by producing vapor, direct contact, pest's stomach poisoning, or through absorption by organism followed by transfer within the system.[20] While one must take into account pesticides besides active ingredients that contain other chemicals such as surfactants, preservatives, or solvents which can also exhibit toxicity.[21] In 1969 the lethal dose (LD50) was established for 98 pesticides and two metabolites to the Sherman strain adult rats.[22] By 2010 it was reported that 250,000–370,000 people die each year because of these chemicals.[23] Moreover, according to the 2004 Garry's

FIG. 2 Scheme representing pesticides contamination of soil, water, and air, along with their influence on human (i.e., nervous, respiratory, digestive, reproductive, excretory, endocrine systems, as well as DNA and fetal).

description,[24] the most reported acute poisoning leading to neurobehavioral toxicity, cancer, and disruption of the endocrine system occurs in children younger than age 5. Awareness in developed countries has since improved, while the constantly increasing number of pesticides led to results for unintentional acute pesticides poisoning (UAPP) presented by Boedeker et al.[25] They estimated 385 million cases of UAPP occurring annually, with 11,000 fatal cases worldwide. Due to the massive number of pesticides and their wide-ranging influence, it is difficult to select individual compounds that cause specific negative effects on human health. Hence, they have been given in groups and in many cases one pesticide can induce diseases differing in symptoms.

Symptoms of pesticide exposure can be specific and occur singly, but they are often mixed, especially in the case of intense subjection to highly toxic chemicals that trigger rapid outcomes. Immediate effects of pesticides activity are general irritation of the skin, respiratory system, mucous membranes in the form of redness, blisters, rash, as well as visual disturbances, nausea, vomiting, diarrhea, dizziness, in extreme cases blindness, loss of consciousness, or even death.[26–28] The most common side effect after unintentional, short-term but high-level exposure to popular organophosphorus (OP) pesticides (e.g., dichlorvos, malathion, diazinon, dimethoate, methamidophos) is respiratory pathology.[29] They inhibit cholinesterase and induce irritation leading to the development of reactive airways dysfunction syndrome (RADS), which is a subset of asthma.[30] Endangering the OP pesticides (e.g., chlorpyrifos, phosalone, ethion, phorate) often manifests in headaches, burning sensation, and eye-watering. While the cumulative exposure cause delayed neurotoxicity manifested by weakness, paralysis, and/or paresthesia.[31] If the poisoning is severe by the high dose of OP pesticide, pulmonary edema and broncho-constriction are observed, often with respiratory muscle paralysis leading to death.[32] The liver is the certain stop in the metabolic path for the toxicities and among hundreds of pesticides the following are particularly hepatotoxic: dimethoate, fenitrothion, diazinon, monocrotophos, phorate, and RPR-II (2-butenoic acid-3-(diethoxyphosphinothioyl methyl ester). They affect the metabolism occurring in the liver's cells specifically in the cytoplasm, mitochondria, and peroxisomes causing oxidative damage.[33] Referring to the metabolism, followed by excretion, the hematological parameter's disturbance and kidneys malfunction have been reported mostly due to the use of cypermethrin, monocrotophos, and dichlorodiphenyldichloroethylene (DDE). They affect bone marrow leading to the increased values for the lymphocyte and eosinophil, whereas the neutrophils and red blood cells indicated the decrease, ultimately causing anemia. In the case of kidneys, they alter creatinine and urea concentrations.[34] Nephrotoxic pesticides are one of the main reasons for CKDu (kidney chronic disease of unknown etiology).[35]

Affecting the human antioxidant defense system results in producing oxygen-free radicals and cell damage, which can also stimulate the development of cancer.[36] Polyhalogenated cyclic hydrocarbons, bipyridyl, and chlorinated acetamide herbicides are highly stimulative toward free radical generation which can lead to lipid, protein, DNA peroxidation.[37,38] The pro-carcinogenic effect of pesticides manifests due to their geno- and cytotoxicity (e.g., endosulfan, simazine, carbaryl, dichlone, lindane, chlorophenol, malathion, dieldrin, paraquat, vinclozolin, fipronil) most often prompting the brain,[39] prostate,[40] ovarian,[41] lung,[42] breast,[43] bladder,[44] colon,[45] and colorectal cancers.[46] Single pesticide such as aflatoxin not only causes liver cancer, but also affects the immune system and impair children's growth development.[47] Most pesticides' cancer-inducing mechanisms are still not clear, as in piperonyl butoxide insecticide. It increases cytochrome *CYP1A1* gene expression toward proteins that are able to catalyze carcinogenic intermediates of polycyclic aromatic hydrocarbons.[48] Another mechanism example is the organochlorines imitation of xenoestrogens, with subsequent affecting the endocrine.[49] Endocrine-disrupting chemicals interfere with human hormones (estrogens, androgens, thyroid) and violate embryonic development, hemostasis, and regular physiology processes.[50,51] As the result, they are responsible for diseases such as neuroblastoma,[52] non-Hodgkin,[53] and Burkitt lymphomas.[54] Also, it is suspected that pesticides cause most childhood cancers, in particularly leukemia.[55,56]

Studies show that sclerosis, dementia, and Alzheimer's disease very often occur in areas contaminated with pesticides such as dichlorodiphenyltrichloroethane (DDT), polychlorinated biphenyls, organochlorines, and organophosphates, i.e., indane, chlorophenol, atrazine, chlorpyrifos, diazinon, malathion, parathion.[57–59] The neuropathological potential is broad and involves deviation in biological redox steady stress, aggregation of amyloid β peptides, and neuroinflammation. Indirect effects are also noted consequently leading to DNA damage and somatic mutations.[57,60] Part of the previous pesticides along with fenpyroximate, trichlorphon, paraquat, rotenone, tebufenpyrad, carbaryl, pendimethalin, and endosulfan affect mitochondria and the performance of ubiquitin-proteasome system. Some of them decrease the amount of intracellular adenosine triphosphate (ATP), and it is proven that paraquat and rotenone cause neuroinflammation and dopaminergic cell loss, which overall promote the pathogenesis of Parkinson's disease.[61]

Noteworthy, pesticides have the potential to create disorders of which origin is difficult to identify. Chlorane, heptachlor, aldrin, dioxins, dioxins-like polychlorinated biphenyls, polybrominated diphenyl ethers, and furans can lead to obesity and diabetes.[62] POPs are especially dangerous for pregnant women, who may be at risk such as insulin resistance, steatosis, hepatitis, dyslipidemia, diabetes mellitus, and vascular morbidity.[63] Moreover, in developing countries, pesticides are major agents in self-poisoning, e.g., endosulfan in India.[64] The popularity of pesticides intentional ingestion is due to their relatively low cost and high availability.[65,66]

2.2 Pesticides removal methods

The outline of a plurality of pesticide-caused illnesses requires the demonstration of currently available, or developing methods of removing/neutralizing pesticides. A general division of such methods is as follows: separation/adsorption, containment-immobilization, and destruction. Examples of methods, materials, and their performance in pesticide removal are presented in Table 1.

Separation techniques are one of the primary and older types of pesticide concentration reduction methods in water and soil.[76] However, these processes do not break down contaminants. They can be proceeded in situ—by using radiofrequency heating of soil, resulting in contaminants (e.g., endrin, isodrin, dieldrin, aldrin) vaporization, with subsequent concentration by vacuum. The second option is ex situ using the following techniques: washing, thermal desorption, and solvent extraction. Washing the soil with various solvents or synthetic surfactant is especially effective for soils with lower organic matter remediation content, which was observed for the removal of the chlorinated pesticides (lindane, heptachlor epoxide, and α-hexachlorocyclohexane).[77,78] Thermal desorption conducted ex situ at temperatures 95–540 °C is used especially toward volatile pesticides (e.g., DDT),[79] which are next condensed, reclaimed, or adsorbent-bonded, whereas in the extraction method, organic solvents or supercritical fluids (highly compressed gas) are used not only for the removal of volatile but also for non-volatile and non-polar pesticides such as carbosulfan.[80–82] Containment-immobilization technologies are one of the broadly used routes, which involve landfilling of hazardous waste with protective methods to prevent leaching and infiltration.[83] It is expected that most of the encapsulated and secured waste will degrade in time. To enhance decomposition, the bioremediation approach involving enzymes is used.[84] The accumulation in further steps involves supplementary methods to increase pesticides removal. More actual technologies must take into account less harmful

TABLE 1 Examples of pesticides, their removal methods and performance.

Pesticide	Removal method	Material/media/ equipment	Initial conc.	(E)/ $(Q_e)/(C_F)$	Reference
2,4-Dichlorophe-noxyacetic acid	Soil washing	5 g/L sodium dodecyl sulphate solution	–	(E): 48.7%	67
Polycyclic aromatic hydrocarbons	Thermal desorption	Horizontal tubular reactor	0.989 ppm in 2-mm soil particle	(C_F): 0.201 ppm	68
Tebuconazole	Supercritical extraction	CO_2/MeOH	–	(E): ~70%	69
Hexachloro-cyclohexane	Adsorption	Nitrogenous porous carbon (OAMPC-700)	700 µg/mL	(Q_e): 751.6 mg/g	70
Fenitrothion	Adsorption	Active-extruded MOF UiO-66	0.028 g/L	(Q_e): 0.37 mmol/g	71
Malathion	Immobilization/ enzymatic degradation	Carboxyleste–rase in SBA-15 silica sieves	–	(E): ~60%	72
Lindane	Gamma radiation	Water/ethanol	0.2 mg/L$_{(aq.)}$ 0.25 g/L$_{(EtOH)}$	(E): 100%/ 88%	73
1,2,4-Trichlorobenzene	Advanced oxidation under electrokinetic technology	$Na_2S_2O_8$ in soil treatment reactor	–	(E): 88.05%	74
3 Phenoxybenzoic acid	Fungi bioremediation	Aspergillus niger YAT	100 mg/L	(E): 100%	75

E, removal efficiency; Q_e, adsorbent sorption capacity toward pollutant; C_F, final concentration of the pesticide after treatment.

for the environment ways and materials, thereby porous adsorbent, especially biochar,[85] porous carbons,[86] polymeric adsorbents,[87] and metal-organic frameworks (MOFs)[88] are more often considered. Porous material with tunable textural parameters, if carefully developed, can act as selective and highly efficient adsorbent, which in addition can be readily reused after desorption of contaminants (e.g., organochlorines and -phosphates).[89]

Less common methods tend to be more complex and costly; nevertheless, their development is necessary especially when conventional methods fail to remove resistant pesticides. Such methods include Fenton advanced oxidation processes toward, e.g., nitrogen-containing pesticides elimination from aquatic systems, for instance, oxamyl, cymoxanil, pyrimethanil, dimethoate, and telone.[90,91] Using various radiation is particularly effective for POPs. Radiolysis of water utilizes a beam of accelerated electrons or gamma-/X-rays which generate free radicals responsible for pesticide degradation (e.g., chlordane, DDT, endrin, dioxins).[73,92] A high increase of the removal efficiency was achieved for electrokinetic technology with the addition of various surfactants, e.g., Tween 90 or dicarboxymethyl glutamic acid tetrasodium.[93] Organochlorine pesticides indicated better decomposition with the use of non-ionic surfactant TritonX-100 in the oxidation process under electrokinetic technology. The biggest limitations in common oxidation technologies are hydrophobicity and low solubility of pollutants, hence applying specific co-solvent resulted in nearly 34% higher elimination efficiency.[74] A final example that is more sympathetic to ecosystems is soil and water remediation with the use of enzymes,[94] bacteria,[95] algae,[96] and fungi.[97] Bioremediation by microorganisms occurs when they are able to proceed biosorption, bioaccumulation, and biodegradation by pesticide-decomposing chemicals (e.g., laccase, peroxidase). Moreover, the bioremoval of contaminants may be further extended in obtaining value products (biochar, biodiesel); therefore, it seems to be an alternative of most worthy consideration.[98]

The strong toxicity of pesticides is a major contributor to the immense number of disorders, often called civilization diseases, which commonly are not associated with the intense development of the modern world. When choosing a suitable pollutant removing method, time and cost, along with soil erosion, nutrient leaching, and soil fertility loss have to be taken into account. Ideally, prior to the process of pest removal, consideration should be given to the least harmful pesticide that will be biodegradable to safe compounds. It also should be easily disposed of as not to significantly affect the quality of ecosystems. However, the use of highly irritating chemicals will almost always have human health consequences, so vividly developing agriculture will contribute negatively to modern civilization if damage-neutralizing methods do not counteract excessive pollution on an ongoing basis. Nevertheless, pesticides high toxicity causing severe and often irreversible diseases, and examples of their use for self-injury must prompt the search for much less harmful means of pest control.

3 Dyes as pollutants

Dyes represent a wide range of organic contaminants that are introduced into water bodies. They come from industries such as textile, pharmaceutical, cosmetic, food, paper, tannery, or photographic, with the textile industry being the main source of water pollution (Fig. 3). Approximately 10,000 various dyes find application in these fields. The annual worldwide production reaches 700,000 t, among which 200,000 t lost to effluents come from the textile industry.[10] Because of the toxicity and carcinogenicity of dyes and their degradation products, which contain, e.g., benzamine and naphthalene, discharging them into water reservoirs causes a lot of undesirable effects. The presence of dyes in water blocks accesses to light

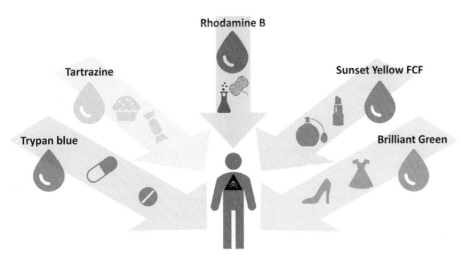

FIG. 3 Scheme showing the negative effect of dyes from different industries on human organism.

affecting the photosynthesis process, and thus reducing the amount of dissolved oxygen. Moreover, due to their non-biodegradability, they accumulate in sediments and living organisms such as algae or fish. Exposure to dyes may lead to skin irritation, allergies, or tissue changes and their absorption through skin, lungs, or digestive tract is a high potential risk for human health.[99–101]

3.1 Impact of dyes on human health

One of the greatest risks carried by dyes is mutagenicity. It is related to inducting permanent and transmissible modifications in the structure or amount of genetic material in the organisms' cells. A particular gene as well as a block of genes and chromosomes can be involved in these changes. A lot of dyes (e.g., Azure B,[102] Trypan blue,[103] Disperse Red 1,[102] Disperse Orange 1,[102] Alizarin Yellow GG,[104] methyl red,[104] methyl orange[104]) are recognized because of their mutagenic activity.[105] Azure B, a dye frequently used in the textiles production, intercalates with DNA double helix structure and duplex RNA. It also shows cytotoxic activity by being a reversible inhibitor of monoamine oxidase A (central nervous system enzyme), which influences human behavior, and glutathione reductase that is crucial in cellular redox homeostasis.[102] Trypan blue is widely used in industries such as food, textile, and paint. Due to the fused aromatic ring structure, it has high resistance to microbial degradation. This dye is known to be a mutagen, hence its release into the environment affects human health.[103]

Mutagenic agents are among factors that promote carcinogenesis. Dyes that exhibit carcinogenic activity are mostly of azo and nitro type (e.g., Sudan I,[102] Basic Red 9,[102] Congo red,[106] Direct Blue 6,[106] Direct Black 38,[106] Direct Brown 95,[106] Solvent Yellow 1,[107] Acid Red 26,[107] Disperse Blue 1,[107] Disperse Orange 37[107]). The Sudan I dye, despite the fact that its application in the food industry is illegal, is still used in products such as paprika. As a result of the action of human gut microbiota, it is enzymatically converted into carcinogenic aromatic amines. These compounds are also breakdown products of Basic Red 9 under anaerobic conditions, hence disposal of this dye in water carries the risk of tumors and sarcomas in the mammary glands, liver, bladder, and hematopoietic system.[102] Carcinogenic effects can be detected in benzidine-containing dyes such as Congo red, Direct Blue 6, Direct Black 38, and Direct Brown 95. Even though their production has declined, they are still in use. Benzidine can induce urinary bladder, lung, liver, pancreatic, bile duct, gallbladder, colorectal, stomach, and genitourinary tract cancer as well as non-Hodgkin lymphoma.[106] Furthermore, metal atoms such as chromium, cobalt, nickel, and copper can be incorporated, mainly into acid dyes molecules, forming metal-complex dyes. This family of dyes exhibits high resistance and their half-lives range from 2 to 13 years. They accumulate in fish gills and are transferred through the food chain to human organs leading to serious diseases considering their carcinogenic activity.[102]

Genotoxicity is considered the biggest possible future threat to human health.[102] Some of the dyes, including Disperse Blue 291,[108] amaranth,[109] erythrosine,[109] and tartrazine[109] are suspected to show genotoxicity. Genotoxic agents cause point mutations as well as damages within DNA molecules in locations like DNA adducts, cross-links, and abasic sites. Severe chromosomal disorders such as aneuploidy, clastogenicity, or chromosomal rearrangements may also develop.[110] One of the dyes whose genotoxic effect was investigated is Disperse Blue 291. The results showed that the contact with the studied dye caused the cell viability to decrease. Disperse Blue 291 leads to DNA fragmentation as well as the increase of apoptotic index in human hepatoma cell lines HepG2.[108]

Skin and eyes itchiness, contact dermatitis, rhinitis, conjunctivitis, asthma, and further allergic reactions may be the result of the oral ingestion of dyes such as tartrazine,[111] Sunset Yellow FCF,[111] Crystal violet,[112] Allura Red AC,[113] Brilliant Blue FCF,[114] or Brilliant Green.[115] The reason is the reaction between the dye molecules and human serum albumin. Thereby, a conjugate acting as the antigen is generated. It binds to immunoglobulin E found on the surface of human mast cells causing the release of histamine.[102] An example of a dye that induces histamine releasing is tartrazine. After the ingestion via the oral route, asthma exacerbation may occur to people suffering from this disease. Tartrazine can cause hay fever, itching around the facial area, rash, and inflammatory skin conditions as well. Another dye influencing human health is Sunset Yellow FCF which is found to trigger hypersensitivity and allergic reactions. It has already been banned in countries such as Finland and Norway.[111] Although triphenylmethane dyes are among the most popular dermatological agents, they can lead to undesirable symptoms. Crystal violet is toxic toward mammalian cells, thus exposure to it may lead to painful light sensitization, eye irritation, or permanent conjunctiva and cornea damage.[112]

Dyes can also be the reason for other unpleasant conditions. The effects of being exposed to tartrazine are insomnia, headaches, depression, and concentration problems.[111] Inhalation, ingestion, or skin absorption of methyl yellow may cause hepatomegaly, hematuria, frequent urination, and disorders in kidney functions.[106] Allura Red AC, a food dye applied in red-colored beverages, is found to induce migraine, diarrhea, and hyperactivity in children's behavior among others. For this reason, the European law allows Allura Red AC to be used only in concentrations not exceeding 100 µg/mL.[116]

3.2   Dyes removal methods

Since using dyes entails numerous disadvantages, the development of effective methods for their removal is a very important issue. Recently, physicochemical methods (such as adsorption, coagulation, flocculation, oxidation, ion exchange) and biological methods (e.g., biological oxidation) have been applied in water treatment. Due to the cost-effectiveness and the fact that it is easy to conduct, adsorption is considered the most efficient technique among all.[117] Karthikeyan et al.[118] used hydroxyapatite-decorated reduced graphene oxide (HA@rGO) in the removal of trypan blue (TB) and Congo red (CR). Maximum adsorption capacities of 146.51 mg/g and 150.09 mg/g were obtained for TB and CR, respectively. Activated carbon is considered one of the most effective adsorbents and is frequently used in the adsorption of diverse dyes (e.g., Sunset Yellow FCF,[119] tartrazine,[120] Azure B,[121] Brilliant Green).[122] Streit et al.[123] described the adsorption of Allura Red AC and Crystal violet. Sorption capacities toward these dyes were as follows: 287.1 mg/g and 640.7 mg/g. The removal of dyes by chemical oxidation is beneficial because dye compounds undergo degradation and no sludge is produced.[124] This method has been applied to dyes such as methylene blue,[125] methyl blue,[125] Rhodamine B,[126] methyl orange,[127] and Direct Blue 71.[128] Coagulation and flocculation are other methods for water decolorization that can be used, e.g., in Congo red,[129,130] Reactive Black 5,[131] or Remazol Brilliant Blue R[132] removal. Different examples of various effective methods for removing dyes such as tartrazine and Sunset Yellow FCF are summarized in Table 2.

TABLE 2 Examples of dyes, their removal methods and performance.

Dye	Removal method	Material	Initial conc.	(E)/(Q$_e$)	Reference
Tartrazine	Adsorption	Ordered mesoporous carbon (5 Ce/C$_{KIT-6}$)	100 mg/L	(Q$_e$): 171.20 mg/g (E): 66%	111
Tartrazine	Adsorption	Zinc-aluminum layered double hydroxide	240 mg/L	(Q$_e$): 282.48 mg/g	133
Tartrazine	Adsorption	Lanthanum enriched aminosilane-grafted mesoporous carbon (La/C$_{KIT-6-A}$)	250 mg/L	(Q$_e$): 210.31 mg/g	134
Sunset Yellow FCF	Adsorption	Ordered mesoporous carbons (5 Ce/C$_{KIT-6}$)	100 mg/L	(Q$_e$): 323.91 mg/g (E): 97%	111
Sunset Yellow FCF	Adsorption	Activated Gbafilo (*Chrysobalanus icaco*) shell	150 mg/L	(Q$_e$): 19.65 mg/g (E): 87.33%	135
Sunset Yellow FCF	Adsorption	N-methyl pyrrolidinium-based Polymeric Ionic Liquid (PILPyr$^+$ AA-TFSI$^-$)	100 mg/L	(Q$_e$): 284.1 mg/g	136
Malachite Green	Adsorption	Magnetic Graphene Oxide decorated with persimmon tannin Fe$_3$O$_4$/PT/GO	150 mg/L	(Q$_e$): 560.58 mg/g (E): 52.2%	137
Malachite Green	Adsorption	Biochar modified with Zero Valent Iron nanoparticles (nZVI/BC)	150 mg/L	(Q$_e$): 515.77 mg/g	138
Malachite Green	Adsorption	Oxidized mesoporous carbon (OMC)	500 mg/L	(Q$_e$): 963.1 mg/g	139
Basic Red 46	Adsorption	Bentonite	80 mg/L	(Q$_e$): 776 mg/g (E): 75%	140
Basic Red 46	Adsorption	Activated Bentonite	25 mg/L	(E): 97%	141
Brilliant Green	Adsorption	Magnetic rice husk ash	200 mg/L	(E): 96.65%	142
Brilliant Green	Adsorption	Hydroxyapatite-chitosan composite (HAp-CS)	80 mg/L	(E): 100%	143

Continued

TABLE 2 Examples of dyes, their removal methods and performance—cont'd

Dye	Removal method	Material	Initial conc.	(E)/(Q$_e$)	Reference
Congo red	Adsorption	MgO-SiO$_2$ composites	1000 mg/L	(Q$_e$): 4000 mg/g	144
Congo red	Adsorption	Coconut residual fiber (CRF)	4000 mg/L	(Q$_e$): 128.94 mg/g (E): 64.42%	145
Rhodamine B	Adsorption	Cobalt nanoparticles-embedded magnetic ordered mesoporous carbon (Co/OMC)	200 mg/L	(Q$_e$): 468 mg/g	146
Rhodamine B	Electro Fenton oxidation	Ferric chloride	50 mg/L	(E): 98%	147
Orange G	Adsorption	N-methyl pyrrolidinium-based Polymeric Ionic Liquid (PILPyr$^+$ AA-TFSI$^-$)	100 mg/L	(Q$_e$): 182.5 mg/g	136
Acid Orange 7	Adsorption	N-methyl pyrrolidinium-based Polymeric Ionic Liquid (PILPyr$^+$ AA-TFSI$^-$)	100 mg/L	(Q$_e$): 247 mg/g	136
Methylene blue	Electrocatalytic degradation	PbO$_2$-ZrO$_2$ nanocomposite electrodes	30 mg/L	(E): 100%	148
Reactive Blue 19	Coagulation	Polyaluminum chloride	100 mg/L	(E): 91%	149

E, removal efficiency; Q_e, adsorbent sorption capacity toward pollutant.

In the present section, the negative effects of synthetic dyes on human health have been highlighted. These compounds exhibit mutagenic and carcinogenic activity, which is the cause of numerous diseases such as liver, bladder, or lung cancer. Moreover, exposure to dyes leads to exacerbation of asthma and allergy symptoms (e.g., hay fever, skin, and eyes itchiness) as well as other undesirable conditions like headaches or problems with concentration. Thus, the second part of this section focuses on emphasizing the importance of the removal processes of dyes from water bodies. The adsorption method is the most commonly used because of its effectiveness, hence it is crucial to look for adsorbents that can reduce environmental pollution and the resulting negative impact on human health.[150]

4 Heavy metals as pollutants

Heavy metals are one of the most common contaminants and the problem of their presence in the environment is widespread.[151] Despite the lack of a specific definition of heavy metal, it is described in the literature as a natural element with a high atomic mass and density, five times higher than the density of water.[152] Between the 35 natural existing metals, 23 are characterized by high specific density above 5 g/cm^3 and atomic weight higher than 40.04 u.[153] The group of heavy metals contains elements required for normal metabolic processes and physiological and biochemical functions in the body as well as elements whose even low concentrations are very dangerous to humans and animals. Iron (Fe), manganese (Mn), copper (Cu), zinc (Zn), molybdenum (Mo) (micronutrients) belong to the first group, whereas elements such as cadmium (Cd), arsenic (As), mercury (Hg), lead (Pb), and nickel (Ni) to the second group.[151] Heavy metals may originate from both natural and anthropogenic sources. In the case of the first one, natural emissions of heavy metals occur, i.e., during volcanic eruptions, sea-salt sprays, forest fires, and rock weathering. Anthropogenic pollutions could derive from agriculture, industries, wastewater, mining, refineries, and metallurgical processes. Automobile exhaust releases lead, smelting—arsenic, zinc and copper, insecticides—arsenic and burning of fossil fuels—nickel, vanadium, mercury, selenium, and tin. They considerably promote heavy metal environmental contamination.[152] The major environmental components affected by heavy metal pollution are soil, water, and air. Mercury, cadmium, and lead are very common pollutants in the air and they originate from different industrial branches, e.g., metal smelters, cement factories, or natural processes such as soil erosion, rock weathering, and volcanic eruptions.[150] The soils were considered major sinks of heavy metals discharged into the environment via leaded gasoline and paints, mine tailings, industrial activities, land application of fertilizers, sewage sludge, pesticides, animal manures, and spillage of petrochemicals. Their total concentration is maintained for a long time after

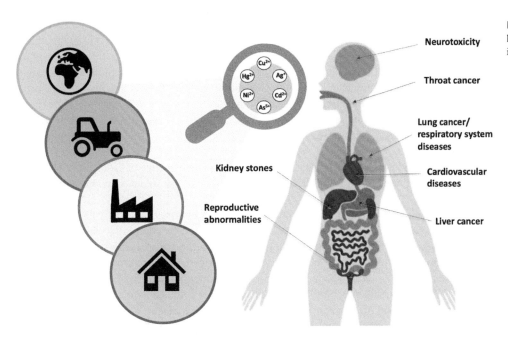

FIG. 4 The major sources of heavy metals pollution and their influence on human health.

release into the environment because they do not degrade.[152,154] Despite a large number of sources of water pollution with heavy metals, their greatest amounts get into the water by runoff from industries and urban areas as well as sediment in the water reservoirs.[152]

Heavy metal ions can accumulate in different parts of plants and contaminate water, which results in harmful consequences on human health (Fig. 4). In addition, they can be found in the body organs, i.e., kidneys, heart, liver, and brain, impairing their normal biological functioning. The most hazardous heavy metals, causing the greatest damage to the body include lead, aluminum, nickel, iron, cadmium, mercury, and arsenic.[155] Among the possible ways by which they can get into the body, skin, inhalation routes, and intake of heavy metals through contaminated water and food should be mentioned.[156] Due to their presence in the body, heavy metals are starting to be used as substitutes for essential elements, an example of which might be zinc substituted by cadmium or calcium substituted by lead. They also have a long biological half-time—17–30 years for cadmium, 25 years for lead, and 30–60 days for mercury. After getting into the body, they may be distributed in the blood throughout the body and accumulate in soft tissues, brain, heart, liver, kidneys, and bones.[156–158]

4.1   Impact of heavy metals on human health

Heavy metals such as cadmium, arsenic, lead, and nickel have a strong influence on the respiratory system. Inhalation of Cd fumes can result in serious damage to the respiratory system what can lead to shortness of breath, lung edema, enlarged pulmonary septa, small areas of atelectasis, and disruption of the mucous membranes.[159] Milton and co-workers'[160] results show that the ingestion of arsenic can contribute to pulmonary effects, manifested by shortness of breath, chest sounds, or coughing. Chronic inhalational exposure to lead is the reason for respiratory system disorders such as pneumoconiosis, bronchitis, and asthma.[159,161] Nickel has a long lung retention time and can accumulate in this area over time.[161]

The presence of heavy metals as cadmium and lead in the environment can attribute to neurological pathologies. Additionally, children are most exposed to this type of disease which may result in learning deficits, lower vocabulary and grammatical-reasoning abilities, or changes in perceptual ability. Moreover, Cd can lead to damage of the central nervous system (CNS) during the early phase of neonatal development. It can cross the placental barrier and affect fetal tissue causing injurious effects including neurotoxicity.[158,159]

Mercury, nickel, lead, cadmium, and arsenic can have an impact on the reproductive system. Cadmium, mercury, and lead decrease density, volume, vigor, and a number of sperms as well as libido and fertility. Exposure to nickel results in a low weight of newborns. Moreover, cadmium adversely affects the ovary, prostate, and placenta.[162] In females, mercury levels were connected with spontaneous abortion, malformations, and enhanced rates of menstrual disorders.[163] High blood lead levels (30–39 μg) may be the reason for a postnatal developmental delay.[164] Lead could also contribute to the decreased weight of testes and accessory sex organs.[165] In turn, Souza et al.[166] showed that exposure to arsenic increases nitrosative stress in the testis and epididymis as well as changes the testosterone level in organs previously injured by diabetes. Additionally, Ni, Zn, and Cu could take part in the pathogenesis of polycystic ovary syndrome (PCOS) connected to the reproductive hormone levels.[167]

Bones are a very susceptible site to the accumulation of heavy metals, i.e., Cd, Co, Pb, and Cr. Chromium and cadmium are able to affect the activity of osteoblasts via calcium accretion. Even low cadmium level results in a high risk of osteomalacia, osteoporosis, and bone fracture due to their increased fragility.[168] Long-term exposure to Cd leads to bone tissue loss.[169] Fleury and co-workers[170] revealed that accumulation of Co may enhance oxidative stress in bone. The presence of metals, such Pb, Cr, and Cd, could promote the evolution of osteoporosis at both cellular/molecular and epigenetic levels.[171]

Heavy metals such as cadmium, arsenic, and mercury accumulate in the kidneys. The presence of cadmium in the renal cortex may be the reason for end-stage renal disease and the formation of kidney stones.[172] It induces apoptosis in vivo in kidneys.[169]

Heavy metals are known as carcinogenic agents to human health. Lim et al.[173] observed associations between the presence of arsenic and zinc and prostate cancer risk. This type of cancer is also caused by cadmium, which can be the reason for malignant transformations in human prostatic epithelial cells.[174] Due to the competition of cadmium with estradiol for binding to estrogen receptors, cadmium could be perceived as estrogen in the uterus. It can affect the androgen-estrogen balance and increase the level of testosterone, which is associated with a high risk for the development of breast cancer. Exposure to cadmium, chromium, arsenic, beryllium, lead, and nickel could be associated with the risk of lung cancer,[175–179] whereas cadmium, mercury, and nickel increase the incidence of liver cancer.[180–182] Skin cancer is the most frequent form of neoplasm associated with arsenic and nickel intake. Nickel chloride in conjunction with UV radiation (UVR) can induce skin tumor incidence.[175,183] Exposure to lead, cadmium, arsenic, and aluminum plays an important role in the growth of malignant tissues in the bladder.[184] It was also observed that even low levels of arsenic, lead, cadmium, and mercury may result in gastric cancer.[181,185] Inhalation of nickel attributes to throat cancer, and increased concentrations of cadmium, cobalt, mercury, and lead in gallstones were associated with cancer in the gallbladder.[186,187]

Various studies confirm that the connection between exposure to arsenic and cadmium could have diabetogenic effects on the pancreas. It could distort glucose metabolism. Schwartz et al.[188] showed an association between fasting glucose level and the presence of cadmium and arsenic in urine.

Nickel, cobalt, and chromium are some of the most common reasons for allergic contact dermatitis. They are commonly used in jewelry, watches, coins, buttons, and zippers. After skin contact with metallic nickel, a rash might be observed.[187] The presence of chromium increases the risk of dermatitis and the severity of hand eczema.[189]

Furthermore, heavy metals can affect the human brain. Khalil and co-workers[190] revealed that even small amounts of lead in bones can lead to decreased levels of cognitive functions. It was also found that it could decrease verbal and non-verbal memory.[191] A higher concentration of arsenic in children who live in areas contaminated by As results in low verbal intelligence and neurobehavioral performance as well as an increased risk of reduced intelligence quotient scores.[192] Exposure to cadmium during childhood has an impact on boys' cognitive functions and lower intelligence.[193,194]

4.2 Heavy metals removal methods

Recently, the removal of heavy metals became a very important issue for scientists. Their elimination could be achieved by using different treatment methods. They include, e.g., adsorption, chemical precipitation, ion exchange as well as membrane separation. The adsorption process is commonly used for the removal of heavy metal ions from wastewater due to its availability, low cost, and eco-friendliness. Cr(VI) ions can be efficiently removed in the adsorption process using graphene sand composite.[195] The sorption capacity of the material was extremely high—2859.38 mg/g. Pb(II), Hg(II), and Cu(II) ions adsorption efficiency achieved 86.4, 85.9, and 87.6%, respectively, using functionalized magnetic graphene oxide (EDTA-mGO) as adsorbent which could be recycled five times.[196] Fe-MoS$_4$ material shows enormous removal abilities of Ag(I), Hg(II), and Pb(II) ions—99.99%. It is also characterized by excellent reusability and selectivity even in attendance of a large concentration of cations.[197] Metal-organic frameworks are promising adsorbents for As ions removal. Bimetallic Fe/Mg metal-organic frameworks (Fe/Mg-MIL-88B) demonstrated high sorption capacity toward arsenic—303.6 mg/g. The material was also characterized by structural stability and good reusability after five adsorption-desorption cycles.[198] Another MOF material used for arsenic adsorption was UiO-66, which achieved a sorption capacity of 303.34 mg/g.[199] Removal of Ni(II) ions from water could be conducted efficiently by oxidized porous carbon (PC-KF)—99.59% of nickel ions was adsorbed after 5 min.[200] In turn, multi-walled carbon nanotubes (MWCNTs) are able to eliminate cadmium ions from water solutions. They exhibit a sorption capacity of 181.8 mg/g.[201] Examples of efficient adsorbents of various heavy metals are presented in Table 3.

TABLE 3 Examples of heavy metals, their removal methods and efficiency.

Heavy metal	Removal method	Material	Initial conc.	(E)/(Q$_e$)	Reference
Cr(VI)	Adsorption	Microalgal biochar	10 mg/L	(Q$_e$): 25.19 mg/g (E): 100%	202
Cr(VI)	Adsorption	Graphene sand composite	1000 mg/L	(Q$_e$): 2859.4 mg/g	195
Cr(VI)	Adsorption	Ppy-Fe$_3$O$_4$/rGO	48.4 mg/L	(Q$_e$): 293.3 mg/g	203
Pb(II)	Adsorption	Functionalized magnetic graphene oxide (EDTA-mGO)	100 mg/L	(Q$_e$): 508.4 mg/g (E): 86.4%	196
Pb(II)	Adsorption	Fe-MoS$_4$	9.94 ppm	(Q$_e$): 345 mg/g (E): 99.99%	197
Pb(II)	Adsorption	Synthetic hydroxyapatite (Sy-HA)	16.5 ppm	(E): 95.52%	204
Hg(II)	Adsorption	Functionalized magnetic graphene oxide (EDTA-mGO)	100 mg/L	(Q$_e$): 268.4 mg/g (E): 85.9%	196
Hg(II)	Adsorption	Fe-MoS$_4$	9.86 ppm	(Q$_e$): 582 mg/g (E): 99.99%	197
Hg(II)	Adsorption	MoS$_2$/MMT	250 mg/L	(Q$_e$): 1836 mg/g	205
Hg(II)	Adsorption	BioMOF	10 ppm	(Q$_e$): 900 mg/g (E): 99.95%	206
Cu(II)	Adsorption	Functionalized magnetic graphene oxide (EDTA-mGO)	100 mg/L	(Q$_e$): 301.2 mg/g (E): 87.6%	196
Cu(II)	Adsorption	ZIF-8	420 mg/L	(Q$_e$): 800 mg/g	207
Ag(I)	Adsorption	Fe-MoS$_4$	10.23 ppm	(Q$_e$): 565 mg/g (E): 99.99%	197
As(V)	Adsorption	Fe/Mg-MIL-88B	10 ppm	(Q$_e$): 303.6 mg/g (E): 97%	198
As(V)	Adsorption	UiO-66	50 ppm	(Q$_e$): 303.34 mg/g	199
Ni(II)	Adsorption	Oxidized porous carbon (PC-KF)	100 mg/L	(E): 99.59%	200
Ni(II)	Adsorption	Fe$_3$O$_4$-NH$_2$	100 mg/L	(Q$_e$): 222.12 mg/g (E): 96%	208
Cd(II)	Adsorption	KIT-6-SH	5 mg/L	(E): 98%	209
Cd(II)	Adsorption	Multi-walled carbon nanotubes (MWCNTs)	100 ppm	(Q$_e$): 181.8 mg/g (E): 94.5%	201
Cd(II)	Adsorption	N-rich COF	80 mg/L	(Q$_e$): 369 mg/g (E): 98%	210
Cd(II)	Adsorption	Synthetic hydroxyapatite (Sy-HA)	15.4 ppm	(E): 90.9%	204

E, removal efficiency; Q_e, adsorbent sorption capacity toward pollutant.

Exposure to heavy metals causes a wide range of toxic effects in humans, which could be both chronic and acute. The long-term influence of heavy metals on the human body may result in neurological and psychical degenerative processes. The illnesses most often induced by heavy metals are neurological, gastric, reproductive, and respiratory system disease, as well as various types of cancer. However, a wide range of water, air, and soil purification techniques allow removing these pollutants from the environment efficiently and safely. Among different methods, adsorption is one of the simplest and most economical techniques for wastewater purification.

5 Drugs as pollutants

The growth of the population results in continuously increasing global drugs production and consumption. Once pharmaceuticals have been used, they are released to the environment in the form of animal and human excretions (feces and urine) and wastes (Fig. 5). Unused or expired medications are very often improperly disposed of and end up in landfills or they are tossed directly into the toilet. Emission of veterinary drugs is also a huge problem since they can enter the environment directly, via feces of animals kept on pasture as well as indirectly, via soil leaching. Excreted drugs are often unchanged or just slightly transformed. Up to 2014, about 631 drugs limits were exceeded in lakes, seas, soil, groundwaters, and rivers. Their sustainability and ability to accumulate are different but even their trace elements could result in changes in the environment due to their biological activity. Hence, pharmaceuticals are perceived as contaminants dangerous for human health.[211–213] Various groups of medications represent hazardous environmental pollutants. Among them, antibiotics, cytostatic drugs, therapeutic pharmaceuticals, and analgesic drugs can be distinguished. Antibiotics can stay in the environment for a long time and they have the potential to accumulate there. Additionally, they are nonbiodegradable. Their worldwide yearly consumption reaches 200,000 t. Cytostatic drugs are a group of medications commonly used in chemotherapy for cancer treatment. The main sources of these drugs are household discharge and hospital effluents. They can be found in drinking water and soil in quantities from 19×10^{-4} mg/L to 2.12×10^{-4} mg/L. They were classified by EU Commission Decision 2000/532/EC251 as dangerous wastes. The concentrations of drugs detected in wastes are affected by various factors such as the type of used drugs, number of patients, sampling storage, as well as excretion speed.[213] The group of analgesic drugs consists of narcotics (opioids), non-narcotics (acetaminophen), and non-steroidal anti-inflammatory

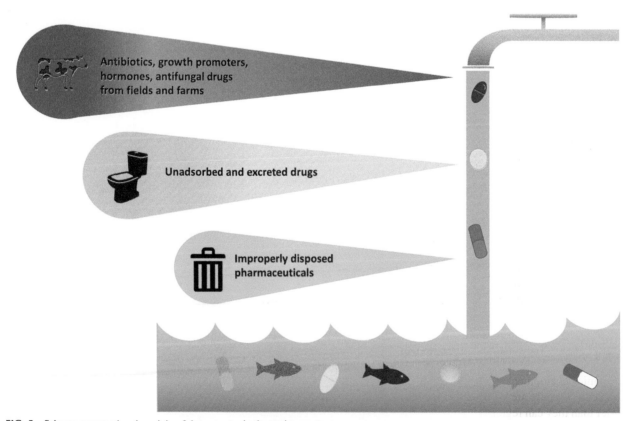

FIG. 5 Scheme representing the origin of drugs wastes in the environment.

drugs (e.g., naproxen, ketoprofen, diclofenac). They are very popular and commonly used for aches, colds, and pain treatment.[214] Hence, the matters of occurrence and impact of drugs on the environment have become the subject of global research.[215]

5.1   Impact of pharmaceuticals on human health

One of the most popular antibiotics are tetracyclines (e.g., tetracycline, chlortetracycline, oxytetracycline, doxycycline). Due to the low metabolism of these drugs in the gastrointestinal tract, over 75% are excreted in their active form in the urine or feces. Once released into the environment, tetracyclines may severely affect human health by accumulating in the food chain. Exposure to them can lead to central nervous system disorders, endocrine dysfunctions, photosensitivity, nephropathy, or problems with joints.[216,217] A group of synthetic antibiotics—fluoroquinolones, among which are ciprofloxacin and norfloxacin, might permeate to aquatic environment indirectly influencing human health. This is due to the fact that bacteria become more resistant to them and thus are more dangerous to people.[218] Carfentanil, a mu-opioid agonist, is used in veterinary medicine. This drug can enter the human body through skin wounds or mucous membranes and cause hypoventilation, respiratory failure, cardiovascular collapse, depression, lethargy, dizziness, and pinpoint pupils.[219] Another medicament that is found in water is acetylsalicylic acid (aspirin). Direct contact can cause irritation of the eyes, skin, or upper respiratory tract, while after chronic ingestion the gastrointestinal bleeding may occur.[220] Acetaminophen (paracetamol) is a medicine with an analgesic effect that is resistant to wastewater treatment methods. Hence, it can get to the human organisms by oral route and lead to chronic hepatitis, centrilobular necrosis, hepatic necrosis, or renal diseases. Moreover, it was reported that acetaminophen may form a covalent bond with DNA and thus trigger chromosomal aberrations.[221] Special focus should be given to steroid estrogens, natural and synthetic hormones, which are discharged into water bodies and may occur in drinking water. Exposure to these compounds can be the reason for breast and testicular cancer as well as fertility decline in men. The presence of anticancer pharmaceuticals in drinking water is also dangerous, mostly to pregnant women. Some cytotoxic agents, e.g., fluorouracil, penetrate the blood-placental barrier and lead to embryotoxic or teratogenic effects.[222]

5.2   Pharmaceuticals removal methods

Various techniques of water treatment were examined in order to improve pharmaceutical uptake from water (Table 4). Among them, flotation, sedimentation, coagulation, membrane technology, adsorption, and filtration are worth to be mentioned. The adsorption process in the removal of pharmaceuticals is characterized by the lack of by-products, low costs, and simple design. Activated carbons were widely used in the uptake of antibiotics. In 2017, Ahmed et al. described this type of adsorbent whose maximum sorption capacities were 1340.8, 638.6, and 570.4 mg/g toward oxytetracycline, norfloxacin, and amoxicillin, respectively.[230] Another activated carbon material obtained from the rice hull was also an efficient adsorbent in the removal of paracetamol, ibuprofen, and aspirin from hospital wastewater.[231] Metal-organic frameworks are considered good materials for pharmaceutical removal as well. MIL-53(Al) exhibits great sorption capacity toward nitroimidazole antibiotic—dimetridazole—467.3 mg/g and maintains about 98% of initial capacity after four cycles of adsorption processes.[232] Sulfachlorpyridazine antibiotic was also successfully removed from water using Cu-based MOF—HKUST-1 (384 mg/g at 25 °C).[233] Similar results were achieved through adsorption of the same drug onto different MOF—UiO-66.[227] Membrane technology became a very interesting topic in wastewater treatment due to its low economical costs and high efficiency. This method was exploited, e.g., for adsorption of cytostatic pharmaceuticals as well as non-steroidal anti-inflammatory drugs and antibiotics.[234,235] A coagulation-flotation method was successfully applied in the removal of tetracycline, fluoroquinolone, and different acidic and basic pharmaceuticals from water (95%–99%).[236]

Drugs discharged into water reservoirs can be the reason for numerous human diseases. Problems with the liver, kidneys, nervous, respiratory, and cardiovascular systems may occur after exposure to them. Hence, it is important to apply various removal methods such as adsorption or membrane technology, which can diminish the number of drugs in water and thereby contribute to the reduction of adverse impact on human health.

6   Conclusions

Dynamic development of industry and fast-growing urbanization are responsible for the rapid pollution of the natural environment. Not only are we producing waste at unsustainable rates, but we are also consuming the planet's natural resources faster than they can regenerate. Various types of hazardous, toxic chemicals, and their degradation products are discharged along with sewage into underground and surface reservoirs, soil, and air. Therefore, it is not surprising that in recent years

TABLE 4 Examples of drugs, their removal methods, and efficiency.

Drug	Removal method	Material	Initial conc.	(E)/(Q$_e$)	Reference
Tetracycline	Adsorption	Alkali-acid modified magnetic biochar (AAMS-biochar)	200 mg/L	(E): 86%	223
Acetylsalicylic acid	Adsorption	Tyre waste adsorbent (TWA)	100 mg/L	(Q$_e$): 40.4 mg/g	224
Acetaminophen	Adsorption	Amine functionalized super para-magnetic silica nanocomposite (NiFe$_2$O$_4$@SiO$_2$@APTS)	12 mg/L	(E): 94%	225
Ciprofloxacin	Adsorption	Bentonite	20 mg/L	(E): 91%	226
Ibuprofen	Adsorption	Amine functionalized superpara-magnetic silica nanocomposite (NiFe$_2$O$_4$@SiO$_2$@APTS)	12 mg/L	(E): 97%	225
Sulfachloropyridazine	Adsorption	UiO-66	45 mg/L	(Q$_e$): 403 mg/g	227
Diclofenac	Adsorption	N-biochar	10 mM	(Q$_e$): 372 mg/g	228
Naproxen	Adsorption	N-biochar	10 mM	(Q$_e$): 290 mg/g	228
Caffeine	Adsorption	Activated carbon	200 mg/L	(Q$_e$): 363.63 mg/g	229
Tannic acid	Adsorption	Activated carbon	200 mg/L	(Q$_e$): 349.65 mg/g	229

E, removal efficiency; *Q$_e$*, adsorbent sorption capacity toward pollutant.

there has been a huge increase in the incidence of cancers, allergies, neurological diseases, asthma, or irritation of skin and eyes in the human population. Diseases that previously did not occur at all or were very rare begin to spread not only in adults but also in children, affecting their development and perceptive skills. Importantly, this phenomenon will continue if the awareness of the need to take care of our planet and the environment does not increase, even in poor and underdeveloping communities.

In many countries, legal regulations on particular forms and sources of emissions are more restrictive from year to year stimulating research on new technologies that can selectively collect and eliminate different contaminants. According to one possible approach, the industry should limit the production of harmful substances by introducing new, so-called pure technologies. Another way to counterpart this phenomenon is directed at the purification of exhaust gases and wastewater by adsorption, absorption, or catalytic decomposition. Collaboration between industry and academia can bring significant benefits in terms of finding the perfect method to remove certain types of toxic chemicals, such as pesticides. Great progress in these fields has been achieved thanks to the development of functional materials with high surface areas, exceptional mechanical strength, and chemical mutability, and sufficient thermal stability. Carbon materials and MOFs are already known to be efficient adsorbents with high selectivity for many contaminants of interest, such as synthetic dyes, heavy metal ions, pesticides, phosphates, arsenates, fluorides, antibiotics, and gases (including CO_2, CH_4, SO_2, and NO_2). There is an ever-growing list of studies showing that these porous materials can be widely used in water and air purification processes. Therefore, their full potential should now be exploited to protect the environment.

References

1. Mudu P, Terracin B, Martuzzi M. *Human Health in Areas with Industrial Contamination*. WHO Regional Office for Europe; 2014.
2. Carré F, Caudeville J, Bonnard R, Bert V, Boucard P, Ramel M. Soil contamination and human health: a major challenge for global soil security. In: *Global Soil Security*. Springer; 2017:275–295. https://doi.org/10.1007/978-3-319-43394-3_25.

3. Steffan JJ, Brevik EC, Burgess LC, Cerdà A. The effect of soil on human health: an overview. *Eur J Soil Sci.* 2018;69(1):159–171. https://doi.org/10.1111/ejss.12451.

4. Manisalidis I, Stavropoulou E, Stavropoulos A, Bezirtzoglou E. Environmental and health impacts of air pollution: a review. *Front Public Health.* 2020;8:14. https://doi.org/10.3389/fpubh.2020.00014.

5. Rani L, Thapa K, Kanojia N, et al. An extensive review on the consequences of chemical pesticides on human health and environment. *J Clean Prod.* 2021;283. https://doi.org/10.1016/j.jclepro.2020.124657, 124657.

6. Tkaczyk A, Mitrowska K, Posyniak A. Synthetic organic dyes as contaminants of the aquatic environment and their implications for ecosystems: a review. *Sci Total Environ.* 2020;717. https://doi.org/10.1016/j.scitotenv.2020.137222, 137222.

7. Mishra S, Bharagava RN, More N, et al. Heavy metal contamination: an alarming threat to environment and human health. In: *Environmental Biotechnology: For Sustainable Future.* Springer; 2019:103–125. https://doi.org/10.1007/978-981-10-7284-0_5.

8. Larsson DGJ. Pollution from drug manufacturing: review and perspectives. *Philos Trans R Soc Lond B Biol Sci.* 2014;369(1656). https://doi.org/10.1098/rstb.2013.0571, 20130571.

9. Domingo-Echaburu S, Dávalos LM, Orive G, Lertxundi U. Drug pollution & sustainable development goals. *Sci Total Environ.* 2021;800. https://doi.org/10.1016/j.scitotenv.2021.149412, 149412.

10. Karri RR, Ravindran G, Dehghani MH. Wastewater—sources, toxicity, and their consequences to human health. In: *Soft Computing Techniques in Solid Waste and Wastewater Management.* Elsevier; 2021:3–33. https://doi.org/10.1016/B978-0-12-824463-0.00001-X.

11. US EPA O of PP. *Types of Pesticides*; 2020. http://www.epa.gov/opp00001/about/types.htm.

12. Rajmohan KS, Chandrasekaran R, Varjani S. A review on occurrence of pesticides in environment and current technologies for their remediation and management. *Indian J Microbiol.* 2020;60(2):125–138. https://doi.org/10.1007/s12088-019-00841-x.

13. Han MA, Kim JH, Song HS. Persistent organic pollutants, pesticides, and the risk of thyroid cancer: systematic review and meta-analysis. *Eur J Cancer Prev.* 2019;28(4):344–349. https://doi.org/10.1097/CEJ.0000000000000481.

14. Dirinck E, Jorens PG, Covaci A, et al. Obesity and persistent organic pollutants: possible obesogenic effect of organochlorine pesticides and polychlorinated biphenyls. *Obesity.* 2011;19(4):709–714. https://doi.org/10.1038/oby.2010.133.

15. Harner T, Shoeib M, Diamond M, Stern G, Rosenberg B. Using passive air samplers to assess urban-rural trends for persistent organic pollutants. 1. Polychlorinated biphenyls and organochlorine pesticides. *Environ Sci Technol.* 2004;38(17):4474–4483. https://doi.org/10.1021/es040302r.

16. Pehkonen SO, Zhang Q. The degradation of organophosphorus pesticides in natural waters: a critical review. *Crit Rev Environ Sci Technol.* 2002;32 (1):17–72. https://doi.org/10.1080/10643380290813444.

17. Van Der Werf HMG. Assessing the impact of pesticides on the environment. *Agric Ecosyst Environ.* 1996;60(2–3):81–96. https://doi.org/10.1016/S0167-8809(96)01096-1.

18. Dehghani MH, Omrani GA, Karri RR. Solid waste—sources, toxicity, and their consequences to human health. In: *Soft Computing Techniques in Solid Waste and Wastewater Management.* Elsevier; 2021:205–213. https://doi.org/10.1016/B978-0-12-824463-0.00013-6.

19. Stoytcheva M. *Pesticides: Formulations, Effects, Fate.* BoD–Books on Demand; 2011.

20. Stenersen J. *Chemical Pesticides Mode of Action and Toxicology.* CRC Press; 2004. https://doi.org/10.1201/9780203646830.

21. Cox C, Surgan M. Unidentified inert ingredients in pesticides: implications for human and environmental health. *Environ Health Perspect.* 2006;114 (12):1803–1806. https://doi.org/10.1289/ehp.9374.

22. Gaines TB. Acute toxicity of pesticides. *Toxicol Appl Pharmacol.* 1969;14(3):515–534. https://doi.org/10.1016/0041-008X(69)90013-1.

23. Dawson AH, Eddleston M, Senarathna L, et al. Acute human lethal toxicity of agricultural pesticides: a prospective cohort study. *PLoS Med.* 2010;7 (10). https://doi.org/10.1371/journal.pmed.1000357.

24. Garry VF. Pesticides and children. *Toxicol Appl Pharmacol.* 2004;198(2):152–163. https://doi.org/10.1016/j.taap.2003.11.027.

25. Boedeker W, Watts M, Clausing P, Marquez E. The global distribution of acute unintentional pesticide poisoning: estimations based on a systematic review. *BMC Public Health.* 2020;20(1):1–19. https://doi.org/10.1186/s12889-020-09939-0.

26. Blair A, Ritz B, Wesseling C, Freeman LB. Pesticides and human health. *Occup Environ Med.* 2015;72(2):81–82. https://doi.org/10.1136/oemed-2014-102454.

27. Kim KH, Kabir E, Jahan SA. Exposure to pesticides and the associated human health effects. *Sci Total Environ.* 2017;575:525–535. https://doi.org/10.1016/j.scitotenv.2016.09.009.

28. Bonner MR, Alavanja MCR. Pesticides, human health, and food security. *Food Energy Secur.* 2017;6(3):89–93. https://doi.org/10.1002/fes3.112.

29. Mamane A, Raherison C, Tessier JF, Baldi I, Bouvier G. Environmental exposure to pesticides and respiratory health. *Eur Respir Rev.* 2015;24 (137):462–473. https://doi.org/10.1183/16000617.00006114.

30. Hernández AF, Parrón T, Alarcón R. Pesticides and asthma. *Curr Opin Allergy Clin Immunol.* 2011;11(2):90–96. https://doi.org/10.1097/ACI.0b013e3283445939.

31. Rastogi SK, Tripathi S, Ravishanker D. A study of neurologic symptoms on exposure to organophosphate pesticides in the children of agricultural workers. *Indian J Occup Environ Med.* 2010;14(2):54–57. https://doi.org/10.4103/0019-5278.72242.

32. Peiris-John RJ, Ruberu DK, Wickremasinghe AR, Van der Hoek W. Low-level exposure to organophosphate pesticides leads to restrictive lung dysfunction. *Respir Med.* 2005;99(10):1319–1324. https://doi.org/10.1016/j.rmed.2005.02.001.

33. Karami-Mohajeri S, Ahmadipour A, Rahimi HR, Abdollahi M. Adverse effects of organophosphorus pesticides on the liver: a brief summary of four decades of research. *Arh Hig Rada Toksikol.* 2017;68(4):261–275. https://doi.org/10.1515/aiht-2017-68-2989.

34. Hassanin NM, Awad OM, El-Fiki S, Abou-Shanab RAI, Abou-Shanab ARA, Amer RA. Association between exposure to pesticides and disorder on hematological parameters and kidney function in male agricultural workers. *Environ Sci Pollut Res.* 2018;25(31):30802–30807. https://doi.org/10.1007/s11356-017-8958-9.

35. Jayasumana C, Gajanayake R, Siribaddana S. Importance of arsenic and pesticides in epidemic chronic kidney disease in Sri Lanka. *BMC Nephrol.* 2014;15(1):1–5. https://doi.org/10.1186/1471-2369-15-124.

36. Dreher D, Junod AF. Role of oxygen free radicals in cancer development. *Eur J Cancer.* 1996;32(1):30–38. https://doi.org/10.1016/0959-8049(95)00531-5.

37. Abdollahi M, Ranjbar A, Shadnia S, Nikfar S, Rezaie A. Pesticides and oxidative stress: a review. *Med Sci Monit.* 2004;10(6):RA141–RA147.

38. Mena S, Ortega A, Estrela JM. Oxidative stress in environmental-induced carcinogenesis. *Mutat Res Genet Toxicol Environ Mutagen.* 2009;674(1–2):36–44. https://doi.org/10.1016/j.mrgentox.2008.09.017.

39. Lombardi C, Thompson S, Ritz B, Cockburn M, Heck JE. Residential proximity to pesticide application as a risk factor for childhood central nervous system tumors. *Environ Res.* 2021;197. https://doi.org/10.1016/j.envres.2021.111078, 111078.

40. Kabir A, Zendehdel R, Tayefeh-Rahimian R. Dioxin exposure in the manufacture of pesticide production as a risk factor for death from prostate cancer: a meta-analysis. *Iran J Public Health.* 2018;47(2):148–155.

41. Shah HK, Bhat MA, Sharma T, Banerjee BD, Guleria K. Delineating potential transcriptomic association with organochlorine pesticides in the etiology of epithelial ovarian cancer. *Open Biochem J.* 2018;12(1):16–28. https://doi.org/10.2174/1874091x01812010016.

42. Shankar A, Dubey A, Saini D, et al. Environmental and occupational determinants of lung cancer. *Transl Lung Cancer Res.* 2019;8(Suppl 1):S31–S49. https://doi.org/10.21037/tlcr.2019.03.05.

43. Høyer AP, Gerdes AM, Jørgensen T, Rank F, Bøggild Hartvig H. Organochlorines, p53 mutations in relation to breast cancer risk and survival. A Danish cohort-nested case-controls study. *Breast Cancer Res Treat.* 2002;71(1):59–65. https://doi.org/10.1023/A:1013340327099.

44. Mortazavi N, Asadikaram G, Ebadzadeh MR, et al. Organochlorine and organophosphorus pesticides and bladder cancer: a case-control study. *J Cell Biochem.* 2019;120(9):14847–14859. https://doi.org/10.1002/jcb.28746.

45. Martin FL, Martinez EZ, Stopper H, Garcia SB, Uyemura SA, Kannen V. Increased exposure to pesticides and colon cancer: early evidence in Brazil. *Chemosphere.* 2018;209:623–631. https://doi.org/10.1016/j.chemosphere.2018.06.118.

46. Abolhassani M, Asadikaram G, Paydar P, et al. Organochlorine and organophosphorous pesticides may induce colorectal cancer; a case-control study. *Ecotoxicol Environ Saf.* 2019;178:168–177. https://doi.org/10.1016/j.ecoenv.2019.04.030.

47. Mahmood I, Imadi SR, Shazadi K, Gul A, Hakeem KR. Effects of pesticides on environment. In: Hakeem KR, Akhtar MS, Abdullah SNA, eds. *Plant, Soil and Microbes: Volume 1: Implications in Crop Science.* Cha Springer International Publishing; 2016:253–269. https://doi.org/10.1007/978-3-319-27455-3_13.

48. Verma H, Sharma T, Gupta S, Banerjee BD. CYP1A1 expression and organochlorine pesticides level in the etiology of bladder cancer in north Indian population. *Hum Exp Toxicol.* 2018;37(8):817–826. https://doi.org/10.1177/0960327117734623.

49. George J, Shukla Y. Pesticides and cancer: insights into toxicoproteomic-based findings. *J Proteome.* 2011;74(12):2713–2722. https://doi.org/10.1016/j.jprot.2011.09.024.

50. Narbonne JF. Pesticides and health. *Sci Aliment.* 2008;28(3):213–221. https://doi.org/10.3166/sda.28.213-221.

51. Jabłońska-Trypuć A, Wołejko E, Wydro U, Butarewicz A. The impact of pesticides on oxidative stress level in human organism and their activity as an endocrine disruptor. *J Environ Sci Health B.* 2017;52(7):483–494. https://doi.org/10.1080/03601234.2017.1303322.

52. Chorfa A, Bétemps D, Morignat E, et al. Specific pesticide-dependent increases in α-synuclein levels in human neuroblastoma (SH-SY5Y) and melanoma (SK-MEL-2) cell lines. *Toxicol Sci.* 2013;133(2):289–297. https://doi.org/10.1093/toxsci/kft076.

53. Hu L, Luo D, Zhou T, Tao Y, Feng J, Mei S. The association between non-Hodgkin lymphoma and organophosphate pesticides exposure: a meta-analysis. *Environ Pollut.* 2017;231:319–328. https://doi.org/10.1016/j.envpol.2017.08.028.

54. Rudant J, Menegaux F, Leverger G, et al. Household exposure to pesticides and risk of childhood hematopoietic malignancies: the ESCALE study (SFCE). *Environ Health Perspect.* 2007;115(12):1787–1793. https://doi.org/10.1289/ehp.10596.

55. Sabarwal A, Kumar K, Singh RP. Hazardous effects of chemical pesticides on human health—cancer and other associated disorders. *Environ Toxicol Pharmacol.* 2018;63:103–114. https://doi.org/10.1016/j.etap.2018.08.018.

56. Varghese JV, Sebastian EM, Iqbal T, Tom AA. Pesticide applicators and cancer: a systematic review. *Rev Environ Health.* 2020. https://doi.org/10.1515/reveh-2020-0121.

57. Tang BL. Neuropathological mechanisms associated with pesticides in Alzheimer's disease. *Toxics.* 2020;8(2):21. https://doi.org/10.3390/TOXICS8020021.

58. Medehouenou TCM, Ayotte P, Carmichael PH, et al. Exposure to polychlorinated biphenyls and organochlorine pesticides and risk of dementia, Alzheimer's disease and cognitive decline in an older population: a prospective analysis from the Canadian Study of Health and Aging. *Environ Health: Glob Access Sci Source.* 2019;18(1):1–11. https://doi.org/10.1186/s12940-019-0494-2.

59. Tsai HH, Yen RF, Lin CL, Kao CH. Increased risk of dementia in patients hospitalized with acute kidney injury: a nationwide population-based cohort study. *PLoS One.* 2017;12(2). https://doi.org/10.1371/journal.pone.0171671.

60. Li Y, Fang R, Liu Z, et al. The association between toxic pesticide environmental exposure and Alzheimer's disease: a scientometric and visualization analysis. *Chemosphere.* 2021;263. https://doi.org/10.1016/j.chemosphere.2020.128238, 128238.

61. Chen T, Tan J, Wan Z, et al. Effects of commonly used pesticides in China on the mitochondria and ubiquitin-proteasome system in Parkinson's disease. *Int J Mol Sci.* 2017;18(12):2507. https://doi.org/10.3390/ijms18122507.

62. Yang C, Kong APS, Cai Z, Chung ACK. Persistent organic pollutants as risk factors for obesity and diabetes. *Curr Diab Rep.* 2017;17(12):1–11. https://doi.org/10.1007/s11892-017-0966-0.

63. Wahlang B. Exposure to persistent organic pollutants: impact on women's health. *Rev Environ Health.* 2018;33(4):331–348. https://doi.org/10.1515/reveh-2018-0018.

64. Bonvoisin T, Utyasheva L, Knipe D, Gunnell D, Eddleston M. Suicide by pesticide poisoning in India: a review of pesticide regulations and their impact on suicide trends. *BMC Public Health*. 2020;20(1):1–16. https://doi.org/10.1186/s12889-020-8339-z.

65. Gunnell DJ, Eddleston M. Suicide by intentional ingestion of pesticides: a continuing tragedy in developing countries. *Int J Epidemiol*. 2003;32 (6):902–909. https://doi.org/10.1093/ije/dyg307.

66. Eddleston M, Phillips MR. Self poisoning with pesticides. *Br Med J*. 2004;328(7430):42–44. https://doi.org/10.1136/bmj.328.7430.42.

67. Bandala ER, Aguilar F, Torres LG. Surfactant-enhanced soil washing for the remediation of sites contaminated with pesticides. *Land Contam Reclam*. 2010;18(2):151–159. https://doi.org/10.2462/09670513.991.

68. Bulmău C, Mărculescu C, Lu S, Qi Z. Analysis of thermal processing applied to contaminated soil for organic pollutants removal. *J Geochem Explor*. 2014;147(PB):298–305. https://doi.org/10.1016/j.gexplo.2014.08.005.

69. Forero-Mendieta JR, Castro-Vargas HI, Parada-Alfonso F, Guerrero-Dallos JA. Extraction of pesticides from soil using supercritical carbon dioxide added with methanol as co-solvent. *J Supercrit Fluids*. 2012;68:64–70. https://doi.org/10.1016/j.supflu.2012.03.017.

70. Guo J, Yang J, Tian H, He J. High efficiency enrichment of organochlorine pesticides from water by nitrogenous porous carbon materials towards their extremely low concentration detection. *Colloids Surf A Physicochem Eng Asp*. 2021;631. https://doi.org/10.1016/j.colsurfa.2021.127728, 127728.

71. Ashouri V, Adib K, Rahimi NM. A new strategy for the adsorption and removal of fenitrothion from real samples by active-extruded MOF (AE-MOF UiO-66) as an adsorbent. *New J Chem*. 2021;45(11):5029–5039. https://doi.org/10.1039/d0nj05693f.

72. Diao J, Zhao G, Li Y, Huang J, Sun Y. Carboxylesterase from *Spodoptera litura*: immobilization and use for the degradation of pesticides. *Procedia Environ Sci*. 2013;18:610–619. https://doi.org/10.1016/j.proenv.2013.04.084.

73. Trojanowicz M. Removal of persistent organic pollutants (POPs) from waters and wastewaters by the use of ionizing radiation. *Sci Total Environ*. 2020;718. https://doi.org/10.1016/j.scitotenv.2019.134425, 134425.

74. Suanon F, Tang L, Sheng H, et al. Organochlorine pesticides contaminated soil decontamination using TritonX-100-enhanced advanced oxidation under electrokinetic remediation. *J Hazard Mater*. 2020;393. https://doi.org/10.1016/j.jhazmat.2020.122388, 122388.

75. Deng W, Lin D, Yao K, et al. Characterization of a novel β-cypermethrin-degrading *Aspergillus niger* YAT strain and the biochemical degradation pathway of β-cypermethrin. *Appl Microbiol Biotechnol*. 2015;99(19):8187–8198. https://doi.org/10.1007/s00253-015-6690-2.

76. Dehghani MH, Hassani AH, Karri RR, et al. Process optimization and enhancement of pesticide adsorption by porous adsorbents by regression analysis and parametric modelling. *Sci Rep*. 2021;11(1):11719. https://doi.org/10.1038/s41598-021-91178-3.

77. Wilson DJ, Clarke AN. Soil surfactant flushing/washing. In: *Hazardous Waste Site Soil Remediation: Theory and Application of Innovative Technologies*. CRC Press; 2017:493–550. https://doi.org/10.1201/9780203752258.

78. Lee JF, Hsu MH, Chao HP, Huang HC, Wang SP. The effect of surfactants on the distribution of organic compounds in the soil solid/water system. *J Hazard Mater*. 2004;114(1–3):123–130. https://doi.org/10.1016/j.jhazmat.2004.07.016.

79. Cong X, Li F, Kelly RM, Xue N. Distribution and removal of organochlorine pesticides in waste clay bricks from an abandoned manufacturing plant using low-temperature thermal desorption technology. *Environ Sci Pollut Res*. 2018;25(12):12119–12126. https://doi.org/10.1007/s11356-018-1422-7.

80. Tolcha T, Gemechu T, Al-Hamimi S, Megersa N, Turner C. High density supercritical carbon dioxide for the extraction of pesticide residues in onion with multivariate response surface methodology. *Molecules*. 2020;25(4):1012. https://doi.org/10.3390/molecules25041012.

81. Eskilsson CS, Mathiasson L. Supercritical fluid extraction of the pesticides carbosulfan and imidacloprid from process dust waste. *J Agric Food Chem*. 2000;48(11):5159–5164. https://doi.org/10.1021/jf000275y.

82. Hsu YC, Wu FC, Chen PS. Review of separation technologies for treating pesticide-contaminated soil. *J Air Waste Manage Assoc*. 1998;48(4):434–440. https://doi.org/10.1080/10473289.1998.10463691.

83. Morillo E, Villaverde J. Advanced technologies for the remediation of pesticide-contaminated soils. *Sci Total Environ*. 2017;586:576–597. https://doi.org/10.1016/j.scitotenv.2017.02.020.

84. Rao MA, Scelza R, Scotti R, Gianfreda L. Role of enzymes in the remediation of polluted environments. *J Soil Sci Plant Nutr*. 2010;10(3):333–353. https://doi.org/10.4067/S0718-95162010000100008.

85. Suo F, You X, Ma Y, Li Y. Rapid removal of triazine pesticides by P doped biochar and the adsorption mechanism. *Chemosphere*. 2019;235:918–925. https://doi.org/10.1016/j.chemosphere.2019.06.158.

86. Qureshi UA, Hameed BH, Ahmed MJ. Adsorption of endocrine disrupting compounds and other emerging contaminants using lignocellulosic biomass-derived porous carbons: a review. *J Water Process Eng*. 2020;38. https://doi.org/10.1016/j.jwpe.2020.101380, 101380.

87. Kyriakopoulos G, Doulia D, Anagnostopoulos E. Adsorption of pesticides on porous polymeric adsorbents. *Chem Eng Sci*. 2005;60(4):1177–1186. https://doi.org/10.1016/j.ces.2004.09.080.

88. Xu Y, Wang H, Li X, et al. Metal–organic framework for the extraction and detection of pesticides from food commodities. *Compr Rev Food Sci Food Saf*. 2021;20(1):1009–1035. https://doi.org/10.1111/1541-4337.12675.

89. Tang J, Ma X, Yang J, Feng DD, Wang XQ. Recent advances in metal-organic frameworks for pesticide detection and adsorption. *Dalton Trans*. 2020;49(41):14361–14372. https://doi.org/10.1039/d0dt02623a.

90. Bhat AP, Gogate PR. Degradation of nitrogen-containing hazardous compounds using advanced oxidation processes: a review on aliphatic and aromatic amines, dyes, and pesticides. *J Hazard Mater*. 2021;403. https://doi.org/10.1016/j.jhazmat.2020.123657, 123657.

91. Oller I, Gernjak W, Maldonado MI, Pérez-Estrada LA, Sánchez-Pérez JA, Malato S. Solar photocatalytic degradation of some hazardous water-soluble pesticides at pilot-plant scale. *J Hazard Mater*. 2006;138(3):507–517. https://doi.org/10.1016/j.jhazmat.2006.05.075.

92. Khedr T, Hammad AA, Elmarsafy AM, Halawa E, Soliman M. Degradation of some organophosphorus pesticides in aqueous solution by gamma irradiation. *J Hazard Mater*. 2019;373:23–28. https://doi.org/10.1016/j.jhazmat.2019.03.011.

93. Suanon F, Tang L, Sheng H, et al. TW80 and GLDA-enhanced oxidation under electrokinetic remediation for aged contaminated-soil: does it worth? *Chem Eng J*. 2020;385. https://doi.org/10.1016/j.cej.2019.123934, 123934.

94. Zhang H, Yuan X, Xiong T, Wang H, Jiang L. Bioremediation of co-contaminated soil with heavy metals and pesticides: influence factors, mechanisms and evaluation methods. *Chem Eng J*. 2020;398. https://doi.org/10.1016/j.cej.2020.125657, 125657.

95. Boudh S, Singh JS. Pesticide contamination: environmental problems and remediation strategies. In: *Emerging and Eco-Friendly Approaches for Waste Management*. Springer; 2018:245–269. https://doi.org/10.1007/978-981-10-8669-4_12.

96. Riaz G, Tabinda AB, Iqbal S, et al. Phytoremediation of organochlorine and pyrethroid pesticides by aquatic macrophytes and algae in freshwater systems. *Int J Phytoremediation*. 2017;19(10):894–898. https://doi.org/10.1080/15226514.2017.1303808.

97. Liapun V. *Fungi as Bioremediators*. Vol. 32. Springer Science & Business Media; 2021. https://doi.org/10.36074/logos-01.10.2021.v1.17.

98. Nie J, Sun Y, Zhou Y, et al. Bioremediation of water containing pesticides by microalgae: mechanisms, methods, and prospects for future research. *Sci Total Environ*. 2020;707. https://doi.org/10.1016/j.scitotenv.2019.136080, 136080.

99. Saini RD. Textile organic dyes: polluting effects and elimination methods from textile waste water. *Int J Chem Eng Res*. 2017;9(1):121–136.

100. Drumond Chequer FM, de Oliveira GAR, Anastacio Ferraz ER, Carvalho J, Boldrin Zanoni MV, de Oliveir DP. Textile dyes: dyeing process and environmental impact. In: *Eco-Friendly Textile Dyeing and Finishing*. InTech; 2013:151–176. https://doi.org/10.5772/53659.

101. Dahiya D, Nigam PS. Waste management by biological approach employing natural substrates and microbial agents for the remediation of dyes' wastewater. *Appl Sci*. 2020;10(8):2958. https://doi.org/10.3390/APP10082958.

102. Lellis B, Fávaro-Polonio CZ, Pamphile JA, Polonio JC. Effects of textile dyes on health and the environment and bioremediation potential of living organisms. *Biotechnol Res Innov*. 2019;3(2):275–290. https://doi.org/10.1016/j.biori.2019.09.001.

103. Lade H, Kadam A, Paul D, Govindwar S. A low-cost wheat bran medium for biodegradation of the benzidine-based carcinogenic dye trypan blue using a microbial consortium. *Int J Environ Res Public Health*. 2015;12(4):3480–3505. https://doi.org/10.3390/ijerph120403480.

104. Chung KT, Cerniglia CE. Mutagenicity of azo dyes: structure-activity relationships. *Mutat Res/Rev Genet Toxicol*. 1992;277(3):201–220. https://doi.org/10.1016/0165-1110(92)90044-A.

105. Sundar R, Jain MR, Valani D. Mutagenicity testing: regulatory guidelines and current needs. In: *Mutagenicity: Assays and Applications*. Elsevier; 2017:191–228. https://doi.org/10.1016/B978-0-12-809252-1.00010-9.

106. Chung KT. Azo dyes and human health: a review. *J Environ Sci Health C Environ Carcinog Ecotoxicol Rev*. 2016;34(4):233–261. https://doi.org/10.1080/10590501.2016.1236602.

107. Ma Q, Bai H, Zhang Q, et al. Determination of carcinogenic and allergenic dyestuffs in toys by LC coupled to UV/Vis spectrometry and tandem mass spectrometry. *Chromatographia*. 2010;72(1–2):85–93. https://doi.org/10.1365/s10337-010-1634-6.

108. Tsuboy MS, Angeli JPF, Mantovani MS, Knasmüller S, Umbuzeiro GA, Ribeiro LR. Genotoxic, mutagenic and cytotoxic effects of the commercial dye CI disperse blue 291 in the human hepatic cell line HepG2. *Toxicol in Vitro*. 2007;21(8):1650–1655. https://doi.org/10.1016/j.tiv.2007.06.020.

109. Mpountoukas P, Pantazaki A, Kostareli E, et al. Cytogenetic evaluation and DNA interaction studies of the food colorants amaranth, erythrosine and tartrazine. *Food Chem Toxicol*. 2010;48(10):2934–2944. https://doi.org/10.1016/j.fct.2010.07.030.

110. Doak SH, Liu Y, Chen C. Genotoxicity and cancer. In: *Adverse Effects of Engineered Nanomaterials: Exposure, Toxicology, and Impact on Human Health*. 2nd ed. Elsevier; 2017:423–445. https://doi.org/10.1016/B978-0-12-809199-9.00018-5.

111. Goscianska J, Marciniak M, Pietrzak R. Ordered mesoporous carbons modified with cerium as effective adsorbents for azo dyes removal. *Sep Purif Technol*. 2015;154:236–245. https://doi.org/10.1016/j.seppur.2015.09.042.

112. Mani S, Bharagava RN. Exposure to crystal violet, its toxic, genotoxic and carcinogenic effects on environment and its degradation and detoxification for environmental safety. *Rev Environ Contam Toxicol*. 2016;237:71–104. https://doi.org/10.1007/978-3-319-23573-8_4.

113. Rovina K, Siddiquee S, Shaarani SM. Extraction, analytical and advanced methods for detection of Allura Red AC (E129) in food and beverages products. *Front Microbiol*. 2016;7(May). https://doi.org/10.3389/fmicb.2016.00798.

114. Gupta VK, Mittal A, Krishnan L, Mittal J. Adsorption treatment and recovery of the hazardous dye, Brilliant Blue FCF, over bottom ash and de-oiled soya. *J Colloid Interface Sci*. 2006;293(1):16–26. https://doi.org/10.1016/j.jcis.2005.06.021.

115. Hatch KL, Maibach HI. Textile dye dermatitis. *J Am Acad Dermatol*. 1995;32(4):631–639. https://doi.org/10.1016/0190-9622(95)90350-X.

116. Diacu E, Ungureanu EM, Ivanov AA, Jurcovan MM. Erratu voltammetric techniques for determination of synthetic pigment allura red AC in beverages. *Rev Chim*. 2015;66(11):1779.

117. Shanker U, Rani M, Jassal V. Degradation of hazardous organic dyes in water by nanomaterials. *Environ Chem Lett*. 2017;15(4):623–642. https://doi.org/10.1007/s10311-017-0650-2.

118. Karthikeyan P, Nikitha M, Pandi K, Meenakshi S, Park CM. Effective and selective removal of organic pollutants from aqueous solutions using 1D hydroxyapatite-decorated 2D reduced graphene oxide nanocomposite. *J Mol Liq*. 2021;331. https://doi.org/10.1016/j.molliq.2021.115795, 115795.

119. Erdogan FO. Comparative study of sunset yellow dye adsorption onto cornelian cherry stones-based activated carbon and carbon nanotubes. *Bulg Chem Commun*. 2018;50(4):592–601.

120. Reck IM, Paixão RM, Bergamasco R, Vieira MF, Vieira AMS. Removal of tartrazine from aqueous solutions using adsorbents based on activated carbon and Moringa oleifera seeds. *J Clean Prod*. 2018;171:85–97. https://doi.org/10.1016/j.jclepro.2017.09.237.

121. Duraisamy R, Yilma B. Adsorption of azure B dye on rice husk activated carbon: equilibrium, kinetic and thermodynamic studies. *Int J Water Res*. 2015;5(2):18–28.

122. Abbas M. Removal of brilliant green (BG) by activated carbon derived from medlar nucleus (ACMN) – kinetic, isotherms and thermodynamic aspects of adsorption. *Adsorpt Sci Technol.* 2020;38(9–10):464–482. https://doi.org/10.1177/0263617420957829.

123. Streit AFM, Côrtes LN, Druzian SP, et al. Development of high quality activated carbon from biological sludge and its application for dyes removal from aqueous solutions. *Sci Total Environ.* 2019;660:277–287. https://doi.org/10.1016/j.scitotenv.2019.01.027.

124. Nidheesh PV, Zhou M, Oturan MA. An overview on the removal of synthetic dyes from water by electrochemical advanced oxidation processes. *Chemosphere.* 2018;197:210–227. https://doi.org/10.1016/j.chemosphere.2017.12.195.

125. Abdel-Aziz MH, Bassyouni M, Zoromba MS, Alshehri AA. Removal of dyes from waste solutions by anodic oxidation on an array of horizontal graphite rods anodes. *Ind Eng Chem Res.* 2019;58(2):1004–1018. https://doi.org/10.1021/acs.iecr.8b05291.

126. Di J, Zhu M, Jamakanga R, Gai X, Li Y, Yang R. Electrochemical activation combined with advanced oxidation on NiCo2O4 nanoarray electrode for decomposition of Rhodamine B. *J Water Process Eng.* 2020;37. https://doi.org/10.1016/j.jwpe.2020.101386, 101386.

127. Mais L, Vacca A, Mascia M, Usai EM, Tronci S, Palmas S. Experimental study on the optimisation of azo-dyes removal by photo-electrochemical oxidation with TiO$_2$ nanotubes. *Chemosphere.* 2020;248. https://doi.org/10.1016/j.chemosphere.2020.125938, 125938.

128. Ertugay N, Acar FN. Removal of COD and color from direct blue 71 azo dye wastewater by Fenton's oxidation: kinetic study. *Arab J Chem.* 2017;10:1158–1163. https://doi.org/10.1016/j.arabjc.2013.02.009.

129. Garvasis J, Prasad AR, Shamsheera KO, Jaseela PK, Joseph A. Efficient removal of Congo red from aqueous solutions using phytogenic aluminum sulfate nano coagulant. *Mater Chem Phys.* 2020;251. https://doi.org/10.1016/j.matchemphys.2020.123040, 123040.

130. Kristianto H, Rahman H, Prasetyo S, Sugih AK. Removal of Congo red aqueous solution using *Leucaena leucocephala* seed's extract as natural coagulant. *Appl Water Sci.* 2019;9(4):88. https://doi.org/10.1007/s13201-019-0972-2.

131. Beluci NDCL, Mateus GAP, Miyashiro CS, et al. Hybrid treatment of coagulation/flocculation process followed by ultrafiltration in TIO$_2$-modified membranes to improve the removal of reactive black 5 dye. *Sci Total Environ.* 2019;664:222–229. https://doi.org/10.1016/j.scitotenv.2019.01.199.

132. Sonal S, Singh A, Mishra BK. Decolorization of reactive dye Remazol Brilliant Blue R by zirconium oxychloride as a novel coagulant: optimization through response surface methodology. *Water Sci Technol.* 2018;78(2):379–389. https://doi.org/10.2166/wst.2018.307.

133. Ouassif H, Moujahid EM, Lahkale R, et al. Zinc-aluminum layered double hydroxide: high efficient removal by adsorption of tartrazine dye from aqueous solution. *Surf Interfaces.* 2020;18. https://doi.org/10.1016/j.surfin.2019.100401, 100401.

134. Goscianska J, Ciesielczyk F. Lanthanum enriched aminosilane-grafted mesoporous carbon material for efficient adsorption of tartrazine azo dye. *Microporous Mesoporous Mater.* 2019;280(January):7–19. https://doi.org/10.1016/j.micromeso.2019.01.033.

135. Ekuma FK, Chukwuemeka-Okorie HO, Okoyeagu A, Chimeziri CC. Studies on the adsorption of tartrazine and sunset yellow dyes from aqueous solution using activated gbafilo (*Chrysobalanus icaco*) shell. *J Chem Soc Niger.* 2019;44(5):937–947.

136. Makrygianni M, Lada ZG, Manousou A, Aggelopoulos CA, Deimede V. Removal of anionic dyes from aqueous solution by novel pyrrolidinium-based polymeric ionic liquid (pil) as adsorbent: investigation of the adsorption kinetics, equilibrium isotherms and the adsorption mechanisms involved. *J Environ Chem Eng.* 2019;7(3). https://doi.org/10.1016/j.jece.2019.103163, 103163.

137. Gao M, Wang Z, Yang C, Ning J, Zhou Z, Li G. Novel magnetic graphene oxide decorated with persimmon tannins for efficient adsorption of malachite green from aqueous solutions. *Colloids Surf A Physicochem Eng Asp.* 2019;566:48–57. https://doi.org/10.1016/j.colsurfa.2019.01.016.

138. Eltaweil AS, Ali Mohamed H, Abd El-Monaem EM, El-Subruiti GM. Mesoporous magnetic biochar composite for enhanced adsorption of malachite green dye: characterization, adsorption kinetics, thermodynamics and isotherms. *Adv Powder Technol.* 2020;31(3):1253–1263. https://doi.org/10.1016/j.apt.2020.01.005.

139. Zhang X, Lin Q, Luo S, Ruan K, Peng K. Preparation of novel oxidized mesoporous carbon with excellent adsorption performance for removal of malachite green and lead ion. *Appl Surf Sci.* 2018;442:322–331. https://doi.org/10.1016/j.apsusc.2018.02.148.

140. Paredes-Quevedo LC, González-Caicedo C, Torres-Luna JA, Carriazo JG. Removal of a textile azo-dye (Basic Red 46) in water by efficient adsorption on a natural clay. *Water Air Soil Pollut.* 2021;232(1):4. https://doi.org/10.1007/s11270-020-04968-2.

141. Elhadj M, Samira A, Mohamed T, Djawad F, Asma A, Djamel N. Removal of Basic Red 46 dye from aqueous solution by adsorption and photo-catalysis: equilibrium, isotherms, kinetics, and thermodynamic studies. *Sep Sci Technol.* 2020;55(5):867–885. https://doi.org/10.1080/01496395.2019.1577896.

142. Dahlan I, Zwain HM, Seman MAO, Baharuddin NH, Othman MR. Adsorption of brilliant green dye in aqueous medium using magnetic adsorbents prepared from rice husk ash. *AIP Conf Proc.* 2019;2124. https://doi.org/10.1063/1.5117077, 20017.

143. Ragab A, Ahmed I, Bader D. The removal of brilliant green dye from aqueous solution using nano hydroxyapatite/chitosan composite as a sorbent. *Molecules.* 2019;24(5):847. https://doi.org/10.3390/molecules24050847.

144. Hu M, Yan X, Hu X, Zhang J, Feng R, Zhou M. Ultra-high adsorption capacity of MgO/SiO$_2$ composites with rough surfaces for Congo red removal from water. *J Colloid Interface Sci.* 2018;510:111–117. https://doi.org/10.1016/j.jcis.2017.09.063.

145. Rani KC, Naik A, Chaurasiya RS, Raghavarao KSMS. Removal of toxic Congo red dye from water employing low-cost coconut residual fiber. *Water Sci Technol.* 2017;75(9):2225–2236. https://doi.org/10.2166/wst.2017.109.

146. Tang L, Cai Y, Yang G, et al. Cobalt nanoparticles-embedded magnetic ordered mesoporous carbon for highly effective adsorption of rhodamine B. *Appl Surf Sci.* 2014;314:746–753. https://doi.org/10.1016/j.apsusc.2014.07.060.

147. Nidheesh PV, Gandhimathi R. Electro Fenton oxidation for the removal of rhodamine B from aqueous solution in a bubble column reactor under continuous mode. *Desalin Water Treat.* 2015;55(1):263–271. https://doi.org/10.1080/19443994.2014.913266.

148. Yao Y, Zhao C, Zhao M, Wang X. Electrocatalytic degradation of methylene blue on PbO$_2$-ZrO$_2$ nanocomposite electrodes prepared by pulse electrodeposition. *J Hazard Mater.* 2013;263:726–734. https://doi.org/10.1016/j.jhazmat.2013.10.038.

149. Nateghi R, Bonyadinejad G, Amin M, Assadi A. Application of coagulation process reactive blue 19 dye removal from textile industry wastewater. *Int J Environ Health Eng.* 2013;2(1):5. https://doi.org/10.4103/2277-9183.107913.

150. Khan FSA, Mubarak NM, Tan YH, et al. A comprehensive review on magnetic carbon nanotubes and carbon nanotube-based buckypaper for removal of heavy metals and dyes. *J Hazard Mater.* 2021;413:125375. https://doi.org/10.1016/j.jhazmat.2021.125375.

151. Zwolak A, Sarzyńska M, Szpyrka E, Stawarczyk K. Sources of soil pollution by heavy metals and their accumulation in vegetables: a review. *Water Air Soil Pollut.* 2019;230(7). https://doi.org/10.1007/s11270-019-4221-y.

152. Masindi V, Muedi KL. Environmental contamination by heavy metals. In: *Heavy Metals.* IntechOpen; 2018. https://doi.org/10.5772/intechopen.76082.

153. Azeh Engwa G, Udoka Ferdinand P, Nweke Nwalo F, N. Unachukwu M. Mechanism and health effects of heavy metal toxicity in humans. In: *Poisoning in the Modern World – New Tricks for an Old Dog?* InTechOpen; 2019. https://doi.org/10.5772/intechopen.82511.

154. Li C, Zhou K, Qin W, et al. A review on heavy metals contamination in soil: effects, sources, and remediation techniques. *Soil Sediment Contam.* 2019;28(4):380–394. https://doi.org/10.1080/15320383.2019.1592108.

155. Rusyniak DE, Arroyo A, Acciani J, Froberg B, Kao L, Furbee B. Heavy metal poisoning: management of intoxication and antidotes. *EXS.* 2010;100:365–396. https://doi.org/10.1007/978-3-7643-8338-1_11.

156. Fu Z, Xi S. The effects of heavy metals on human metabolism. *Toxicol Mech Methods.* 2020;30(3):167–176. https://doi.org/10.1080/15376516.2019.1701594.

157. Park JD, Zheng W. Human exposure and health effects of inorganic and elemental mercury. *J Prev Med Public Health.* 2012;45(6):344–352. https://doi.org/10.3961/jpmph.2012.45.6.344.

158. Zhang R, Wilson VL, Hou A, Meng G. Source of lead pollution, its influence on public health and the countermeasures. *Int J Health.* 2015;2(1):1. https://doi.org/10.13130/2283-3927/4785.

159. Rehman K, Fatima F, Waheed I, Akash MSH. Prevalence of exposure of heavy metals and their impact on health consequences. *J Cell Biochem.* 2018;119(1):157–184. https://doi.org/10.1002/jcb.26234.

160. Milton AH, Hasan Z, Rahman A, Rahman M. Chronic arsenic poisoning and respiratory effects in Bangladesh. *J Occup Health.* 2001;43(3):136–140. https://doi.org/10.1539/joh.43.136.

161. Buxton S, Garman E, Heim KE, et al. Concise review of nickel human health toxicology and ecotoxicology. *Inorganics.* 2019;7(7):89. https://doi.org/10.3390/inorganics7070089.

162. Aprioku JS, Ebenezer B, Ijomah MA. Toxicological effects of cadmium during pregnancy in Wistar albino rats. *Toxicol Environ Heal Sci.* 2014;6(1):16–24. https://doi.org/10.1007/s13530-014-0183-z.

163. Henriques MC, Loureiro S, Fardilha M, Herdeiro MT. Exposure to mercury and human reproductive health: a systematic review. *Reprod Toxicol.* 2019;85(2018):93–103. https://doi.org/10.1016/j.reprotox.2019.02.012.

164. Sachdeva C, Thakur K, Sharma A, Sharma KK. Lead: tiny but mighty poison. *Indian J Clin Biochem.* 2018;33(2):132–146. https://doi.org/10.1007/s12291-017-0680-3.

165. Reshma Anjum M, Madhu P, Pratap Reddy K, Sreenivasula Reddy P. The protective effects of zinc in lead-induced testicular and epididymal toxicity in Wistar rats. *Toxicol Ind Health.* 2017;33(3):265–276. https://doi.org/10.1177/0748233716637543.

166. Souza ACF, Bastos DSS, Sertorio MN, et al. Combined effects of arsenic exposure and diabetes on male reproductive functions. *Andrology.* 2019;7(5):730–740. https://doi.org/10.1111/andr.12613.

167. Zheng G, Wang L, Guo Z, et al. Association of serum heavy metals and trace element concentrations with reproductive hormone levels and polycystic ovary syndrome in a Chinese population. *Biol Trace Elem Res.* 2015;167(1). https://doi.org/10.1007/s12011-015-0294-7.

168. Wallin M, Barregard L, Sallsten G, et al. Low-level cadmium exposure is associated with decreased bone mineral density and increased risk of incident fractures in elderly men: the MrOS Sweden study. *J Bone Miner Res.* 2016;31(4):732–741. https://doi.org/10.1002/jbmr.2743.

169. Genchi G, Sinicropi MS, Lauria G, Carocci A, Catalano A. The effects of cadmium toxicity. *Int J Environ Res Public Health.* 2020;17(11):3782. https://doi.org/10.3390/ijerph17113782.

170. Fleury C, Petit A, Mwale F, et al. Effect of cobalt and chromium ions on human MG-63 osteoblasts in vitro: morphology, cytotoxicity, and oxidative stress. *Biomaterials.* 2006;27(18):3351–3360. https://doi.org/10.1016/j.biomaterials.2006.01.035.

171. Sundukov YN. First record of the ground beetle Trechoblemus postilenatus (Coleoptera, Carabidae) in Primorskii krai. *Far East Entomol.* 2006;165(April):16. https://doi.org/10.1002/tox.

172. Prozialeck WC, Edwards JR. Mechanisms of cadmium-induced proximal tubule injury: new insights with implications for biomonitoring and therapeutic interventions. *J Pharmacol Exp Ther.* 2012;343(1):2–12. https://doi.org/10.1124/jpet.110.166769.

173. Lim JT, Tan YQ, Valeri L, et al. Association between serum heavy metals and prostate cancer risk – a multiple metal analysis. *Environ Int.* 2019;132. https://doi.org/10.1016/j.envint.2019.105109, 105109.

174. Zimta AA, Schitcu V, Gurzau E, et al. Biological and molecular modifications induced by cadmium and arsenic during breast and prostate cancer development. *Environ Res.* 2019;178. https://doi.org/10.1016/j.envres.2019.108700, 108700.

175. Martinez VD, Vucic EA, Becker-Santos DD, Gil L, Lam WL. Arsenic exposure and the induction of human cancers. *J Toxicol.* 2011;2011:13. https://doi.org/10.1155/2011/431287.

176. Cui J, Xu W, Chen J, et al. M2 polarization of macrophages facilitates arsenic-induced cell transformation of lung epithelial cells. *Oncotarget.* 2017;8(13):21398–21409. https://doi.org/10.18632/oncotarget.15232.

177. Shay E, De Gandiaga E, Madl AK. Considerations for the development of health-based surface dust cleanup criteria for beryllium. *Crit Rev Toxicol.* 2013;43(3):220–243. https://doi.org/10.3109/10408444.2013.767308.

178. Steenland K, Barry V, Anttila A, et al. A cohort mortality study of lead-exposed workers in the USA, Finland and the UK. *Occup Environ Med.* 2017;74(11):785–791. https://doi.org/10.1136/oemed-2017-104311.

179. Silvera SAN, Rohan TE. Trace elements and cancer risk: a review of the epidemiologic evidence. *Cancer Causes Control.* 2007;18(1):7–27. https://doi.org/10.1007/s10552-006-0057-z.

180. Yuan W, Yang N, Li X. Advances in understanding how heavy metal pollution triggers gastric cancer. *Biomed Res Int.* 2016;2016:1–10. https://doi.org/10.1155/2016/7825432.

181. Chen K, Liao QL, Ma ZW, et al. Association of soil arsenic and nickel exposure with cancer mortality rates, a town-scale ecological study in Suzhou, China. *Environ Sci Pollut Res.* 2015;22(7):5395–5404. https://doi.org/10.1007/s11356-014-3790-y.

182. Cartularo L, Laulicht F, Sun H, Kluz T, Freedman JH, Costa M. Gene expression and pathway analysis of human hepatocellular carcinoma cells treated with cadmium. *Toxicol Appl Pharmacol.* 2015;288(3):399–408. https://doi.org/10.1016/j.taap.2015.08.011.

183. Uddin AN, Burns FJ, Rossman TG, Chen H, Kluz T, Costa M. Dietary chromium and nickel enhance UV-carcinogenesis in skin of hairless mice. *Toxicol Appl Pharmacol.* 2007;221(3):329–338. https://doi.org/10.1016/j.taap.2007.03.030.

184. Abdel-Gawad M, Elsobky E, Shalaby MM, Abd-Elhameed M, Abdel-Rahim M, Ali-El-Dein B. Quantitative evaluation of heavy metals and trace elements in the urinary bladder: comparison between cancerous, adjacent non-cancerous and normal cadaveric tissue. *Biol Trace Elem Res.* 2016;174 (2):280–286. https://doi.org/10.1007/s12011-016-0724-1.

185. Yuan W, Yang N, Li X. Advances in understanding how heavy metal pollution triggers gastric cancer. *Biomed Res Int.* 2016;2016. https://doi.org/10.1155/2016/7825432.

186. Mondal B, Maulik D, Mandal M, Sarkar GN, Sengupta S, Ghosh D. Analysis of carcinogenic heavy metals in gallstones and its role in gallbladder carcinogenesis. *J Gastrointest Cancer.* 2017;48(4):361–368. https://doi.org/10.1007/s12029-016-9898-1.

187. Shukla VV. Secondary hypertension manifests renal artery stenosis and weakened kidney. *J Mech Med Biol.* 2011;11(1):73–100. https://doi.org/10.1142/S021951941000371X.

188. Schwartz GG, Il'Yasova D, Ivanova A. Urinary cadmium, impaired fasting glucose, and diabetes in the NHANES III. *Diabetes Care.* 2003;26 (2):468–470. https://doi.org/10.2337/diacare.26.2.468.

189. Kanto H. Metal allergy. *Otolaryngol Head Neck Surg.* 2015;87(9):684–690. https://doi.org/10.30895/2221-996x-2019-19-2-88-93.

190. Khalil N, Morrow LA, Needleman H, Talbott EO, Wilson JW, Cauley JA. Association of cumulative lead and neurocognitive function in an occupational cohort. *Neuropsychology.* 2009;23(1):10–19. https://doi.org/10.1037/a0013757.

191. Wang T, Guan RL, Liu MC, et al. Lead exposure impairs hippocampus related learning and memory by altering synaptic plasticity and morphology during juvenile period. *Mol Neurobiol.* 2016;53(6):3740–3752. https://doi.org/10.1007/s12035-015-9312-1.

192. Mondal P, Chattopadhyay A. Environmental exposure of arsenic and fluoride and their combined toxicity: a recent update. *J Appl Toxicol.* 2020;40 (5):552–566. https://doi.org/10.1002/jat.3931.

193. Gustin K, Tofail F, Vahter M, Kippler M. Cadmium exposure and cognitive abilities and behavior at 10 years of age: a prospective cohort study. *Environ Int.* 2018;113(December 2017):259–268. https://doi.org/10.1016/j.envint.2018.02.020.

194. Sanders AP, Claus Henn B, Wright RO. Perinatal and childhood exposure to cadmium, manganese, and metal mixtures and effects on cognition and behavior: a review of recent literature. *Curr Environ Health Rep.* 2015;2(3):284–294. https://doi.org/10.1007/s40572-015-0058-8.

195. Dubey R, Bajpai J, Bajpai AK. Green synthesis of graphene sand composite (GSC) as novel adsorbent for efficient removal of Cr (VI) ions from aqueous solution. *J Water Process Eng.* 2015;5:83–94. https://doi.org/10.1016/j.jwpe.2015.01.004.

196. Cui L, Wang Y, Gao L, et al. EDTA functionalized magnetic graphene oxide for removal of Pb(II), Hg(II) and Cu(II) in water treatment: adsorption mechanism and separation property. *Chem Eng J.* 2015;281:1–10. https://doi.org/10.1016/j.cej.2015.06.043.

197. Chen Z, Jawad A, Liao Z, et al. Fe-MoS4: an effective and stable LDH-based adsorbent for selective removal of heavy metals. *ACS Appl Mater Interfaces.* 2017;9(34):28451–28463. https://doi.org/10.1021/acsami.7b07208.

198. Gu Y, Xie D, Wang Y, et al. Facile fabrication of composition-tunable Fe/Mg bimetal-organic frameworks for exceptional arsenate removal. *Chem Eng J.* 2019;357(September 2018):579–588. https://doi.org/10.1016/j.cej.2018.09.174.

199. Wang C, Liu X, Chen JP, Li K. Superior removal of arsenic from water with zirconium metal-organic framework UiO-66. *Sci Rep.* 2015;5:1–10. https://doi.org/10.1038/srep16613.

200. Siddiqui MN, Ali I, Asim M, Chanbasha B. Quick removal of nickel metal ions in water using asphalt-based porous carbon. *J Mol Liq.* 2020;308. https://doi.org/10.1016/j.molliq.2020.113078, 113078.

201. Bhanjana G, Dilbaghi N, Kim KH, Kumar S. Carbon nanotubes as sorbent material for removal of cadmium. *J Mol Liq.* 2017;242:966–970. https://doi.org/10.1016/j.molliq.2017.07.072.

202. Daneshvar E, Zarrinmehr MJ, Kousha M, et al. Hexavalent chromium removal from water by microalgal-based materials: adsorption, desorption and recovery studies. *Bioresour Technol.* 2019;293(August). https://doi.org/10.1016/j.biortech.2019.122064, 122064.

203. Wang H, Yuan X, Wu Y, et al. Facile synthesis of polypyrrole decorated reduced graphene oxide-Fe$_3$O$_4$ magnetic composites and its application for the Cr(VI) removal. *Chem Eng J.* 2015;262(VI):597–606. https://doi.org/10.1016/j.cej.2014.10.020.

204. Ramdani A, Kadeche A, Adjdir M, et al. Lead and cadmium removal by adsorption process using hydroxyapatite porous materials. *Water Pract Technol.* 2020;15(1):130–141. https://doi.org/10.2166/wpt.2020.003.

205. Mário EDA, Liu C, Ezugwu CI, Mao S, Jia F, Song S. Molybdenum disulfide/montmorillonite composite as a highly efficient adsorbent for mercury removal from wastewater. *Appl Clay Sci.* 2020;184(November). https://doi.org/10.1016/j.clay.2019.105370, 105370.

206. Mon M, Lloret F, Ferrando-Soria J, Martí-Gastaldo C, Armentano D, Pardo E. Selective and efficient removal of mercury from aqueous media with the highly flexible arms of a BioMOF. *Angew Chem Int Ed.* 2016;55(37):11167–11172. https://doi.org/10.1002/anie.201606015.

207. Zhang Y, Xie Z, Wang Z, Feng X, Wang Y, Wu A. Unveiling the adsorption mechanism of zeolitic imidazolate framework-8 with high efficiency for removal of copper ions from aqueous solutions. *Dalton Trans.* 2016;45(32):12653–12660. https://doi.org/10.1039/c6dt01827k.

208. Norouzian Baghani A, Mahvi AH, Gholami M, Rastkari N, Delikhoon M. One-pot synthesis, characterization and adsorption studies of amine-functionalized magnetite nanoparticles for removal of Cr (VI) and Ni (II) ions from aqueous solution: kinetic, isotherm and thermodynamic studies. *J Environ Health Sci Eng.* 2016;14(1). https://doi.org/10.1186/s40201-016-0252-0.

209. Bagheri S, Amini MM, Behbahani M, Rabiee G. Low cost thiol-functionalized mesoporous silica, KIT-6-SH, as a useful adsorbent for cadmium ions removal: a study on the adsorption isotherms and kinetics of KIT-6-SH. *Microchem J.* 2019;145:460–469. https://doi.org/10.1016/j.microc.2018.11.006.

210. Dinari M, Hatami M. Novel N-riched crystalline covalent organic framework as a highly porous adsorbent for effective cadmium removal. *J Environ Chem Eng.* 2019;7(1). https://doi.org/10.1016/j.jece.2019.102907, 102907.

211. Baran W, Adamek E, Ziemiańska J, Sobczak A. Effects of the presence of sulfonamides in the environment and their influence on human health. *J Hazard Mater.* 2011;196:1–15. https://doi.org/10.1016/j.jhazmat.2011.08.082.

212. Reddersen K, Heberer T, Dünnbier U. Identification and significance of phenazone drugs and their metabolites in ground- and drinking water. *Chemosphere.* 2002;49(6):539–544. https://doi.org/10.1016/S0045-6535(02)00387-9.

213. Jureczko M, Kalka J. Cytostatic pharmaceuticals as water contaminants. *Eur J Pharmacol.* 2020;866. https://doi.org/10.1016/j.ejphar.2019.172816, 172816.

214. Felis E, Miksch K. Removal of analgesic drugs from the aquatic environment using photochemical methods. *Water Sci Technol.* 2009;60(9):2253–2259. https://doi.org/10.2166/wst.2009.668.

215. Rezka P, Balcerzak W. *Występowanie Antybiotyków W Środowisku*; 2016. https://doi.org/10.4467/2353737XCT.16.203.5952.

216. Xu L, Zhang H, Xiong P, Zhu Q, Liao C, Jiang G. Occurrence, fate, and risk assessment of typical tetracycline antibiotics in the aquatic environment: a review. *Sci Total Environ.* 2021;753. https://doi.org/10.1016/j.scitotenv.2020.141975, 141975.

217. Dehghan A, Zarei A, Jaafari J, Shams M, Mousavi Khaneghah A. Tetracycline removal from aqueous solutions using zeolitic imidazolate frameworks with different morphologies: a mathematical modeling. *Chemosphere.* 2019;217:250–260. https://doi.org/10.1016/j.chemosphere.2018.10.166.

218. Zhao J, Liang G, Zhang X, et al. Coating magnetic biochar with humic acid for high efficient removal of fluoroquinolone antibiotics in water. *Sci Total Environ.* 2019;688:1205–1215. https://doi.org/10.1016/j.scitotenv.2019.06.287.

219. Lust EB, Barthold C, Malesker MA, Wichman TO. Human health hazards of veterinary medications: information for emergency departments. *J Emerg Med.* 2011;40(2):198–207. https://doi.org/10.1016/j.jemermed.2009.09.026.

220. Schulman LJ, Sargent EV, Naumann BD, Faria EC, Dolan DG, Wargo JP. A human health risk assessment of pharmaceuticals in the aquatic environment. *Hum Ecol Risk Assess Int J.* 2002;8(4):657–680. https://doi.org/10.1080/20028091057141.

221. Blanset DL, Zhang J, Robson MG. Probabilistic estimates of lifetime daily doses from consumption of drinking water containing trace levels of N, N-diethyl-meta-toluamide (DEET), triclosan, or acetaminophen and the associated risk to human health. *Hum Ecol Risk Assess.* 2007;13(3):615–631. https://doi.org/10.1080/10807030701341209.

222. Szymonik A, Lach J, Malińska K. Fate and removal of pharmaceuticals and illegal drugs present in drinking water and wastewater. *Ecol Chem Eng S.* 2017;24(1):65–85. https://doi.org/10.1515/eces-2017-0006.

223. Zhou Y, He Y, He Y, et al. Analyses of tetracycline adsorption on alkali-acid modified magnetic biochar: site energy distribution consideration. *Sci Total Environ.* 2019;650:2260–2266. https://doi.org/10.1016/j.scitotenv.2018.09.393.

224. Azman A, Ngadi N, Zaini DKA, Jusoh M, Mohamad Z, Arsad A. Effect of adsorption parameter on the removal of aspirin using Tyre waste adsorbent. *Chem Eng Trans.* 2019;72(August 2018):157–162. https://doi.org/10.3303/CET1972027.

225. Chandrashekar Kollarahithlu S, Balakrishnan RM. Adsorption of pharmaceuticals pollutants, ibuprofen, acetaminophen, and streptomycin from the aqueous phase using amine functionalized superparamagnetic silica nanocomposite. *J Clean Prod.* 2021;294. https://doi.org/10.1016/j.jclepro.2021.126155, 126155.

226. Genç N, Dogan EC. Adsorption kinetics of the antibiotic ciprofloxacin on bentonite, activated carbon, zeolite, and pumice. *Desalin Water Treat.* 2015;53(3):785–793. https://doi.org/10.1080/19443994.2013.842504.

227. Azhar MR, Abid HR, Periasamy V, Sun H, Tade MO, Wang S. Adsorptive removal of antibiotic sulfonamide by UiO-66 and ZIF-67 for wastewater treatment. *J Colloid Interface Sci.* 2017;500:88–95. https://doi.org/10.1016/j.jcis.2017.04.001.

228. Jung C, Boateng LK, Flora JRV, et al. Competitive adsorption of selected non-steroidal anti-inflammatory drugs on activated biochars: experimental and molecular modeling study. *Chem Eng J.* 2015;264:1–9. https://doi.org/10.1016/j.cej.2014.11.076.

229. Sarici-Özdemir Ç, Önal Y. Study to observe the applicability of the adsorption isotherms used for the adsorption of medicine organics onto activated carbon. *Part Sci Technol.* 2018;36(2):254–261. https://doi.org/10.1080/02726351.2016.1246497.

230. Ahmed MJ. Adsorption of quinolone, tetracycline, and penicillin antibiotics from aqueous solution using activated carbons: review. *Environ Toxicol Pharmacol.* 2017;50:1–10. https://doi.org/10.1016/j.etap.2017.01.004.

231. Mupa MMT. Preparation of rice hull activated carbon for the removal of selected pharmaceutical waste compounds in hospital effluent. *J Environ Anal Toxicol.* 2015. https://doi.org/10.4172/2161-0525.s7-008, s7-008.

232. Peng Y, Zhang Y, Huang H, Zhong C. Flexibility induced high-performance MOF-based adsorbent for nitroimidazole antibiotics capture. *Chem Eng J.* 2018;333:678–685. https://doi.org/10.1016/j.cej.2017.09.138.

233. Azhar MR, Abid HR, Sun H, Periasamy V, Tadé MO, Wang S. Excellent performance of copper based metal organic framework in adsorptive removal of toxic sulfonamide antibiotics from wastewater. *J Colloid Interface Sci.* 2016;478:344–352. https://doi.org/10.1016/j.jcis.2016.06.032.

234. Wang X, Zhang J, Chang VWC, She Q, Tang CY. Removal of cytostatic drugs from wastewater by an anaerobic osmotic membrane bioreactor. *Chem Eng J*. 2018;339(January):153–161. https://doi.org/10.1016/j.cej.2018.01.125.

235. Schröder HF, Tambosi JL, Sena RF, Moreira RFPM, José HJ, Pinnekamp J. The removal and degradation of pharmaceutical compounds during membrane bioreactor treatment. *Water Sci Technol*. 2012;65(5):833–839. https://doi.org/10.2166/wst.2012.828.

236. Saitoh T, Shibata K, Fujimori K, Ohtani Y. Rapid removal of tetracycline antibiotics from water by coagulation-flotation of sodium dodecyl sulfate and poly(allylamine hydrochloride) in the presence of Al(III) ions. *Sep Purif Technol*. 2017;187(III):76–83. https://doi.org/10.1016/j.seppur.2017.06.036.

Chapter 3

Agriculture risks of pollutants in water and their benefits after purification

Ali Sohani[a,*], Sandra Ricart[b], Hadi Omidinasab[c], Ala Sadooghi[a], Hüseyin Yagli[d], and Hitesh Panchal[e]

[a]*Optimization of Energy Systems' Installations Lab., Faculty of Mechanical Engineering-Energy Division, K.N. Toosi University of Technology, Tehran, Iran,* [b]*Water and Territory Research Group, Interuniversity Institute of Geography, University of Alicante, Alicante, Spain,* [c]*Department of Mechanical Engineering, South Tehran Branch, Islamic Azad University, Tehran, Iran,* [d]*Gaziantep University, Gaziantep, Turkey,* [e]*Mechanical Engineering Department, Government Engineering College Patan, Patan, Gujarat, India*

[*]*Corresponding author.*

1 Introduction

Treatment of wastewater is one of the most practical solutions to exit the water crises, including lack of water resources and the environmental effect of wastewater pollutants.[1,2] Wastewater could be generally described as a mixture of liquid or water-carried waste from residential or industrial establishments.[2–4] Generally speaking, it contains a high load of waste-requiring oxygen, inorganic chemicals, pathogens, organic matter, plant growth-promoting nutrients, sediments, and minerals. It may also contain toxic compounds.[5]

The wastewater that comes from the textile industry could be given as an example for more detailed explanations.[6,7] Based on the reports published by the World Bank, the textile industry is responsible for almost 20% of global water pollution. The effluents in the utilized water in the textile industry have high levels of phosphates, suspended solids (SS), salts, dyes, non-biodegradable organics, organo-pesticides, and heavy metals.[6,8] By proper treatment, wastewater could reach the standard for the considered usage.[9]

The treated water could be utilized for various applications from which corps farming is one of the most popular ones. For example, China is one of the leading countries in wastewater treatment due to some limited resources and its population.[10]

On the one hand, since it contains more nutrients and organic materials, using treated wastewater has several benefits compared to the other possible alternatives.[11] On the other hand, it has serious challenges, including acceptance among the people and farmers. The economic condition, the used treatment technologies, and public health are among the most important factors that impact the acceptance of water treatment projects in different applications, including agriculture.

A simple schematic describing the sanitation history is provided in Fig. 1. Despite considerable progress during the past years, acceptance has been considered one of the biggest challenges to developing wastewater treatment plans. Considering this point and the high popularity in agriculture, this book chapter is prepared. It provides a framework to describe the agricultural risks of pollutants in water, their benefits after purification, and the working principle of a wastewater treatment plant. Acquiring knowledge about the indicated issues will help achieve higher progress in water treatment plans.

This chapter starts with the introduction part, i.e., Section 1. Then, the risks and benefits of using treated wastewater are explained in Section 2. Section 3 provides the details about the acceptability of wastewater treatment in society. The working principle of a typical wastewater treatment plant is introduced in Section 4, while Section 5 discusses the quantitative measures for risk assessment.

2 Deepening on farmers' perception: Risks and benefits of using treated wastewater

Water pollution and scarcity are worldwide phenomena, and two of the biggest challenges society is currently dealing with.[12] Treated wastewater is considered a practical solution for coping with this changing scenario as an alternative way for water supply and trustable water resources even in severe conditions. Additionally, reusing treated wastewater improves water management strategies twofold: increasing water resources availability and decreasing the environmental

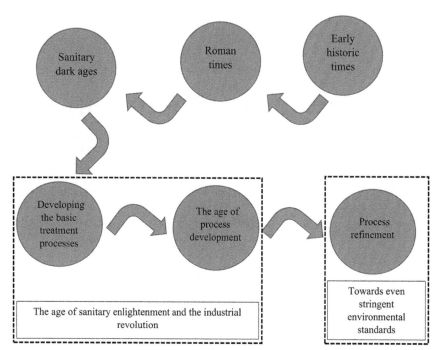

FIG. 1 A simple schematic describing the sanitation history. *(Figure was author's original, while data are from Lofrano G, Brown J. Wastewater management through the ages: a history of mankind. Sci Total Environ 2010;408(22):5254–5264. https://doi.org/10.1016/j.scitotenv.2010.07.062[2010/ 10/15].)*

effects of untreated wastewater disposal.[13] Partially integrated into water management schemes to guarantee sustainability goals, many regions (e.g., Africa, Central America, Southern Asia, Southern Europe) are exploring this non-conventional water resource in response to emerging water scarcity issues, water shortages now and in the future, and increasing pressures on water sources worldwide.[14] Treated wastewater used for crop irrigation globally increases by 10–30% per year in Europe, the United States, and China, and up to 40% in Australia.[15] Since almost 50% of the water bodies worldwide have been seriously polluted,[16] treatment of wastewater and reuse of that can boost environmental security by alleviating freshwater resources' pollution and providing water supply for irrigation, contributing to (a) reducing pressure on overstressed aquifers; (b) increase groundwater recharge, and (c) minimize the application of fertilizers due to remaining nutrients.[17] Additional advantages for agricultural use include the following: (1) a trustable nutrient's source (potassium, phosphorus, and nitrogen), and supplying organic compounds to be used in maintaining the fertility of the soil (fertigation) and making that productive; (2) a sustainable source for water provision using lower cost and energy than other alternatives (e.g.,

TABLE 1 Main benefits and risks of treated wastewater for agricultural use and the way for minimizing the risks.

Benefits	Risks	The way for minimizing
Regular availability in space and time (no seasonal or geographical affection)	Microbial pathogens	Controlling the temperature Clean water and ice utilization
Less expensive and controversial than water transfers or desalination	Micropollutants	Advanced treatment ways
Reduced over-exploitation of surface and groundwater resources	Higher soil salinity	Irradiation channels efficiency enhancement Salty drainage water treatment
Ecological flow and landscaping	Higher heavy metals concentration	Advanced treatment ways
Fertigation through LCA	Limited regulation	Correct policymaking
Crop yield productivity	Yuck factor	Correct policymaking

desalination or importing water); (3) a resource for avoiding the effects of new equipment for water supply (like dams); and (4) the decrease in the amount of contaminants discharged to river systems and the environment[18] (Table 1).

Even with these contrasted benefits, the acceptance of treated wastewater and successful implementation is not ensured among farmers due to the diverse challenges of putting this concept into practice.[19] These barriers usually come from institutional and economic deficiencies, public health, unavailability of high treatment technologies, and risks.[20] According to Chfadi et al.[21] two main reasons are explaining the attitudes toward treated wastewater reuse: the first factor is mainly rational, and it stems from the aversion to the risk of contamination, while the second factor is more psychological and prompted by the disgust toward wastewater, and repulsion toward it, the so-called "yuck factor."[22] Among the financial constraints, inadequate regulatory frameworks, or engineering issues, several drawbacks mainly associated with health risks and significant sources of pollution to natural water bodies and the environment have increased the yuck factor.[23]

Consequently, treated wastewater effluent quality standards become very important in terms of its performance to groundwater, soil, and plant, being important to adhere to the lowest quality norms to diminish public health and other environmental risks. While the feasibility of wastewater recycling has long been demonstrated and technological progress mainly ensured that water recycling is safe,[24] perceived risk to human health and the environment together with low levels of water quality standards are conditioning farmers' acceptance and willingness to support treated wastewater reuse.[25] Investigations have recommended that interventions are more likely to be successful if they are designed to incorporate the target group's attitudes, perceptions, constraints, and suggestions/knowledge.[26] In this line, some studies analyzed treated wastewater users' perception, identifying significantly higher positive attitudes toward treated wastewater application's economic, social, and environmental sustainability. However, one primary concern associated with reusing wastewater in agricultural irrigation is the high concentration of heavy metals (e.g., Cd, Ca, Mg, and Na) and their affection to irrigated crops compared with freshwater irrigated plants.[27] In most cases, the lack of local monitoring and oversight of these chemical parameters determines farmers' predisposition to use treated wastewater.[28] For example, it was estimated that more than 10% of the total cultivated area worldwide is irrigated by treated wastewater with significantly limited treatment processes,[29] while around one in 10 citizens globally consumes crops and vegetables produced with untreated wastewater.[30]

Specific socio-economic conditions and cultural factors influence farmers' rejection or acceptance according to personal experience[31] (Fig. 2). For example, Alhumoud and Madzikanda[32] surveyed public willingness to use treated wastewater for a variety of purposes, highlighting that the main accepted uses are in agriculture and landscaping as well as that education level and knowledge of wastewater reuse are the main factors affecting the respondents' choices, other factors include age, nationality, and gender. Likewise, Al-Shenaifi et al.[33] concluded that young farmers had more positive attitudes toward treated wastewater than older ones, while Al-Najar and El Hamarneh[34] confirmed that educated people reflected elevated awareness of the health effects of using treated wastewater. Other studies concluded that cost is a key parameter having impacts on farmers' decision to employ treated wastewater since farmers assume that treated wastewater has the cheaper quality and, therefore, it must have a lower cost than freshwater.[35] However, the main reason for the negative environmental attitudes toward treated wastewater can be related to insufficient or abstract knowledge and information in addition to the weak lack of social trust in information sources,[36] together with insufficient or inadequate fair procedures by the water authorities.[37] These weaknesses motivated the request for effective management of treated wastewater to mitigate public health and environmental consequences to give the confidence that reuse of wastewater is a sustainable practice or at least an acceptable risk.[38]

As public and farmers' acceptance has a vital role in the uptake and implementation of treated wastewater for agricultural use, facing main barriers and enabling factors concerning wastewater reuse schemes should be one of the main issues

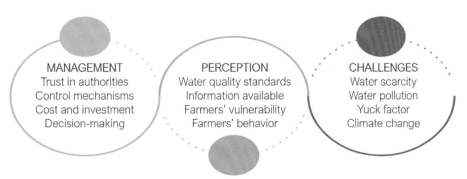

FIG. 2 Driving factors having impacts on the treated wastewater reuse.

for research works in the future. Therefore, building confidence, ensuring accountability and transparency, and encouraging farmers' participation via proper channels for communication are fundamental.[39] Likewise, to improve farmers' acceptance of treated wastewater use, some specific strategies should be addressed, including applying advanced technologies to boost wastewater quality and decrease environmental and health risks, promoting agricultural extension services, and ensuring better farm management and constant supervision.

3 Acceptability of wastewater treatment in the society

There are always two approaches to wastewater in society; the approach of researchers and environmentalists and the approach of other people. These two approaches are mainly in conflict with each other. On the one hand, the people's approach in society is that the reuse of sewage is not hygienic and contains germs and bacteria. On the other hand, researchers and environmentalists show that based on their scientific studies, the treated wastewater could be safely used in various applications.

Therefore, since the mid-19th century, cutting-edge social orders have raised the open wellbeing standard by first collecting and treating sewage water. In the past years, administrative specialists in industrialized districts have moreover tried to make strides in neighborhood accepting endeavored to improve local receiving water quality by more advanced forms of wastewater treatment, including organic supplement evacuation, etc.[40] The effluent is pre-treated at different levels to evacuate pathogens, natural matter, and supplement components, with controls and rules based on open wellbeing affirmation criteria whereas too depending on financial and innovative capabilities.[41–43] In more innovative nations, the properties and compositions of the wastewater items (emanating and slime) will decide their utilization alternatives for typical rural scenarios.[44–46]

4 The working principle of a typical wastewater treatment plant

There are several issues to consider in a wastewater treatment process. The most basic is that the pollution in the wastewater comes from two parts, and a good water treatment plant should have solutions for removing both of them. One part appears in liquid form, and another appears in the form of solid and sludge. Considering this point, a typical wastewater treatment plant is introduced here.

A typical water treatment plant is composed of different steps in two main categories; a wastewater line and a sludge collector line. They are shown in Figs. 3 and 4.[47]

The wastewater enters the treatment plant. It goes into the preliminary section and stays for a while. Then, wastewater enters some mechanical filters such as a bar screen for filtering the bigger particles and a filter for segregating grease from wastewater.

After that, wastewater enters the primary sludge area, including a huge clarifier. In the clarifier, with the aid of gravity, some solid particles settle down to the bottom. On the one hand, the collected sludge moves to the other part of the plant for sludge treatment. On the other hand, from the top of the clarifier, small particles are collected with the aid of low-velocity drag.[47]

Next, the wastewater, which includes only some chemical germs and microbial particles, remains. It goes to the secondary and the most important part of a wastewater plant to eliminate chemical compounds. This part of the plant could be named the biological treatment section. In this part, wastewater is deeply clarified with the aid of micro-organisms and their chemical reaction with bacteria and germs.

There are two subsections in the biological part. One is the aeration part which is responsible for adding air and bubbles with micro-organisms into the wastewater to incite little bacteria to start a chemical reaction and get bigger or eliminate. Another is a clarifier with the same working principle as the previously introduced clarifiers.

The third part of a wastewater treatment plant is the chemical part. It is responsible for eliminating the micro-organisms which are added in the second part due to the fact that micro-organisms are dangerous for human health. In the chemical section of the plant, with the aid of ultraviolet (UV) disinfection and chlorination/dichlorination, the goal of elimination of micro-organisms is achieved. Furthermore, the different phosphor and nitrogen which are added in the second part are removed.

So far, what happens in the wastewater line has been described. The sludge collector line is another line in a water treatment plant that collects and reclaims the sludge. At first, it seems the collected sludge is sewage or rubbish products. Nonetheless, after the proper further treatment, it could be used for different applications.[47] For example, the final product could be utilized as fuel and an energy source to overcome the existing energy crises.[48] It is used as the final product in improving soil quality and could be given as another example of applications of sludge line products.

Sludge Treatment Process Flow Diagram

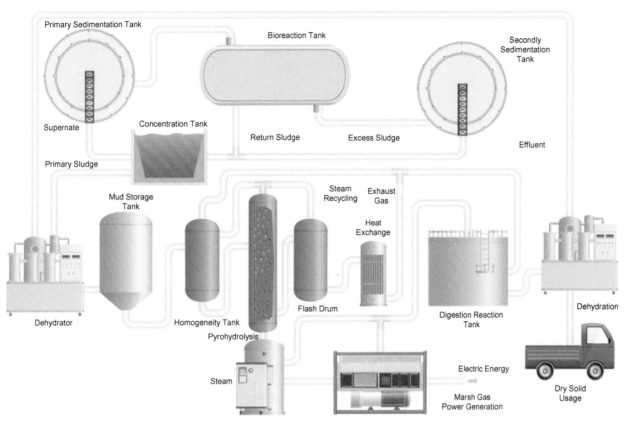

FIG. 3 Schematic of components and their connections in a treatment process plant.

5 The quantitative measures for risk assessment

The quantitative indices are required for different applications, including engineering purposes, and in wastewater treatment literature, Quantitative Microbial Risk Assessment (QMRA) is considered a powerful quantitative tool for risk assessment of treated water.[49]

Here are the four main steps to perform QMRA[50]:

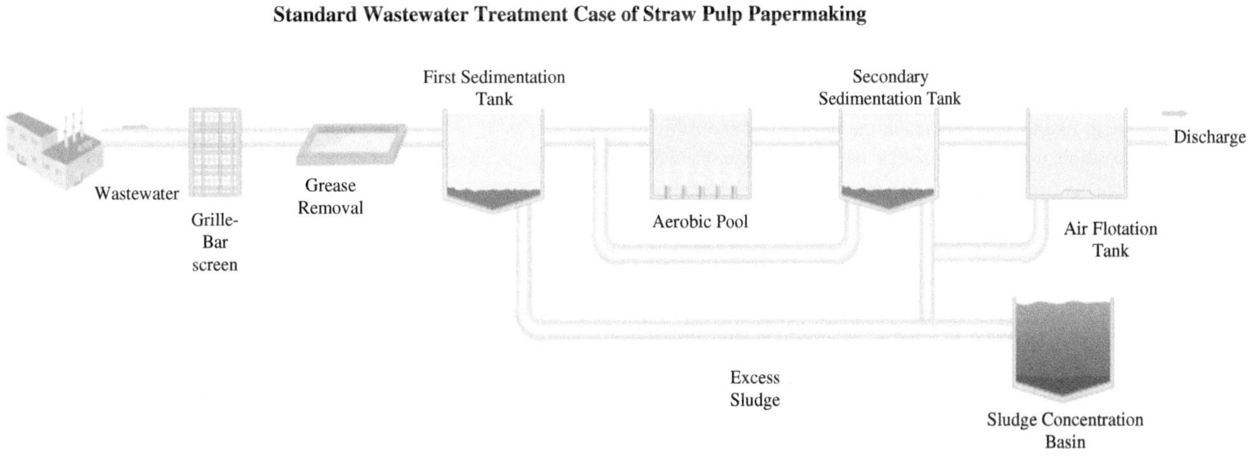

Standard Wastewater Treatment Case of Straw Pulp Papermaking

FIG. 4 Schematic of processes in a treatment process plant.

TABLE 2 The regulation limits for several micro-organisms and compounds; the values are in percent.

Scenario	Station	Pathogens	
		G. lamblia	Coli O157:H7
Farming activities	One	0.64	3.40
	Two	1.83	2.85
	Three	2.66	4.57
	Four	0.95	0.22
Consumption of salad	One	0.00	23.25
	Two	0.00	0.00
	Three	1.01	21.25
	Four	0.00	12.25

More information about QMRA could be found in the literature including the studies such as Zimmer-Faust et al.,[52] Owens et al.,[53] He and Huang,[54] and so on.

1. Hazard identification: This stage is composed of two parts. Initially, and in the first part, the investigated micro-organisms and compounds are determined. Then, and in the second part, the diseases from each micro-organism and compound are listed.
2. Exposure assessment: The amount of each micro-organism and compound in the medium (water, soil, etc.), which individuals are exposed to, is determined.
3. Dose-response: A proper dose-response model, which researchers have previously developed, is chosen for further analysis.
4. Risk characterization: Using the governing equations for the dose-response model for each case, the amount of risk is specified for the investigated dose of each micro-organism and compound.

One of the best studies in the field has been the work done by Kouamé et al.[51] The study was done for Côte d'Ivoire in the western part of Africa. The study was carried out by considering the vegetables of green salad, while sampling took place at four stations. Stations 1, 2, and 4 were the lake water, whereas, at station 3, the wastewater drainage was used. According to the details found in Ref. 51, Table 2 is obtained about the risk assessment of four stations. In this table, the probability is reported for the diarrhoeal problems. It is worth mentioning that *E. coli* O157:H7 is a bacterium, while *G. lamblia* is a micro-organism.

The provided discussions have highlighted the necessity of working more on ways to decrease the agricultural risks of pollutants in water, especially by the growing concerns that come from water scarcity and water pollution issues. Future works could find more efficient, economic, and environmentally friendly water treatment ways and find proper strategies for more acceptance of using treated water for policymakers, investors, and people in society.

Acknowledgment

The authors would like to sincerely thank the Book Editors, Professor Inderjeet Tyagi, and Professor Rama Rao Karri, for their kind support during the preparation process of the book chapter. Moreover, we would like to express our sincere gratitude to all the production team members for their great efforts to prepare the book chapter in a high-quality format.

References

1. Safari M, Sohani A, Sayyaadi H. A higher performance optimum design for a tri-generation system by taking the advantage of water-energy nexus. *J Clean Prod.* 2021;284:124704. https://doi.org/10.1016/j.jclepro.2020.124704 [2021/02/15].
2. Karri RR, Ravindran G, Dehghani MH. Wastewater—sources, toxicity, and their consequences to human health. In: *Soft Computing Techniques in Solid Waste and Wastewater Management.* Elsevier; 2021:3–33. https://doi.org/10.1016/B978-0-12-824463-0.00001-X.
3. Boukhari S, Djebbar Y, Amarchi H, Sohani A. Application of the analytic hierarchy process to sustainability of water supply and sanitation services: the case of Algeria. *Water Supply.* 2017;18(4):1282–1293. https://doi.org/10.2166/ws.2017.194.

4. Dehghani MH, Omrani GA, Karri RR. Solid waste—sources, toxicity, and their consequences to human health. In: *Soft Computing Techniques in Solid Waste and Wastewater Management*. Elsevier; 2021:205–213. https://doi.org/10.1016/B978-0-12-824463-0.00013-6.

5. Shahverdian MH, Sohani A, Sayyaadi H. Water-energy nexus performance investigation of water flow cooling as a clean way to enhance the productivity of solar photovoltaic modules. *J Clean Prod*. 2021;312:127641. https://doi.org/10.1016/j.jclepro.2021.127641 [2021/08/20].

6. Khan FSA, Mubarak NM, Tan YH, et al. A comprehensive review on magnetic carbon nanotubes and carbon nanotube-based buckypaper for removal of heavy metals and dyes. Review. *J Hazard Mater*. 2021;413:125375. https://doi.org/10.1016/j.jhazmat.2021.125375.

7. Khan FSA, Mubarak NM, Tan YH, et al. Magnetic nanoparticles incorporation into different substrates for dyes and heavy metals removal—a review. *Environ Sci Pollut Res*. 2020;27(35):43526–43541. https://doi.org/10.1007/s11356-020-10482-z.

8. Dehghani MH, Hassani AH, Karri RR, et al. Process optimization and enhancement of pesticide adsorption by porous adsorbents by regression analysis and parametric modelling. *Sci Rep*. 2021;11(1):11719. https://doi.org/10.1038/s41598-021-91178-3.

9. Donkadokula NY, Kola AK, Naz I, Saroj D. A review on advanced physico-chemical and biological textile dye wastewater treatment techniques. *Rev Environ Sci Biotechnol*. 2020;19(3):543–560. https://doi.org/10.1007/s11157-020-09543-z.

10. Xu A, Wu Y-H, Chen Z, et al. Towards the new era of wastewater treatment of China: development history, current status, and future directions. *Water Cycle*. 2020;1:80–87. https://doi.org/10.1016/j.watcyc.2020.06.004 [2020/01/01].

11. Mehmood A, Khan FSA, Mubarak NM, et al. Magnetic nanocomposites for sustainable water purification—a comprehensive review. *Environ Sci Pollut Res*. 2021;28(16):19563–19588. https://doi.org/10.1007/s11356-021-12589-3.

12. Liu J, Liu Q, Yang H. Assessing water scarcity by simultaneously considering environmental flow requirements, water quantity, and water quality. *Ecol Indic*. 2016;60:434–441. https://doi.org/10.1016/j.ecolind.2015.07.019.

13. Goyal K, Kumar A. Development of water reuse: a global review with the focus on India. *Water Sci Technol*. 2021. https://doi.org/10.2166/wst.2021.359.

14. Aleisa E, Al-Zubari W. Wastewater reuse in the countries of the Gulf Cooperation Council (GCC): the lost opportunity. *Environ Monit Assess*. 2017;189(11):1–15. https://doi.org/10.1007/s10661-017-6269-8.

15. Lissaneddine A, Mandi L, El Achaby M, et al. Performance and dynamic modeling of a continuously operated pomace olive packed bed for olive mill wastewater treatment and phenol recovery. *Chemosphere*. 2021;280:130797. https://doi.org/10.1016/j.chemosphere.2021.130797 [2021/10/01].

16. Garrone P, Grilli L, Groppi A, Marzano R. Barriers and drivers in the adoption of advanced wastewater treatment technologies: a comparative analysis of Italian utilities. *J Clean Prod*. 2018;171:S69–S78. https://doi.org/10.1016/j.jclepro.2016.02.018.

17. Jaramillo MF, Restrepo I. Wastewater reuse in agriculture: a review about its limitations and benefits. *Sustainability*. 2017;9(10):1734. https://doi.org/10.3390/su9101734.

18. Sathaiah M, Chandrasekaran M. A bio-physical and socio-economic impact analysis of using industrial treated wastewater in agriculture in Tamil Nadu, India. *Agric Water Manag*. 2020;241:106394. https://doi.org/10.1016/j.agwat.2020.106394.

19. Mohsenpour SF, Hennige S, Willoughby N, Adeloye A, Gutierrez T. Integrating micro-algae into wastewater treatment: a review. *Sci Total Environ*. 2021;752:142168. https://doi.org/10.1016/j.scitotenv.2020.142168 [2021/01/15].

20. Chai Q, Hu A, Qian Y, et al. A comparison of genotoxicity change in reclaimed wastewater from different disinfection processes. *Chemosphere*. 2018;191:335–341. https://doi.org/10.1016/j.chemosphere.2017.10.024 [2018/01/01].

21. Chfadi T, Gheblawi M, Thaha R. Public acceptance of wastewater reuse: new evidence from factor and regression analyses. *Water*. 2021;13(10):1391. https://doi.org/10.3390/w13101391.

22. Nemeroff C, Rozin P, Haddad B, Slovic P. Psychological barriers to urban recycled water acceptance: a review of relevant principles in decision psychology. *Int J Water Resour Dev*. 2020;36(6):956–971. https://doi.org/10.1080/07900627.2020.1804841.

23. Inyinbor AA, Bello OS, Oluyori AP, Inyinbor HE, Fadiji AE. Wastewater conservation and reuse in quality vegetable cultivation: overview, challenges and future prospects. *Food Control*. 2019;98:489–500. https://doi.org/10.1016/j.foodcont.2018.12.008.

24. Gatta G, Gagliardi A, Disciglio G, et al. Irrigation with treated municipal wastewater on artichoke crop: assessment of soil and yield heavy metal content and human risk. *Water*. 2018;10(3):255. https://doi.org/10.3390/w10030255.

25. Ricart S, Rico AM, Ribas A. Risk-yuck factor nexus in reclaimed wastewater for irrigation: comparing farmers' attitudes and public perception. *Water*. 2019;11(2):187. https://doi.org/10.3390/w11020187.

26. Antwi-Agyei P, Peasey A, Biran A, Bruce J, Ensink J. Risk perceptions of wastewater use for urban agriculture in Accra, Ghana. *PLoS One*. 2016;11(3):e0150603. https://doi.org/10.1371/journal.pone.0150603.

27. Sheidaei F, Karami E, Keshavarz M. Farmers' attitude towards wastewater use in Fars Province, Iran. *Water Policy*. 2016;18(2):355–367. https://doi.org/10.2166/wp.2015.045.

28. Dare A, Mohtar RH. Farmer perceptions regarding irrigation with treated wastewater in the West Bank, Tunisia, and Qatar. *Water Int*. 2018;43(3):460–471. https://doi.org/10.1080/02508060.2018.1453012.

29. Thebo AL, Drechsel P, Lambin EF, Nelson KL. A global, spatially-explicit assessment of irrigated croplands influenced by urban wastewater flows. *Environ Res Lett*. 2017;12(7):074008. https://doi.org/10.1088/1748-9326/aa75d1.

30. Khalid S, Shahid M, Bibi I, Sarwar T, Shah AH, Niazi NK. A review of environmental contamination and health risk assessment of wastewater use for crop irrigation with a focus on low and high-income countries. *Int J Environ Res Public Health*. 2018;15(5):895. https://doi.org/10.3390/ijerph15050895.

31. Ganoulis J. Risk analysis of wastewater reuse in agriculture. *Int J Recycl Org Waste Agric*. 2012;1(1):1–9. https://doi.org/10.1186/2251-7715-1-3.

32. Alhumoud JM, Madzikanda D. Public perceptions on water reuse options: the case of Sulaibiya wastewater treatment plant in Kuwait. *Int Bus Econ Res J*. 2010;9(1). https://doi.org/10.19030/iber.v9i1.515.

33. Al-Shenaifi M, Al-Shayaa M, Alharbi M. Perception and attitudes of farmers toward the uses of treated sewage water in palm trees irrigation. *Jordan J Agric Sci.* 2015;11(3):693–704.

34. Al-Najar H, El Hamarneh B. The effect of education level on accepting the reuse of treated effluent in irrigation. *Indones J Sci Technol.* 2019;4(1):28–38. https://doi.org/10.17509/ijost.v4i1.14881.

35. Deh-Haghi Z, Bagheri A, Fotourehchi Z, Damalas CA. Farmers' acceptance and willingness to pay for using treated wastewater in crop irrigation: a survey in western Iran. *Agric Water Manag.* 2020;239:106262. https://doi.org/10.1016/j.agwat.2020.106262.

36. Furlong C, Jegatheesan J, Currell M, Iyer-Raniga U, Khan T, Ball AS. Is the global public willing to drink recycled water? A review for researchers and practitioners. *Util Policy.* 2019;56:53–61. https://doi.org/10.1016/j.jup.2018.11.003 [2019/02/01].

37. Greenaway T, Fielding KS. Positive affective framing of information reduces risk perceptions and increases acceptance of recycled water. *Environ Commun.* 2020;14(3):391–402. https://doi.org/10.1080/17524032.2019.1680408.

38. Woldetsadik D, Drechsel P, Keraita B, Itanna F, Gebrekidan H. Farmers' perceptions on irrigation water contamination, health risks and risk management measures in prominent wastewater-irrigated vegetable farming sites of Addis Ababa, Ethiopia. *Environ Syst Decis.* 2018;38(1):52–64. https://doi.org/10.1007/s10669-017-9665-2.

39. Massoud MA, Terkawi M, Nakkash R. Water reuse as an incentive to promote sustainable agriculture in Lebanon: stakeholders' perspectives. *Integr Environ Assess Manag.* 2019;15(3):412–421. https://doi.org/10.1002/ieam.4131.

40. Saravanan A, Kumar PS, Varjani S, et al. A review on algal-bacterial symbiotic system for effective treatment of wastewater. *Chemosphere.* 2021;271:129540. https://doi.org/10.1016/j.chemosphere.2021.129540 [2021/05/01].

41. Ofori S, Puškáčová A, Růžičková I, Wanner J. Treated wastewater reuse for irrigation: pros and cons. *Sci Total Environ.* 2021;760:144026. https://doi.org/10.1016/j.scitotenv.2020.144026 [2021/03/15].

42. Melo W, Delarica D, Guedes A, et al. Ten years of application of sewage sludge on tropical soil. A balance sheet on agricultural crops and environmental quality. *Sci Total Environ.* 2018;643:1493–1501. https://doi.org/10.1016/j.scitotenv.2018.06.254 [2018/12/01].

43. Moretti M, Van Passel S, Camposeo S, et al. Modelling environmental impacts of treated municipal wastewater reuse for tree crops irrigation in the Mediterranean coastal region. *Sci Total Environ.* 2019;660:1513–1521. https://doi.org/10.1016/j.scitotenv.2019.01.043 [2019/04/10].

44. López A, Baguer B, Goñi P, et al. Assessment of the methodologies used in microbiological control of sewage sludge. *Waste Manag.* 2019;96:168–174. https://doi.org/10.1016/j.wasman.2019.07.024 [2019/08/01].

45. Jafari M, Botte GG. Electrochemical treatment of sewage sludge and pathogen inactivation. *J Appl Electrochem.* 2021;51(1):119–130. https://doi.org/10.1007/s10800-020-01481-6 [2021/01/01].

46. Bertanza G, Canato M, Laera G. Towards energy self-sufficiency and integral material recovery in waste water treatment plants: assessment of upgrading options. *J Clean Prod.* 2018;170:1206–1218. https://doi.org/10.1016/j.jclepro.2017.09.228 [2018/01/01].

47. Klingensmith N, Sridhar A, LaVallee Z, Banerjee S. Water or slime? A platform for automating water treatment systems. In: *Proceedings of the 2nd ACM International Conference on Embedded Systems for Energy-Efficient Built Environments*; 2015:75–84.

48. Delfani F, Rahbar N, Aghanajafi C, Heydari A, Khalesi DA. Utilization of thermoelectric technology in converting waste heat into electrical power required by an impressed current cathodic protection system. *Appl Energy.* 2021;302:117561. https://doi.org/10.1016/j.apenergy.2021.117561 [2021/11/15].

49. Gerba CP. Risk assessment. In: Pepper IL, Gerba CP, Gentry TJ, eds. *Environmental Microbiology.* 3rd ed. Academic Press; 2015:565–579. https://doi.org/10.1016/B978-0-12-394626-3.00024-7 [chapter 24].

50. *Department of Health. Quantitative Microbial Risk Assessment Basics.* https://www.health.state.mn.us/communities/environment/risk/guidance/dwec/basics.html;. Accessed 10 November 2021.

51. Kouamé PK, Nguyen-Viet H, Dongo K, Zurbrügg C, Biémi J, Bonfoh B. Microbiological risk infection assessment using QMRA in agriculture systems in Côte d'Ivoire, West Africa. *Environ Monit Assess.* 2017;189(11):1–11. https://doi.org/10.1007/s10661-017-6279-6.

52. Zimmer-Faust AG, Steele JA, Griffith JF, Schiff K. The challenges of microbial source tracking at urban beaches for quantitative microbial risk assessment (QMRA). *Mar Pollut Bull.* 2020;160:111546. https://doi.org/10.1016/j.marpolbul.2020.111546 [2020/11/01].

53. Owens CEL, Angles ML, Cox PT, Byleveld PM, Osborne NJ, Rahman MB. Implementation of quantitative microbial risk assessment (QMRA) for public drinking water supplies: systematic review. *Water Res.* 2020;174:115614. https://doi.org/10.1016/j.watres.2020.115614 [2020/05/01].

54. He X, Huang K. Assessment technologies for hazards/risks of wastewater. In: Ren H, Zhang X, eds. *High-Risk Pollutants in Wastewater.* Elsevier; 2020:141–167. https://doi.org/10.1016/B978-0-12-816448-8.00007-1 [chapter 7].

Chapter 4

Effectiveness of metal-organic framework as sensors: Comprehensive review

Diksha Praveen Pathak[a,*], Yogendra Kumar[b], and Shalu Yadav[c]

[a]*Department of Chemical and Biochemical Engineering, Rajiv Gandhi Institute of Petroleum Technology, Jais, Uttar Pradesh, India,* [b]*Department of Chemical Engineering, Indian Institute of Technology, Madras, Tamil Nadu, India,* [c]*Department of Basic Science and Humanity, Rajiv Gandhi Institute of Petroleum Technology, Jais, Uttar Pradesh, India*

[*]*Corresponding author.*

1 Introduction

Surplus emissions of harmful pollutants have become a serious and worldwide concern. These pollutants, which have high mobility and solubility, may easily disperse into the air, water, and soil, presenting a significant threat to the ecosystem, and also pollutants like heavy metal ions and organic debris account for approximately 20% of terminal illnesses in living beings.[1,2] A variety of nanomaterials and nanoframeworks have been explored in the quest for developing sustainable technologies to combat the adverse effects of environmental pollution and climate change.[3] In this regard, an efficacious system for monitoring and sensing pollutants is crucial.

Sensitivity is an important characteristic that is defined as the smallest change in an external condition that causes an observable signal alteration.[4] The construction of the sensor, associated electronics, and signal processing unit,[5] as well as the detecting material, all have an impact on sensitivity. A sensor mainly constitutes a detecting device and a transduction unit that converts the perceived data into an optical or electrical signal. A sensor works on the principle of transduction which is driven by changes in the electrical, mechanical, optical, or photophysical characteristics of the sensing material of the sensor as it contacts with the samples.[6,7] Some of the essential characteristics of a good sensor are good sensitivity, high specificity, low noise, reusability, fast response, durability, and being economically viable.[8] Generally, micro- and nanomaterials are utilized in sensing environmental pollutants[9] such as metals and their oxides,[6] graphene and carbon nanotubes,[10,11] polymers,[12] quantum dots (QDs),[13] and semiconductors.[14]

One of the potential options for monitoring pollutants is sensors based on MOFs. MOFs are a unique type of porous structure made up of metal nodes and organic linkers. MOFs are now a rapidly developing area as their structures and operations can be easily tuned according to the desired applications. The structural properties and characteristics of MOFs can be prognosticated utilizing an amalgamation of specific metal ions/clusters and customized organic ligands.[15] As a result, MOFs might serve as a versatile platform for a variety of applications, such as catalysis,[16] luminescence,[17] gas adsorption and separation,[18] drug delivery,[19] optics,[20] proton conduction,[21] and magnetism.[22] MOFs are also highly switchable toward external stimuli. The modular characteristic and high porosity of MOFs make them a promising candidate for their utilization in devices used for sensing. MOFs are sensitive even when there is a minor variation in concentration to impact the host–guest relationship. MOFs can function as molecular sieves in sensors as their porosity can be easily tuned.[23] Depending on the size, shape, and structure of MOFs, nanoscale pores in MOFs can enable the passage of small molecules into the pore or across the MOFs layer while blocking the diffusion of bigger molecules. In the instance of IRMOF-16, the free volume inside MOFs reaches 91.1% of the total volume, while for most of the open zeolites, it would not exceed 30%–40%.[24] Unlike zeolites, which usually have a spherical slit, and elliptical pores, MOFs also have square, rectangular, and triangular pores.[25,26] Due to such a variety of pore geometries, specific adsorption and monitoring may be achieved more efficiently.[27] Many pollutants like heavy metals, organic components, and poisonous gases have been detected by careful observations of the different physical and chemical changes that occur in a MOF when a guest molecule is adsorbing onto it.[28] In current years, numerous MOFs have been created based on the diverse principle of operation for prospective application as magnetic sensors, ferroelectric sensors, chemiresistive sensors, luminous sensors, and colorimetric sensors.[2,29] Recent works are more focused on new sensing techniques like turn-on mechanisms rather than conventional turn-off mechanism for luminescent sensors, and multifunctional MOF-based sensors due to their improved sensitivity and selectivity.

In this chapter, the detailed working mechanism of MOF-based sensors and specific properties of MOFs that are utilized in a particular type of sensor are discussed. Different synthesis methods of MOFs with their advantages and limitations, the recent trends and future perspectives in this field have been discussed in sections below. The problems and potential of this research area are also examined in depth, providing deep insight into the future development of improved MOF materials for next-generation MOF-predicated sensors.

2 MOF-based sensors & sensing techniques: Working process and mechanism

2.1 Electrochemical sensors and sensing techniques

The redox reactions of the sample matrix in an electrochemical process are the prime driving force in an electrochemical sensor. Due to the enhanced catalytic activity, large surface area, high porosity, and superior absorbability of surface modifiers, they are commonly used for electrochemical sensing. It is difficult to construct electrochemical MOF-based sensing devices since a majority of MOFs are insulators. The standard Zn-MOF-5 (terephthalate) has shown semiconducting properties in one of the studies.[30] Certain MOFs with significant electroconductivity have also been discovered in recent studies.[31,32] MOFs which are less conductive may also be utilized in sensing the changes in impedance,[33] work function,[34] and capacitance.[35] Some MOFs having high capacitance can be used to detect compounds that affect this parameter like alcohols. When it comes to the sensing of VOCs, the Cu-BTC MOF-199 material has demonstrated encouraging results.[36] Also, H_2S[37] and ammonia[38] can be effectively controlled using a strategy similar to this one.

MOFs can also effectively function as chemiresistive sensors.[39] The chemiresistive sensor's sensing process is mostly ascribed to the movement of electrons or holes induced by interactions and adsorptions of analyte gas with the surface of the sensor. As a result, the electronic properties of the chemiresistive sensor change on its reaction with the adsorbed gases. The selective adsorbing and penetrating characteristics of MOFs provide a viable answer to the selectivity problem in chemiresistive sensors.

Field-effect transistors (FETs) are also excellent sensing devices. FETs produced using MOFs offer fascinating properties, including scalability, resilience, and flexibility with scalable manufacturing techniques.[40] MOFs lacking substantial conductivity can be used for vapor analysis, and the Kelvin probe method may be used for detecting variations in the work function.

2.2 Luminescent sensors and sensing techniques

Luminescence property of MOFs can be attributed to the following factors: luminescence due to organic ligands having expanded π-clouds, emission from metal-centers of transition metal (d^{10}), lanthanides and clusters of silver metal, luminescence due to dyes and lanthanides trapped inside the porous structure of MOFs and emission from charge transfer.[15] The phenomenon of luminescence happens when electrons from excited singlet levels emit photons and come back to the lowest energy state. This phenomenon of luminescence by MOF is quenched upon analyte absorption and is termed the "turn-off" technique.[41] Luminescent MOFs have been the most frequently researched form of MOF sensor due to their distinct features. The porosity guarantees analytes adsorption inside the MOF cavities, and effective contact can take place between host framework and guest analytes, including hydrogen bonding, van der Waals force, π-π interactions, halogen bonding, and the controlled porosity improves the selectivity of MOFs. MOF luminescence is highly reactive to the variety and quantity of guest samples, as well as subtle alterations in the surroundings. The sensing methods in luminescent MOFs comprise selective adsorption, charge transfer, photoelectron transfer, etc. MOFs have the potential to be multipurpose luminous substances because they may create a luminous signal from both inorganic and organic species. The two-fold diffused MOF ([Zn$_2$(bdc)$_2$(dpNDI)]n, bdc = 1,4-benzene dicarboxylate, dpNDI = N, N-di(4-pyridyl)-1,4,5,8-naphthalene diimide) may alter its shape when it interacts with chemical species, choosing an orientation that enhanced the total binding affinity. This MOF showed significant selectivity for bulkier methylated aromatic molecules like p-xylene and toluene compared to benzene. The aromatic compounds form an excited complex with the dpNDI unit. The intensity of the charge transfer on the interaction between host and guest determines the location of the emission band.[42] Regarding MOF materials, the luminous characteristics of Cd(II) are remarkably comparable to those of Zn(II). The fluorescence-based titration and quenching techniques were used to detect nitroaromatic chemicals in ethanol using MOF ([Cd(atc)(H$_2$O)$_2$]n, atc = 2-aminoterephthalic acid).[43] Overall, possibilities for explosive detection by MOFs are extremely promising, and fresh studies on such uses of MOFs have been conducted. A highly selective method to differentiate and identify 2,4,6-trinitrophenol from other nitroaromatic chemicals has been reported. The MOF utilized was ([(CH$_3$)$_2$NH$_2$]

+$[Zn_4(\mu_4\text{-}O)(NTB)_2(NO_2\text{-}bdc)_{0.5}]\cdot 3DMA$, NTB = $4,4',4''$-nitrilotribenzoic acid, NO_2-bdc = 2-nitro-4-benzene dicarboxylic acid) which quenched luminescence even at a very low explosive concentration.[44] Luminescent MOFs are widely used in optical telecommunication, chemical sensors, biomedical sensors, and many electronic devices.[45]

2.3 Magnetic sensors and sensing techniques

Analyzing the magnetic characteristics of MOFs is yet another technique of sensing. These kinds of materials have distinct magnetic properties that are affected by temperature, adsorbed species, and the host-guest relation. The MOF MOROF-1, with large pores (2.8–3.1 nm) displayed a selective and reversible solvent evoked "shrinking-breathing" behavior that directly impacts its magnetism, making this magnetic sponge-like action an encouraging path toward magnetic detectors.[46] The spin crossover (SCO) MOFs and single ion or single-molecule magnet (SIM/SMM) MOFs are the two kinds of magnetic MOFs that are used to make magnetic sensing devices. Among others, one particular topic of study concerning molecular functional structures is the spin cross over MOFs. Since the ions of transition metals (d^4–d^7) can persist in both low spin and high spin states, the metal sites in SCO MOFs usually constitute these metal ions. SCO MOFs are very sensitive toward any type of chemical or physical change in its surrounding and rapidly switches between low spin and high spin state, which leads to change in their physical parameters like magnetism, structure, luminescence, color, diamagnetism, etc.[47,48] SCO MOFs are usually subdivided into two categories as follows.

2.3.1 SCO MOFs modulated through guest adsorption

The ($Fe_2(azpy)_4(NCS)_4$, azpy = trans-4,40-azopyridine) porous MOF exhibits reversible absorption and removal of adsorbates such as methanol, ethanol, and propanol that make H-bonds with the system. With the absorption and removal of guests, the framework exhibits remarkable flexibility, generating a significant impact on the original geometry of the Fe(II) sites. As a result, the adsorbed parts experience crossovers of half-spin, but the desorbed parts do not.[49]

2.3.2 SCO MOFs are induced by the change in the metal's oxidation state

A layered MOF [$Co^{III}La^{III}$-(notp)(H_2O)$_4$]nH_2O, (notp)$^{6-}$ = hexaanion of the 1,4,7-triazacyclononane1,4,7-triyl-tris(methylene phosphonic acid)] is an excellent illustration of how heating a MOF initiates the change of metal's oxidation state that further gives rise to a controlled regulation of the magnetism.[50]

The magnetic properties of (SMM)/(SIM) MOFs (like hysteresis loop, magnetic ordering, SCO, magnetic susceptibility, and so on) may be conveniently modified using the single-crystal-to-single-crystal (SCSC) transformation, implying that MOFs are potential magnetic sensing materials.[51]

2.4 Few other sensors and sensing techniques

Calorimetric sensors are based on chromism. The most recognizable and simple signal obtained using gas sensors is the color shift of MOFs. Generally, the color shift is accomplished by one of these two processes. The first one is the reversible elimination/resorption of coordinated subunits which changes the coordination geometry of the metal. When coordinated subunits are lost due to variation in pressure, light, or temperature, causing the metal center's coordination states to shift, a d-d transition changes, and the MOFs change their absorption spectrum.[52] The second one is that contact among a large number of guest and host entities can change the process of charge transfer between electronic states, altering the absorption spectrum.[53] The constituents and quantities of analyte can be measured in calorimetric sensors by detecting the variation in color of the sample. Gu et al. investigated the performance of chromophoric Ru-MOFs (RuUiO-67) as sensing devices for sensing lead in an aqueous medium.[54]

Ferroelectric sensors and sensing techniques have also been explored for their application in sensing devices. For ferroelectric devices, crystallinity is required. Due to its excellent properties, such as adaptable functionality, ultra-high porosity, and adjustable pore volume, crystal MOF substances may be supplemented with ferroelectric entities. MOFs' ferroelectric characteristics are mostly obtained from MOF-polar guest molecule H-bond interactions between the polar guest species and MOF surface that provide them their ferroelectric properties.[55] The vapors of CH_3OH, CH_2OH, and H_2O were detected utilizing HKUST-1 MOF on the microcantilever, while carbon dioxide was detected with the help of piezoresistive cantilevers comprising MOF layers.[56] Along with microcantilever sensing techniques, other mechanical sensing technologies, like microresonator, surface acoustic wave (SAW) detectors, and quartz microbalance coated using a thin MOF coating capable of increasing the target gas adsorption, are also available.[57,58] All of these aforementioned sensors and sensing techniques as well as their application in detecting different pollutants and harmful substances have been summarized in Fig. 1.

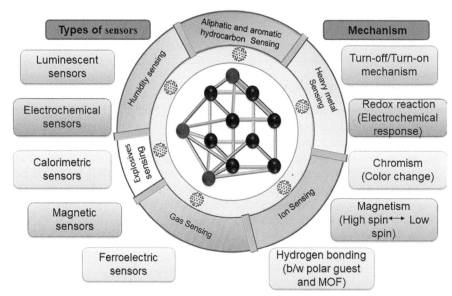

FIG. 1 Different MOF-based sensors and sensing mechanism along with their sensing applications.

3   Application of MOFs in sensing volatile organic compounds (VOCs) and aqueous pollutants

Volatile organic compounds, commonly known as VOCs, have been classified as constantly and pervasively harmful vapors for both human and animal health. The long-term exposure of these VOCs may cause both chronic and acute illnesses. Humans are always in danger of being exposed to these vapors through inhalation. VOCs are often found in groundwater and soil, workplaces, the atmosphere, chemical industries, and research labs. The VOCs are low boiling substances so the rapid rate of evaporation makes the situation very critical and concentration raises quickly to harmful levels within a few seconds that may be fatal in enclosed surroundings. There are different routes of VOCs exposure to humans such as injection (with a contaminated surface), ingestion (eating and drinking), and inhalation (breathing filthy air). Gas chromatography and mass spectrometry are two reliable and accurate techniques used for VOCs quantification. However, GC/MS requires costly chemicals and equipment for sample preparation. Furthermore, GC/MS is a time-consuming process as it requires time for sample preparation and collection; thus, real-time monitoring is difficult in this procedure. These aforementioned constraints make GC/MS procedure troublesome, and in the end, it limits the testing quality, frequency, and amount. As a result, stable, robust, rapid, nondestructive, cost-effective, and reliable sensors for VOC detection are in high demand. Some other techniques have also been tested in the last decade such as quartz crystal microbalances (QCMs) and metal-organic frameworks (MOFs) based techniques. The QCMs are highly reliable and cost-effective due to in-situ monitoring capabilities and high sensitivity. The QCMs come with usual benefits like low-temperature operability, low power consumption, longer battery life, and no sample preparation requirement. These nanoporous materials have excellent mechanical, thermal, and chemical stability. MOFs have high molecular transport rates as well as large interfacial surface areas due to regularity and uniformity of pore size. Many of the mechanical, morphological, and chemical features of MOFs can also be accurately modified due to their high modularity of synthesis and a large variety of inorganic-organic compositions. Nanoporosity and high interfacial area of MOFs make them intriguing for sensing applications, especially in gas, heavy metal, water vapor, hydrocarbon, and ion sensing. Less work has been done in the field of MOF gas and VOC sensing applications so far. MOF-based film sensors are prepared over the surface of quartz crystals using drop-casting and electrospinning. The thin MOF film has an aesthetically pleasing structure.

Water vapor, oxygen, and dangerous industrial gases (such as sulfur dioxide, carbon dioxides, carbon disulfide, hydrogen sulfide, and nitrogen oxide) can be detected using switching materials and MOF-based sensing. In addition, MOFs with different functional groups and large surface areas can interact strongly with guest molecules, resulting in alterations in electrical, mechanical, and optical signals. When it comes to gas sensing, the most common techniques used include: optical (such as vaporchromism), mechanical (such as SAW devices, microcantilevers, and QCM), magnetic, and electrical sensors (chemiresistive, ferroelectric, impedance, and chemicapacitive). The attractive thing with MOFs is that topologically/morphologically diverse MOFs could be synthesized with metal nodes using the aforementioned building blocks and different coordination number organic linkers. Unique features like ultra-high porosity and large BET surface areas may be obtained. Several interactions between guest analytes and host framework attribute to selective adsorption through halogen-hydrogen bonds, van der Waals interactions, and pore size management, which could be the main reason that MOFs can be used to selectively target specific analytes.[45]

VOCs are the newer class of organic contaminant that easily volatilizes in the air under typical meteorological conditions. In addition to alcohols (butanol, ethanol, methanol, etc.), aliphatic hydrocarbons (ethylene, propylene, and butanone), (formaldehyde and acetaldehydes), and aromatic hydrocarbons (toluene, xylene, and benzene), and aldehydes, VOCs also contain chlorinated hydrocarbons. Among these, chlorinated hydrocarbons are the most common (chloroform, dichloromethane, etc.).

3.1 Aromatic and aliphatic hydrocarbons

Numerous MOFs as sensors for aromatic and aliphatic hydrocarbons have been extensively investigated in the past few years, including magnetism (magnetization), luminescence (luminescence), mass change (QCMs),[59,60] chemiresistance (chemical resistance), and capacitance-based system.[61] A set of 2D MOFs with structurally similar but chemically heterogeneous properties was chosen by Dinca and co-workers[62] for a study on how chemical alterations affect the sensing responses. To create cross-reactive chemiresistive sensor arrays, the researchers employed three materials: $Cu_3(HHTP)$, $Cu_3(HITP)$, and $Ni_3(HITP)$. Each of the MOF-based sensor arrays has a varied reaction to five characteristic VOCs (aromatic hydrocarbons, amines, aliphatic hydrocarbon, ethers/ketones, and alcohols). The common sensing processes in chemiresistive devices include Schottky barrier modulation, charge transfer, and swelling/solvation. The benefits of MOF-based chemiresistors include low swelling, ohmic contacts, low Schottky barrier modulation due to rigid MOF structure. Furthermore, the MOF's charge density plays an important role in charge transfer sensing. Hydrogen bonding could also be a reason for the reported sensing response.

When one particular MOF is insufficient to match the sensing device's needs, Xu and co-workers[63] created heterostructured MOF over thin MOF film utilizing the van der Waals (vdW) integration approach to combine the varied features of several MOF layers.

Cu-HHTP and Cu-TCPP (TCPP = 5,10,15,20-tetrakis-(4-carboxyphenyl) porphyrin) are two lattice-mismatched MOF layers, which were selected to create high-quality MOF over MOF thin films. The high-quality MOF heterostructure is created by combining a Cu-TCPP layer (molecular sieving) with a tunable thickness over the sensing Cu-HHTP layer (chemiresistive). The MOF Cu-TCPP over Cu-HHTP-based sensor had very high responsiveness and selectivity to benzene. It is also able to detect and differentiate five properties of breath biomarkers, including acetone, ammonia, hexane, benzene, and carbon monoxide.

The most common raw materials in the majority of the industrial settings are aliphatic hydrocarbons that have a significant environmental and toxicological impact on the environment. Long and group[64] studied the previously reported conductive 3D MOF that is $Cu[Ni(pdt)_2]$ (pdt^{2-} = pyrazine2,3-dithiolate), the conductivity of which is extremely responsive to the state of solvation. Furthermore, the impressive adsorption capacity of $Cu[Ni(pdt)_2]$ toward hydrocarbons exhibits excellent chemiresistive sensing properties for VOCs and hydrocarbons (acetylene, propylene, ethane, cis-2-butene, propane, and ethylene), and the change of conductivity will depend on hydrocarbon. Furthermore, $Cu[M(pdt)_2]$ comprises a large number of active metal sites, which is advantageous for sensing applications. $Cu[M(pdt)_2]$ material is highly selective toward acetylene that can be a result of high affinity between active metal site and guest. Moreover, the π-π interactions and hydrogen bonding between ligands and acetylene can also be responsible for the high selectivity of $Cu[M(pdt)_2]$ toward acetylene.[65]

3.2 Inorganic anions

Nutrients in water, such as phosphate ions (PO_4^{3-}), cause eutrophication and reduce or eliminate dissolved oxygen, having a deleterious impact on the aquatic ecosystem.[66] Several luminescent-based MOF sensors were used in a decade to detect inorganic anions and demonstrated high selectivity toward inorganic ions. Qian and co-workers[67] described a PO_4^{3-} sensor based on photoluminescence quenching having high selectivity and based on the TbNTA1 (NTA = nitrilotriacetate) MOF. Several studies demonstrate that inorganic ions (NO^{3-}, NO^{2-}, Br^-, Cl^-, F^-, I^-, CO_3^{2-}, SO_4^{2-}, and HCO_3^{3-}) had no effect on TbNTA1 fluorescence intensity, but only PO_4^{3-} adsorbed TbNTA1 had a significant quenching effect that can be defined by TbNTA1's affinity for anions. The luminescence quenching effect is the result of nonradioactive relaxation. When

TbNTA1 is combined with phosphate ion that has a tetrahedral structure, the Tb-O ion reduces energy transmitted to Tb^{3+} ions.

In another MOF-based sensor, 3-aminopropyl, trimethoxysilane capped ZnO QDs (zinc oxide quantum dots) commonly known as APTMS-ZnO QDs were used. The APTMS-ZnO QDs are connected to MOFs with negative charge via electrostatic and amine zinc interaction. The inclusion of phosphate in the QD-MOF system was found to decrease the quenching and restore the fluorescence of ZnO QDs. The fluorescence intensity of ZnO QDs is unaffected by other species and has been determined by phosphate concentration.[68]

The cyanide ion is another inorganic anion that is highly poisonous and fatal present in water. Ghosh and co-workers[69] studied a fluorescent-based MOF device that is highly selective and sensitive for cyanide ion detection in the aqueous phase.

The detection of chloride and free ClO ions in drinking water is critical to analyze as high levels of these are highly hazardous to human health, and it produces a plethora of disinfection byproducts. Although lower concentrations of these ions are not acute toxic for humans, they may cause illness over a longer period. The amine-based NH_2-MIL-53(Al) MOF sensor was prepared using $AlCl_3.6H_2O$ and NH_2-H_2BDC precursors in the aqueous phase of urea as a modulator in a single-step hydrothermal reactor. The water solubility and stability of the as-synthesized Al nanoplates were outstanding. The high fluorescence of Al nanoplates was dramatically reduced by adding free chlorine. The sensor platform exhibited an excellent LOD (0.04 M) and a broad detection range (0.05–15 µM). The hydrogen bonds of N-HO-Cl and NH_2-ClO surface interactions play in the suppression of fluorescence.[70]

Novel MOF ($[Y(tp)(ox)_{0.5}(H_2O)_2]\cdot H_2O$, $[Y_{0.8}Tb_{0.2}(tp)(ox)_{0.5}(H_2O)_2]\cdot H_2O$) showed good identification capacity for five hazardous oxo-anions in aqueous medium: phosphate (PO_4^{3-}), chromate (CrO_4^{2-}), arsenate ($HAsO_4^{2-}$), dichromate ($Cr_2O_7^{2-}$), and permanganate (MnO_4^-) in UV radiation. Anions $HAsO_4^{2-}$, MnO_4^-, and PO_4^{3-} entered the MOF channel and interacted at a molecular level with Y^{3+}/Tb^{3+} centers. By weakening Tb-O interactions in carboxylates, it inhibits energy transfer from the ligands to metal centers, resulting in a quenching of the Tb site emission and an increase in ligand center emission. The flow of energy from ligands to metal sites is prevented by weakening carboxylates Tb-O bonds, which resulted in Tb center quenching and simultaneous increase in ligand center emission. The inner-filter effect contributes majorly to Tb center emission damping with a simultaneous decrease in luminescence intensity in CrO_4^{2-} and $Cr_2O_7^{2-}$ anions.[71]

3.3 Heavy metal ions

Heavy metal ions are nonbiodegradable water contaminants highly poisonous and hazardous for human health, even in trace amounts. Heavy metal ions (As, Cr, Pb, Cd, Cu, Hg) can be detected using MOF-based sensors. Few recently synthesized MOFs with their limit of detection (LOD) for sensing heavy metal ions are listed in Table 1. Lin et al. studied a

TABLE 1 MOF-based sensors for heavy metal ion detection.

Heavy metal ions	MOFs for sensing	LOD	References
Pb(II) (lead ion)	MOF-177	0.004×10^{-6} M	72
Pb(II) (lead ion)	Fe-MOFs/PdPt nano particles hybrid	2×10^{-12} M	73
Pb(II) (lead ion)	NH_2-$Cu_3(BTC)_2$;BTC = benzene-1,3,5-tricarboxylate	5×10^{-9} M	74
Pb(II) (lead ion)	GR–5/(Fe-P)$_n$-MOF	0.034×10^{-9} M	75
Hg(II) (mercury ion)	Adenine-La-MOF	0.2×10^{-9} M	76
Hg(II) (mercury ion)	Porphyrinic-MOF	6×10^{-9} M	77
Hg(II) (mercury ion)	ZnAPA (H_2APA = 5-aminoisophthalic acid)	0.124×10^{-6} M	78
Cd(II) (cadmium ion)	Eu^{3+}@UiO-66(Zr)-COOH	0.06×10^{-6} M	79
Cd(II) (cadmium ion)	PCN-224	0.002×10^{-6} M	80
Cd(II) (cadmium ion)	MOF-177	0.03×10^{-6} M	72
Ag(I) (silver ion)	Eu^{3+}@MIL-82	0.09×10^{-6} M	81

copper ion (Cu^{2+}) detection probe based on florescent MOF, highly selective and sensitive for copper ion detection. The zeolitic imidazolate framework (ZIF-8) was encapsulated with branched polyethylenimine (BPEI) capped with CQDs (Carbon-QDs) for robust fluorescence activity (quantum yield >40%) and outstanding Cu^{2+} selectivity. The ZIF-8/BPEI-CQDs composites' material was employed for copper ion sensing that was ultrasensitive and very selective. ZIF-8, in combination with CQDs, has a high absorption capacity that is attributed to the accumulation of Cu^{2+} ions and demonstrates high fluorescent activity and ion selectivity. Using ZIF-8/BPEI-CQDs sensors with a LOD of 80 pM, detection of 2–1000 nM concentration of Cu^{2+} ions is possible. These sensors have two orders lower magnitude as compared to other fluorescent sensors without amplification in guest luminophore inclusion.[82]

The Chi and co-workers[83] developed a Ru-MOF-based quick and selective sensing approach for hazardous Hg^{2+} detection. The luminous $Ru(bpy)_3^{2+}$ was encapsulated in MOFs pore after being doped in a Ru-MOF framework. The Ru-MOFs form a yellow precipitate in an aqueous solution and emit red color light in UV radiation. Hg^{++} ions rapidly react with Ru-MOFs and produce a significant amount of luminous $Ru(bpy)_3^{2+}$ guest inclusions in an aqueous phase, which results in electrochemiluminescence or intense fluorescence signals. The water solution turned yellow as the $Ru(bpy)_3^{2+}$ was liberated, and UV light revealed red light emission.

Chen and co-workers[84] developed a lanthanide MOF nanoparticle-based turn-on fluorescence sensor. An inner-filter effect allows this sensor to detect Hg^{2+}. The synthesized Eu-isophthalate MOF was subjected to imidazole-4,5-dicarboxylic acid (IDA). IDA may interact with MOF particles and quench its emission due to IDA significant absorption in the MOF excitation region. When Hg^{2+} was added, the emissions were restored as it formed a strong bond with imidazole-4,5-dicarboxylic acid and mercury ions freed imidazole-4,5-dicarboxylic acid molecules from the MOF surface.

A highly efficient Eu^{3+} ion-based MOF (CPM-17-Zn) has shown excellent sensitivity and selectivity for sensing cd^{2+} in solution.[85]

4 Application of MOFs in sensing atmospheric pollutants

4.1 Carbon dioxide (CO_2)

CO_2 gas detection using a microwave sensor was presented by Mohammad et al. A commercially available porous bed (Zeolite 13X) and two synthesized MOF-199-M1 and MOF-199-M2 were used to compare sensitivity improvements. MOF-199-M2 had a maximum sensitivity of 24 kHz/percent CO_2, while the Zeolite 13X had a minimum sensitivity of 10 kHz/percent CO_2, allowing MOF-199-M2 to be used as a sensor even at high CO_2 concentrations (>45 vol.%).[86] SAW systems that use nanoporous MOFs sensors are gaining popularity. Similarly, a zeolitic imidazolate framework-8 (ZIF-8) MOF-coated SAW mass detector for sensitive methane (CH_4) and carbon dioxide detection (CO_2) was proposed. They found that identical devices with detection limits of 7.01 vol.% and 0.91 vol.% for CH_4 and CO_2 had four times improved sensitivity at 860 MHz compared to 430 MHz.[87] Stassen et al. integrated imino-semiquinonate moieties, i.e., well-defined N-heteroatom functionalization, into an electrically conducting, nanoporous two-dimensional MOF. Cu_3(hexaaminobenzene)$_2$ (Cu_3HIB_2) showed the highest performance, with selective, robust and ambient CO_2 detecting properties in the 400–2500 ppm (ppm) range. In the 10%–80% RH range, the measured ambient CO_2 sensitivity is almost RH-independent. Cu_3HIB_2 was shown to be more sensitive than those of any other existing chemiresistor over a wider RH range.[88] A Ni(II) MOF **CSMCRI-3**, where CSMCRI stands for Central Salt and Marine Research Institute, was synthesized solvothermally using nitrogen-rich, optically fluorescent organic supports by Ranadip et al.,[89] and similarly, a Cd(II) MOF **CSMCRI-5**[90] was also synthesized. The activated MOFs showed exceptionally selective CO_2 adsorption against N_2 (292.5) and CH_4 (11.7) and N_2 (259.94) and CH_4 (14.34), respectively, as a result of enhanced pore functionality and stability. A moisture-stable 3D-MOF, {(Me$_2$NH$_2$)-[Zn$_2$(bpydb)$_2$(ATZ)](DMA)(NMF)$_2$}n (where bpydb = 4,4'-(4,4'-bipyridine-2,6-diyl)dibenzoate, ATZ = deprotonated 5-aminotetrazole, DMA = N,N-dimethylacetamide, and NMF = N-methylformamide), with uncoordinated N-donor regions and charged framework was fabricated. This MOF has an intriguing structural dynamic with high CO_2/N_2 (127) and CO_2/CH_4 (131) sorption.[91] Hromatka et al. used in-situ crystallization to create a 2-m-long period grating CO_2 sensor coated using HKUST-1 in an optical fiber. With 401 ppm of gas concentration, a 40-layer film showed the best sensitivity.[92]

4.2 Sulfur dioxide (SO_2)

Due to the obvious environmental and health risks posed by sulfur dioxide (SO_2) gas, rapid and selective sensing of this gas is gaining popularity. A study was conducted on the sensing property of an amino-functionalized luminescent MOF-5-NH_2 for SO_2 gas. The results show that such MOF probe can detect SO_2 derivatives like SO_3^{2-} selectively and sensitively in real-time, with a detection limit of 0.168 ppm and a response time of 15 s. Furthermore, MOF-5-NH$_2$ also showed good anti-interference and selectivity.[93] In a work by Janiak et al., three different MOFs such as MOF-177, MIL-160, and NH$_2$-MIL-125(Ti) were prepared for SO$_2$ sensing. MOF-177 had the maximum SO$_2$ uptake of 25.7 mmol/g among the contenders, whereas the others were judged to be suitable for removing levels below 500 ppm.[94] A doped capacitive interdigitated electrode was coated solvothermally with an In-based MFM-300. The resulting sensor has a lower sensitivity limit of roughly 5 parts per billion (ppb) down to 75 parts per billion (ppb).[95]

Similarly, the iridium organometallic fragment [Ir{κ^3(P, Si, Si)PhP(o-C$_6$H$_4$CH$_2$Si^{i}Pr$_2$)$_2$}] was incorporated into NU-1000 to produce a novel Ir-based MOF-type [Ir]@NU-1000. The new material showed exceptional SO$_2$ uptake at room temperature and hence it is a promising adsorbent for SO$_2$ capture.[96] Further, the MOFs MIL-53(Al)-TDC and MIL-53 (A)-BDC were proposed as promising MOFs for sensing and SO$_2$ storage, respectively.[97] Cobalt and zinc co-doping were used to create carbon nanotube networks based on zinc-doped zeolitic imidazolate frameworks (ZIF-67) (bimetallic MOFs). Pyrolysis yielded good SO$_2$ sensitivity, cross-selectivity, and durability, combining the benefits of MOFs and CNTs.[98]

4.3   Nitrogen dioxide (NO$_2$)

NOx is a dangerous air pollutant that acts as a precursor to acid rain. The electrochemical principles can be utilized to assess the analytical performance of various MOFs, as well as the benefits and drawbacks of their application in NO$_2$ detection. For enhanced gas detection, the flexible ZIF-MOFs are produced utilizing a solvothermal technique. Polyhedral ZIF-8 nanostructures with high specific surface area, selectivity, porosity, and high sensitivity of 118.5 were produced in this line.[99] Eddaoudi et al. have successfully shown an ultrasensitive and selective OFET sensor for NO$_2$ gas known as MOF-A ([Ni(TiF$_6$)(TPyP)]n). It is a first-of-its-kind combination of an isoreticular fluorinated 3D MOF and PDVT-10, with a sensitivity of 680 nA/ppb. From 8 ppb to 100 ppm, the device displayed a repeatable performance that was unaffected by humidity or ambient conditions.[100] Xiaohui et al. used the hydrothermal technique to produce Cu-based MOF precursors as a template. Cu$_2$O-CuO octahedrons were shown to be substantially better at sensing NO$_2$ than commercially available CuO powder, with detection limits of up to 8.25–500 ppb NO$_2$ at RT compared to 2.88 ppb at ambient temperature for CuO powder.[101] Porous hollow SmFeO$_3$ microspheres with a diameter of about 2 μm were prepared from a simple combination of a MOF and a templating precursor. These chemiresistive gas sensors showed an excellent sensitivity response of 10.2 under 200 ppb with a lower detection limit of 50 ppb.[102]

4.4   Ammonia (NH$_3$)

Yang Shi et al. demonstrated excellent catalytic activity of the produced crystalline Ni-MOF for NH$_3$-SCR. This Ni-MOF SCR catalyst had superior thermal stability (440°C) than other MIL-100(Fe) or Cu-BTC SCR catalysts previously described. Furthermore, its action was observed to be improved by pretreatment with N$_2$.[103] The hydrothermal methodology was used to successfully produce Co/Mn-MOF-74. The influence of the ratio of Co/Mn ratio on the framework and catalytic properties of Co/Mn-MOF-74 solids for low-temperature NH$_3$-SCR of NOx was examined. The Mn$_{0.66}$Co$_{0.34}$-MOF-74 chemical composition produced extremely large NO conversions (>96.0%) around 180–240°C. Mn$_{0.66}$Co$_{0.34}$-MOF-74 also demonstrated exceptional resistance to SO$_2$ poisoning. Co inclusion was found to diminish SO$_2$ adsorption strength on the catalyst surface, increasing the catalyst's SO$_2$ resistance to poisoning. The results of this study imply that Mn$_{0.66}$Co$_{0.34}$-MOF-74 could be a promising industrial NH$_3$-SCR NOx control catalyst.[104] Following NH$_3$ adsorption, a series of M(BDC) (M = Cu, Zn, Cd)(1,4-benzene dicarboxylic acid) materials were developed into one-dimensional M(BDC)(NH$_3$)$_2$. The values of adsorption capacities for Cu, Zn, and Cd, were found to be 17.2, 14.1, and 7.4 mmol/g, respectively. These MOFs possess good regeneration capacity at higher temperatures (250°C). Cu(BDC) and Zn(BDC) showed better results in removing lower concentrations of NH$_3$.[105] Combination of two or more active components gives hybrid materials with enhanced properties. In this regard, Shunxing et al. used a combination of organic single crystals and MOFs to produce a novel 1D core-shell heterostructure with a mix of single copper phthalocyanine (CuPc) ribbon cores and an isoreticular MOF-3 (IRMOF-3) as the shell. The detection limit of the heterostructured, CuPc@IRMOF-3 sensors was estimated to be 52 ppb which showed good stability.[106] New Cu-MOF-74 material was also synthesized solvothermally and then employed as a catalyst for NO elimination for selective catalytic reduction (SCR) at low temperatures with ammonia. It was found that the Cu-MOF-74-iso-80 catalyst showed the maximum NH$_3$-SCR activity, giving 100% N$_2$ selectivity and 97.8% NO conversion at 230°C with the plus water resistance performance.[107] A composite of reduced graphene oxide (rGO) and zeolite (ZIF-67) was made by hydrothermal method for the detection of ammonia. A porous and high surface area (1080 m^2 g^{-1}) composite was obtained and drop cast on interdigitated gold electrodes. Sensor responses of 1.22 ± 0.02 and 4.77 ± 0.15 were obtained for 20 and 50 ppm of NH_3, which was stable and tolerable against humidity. The overall detection limit of the sensor was 74 ppb.[108]

4.5 Hydrogen sulfide (H_2S)

Hydrogen sulfide (H_2S) is indeed a poisonous, combustible, colorless, and corrosive gas with a foul-smelling odor.[109] Salama et al. developed an enhanced sensor that can detect H_2S at ambient temperature. They employed rare-earth element-based MOF thin films with a fundamental fcu architecture.[110] This unique sensor was made by in-situ growth on an electrode. The sensors may detect H_2S selectively in the mixture of other gases within a range of 5–100 ppb. A MOF having a flexible mixed-matrix polymer membrane was designed. MOF-5 microparticles were implanted on a conductivity-regulated chitosan membrane via mixing it with the ionic liquid of glycerol with variable concentrations to create a gas sensor that might detect even 1 ppm at ambient temperature. At 50 ppm H_2S, the MOF-5/CS/IL sensor demonstrated good selectivity, little recovery, a rapid reaction time (8 s), and remarkable detecting stability of about 97%.[111] Because of its high toxicity, H_2S should be recognized immediately at quantities exceeding 10 ppm. MOFs coated with silver oxide nanoparticles have been used as the detecting material in this study. Impregnation was used to deposit nanoparticles over three MOFs (BDC-N3, UiO-66(Zr), UiO-66(Zr) BDC). The UiO-66(Zr) BDC-NO_2 packed with Ag_2O has the highest sensitivity to H_2S of all of them.[112] The monitoring of trace (11.2;1 ppm) amounts of H_2S using 3D inverse opal ZnO sensor enhanced with uniformly dispersed Pt nanoparticles had a low LOD (25 ppb), good selectivity, and stability.[113]

4.6 Carbon monoxide (CO)

Haitao et al. developed MOF crystals of Ni-MOF-74 as gravimetric sensing components on resonant microcantilever detectors for high-sensitivity monitoring of 10 parts per billion CO. For particulates smaller than 10 ppb, the particular reaction between CO and the reactive sites of Ni^{2+} in MOF crystal resulted in better sensitivity. The sensor's reproducibility and stability were likewise impressive. For a better outcome, eight types of interfering gases were examined.[114] The hydrothermal technique was used to create a new MOF-derived nanocomposite of molybdenum diselenide ($MoSe_2$) and tin dioxide (SnO_2). Because of the n-n heterojunction generated at the junction of SnO_2 nanoparticles and $MoSe_2$ nanoflowers,

TABLE 2 A summary of MOF-based sensors for sensing toxic gases in the atmosphere.

Year	MOF	Target gas	References
2019	Cu_3(hexaiminobenzene)$_2$	CO_2	88
2019	[$Ni_2(\mu_2$-OH)(azdc)(tpim)](NO$_3$)·6DMA·6MeOH (CSMCRI-3)	CO_2	89
2016	{(Me$_2$NH$_2$)-[Zn$_2$(bpydb)$_2$(ATZ)](DMA)(NMF)$_2$}n	CO_2	91
2020	[Cd1.5(L)$_2$(bpy)(NO$_3$)]·2DMF·2H$_2$O (CSMCRI-5)	CO_2	90
2018	nanoZIF-8	CO_2	116
2018	[Cd(L)$_2$]·(DMF)$_{0.92}$	CO_2	117
2019	Ni@ZrOF	CO_2	118
2018	HKUST-1	CO_2	92
2021	PET@NH$_2$-UiO-66	CO_2	119
2017	MOF-199-M1 and MOF-199-M2	CO_2	86
2018	MOF-5-NH$_2$	SO_2	93
2018	MFM-300-MOF (In)	SO_2	95
2019	MOF-177, NH$_2$-MIL-125(Ti), and MIL-160	SO_2	94
2020	[Ir]@NU-1000	SO_2	96
2021	MIL-53(Al)-TDC & MIL-53(Al)-BDC	SO_2	97
2021	Zr- and Al-MOFs	SO_2	120

Continued

TABLE 2 A summary of MOF-based sensors for sensing toxic gases in the atmosphere—cont'd

Year	MOF	Target gas	References
2018	ZIF-67	SO_2	98
2020	ZIF-8	NO_2	99
2020	In(acac)$_3$@ZIF-8	NO_2	121
2020	[Ni(TiF$_6$)(TPyP)]$_n$ (MOF-A)	NO_2	100
2021	Cu_2O-CuO MOFs	NO_2	101
2018	[Zr$_6$O$_4$ (OH)$_4$ (FA)$_6$]$_2$ (cal)$_3$]	NO_2	122
2018	$SmFeO_3$	NO_2	102
2020	$Co_3V_2O_8$ MOFs	NO_2	123
2019	In_2O_3/ZIF-8	NO_2	124
2017	PdO-Co$_3$O$_4$ HNCs	NO_2	125
2019	Co_3O_4/$NiCo_2O_4$	H_2S	126
2016	Fe^{III}-MIL-88-NH$_2$	H_2S	127
2020	3DIO ZnO	H_2S	113
2020	CuPc@IRMOF-3	NH_3	106
2018	NiPc-M and NiNPc-M	NH_3, H_2S, CO	128
2018	Ni-MOF-274	CO	114
2019	SnO_2/$MoSe_2$	CO	115

these two had high responsiveness, anti-humidity, and reproducibility for CO-sensing applications at ambient temperature.[115] A summary of MOFs-based sensors for detecting toxic gases in the atmosphere is presented in Table 2.

5   Synthesis

MOFs are frequently prepared via a one-stage synthesis inside a solution. In a nutshell, organic linkers and metal ions are combined with a suitable solvent to self-assemble stable and organized frameworks.[129] The reaction is usually done thermally or at room temperature, where the temperature is usually below 250°C.[130]

Following are some of the most common or conventional methods of MOF synthesis:

5.1   Crystallization synthesis

Gradually adding reactive components to decelerate crystal nucleation in MOFs helps in synthesizing a framework having good crystallinity.[41] Slow evaporation process at ambient temperature or diffusion across a reaction contact between two constituents are the two ways of crystal synthesis. Diffusion is frequently done in basic conditions, which removes hydrogen from organic ligands while also preventing competitive binding between organic linkers and unreacted metal ions. The method's benefits include the use of mild conditions and the creation of single crystals. However, it takes a long time to synthesize (10 days or more) and depends on the reactant's solubility at ambient temperature.[131]

5.2   Solvothermal synthesis

A majority of MOFs are synthesized via solvothermal technique since it produces excellent crystal formation and requires less equipment.[132] Not only it successfully addresses the problem of organic ligand solubility, but it also reduces reaction time. The organic binders, metal ions, and the solvent are mixed in a specific ratio inside a Teflon reactor and reacts at high pressure as well as high-temperature conditions to create a crystalline MOF in hydrothermal synthesis. Usually, polar solvents which boil at high temperatures, including ethanol formamide and water, are used. Following the thermal treatment to reveal an unsaturated metal site, these polar solvents readily couple with that unsaturated metal site and may be removed later. The revealed metal site may adsorb and detach gases by coordination, and it can also link with carboxyl or amino group-containing materials.[133,134]

5.3 Microwave-assisted synthesis

The solvothermal synthesis of MOFs could be aided by ultrasonic or microwave radiation. The solvent may reach extremely high temperatures in the presence of microwaves in just a brief period. By changing the temperature and time of the reaction, a material having a high surface area can be created as compared to hydrothermal synthesis. Ultrasonic waves accelerate the development and destruction of solvent bubbles, create cavitation, and form air and voids in solutions. This results in extraordinarily elevated local pressures and temperatures and allows MOFs to crystallize and precipitate quickly.[135] Microwave radiation has several benefits, including consistent and fast heating, rapid response time, and no temperature drop.

5.4 Direct synthesis

Only particular reaction methods are suitable for direct synthesis. Initially, a metal ion and an organic binder are added in a suitable solvent and afterwards mechanically agitated around 100°C.[136] Mixing solvents such as H_2O or CH_3OH, with Co or Zn metal ions and 2-methylimidazole can produce ZIF-8 or ZIF-67.[137]

Four unique synthesis strategies have been developed by the modification of mechanochemistry and reaction media (deep eutectic solvents, ionic liquids(ILs), surfactants) in the ongoing quest to make MOF materials in a practical and environmentally acceptable manner.[138–141] These strategies have been briefly summarized in Fig. 2.

Ionic liquid microemulsion (ILMEs) is a noble technique for preparing MOFs of controlled size and microstructure. The elementary substance of MOFs is first dissolved in an aqueous solution. Ionic liquids are used to disperse these building blocks in the form of nanodroplets. These nanodroplets are then subjected to the solvothermal process. Fresh as well as previously used ionic liquid can be utilized that make this process environment friendly due to easy recovery of ionic liquids. The outstanding and distinct properties of ionic liquids are high dissolution, low vapor pressure, cation and anion selection versatility, and thermal stability. Moreover, cation and anion presence in a solvent can help in charge compensation during synthesis. The collision of ionic liquid surrounded complex (nanodroplets) that consists of metal ions and

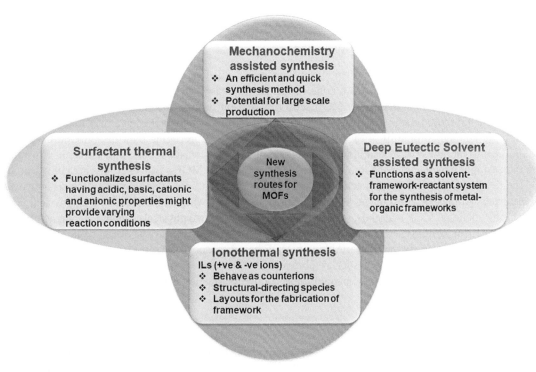

FIG. 2 A brief description of new routes of synthesis for MOFs.

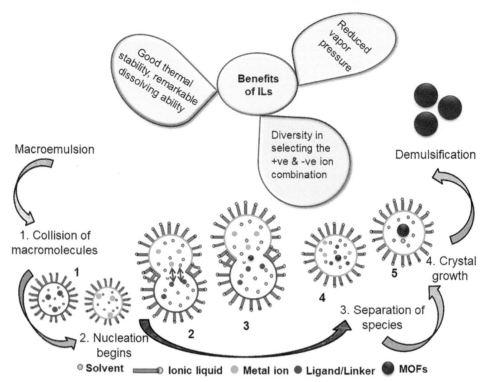

FIG. 3 Synthesis and downsizing of MOFs using the bottom-up macroemulsion technique in ionic liquids along with benefits of using ionic liquids.[142] *(Modified with permission from Usman et al. under creative commons attribution 4.0 international license.)*

ligand separately occurs in microemulsion. IL's complex (nanodroplets) also consists of a solvent in which metal ions and ligands are dispersed. Due to the collision of two ionic liquids, complex charge transfer occurs and nuclei of MOFs are formed. The nuclei of MOF gradually start growing by Ostwald ripening to the desired size. The ILs and solvent are demulsified and MOF is collected. This technique (IELMs) and various advantages of using ionic liquids are summarized in Fig. 3.

6 Recent advances and future perspective

Luminescent MOF sensors are well known as potent sensing devices due to their unique structural features and controlled luminescence characteristics. Most of the work done in this field is based on the turn-off mechanism of quenching the fluorescence; however, in recent years, many MOF sensors based on turn-on mechanism have also been studied. The turn-on mechanism is far more sensitive and has lower detection limits since the naked eye may immediately distinguish the significant emission and color shift of the visible light region from a preceding dark background or previously low emitting MOF. Due to the low cost of manufacturing and simplicity of downsizing, chemiresistive monitoring of MOFs has become increasingly popular since the introduction of electrochemical MOFs. However, this area of research is still in its infancy. The development of new redox-active linkers or guest species might greatly improve the sensing capability of such MOF materials. At present, most of the VOC and gas MOF sensors with distinct chemical or physical reactions are independent. Multifunctional MOF materials, wherein two and more than two physical or chemical characteristics, such as luminescence, magnetism, chirality, conductivity, ferroelectricity, porosity, etc., coexist or can be linked, might aid in the creation of novel sensing devices in the future. Although MOF-based sensors have already shown excellent sensing capability, a few constraints need to be addressed such as fabricating MOF-based sensing elements into practical devices, maintaining their efficiency over a long time, and making their commercialization economically viable.

7 Conclusion

MOF-predicated sensing devices have demonstrated excellent sensing performance that can be further enhanced by mixing them with some other functional entities. The present state of MOF-predicated sensors for monitoring environmental pollutants has been reviewed in this chapter. Owing to their distinctive structure and characteristics, including an immensely colossal surface area, high catalytic activity, customizable functionalization, good specificity, and selectivity, MOFs have huge potential for their application in sensors. This chapter thoroughly covered the following crucial aspects of MOF-predicated sensors:

- Different sensing techniques (electrochemical, luminescent, magnetic, calorimetric, ferroelectric, chemiresistive) along with their working principles and methodology.
- MOFs have been utilized as sensors for monitoring gases (CO, CO_2, NO_2, SO_2, NH_3, H_2S), VOCs, heavy metal ions (Ag^+, Cd^{2+}, Cu^{2+}, Hg^{2+}, Pb^{2+}), and other toxic pollutants.
- Both conventional (crystallization, solvothermal, microwave-assisted, direct) and new synthesis (ionothermal, deep eutectic solvent assisted, surfactant thermal, mechanochemistry assisted) strategies for the preparation of MOFs.

Overall, this analysis is contemporary, informative, and delivers insight into the future development of improved MOF materials as the next-generation sensors for monitoring and controlling environmental pollution. However, a majority of the sensor trials were done with samples manufactured in the lab. As a result, sensor execution in a complicated environmental condition should be assessed, and the sensor sensitivity and durability must be enhanced, as these are the two most important criteria for the practical application of the sensors. Despite the potential of MOF-based materials, further research is needed to enhance sensor performance in terms of selectivity, reusability, sensitivity, and stability. Undoubtedly, massive and coordinated efforts are required to overcome these issues. However, MOF-based sensors have a huge potential for monitoring and controlling environmental contaminants.

References

1. Yang GL, Jiang XL, Xu H, Zhao B. Applications of MOFs as luminescent sensors for environmental pollutants. *Small*. 2021;17(22). https://doi.org/10.1002/smll.202005327.
2. Lustig WP, Mukherjee S, Rudd ND, Desai AV, Li J, Ghosh SK. Metal-organic frameworks: functional luminescent and photonic materials for sensing applications. *Chem Soc Rev*. 2017;46(11):3242–3285. https://doi.org/10.1039/c6cs00930a.
3. Kumar Y, Yogeshwar P, Bajpai S, et al. Nanomaterials: stimulants for biofuels and renewables, yield and energy optimization. *Mater Adv*. 2021;2(16):5318–5343. https://doi.org/10.1039/D1MA00538C.
4. Meng Z, Stolz RM, Mendecki L, Mirica KA. Electrically-transduced chemical sensors based on two-dimensional nanomaterials. *Chem Rev*. 2019;119(1):478–598. https://doi.org/10.1021/acs.chemrev.8b00311.
5. Stassen I, Burtch N, Talin A, Falcaro P, Allendorf M, Ameloot R. An updated roadmap for the integration of metal-organic frameworks with electronic devices and chemical sensors. *Chem Soc Rev*. 2017;46(11):3185–3241. https://doi.org/10.1039/C7CS00122C.
6. Falcaro P, Ricco R, Yazdi A, et al. Application of metal and metal oxide nanoparticles at MOFs. *Coord Chem Rev*. 2016;307:237–254. https://doi.org/10.1016/j.ccr.2015.08.002.
7. Wang J, Jiu J, Araki T, et al. Silver nanowire electrodes: conductivity improvement without post-treatment and application in capacitive pressure sensors. *Nanomicro Lett*. 2015;7(1):51–58. https://doi.org/10.1007/s40820-014-0018-0.
8. Moldovan O, Iñiguez B, Deen MJ, Marsal LF. Graphene electronic sensors—review of recent developments and future challenges. *IET Circ Devices Syst*. 2015;9(6):446–453. https://doi.org/10.1049/iet-cds.2015.0259.
9. Chen X, Zhou G, Mao S, Chen J. Rapid detection of nutrients with electronic sensors: a review. *Environ Sci: Nano*. 2018;5(4):837–862. https://doi.org/10.1039/c7en01160a.
10. Bo Z, Yuan M, Mao S, Chen X, Yan J, Cen K. Decoration of vertical graphene with tin dioxide nanoparticles for highly sensitive room temperature formaldehyde sensing. *Sensors Actuat B Chem*. 2018;256:1011–1020. https://doi.org/10.1016/j.snb.2017.10.043.
11. Mao S, Pu H, Chang J, et al. Ultrasensitive detection of orthophosphate ions with reduced graphene oxide/ferritin field-effect transistor sensors. *Environ Sci: Nano*. 2017;4(4):856–863. https://doi.org/10.1039/c6en00661b.
12. Kumar H, Kumari N, Sharma R. Nanocomposites (conducting polymer and nanoparticles) based electrochemical biosensor for the detection of environment pollutant: its issues and challenges. *Environ Impact Assess Rev*. 2020;85. https://doi.org/10.1016/j.eiar.2020.106438, 106438.
13. Fan L, Wang Y, Li L, Zhou J. Carbon quantum dots activated metal organic frameworks for selective detection of Cu(II) and Fe(III). *Colloids Surf A Physicochem Eng Asp*. 2020;588. https://doi.org/10.1016/j.colsurfa.2019.124378, 124378.
14. Huo X, Liu P, Zhu J, Liu X, Ju H. Electrochemical immunosensor constructed using TiO₂ nanotubes as immobilization scaffold and tracing tag. *Biosens Bioelectron*. 2016;85:698–706. https://doi.org/10.1016/j.bios.2016.05.053.
15. Li HY, Zhao SN, Zang SQ, Li J. Functional metal-organic frameworks as effective sensors of gases and volatile compounds. *Chem Soc Rev*. 2020;49(17):6364–6401. https://doi.org/10.1039/c9cs00778d.
16. Zhao M, Yuan K, Wang Y, et al. Metal-organic frameworks as selectivity regulators for hydrogenation reactions. *Nature*. 2016;539(7627):76–80. https://doi.org/10.1038/nature19763.
17. Qu X-L, Gui D, Zheng X-L, et al. A Cd(II)-based metal-organic framework as a luminance sensor to nitrobenzene and Tb(III) ion. *Dalton Trans*. 2016;45(16):6983–6989. https://doi.org/10.1039/C6DT00162A.
18. Qian Q, Asinger PA, Lee MJ, et al. MOF-based membranes for gas separations. *Chem Rev*. 2020;120(16):8161–8266. https://doi.org/10.1021/acs.chemrev.0c00119.

19. Abánades Lázaro I, Forgan RS. Application of zirconium MOFs in drug delivery and biomedicine. *Coord Chem Rev.* 2019;380:230–259. https://doi.org/10.1016/j.ccr.2018.09.009.

20. Arora H, Park S, Dong R, Erbe A. 2D MOFs: a new platform for optics? *Opt Photonics News.* 2020;31(10):36–43. https://doi.org/10.1364/OPN.31.10.000036.

21. Yang F, Xu G, Dou Y, et al. A flexible metal-organic framework with a high density of sulfonic acid sites for proton conduction. *Nat Energy.* 2017;2 (11):877–883. https://doi.org/10.1038/s41560-017-0018-7.

22. Yang C, Dong R, Wang M, et al. A semiconducting layered metal-organic framework magnet. *Nat Commun.* 2019;10(1):3260. https://doi.org/10.1038/s41467-019-11267-w.

23. Kustov LM, Isaeva VI, Přech J, Bisht KK. Metal-organic frameworks as materials for applications in sensors. *Mendeleev Commun.* 2019;29(4):361–368. https://doi.org/10.1016/j.mencom.2019.07.001.

24. Lin CL, Chen YF, Qiu LJ, et al. Synthesis, structure and photocatalytic properties of coordination polymers based on pyrazole carboxylic acid ligands. *CrystEngComm.* 2020;22(41):6847–6855. https://doi.org/10.1039/d0ce01054e.

25. Zheng J, Vemuri RS, Estevez L, et al. Pore-engineered metal-organic frameworks with excellent adsorption of water and fluorocarbon refrigerant for cooling applications. *J Am Chem Soc.* 2017;139(31):10601–10604. https://doi.org/10.1021/jacs.7b04872.

26. Yang J, Zhang YB, Liu Q, et al. Principles of designing extra-large pore openings and cages in zeolitic imidazolate frameworks. *J Am Chem Soc.* 2017;139(18):6448–6455. https://doi.org/10.1021/jacs.7b02272.

27. Pribylov AA, Murdmaa KO, Solovtsova OV, Knyazeva MK. Methane adsorption on various metal-organic frameworks and determination of the average adsorption heats at supercritical temperatures and pressures. *Russ Chem Bull.* 2018;67(10):1807–1813. https://doi.org/10.1007/s11172-018-2293-2.

28. Kumar V, Kim KH, Kumar P, Jeon BH, Kim JC. Functional hybrid nanostructure materials: advanced strategies for sensing applications toward volatile organic compounds. *Coord Chem Rev.* 2017;342:80–105. https://doi.org/10.1016/j.ccr.2017.04.006.

29. Bobbitt NS, Mendonca ML, Howarth AJ, et al. Metal-organic frameworks for the removal of toxic industrial chemicals and chemical warfare agents. *Chem Soc Rev.* 2017;46(11):3357–3385. https://doi.org/10.1039/c7cs00108h.

30. Llabrés i Xamena FX, Corma A, Garcia H. Applications for metal-organic frameworks (MOFs) as quantum dot semiconductors. *J Phys Chem C.* 2007;111(1):80–85. https://doi.org/10.1021/jp063600e.

31. Sun L, Campbell MG, Dincǝ M. Electrically conductive porous metal-organic frameworks. *Angew Chem Int Ed.* 2016;55(11):3566–3579. https://doi.org/10.1002/anie.201506219.

32. Sun L, Park SS, Sheberla D, Dincă M. Measuring and reporting electrical conductivity in metal–organic frameworks: Cd^2 (TTFTB) as a case study. *J Am Chem Soc.* 2016;138(44):14772–14782. https://doi.org/10.1021/jacs.6b09345.

33. Liu R, Liu Y, Yu S, Yang C, Li Z, Li G. A highly proton-conductive 3D ionic cadmium-organic framework for ammonia and amines impedance sensing. *ACS Appl Mater Interfaces.* 2019;11(1):1713–1722. https://doi.org/10.1021/acsami.8b18891.

34. Davydovskaya P, Ranft A, Lotsch BV, Pohle R. Analyte detection with Cu-BTC metal-organic framework thin films by means of mass-sensitive and work-function-based readout. *Anal Chem.* 2014;86(14):6948–6958. https://doi.org/10.1021/ac500759n.

35. Hosseini MS, Zeinali S, Sheikhi MH. Fabrication of capacitive sensor based on Cu-BTC (MOF-199) nanoporous film for detection of ethanol and methanol vapors. *Sensors Actuat B Chem.* 2016;230:9–16. https://doi.org/10.1016/j.snb.2016.02.008.

36. Homayoonnia S, Zeinali S. Design and fabrication of capacitive nanosensor based on MOF nanoparticles as sensing layer for VOCs detection. *Sensors Actuat B Chem.* 2016;237:776–786. https://doi.org/10.1016/j.snb.2016.06.152.

37. Yassine O, Shekhah O, Assen AH, Belmabkhout Y, Salama KN, Eddaoudi M. H$_2$S sensors: fumarate-based fcu-MOF thin film grown on a capacitive interdigitated electrode. *Angew Chem Int Ed.* 2016;55(51):15879–15883. https://doi.org/10.1002/anie.201608780.

38. Assen AH, Yassine O, Shekhah O, Eddaoudi M, Salama KN. MOFs for the sensitive detection of ammonia: deployment of fcu-MOF thin films as effective chemical capacitive sensors. *ACS Sens.* 2017;2(9):1294–1301. https://doi.org/10.1021/acssensors.7b00304.

39. Campbell MG, Dincă M. Metal-organic frameworks as active materials in electronic sensor devices. *Sensors (Switzerland).* 2017;17(5). https://doi.org/10.3390/s17051108.

40. Wu G, Huang J, Zang Y, He J, Xu G. Porous field-effect transistors based on a semiconductive metal-organic framework. *J Am Chem Soc.* 2017;139 (4):1360–1363. https://doi.org/10.1021/jacs.6b08511.

41. Li X, Yang L, Zhao L, Wang XL, Shao KZ, Su ZM. Luminescent metal-organic frameworks with anthracene chromophores: small-molecule sensing and highly selective sensing for nitro explosives. *Cryst Growth Des.* 2016;16(8):4374–4382. https://doi.org/10.1021/acs.cgd.6b00482.

42. Myers M, Podolska A, Heath C, Baker MV, Pejcic B. Pore size dynamics in interpenetrated metal organic frameworks for selective sensing of aromatic compounds. *Anal Chim Acta.* 2014;819:78–81. https://doi.org/10.1016/j.aca.2014.02.004.

43. Wang Y-P, Wang F, Luo D-F, Zhou L, Wen L-L. A luminescent nanocrystal metal-organic framework for sensing of nitroaromatic compounds. *Inorg Chem Commun.* 2012;19:43–46. https://doi.org/10.1016/j.inoche.2012.01.033.

44. Nagarkar SS, Joarder B, Chaudhari AK, Mukherjee S, Ghosh SK. Highly selective detection of nitro explosives by a luminescent metal-organic framework. *Angew Chem Int Ed.* 2013;52(10):2881–2885. https://doi.org/10.1002/anie.201208885.

45. Li HY, Zhao SN, Zang SQ, Li J. Functional metal-organic frameworks as effective sensors of gases and volatile compounds. *Chem Soc Rev.* 2020;49 (17):6364–6401. https://doi.org/10.1039/c9cs00778d.

46. Maspoch D, Ruiz-Molina D, Wurst K, et al. A nanoporous molecular magnet with reversible solvent-induced mechanical and magnetic properties. *Nat Mater.* 2003;2(3):190–195. https://doi.org/10.1038/nmat834.

47. Clements JE, Airey PR, Ragon F, Shang V, Kepert CJ, Neville SM. Guest-adaptable spin crossover properties in a dinuclear species underpinned by supramolecular interactions. *Inorg Chem.* 2018;57(23):14930–14938. https://doi.org/10.1021/acs.inorgchem.8b02625.

48. Wang L-F, Zhuang W-M, Huang G-Z, et al. Spin-crossover modulation via single-crystal to single-crystal photochemical [2 + 2] reaction in Hofmann-type frameworks. *Chem Sci.* 2019;10(32):7496–7502. https://doi.org/10.1039/C9SC02274K.

49. Halder GJ, Kepert CJ, Moubaraki B, Murray KS, Cashion JD. Guest-dependent spin crossover in a nanoporous molecular framework material. *Science.* 2002;298(5599):1762–1765. https://doi.org/10.1126/science.1075948.

50. Bao S-S, Liao Y, Su Y-H, et al. Tuning the spin state of cobalt in a Co-La heterometallic complex through controllable coordination sphere of La. *Angew Chem.* 2011;123(24):5618–5622. https://doi.org/10.1002/ange.201007872.

51. Ni Z-P, Liu J-L, Hoque MN, et al. Recent advances in guest effects on spin-crossover behavior in Hofmann-type metal-organic frameworks. *Coord Chem Rev.* 2017;335:28–43. https://doi.org/10.1016/j.ccr.2016.12.002.

52. Ullman AM, Jones CG, Doty FP, Stavila V, Talin AA, Allendorf MD. Hybrid polymer/metal-organic framework films for colorimetric water sensing over a wide concentration range. *ACS Appl Mater Interfaces.* 2018;10(28):24201–24208. https://doi.org/10.1021/acsami.8b07377.

53. Gładysiak A, Nguyen TN, Navarro JAR, Rosseinsky MJ, Stylianou KC. A recyclable metal-organic framework as a dual detector and adsorbent for ammonia. *Chem Eur J.* 2017;23(55):13602–13606. https://doi.org/10.1002/chem.201703510.

54. Wang Z, Yang J, Li Y, Zhuang Q, Gu J. Zr-based MOFs integrated with a chromophoric ruthenium complex for specific and reversible Hg^{2+} sensing. *Dalton Trans.* 2018;47(16):5570–5574. https://doi.org/10.1039/c8dt00569a.

55. Zhang W, Xiong R-G. Ferroelectric metal-organic frameworks. *Chem Rev.* 2012;112(2):1163–1195. https://doi.org/10.1021/cr200174w.

56. Venkatasubramanian A, Lee J-H, Houk RJ, Allendorf MD, Nair S, Hesketh PJ. Characterization of HKUST-1 crystals and their application to MEMS microcantilever array sensors. *ECS Trans.* 2019;33(8):229–238. https://doi.org/10.1149/1.3484126.

57. Tu M, Wannapaiboon S, Khaletskaya K, Fischer RA. Engineering zeolitic-imidazolate framework (ZIF) thin film devices for selective detection of volatile organic compounds. *Adv Funct Mater.* 2015;25(28):4470–4479. https://doi.org/10.1002/adfm.201500760.

58. Khoshaman AH, Bahreyni B. Application of metal organic framework crystals for sensing of volatile organic gases. *Sensors Actuat B Chem.* 2012;162(1):114–119. https://doi.org/10.1016/j.snb.2011.12.046.

59. Xu F, Sun L, Huang P, et al. A pyridine vapor sensor based on metal-organic framework-modified quartz crystal microbalance. *Sensors Actuat B Chem.* 2018;254:872–877. https://doi.org/10.1016/j.snb.2017.07.026.

60. Jiang J, Ma Z-H, Liu R, He X. Structural transformation of two copper coordination polymers and their enhanced benzene vapor selective detection. *Inorg Chim Acta.* 2020;501. https://doi.org/10.1016/j.ica.2019.119241, 119241.

61. Yuan H, Tao J, Li N, et al. On-chip tailorability of capacitive gas sensors integrated with metal-organic framework films. *Angew Chem Int Ed.* 2019;58(40):14089–14094. https://doi.org/10.1002/anie.201906222.

62. Campbell MG, Liu SF, Swager TM, Dincă M. Chemiresistive sensor arrays from conductive 2D metal-organic frameworks. *J Am Chem Soc.* 2015;137(43):13780–13783. https://doi.org/10.1021/jacs.5b09600.

63. Yao MS, Xiu JW, Huang QQ, et al. Van der waals heterostructured MOF-on-MOF thin films: cascading functionality to realize advanced chemiresistive sensing. *Angew Chem Int Ed.* 2019;58(42):14915–14919. https://doi.org/10.1002/anie.201907772.

64. Aubrey ML, Kapelewski MT, Melville JF, et al. Chemiresistive detection of gaseous hydrocarbons and interrogation of charge transport in Cu[Ni (2,3-pyrazinedithiolate)2] by gas adsorption. *J Am Chem Soc.* 2019;141(12):5005–5013. https://doi.org/10.1021/jacs.9b00654.

65. Peng YL, Pham T, Li P, et al. Robust ultramicroporous metal-organic frameworks with benchmark affinity for acetylene. *Angew Chem Int Ed.* 2018;57(34):10971–10975. https://doi.org/10.1002/anie.201806732.

66. Kolliopoulos AV, Kampouris DK, Banks CE. Rapid and portable electrochemical quantification of phosphorus. *Anal Chem.* 2015;87(8):4269–4274. https://doi.org/10.1021/ac504602a.

67. Xu H, Xiao Y, Rao X, et al. A metal-organic framework for selectively sensing of PO_4^{3-} anion in aqueous solution. *J Alloys Compd.* 2011;509:2552–2554. https://doi.org/10.1016/j.jallcom.2010.11.087.

68. Fang X, Zong B, Mao S. Metal-organic framework-based sensors for environmental contaminant sensing. *Nanomicro Lett.* 2018;10(4):64. https://doi.org/10.1007/s40820-018-0218-0.

69. Karmakar A, Kumar N, Samanta P, Desai AV, Ghosh SK. A post-synthetically modified MOF for selective and sensitive aqueous-phase detection of highly toxic cyanide ions. *Chem Eur J.* 2016;22(3):864–868. https://doi.org/10.1002/chem.201503323.

70. Lu T, Zhang L, Sun M, Deng D, Su Y, Lv Y. Amino-functionalized metal-organic frameworks nanoplates-based energy transfer probe for highly selective fluorescence detection of free chlorine. *Anal Chem.* 2016;88(6):3413–3420. https://doi.org/10.1021/acs.analchem.6b00253.

71. Daga P, Sarkar S, Majee P, et al. A selective detection of nanomolar-range noxious anions in water by a luminescent metal-organic framework. *Mater Adv.* 2021;2(3):985–995. https://doi.org/10.1039/D0MA00811G.

72. Sangeetha S, Krishnamurthy G. Fabrication of MOF-177 for electrochemical detection of toxic Pb^{2+} and Cd^{2+} ions. *Bull Mater Sci.* 2019;43(1):29. https://doi.org/10.1007/s12034-019-1979-x.

73. Yu Y, Yu C, Niu Y, et al. Target triggered cleavage effect of DNAzyme: relying on Pd-Pt alloys functionalized Fe-MOFs for amplified detection of Pb^{2+}. *Biosens Bioelectron.* 2018;101:297–303. https://doi.org/10.1016/j.bios.2017.10.006.

74. Wang Y, Ge H, Wu Y, Ye G, Chen H, Hu X. Construction of an electrochemical sensor based on amino-functionalized metal-organic frameworks for differential pulse anodic stripping voltammetric determination of lead. *Talanta.* 2014;129:100–105. https://doi.org/10.1016/j.talanta.2014.05.014.

75. Cui L, Wu J, Li J, Ju H. Electrochemical sensor for lead cation sensitized with a DNA functionalized porphyrinic metal-organic framework. *Anal Chem.* 2015;87(20):10635–10641. https://doi.org/10.1021/acs.analchem.5b03287.

76. Tan H, Liu B, Chen Y. Lanthanide coordination polymer nanoparticles for sensing of mercury(II) by photoinduced electron transfer. *ACS Nano.* 2012;6(12):10505–10511. https://doi.org/10.1021/nn304469j.

77. Yang J, Wang Z, Li Y, Zhuang Q, Zhao W, Gu J. Porphyrinic MOFs for reversible fluorescent and colorimetric sensing of mercury(II) ions in aqueous phase. *RSC Adv.* 2016;6(74):69807–69814. https://doi.org/10.1039/C6RA13766K.

78. Jiang J, Lu Y, Liu J, Zhou Y, Zhao D, Li C. An acid-base resistant Zn-based metal-organic framework as a luminescent sensor for mercury(II). *J Solid State Chem.* 2020;283. https://doi.org/10.1016/j.jssc.2019.121153, 121153.

79. Hao J-N, Yan B. A water-stable lanthanide-functionalized MOF as a highly selective and sensitive fluorescent probe for Cd^{2+}. *Chem Commun.* 2015;51(36):7737–7740. https://doi.org/10.1039/C5CC01430A.

80. Moradi E, Rahimi R, Farahani YD, Safarifard V. Porphyrinic zirconium-based MOF with exposed pyrrole Lewis base site as a luminescent sensor for highly selective sensing of Cd^{2+} and Br^- ions and THF small molecule. *J Solid State Chem.* 2020;282. https://doi.org/10.1016/j.jssc.2019.121103, 121103.

81. Ge K-M, Wang D, Xu Z-J, Chu R-Q. A luminescent Eu(III)-MOF for selective sensing of Ag^+ in aqueous solution. *J Mol Struct.* 2020;1208. https://doi.org/10.1016/j.molstruc.2020.127862, 127862.

82. Lin X, Gao G, Zheng L, Chi Y, Chen G. Encapsulation of strongly fluorescent carbon quantum dots in metal-organic frameworks for enhancing chemical sensing. *Anal Chem.* 2014;86(2):1223–1228. https://doi.org/10.1021/ac403536a.

83. Lin X, Luo F, Zheng L, Gao G, Chi Y. Fast, sensitive, and selective ion-triggered disassembly and release based on tris(bipyridine)ruthenium(II)-functionalized metal-organic frameworks. *Anal Chem.* 2015;87(9):4864–4870. https://doi.org/10.1021/acs.analchem.5b00391.

84. Li Q, Wang C, Tan H, Tang G, Gao J, Chen C-H. A turn on fluorescent sensor based on lanthanide coordination polymer nanoparticles for the detection of mercury(II) in biological fluids. *RSC Adv.* 2016;6(22):17811–17817. https://doi.org/10.1039/C5RA26849D.

85. Liu C, Yan B. Zeolite-type metal organic frameworks immobilized Eu^{3+} for cation sensing in aqueous environment. *J Colloid Interface Sci.* 2015;459:206–211. https://doi.org/10.1016/j.jcis.2015.08.025.

86. Zarifi MH, Gholidoust A, Abdolrazzaghi M, Shariaty P, Hashisho Z, Daneshmand M. Sensitivity enhancement in planar microwave active-resonator using metal organic framework for CO_2 detection. *Sensors Actuat B Chem.* 2018;255:1561–1568. https://doi.org/10.1016/j.snb.2017.08.169.

87. Devkota J, Greve DW, Hong T, Kim KJ, Ohodnicki PR. An 860 MHz wireless surface acoustic wave sensor with a metal-organic framework sensing layer for CO_2 and CH_4. *IEEE Sensors J.* 2020;20(17):9740–9747. https://doi.org/10.1109/JSEN.2020.2990997.

88. Stassen I, Dou JH, Hendon C, Dincǎ M. Chemiresistive sensing of ambient CO_2 by an autogenously hydrated Cu_3(hexaiminobenzene)$_2$ framework. *ACS Cent Sci.* 2019;5(8):1425–1431. https://doi.org/10.1021/acscentsci.9b00482.

89. Goswami R, Seal N, Dash SR, Tyagi A, Neogi S. Devising chemically robust and cationic Ni(II)-MOF with nitrogen-rich micropores for moisture-tolerant CO_2 capture: highly regenerative and ultrafast colorimetric sensor for tnp and multiple oxo-anions in water with theoretical revelation. *ACS Appl Mater Interfaces.* 2019;11(43):40134–40150. https://doi.org/10.1021/acsami.9b15179.

90. Singh M, Senthilkumar S, Rajput S, Neogi S. Pore-functionalized and hydrolytically robust Cd(II)-metal-organic framework for highly selective, multicyclic CO_2 adsorption and fast-responsive luminescent monitoring of Fe(III) and Cr(VI) ions with notable sensitivity and reusability. *Inorg Chem.* 2020;59(5):3012–3025. https://doi.org/10.1021/acs.inorgchem.9b03368.

91. Chen DM, Tian JY, Chen M, Sen LC, Du M. Moisture-stable Zn(II) metal-organic framework as a multifunctional platform for highly efficient CO_2 capture and nitro pollutant vapor detection. *ACS Appl Mater Interfaces.* 2016;8(28):18043–18050. https://doi.org/10.1021/acsami.6b04611.

92. Hromadka J, Tokay B, Correia R, Morgan SP, Korposh S. Carbon dioxide measurements using long period grating optical fibre sensor coated with metal organic framework HKUST-1. *Sensors Actuat B Chem.* 2018;255:2483–2494. https://doi.org/10.1016/j.snb.2017.09.041.

93. Wang M, Guo L, Cao D. Amino-functionalized luminescent metal-organic framework test paper for rapid and selective sensing of SO_2 gas and its derivatives by luminescence turn-on effect. *Anal Chem.* 2018;90(5):3608–3614. https://doi.org/10.1021/acs.analchem.8b00146.

94. Brandt P, Nuhnen A, Lange M, Möllmer J, Weingart O, Janiak C. Metal-organic frameworks with potential application for SO_2 separation and flue gas desulfurization. *ACS Appl Mater Interfaces.* 2019;11(19):17350–17358. https://doi.org/10.1021/acsami.9b00029.

95. Chernikova V, Yassine O, Shekhah O, Eddaoudi M, Salama KN. Highly sensitive and selective SO_2 MOF sensor: the integration of MFM-300 MOF as a sensitive layer on a capacitive interdigitated electrode. *J Mater Chem A.* 2018;6(14):5550–5554. https://doi.org/10.1039/c7ta10538j.

96. Gorla S, Díaz-Ramírez ML, Abeynayake NS, et al. Functionalized NU-1000 with an iridium organometallic fragment: SO_2 capture enhancement. *ACS Appl Mater Interfaces.* 2020;12(37):41758–41764. https://doi.org/10.1021/acsami.0c11615.

97. López-Olvera A, Zárate JA, Martínez-Ahumada E, et al. SO_2 capture by two aluminum-based MOFs: rigid-like MIL-53(Al)-TDC versus breathing MIL-53(Al)-BDC. *ACS Appl Mater Interfaces.* 2021;13(33):39363–39370. https://doi.org/10.1021/acsami.1c09944.

98. Li Q, Wu J, Huang L, et al. Sulfur dioxide gas-sensitive materials based on zeolitic imidazolate framework-derived carbon nanotubes. *J Mater Chem A.* 2018;6(25):12115–12124. https://doi.org/10.1039/c8ta02036a.

99. Zhan M, Hussain S, AlGarni TS, et al. Facet controlled polyhedral ZIF-8 MOF nanostructures for excellent NO_2 gas-sensing applications. *Mater Res Bull.* 2021;136. https://doi.org/10.1016/j.materresbull.2020.111133.

100. Yuvaraja S, Surya SG, Chernikova V, et al. Realization of an ultrasensitive and highly selective OFET NO_2 sensor: the synergistic combination of PDVT-10 polymer and porphyrin-MOF. *ACS Appl Mater Interfaces.* 2020;12(16):18748–18760. https://doi.org/10.1021/acsami.0c00803.

101. Wang W, Zhang Y, Zhang J, et al. Metal-organic framework-derived Cu_2O-CuO octahedrons for sensitive and selective detection of ppb-level NO_2 at room temperature. *Sensors Actuat B Chem.* 2021;328. https://doi.org/10.1016/j.snb.2020.129045.

102. Huang HT, Zhang WL, Zhang XD, Guo X. NO_2 sensing properties of $SmFeO_3$ porous hollow microspheres. *Sensors Actuat B Chem.* 2018;265:443–451. https://doi.org/10.1016/j.snb.2018.03.073.

103. Sun X, Shi Y, Zhang W, et al. A new type Ni-MOF catalyst with high stability for selective catalytic reduction of NOx with NH_3. *Catal Commun.* 2018;114:104–108. https://doi.org/10.1016/j.catcom.2018.06.012.

104. Jiang H, Niu Y, Wang Q, Chen Y, Zhang M. Single-phase SO_2-resistant to poisoning Co/Mn-MOF-74 catalysts for NH_3-SCR. *Catal Commun.* 2018;113:46–50. https://doi.org/10.1016/j.catcom.2018.05.017.

105. Chen Y, Du Y, Liu P, Yang J, Li L, Li J. Removal of ammonia emissions via reversible structural transformation in M(BDC) (M = Cu, Zn, Cd) metal-organic frameworks. *Environ Sci Technol.* 2020;54(6):3636–3642. https://doi.org/10.1021/acs.est.9b06866.

106. Zheng J, Pang K, Liu X, et al. Integration and synergy of organic single crystals and metal-organic frameworks in core-shell heterostructures enables outstanding gas selectivity for detection. *Adv Funct Mater.* 2020;30(52). https://doi.org/10.1002/adfm.202005727.

107. Jiang H, Zhou J, Wang C, Li Y, Chen Y, Zhang M. Effect of cosolvent and temperature on the structures and properties of Cu-MOF-74 in low-temperature NH_3-SCR. *Ind Eng Chem Res.* 2017;56(13):3542–3550. https://doi.org/10.1021/acs.iecr.6b03568.

108. Garg N, Kumar M, Kumari N, Deep A, Sharma AL. Chemoresistive room-temperature sensing of ammonia using zeolite imidazole framework and reduced graphene oxide (ZIF-67/RGO) composite. *ACS Omega.* 2020;5(42):27492–27501. https://doi.org/10.1021/acsomega.0c03981.

109. Vikrant K, Kumar V, Ok YS, Kim KH, Deep A. Metal-organic framework (MOF)-based advanced sensing platforms for the detection of hydrogen sulfide. *TrAC, Trends Anal Chem.* 2018;105:263–281. https://doi.org/10.1016/j.trac.2018.05.013.

110. Yassine O, Shekhah O, Assen AH, Belmabkhout Y, Salama KN, Eddaoudi M. H_2S sensors: fumarate-based fcu-MOF thin film grown on a capacitive interdigitated electrode. *Angew Chem.* 2016;128(51):16111–16115. https://doi.org/10.1002/ange.201608780.

111. Ali A, Alzamly A, Greish YE, Bakiro M, Nguyen HL, Mahmoud ST. A highly sensitive and flexible metal-organic framework polymer-based H_2S gas sensor. *ACS Omega.* 2021;6(27):17690–17697. https://doi.org/10.1021/acsomega.1c02295.

112. Surya SG, Bhanoth S, Majhi SM, More YD, Teja VM, Chappanda KN. A silver nanoparticle-anchored UiO-66(Zr) metal-organic framework (MOF)-based capacitive H_2S gas sensor. *CrystEngComm.* 2019;21(47):7303–7312. https://doi.org/10.1039/c9ce01323g.

113. Zhou X, Lin X, Yang S, et al. Highly dispersed metal-organic-framework-derived Pt nanoparticles on three-dimensional macroporous ZnO for trace-level H_2S sensing. *Sensors Actuat B Chem.* 2020;309. https://doi.org/10.1016/j.snb.2020.127802.

114. Lv Y, Xu P, Yu H, Xu J, Li X. Ni-MOF-74 as sensing material for resonant-gravimetric detection of ppb-level CO. *Sensors Actuat B Chem.* 2018;262:562–569. https://doi.org/10.1016/j.snb.2018.02.058.

115. Yang Z, Zhang D, Wang D. Carbon monoxide gas sensing properties of metal-organic frameworks-derived tin dioxide nanoparticles/molybdenum diselenide nanoflowers. *Sensors Actuat B Chem.* 2020;304. https://doi.org/10.1016/j.snb.2019.127369.

116. Chocarro-Ruiz B, Pérez-Carvajal J, Avci C, et al. A CO_2 optical sensor based on self-assembled metal-organic framework nanoparticles. *J Mater Chem A.* 2018;6(27):13171–13177. https://doi.org/10.1039/c8ta02767f.

117. Senthilkumar S, Goswami R, Smith VJ, Bajaj HC, Neogi S. Pore wall-functionalized luminescent Cd(II) framework for selective CO_2 adsorption, highly specific 2,4,6-trinitrophenol detection, and colorimetric sensing of Cu^{2+} ions. *ACS Sustain Chem Eng.* 2018;6(8):10295–10306. https://doi.org/10.1021/acssuschemeng.8b01646.

118. Singh M, Solanki P, Patel P, Mondal A, Neogi S. Highly active ultrasmall ni nanoparticle embedded inside a robust metal-organic framework: remarkably improved adsorption, selectivity, and solvent-free efficient fixation of CO_2. *Inorg Chem.* 2019;58(12):8100–8110. https://doi.org/10.1021/acs.inorgchem.9b00833.

119. Li PX, Yan XY, Song XM, et al. Zirconium-based metal-organic framework particle films for visible-light-driven efficient photoreduction of CO_2. *ACS Sustain Chem Eng.* 2021;9(5):2319–2325. https://doi.org/10.1021/acssuschemeng.0c08559.

120. Brandt P, Xing SH, Liang J, et al. Zirconium and aluminum MOFs for low-pressure SO_2 adsorption and potential separation: elucidating the effect of small pores and NH2 groups. *ACS Appl Mater Interfaces.* 2021;13(24):29137–29149. https://doi.org/10.1021/acsami.1c06003.

121. Li Z, Zhang Y, Zhang H, Jiang Y, Yi J. Superior NO_2 sensing of MOF-derived indium-doped ZnO porous hollow cages. *ACS Appl Mater Interfaces.* 2020;12(33):37489–37498. https://doi.org/10.1021/acsami.0c10420.

122. Schulz M, Gehl A, Schlenkrich J, Schulze HA, Zimmermann S, Schaate A. A calixarene-based metal-organic framework for highly selective NO_2 detection. *Angew Chem Int Ed.* 2018;57(39):12961–12965. https://doi.org/10.1002/anie.201805355.

123. Li C, Meng W, Xu X, et al. High performance solid electrolyte-based NO_2 sensor based on $Co_3V_2O_8$ derived from metal-organic framework. *Sensors Actuat B Chem.* 2020;302. https://doi.org/10.1016/j.snb.2019.127173.

124. Liu Y, Wang R, Zhang T, Liu S, Fei T. Zeolitic imidazolate framework-8 (ZIF-8)-coated In_2O_3 nanofibers as an efficient sensing material for ppb-level NO 2 detection. *J Colloid Interface Sci.* 2019;541:249–257. https://doi.org/10.1016/j.jcis.2019.01.052.

125. Choi SJ, Choi HJ, Koo WT, Huh D, Lee H, Kim ID. Metal-organic framework-templated PdO-Co_3O_4 nanocubes functionalized by SWCNTs: improved NO_2 reaction kinetics on flexible heating film. *ACS Appl Mater Interfaces.* 2017;9(46):40593–40603. https://doi.org/10.1021/acsami.7b11317.

126. Tan J, Hussain S, Ge C, et al. ZIF-67 MOF-derived unique double-shelled Co_3O_4/$NiCo_2O_4$ nanocages for superior gas-sensing performances. *Sensors Actuat B Chem.* 2020;303. https://doi.org/10.1016/j.snb.2019.127251.

127. Cao YY, Guo XF, Wang H. High sensitive luminescence metal-organic framework sensor for hydrogen sulfide in aqueous solution: a trial of novel turn-on mechanism. *Sensors Actuat B Chem.* 2017;243:8–13. https://doi.org/10.1016/j.snb.2016.11.085.

128. Meng Z, Aykanat A, Mirica KA. Welding metallophthalocyanines into bimetallic molecular meshes for ultrasensitive, low-power chemiresistive detection of gases. *J Am Chem Soc.* 2019;141(5):2046–2053. https://doi.org/10.1021/jacs.8b11257.

129. Yilmaz G, Yam KM, Zhang C, Fan HJ, Ho GW. In situ transformation of MOFs into layered double hydroxide embedded metal sulfides for improved electrocatalytic and supercapacitive performance. *Adv Mater.* 2017;29(26). https://doi.org/10.1002/adma.201606814.

130. Chen K, Sun Z, Fang R, Shi Y, Cheng HM, Li F. Metal-organic frameworks (MOFs)-derived nitrogen-doped porous carbon anchored on graphene with multifunctional effects for lithium-sulfur batteries. *Adv Funct Mater*. 2018;28(38). https://doi.org/10.1002/adfm.201707592.

131. Yin N, Wang K, Wang L, Li Z. Amino-functionalized MOFs combining ceramic membrane ultrafiltration for Pb(II) removal. *Chem Eng J*. 2016;306:619–628. https://doi.org/10.1016/j.cej.2016.07.064.

132. Wu JX, Yan B. Lanthanides post-functionalized indium metal-organic frameworks (MOFs) for luminescence tuning, polymer film preparation and near-UV white LED assembly. *Dalton Trans*. 2016;45(46):18585–18590. https://doi.org/10.1039/c6dt03738k.

133. Armstrong M, Sirous P, Shan B, et al. Prolonged HKUST-1 functionality under extreme hydrothermal conditions by electrospinning polystyrene fibers as a new coating method. *Microporous Mesoporous Mater*. 2018;270:34–39. https://doi.org/10.1016/j.micromeso.2018.05.004.

134. Hu Z, Mahdi EM, Peng Y, et al. Kinetically controlled synthesis of two-dimensional Zr/Hf metal-organic framework nanosheets via a modulated hydrothermal approach. *J Mater Chem A*. 2017;5(19):8954–8963. https://doi.org/10.1039/c7ta00413c.

135. Armstrong MR, Senthilnathan S, Balzer CJ, Shan B, Chen L, Mu B. Particle size studies to reveal crystallization mechanisms of the metal organic framework HKUST-1 during sonochemical synthesis. *Ultrason Sonochem*. 2017;34:365–370. https://doi.org/10.1016/j.ultsonch.2016.06.011.

136. Martín-Jimeno FJ, Suárez-García F, Paredes JI, et al. A "nanopore lithography" strategy for synthesizing hierarchically micro/mesoporous carbons from ZIF-8/graphene oxide hybrids for electrochemical energy storage. *ACS Appl Mater Interfaces*. 2017;9(51):44740–44755. https://doi.org/10.1021/acsami.7b16567.

137. Zhou K, Mousavi B, Luo Z, Phatanasri S, Chaemchuen S, Verpoort F. Characterization and properties of Zn/Co zeolitic imidazolate frameworks vs. ZIF-8 and ZIF-67. *J Mater Chem A*. 2017;5(3):952–957. https://doi.org/10.1039/C6TA07860E.

138. Khosravi A, Mokhtari J, Naimi-Jamal MR, Tahmasebi S, Panahi L. Cu$_2$(BDC)$_2$(BPY)-MOF: an efficient and reusable heterogeneous catalyst for the aerobic Chan-Lam coupling prepared via ball-milling strategy. *RSC Adv*. 2017;7(73):46022–46027. https://doi.org/10.1039/C7RA09772G.

139. Kang P, Jung S, Lee J, Kang HJ, Lee H, Choi M-G. Anion induced structural transformation in silver-(3,6-dimethoxy-1,2,4,5-tetrazine) coordination polymers under mechanochemical conditions. *Dalton Trans*. 2016;45(30):11949–11952. https://doi.org/10.1039/C6DT01834C.

140. Li P, Cheng F-F, Xiong W-W, Zhang Q. New synthetic strategies to prepare metal–organic frameworks. *Inorg Chem Front*. 2018;5(11):2693–2708. https://doi.org/10.1039/C8QI00543E.

141. Tella AC, Mehlana G, Alimi LO, Bourne SA. Solvent-free synthesis, characterization and solvent-vapor interaction of zinc (II) and copper (II) coordination polymers containing nitrogen-donor ligands. *Zeitschrift für anorganische und allgemeine Chemie*. 2017;643(8):523–530. https://doi.org/10.1002/zaac.201600460.

142. Usman KAS, Maina JW, Seyedin S, et al. Downsizing metal-organic frameworks by bottom-up and top-down methods. *NPG Asia Mater*. 2020;12(1). https://doi.org/10.1038/s41427-020-00240-5.

Chapter 5

Graphene oxides and its composites as new generation adsorbents for remediation of toxic pollutants from water: An overview

Arunima Nayak*, Brij Bhushan, Nupur Kukretee, and Shreya Kotnala
Department of Chemistry, Graphic Era University, Dehradun, India
*Corresponding author.

1 Introduction

Because of rapid industrialization, urbanization, and modern lifestyle, human activities have increased levels of noxious pollutants in aqueous streams.[1,2] The voluminous generation of wastewater has not only increased, but also such wastewaters have shown the presence of diversified pollutants in the form of inorganic contaminants (heavy metal ions, inorganic ions), organic contaminates (dyes, pesticides, herbicides, pharmaceuticals, personal care products, polyhydroxyalkanoates, phenols); all of these are the cause of various diseases.[3,4] Many of them may also be carcinogenic.[5,6] Many of such pollutants are persistent and are bio-accumulative in nature. Although the fresh water on Earth is a meagre 3% but only 0.5% of which is available for drinking purposes, thereby making it a limiting natural resource.[7] Rapid environmental degradation and pollution have further reduced the availability of fresh water. This has raised concerns among environmentalists, and there is the need for low-cost water recycling. Water is a limiting natural resource, and hence it is recycling by treatment technologies for its reuse has been a major environmental research area. Among the treatment technologies currently being employed to remove pollutants, adsorption is a cost-effective and environmentally friendly alternative that has demonstrated high efficiency irrespective of the nature of the pollutant type.[8–11]

The characteristics of the adsorption technology depend on the physico-chemical quality of the adsorbent.[9] Because of the diversified presence of contaminants and because of the complexity of wastewater, the efficiency of the adsorption technology is critically dependent on the quality of the adsorbents. The high quality of the adsorbents depends on superior characteristics like high surface-to-volume ratio, well-distributed porosity, and surface functionality.[12,13] With the advent of nanotechnology, different nano-sized adsorbents like the carbon nanotubes (CNTs), graphite, metal nanoparticles, nanocomposites, etc., have been used in wastewater purification.[14–16] Graphene has recently captured the attention of chemists, environmentalists, metallurgists, engineers, etc., as an advanced material on account of its properties like high mechanical strength (Young's modulus of about 1000 GPa),[17] high electrical conductivity (zero bandgaps),[18] and high optical transparency (absorption of approximately 2.3% toward visible light).[19,20] Some of its other unique properties are enhanced physicochemical properties demonstrating excellent chemical and thermal stability (2000 and 5000 W/m K) and exhibiting zero thickness.[21] Graphene consists of sp^2 carbon atoms bonded to each other in a honeycomb-like lattice nanostructure and arranged in a single layer (two dimensional). Since its discovery in 2004,[22] and because of its unique properties, graphene has been used as a base material in various fields of application. More recently, characteristics like its high surface area (2360 m²/g), large surface-binding active sites, and a great delocalized π-electron system have enabled graphene to be used as advanced adsorbents for wastewater purification.[23] Graphene has demonstrated superior binding capacity for trapping diverse types of both organic and inorganic contaminants. Many studies have shown that the extended delocalized π-electron system in graphene could trap various aromatic moieties in organic-based contaminates via a strong hydrophobic effect or via π-π interactions; such interactions have helped in the separation of the contaminants from water bodies.

Despite the better properties of graphene as an adsorbent for water purification, its physical handling is a significant problem in the sense that graphene sheets are insoluble in most solvents.[14] In an aqueous medium, graphene sheets form

Sustainable Materials for Sensing and Remediation of Noxious Pollutants. https://doi.org/10.1016/B978-0-323-99425-5.00014-1

irreversible agglomerates or roll back into graphitic sheets via their π-π stacking and van der Waals interaction. This has resulted in a decreased surface area and loss of properties, which has hindered its use in wastewater treatment, and various other applications.[24] Compared to pristine graphene, their oxides (GO) and their reduced form (rGO) have exhibited greater stability and better handling features. This is because of the higher proportions of oxygenated functional groups in GO and rGO.[25–27] These features also make them favorable for further tuning to other graphene-based derivatives or composites. Both GO and rGO have exhibited lower production cost which is again an added advantage over pristine graphene. Besides graphene oxides (GO and rGO), graphene-based composites [prepared via incorporation of various functionalities of metal nanoparticles, organic monomers, metal-organic frameworks (MOFs), polymers, etc.] have exhibited not only better stability but also have demonstrated better binding to different aqueous phase pollutants as compared to that of pristine graphene-based adsorbents. New functionalities added to graphene oxide (GO, rGO) because of the incorporation of various organic and or inorganic moieties via covalent bond interactions have caused enhanced solubility of graphene-based composites in both organic and inorganic solvents.[28]

Various reviews have been published highlighting the importance of GO as adsorbents. Discussions have been placed on GO for removal of rare-earth ions[29]; GO-chitosan composite for broad-spectrum adsorbents in water purification[30]; graphene family nano-adsorbents for removal of toxic pollutants[31,32]; GOs for removal of heavy metal, and precious metal ions[33]; graphene nanocomposites for cadmium ion removal[34]; graphene-based adsorbents for decontamination of wastewater[35]; and more lately, graphene-based adsorbents for removal of noxious pollutants.[7,25–27] The outline of the present chapter is to put forth updated and latest information and put forth discussions on the (Fig. 1)

→ synthetic routes applied for fabrication of the graphene and its derivatives,
→ application of GO-based nanocomposites as advanced adsorbents for remediation of contaminates from water bodies, and finally,
→ the management of the spent adsorbents.

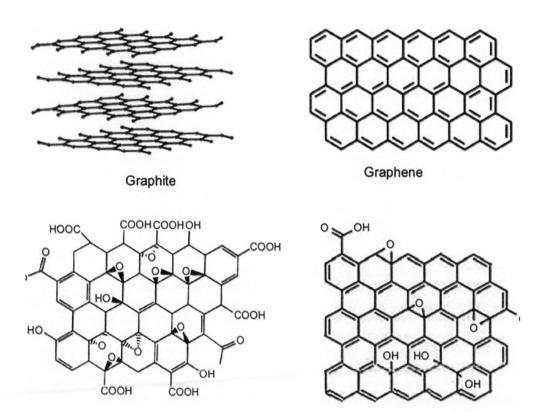

FIG. 1 Structure of graphite, graphene, graphene oxide, and reduced graphene oxide. *Adapted from Bai R.G., Muthoosamy K., Manickam S. Graphene-based 3D scaffolds in tissue engineering: fabrication, applications, and future scope in liver tissue engineering. Int J Nanomedicine. 2019;14:5753–5783 permission for Elsevier.*

2   Fabrication of graphene oxides/its derivatives and their properties

Graphene is an allotrope of carbon in which the sp^2 hybrid carbon atoms are covalently bonded in a single layer to form a hexagonal honeycomb-like 2D structure.[36] Its unique properties as demonstrated in its high mobility (2×10^5 cm^2/v/s), high mechanical strength (Young's modulus of $\sim$100GPa), high electrical conductivity (due to zero bandgap), thermal conductivity ($\sim$5000 W/m/K), as well as high optical transmittance ($\sim$98%) have rendered its widespread use in electronics and energy. Graphene has a large aspect ratio, and its surface area of 2630 m^2/g makes it suitable as an advanced adsorbent for use in medicine, drug delivery, biotechnology, and pollution abatement of air and water. Graphene can be wrapped into fullerenes, rolled into CNTs, or stacked into graphite.

The properties of graphene depend on the synthesis routes adopted for its fabrication, which ultimately bears its size, shape, and defects.[31] The synthesis methods used are top-down (extracting graphene from graphite or other carbon sources) and bottom-up (fabricating graphene from simple carbon sources) approaches.[37] Among the top-down synthesis methods adopted are the exfoliation of natural or synthetic graphite, liquid-phase exfoliation, graphite intercalation compounds (GICs), and electrochemical exfoliation. The top-down approaches have resulted in high yields of low-cost graphene. But the quality of such graphene sheets is underrated because of defects introduced by exfoliation.[38] Bottom-up approaches for the synthesis of graphene involve direct chemical methods,[39] epitaxial growth, and chemical vapor deposition methods. Such methods result in the fabrication of defect-free graphene with superior physical properties but suffer from the disadvantages of higher fabrication cost, more time-consuming, and scalability issues.[38] Top-down synthesis methods are more favorable for fabricating GO, rGO, graphene, and its nanocomposite. Exfoliation methods used to fabric graphene from aqueous solutions of surfactants or organic solvents/ionic liquids, etc., using ultrasonication have been reported in various studies.[40] But mass production and application of single-layer graphene are not cost-efficient. Other technologies used during exfoliation of graphite are via using mechanical and/or chemical methods.

Poor solubility of pristine graphene and its agglomeration tendency in the solution phase has resulted in hurdles faced during its handling and application.[41–43]

GO fabricated from oxidation of graphite in protonated solvents has a similar hexagonal single-layered 2D structure to graphene but has a rich presence of oxygen functionalities (hydroxyl, carboxyl, carbonyl, alkoxy, etc.).[44] Studies have shown that GO can be synthesized via either the Brodie[45] or Staudenmaier[46] or Hummers method[47] or with some variation of these methods. While all methods involve the oxidation of graphite in the presence of strong acids, the difference lies in the oxidizing agent and acid used. For example, both Brodie's and Staudenmaier's methods involve the use of potassium chlorate and nitric acid as an oxidizing agent and acid, respectively. On the other hand, Hummer's method involves the oxidation of graphite in the presence of potassium permanganate and sulfuric acid.

Irrespective of the methods used for fabrication, GO has demonstrated many advantages over pristine graphene-like ease of fabrication, better solubility, enhanced stability, and greater tendency to form multi-functional nanocomposite materials due to the presence of rich surface functionalities.[48] Although GO has proved to be a better and more stable adsorbent than pristine graphene, the rich oxygen functionality in GO has resulted in weak interactions with different water-based anionic contaminates because of the electrostatic repulsion between them.[49] Also, the difficulty faced in separating GO-based adsorbents from treated water has caused concerns of water recontamination.[41] Besides the lower adsorption efficiency, GO has demonstrated weaker conductivity because of the disruption of the sp^2 bonding network, mainly because of the creation of oxygen-rich functionality. Reduced GO, which is fabricated via thermal, chemical, or electrochemical reduction of GO, has a close structural resemblance to pristine graphene and lesser oxygen-rich functionalities than GO.[44] rGO has attracted the research community's attention mainly on account of its effectiveness for the cost-effective and mass production of graphene-based materials. Conductivity is partially recovered in rGO because of the close structural resemblance to graphene. rGO has demonstrated a greater tendency to be modified via non-covalent physisorption of polymers and/or nano-sized molecules like metals or metal oxides via either π-π stacking or van der Waals interactions.[42] The resulting nanocomposites have demonstrated new functionalities and new properties.[50] The anchoring of nano-sized metals/metal oxides or polymers onto the two-dimensional graphene structure has resulted in reduced agglomeration in the aqueous phase and demonstrated increased surface area, resulting in enhanced adsorption capacity more excellent selectivity to bind to various organic or inorganic contaminants as compared to pristine graphene.

The fabrication methods used for graphene nanocomposites derived from inorganic nano-moieties, as evident from literature, are ex situ hybridization[51] and in situ crystallization.[52] The ex situ process involves mixing GO/rGO nanosheets and inorganic nanoparticles that are either synthesized in the laboratory or commercially purchased. In situ crystallization methods are chemical reduction, electroless deposition, sol-gel, hydrothermal, electrochemical deposition, thermal evaporation, and in situ self-assembly. A detailed discussion on the unique fabrication method is put forth in a review published by.[42] The different fabrication methods used for graphene-filled polymer composites are solution mixing, melt

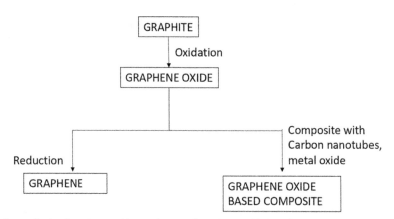

FIG. 2 General procedure for synthesis of graphene oxide, graphene, and graphene-oxide-based composites.

compounding, and in situ polymerization. Graphene and its derivatives are known to enhance the polymeric matrix's mechanical, electrical, and thermal properties.[15] The dispersibility and bonding of graphene-based filler materials onto the polymer matrix are essential factors that affect the properties of the graphene nanocomposites. Polymer functionalized graphene nanosheets are fabricated using graphene derivatives which act as templates for polymer impregnation via either covalent or non-covalent functionalization. The functional groups of polymers react with the oxygenated functionalities of the GO/rGO surface via covalent interactions.[53] Non-covalent functionalization, on the other hand, involves interactions via van der Waals force, electrostatic interaction, or π-π stacking. Such a type of functionalization is easy for the fabrication of graphene-polymer nanocomposites as it does not alter the structural aspects of the GO/rGO. Also, this method favors the tailoring of the various properties of graphene-based nanosheets.[15] Fig. 2 represents a general procedure for the synthesis of graphene, GO, and GO-based composites.

3 Application of graphene oxide-based nanocomposite adsorbents for remediation of organic pollutants

The advancement in the design of novel graphene-based adsorbents is at the leading edge among the most recent exploration in material designing. Toxic contaminants in water are mainly organic contaminants, including dyes, drugs, antibiotics, pesticides, and oils. Once delivered into the aqueous environment, such contaminates might cause an extreme threat to humans and the encompassing plants and animals. Thus, there is an urgent need to eliminate organic-based pollutants from the aqueous bodies. Many advancements have been carried out to fabricate novel GO-based nanocomposite adsorbents for wastewater remediation. This is mainly on account of its novel characteristics.

Yang et al.[54] fabricated an initial GO by a modified Hummers process followed by its crosslinking with β-cyclodextrin to develop a composite of β-cyclodextrin/graphene oxide (β-CD/GO). β-cyclodextrin was used as a crosslinking agent as it is known to form a host-guest complex by encapsulating the pollutants. β-Cyclodextrin, along with the GO nanosheets, resulted in the formation of plywood-like structures with cages of nano-scale that made the composite a more efficient adsorbent for the organic dye methylene blue (MB). The adsorption potential of the as-synthesized composite was evaluated for MB dye. An optimized adsorbent dosage of 0.04 g/L was reported to bind to MB at a maximum capacity of 76.4 mg/g under optimum operating conditions (time of 60 min, temperature of 70°C).

To remove Acid blue 92 dye from an aqueous solution, Hoseinzadeh et al.[55] developed a green adsorbent using zeolitic imidazolate framework (ZIF-8) with GO. They functionalized it by 3-amiopropyltrimethoxy silane (ATPMS) to produce a ZIF-8/GO/APTMS composite. The developed nanocomposite displayed an excellent adsorption capacity of 371.8 mg/g for Acid blue 92 dye under the following optimum conditions: pH 2.2, temp 25°C, an adsorbent dosage of 4 mg, contact time 60 min, adsorbate concentration of 20 mg/L. Experimental data well suited the pseudo-second-order kinetic model, and the major mechanisms governing the adsorption process were π-π stacking interaction, electrostatic interaction, and hydrogen bonding.

The solution combustion method was employed by Bayantong et al.[56] to fabricate different metal (M = Cu, Co, and Ni) ferrite-based GO nanocomposites (MFe$_2$O$_4$@GO). The fabricated composites of CuFe$_2$O4@GO, CoFe$_2$O$_4$@GO, and NiFe$_2$O$_4$@GO exhibited maximum binding to MB dye (of an initial concentration of 16 mg/L) at temperatures of 25°C. The maximum binding capacities reported were 25.81, 50.15, and 76.34 mg/g, respectively. Because of the high surface area and mesoporosity, $NiFe_2O_4$@GO adsorbent displayed the highest adsorption capacity of 76.34 mg/g. The binding mechanism reported was that of chemical adsorption. A nanocomposite from the zirconium-based MOF and GO (Uio-66-$(OH)_2$GO was fabricated via hydrothermal strategy, and their adsorption consequences for the dye (MB) and antibiotic (tetracycline-TC) were reported by Sun et al.[57] as 96.69 and 37.96 mg/g, respectively, at a pH of 11. Electrostatic interactions, π-π stacking interactions, and hydrogen bonding were the key mechanisms reported for the adsorption of MB and TC pollutants onto the composite adsorbent.

A biocompatible, biodegradable nanoporous composite hydrogel (NHA) was fabricated via the incorporation of an organic copolymer (starch grafted poly-acryl amide) and a mineral (hydroxyapatite) onto GO.[58] A free radical graft co-polymerization of acrylamide monomer onto the starch polymer was adopted during the fabrication. The as-synthesized adsorbent showed high removal potential for a model Malachite Green (MG) dye (adsorption capacity of 297 mg/g) under the optimized agitation time of 60 min, pH of 10, and taking an initial dye concentration of 100 mg/L. The adsorbent demonstrated good reusability for a consecutive five cycles of adsorption-desorption. A maximum of 27% of the adsorbent MG dye was found to be liberated from the adsorbent after the fifth cycle. Results thus have indicated the good potential of NHA as an advanced adsorbent for the model MG dye.

Sadegh et al.[59] fabricated a magnetic reduced GO nanocomposite (RGO-Fe_3O_4) via chemical precipitation with Fe_3O_4 for simultaneous removal of two cationic dyes viz. Rhodamine B(Rh-B) and MG from aqueous solution. The objective was to test the efficacy of RGO-Fe_3O_4 in an industrial effluent containing diverse types of dyes. Batch studies confirmed the optimum conditions for maximum adsorption performance with a pH of 8, mixture adsorbate concentration of each dye of 16 mg/L, a contact time of 20 min, and adsorbent dosage of 0.028 g. The dyes binding was found to be in the form of a monolayer, while kinetic studies revealed the applicability of pseudo-second-order kinetics. The adsorbent further demonstrated good reusability up to a maximum of four consecutive adsorption-desorption cycles. Because of its magnetic properties, the adsorbent demonstrated good separability after its use from the aqueous medium.

In another study, Jinendra et al.[60] fabricated a reduced graphene oxide-nickel (RGO-Ni) nanocomposite to estimate the adsorption potential for removing a model Rhodamine B dye from an aqueous solution. The maximum uptake of dye (65.31 mg/g) by the composite adsorbent was reported under optimum pH 8, contact time of 60 min, adsorbent dose of 0.5 mg, and adsorbate concentration of 50 mg/100 mL. A sodium alginate/graphene oxide (Alg/GO) composite bead was fabricated by Ajeel et al.,[61] and the reported performance for a model MB dye from aqueous solution was 91.26% at pH 7.78, contact time of 12 h, and temperatures of 25°C.

Noreen et al.[62] fabricated a series of polymeric GO nanocomposites like GO/polyaniline (PAN), GO/polypyrrole (PPy), and GO/polystyrene (PS) for studying the removal efficiency for a model antacid orange RL (AO-RL) dye. Maximum adsorption of the AO-RL dye was reported as follows: 39.54, 44.76, and 21.43 mg/g for GO/PAN, GO/PPy, and GO/PS at a pH of 2, 6, and 3, respectively. In all cases, the optimum adsorbent dosage was 0.01 g/50 mg/L of the dye, and the temperature maintained was 35°C. Irrespective of the nature of the polymer used, the binding of the dye was monolayer. Desorption studies demonstrated the use of 0.05 N NaOH brought about maximum desorption of the dye.

In a study by Verma et al.,[63] graphene oxide-manganese dioxide (GO-MnO_2) nanocomposite fabricated via hydrothermal method was used for the removal of two model dyes [MB and methyl orange (MO) dyes]. Maximum binding of 149.25 and 178.25 mg/g was reported for the MO and MB dyes, respectively, with the binding affinity being a single layer (Langmuir isotherm model). Ahlawat et al.[64] initially prepared a GO via a modified Hummer's process followed by its fabrication with activated carbon to form a composite (GO@AC). To investigate different adsorption factors, batch tests were performed to expel MB and MO. A comparative study revealed the higher potential of GO@AC nanocomposite as an adsorbent for both dyes (adsorption efficiency of GO: 276.5, and of GO@AC: 423.15 mg/g for MB dye, and of GO: 220, and of GO@AC: 346.51 mg/g for MO, respectively). Irrespective of the nature of the adsorbent type, adsorption kinetics was best described by the pseudo-second-order kinetic model. A multi-layered adsorption mechanism was found to be operative. The possibility of its reuse was demonstrated in GO@AC, with its efficiency for desorption being 92% and 85% for MB and MO dyes, respectively. Composite aerogels comprising GO and recovered cellulose (RCE/GO) were fabricated by Ren et al.[65] via a solution mixing-regeneration and freeze-drying method. The RCE/GO nanocomposite aerogel displayed a 3D organization of thin-walled porous structure with an enormous explicit surface region. The maximum adsorption capacity demonstrated for a model MB dye was reported to be 68 mg/g at a pH of 6 and under a contact time of 30 min. The binding mechanism was electrostatic.

Certain noxious dyes like Trypan blue and Congo red are highly water-soluble and have low biodegradability. Thus, Karthikeyan et al.[66] fabricated a nanocomposite-based adsorbent via hydroxyapatite-decorated reduced graphene oxide (HA@rGO) through a hydrothermal method and utilized them for the expulsion of Trypan blue and Congo red dyes from an aqueous medium. In the batch tests, the resultant nanocomposites demonstrated high removal rates in a wide pH range. They displayed good adsorption efficiencies of 146.51 mg/g (Trypan blue) and 150.09 mg/g (Congo red) individually. A few portrayal examination results showed that the expulsion execution of the HA@rGO nanocomposites for dyes was mainly determined by electrostatic, hydrophobic, π-π collaborations, and H-bonding between the adsorbents and dye molecules. After adsorption, the depleted HA@rGO nanocomposites could be recovered in antacid arrangement and reused all through five progressive adsorption-desorption cycles.

In another study,[67] alginate supported reduced graphene oxide@hydroxyapatite (rGO@HAP-Alg) hybrids were fabricated via the co-precipitation method for the removal of Reactive Blue 4 (RB4), Indigo Carmine (IC), and Acid Blue 158 (AB158) azo colors; the demonstrated efficiencies were reported as 45.56, 47.16, and 48.26 mg/g, respectively. The thermodynamic parameters showed the endothermic nature of the adsorption cycle. The adsorption framework was pH-subordinate and showed the highest dye expulsion at pH 6–7. The mechanism reported was electrostatic interactions, surface complexation, and the hydrogen bonding between the adsorbate and adsorbent during the adsorption cycle.

A free radical method was employed to fabricate acrylamide/graphene oxide-bonded sodium alginate (AM-GO-SA) hydrogels to remove crystal violet dye by Pashaei-Fakhri et al.[68] The adsorption interaction tests revealed its dependence on pH. AM-GO-SA hydrogels with a surface area of $6.983 m^2/g$ demonstrated a maximum adsorption capacity of 100.30 mg/g for crystal violet dye. Another adsorbent prepared from reduced graphene oxide functionalized with L-cysteine and polyaniline (L-Cys/rGO//PANI) was viably synthesized by a straightforward and easy strategy for the removal of Congo red dye.[69] The ideal conditions for dye removal were the contact time of 10 min and 0.025 g of adsorbent dosage, and maximum adsorption was observed to be 56.57 mg/g.

Novel nanostructures of 3D GOs and CNTs in a weight proportion of 1:5 were prepared by Hu et al.[70] via a facile freeze-drying technique. The nanostructure demonstrated a high surface area of $257.6 m^2/g$, and the adsorption capacity was 248.48 mg/g for a cationic dye (Rhodamine B, RhB) and 66.96 mg/g for an anionic dye (MO).

Besides the dyes, another important class of organic pollutants showing their prevalence in water bodies is the pharmaceuticals. It is because of the rapid increase in the use of pharmaceuticals that have resulted in their presence in water bodies. Passive intake of such contaminates caused harm to the aquatic and plant community. It is a matter of concern to the oceanic climate and human wellbeing as well. Measures have been adopted on a laboratory scale to fabricate GO-based nanocomposite adsorbents for the removal of such toxic water pollutants. A graphene nanoplatelet/boron nitride composite aerogel (GNP/BNA) was fabricated by Han et al.[71] via a one-pot foam gel casting/nitridation. The same was used as an adsorbent for removing a model drug, ciprofloxacin (CIP), from its aqueous solution. Approximately 99% of CIP was removed at a near-neutral pH and under room temperature conditions via the application of a low adsorbent dosage (250 mg/L). In yet another study, GNPs have demonstrated high adsorption efficiency of 210.08 mg/g for model antibiotics sulfamethoxazole (SMX) and 56.21 mg/g for model analgesic acetaminophen (ACM).[72] A sustainable adsorbent based on GO [NiZrAl-layered double hydroxide-graphene oxide-chitosan (NiZrAl-LDH-GO-CS NC)] was fabricated and used for the removal of nalidixic acid (NA).[73] The adsorbent demonstrated fast kinetics and high adsorption of 277.79 mg/g for NA. The nature of the binding process was chemisorption. A one-step hydrothermal method was used by Feng et al.[74] for the fabrication of a β-cyclodextrin-immobilized reduced graphene oxide composite (β-CD/rGO) for the removal of naproxen from its aqueous suspension. The adsorbent showed high removal of 361.85 mg/g under optimum pH and temperature conditions of 6.5°C and 40°C, respectively. The adsorbent dosage requirement was as little as 0.1 g/L, and the results were highly promising. Isotherm and kinetic studies revealed that the major binding mechanisms of naproxen onto β-CD/rGO composite were hydrogen bonding and π-π interactions.

Segovia et al.[75] synthesized a progression of carbon xerogel/graphene hybrids (GO/CX) by adding an increasing amount of GO solution. The obtained samples were functionalized with urea. Carbon xerogel/graphene hybrids (GO/CX) were evaluated as adsorbents for metronidazole (MNZ) expulsion from aqueous solutions under various functional conditions to decide the mechanism of adsorption. The maximum adsorption capacities displayed by GO/CX composites were in the range of 110–166 mg/g for MNZ at pH 5 and 298 K. The binding mechanism was concluded to be due to π-π dispersive interactions, electrostatic interactions, and hydrogen bonding. A nanocomposite was fabricated from chemically reduced GO via the incorporation of magnesium ascorbyl phosphate (GBM-MAP).[76] The adsorbent demonstrated high removal potential for a model Bisphenol A (BPA) (adsorption capacity of 324 mg/g at 20°C). The major binding mechanisms determined were hydrogen bonding, electrostatic interaction, hydrophobic interaction, and π-π stacking interactions.

Adsorption of sodium diclofenac (DCF) on graphene oxide nanosheets (GON) was assessed by Guerra et al.[77] The best working condition was observed at the normal pH of 6.2 and 0.25 g/L adsorbent dosage. The most extreme adsorption limit of DCF at 25°C was 128.74 mg/g with a removal rate of 74% in 300 min. The cycle was ideal and spontaneous with adsorptive limit diminishing with increasing temperature.

Zhang et al.[78,79] fabricated graphene aerogels (S-rGA) by a simple $NaHSO_3$ chemical reduction method for the expulsion of tetrabromobisphenol A (TBBA) from aqueous solutions. The batch results best fitted the Langmuir model with maximum capacity found to be 128.37 mg/g under optimum pH and temperature conditions (pH of 7 and temperature of 25°C). The study concluded the high removal efficiency was undertaken at very low adsorbent dosage conditions. The major binding mechanism was that of π-π interaction. GON were demonstrated to be proficient adsorbents for the removal of 17β-Estradiol (E2) from an aqueous solution.[80] Batch adsorption studies showed that the maximum binding was 149.4 mg/g, which took place at 298 K, and the binding followed the monolayer mechanism. The binding interaction of E2 for GO was π-π interactions as well as hydrogen bonding. Al-Khateeb et al.,[81] in a study conducted in 2014, carried out studies testing the efficiency of GNPs for removal of model drugs aspirin (ASA), acetaminophen (APAP), and caffeine (CAF) from aqueous solution. Characterization studies demonstrated that the GNP had a high specific surface area of 635.2 m²/g because of which the maximum adsorption capacities as exhibited for ASA, CAF, and APAP were 12.98, 19.72, and 18.07 mg/g, respectively, under optimum conditions pH of 8, contact time of 10 min, and temperature 23°C.

Another class of organic-based pollutants is pesticides/insecticides and herbicides used to enhance agricultural productivity. But their toxicity to the aquatic community has raised alarms among environmentalists. In this respect, GO-based nanocomposite adsorbents have shown good removal potential under laboratory conditions for model pesticides. In a study conducted by Nikou et al.,[82] a nanocomposite of ZIF-8 MOF and graphene oxide (GO/ZIF-8) was fabricated by a simple solvothermal method and was utilized for the removal of diazinon (DIZ) and chlorpyrifos (CLP) pesticides from the aqueous environment. The study's outcome showed that the maximum removal efficiencies of the nanocomposite for CLP and DIZ were 83% and 73%, respectively. In another study, Mahdavi et al.[83] fabricated a GO-based nano-adsorbent via the functionalization of GO by aminoguanidine (AGu@mGO(R)). Rice husk was used to fabricate the graphene-oxide (mGO) via a modified Hummer's technique which was later functionalized by aminoguanidine to develop AGu@mGO(R) nanoparticles. AGu@mGO(R) proficiency was assessed for the elimination of chlorpyrifos. Under the optimum conditions of pH 6.5, the adsorbent dosage of 10 mg, and contact time of 30 min, the nanocomposite displayed a maximum uptake capacity of 85.47 mg/g for chlorpyrifos.

Atrazine (an endocrine-disrupting herbicide) is a noxious pollutant whose removal from water bodies has demanded attention by researchers. A fabricated nano-adsorbent removed atrazine by Khawaja et al.[84] The nano-adsorbent was fabricated from magnetic graphene oxide crosslinked with carboxymethylcellulose (GO-CMC-Fe). The developed composite presented a high adsorption capacity of 193.8 mg/g for atrazine when the adsorbent dosage of 100 mg was used. The process of adsorption was endothermic and spontaneous.

In another study, Muthusaravanan et al.[85] fabricated GO nanosheets using Hummers' strategy with a minor adjustment for atrazine expulsion from simulated wastewater. The response surface methodology (RSM) utilizing Box Behnken plan (BBD) was used effectively for determining the optimum environmental conditions like an adsorbent dosage of 121.45 mg/L, initial adsorbate concentrate of 27.03 mg/L, temperature 27.69°C, pH 5.37, and time 180 min. The atrazine adsorption onto GO was observed to be higher in acidic pH and at a lower temperature. Major mechanisms responsible for atrazine adsorption onto the surface of GON were the π-π interactions and hydrogen bonding. The GO nanosheets displayed high atrazine uptake of 138.19 mg/g at 318 K. The reliability investigation showed that GO nanosheets could be successfully reused utilizing 0.01 N NaOH up to six cycles adsorption-desorption.

A summary of adsorption capacities of various graphene-based nano-adsorbents for removing organic-based adsorbates like the dyes and other hazardous organic pollutants (pharmaceuticals, pesticides, phenolic compounds, etc.) are depicted in Tables 1 and 2, respectively.

4 Adsorption of graphene-based adsorbents for remediation of inorganic pollutants

The pollution of water bodies by inorganic components such as copper, chromium, cadmium, arsenic, lead, iron, mercury, silver, zinc, nickel, cobalt, and manganese has increased over the recent years because of the uncontrolled release of waste effluents from numerous businesses, like metal industries, painting, mining, purifying, tanneries, printing, vehicle assembling, and oil refining, overflow from an agrarian, and backwoods lands.[86,87] Such inorganic contaminants accumulate in living systems because of their unmanageable and perpetual nature in the environment, posing detrimental effects on human health and other organic frameworks.[87,88] At the front line of graphene upheaval, various examinations have consequently been attempted to assess the metal expulsion limit of graphene-based nano-adsorbents from water frameworks.

A solvothermal method was used to fabricate magnetic GO/MgAl-layered (MGO/Mg-Al) double hydroxide by Huang et al.[89] Batch studies revealed high uptake capacities of the fabricated adsorbent for Cd(II) (45.05 mg/g), Pb(II) (192.3 mg/g), and Cu(II) (23.04 mg/g). The major mechanism involved in the adsorption process was surface complexation.[90] developed polyamidoamine dendrimer magnetic graphene oxide (PAMD/MGO) to eliminate toxic metal ions from contaminated water. The developed adsorbent exhibited maximum adsorption capacities of 326.729, 435.85, and 353.59, and mg/g for Pb(II), Cd(II), and Cu(II) ions, respectively, at a pH of 7.2. The ability of zinc oxide graphene reduced (ZnO-G) nanomaterials for the expulsion of Ni(II) from wastewater was demonstrated by Hadadian et al.[91], and the

TABLE 1 Adsorption capacities of graphene oxide-based nano-adsorbents for the removal of organic dyes.

Adsorbent	Method of synthesis	Adsorbate	Experimental conditions	Maximum adsorption capacity (mg/g)	References
β-CD/GO	Modified Hummers method followed by crosslinking	Methylene blue	Contact time 60 min, adsorbent dosage 0.04 g/L, temp 70°C	76.4	54
ZIF-8/GO/APTMS	NA	Acid blue 92	Adsorbate conc 20 mg/L, pH 2.2, contact time 60 min, adsorbent dosage 4 mg, temp 25°C	371.8	55
$CuFe_2O_4$@GO $CoFe_2O_4$@GO $NiFe_2O_4$@GO	Solution combustion method	Methylene blue Methylene blue Methylene blue	Adsorbate conc 16 mg/L, adsorbent dosage 0.25 mg, temp 25°C	25.81 50.15 76.34	56
Uio-66-$(OH)_2$GO	Hydrothermal method	Methylene blue	pH 11	96.69	57
RGO-Ni	NA	Rhodamine B	Adsorbate conc 50 mg/l00 mL, adsorbent dosage 0.5 mg, pH 8, contact time 60 min	65.31	60
Alg/GO	NA	Methylene blue	pH 7.78, contact time 12 h, temp 25°C	NA	61
GO-MnO_2	Hydrothermal method	Methylene blue Methyl orange	–	178.253 149.253	63
GO GO@AC RCE/GO	Modified Hummers process NA Solution mixing-regeneration and freeze-drying measure	Methylene blue Methyl orange Methylene blue Methyl orange Methylene blue	– – pH 6, contact time 30 min	276.5 220 423.15 346.51 68	64 65
HA@rGO	Hydrothermal method	Trypan blue Congo red	Adsorbate conc 50 mg/L, adsorbent dosage 100 mg, temp 25°C	146.51 150.09	66
rGO@HAP-Alg	Co-precipitation method	Reactive blue 4 Indigo carmine Acid blue 158	pH 6–7 pH 6–7 pH 6–7	45.56 47.16 48.26	67
AM-GO-SA	Free radical method	Crystal violet	Adsorbate conc 10 mg/L, pH 8, temp 30°C	100.30	68
L-Cys/rGO/PANI	NA	Congo red	Adsorbent dosage 0.025 g, contact time 10 min	56.57	69

TABLE 1 Adsorption capacities of graphene oxide-based nano-adsorbents for the removal of organic dyes—cont'd

Adsorbent	Method of synthesis	Adsorbate	Experimental conditions	Maximum adsorption capacity (mg/g)	References
GO/CNTs	Freeze-drying method	Rhodamine B Methyl orange	– –	248.48 66.96	70
Nanoporous composite hydrogel (NHA)	Fabrication via incorporation of an organic copolymer (starch grafted poly-acryl amide) and a mineral (hydroxyapatite) onto graphene oxide	Malachite Green	Time of 60 min, pH of 10 and initial dye concentration of 100 mg/L	297	58
Magnetic reduced graphene oxide nanocomposite (RGO-Fe$_3$O$_4$)	Fabrication via chemical precipitation	Rhodamine B(Rh-B) and Malachite Green (MG)	pH of 8, mixture adsorbate concentration of each dye of 16 mg/L, a contact time of 20 min and an adsorbent dosage of 0.028 g	NA	59
Series of polymeric GO nanocomposites GO/polyaniline (PAN), GO/polypyrrole (PPy), and GO/polystyrene (PS)	NA	Antacid orange RL (AO-RL)	pH of 2, 6, and 3 for GO/PAN, GO/PPy, and GO/PS. Adsorbent dosage in all cases was 0.01 g/50 mg/L (dye) and temperature of 35°C	39.54, 44.76, and 21.43 mg/g for GO/PAN, GO/PPy, and GO/PS	62

TABLE 2 Adsorption capacities of graphene-oxide-based nano-adsorbents for the removal of pharmaceuticals, pesticides, and other harmful organic pollutants.

Adsorbent	Method of synthesis	Adsorbate	Experimental conditions	Maximum adsorption capacity (mg/g)	References
GNP/BNA	One-pot foam gel casting/nitridation method	Ciprofloxacin	Adsorbate conc 10.5 mg/L, adsorbent dosage 250 mg/L, pH 7, temp 22 ± 3°C pH 4 pH 8	185	71
GNPs(C300)	NA	Sulfamethoxazole Acetaminophen	–	210.08 56.21	72
NiZrAl-LDH-GO-CS NC	NA	Nalidixic acid	pH 6.5, contact time 6 h, adsorbent conc 0.1 g/L, temp 40°C	277.79	73
β-CD/rGO	One-pot hydrothermal method	Naproxen	pH 5, temp 25°C	361.85	74

Continued

TABLE 2 Adsorption capacities of graphene-oxide-based nano-adsorbents for the removal of pharmaceuticals, pesticides, and other harmful organic pollutants—cont'd

Adsorbent	Method of synthesis	Adsorbate	Experimental conditions	Maximum adsorption capacity (mg/g)	References
GO/CX	NA	Metronidazole	Temp 20°C	110–166	75
GBM-MAP	NA	Bisphenol A	pH 7, adsorbate conc 1 mg/L, adsorbent dosage 3 mg, temp 25°C	324	76
S-rGA	Chemical reduction method	Tetrabromo-bisphenol A	pH 7, temp 25°C	128.37	78
Few layered GO	Modified Hummers method	17β-Estradiol	pH 8, contact time 10 min, 10 mg of adsorbent dosage, temp 23°C	149.4	80
GNPs	NA	Aspirin Caffeine Acetaminophen	pH 7, contact time 24 min, adsorbent dosage 24 mg –	12.98 19.72 18.07	81
GO/ZIF-8	Solvothermal method	Chlorpyrifos diazinon		103.72 90.17	82
Uio-66-(OH)$_2$GO	Hydrothermal method	Tetracycline		37.96	57
AGu@mGO (R)	Magnetic solid phase extraction	Chlorpyrifos	Adsorbent dosage 10 mg, pH 6.5, contact time 30 min	85.47	83
GO-CMC-Fe	NA	Atrazine	Adsorbent dosage 100 mg, adsorbate conc 1200 mg/L	193.8	84
GO	Modified Hummers process	Atrazine	Adsorbent dosage 121.45 mg/L, adsorbate conc 27.03 mg/L, pH 5.37, contact time 180 min, temp 27.69°C	138.19	85
GO nanosheets	Modified Hummers process	Sodium diclofenac	pH 6.2, contact time 300 min, adsorbent dosage 0.25 g/L, temp 25°C	128.74	77

composite exhibited better uptake capacity (66.7 mg/g) at pH 3.6. The spontaneity and feasibility of the adsorption process were predicted from thermodynamic parameters. An oxidative polymerization technique was applied to fabricate carbon nanotube-graphene oxide composite combined with polyaniline (CNT-GO/Polyaniline) by Ansari et al.[92] to eliminate Cr(VI) ions. The developed nanocomposite displayed a high adsorption capacity of 139.9 mg/g at acidic pH 2 and 30°C. The adsorption process was exothermic, as evident from thermodynamic studies. Graphene oxide-based nickel ferrite (GONF) was synthesized via the co-precipitation method for the expulsion of U(VI) and Th(IV) from wastewater by Lingamdinne et al.[93] SEM images revealed the porous structure of the composite adsorbent. Maximum adsorption capacities demonstrated by adsorbent were 45.49 and 34.57 mg/g for U(VI) and Th(IV), respectively, under acidic conditions. The adsorption process was endothermic, as evident from thermodynamic studies. The tendency of iron oxide-graphene oxide (FeO-GO) nanocomposite for the eviction of As(V) and As(III) was reported by Su et al.[94] from industrial wastewater. The nanocomposite exhibited a high surface area of about 341 m^2/g with a mesoporous structure. The nanocomposite displayed maximum uptake for As(III) (147 mg/g) and As(V) (113 mg/g). In another work, Zhang et al.[78,79] synthesized reduced graphene oxide/nickel oxide nanocomposite (rGO/NiO) and evaluated its adsorption performance for the eviction of Cr(VI) from wastewater. The adsorption capacity attained was maximum (198 mg/g) under optimized conditions (pH 4 and temperature 25°C). Synergistic adsorption was considered to be the phenomenon governing the process of adsorption.

A sandwiched nanocomposite of graphene oxide, manganese oxide, iron oxide, and polypyrrole (GO/MnO$_2$/Fe$_3$O$_4$/PPy) was fabricated by Liu et al.[95] for the expulsion of Cr(VI) from water. The binding mechanism, as reported in the study, was via adsorption-reduction. The nanocomposite displayed a high adsorption capacity of 374.53 mg/g at pH 2. The electrostatic attraction was the major mechanism that was responsible for adsorption. It was inferred that the nanocomposite possessed immense perspective as a low-cost adsorbent material to eradicate Cr(VI) from wastewater.

Leng et al.[96] demonstrated the good efficiency of graphene as an advanced adsorbent for removing Sb(II) from an aqueous solution. Under the favorable conditions of pH, temperature, adsorbent dose, adsorbate concentration, and contact time, graphene displayed high removal capacity of 10.92 mg/g for Sb(III) ions. The adsorbent was regenerated by using 0.1 mol/l EDTA. Wu et al.[97] demonstrated the adsorption of Cu(II) onto GO. Maximum removal of Cu(II) recorded was 117.5 mg/g for Cu(II), and the optimum conditions were a pH of 5.3, adsorbent dosage of 1 mg/mL, and contact time of 150 min. The major mechanisms governing the process of adsorption were electrostatic interaction, surface complexation, and ion exchange. The adsorbent loaded with Cu(II) was regenerated using HCl as a desorbing agent. A sacrificial-template polymerization method was employed to fabricate the composite nanosheets of polypyrrole and graphene oxide (PPy/GO) by Li et al.[98] for the adsorption of Cr(VI) from an aqueous environment. The batch studies concluded that the composite displayed a high removal efficiency of 497.1 mg/g for Cr(VI) at pH 3 and a contact time of 24 h.

A nanocomposite of Fe-Al layered hydroxide coated with sodium alginate and reduced graphene oxide (FAH-rGO/SA) was fabricated via hydrothermal and ex situ polymerization strategy for the eradication of As(V) ions by.[88] The composite with the highest weight percentage of sodium alginate demonstrated a great adsorption capacity of 190.84 mg/g for As(V) at room temperature.[87] designed a nanocomposite of magnetic oak wood ash and graphene oxide (Ash/GO/Fe$_3$O$_4$) for the eviction of toxic metals like Pb(II) and Cd(II) from the aquatic medium. The composite presented high adsorption capacities for Pb(II) (47.16 mg/g) and for Cd(II) (43.66 mg/g). The designed composite displayed high reusability. In another study, the one-pot hydrothermal method was used to fabricate carbon layer encapsulated magnetic nanoparticles and graphene oxide (Fe$_3$O$_4$@C-GO) nanocomposite by Wang et al.[99] The capacity of nanocomposite to remove toxic metal ions like Ag(I), Pb(II), Cr(VI), and Al(III) from an aqueous solution was assessed. Fe$_3$O$_4$@C-GO nanocomposite presented high surface area and pore volume of 160 m^2/g, and 0.283 cm^3/g, respectively. VSM studies indicated the superparamagnetic nature of composite. The maximum adsorption capacities displayed by the nanocomposites were 162.9 mg/g for Ag(I), 125.8 mg/g for Pb(II), 158.2 mg/g for Cr(VI), and 173.9 mg/g for Al(III).

The adsorption ability of graphene oxide with manganese ferrite (GO-MnFe$_2$O$_4$) nanocomposite for Pb(II) ions was assessed.[100] A one-pot hydrothermal method was employed for the fabrication of the nanocomposite adsorbent. The fabricated nanocomposite displayed a high adsorption capacity of 621.11 mg/g for Pb(II) ions via electrostatic interactions. A few adsorption-desorption cycles were done to gauge the reusability of the composite material that exhibited its strength and cost adequacy to be utilized as a powerful adsorbent for the expulsion of harmful lead particles from wastewater.

Cellulose acetate/graphene oxide (CA-GO) nanocomposite films were prepared, and its potential as an adsorbent for Ni(II) ions was demonstrated by Aldalbahi et al.[101] The composite presented a high surface area with a porous structure. Batch studies were performed to elucidate the effect of several adsorption parameters on the adsorption capacity of the CA-GO nanocomposite films. Langmuir's model best described the experimental data, and the removal efficiency of about 96.77% was obtained for Ni(II) ions.

Anush et al.[102] fabricated a composite consisting of graphene oxide and chitosan-magnetite (GO/CT-Fe$_3$O$_4$) for the adsorption of Cu(II) and Cr(VI) from an aqueous environment. The prepared nanocomposite displayed high adsorption capacities of 111.11 for Cu(II) and 142.85 mg/g for Cr(VI). From the thermodynamic examinations, it was uncovered that the adsorption interaction was endothermic and unconstrained. The hydrothermal method was used to fabricate cobalt oxide graphene nanocomposite functionalized with polypyrrole(COPYGO) for the eradication of Pb(II) and Cd(II) from an aqueous solution by Anuma et al.[103] The nanocomposite exhibited a high surface area of 133 m^2/g. The observed maximum adsorption capacities of COPYGO nanocomposite were 780.363 and 794.188 mg/g for Pb(II) and Cu(II) ions, respectively. The adsorption process was endothermic, and spontaneous which was evident from thermodynamic parameters.

Archana et al.[104] synthesized p-phenylenediamine graphene oxide nanocomposite via the hydrothermal method (PPDG). The fabricated adsorbent was characterized by several techniques like FESEM, XRD, FTIR, TEM, and EDX. SEM and TEM analyses revealed that the composite adsorbent consisted of flaky nanoparticles of size ranging from 100 to 200 nm. The adsorption capacity of composite material evaluated for Pb(II) ions was 800 mg/g. To establish the effectiveness and cost adequacy of the PPDG nanocomposite, regeneration studies were carried out by using 0.5 NaOH solutions as a desorbing agent. It was observed that 30% evacuation effectiveness for Pb(II) ions could be accomplished even after three cycles.

A recent study used an in situ emulsion polymerization method to fabricate polyacrylonitrile grafted ethylene diamine partially reduced graphene oxide (PAN-PRGO) nanocomposite.[105] The composite was adjusted by fusing a few functionalities onto the outer layer of graphene nanosheets via incomplete hydrolysis to obtain HPAN-PRGO. The maximum uptake capacity of the nanocomposite adsorbent was investigated for Hg(II) ions, and it was obtained to be 324 mg/g at pH 5. The major mechanism depicted from XPS analysis was chelation between the functional groups (amine, amide, carboxylic, and amidoxime) of adsorbent and Hg(II) ions. The regeneration studies concluded that the composite could retain 90% adsorption efficiency even after six adsorption-desorption cycles.

Khan et al.[15] developed polyethylene glycol/graphene/carbon nanotube (PEG/G/CNT) via dissolution process along with ultrasonication. The uptake capacity of the developed adsorbent was tested for Au(III) ions. High uptake capacity of 80.80 mg/g of Au(III) was reported for PEG/G/CNT. It was concluded that the fabricated nanocomposite could be used on a large scale in industries for gold extraction to ensure better productivity.

A nano adsorbent of graphene oxide and hydroxyapatite (GO/HAP) was fabricated via the hydrothermal method for the eviction of U(VI) ions from an aqueous solution[106] At a pH of 3 and a contact time of 5 min, the GO/HAP composite demonstrated high adsorption limit of 373 mg/g for U(VI). The boundaries from thermodynamic examination demonstrated that the composite assimilated U(VI) via a course of unconstrained and exothermic adsorption. According to different characterization techniques like FTIR, XRD, and XPS studies, the phosphate bunch in the composite was predominantly used for U(VI) maintenance and consolidation. The GO/HAP composite's improved U(VI) adsorption limit is doubtlessly attributed to the synergistic impact in the wake of functionalizing with nano HAP.

The adsorptive potential of EDTA-modified graphene oxide (MGO-EDTA) nano-adsorbents for the eviction of Pb(II) ions from aqueous solution was demonstrated by Xing et al.[107] The adsorption capacity of the adsorbent was enhanced due to the chelating tendency of EDTA. The maximum adsorption capacity of MGO-EDTA for Pb(II) ions was 211.3 mg/g at pH 6.6, and contact time of 50 min. Freundlich model well described the adsorption data.

In another study, the elimination of Pb(II) ions from wastewater by zinc oxide modified graphene oxide (ZnO/GO) nano-adsorbents was carried out by Ahmad et al.[108] The ZnO/GO was fabricated by the solvothermal method and was characterized using XRD, SEM, EDX, and FTIR techniques. The composite presented a high adsorption capacity of 909.09 mg/g for Pb(II) at pH 5. The experimental data well followed the Langmuir and pseudo-first-order kinetic model.

Graphene oxide coated with activated sawdust (GO/ASD) was synthesized, and its potential as an adsorbent for the removal of strontium was investigated by Chakraborty et al.[109] The developed nano adsorbents displayed a good adsorption capacity of 4.09 mg/g for strontium ions at pH of 6–8, temperature 45°C, and an initial metal ion concentration of 50 mg/L. The results were promising as compared to other reported adsorbents.

The removal of the uncommon earth components in water and hydrogen systems is a pressing issue nowadays. Hence, graphene oxide/polyhydroquinone nanocomposite was utilized to eliminate samarium particles (Sm(III)) from an aqueous solution by Ali et al.[110] The sorption and desorption of Sm^{3+} particles on the graphene oxide/polyhydroquinone nanocomposite were examined with contact time (15.0 min), initial metal ion concentration (100.0 mg/L), adsorbent dosage (0.3 g/L), pH (6.0), and temperature (25°C). Under such optimized conditions, Langmuir adsorption capacity was reported to be 357.14 mg/g. Husein et al.[111] fabricated a nitrogen-doped MgO/graphene nanocomposite (N-MgO/G). For testing its efficacy for the removal of lead metal particles (Pb^{2+}) from an aqueous environment. Batch studies revealed the adsorption capacity of 294.12 mg/g for Pb^{2+} under the following ideal conditions: pH 5.13, contact time 35 min, adsorbent dosage 0.025 g, initial metal ion concentration of 400 mg/L, and a temperature of 36°C. Adsorption data fitted well in the pseudo-second-order model, and thermodynamics revealed that Pb^{2+} adsorption was endothermic.

In another study, a green and high-proficient adsorbent magnetic cellulose supported by graphene oxide (Fe_3O_4/ CMC/GO) was fabricated by Chen et al.[112] for the removal of Cu(II) ions from an aqueous solution. Batch adsorption studies were conducted to elucidate the effect of operating parameters on the maximum adsorption capacity of the adsorbent. Langmuir model fitted well the adsorption data, and the composite adsorbent displayed a high adsorption limit of 199.98 mg/g under the ideal conditions of pH 6–7 and temperature 45°C. An in situ strategy was applied to develop manganese ferrite/graphene oxide magnetic nanocomposite ($MnFe_2O_4$/GO) for the removal of Pb^{2+} ions from aqueous solution.[113] Ideal conditions for Pb^{2+} adsorption over $MnFe_2O_4$/GO nanocomposite were controlled by focusing on operating parameters such as contact time, pH, temperature, adsorbent, and adsorbate dosage. As per the Langmuir isotherm model, the maximum adsorption capacity obtained for Pb^{2+} ions was 636.94 mg/g at a contact time of 120 min and pH of 6. The developed adsorbent displayed reusability for five adsorption desorption cycles that demonstrated its stability and reusability.

Method of fabrication, experimental conditions, and the maximum adsorption capacities of various graphene-based nano-adsorbents for the removal of toxic metal ions are depicted in Table 3.

TABLE 3 Adsorption capacities of graphene-oxide-based adsorbents for the removal of inorganic pollutants.

Adsorbent	Method of synthesis	Adsorbate	Experimental conditions	Maximum adsorption capacity (mg/g)	References
MGO/Mg-Al	One-pot solvothermal method	Pb(II) Cu(II) Cd(II)	Conc 0.16 mg/L, contact time 240 min	192.3 23.04 45.05	89
ZnO-G	Microwave-assisted hydrothermal method	Ni(II)	Contact time 800 min, pH 3.6, temp 25°C	66.7	91
PAMD/MGO	NA	Cd(II) Pb(II) Cu(II)	Conc 2–10 mg/L, contact time 160 min, temp 500°C, pH 7.2	435.85 326.729 353.59	90
CNT-GO/ Polyaniline	Oxidative polymerization method followed by doping with para toluene sulphonic acid	Cr(VI)	Conc 25–200 mg/L, contact time 650 min, temp 30°C, pH 2	139.9	92
GONF	Co-precipitation method	U(VI) Th(IV)	Conc 4–7 mg/L, contact time 350 min, temp 60°C, pH 3.5	44.49 34.57	93
FeO$_x$-GO	NA	As(III) As(V)	Contact time 1500 min, temp 250°C, pH 7	147 113	94
rGO/NiO	Hummers process followed by co-precipitation	Cr(VI)	Contact time 1500 min, temp 25°C, pH 4	198	78
GO/MnO$_2$/Fe$_3$O$_4$/ PPy	NA	Cr(VI)	Contact time 1400 min, pH 2	374.53	95
Graphene	Modified Hummers process	Sb(III)	Conc 1–10 mg/L, contact time 240 min, temp 30°C, pH 11	10.919	96
GO	Modified Hummers process	Cu(II)	Conc 25–250 mg/L, contact time 150 min, pH 5.3	117.5	97
PPy/GO	Sacrificial-template polymerization method	Cr(VI)	Contact time 1440 min, pH 3	497.1	98
FAH-rGO/SA	Hydrothermal and ex situ polymerization method	As(V)	Adsorbent dosage 0.07 g/L, contact time 360 min, pH 7	190.84	88
Ash/GO/Fe$_3$O$_4$	Improved Hummers process followed by sonication	Pb(II) Cd(II)	Adsorbent dosage 1 g/L, pH 6, temp 25°C	47.16 43.66	87
Fe$_3$O$_4$@C-GO	One pot hydrothermal method	Ag(I) Pb(II) Al(III) Cr(VI)	Contact time 20 min, temp 25°C, pH 6 Contact time 20 min, temp 25°C, pH 2	162.9 125.8 173.9 158.2	99
GO-MnFe$_2$O$_4$	One-pot hydrothermal method	Pb(II)	Contact time 30 min, pH 6	621.11	100
CA-GO	NA	Ni(II)	Contact time 30 min, temp 25°C	NA	101
GO/CT-Fe$_3$O$_4$	NA	Cu(II) Cr(VI)	Adsorbent dosage 30 mg, pH 6.8 Adsorbent dosage 30 mg, pH 3	111.11 142.85	102

Continued

TABLE 3 Adsorption capacities of graphene-oxide-based adsorbents for the removal of inorganic pollutants—cont'd

Adsorbent	Method of synthesis	Adsorbate	Experimental conditions	Maximum adsorption capacity (mg/g)	References
COPYGO	Hydrothermal method	Pb(II) Cd(II)	pH 5.5 pH 6.1	780.363 794.188	103
PPDG	Hydrothermal method	Pb(II)	–	800	104
HPAN-PRGO	In situ emulsion polymerization method	Hg(II)	pH 5	324	105
PEG/G/CNT	Dissolution process and ultrasonication	Au(III)	Temp 25°C, pH 1	80.80	15
GO/HAP	Hydrothermal method	U(VI)	Contact time 5 min, pH 3	373	106
MGO-EDTA	NA	Pb(II)	Contact time 50 min, pH 6.6	211.3	107
ZnO/GO	Solvothermal method	Pb(II)	pH 5	909.09	108
Graphene oxide/polyhydroquinone	NA	Sm(III)	Adsorbent dosage 0.3 g/L, adsorbate conc 100 mg/L, contact time 15 min, pH 6, temp 25°C	357.14	110
N-MgO/G	Green synthesis via electrochemical exfoliation process	Pb(II)	Adsorbent dosage 0.025 g, adsorbate conc 400 mg/L, pH 5.3, contact time 35 min, temp 36°C	294.12	111
Fe_3O_4/CMC/GO	NA	Cu(II)	Adsorbate conc 200 mg/L, pH 6–7, temp 45°C	199.98	112
$MnFe_2O_4$/GO	In situ method	Pb(II)	pH 6, contact time 120 min	636.94	113
GO/ASD	NA	Sr(II)	Adsorbate conc 50 mg/L, pH 6–8, temp 45°C	4.09	109

5 Management of spent graphene-oxide-based adsorbents

The regeneration and recycling of spent graphene-based adsorbents promise the commercialization and economic feasibility of the wastewater treatment process. Desorption efficiency depends on the binding mechanism. Metal ions desorbed from spent graphene-based adsorbents is pH sensitive, with desorption efficiency increasing with reduced pH values of the desorption medium. For example, a highly acidic medium (0.1 M HCl) resulted in maximum desorption of Zn(II) metal ions to the tune of 91.6%. In another study conducted by Wu et al.,[97] more than 74% of Cu(II) could be recovered using strong hydrochloric acid as the desorption medium. But desorption efficiency decreased with increased pH of the aqueous medium. Binding efficiency could be retained up to 95% after five sorption-desorption cycles and up to 90% after 10 such cycles. Experiments thus suggested the high efficiency of the reusability of the spent GO. Besides a highly acidic medium, EDTA and thiourea-HCl have served as efficient desorbing mediums.[114] As per a study conducted by Tan et al.,[115] GO membrane could retain its adsorption potential for Cu(II), Cd(II), and Ni(II) for up to six cycles of sorption-desorption. After six cycles, the adsorption capacity of the GO membrane decreased to 10%, 12%, and 21% for Cu(II), Cd(II), and Ni(II), respectively. The decreased efficiency was attributed to a loss in the binding sites with the repeated sorption-desorption process. Biobased nanocomposite fabricated from GO, biopolymer tragacanth, and polyvinyl alcohol demonstrated high removal efficiency for metal ions like Pb(II), Cu(II), as well as organic dyes like crystal violet (CV) and Congo red (CR). The spent adsorbent showed good reusability with no significant activity loss on the completion of three sorption-desorption cycles. 0.1 N HNO_3 was used as the desorbing medium.[116] After three cycles of desorption, the adsorption efficiency of the biobased magnetic beads decreased to 2.75% for Pb(II) and 6.5% for Cu(II). While the same for the organic dyes, the adsorption efficiency decreased to 7.5% for CV and 4.4% for CR, respectively. Desorption efficiency decreased to approximately 7% for Pb(II) and Cu(II), 9% for CV, and 5.5% for CR after three cycles. The study thus demonstrated that the magnetic beads retained their efficiency as an adsorbent even under harsh acidic conditions and after repeated use for three cycles of sorption and desorption.

Chen et al.[117] fabricated a GO-cellulose hydrogel and demonstrated its high binding affinity for the Cu(II) from an aqueous medium. 1 M HCl showed the highest desorption of the metal ions, and the adsorbent retained its efficacy even after five continuous cycles of sorption-desorption. Oak-wood ash/GO/Fe$_3$O$_4$ nanocomposites demonstrated high reusability with the negligible decrease in adsorption and also demonstrated high stability for approximately eight cycles of sorption-desorption.[87] GO and GO functionalized with silver nanoparticles as fabricated by Martinez et al.[118] could be regenerated using an acidic medium (0.001 M HNO$_3$) with phosphate recovery of 98% and 80%, respectively. The study further confirmed that there were no significant alterations in the structural characteristics of both adsorbents.

6 Comparative performance analysis of graphene-oxide-based adsorbents vis-a-vis other reported adsorbents

A comparison of the performance of different adsorbents was conducted and the results are depicted in Table 4. The adsorbates selected for the comparative study were methylene blue (dyes), trinitrophenol (organic-based pollutants other than dyes), and copper ions (metal ion pollutants). The adsorptive removal of pollutants from aqueous medium depends not only on the features of adsorbent and the adsorbate but also on the environmental factors like the pH, temperature, contact time, etc. Thus, a comparison of the performance of an adsorbent requires the consideration of various such parameters. Table 4 reveals the efficacy of GO-based adsorbents vis-a-vis other reported adsorbents like CNTs, activated carbon, and various other polymer-based nanocomposites. In fact, for graphene oxide modified with polymers, metal oxides have demonstrated

TABLE 4 Performance of graphene-oxide-based adsorbents vis-à-vis other reported adsorbents.

Adsorbent	Adsorbate	Experimental conditions	Maximum adsorption capacity (mg/g)	References
Chitosan-activated carbon clay	Methylene blue	Adsorbent dosage: 2 g/L; pH: 7; contact time: 60 min; temp: 30°C	86.08	119
Activated carbon from spent bleach sorbent	Methylene blue	pH: 9; contact time: 300 min; temp: 30°C	178.64	120
MWCNT-Ca-Alginate COOH	Methylene blue	Temp: 22°C; pH; adsorbate concentration: 50 mg/L; contact time: 8 h; adsorbent dosage: 250 mg/L	100.7	121
MWCNT	Methylene blue	Adsorbent concentration: 400 mg/L; adsorbate concentration: 10 mg/L; contact time: 120 min; pH: 4; temp: 310 K	132.6	122
Graphene oxide/Ca-alginate composites	Methylene blue	pH: 5.4; temp: 25°C; adsorbent dose: 0.5 g/L; adsorbate concentration: 20–70 mg/L	181.81	123
Graphene oxide/TiO$_2$	Methylene blue	Adsorbate concentration: 150 mg/L	407.6	124
Graphene oxide/MgO	Methylene blue	pH: 11; adsorbate concentration: 20 mg/L; contact time: 20 min	833	125
Graphene oxide	Methylene blue	pH: 5.4; temp 25°C; adsorbent dose: 0.5 g/L; adsorbate concentration: 20–70 mg/L	144.92	123
Fe3O4-MCNT (*Pseudomonas aeruginosa* immobilized)	Trinitrophenol	Contact time: 45 min; pH: 2; adsorbent dosage: 25 mg; temp: 15°C	100	126

Continued

TABLE 4 Performance of graphene-oxide-based adsorbents vis-à-vis other reported adsorbents—cont'd

Adsorbent	Adsorbate	Experimental conditions	Maximum adsorption capacity (mg/g)	References
Fe_3O_4-AC from almond shell	Trinitrophenol	Adsorbent dosage: 2 g/L; contact time: 25 min; pH: 2	73.96	127
Coconut shell-derived AC-70	Trinitrophenol	Temp: 30°C	269	128
Graphene-oxide-Chitosan	Trinitrophenol	pH: 2; temp: 30°	263.1	129
MWCNT-Chitosan	Trinitrophenol	pH: 2; temp: 30°		130
MWCNT-polymer	Cu^{2+}	Temp: 25°C; pH: 7; contact time: 180 min; adsorbate concentration: 100 mg/L	189	131
GO/CT-Fe_3O_4	Cu^{2+}	Adsorbent dosage: 30 mg, pH: 6.8	111.11	102

better adsorptive removal for pollutants (irrespective of the nature of pollutants). The unique characteristics of GO-based nanocomposites that may have contributed to their better adsorptive performance are demonstrated in their rich surface-binding sites, high surface area, and high aspect ratio.

7 Future perspectives and conclusion

The chapter has discussed the favorable properties of graphene and graphene-based materials such as GO, rGO, graphene-nano metal composites, graphene-polymer nanocomposites, etc. The different fabrication methods employed till date have a contributing effect on graphene-based materials' physicochemical, mechanical, optical, and conducting properties. Besides the strong mechanical properties of graphene, the rich surface-binding sites (due to oxygen-based functionalities, rich π electron system, and properties exhibited in incorporated nano-metal/metal oxide or polymer system), high surface area, high surface to volume or aspect ratio in GO, rGO, and GO/rGO-based nanocomposites are responsible for advanced adsorbent. Enhanced adsorption properties in graphene-derived materials as compared to as-reported carbon or other inorganic-based adsorbents. Such properties have resulted in high binding efficiency of GO/rGO/GO nanocomposites toward organic (hazardous dyes, pesticides, pharmaceuticals, phenolics, PAHs, etc.) and various inorganic-based (noxious heavy metal ions) water pollutants. The good adsorption performance, along with the ability to be regenerated, and reused without significant loss of performance, can enable GO-based materials to solve the crisis of water pollution via adsorption technology. Such GO-based adsorbents can easily replace the costly commercial activated carbons used for use as water filters. But certain limitation exists, which creates hurdles in the path of cost-efficiency and scalability, thereby hindering its application on a commercial scale. Firstly, GO-based adsorbents may pose toxicity because of their nano-scale dimensions, thereby influencing adverse effects on health and the environment. Secondly, there are issues related to cost-efficient and scalable manufacturing of the GO-based adsorbents for large-scale water treatment. Thirdly, the disposal of the spent GO-based adsorbents is an issue of concern.

Because of the favorable outcomes like enhanced, safe, and cost-efficient use for water treatments via application of GO-based adsorbents under low dosages, more research efforts are required to design and/or fabricate such adsorbents under sustainable conditions to nullify the limitations to a great extent.

References

1. Huber M, Welker A, Helmreich B. Critical review of heavy metal pollution of traffic area runoff: occurrence, influencing factors, and partitioning. *Sci Total Environ.* 2016;541:895–919. https://doi.org/10.1016/j.scitotenv.2015.09.033.
2. Nayak A, Bhushan B, Kotnala S. Fabrication of chitosan-hydroxyapatite nano-adsorbent for removal of norfloxacin from water: isotherm and kinetic studies. *Mater Today Proc.* 2021. https://doi.org/10.1016/j.matpr.2021.07.356.
3. Karri RR, Ravindran G, Dehghani MH. Wastewater—sources, toxicity, and their consequences to human health. In: *Soft Computing Techniques in Solid Waste and Wastewater Management.* Elsevier; 2021:3–33.

4. Dehghani MH, Omrani GA, Karri RR. Solid waste—sources, toxicity, and their consequences to human health. In: *Soft Computing Techniques in Solid Waste and Wastewater Management.* Elsevier; 2021:205–213.

5. Awad AM, Jalab R, Benamor A, et al. Adsorption of organic pollutants by nanomaterial-based adsorbents: an overview. *J Mol Liq.* 2020;301:112335. https://doi.org/10.1016/j.molliq.2019.112335.

6. Nayak A, Bhushan B, Gupta V, Kotnala S. Fabrication of microwave assisted biogenic magnetite-biochar nanocomposite: a green adsorbent from jackfruit peel for removal and recovery of nutrients in water sample. *J Ind Eng Chem.* 2021;100:134–148. https://doi.org/10.1016/j.jiec.2021.05.028.

7. Ali I, Basheer AA, Mbianda XY, et al. Graphene based adsorbents for remediation of noxious pollutants from wastewater. *Environ Int.* 2019;127:160–180. https://doi.org/10.1016/j.envint.2019.03.029.

8. Ighalo JO, Adeniyi AG. Adsorption of pollutants by plant bark derived adsorbents: an empirical review. *J Wat Proc Eng.* 2020;35:101228.

9. Lima ÉC, Dehghani MH, Guleria A, et al. Adsorption: fundamental aspects and applications of adsorption for effluent treatment. In: *Green Technologies for the Defluoridation of Water.* Elsevier; 2021:41–88.

10. Awad AM, Shaikh SM, Jalab R, et al. Adsorption of organic pollutants by natural and modified clays: a comprehensive review. *Sep Purif Technol.* 2019;228:115719.

11. Khan FSA, Mubarak NM, Khalid M, et al. Magnetic nanoadsorbents' potential route for heavy metals removal—a review. *Environ Sci Poll Res.* 2020;27(19):24342–24356.

12. Nayak A, Bhushan B, Gupta V, Sharma P. Chemically activated carbon from lignocellulosic wastes for heavy metal wastewater remediation: effect of activation conditions. *J Colloid Interface Sci.* 2017;493:228–240. https://doi.org/10.1016/j.jcis.2017.01.031.

13. Tripathi A, Ranjan MR. Heavy metal removal from wastewater using low cost adsorbents. *J Bioremed Biodegr.* 2015;06:06. https://doi.org/10.4172/2155-6199.1000315.

14. Faysal HM, Akther N, Zhou Y. Recent advancements in graphene adsorbents for wastewater treatment: current status and challenges. *Chin Chem Lett.* 2020;31(10):2525–2538. https://doi.org/10.1016/j.cclet.2020.05.011.

15. Khan A, Hussein MA, Alsheri FM, Alamry KA, Asiri AM. A new class of polyethylene glycol-grafted graphene carbon nanotube composite as a selective adsorbent for Au (III). *Waste Biomass Valoriz.* 2021;12(2):937–946. https://doi.org/10.1007/s12649-020-01053-x.

16. Mehmood A, Khan FSA, Mubarak NM, et al. Magnetic nanocomposites for sustainable water purification—a comprehensive review. *Environ Sci Poll Res.* 2021;28(16):19563–19588.

17. Vadukumpully S, Paul J, Mahanta N, Valiyaveettil S. Flexible conductive graphene/poly(vinyl chloride) composite thin films with high mechanical strength and thermal stability. *Carbon.* 2011;49(1):198–205. https://doi.org/10.1016/j.carbon.2010.09.004.

18. Jorgensen JH, Cabo AG, Balog R. Symmetry-driven band gap engineering in hydrogenfunctionalized graphene. *ACS Nano.* 2016;10(12):10798–10807. https://doi.org/10.1021/acsnano.6b04671.

19. Mak KF, Ju L, Wang F, Heinz TF. Optical spectroscopy of graphene: from the far infrared to the ultraviolet. *Solid State Commun.* 2012;152(15):1341–1349. https://doi.org/10.1016/j.ssc.2012.04.064.

20. Pirruccio G, Martín ML, Lozano G, Gómez RJ. Coherent andbroadband enhanced optical absorption in graphene. *ACS Nano.* 2013;7(6):4810–4817. https://doi.org/10.1021/nn4012253.

21. Renteria J, Nika D, Balandin AA. Graphene thermal properties: applications in thermal management and energy storage. *Appl Sci.* 2014;4(4):525. https://doi.org/10.3390/app4040525.

22. Novoselov KS, Geim AK, Morozov SV, et al. Electric field effect in atomically thin carbon films. *Science.* 2004;306(5696):666–669. https://doi.org/10.1126/science.110.

23. Gandhi MR, Vasudevan S, Shibayama A, Yamada M. Graphene and graphene-based composites: a rising star in water purification—a comprehensive overview. *ChemistrySelect.* 2016;1(15):4358–4385. https://doi.org/10.1002/SLCT.201600693.

24. Mohan VB, Lau K-T, Hui D, Bhattacharyya D. Graphene-based materials and their composites: a review on production, applications and product limitations. *Compos B Eng.* 2018;142:200–220. https://doi.org/10.1016/j.compositesb.2018.01.013.

25. Lingamdinne LP, Koduru JR, Chang YY, Karri RR. Process optimization and adsorption modeling of Pb (II) on nickel ferrite-reduced graphene oxide nano-composite. *J Mol Liq.* 2018;250:202–211.

26. Lingamdinne LP, Koduru JR, Karri RR. A comprehensive review of applications of magnetic graphene oxide based nanocomposites for sustainable water purification. *J Environ Manag.* 2019;231:622–634.

27. Narayana PL, Lingamdinne LP, Karri RR, et al. Predictive capability evaluation and optimization of Pb (II) removal by reduced graphene oxide-based inverse spinel nickel ferrite nanocomposite. *Environ Res.* 2022;204:112029.

28. Haseen U, Umar K, Ahmad H, Parveen T, Mohamad Ibrahim MN. Graphene and its composites: applications in environmental remediation. In: *Modern Age Wastewater Problems*; 2019:85–91. https://doi.org/10.1007/978-3-030-08283-3_5.

29. Ashour RW, Abdelhamid HN, Abdel-Magied AF, et al. Rare earth ions adsorption onto graphene oxide nanosheets. *Solvent Extr Ion Exch.* 2017;35:91–103. https://doi.org/10.1080/07366299.2017.1287509.

30. Chen Y, Chen L, Bai H, Li L. Graphene oxide-chitosan composite hydrogels asbroad-spectrum adsorbents for water purification. *J Mater Chem A.* 2013;1:1992–2001. https://doi.org/10.1039/C2TA00406B.

31. Choudhury S, Balasubramanian R. Recent advances in the use of graphene-family nanoadsorbents for removal of toxic pollutants from wastewater. *Adv Colloid Interf Sci.* 2014;204:35–56. https://doi.org/10.1016/j.cis.2013.12.005.

32. Lingamdinne LP, Choi JS, Choi YL, et al. Process modeling and optimization of an iron oxide immobilized graphene oxide gadolinium nanocomposite for arsenic adsorption. *J Mol Liq.* 2020;299:112261.

33. Duru I, Ege D, Kamali AR. Graphene oxides for removal of heavy and precious metals from wastewater. *J Mater Sci.* 2016;51:6097–6116. https://doi.org/10.1007/s10853-016-9913-8.

34. Kumar R, Chawla J, Kaur I. Removal of cadmium ion from wastewater by carbon-based nanosorbents, a review. *J Water Health.* 2015;13:19–33. https://doi.org/10.2166/wh.2014.024.

35. Koduru JR, Karri RR, Mubarak NM. Smart materials, magnetic graphene oxide-based nanocomposites for sustainable water purification. *Sustainable Polymer Composites and Nanocomposites.* Springer; 2019:759–781.

36. Eigler S, Hirsch A. Chemistry with graphene and graphene oxide-challenges for synthetic chemists. *Angew Chem Int Ed.* 2014;53 (30):7720–7738. https://doi.org/10.1002/anie.201402780.

37. Choi HJ, Jung SM, Seo JM, Chang DW, Dai L, Baek JB. Graphene for energy conversion and storage in fuel cells and supercapacitors. *Nano Energy.* 2012;1(4):534–551. https://doi.org/10.1016/j.nanoen.2012.05.001.

38. Iwan A, Chuchmala A. Perspectives of applied graphene: polymer solar cells. *Prog Polym Sci.* 2012;37(12):1805–1828. https://doi.org/10.1016/j.progpolymsci.2012.08.001.

39. Allen MJ, Tung VC, Kaner RB. Honeycomb carbon: a review of graphene. *Chem Rev.* 2010;110(1):132–145. https://doi.org/10.1021/cr900070d.

40. Melezhyk AV, Pershin VF, Memtov NR, Tkachev AG. Mechanochemical synthesis of graphene nanoplatelets from expanded graphene compound. *Nanotechnol Russ.* 2016;11:421–429. https://doi.org/10.1134/S1995078016040121.

41. Cheng JS, Du J, Zhu W. Facile synthesis of three dimensional chitosan-graphene meso-structures for reactive black 5 removal. *Carbohydr Polym.* 2012;88(1):61–67. https://doi.org/10.1016/j.carbpol.2011.11.065.

42. Huang X, Qi X, Boey F, Zhang H. Graphene-based composites. *Chem Soc Rev.* 2012;41(2):666. https://doi.org/10.1039/C1CS15078B.

43. Zhu Y, Murali S, Cai W, et al. Graphene and graphene oxide: synthesis, properties, and applications. *Adv Mater.* 2010;22(35):3906–3924. https://doi.org/10.1002/adma.201001068.

44. Dreyer DR, Park S, Bielawski CW, Ruoff RS. The chemistry of graphene oxide. *Chem Soc Rev.* 2009;39(1):228–240. https://doi.org/10.1039/b917103g.

45. Brodie BC. On the atomic weight of graphite. *Philos Trans R Soc Lond A.* 1859;149:249–259. https://doi.org/10.1098/rstl.1859.0013.

46. Staudenmaier L. VerfahrenzurDarstellung der Graphitsäure. *Chem Europe.* 1898;31(2):1481–1487. https://doi.org/10.1002/cber.18980310237.

47. Hummers WS, Offeman RE. Preparation of graphitic oxide. *J Am Chem Soc.* 1958;80(6):1339. https://doi.org/10.1021/ja01539a017.

48. Krishnan D, Kim F, Luo J, et al. Energetic graphene oxide: challenges and opportunities. *NanoToday.* 2012;7(2):137–152. https://doi.org/10.1016/j.nantod.2012.02.003.

49. Singh V, Joung D, Zhai L, Das S, Khondaker SI, Seal S. Graphene based materials: past, present and future. *Prog Mater Sci.* 2011;56 (8):1178–1271. https://doi.org/10.1016/j.pmatsci.2011.03.003.

50. Kuilla T, Bhadra S, Yao D, Kim NH, Bose S, Lee JH. Recent advances in graphene based polymer composites. *Prog Polym Sci.* 2010;35 (11):1350–1375. https://doi.org/10.1016/j.progpolymsci.2010.07.005.

51. Zhou X, Huang X, Qi X, et al. In situ synthesis of metal nanoparticles on single-layer graphene oxide and reduced graphene oxide surfaces. *J Phys Chem C.* 2009;113:10842–10846. https://doi.org/10.1021/jp903821n.

52. Huang J, Zhang L, Chen B, et al. Nanocomposites of size-controlled gold nanoparticles and graphene oxide: formation and applications in SERS and catalysis. *Nanoscale.* 2010;2:2733–2738. https://doi.org/10.1039/C0NR00473A.

53. Song M. Graphene functionalization and its application to polymer composite materials. *Nano Energy.* 2013;2(2):97–111. https://doi.org/10.1680/nme.12.00035.

54. Yang Z, Liu X, Liu X, et al. Preparation of β-cyclodextrin/graphene oxide and its adsorption properties for methylene blue. *Colloids Surf B: Biointerfaces.* 2021;200. https://doi.org/10.1016/j.colsurfb.2021.111605, 111605.

55. Hoseinzadeh H, Hayati B, Ghaheh FS, Seifpanahi-Shabani K, Mahmoodi NM. Development of room temperature synthesized and functionalized metal-organic framework/graphene oxide composite and pollutant adsorption ability. *Mater Res Bull.* 2021;142. https://doi.org/10.1016/j.materresbull.2021.111408, 111408.

56. Bayantong ARB, Shih YJ, Ong DC, Abarca RRM, Dong CD, de Luna MDG. Adsorptive removal of dye in wastewater by metal ferrite-enabled graphene oxide nanocomposites. *Chemosphere.* 2021;274. https://doi.org/10.1016/j.chemosphere.2020.129518, 129518.

57. Sun Y, Chen M, Liu H, Zhu Y, Wang D, Yan M. Adsorptive removal of dye and antibiotic from water with functionalized zirconium-based metal organic framework and graphene oxide composite nanomaterial Uio-66-(OH)2/GO. *Appl Surf Sci.* 2020;525. https://doi.org/10.1016/j.apsusc.2020.146614, 146614.

58. Hosseinzadeh H, Ramin S. Fabrication of starch- graft -poly(acrylamide)/graphene oxide/hydroxyapatite nanocomposite hydrogel adsorbent for removal of Malachite Green dye from aqueous solution. *Int J Biol Macromol.* 2018;106:101–115. https://doi.org/10.1016/j.ijbiomac.2017.07.182.

59. Sadegh N, Haddadi H, Arabkhani P, Asfaram A, Sadegh F. Simultaneous elimination of Rhodamine B and Malachite Green dyes from the aqueous sample with magnetic reduced graphene oxide nanocomposite: optimization using experimental design. *J Mol Liq.* 2021;343. https://doi.org/10.1016/j.molliq.2021.117710, 117710.

60. Jinendra U, Bilehal D, Nagabhushana BM, Kumar AP. Adsorptive removal of Rhodamine B dye from aqueous solution by using graphene-based nickel nanocomposite. *Heliyon.* 2021;7(4). https://doi.org/10.1016/j.heliyon.2021.e06851, e06851.

61. Ajeel SJ, Beddai AA, Almohaisen AMN. Preparation of alginate/graphene oxide composite for methylene blue removal. *Mater Today Proc.* 2021. https://doi.org/10.1016/j.matpr.2021.05.331.

62. Noreen S, Tahira M, Ghamkhar M, et al. Treatment of textile wastewater containing acid dye using novel polymeric graphene oxide nanocomposites (GO/PAN, GO/PPy, GO/PSty). *J Mater Res Technol.* 2021;14:25–35. https://doi.org/10.1016/j.jmrt.2021.06.007.

63. Verma M, Tyagi I, Kumar V, Goel S, Vaya D, Kim H. Fabrication of GO-MnO₂ nanocomposite using hydrothermal process for cationic and anionic dyes adsorption: kinetics, isotherm, and reusability. *J Environ Chem Eng.* 2021;9(5). https://doi.org/10.1016/j.jece.2021.106045, 106045.

64. Ahlawat W, Dilbaghi N, Kumar S. Evaluation of graphene oxide and its composite as potential sorbent for removal of cationic and anionic dyes. *Mater Today: Proc.* 2021;45(6):5500–5505. https://doi.org/10.1016/j.matpr.2021.02.215.

65. Ren F, Li Z, Tan WZ, et al. Facile preparation of 3D regenerated cellulose/graphene oxide composite aerogel with high-efficiency adsorption towards methylene blue. *J Colloid Interface Sci.* 2018;532:58–67. https://doi.org/10.1016/j.jcis.2018.07.101.

66. Karthikeyan P, Nikitha M, Pandi K, Meenakshi S, Park CM. Effective and selective removal of organic pollutants from aqueous solutions using 1D hydroxyapatite-decorated 2D reduced graphene oxide nanocomposite. *J Mol Liq.* 2021;331, 115795.

67. Sirajudheen P, Karthikeyan P, Vigneshwaran S, Meenakshi S. Complex interior and surface modified alginate reinforced reduced graphene oxide-hydroxyapatite hybrids: removal of toxic azo dyes from the aqueous solution. *Int J Biol Macromol.* 2021;175:361–371.

68. Pashaei-Fakhri S, Peighambardoust SJ, Foroutan R, Arsalani N, Ramavandi B. Crystal violet dye sorption over acrylamide/graphene oxide bonded sodium alginate nanocomposite hydrogel. *Chemosphere.* 2021;270, 129419.

69. Razzaq S, Akhtar M, Zulfiqar S, et al. Adsorption removal of Congo red onto L-cysteine/rGO/PANI nanocomposite: equilibrium, kinetics and thermodynamic studies. *J Taibah Univ Sci.* 2021;15(1):50–62.

70. Hu C, Grant D, Hou X, Xu F. High rhodamine B and methyl orange removal performance of graphene oxide/carbon nanotube nanostructures. *Mater Today: Proc.* 2021;34:184–193.

71. Han L, Khalil AM, Wang J, et al. Graphene-boron nitride composite aerogel: a high efficiency adsorbent for ciprofloxacin removal from water. *Sep Purif Technol.* 2021;278. https://doi.org/10.1016/j.seppur.2021.119605, 119605.

72. Rosli FA, Ahmad H, Jumbri K, Abdullah AH, Kamaruzaman S, Fathihah Abdullah NA. Efficient removal of pharmaceuticals from water using graphene nanoplatelets as adsorbent. *R Soc Open Sci.* 2021;8(1). https://doi.org/10.1098/rsos.201076, 201076.

73. Radmehr S, Sabzevari MH, Ghaedi M, Azqhandi MHA, Marahel F. Adsorption of nalidixic acid antibiotic using a renewable adsorbent based on graphene oxide from simulated wastewater. *J Environ Chem Eng.* 2021;9(5). https://doi.org/10.1016/j.jece.2021.105975, 105975.

74. Feng X, Qiu B, Dang Y, Sun D. Enhanced adsorption of naproxen from aquatic environments by β-cyclodextrin-immobilized reduced graphene oxide. *Chem Eng J.* 2021;412. https://doi.org/10.1016/j.cej.2021.128710, 128710.

75. Segovia SSJ, Pastrana MLM, Ocampo PR, Morales TS, Berber MMS, Carrasco MF. Synthesis and characterization of carbon xerogel/graphene hybrids as adsorbents for metronidazole pharmaceutical removal: effect of operating parameters. *Sep Purif Technol.* 2020;237. https://doi.org/10.1016/j.seppur.2019.116341, 116341.

76. Fang Z, Hu Y, Wu X, et al. A novel magnesium ascorbyl phosphate graphene-based monolith and its superior adsorption capability for bisphenol A. *Chem Eng J.* 2018;334:948–956. https://doi.org/10.1016/j.cej.2017.10.067.

77. Guerra ACS, de Andrade MB, Tonial dos Santos TR, Bergamasco R. Adsorption of sodium diclofenac in aqueous medium using graphene oxide nanosheets. *Environ Technol.* 2021;42(16):2599–2609.

78. Zhang K, Li H, Xu X, Yu H. Synthesis of reduced graphene oxide/NiO nanocomposites for the removal of Cr (VI) from aqueous water by adsorption. *Microporous Mesoporous Mater.* 2018;255:7–14. https://doi.org/10.1016/j.micromeso.2017.07.037.

79. Zhang W, Chen J, Hu Y, Fang Z, Cheng J, Chen Y. Adsorption characteristics of tetrabromobisphenol A onto sodium bisulfite reduced graphene oxide aerogels. *Colloids Surf A Physicochem Eng Asp.* 2018;538:781–788. https://doi.org/10.1016/j.colsurfa.2017.11.070.

80. Jiang LH, Liu YG, Zeng GM, et al. Removal of 17β-estradiol by few-layered graphene oxide nanosheets from aqueous solutions: external influence and adsorption mechanism. *Chem Eng J.* 2016;284:93–102. https://doi.org/10.1016/j.cej.2015.08.139.

81. Al-Khateeb LA, Almotiry S, Salam MA. Adsorption of pharmaceutical pollutants onto graphene nanoplatelets. *Chem Eng J.* 2014;248:191–199. https://doi.org/10.1016/j.cej.2014.03.023.

82. Nikou M, Samadi-Maybodi A, Yasrebi K, Sedighi-Pashaki E. Simultaneous monitoring of the adsorption process of two organophosphorus pesticides by employing GO/ZIF-8 composite as an adsorbent. *Environ Technol Innov.* 2021;23. https://doi.org/10.1016/j.eti.2021.101590, 101590.

83. Mahdavi V, Taghadosi F, Dashtestani F, et al. Aminoguanidine modified magnetic graphene oxide as a robust nanoadsorbent for efficient removal and extraction of chlorpyrifos residue from water. *J Environ Chem Eng.* 2021;9(5), 106117.

84. Khawaja H, Zahir E, Asghar MA, Rafique K, Asghar MA. Synthesis and application of covalently grafted magnetic graphene oxide carboxymethyl cellulose nanocomposite for the removal of atrazine from an aqueous phase. *J Macromol Sci B: Phys.* 2021;1–20.

85. Muthusaravanan S, Balasubramani K, Suresh R, et al. Adsorptive removal of noxious atrazine using graphene oxide nanosheets: insights to process optimization, equilibrium, kinetics, and density functional theory calculations. *Environ Res.* 2021;200. https://doi.org/10.1016/j.envres.2021.111428, 111428.

86. Ahmaruzzaman M, Gupta VK. Rice husk and its ash as low-cost adsorbents in water and wastewater treatment. *Ind Eng Chem Res.* 2011;50 (24):13589–13613. https://doi.org/10.1021/ie201477c.

87. Pelalak R, Heidari Z, Khatami SM, Kurniawan TA, Marjani A, Shirazian S. Oak wood ash/GO/Fe₃O₄ adsorption efficiencies for cadmium and lead removal from aqueous solution: kinetics, equilibrium and thermodynamic evaluation. *Arab J Chem.* 2021;14(3). https://doi.org/10.1016/j.arabjc.2021.102991, 102991.

88. Priya VN, Rajkumar M, Mobika J, Sibi SL. Alginate coated layered double hydroxide/reduced graphene oxide nanocomposites for removal of toxic As (V) from wastewater. *Phys E.* 2021;127. https://doi.org/10.1016/j.physe.2020.114527, 114527.

89. Huang Q, Chen Y, Yu H, et al. Magnetic graphene oxide/MgAl-layered double hydroxide nanocomposite: one-pot solvothermal synthesis, adsorption performance and mechanisms for Pb²⁺, Cd²⁺, and Cu²⁺. *Chem Eng J.* 2018;341:1–9. https://doi.org/10.1016/j.cej.2018.01.156.

90. Peer FE, Bahramifar N, Younesi H. Removal of Cd (II), Pb (II) and Cu (II) ions from aqueous solution by polyamidoamine dendrimer grafted magnetic graphene oxide nanosheets. *J Taiwan Inst Chem Eng.* 2018;87:225–240. https://doi.org/10.1016/J.JTICE.2018.03.039.

91. Hadadian M, Goharshadi EK, Fard MM, Ahmadzadeh H. Synergistic effect of graphene nanosheets and zinc oxide nanoparticles for effective adsorption of Ni (II) ions from aqueous solutions. *Appl Phys A Mater Sci Process.* 2018;124(3):1–10. https://doi.org/10.1007/s00339-018-1664-8.

92. Ansari MO, Kumar R, Ansari SA, et al. Anion selective pTSA doped polyaniline@ graphene oxide-multiwalled carbon nanotube composite for Cr (VI) and Congo red adsorption. *J Colloid Interface Sci.* 2017;496:407–415. https://doi.org/10.1016/j.jcis.2017.02.034.

93. Lingamdinne LP, Choi YL, Kim IS, Yang JK, Koduru JR, Chang YY. Preparation and characterization of porous reduced graphene oxide based inverse spinel nickel ferrite nanocomposite for adsorption removal of radionuclides. *J Hazard Mater.* 2017;326:145–156. https://doi.org/10.1016/j.jhazmat.2016.12.035.

94. Su H, Ye Z, Hmidi N. High-performance iron oxide-graphene oxide nanocomposite adsorbents for arsenic removal. *Colloids Surf A Physicochem Eng Asp.* 2017;522:161–172. https://doi.org/10.1016/j.colsurfa.2017.02.065.

95. Liu W, Yang L, Xu S, et al. Efficient removal of hexavalent chromium from water by an adsorption-reduction mechanism with sandwiched nano-composites. *RSC Adv.* 2018;8(27):15087–15093. https://doi.org/10.1039/C8RA01805G.

96. Leng Y, Guo W, Su S, Yi C, Xing L. Removal of antimony (III) from aqueous solution by graphene as an adsorbent. *Chem Eng J.* 2012;211:406–411. https://doi.org/10.1016/j.cej.2012.09.078.

97. Wu W, Yang Y, Zhou H, et al. Highly efficient removal of cu (II) from aqueous solution by using graphene oxide. *Water Air Soil Pollut.* 2013;224 (1):1–8. https://doi.org/10.1007/s11270-012-1372-5.

98. Li S, Lu X, Xue Y, Lei J, Zheng T, Wang C. Fabrication of polypyrrole/graphene oxide composite nanosheets and their applications for Cr (VI) removal in aqueous solution. *PLoS One.* 2012;7(8). https://doi.org/10.1371/journal.pone.0043328, e43328.

99. Wang Y, Ding G, Lin K, Liu Y, Deng X, Li Q. Facile one-pot synthesis of ultrathin carbon layer encapsulated magnetite nanoparticle and graphene oxide nanocomposite for efficient removal of metal ions. *Sep Purif Technol.* 2021;266. https://doi.org/10.1016/j.seppur.2021.118550, 118550.

100. Verma M, Kumar A, Singh KP, et al. Graphene oxide-manganese ferrite (GO-MnFe$_2$O$_4$) nanocomposite: one-pot hydrothermal synthesis and its use for adsorptive removal of Pb^{2+} ions from aqueous medium. *J Mol Liq.* 2020;315. https://doi.org/10.1016/j.molliq.2020.113769, 113769.

101. Aldalbahi A, El-Naggar M, Khattab T, et al. Development of green and sustainable cellulose acetate/graphene oxide nanocomposite films as efficient adsorbents for wastewater treatment. *Polymers.* 2020;12(11):2501. https://doi.org/10.3390/polym12112501.

102. Anush SM, Chandan HR, Gayathri BH, Manju N, Vishalakshi B, Kalluraya B. Graphene oxide functionalized chitosan-magnetite nanocomposite for removal of Cu (II) and Cr (VI) from waste water. *Int J Biol Macromol.* 2020;164:4391–4402. https://doi.org/10.1016/j.ijbiomac.2020.09.059.

103. Anuma S, Mishra P, Bhat BR. Polypyrrole functionalized cobalt oxide graphene (COPYGO) nanocomposite for the efficient removal of dyes and heavy metal pollutants from aqueous effluents. *J Hazard Mater.* 2021;416. https://doi.org/10.1016/j.jhazmat.2021.125929, 125929.

104. Archana S, Radhika D, Jayanna BK, Kannan K, Kumar KY, Muralidhara HB. Functionalization and partial grafting of the reduced graphene oxide with p-phenylenediamine: an adsorption and photodegradation studies. *FlatChem.* 2021;26. https://doi.org/10.1016/j.flatc.2020.100210, 100210.

105. Awad FS, AbouZied KM, Bakry AM, Abou El-Maaty WM, El-Wakil AM, El-Shall MS. Polyacrylonitrile modified partially reduced graphene oxide composites for the extraction of Hg (II) ions from polluted water. *J Mater Sci.* 2021;56(13):7982–7999. https://doi.org/10.1007/s10853-021-05797-2.

106. Su M, Liu Z, Wu Y, et al. Graphene oxide functionalized with nano hydroxyapatite for the efficient removal of U (VI) from aqueous solution. *Environ Pollut.* 2021;268. https://doi.org/10.1016/j.envpol.2020.115786, 115786.

107. Xing C, Xia A, Yu L, Dong L, Hao Y, Qi X. Enhanced removal of Pb (II) from aqueous solution using EDTA-modified magnetic graphene oxide. *Clean-Soil Air Water.* 2021;49(4):2000272. https://doi.org/10.1002/clen.202000272.

108. Ahmad SZN, Salleh WNW, Yusof N, et al. Pb (II) removal and its adsorption from aqueous solution using zinc oxide/graphene oxide composite. *Chem Eng Commun.* 2021;208(5):646–660. https://doi.org/10.1080/00986445.2020.1715957.

109. Chakraborty V, Das P, Roy PK. Synthesis and application of graphene oxide-coated biochar composite for treatment of strontium-containing solution. *Int J Environ Sci Technol.* 2021;18(7):1953–1966.

110. Ali I, Babkin AV, Burakova IV, et al. Fast removal of samarium ions in water on highly efficient nanocomposite based graphene oxide modified with polyhydroquinone: isotherms, kinetics, thermodynamics and desorption. *J Mol Liq.* 2021;329, 115584.

111. Husein DZ, Uddin MK, Ansari MO, Ahmed SS. Green synthesis, characterization, application and functionality of nitrogen-doped MgO/graphene nanocomposite. *Environ Sci Pollut Res.* 2021;28(22):28014–28023.

112. Chen Y, Cui J, Liang Y, Chen X, Li Y. Synthesis of magnetic carboxymethyl cellulose/graphene oxide nanocomposites for adsorption of copper from aqueous solution. *Int J Energy Res.* 2021;45(3):3988–3998.

113. Katubi KMM, Alsaiari NS, Alzahrani FM, Siddeeg MS, Tahoon AM. Synthesis of manganese ferrite/graphene oxide magnetic nanocomposite for pollutants removal from water. *Processes.* 2021;9(4):589.

114. Yang X, Chen C, Li J, Zhao G, Ren X, Wang X. Graphene oxide-iron oxide and reduced graphene oxide-iron oxide hybrid materials for the removal of organic and inorganic pollutants. *RSC Adv.* 2012;2(23):8821. https://doi.org/10.1039/c2ra20885g.

115. Tan P, Sun J, Hu Y, et al. Adsorption of Cu^{2+}, Cd^{2+} and Ni^{2+} from aqueous single metal solutions on graphene oxide membranes. *J Hazard Mater.* 2015;297:251–260. https://doi.org/10.1016/j.jhazmat.2015.04.068.

116. Sahraei R, Pour ZS, Ghaemy M. Novel magnetic bio-sorbent hydrogel beads based on modified gum tragacanth/graphene oxide: removal of heavy metals and dyes from water. *J Clean Prod.* 2016;142:2973–2984. https://doi.org/10.1016/j.jclepro.2016.10.170.

117. Chen X, Zhou S, Zhang L, You T, Xu F. Adsorption of heavy metals by graphene oxide/cellulose hydrogel prepared from NaOH/urea aqueous solution. *Materials.* 2016;9(7):582.

118. Vicente-Martínez Y, Caravaca M, Soto-Meca A, Francisco-Ortiz D, Gimeno F. Graphene oxide and graphene oxide functionalized with silver nano-particles as adsorbents of phosphates in waters. A comparative study. *Sci Total Environ.* 2020;709:136111. https://doi.org/10.1016/j.scitotenv.2019.136111.

119. Marrakchi F, Hameed BH, Hummadi EH. Mesoporous biohybrid epichlorohydrin crosslinked chitosan/carbonaceous clay adsorbent for effective cationic and anionic dyes adsorption. *Int J Biol Macromol.* 2020;163:1079–1086. https://doi.org/10.1016/j.ijbiomac.2020.07.032.

120. Marrakchi F, Bouaziz M, Hameed BH. Activated carbon–clay composite as an effective adsorbent from the spent bleaching sorbent of olive pomace oil: process optimization and adsorption of acid blue 29 and methylene blue. *Chem Eng Res Des.* 2017;128:221–230.

121. Wang B, Gao B, Zimmerman AR, Lee X. Impregnation of multiwall carbon nanotubes in alginate beads dramatically enhances their adsorptive ability to aqueous methylene blue. *Chem Eng Res Des.* 2018;133:235–242. https://doi.org/10.1016/j.cherd.2018.03.026.

122. Zohre S, Ataallah SG, Mehdi A. Experimental study of methylene blue adsorption from aqueous solutions onto carbon nano tubes. *Int J Water Resour Environ Eng.* 2010;2(2):016–028. https://doi.org/10.5897/IJWREE.

123. Li Y, Du Q, Liu T, et al. Methylene blue adsorption on graphene oxide/calcium alginate composites. *Carbohydr Polym.* 2013;95(1):501–507. https://doi.org/10.1016/j.carbpol.2013.01.094.

124. Wang H, Gao H, Chen M, et al. Microwave-assisted synthesis of reduced graphene oxide/titania nanocomposites as an adsorbent for methylene blue adsorption. *Appl Surf Sci.* 2015;360(2016):840–848. https://doi.org/10.1016/j.apsusc.2015.11.075.

125. Heidarizad M, Şengör SS. Synthesis of graphene oxide/magnesium oxide nanocomposites with high-rate adsorption of methylene blue. *J Mol Liq.* 2016;607–617. https://doi.org/10.1016/j.molliq.2016.09.049.

126. Yousefi N, Emtyazjoo M, Sepehr MN, Darzi SJ, Sepahy AA. Green synthesis of *Pseudomonas aeruginosa* immobilized Fe_3O_4-multiwalled carbon nanotubes bio-adsorbent for the removal of 2,4,6-trinitrophenol from aqueous solution. *Environ Technol Innov.* 2020;20:101071. https://doi.org/10.1016/j.eti.2020.101071.

127. Mohan D, Sarswat A, Singh VK, Alexandre-Franco M, Pittman Jr CU. Development of magnetic activated carbon from almond shells for trinitrophenol removal from water. *Chem Eng J.* 2011;172(2–3):1111–1125. https://doi.org/10.1016/j.cej.2011.06.054.

128. Aggarwal P, Misra K, Kapoor S, Bhalla A, Bansal R. Effect of surface oxygen complexes of activated carbon on the adsorption of 2,4,6-trinitrophenol. *Def Sci J.* 2013;48(2):219–222. https://doi.org/10.14429/dsj.48.3902.

129. Kafshgari MM, Tahermansouri H. Development of a graphene oxide/chitosan nanocomposite for the removal of picric acid from aqueous solutions: study of sorption parameters. *Colloids Surf B: Biointerfaces.* 2017;160:671–681. https://doi.org/10.1016/j.colsurfb.2017.10.019.

130. Khakpour R, Tahermansouri H. Synthesis, characterization and study of sorption parameters of multi-walled carbon nanotubes/chitosan nanocomposite for the removal of picric acid from aqueous solutions. *Int J Biol Macromol.* 2018;109:598–610. https://doi.org/10.1016/j.ijbiomac.2017.12.105.

131. Hosseinzadeh H, Pashaei S, Hosseinzadeh S, Khodaparast Z, Ramin S, Saadat Y. Preparation of novel multi-walled carbon nanotubes nanocomposite adsorbent via RAFT technique for the adsorption of toxic copper ions. *Sci Total Environ.* 2018;640–641:303–314. https://doi.org/10.1016/j.scitotenv.2018.05.326.

Chapter 6

Low-cost adsorbent biomaterials for the remediation of inorganic and organic pollutants from industrial wastewater: Eco-friendly approach

Kajol Goria[a], Anu Bharti[a], Shubham Raina[a], Richa Kothari[a,*], V.V. Tyagi[b], Har Mohan Singh[b], and Gagandeep Kour[a]

[a]*Department of Environmental Sciences, Central University of Jammu, Rahya Suchani (Bagla) Samba, Jammu, Jammu and Kashmir, India,*
[b]*School of Energy Management, Shri Mata Vaishno Devi University, Katra, Jammu and Kashmir, India*
*Corresponding author.

1 Introduction

Water is known to be the elixir of life, but rampant population explosion followed by intense industrialization has degraded water quality to great extent.[1] Rapid industrialization has led to the generation of industrial effluents consisting of numerous organic as well as inorganic pollutants. Industries without proper treatment or after inadequate treatment mainly discharged off their toxic effluents directly into the environment due to which millions of people get affected by unhygienic water.[2] Annual wastewater generated globally is reported about 359 billion cubic meters, out of which 48% discharged untreated.[3] However, some reports indicated that wastewater generation in India constitutes approximately 44 million m^3 per day out of which industries are reported to generate around 6.2 million m^3 of untreated effluents per day.[4] In addition to industrial wastewater, a lot of other kinds of wastewater have been generated due to man-made activities like domestic wastewater, agricultural runoff, etc., as these sectors also utilize water. The microbial population occurring in wastewater comprises bacteria, protozoa, viruses, fungi, algae, etc., whose presence can lead to the spread of several deadly diseases. Salmonellosis, diarrhea, typhoid, dysentery, cholera, skin and tissue infections, respiratory infection, meningitis, pneumonia, hepatitis, gastroenteritis, fever, immune suppression, reproductive abnormalities, carcinogenic, dehydration, gastrointestinal disorders, ascariasis, coughing, chest pain, sleeplessness, fever and hookworm disease are few of the commonly occurring waterborne diseases caused by polluted water contaminated with organic, inorganic, or biological pollutants described in proceeding sections.[5–7] In view of the severe harmful and toxic impacts of wastewater, several regulatory guidelines and legislations for wastewater discharge have been proposed globally as well as locally to reduce the risk of wastewater to man and the environment. It is the need of the hour to search for cost-efficient, economic, and eco-friendly methods and their implementation to ensure sustainable management of wastewaters.[8] To comply with these regulations, proper wastewater treatment technologies are recommended from time to time. Few of the traditionally used wastewater remediation technology has been described in the proceeding text. Some of the available physical methods of wastewater treatment comprise filtration methods including microfiltration, ultrafiltration and reverse osmosis, other methods like floatation, sedimentation, adsorption as well as nano-adsorption.[9] On the other hand, there are few chemical methods being utilized for depolluting the polluted water including flocculation, coagulation, chemical adsorption, neutralization, disinfection via ozonation, chlorination or dichlorination and many others. Similarly, biological ways of pollutants remediation called bioremediation techniques are also available including aerobic and anaerobic methods. Examples of aerobic methods are activated sludge technique, oxidation ponds, aeration lagoon, oxygenated bioreactors, trickling or biological filters, microbial treatment, etc., whereas anaerobic methods comprise lagoons and bioreactors devoid of oxygen.[10] Each of the treatment approaches mentioned above offers several limitations in aspects of environmental pollution, sewage accumulation, high expenditure and numerous technical challenges. Therefore, eco-friendly treatment options for treating wastewater through biological means have gained a lot of attention these days.[11] Adsorption through biological materials like fruit and

Sustainable Materials for Sensing and Remediation of Noxious Pollutants. https://doi.org/10.1016/B978-0-323-99425-5.00004-9

vegetable wastes, agricultural waste products including biochar, agro-derived activated carbon, and many other compounds available in nature, provides a prominent treatment technology to remediate industrial as well as other effluents.[12] In this article, different ways to achieve wastewater remediation through green and eco-friendly adsorption approaches are discussed and highly endorsed.

2 Low-cost adsorbent biomaterials

Adsorbing materials that are readily available in nature at a low or even no cost and acquire the property to separate pollutants selectively from wastewater and bind them to their exterior and interior surfaces are termed as low-cost adsorbent biomaterials. There is a wide variety of such low-cost adsorbents developed and commercialized in order to alleviate the serious concern of wastewater remediation. Due to less effectiveness and low economic viability of traditionally used wastewater treatment technologies, biological adsorbing materials are gaining great attention. These materials are thought to be highly potent in the remediation of all kinds of pollutants, whether organic or inorganic including heavy metals. Low-cost biomaterials like activated carbon prepared from natural products such as agricultural biomass waste, fruit and vegetable wastes, several industrial wastes and many others are highly recognized for their excellent adsorption capacity that has been described in the proceeding text. Some of the major biomaterials are discussed here below.

2.1 Biomaterials

Biomaterials are naturally occurring organic matter that is present in abundance on earth. Biomaterials are capable of exhibiting sorption. Apart from this, different biomaterials in the form of organic waste such as egg shells, vegetable and fruit peels, shells of walnuts, almonds, groundnut, sugarcane bagasse, etc., can be easily used as the source of adsorbents.[13–15] These types of materials are easily available at negligible cost. Biomaterials in the form of waste serve as an excellent feedstock for the synthesis of potential sorbents. They also serve as environmentally friendly materials that are renewable, sustainable, biodegradable, and non-toxic.[16] In general, biosorbents are prepared via a two-stage chemical pre-treatment process, after which it is dried, pulverized, and then sieved or sometimes calcined. A potential biosorbent should possess certain general characteristics other than specific ones. These general characteristics include low-cost availability, easily portable, chemically and mechanically stable, and the high surface area along with some other textural and physiochemical properties. But it is difficult to have one biosorbent having all these characteristics; however, the more of these properties possessed by the biosorbent the more potentially active is the biosorbent enhancing the chances of the concentration gradient between the solution and the biosorbent.[17] There are several types of biosorbents which are being used at present and many more are still being investigated in terms of potential, economical aspect, transportability, and ease of modification. These include algal biomass, fungal biomass, plant materials, agricultural waste, magnetic bioadsorbents (magnetic nanoparticles impregnated onto them), fish scales, egg shells, and waste.[18–21] Some microbes are widely used as biosorbent because they have the ability to grow easily in different concentrations, variable pH, a wider range of temperature and they have a good efficiency when it comes to harmful pollutant uptake such as heavy metals dyes and phenols.[22] Plant material's agricultural waste consist of high cellulose content which is very effective in adsorbing pollutants such as heavy metals and dyes. These biomaterials are economic, eco-friendly, renewable, and plentiful and therefore are potential biosorbents in their natural form or chemically modified form. Examples include seeds, roots, bark, peels such as banana and orange, husk such as rice and sesame, and leaves.[23–25]

2.2 Activated carbon prepared from an agricultural waste

Activated carbon is one of the best adsorbents that can be used for the treatment of various kinds of wastewaters originating from households, municipal discharge, agricultural processing units and industries. This material is extensively used due to its high surface area that is suitable for the exchange of various solutes present in water.[26] Activated carbon is a carbonaceous substance that is amorphous in nature and has a large network of pores present on its surface and extensively distributed oxygenated functional groups such as carboxylic acids, phenols, etc.[27,28] Three different classes of pores are present, namely micropores, mesopores, and macropores. Micro and mesopores contribute maximum to the phenomenon of adsorption. Activated carbon is the product of pyrolysis. In this process, suitable organic materials are subjected to heating in the absence of oxygen to produce solid char, bio-oil, and different gases.[29] This solid product is activated using different types of physical and chemical activation processes that will be later discussed in this subsection. This process is different from combustion as pyrolysis is an endothermic process involving thermal degradation of organic matter, while combustion is an exothermic and oxidative process taking place in the presence of oxygen that produces light and heat. Activated carbon can also be

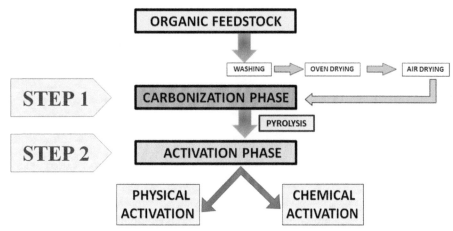

FIG. 1 Preparation of activated carbon.

prepared through gasification, but this method is not preferred due to the poor yield of char. The preparation of activated carbon can be subdivided into two steps as shown in Fig. 1. The first step involved in the preparation of activated carbon is the carbonization of feedstock. Agricultural wastes such as crop residues, plant litter such as twigs and leaves, egg shells, the solid proportion of kitchen waste, etc., can be used for this purpose. On the basis of temperature range, the process of carbonization can be divided into different phases. The first phase having a temperature range below 200°C remains endo-thermic which is useful in the removal of moisture from the feedstock. The second phase or the pre-carbonization phase, having a temperature range below 300°C, is also endothermic; in this phase, few acids such as acetic acid few alcohols like methanol are produced along with gases such as CO_2 and CO, also a light fraction of tar is generated during the process. Temperature above 300°C is the third exothermic phase which is responsible for the escape of different volatiles and fixing of the carbon content of the char.[29] The second step involves the activation of char through physical/ chemical activation. Physical activation includes steam activation, CO_2 activation, activation using high temperature, etc. In chemical activation, different chemicals can be used for the impregnation including NaOH, KOH, ZnCl, H_3PO_4, etc.[30] During the chemical acti-vation of carbon, activating agents are mixed/impregnated in the char that helps in the removal of tar, thereby opening surface pores. After that, carbon particles are oxidized through the burning of char in the presence of activating agents. For well-developed pores, activation time and temperature during the process serve as crucial parameters. The activated carbon thus produced has a highly porous structure that enhances the adsorption of substances to many folds.

2.3 Graphene-based adsorbents and their derivatives

Graphene has gained curiosity due to its adsorption behavior. It has a unique structure, extraordinary properties, and a wide range of applications of which its adsorption behavior is prominent. It has a large surface area, a wide range of functional groups, is easy to prepare, and low-production cost. Graphene and its derivates are being used for the treatment of heavy metals, pesticides, oils, different dyes, etc. Graphene and its corresponding derivatives have various natural advantages that make it a brilliant material for adsorption purposes. It has a single-layered structure; due to this, all the molecules remain in direct contact with the contaminated solution. Its surface area even surpasses Carbon nanotubes (CNT's) that are considered as good adsorbing surfaces.[31,32] Pollutant adsorption patterns of graphene-based adsorbents have also shown higher kinetic rates in comparison to traditional adsorbents. It has an ideal porous structure permitting higher diffusion and adsorption of pollutants. Another interesting property of graphene is that it can handle different types of pollutants simultaneously.[33]

2.3.1 Graphene

Structurally graphene is a two-dimensional single layer of hexagonally packed and sp^2 hybridized carbon atoms held together through σ and π-bonds. Its structure comprises of a honeycomb lattice and building block for various allotropic forms of carbon such as graphite, fullerenes, and carbon nanotubes as shown in Fig. 2. When graphene is arranged in the form of multiple layers, then this three-dimensional (3D) is referred to as graphite. Graphene rolled up in a cylindrical fashion is regarded as a carbon nanotube. Considering the physiochemical properties of graphene, it has high strength, high electrical and thermal conductivity, big specific surface area, electron mobility, and flexible nature. These following

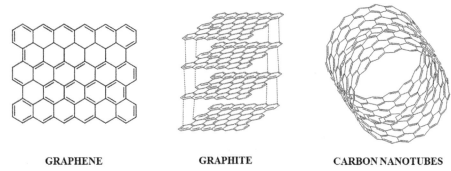

GRAPHENE   GRAPHITE   CARBON NANOTUBES

FIG. 2 Representation of chemical structures of graphene, graphite and carbon nanotubes.

properties allow graphene to be used as a promising adsorbent for various environmental remediations including treatment of wastewaters contaminated with heavy metals, pesticides, dyes, antibiotics, etc.[34]

2.3.2 Graphene oxide

Graphene oxide is one of the important derivatives of graphene for adsorption. It is an oxidized form of graphene having oxygen-containing functional groups. Production of graphene oxide (GO) is generally achieved through the high degree oxidation of graphite. It has poor electrical conductivity, cost-effective material, easily produced, and shows high adsorptive performance.[33]

2.3.3 Reduced graphene oxide

Reduced graphene oxide is also a derivative of graphene. In comparison to graphene oxide, it has a less negative charge and is low oxygen content.[35,36] Chemically, it can be either prepared via reduction of graphene oxide or via direct growth method, as illustrated in Fig. 3. Reduction of graphene oxide is an easy and cost-effective approach; therefore, the chemical reduction of graphene oxide is a widely accepted method. Hydrazine hydrate is largely used as a reducing agent as it is capable of reducing most of the functional groups having oxygen as an element. Other reductants involved in the preparation include $NaBH_4$, $Na_2S_2O_3$, and vitamin C.[33]

2.3.4 Graphene sponge

A graphene sponge is the derivative of graphene having a sponge resembling structure. This structure shows robust performance for the diffusion of pollutants into it. It has the capability for adsorption and desorption of different contaminants including dyes, oils, and various organic solvents. The graphene sponge can be reused without a significant loss in its adsorption-desorption capacity.[37] The sponge-like structure has various practical advantages such as easy handling and usage. Its surface properties and pore structure can be adjusted during the preparation stage as per requirement. Graphene sponges can be prepared from various routes, the key factor that is kept into consideration is to maintain its porous structure.

2.3.5 Functional graphene

Graphene can be further functionalized by the addition of different functional groups. The aim of functionalization is to enhance adsorptive strength. Functionalization of graphene through the covalent route is marked by functional groups attached to its basal plane through intense intermolecular forces having high-binding energy. Oxygen-containing functional groups such as COO, CO, OH, and COOH[38] are among common groups that are covalently functionalized to the graphene.

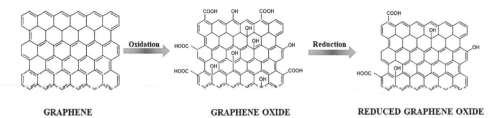

GRAPHENE   GRAPHENE OXIDE   REDUCED GRAPHENE OXIDE

FIG. 3 Conversion of graphene into graphene oxide and reduced graphene oxide.

Functionalization of graphene/graphene oxide through non-covalent method involves the use of functional species such as those that have weak forces of attraction between them and graphene. Compounds such as metallic, organometallic and ionic can be used for the non-covalent functionalization of graphene.

3 Wastewater treatment remediation techniques

In today's world, fresh water, which is safe for consumption, has become a necessity for the survival of every living organism present on earth.[39] It has major impacts on human health, hygiene, plants, and animals. However, during the last few decades, the world has seen tremendous industrialization which has directly or indirectly contributed to the problem of various types of pollutions especially water pollution. Moreover, most of the world's water reservoirs are being polluted by various types of contaminants such as pharmaceuticals, heavy metals, pesticides, insecticides, and so on.[40–42] Water scarcity coupled with water pollution has become a universal problem nowadays. There are numerous ways in which wastewater is generated such as industrial, domestic, and other activities. Due to the problem of water crisis the world is facing today or is going to face in the near future, wastewater reuse is being looked at as a potential option. There are different ways in which water sources can get contaminated thus leading to various types of wastewater remediation techniques. For example, wastewater generated from domestic activities will have different types of contaminants as compared to the wastewater generated from industrial processes. The remediated wastewater can be used for irrigational purposes, manufacturing, and environmental practices. The basic principle involved in wastewater remediation remains the same, however, the process may vary according to the location and source of contamination.

3.1 Water pollutants

There are a number of contaminants present in wastewater but depending upon the nature of the contaminants, which are broadly categorized as inorganic, organic pollutants and biological pollutants. Mainly inorganic pollutants comprise heavy metal, ions, and acids, while organic one includes toxic chemicals such as dyes, pesticides, and phenols and the remaining biological pollutants consist of disease-causing protozoans, viruses, and bacteria.

3.1.1 Inorganic pollutants

Inorganic water pollutants are characterized for their longer persistence time and their potential of not degrading straightforwardly, thus altogether deteriorating the aqueous system. In India, there are several regions where these pollutants have crossed threshold limits hence making the water not fit for drinking purposes.[43] Some of the common inorganic water pollutants are heavy metals, anions, acids, and alkalis.

3.1.1.1 Heavy metals

Metals having a specific density greater than $5\,g/cm^3$ are considered heavy metals. The widely studied heavy metals in wastewater generated by several industrial activities include metals such as arsenic, mercury, chromium, lead, nickel, cadmium, and zinc.[44] Generally, the sources of heavy metals in wastewater include the mining and battery industries, factories producing dyes, electronics, and sewage. They can have both natural as well as anthropogenic sources which have the potential to contaminate the water. Some of the natural sources include volcanic activities, rock breakdown and soil erosion, whereas anthropogenic sources include excessive mining activities, mineral extraction, industrial and agricultural discharge. Anthropogenic sources are more responsible for contaminating the water bodies.[45] Contamination caused by them can pose serious threats to the environment as well as to the organisms residing in it because they bioaccumulate (a process in which a chemical substance gets gradually accumulated in an organism's body and over some time period, the concentration of the chemical becomes more than it is in the environment). These chemicals are accumulated and stored quickly before being broken down and finally excreted out. World Health Organization (WHO) has set up certain permissible limits for heavy metals in water bodies and in case these limits are exceeded, then it can seriously affect the soil and aqueous system.[34] This ultimately leads to an impact on the health of living organisms depending upon these systems. Heavy metals in excess quantity can be lethal, can be carcinogenic sometimes, causing several health issues to living beings.

3.1.1.2 Anions

Harmful anions are another type of water pollutant present in wastewater other than heavy metals, organic or biological pollutants. Hazardous anions such as nitrates, fluorides, phosphates, arsenites, and sulfates are present in wastewater in variable concentrations. These anions have shown concerns in the treatment of wastewaters. Untreated wastewater contains nitrogen in both organic and inorganic (ammonia, nitrite, and nitrate) forms. Among these, all except nitrate is present in gaseous form and is responsible for causing degradation to groundwater and also has serious health effects on almost all living organisms.[46] Excess of nitrogen can be very lethal as it can have carcinogenic properties. Its impacts can be seen by the large cases of methemoglobinemia and cancer.[47] Fluoride is present as fluoride ions in natural waters. The natural sources of fluoride ions include discharge from mineral sediments and anthropogenic sources include industries such as phosphate fertilizers, pesticides, brick manufacturing, mineral extraction, and coal-based thermal power plants.[48] According to standards prescribed by WHO, the permissible limit for fluoride is 1.5 mg/L above which becomes very harmful. Excess of fluoride ions in the body of living beings can cause problems such as dental and skeleton fluorosis.[49,50] The fluoride concentration in the wastewater of countries like India, Africa, and China is more or less similar.[51] Phosphate ions are released into the water bodies via natural as well as anthropogenic activities. Natural sources include the phosphorus cycle that occurs in rocks and mineral sediments during which phosphorus is released as phosphate ions that are water-soluble. Phosphate is present in three forms: orthophosphate, metaphosphate, and organically bound phosphate. They are present as free ions (soluble orthophosphate ion), weakly chemically bounded phosphorus or other forms of phosphorus/oxygen in wastewater.[52,53] Excess amounts of phosphorus compounds in the water bodies cause a phenomenon called eutrophication (gradual increase in the availability of one or more limiting factors like CO_2, nutrients, sunlight leading to aging of aquatic systems). To combat this problem, we need to understand and identify the sources of the phosphorus compounds reaching the aquatic systems. Other than natural sources, anthropogenic sources such as industrial and drainage runoff contribute to an increase in the concentration of phosphorus compounds in the wastewater[54] found in their study that the main source of phosphorus in the river system of the United Kingdom was mainly sewage effluent and agricultural discharge. Although sulfate ions are among the common and non-toxic constituents of several types of water bodies, the excess concentration of these can pose serious threats to the environment such as mineralization of water, corrosion of metal pipes, imbalance in the natural sulfur level.[55,56] Sources of these ions are mainly the effluent discharge from the industries especially steel and iron plants. In these industries, a large quantity of wastewater is generated because a number of processes such as mining, smelting, and rolling are being carried out.

3.1.1.3 Acids

In different types of industrial applications and chemical labs, acids are used. In fewer quantities, they are not harmful because they can be easily neutralized. However, if they are present in larger quantities in the wastewater, they can cause some serious environmental and human health problems. Due to rapid industrialization, the emission of harmful gases such as sulfur dioxide and nitrogen oxides has also increased. These gases get converted into acids (sulfuric acid and nitric acid) and come down in the form of precipitation (acid rain) and finally changing some parameters such as the pH of natural water systems.

3.1.2 Organic pollutants

The environment also contains a large number of organic pollutants, most of which ultimately find their way into wastewater and finally to the natural water bodies. Mostly, these pollutants come from plants, animal/human waste, or artificial sources. Their entry into the wastewater comes from sources such as human waste, excessive use of pesticides, detergents, pharmaceuticals, insecticides, weedicides, fungicides, herbicides, artificial flavors, dyes and personal care products.[38,57] A load of organic waste in the water bodies will lead to a decrease in dissolved oxygen that will have some serious impacts on the organisms residing there. Different types of combinations can be formed by various organic pollutants such as phenols and their derivatives present together, resulting in the production of bad taste and displeasing odor of water. This adds to the problems in the environment. Mostly, these substances are very harmful making the water unfit for any kind of reuse.

3.1.2.1 Phenols

Phenols, along with their derivatives, are the organic chemicals that are used in the manufacturing of materials such as synthetic fibers, thermoplastic polymers, and polycaprolactam. These are categorized as priority pollutants because they are well known for their acute toxicity, genotoxicity, corrosivity, and endocrine-disrupting potency. There has been an increase in the concern of public health due to the widespread presence of phenols in wastewater in the present times. Due to the rise in anthropogenic activities, their concentration in wastewater has also increased. Their variability, toxicity, and persistency can affect the health of the environment and pose a threat to the living organisms by contaminating the aquatic ecosystem.[57]

3.1.2.2 Dyes

Organic dyes are defined as molecules that are complex in nature and are used to impart color to materials such as fabrics. These hold an essential role in industrial processes for producing a variable range of products. They are mostly water-soluble, poor in terms of biodegradability, hard to detect when present in low concentrations, resistant to detergents and heat. Synthetic dyes are among those types of pollutants that have shown negative effects on the health of the environment including the terrestrial as well as aquatic ecosystems.[58] Their discharge in water bodies imparts undesirable color to it and restricts the solar penetration, which affects the biological and other photochemical activities in the aquatic ecosystem. According to Yao et al.[59] throughout the world, approximately 7×10^5 metric tons of dyes are generated annually and out of this, 10% is discharged directly in wastewater. The main sources of dyes in wastewater are textile industries, leather manufacturing units, paper mills, and food processing units. Harmful impacts of dyes as pollutants include low visibility at low concentration, disturbed sunlight penetration, carcinogenic, mutagenic, and disturbing COD/BOD levels.

3.1.2.3 Pesticide

Increased anthropogenic activities have increased the number of chemical pollutants in wastewater effluents. Among these commonly found chemical pollutants are pesticides which have been used extensively in recent times in agriculture in order to get a good crop yield both quality as well as quantity wise. These, along with phenols and others, are categorized under priority pollutants by EU legislation. It can pollute groundwater and surface water through leaching, erosion, runoff, and hence can have very severe impacts on the living systems such as immune suppression, carcinogenicity, reproduction abnormalities, and reduced intelligence.[60] Depending upon the type of pesticide such as herbicide, rodenticide, fungicide, defoliants, and nematicides, the characteristics of wastewater vary.[57] A majority of these types are chemically stable and are very difficult to degrade under normal circumstances. Conventional methods for wastewater treatment are not so efficient in eliminating the pesticides completely from the wastewater effluents; therefore, some advanced treatment technologies are required.

3.1.3 Biological pollutants

The contamination of water bodies by pathogenic waterborne microbes has gained attention as it has degraded the water quality all over the world.[61] The sources of pathogenic microbes are mostly human, animal fecal wastes, hospital effluents, and discharge of untreated or poorly treated sewage effluents directly into the main water bodies. This, in turn, through leaching or runoff, result in polluting the ground water as well as surface water causing a number of health problems. The contamination of drinking water by these pathogenic microbes can have more fatal impacts in undeveloped or low-income countries hence there is a need to understand their effects on every type of aquatic ecosystem. These are the regions where infants, pregnant women, and elderlies with the weak immune system are infected very readily. Bacteria, viruses, helminths, and protozoans are the most commonly found pathogens present in wastewater, drinking water, or another type of recreational water and are responsible for causing long-term as well as short-term health problems such as heart problems, typhoid, cholera, shigellosis, ulcers, salmonellosis, campylobacteriosis, giardiasis, intestinal discomfort, and Hepatitis. Viruses are potentially considered more harmful pathogens present in wastewater as these are highly resistant, infectious, and unable to detect easily. However, bacteria are among the most commonly found pathogens in wastewater[62] responsible for causing a number of diseases, and protozoans are mostly found in wastewater than any other system. If helminth parasites or helminth eggs are present along with these pathogens, it is of great concern with regard to human health as a very small number of these protozoans can cause severe infections. Every year, approximately 3.4 million people die due to the consumption of contaminated water and associated waterborne diseases and children's deaths are recorded more out of this total number.[63] Some of the highly concerned and harmful impacts of a variety of pollutants have been compiled in Table 1.

3.2 Current remediation methods

As the amount of wastewater has increased tremendously in a few decades, it is necessary to have some cost-efficient wastewater treatment technologies to ensure safe disposal or for reuse purposes. At present, there are a number of wastewater remediation methods that are used to get safer and cleaner water. Some of these are conventionally utilized as physical, chemical, and biological wastewater treatment methods as represented in Fig. 4.

TABLE 1 Harmful impacts of some commonly occurring found pollutants in wastewaters.

Water pollutants		Sources	Harmful impacts	References
Inorganic pollutants				
Heavy metals	Arsenic (Ar)	Discharge from industries	Carcinogen, skin problems, immunotoxin	64
	Cadmium (Cd)	Corrosion of galvanized pipes, metallurgical refineries, runoff from paints, batteries	Nephrological problems Kidney damage	65
	Nickel (Ni)	Wastewater from industries such as battery manufacturing, electroplating, alloying etc.	Respiratory implications, cardiological disorders DNA damage, skin problems, birth defects, carcinogenic	66
	Copper (Cu)	Corroded pipes, wood preservatory chemicals	Mucosal irritation, stomach and liver problems, effect on the central nervous system	67
	Zinc (Zn)	Industrial effluent, smelting, batteries, brass, rubber production, fossil fuel combustion	Anemia, pancreatic degradation, irritations, respiratory disorders, infertility	66
	Chromium (Cr)	Leather and textile industries, painting, steel, paper and pulp factories	Chromosomal aberrations, immunity reduction, skin problems, gastrointestinal problems	66
	Lead (Pb)	Corrosion of pipelines	Hypertension, brain damage, retarded physical and mental growth, kidney Problems	65
	Selenium (Se)	Agricultural runoff, mining, industrial effluent	Dysfunction of arms and legs, gastrointestinal problems, liver and spleen damage	68,69
	Mercury (Hg)	Refineries discharge, natural deposits erosion, landfills runoff	Spinal cord disorders, poisonous, mutagenic, kidney disorders	65
Ions	Nitrate ions	Anthropogenic-wastewater treatment plants, fertilizer manufacturing factories	Eutrophication, methemoglobinemia, carcinogen, water quality problems	70
	Fluoride ions	Natural-dissolution of minerals Anthropogenic-industrial effluents, coal-based thermal power plants	Dental and skeletal fluorosis, infertility, thyroid malfunction	71
	Phosphate ions	Natural-phosphorus cycle Anthropogenic-industrial effluent, domestic effluent including human excreta, detergents and food residues	Eutrophication	72
	Sulfate ions	Natural-rocks and minerals dissolution, fertilizer industries Anthropogenic-industrial effluents, sewage infiltration	Diarrhea, dehydration, gastrointestinal disorders, imbalance in the natural sulfur cycle	6

TABLE 1 Harmful impacts of some commonly occurring found pollutants in wastewaters—cont'd

Water pollutants		Sources	Harmful impacts	References
Organic pollutants				
Phenols		Industrial wastewater generated by textile industries, dye manufacturing units, paper mills, pharmaceutical industries and resin factories	Acute toxicity, genotoxicity, corrosivity, endocrine-disrupting potency	73
Dyes		Textile industries, paper mills, leather manufacturing units, pharmaceuticals, and food processing plants	Distorted solar penetration, carcinogenic, skin problems, mutagenic, toxic even in low concentration, raise COD/BOD of aquatic ecosystem	74
Pesticides		Agricultural runoff, leaching, erosion and drainage from pesticide manufacturing plants	Neurotoxicity, reproductive abnormalities, carcinogenic	33
Biological pollutants				
Viruses	Hepatitis A virus, noroviruses, adenovirus, astrovirus, rotavirus and reovirus	Slaughterhouse wastewater	Respiratory infection, meningitis, pneumonia, hepatitis, gastroenteritis and fever	75
Bacteria	*E. coli, Salmonella* sp., *P. aeruginosa, Klebsiella* sp., *Vibrio* sp., *S. aureus* and *Shigella* sp., *Yersinia* sp.	Human and animal fecal waste and food processing units	Salmonellosis, diarrhea, typhoid, gastroenteritis dysentery, cholera, skin and tissue infections	5
Protozoan	*Balantidium coli, Cryptosporidium* sp., *Giardia duodenalis, Toxoplasma gondii* and *Entamoeba* sp.	Sewage effluent, fecal matter	Balantidiasis, amoebic dysentery, toxoplasmosis giardia and cryptosporidium	76
Helminths	*Ascaris lumbricoides, Ascaris suum, Necator Americanus* and *Trichuris trichiura*	Fecal-oral route	Ascariasis, coughing, chest pain, sleeplessness, fever and hookworm disease	7

3.2.1 Chemical precipitation

This process involves the materials dissolved in water to change their form into undissolved solid by altering the ionic equilibrium in the presence of chemical reactions. The solid thus formed acts as a precipitate and is compressed as a pellet by centrifuge. The remaining liquid residue left is called supernatant. This method is commonly applied to eliminate the heavy metal ions by adding counter-ions in order to decrease the solubility because it is a cost-efficient and simple process. Most commonly, it is used to eliminate the metal cations but can also be used to remove metal anions as well as organic pollutants such as dyes, phenols, and aromatic enzymes. The chemical precipitation method, in general, involves the following steps:

- Stage 1—Chemical material such as magnesium hydroxide, calcium sulfide, or soda ash is added to precipitate the metal ions as hydroxides, sulfides, or carbonates.
- Stage 2—Elimination of the precipitated metal as hydroxides or carbonates by methods such as gravity separation or other methods depending upon the type of wastewater to be treated.
- Stage 3—Metal ions that are still not precipitated out in earlier stages are then precipitated by the addition of organic or inorganic sulfides.
- Stage 4—Removal of precipitates formed in stage 3 by suitable separation methods.

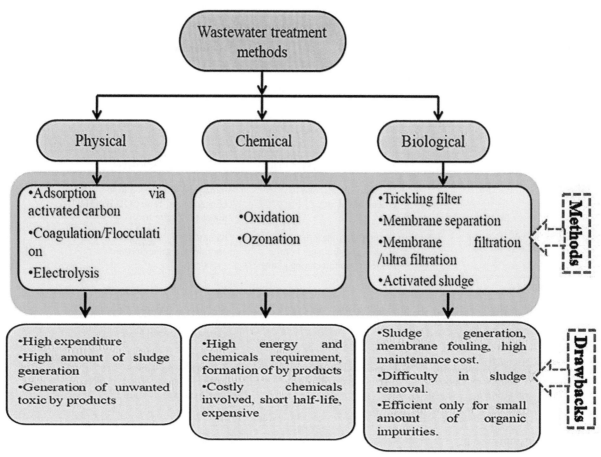

FIG. 4 Current wastewater remediation methods with their possible drawbacks.[77]

3.2.2 Coagulation and flocculation

Coagulation and flocculation methods can be considered as a necessary tool as well as the backbone of most of the water remediation methods as it aids in enhancing the speed of the separation process.[78] Wastewater effluents consist of particles having variable sizes, and hence coagulant and flocculant treatment is used widely to eliminate these variable sized particles from wastewater. The coagulation process involves the addition of chemical material known as a coagulant which helps in the easy removal of suspended particles by binding with them and forming tiny aggregates called flocs. The coagulants that are used in wastewater remediation should be trivalent metal cations or polymers, and they should be non-toxic and should be insoluble or have low solubility in the pH range commonly used for wastewater remediation. There are two main types of coagulants that are widely used viz., organic and inorganic coagulants. Examples of inorganic coagulants involve aluminum sulfate, ferric sulfate, sodium aluminate, polyaluminum chloride, and aluminum chloride.[79] These are cost-efficient as well as can be used in several activities. However, their disadvantage is that they add to the sludge volume. Organic coagulants, on the other hand, produce less sludge, are more effective in low dosage, and do not alter the pH of the water that has been treated. Examples include polyamines, tannins, and melamine formaldehyde. Flocculation is a process that involves slow and gentle optimization of the rate of contact among the particles so as to increase the size of flocs. Hence, in the flocculation process, the particles come closer and get attached to one another building larger molecules which are easier to segregate by settling down or filtration. Sometimes, separate flocculants are used along with coagulants depending upon the composition of wastewater. These flocculants can be naturally occurring such as activated silica and polysaccharides or synthetic flocculants such as polyacrylamide.

3.2.3 Ion exchange

It is a sorption method in which reverse chemical reaction occurs when the ions present in the solution are swapped by other ions having similar charges but are of different species. They are widely used for water softening, heavy metal cation removal, demineralization, or nitrate removal. These particles can be naturally occurring, such as inorganic zeolites or synthetically manufactured such as organic resins. Organic resins are widely used at present because their characteristics can be altered as per requirements. There are two types of resins ion exchange systems: resin anion exchange system and resin cation exchange system. Both the systems are very sensitive to fouling which is caused by organic matter present in the wastewater. Depending upon the type of cation and anions, i.e., Strong Acid Cation (SAC), Weak Acid Cation (WAC), Strong Base Anion (SBA), Weak Base Anion (WBA), the material can further be broken down into several small groups.[80] In the resin ion exchange system, wastewater is made to pass through the cation exchange system. This will cause loss of ions making the metal ions acids and ions lost are swapped by the equal number of hydrogen ions. The acids which are produced are further removed by anionic exchange resin which is alkaline in nature. Here, anions are swapped by an equal number of hydroxides. This exchange bed has a fixed limit so the resins will get exhausted and will undergo the process of regeneration. The cation exchange resin is regenerated by HCl or H_2SO_4 on the other hand, the anion exchange resin is regenerated by NaOH.[81] Advantages of this method involve its cost-effectiveness, low-energy requirements, easier regeneration of the resin. However, there are several disadvantages attached to this method also such as fouling (calcium/iron), bacterial contamination, organic contamination, and chlorine contamination. Applications include softening, demineralization, color removal in sugar cane syrups as a catalyst, and in pharmaceutical industries.

3.2.4 Membrane filtration

The technology widely used to treat wastewater in the place of conventional methods has gained attention in the past few decades as it is capable of enhancing water quality. It has become a well-recognized technology that can help in the generation of safe and clean water out of wastewater. Different types of membranes filtration methods have been used in wastewater treatment or water reuse applications. Microfiltration and ultrafiltration are commonly used methods because they are capable of producing water with low suspended solids.[82,83] Ultrafiltration makes use of membranes having a pore size in the range from 0.1 to 0.001 μm. This type of membrane-based method is very useful for removing molecules with high molecular weight and colloidal substances. For direct wastewater treatment, membrane bio-reactor (MBR) is used as they are capable of producing water that is low in effluent nutrients. Nano filtration and reverse osmosis are the advanced methods used for wastewater treatment.

3.2.5 Adsorption

It is a surface phenomenon that is capable of removing a wide range of pollutants from wastewater. This process involves the binding of the molecules to the porous surface of a solid substance, i.e., it involves the interaction between a solute that is adsorbable in nature and a solid adsorbent. The solute molecules get concentrated on the surface of the adsorbent due to the acting attractive forces between the two and gradually, over some time the concentration of adsorbate on the adsorbent increases. There are several types of adsorbents used such as naturally occurring or synthetically producing zeolites, nano-adsorbents, natural clay minerals, active carbon, silica gel, phytoadsorbents, and activated aluminum. Zeolites are substances having cavities that range below 2 nm. The structure of various zeolites is dependent upon 3D interlinkage between tetrahedral units, i.e., SiO_4^{4-} and AlO_4^{4-} which stay interlinked to each other by shared oxygen atoms.[84] Natural zeolites such as Analcime, chabazite, clinoptilolite, etc., are among commonly used. Clinoptilolite among these has been highly studied zeolite as a potent adsorbent that can address the wastewater problems such as removing pollutants like heavy metals, dyes, phenols, etc.[85] studied these naturally present zeolites in order to get an alternative for a low-cost adsorbents that have a brilliant selectivity range for a large range of cations as well as having better ion exchange capacity. There are non-toxic exchangeable cations present within their like sodium ions, potassium ions, magnesium ions, etc., making them perfect adsorbent for wastewater treatment. Other commonly used adsorbent is active carbon which is suitable for removing the polar contaminants. Depending upon the nature of surface attachment, the adsorption process has been broadly divided into three categories: physical, chemical, or exchange adsorption. Physical adsorption also known as physisorption is a reversible process involving the weak van der Waal forces which are easy to separate without sharing or transferring electrons. Chemical adsorption or chemisorption is an irreversible process involving the formation of chemical bonds between adsorbate and adsorbent. It is difficult to remove the materials which are chemisorbed on the solid surface. Exchange adsorption involves the charge attraction between the adsorbent and adsorbate. Adsorption is considered a very effective and efficient treatment method among all other treatment methods because it is economical, versatile, highly efficient, can remove pollutants in trace concentration, is not affected by toxicity and is easily controllable. However, like all other wastewater treatment methods, it too has some limitations such as producing some materials that can prove to be harmful to the environment.

3.3 Green and eco-friendly adsorption technology for organic and inorganic pollutants remediation

Contamination of water by various pollutants such as inorganic, inorganic, and biological pollutants have escalated several types of a serious problem which has led to the demand for new methods to treat this huge amount of wastewater. However, conventional methods being economical are not environmentally friendly as many of these methods generate sludge or some harmful materials that will again have harmful impacts on the environment as well as living beings. Hence, there is a need to develop green and eco-friendly methods to treat and eliminate pollutants such as heavy metals, dyes, phenols, and their associated products.[86,87] Phytoremediation has also been considered an emerging technique for the elimination of wastewater pollutants. This involves the employment of living plants whose roots absorb toxic contaminants from wastewater. Generally, a detailed mechanism is associated with phytoremediation technology involving phytoextraction or phytoaccumulation, phytodegradation, phytovolatilization, phytostabilization, phytofiltration and phytomining.[88] Numerous plants are recognized for their bioremediation potential. Plant species, e.g., *Typha angustifolia*, has been reported as a promising candidate for remediation of poisonous heavy metals like lead, nickel, zinc, iron, manganese, and copper from contaminated aqueous solution.[89,90] Vetiver grass (*Chrysopogon zizanioides* L.) has also been evaluated for phytoremediation capability which proved it as a good phytoremediation agent for chromium and nickel.[91] Another fascinating approach for wastewater remediation is microbial remediation that makes the use of microbes like algae or bacteria to decontaminate wastewater. Waste water remediation via algae is referred to as phytoremediation. Algal biomass has been proved as a good adsorbent having a great capacity to absorb contaminants including heavy metals, phenols, etc. Microalgal consortia such as *Chlorella* sp., *Scenedesmus* sp., and *C. zofingiensis* have been reported as excellent agents for remediating chemical oxygen demand (COD) and phosphorus from dairy wastewater[92,93] studied the capacity of adsorbents obtained out of green algae *Nannochloropsis* sp. that has been modified using silica and a coat of magnetite particles (ASN-MPs) after that so as to absorb Methylene Blue (ME) as well as Cu(II) cations present in the solution. They found that the values of adsorption capacity (q_m) of the ME and Cu(II) cation mixture present in ASN and ASN-MPs were around 1.39×10^{-1} and 5.32×10^{-1} mmol g^{-1}, respectively. However, these modified adsorbents showed good efficiency for the sorption of ME and Cu(II) cations. Similarly, bacterial consortia of *Bacillus* sp., *Pseudomonas* sp., *Nitrosomonas*, *Nitrobacter*, *Aspergillus niger* and *Rhodococcus* sp. have been evaluated for excellent removal of COD and sulfate impurities.[94] Hence, these can be utilized as a good adsorbent for adsorbing cationic mixtures contaminants such as heavy metals as well as organic dyes. Some of the recent green and eco-friendly technologies used for remediating pollutants are discussed below.

3.3.1 Biosorbents

Biosorbents are materials having biological origins that are used to eliminate pollutants (organic as well as inorganic) from the solution. The process by which this type of remediation is achieved is known as biosorption. This eco-friendly process has been widely accepted as an alternative to conventional technologies due to its practical application as well as economic aspect.[95] Few of the biosorbents involved in heavy metal removal are listed in Table 2 along with their removal efficiency and adsorption capacity in wastewater treatment.

3.3.2 Graphene-based composites

Carbon-based nanoparticles are widely used for eliminating inorganic as well as organic pollutant via adsorption.[114] Graphene along with its derivatives due to its unique characteristics, is attracting much attention in adsorption applications. Graphene is also known as the building block of graphite, is a two-dimensional (2D) single layer of carbon atoms (sp^2 hybridized) that is packed in a honeycomb-like structure. It is gaining the attention of researchers because of its remarkable characteristics such as high hydrophobic surface area, distinct morphology, high thermal conductivity, high optical transparency, high electron mobility, high mechanical strength, ease to modification, and unique chemical structure. The large delocalized network of graphene allows it to form stronger bonds with other harmful pollutants easily hence making graphene more stable as compared to its other carbon nano counter systems.[115] As the properties of graphene are highly dependent upon its shape, size and unique morphology therefore, its synthesis has to be done in a controlled manner in order to obtain graphene having desirable properties. Graphene can be modified and combined with other materials which give rise to new composites possessing increased properties. Extensive research has been done on graphene-based composites and their derivatives such as graphene oxide (GO), reduced graphene oxide (rGO) as adsorbents to be used in wastewater remediation.

TABLE 2 Heavy metal removal tendency and adsorption capacity of low-cost biosorbents.

Low-cost biosorbents	Effluent treated	Heavy metal removal efficiency (%) or adsorption capacity (mg/g)	References
Acid treated carrot residue	Industrial effluent	Zn 29.61 mg/g Cu 32.74 mg/g Cr 45.09 mg/g	96
HCl activated rice husk	Car battery waste water	Pb 54.85% or 0.56731 mg/g	97
Palm kernel cake residue biochar	Industrial effluent	Pb 49.64 mg/g Cd 18.60 mg/g Hg 13.69 mg/g Cr 19.92 mg/g	98
Combined waste rind of *Citrullus lanatus* and waste peel of *Citrus sinensis*	Tannery wastewater	Pb 86% Cr 85%	99
Activated carbon prepared from banana peel	Industrial effluent	Cu 99.84% or 49.5 mg/g Pb 97.53% or 45.6 mg/g Cd 94.69 or 30.7 mg/g Cr 88.02% or 25.2 mg/g	100
Cobalt ferrite-activated carbon from the almond husk	Tannery wastewater	Cr 98.2% or 23.6 mg/g Pb 96.4% or 6.27 mg/g	101
Activated carbon derived from rice husk	Industrial effluent	Hg 55.87 mg/g	102
Orange peel	Industrial effluent	Zn 86.6% or 56.18 mg/g Cu 70.73 mg/g Pb 209.8 mg/g	103
Eggshell	Handloom-dyeing effluents	Fe 93.43% Cu 100% Cr 100% Mn 93.05% Zn 99.82%	104
Sawdust	Handloom-dyeing effluents	Fe 87.22% Cu 100% Cr 100% Mn 51.18% Zn 97.46%	104
Rice husk	Handloom-dyeing effluents	Fe 66.38% Cu 100% Cr 100% Mn 76.50%, Zn 96.53%	104
Lemon peel	Handloom-dyeing effluents	Fe 65.80% Cu 100% Cr 100% Mn 58.57% Zn 87.43%	104
Activated carbon derived from sugarcane bagasse	Textile wastewater	Fe 91% Zn 89%	105
Cyanobacterial strain	Sugar mill effluent	Cu 87.50% Cr 82.96% Pb 86.44%	106
Activated carbon prepared from banana peels	Industrial effluent	Cu 14.3 mg/g Ni 27.4 mg/g Pb 34.5 mg/g	107

Continued

TABLE 2 Heavy metal removal tendency and adsorption capacity of low-cost biosorbents—cont'd

Low-cost biosorbents	Effluent treated	Heavy metal removal efficiency (%) or adsorption capacity (mg/g)	References
Sugarcane bagasse	Industrial effluent	Pb 99.9% or 19.3 mg/g Cu 90% or 13.24 mg/g Ni 66% or 2.99 mg/g	108
Phosphoric acid-activated rice husk	Industrial effluent	Cu 88.9% or 17.0358 mg/g	109
Activated carbon derived from sugarcane bagasse	Industrial effluent	Hg 82.88% Pb 100% Cd 79.5%	110
Peanut shells	Industrial effluent	Cd 55.43 ± 6.82 mg/g	111
Hazelnut shell activated carbon	Industrial effluent	Cr 170 mg/g	112
Orange peel		Ni 158 mg/g	
Soybean hull		Cu 154.9 mg/g	
Jackfruit		Cd 52.08 mg/g	
Banana peels	Domestic/synthetic greywater	Cr 68% Pb 94.5%	113
Orange peels		Cr 55% Pb 87.5%	
Sapodilla peels		Cr 54% Pb 81%	

3.3.2.1 Removal of heavy metals by graphene-based composites

Graphene-based nanosorbent composites can be used broadly to remove heavy metals through the process of adsorption. It can be further enhanced by adding a particular design employing functionalization that includes the addition of high-binding functional groups, adding dopants such as nitrogen or sulfur atom[116] or inserting defects to the basic structure to increase the adsorption performance.[117] Among these above-mentioned methods, functional groups having oxygen, nitrogen, sulfur, and other groups are considered more reliable for the removal of heavy metals as compared to the defects or dopants introduced on graphene structure. Several studies have shown that the inserted functional groups on the graphene network are capable of immobilizing particular heavy metals to a larger extent. Graphene oxide (GO), for example, is usually obtained from graphite by Hummer's and Offeman method. It contains plentiful different functional groups such as carboxyl (—COOH), carbonyl (C=O) and hydroxyl (—OH) groups, which makes GO a potential sorbent as it can adsorb heavy metals more readily. Hence, it has led to several efforts by researchers in the recent past in order to obtain advanced GO-based adsorbent composites that are easy to separate from the wastewater. Magnetic-graphene-based nanocomposites are one such example that is being extensively used for wastewater treatment.[118–120] Hence, a lot of work has been done on designing the graphene network with the addition of some functional groups such as oxygen groups like alginate and cellulose, nitrogen groups such as 3-aminopyrazole, 2,2-dipyridylamine and sulfur groups such as thiacalix-arene tetrasulfonate in order to enhance the efficiency of adsorption for heavy metals.[100,121]

3.3.2.2 Removal of dyes by graphene-based composites

Dyes, in general, are defined as colored organic particles and they can be categorized according to the properties present in their structures such as molecular makeup, charge, and solubility. They can be water-soluble or water-insoluble. The water-soluble dyes include acid, basic, substantive, reactive, and natural dyes[122] whereas water-insoluble dyes include examples like vat, disperse, sulfur.[123] Their structure consists of an aromatic ring that allows the interaction between the delocalized electrons present in their benzene ring and the delocalized π electrons that are present in the graphene. Organic dyes carry certain charged groups on them; hence the interaction between these organic dyes and graphene-based composites having high-binding functional groups will lead to the removal of dyes to a larger extent because the functionalized graphene molecules will show more pronounced adsorption. Graphene derivatives that possess negatively charged groups on their surface will adsorb cationic dyes more readily. For example, GO being a derivative of graphene, carries on its surface negatively charged groups, and it has shown enhanced performance in adsorbing cationic dyes wastewater.[124] Similarly, anionic dyes are adsorbed by graphene derivatives having positively charged functional groups like chitosan and amine.[125] But GO sheets show large dispersibility in water, which restricts the effective segregation of GO sheets that has adsorbed dyes from wastewater. Recently, the option of wastewater remediating materials has been taken to one step further by focusing on the fabrication of three-dimensional macrostructures which are more porous and have a lower density from two-dimensional graphene nanosheets in order to enhance the adsorbing capacity toward different types of pollutants (heavy metal, dyes, and other organic and inorganic pollutants).[126–128] The use of advanced graphene-based composites is expected to lessen water pollution which has taken a frightening shape in recent years and deliver cleaner and safer water by eliminating the multiple pollutants present in wastewater.[129]

3.3.3 Metal oxides

Metal oxide nanoparticles in the past were mostly used for a wider range of applications that include energy-storing devices, catalysis, sensors, environmental remediation, and many more. However, these metal oxide nanoparticles possess extraordinary surface features and microstructural characteristics that make them important sources for the adsorbing pollutants. They possess some active sites along with a high surface area which enables them to perform adsorption with more ease. Most commonly used metal oxides utilized for wastewater remediation include ferric oxides,[90] magnesium oxides,[130] aluminum oxides,[131] titanium oxides,[132,133] cerium oxide,[134] and zinc oxide.[135] Metal oxides nanoparticles have their size in the nano-range that leads to the enhancement of surface-to-volume ratio in an exponential manner.[136] Hence, the surface energy of adsorbent is enhanced by decreasing the size and introduction of some active sites on their surface to interact with pollutants. Thus, nanomaterials are capable of showing pronounced adsorption capacity as compared to the larger counterpart for removing the pollutant present in the wastewater.

4 Application of nanotechnology in wastewater treatment

Nanotechnology has been emerged as an advancing technology in wastewater cleaning by making use of nano-sized (1–100 nm) particles in numerous materials. Such materials incorporating nanoparticles are commonly called nanomaterials. In this modern era of technical advancement, nanotechnology is being employed for the detection and removal of different types of pollutants that are present in wastewater. Few of the characteristics of nanomaterials like little space and energy requirements, low-cost expenditure depending upon the extent of contamination, magnetic nature, and many others make them highly potential to clean wastewater. Nanotechnology is clearly recognized as among one of the most advanced technologies for potential decontamination of industrial wastewaters.[137] A major challenge with wastewater discharge is its efficient yet cost-effective treatment. Pollutants such as dyes, oils and heavy metals present in wastewater pose a serious threat to all life forms. Heavy metals, as we know, are non-biodegradable, are a well-known source of water pollution that can cause various diseases in human beings and disrupt or damage the functioning of vital organs such as the spleen, kidney, liver, heart and may also cause numerous neurological disorders. Nanoparticles have various adsorptive properties that make them excellent materials for the removal of contaminants. Nanoparticles can also be used for chemical and photochemical oxidation-based catalyzes for the destruction of different contaminants.[138] As the nanoparticle occupies magnetic characteristics so can perform quick separation of pollutants as well as quickly recover after treatment under the influence of a magnetic field. Nanomaterials can be classified into three different classes on the basis of their nature. First class includes nano-adsorbents that are made of materials having high adsorptive capacity. These include activated carbon (AC), silica, clay, various metal oxides and composites. The second category involves nano-catalysts such including metal oxides and semiconductor materials. The third class involves the use of nanomembranes. Various materials including nano-metal particles and nanocarbon tubes, are being used for the preparation of nanomembranes.[139] Some of the nanoclusters used in cleaning wastewater might be considered essential micronutrients for plants after treatment as it results from the release of microlevel particles. A few of the nanoparticles explored for wastewater remediation are listed in Table 3.

4.1 Hybrid nanomaterials

Adsorbing materials are formed by the coupling of any two or more than two individual components of organic or inorganic compounds with the incorporation of nanotechnology. Hybrid nanomaterials are generally binary or tertiary hybrid materials and mainly include hybrids based on metal oxides, colloidal compounds, chalcogens, and iron oxide-based magnetic hybrids.[151] Metal oxide-based hybrid nanomaterials like Fe_2O_3, WO_3, TiO_2, and ZnO are widely investigated for pollutants remediation through photocatalytic activities. An investigation carried out by Zhou et al.[152] demonstrated that coupling of Ti_2O and boron nitride results in organic pollutants remediations with greater efficiency. The coupling technique resulted in

TABLE 3 Different nanoparticles and their targeted pollutants investigated previously.

Nanoparticles	Example	Target pollutant	References
Ferric oxide	Goethite (α-FeOOH)	Cu (II)	140
	Hematite	Cu (II)	140
	Hydrous amorphous iron oxide	Pb (II)	141
	Fe nanoparticles doped mesoporous carbon	Organic dye (Methylene blue)	90
	Fe3O4/Al2O3/chitosan composite	Organic dye (Methyl orange)	142
Magnesium oxide	Nano magnesium oxide	Cr (VI)	130
Aluminum oxide	Sodium dodecyl sulfate coated nano α-Al$_2$O$_3$ modified with 2,4-dinitrophenylhydrazine (DNPH)	Pb, Cr, Ni, Co and Mn	143
	Carbon-Al$_2$O$_3$ core-shell spheres	Orange II	131
Titanium oxide	Nano TiO$_2$	Pb, Zn, Hg and Cd	132
	Sodium titanate	Organic dye (Methylene blue)	144
	CNT-loaded TiO$_2$	Methyl orange	145
	Titania nanotubes functionalized with hexadecyltrimethylammonium chloride	Anionic dyes such as acid red 1 and acid blue 9	146
Cerium oxide	CeO$_2$	Pb, Zn and Cd	147,148
	Hydroxyapatite-doped CeO2 nanoparticles	Eriochrome Black T dye	134
	Porous CeO$_2$	Congo red dye	134
Zinc oxide	Plate with nanostructured ZnO	Cu	149
	Chitosan-modified ZnO nanoparticles	Direct Blue 78 and Acid Black 26 dyes	150
	Incorporation of NiFe2O4 in the ZnO nanoparticles	Congo red dye	135

such adsorption enhancement due to boron nitride that promoted charge separation and enhanced sorption capacity of hybrid nanomaterial Ti$_2$O. Similarly, chalcogens, e.g., ZnS, are known to degrade methylene blue pollutants through photocatalysis.[153] Other chalcogens explored for photocatalytic remediation of pollutants such as dyes and heavy metals are Bi$_2$S$_3$, MoS$_2$, etc.[154]

4.2 Nanomembranes

Nanomaterials are regarded as one of the notable developments that took place in the wastewater treatment sector. One such nanomaterial is the nanomembranes which, because of being a simple, eco-friendly, reliable, straightforward process, less energy requirement, and not use any harmful and hazardous chemicals, has gained wide spread acceptability.[155] Any kind of wastewater weather surface water, groundwater or any industrial effluent, can be treated by the application of nanomembranes.[156,157] The number of membranes by using nanoparticles has been developed for wastewater filtration commonly referred nanofiltration membranes. These structures do not require multiple tanks but an only single tank that makes the use of filtering membranes. Wastewater simply passed through the nanomembranes and the clean filtrate gets separated out after pollutants adsorption is performed by the nanomembranes. The adsorption capacity of all kinds of nanomaterials increases with the decreased particle size. Preparation of nanomembranes requires careful considerations of minute details of the materials as well as deposition techniques. Widely used deposition techniques are ultrathin film techniques individually or in coupling with sacrificial layer etching technology for the preparation of nanomembranes.[158] Generally, there are physical and chemical methods for dispositioning nanomembranes. Physical methods comprise physical vapor deposition (PVD), radiofrequency sputtering, the spin coating also called spinning or spin casting, evaporation techniques like thermal evaporation and laser-assisted evaporation, epitaxial growth methods like homoepitaxy or heteroepitaxy while chemical methods incorporate chemical vapor deposition (CVD) like plasma-enhanced CVD, atomic layer deposition (ALD), plating techniques like electroplating, the sol-gel method including chemical bath deposition, spin coating method in case of chemical polymerization, molecular beam epitaxy, etc.[57,58]

5 Challenges and future perspectives

Naturally available low-cost adsorbing biomaterials emerged as a prominent technological option for wastewater remediation due to their high sorption ability, easy accessibility, reliability, environment friendliness, and many other advantages. Still, there are a few barriers that in the path of their commercialization or pilot/large-scale utilization that are required to be removed. Such barriers can restrict them from working efficiently. A few of those challenging barriers are discussed below. However, a few of the recommendations are also proposed for the effective removal of such hurdles that can significantly reflect the future perspectives of these low-cost adsorbing biomaterials.

5.1 Barriers associated with nanotechnology

Nanotechnology, no doubt, is regarded as a highly advanced and effective approach in today's world contributing in infinite sectors. Among them, wastewater treatment has been considered as one of its prominent applications. Although nanotechnology effectively remediates very minute pollutants incapable of remediation by means other than nanotechnology, yet carries few of the disadvantages. Widely used nanofiltration membranes show limitations such as membrane clogging so could be thought of as inefficient in depolluting the polluted water. Similarly, carbon nanotubes due to their great specific surface area, are identified as excellent nanomaterials for fast adsorption, therefore, they are highly recommended for heavy metal removal.[159,160] Heavy metals effectively adsorbed by carbon nanotubes reported are manganese (Mn^{2+}), copper (Cu^{2+}), chromium (Cr^{2+}) and lead (Pb^{2+}).[161] However, these nanomaterials offer a few challenges also that impede their commercialization including the high cost of their synthesis, difficulty in recovery after treatment and risk of secondary pollution.[160]

5.2 Environmental and economic challenges

Adsorbents synthesized from naturally obtained low-cost biomaterials carry several advantages in wastewater remediation. Along with the technique offer a few limitations. Environmental pollution can be considered as one of the key challenges in this regard due to after remains of biomaterials within the treatment system. Few nanomaterials like carbon nanotubes do not easily recover from the treated water and increase the risk of secondary pollution. Similarly, graphene and its derivatives, namely graphene oxides, are difficult to be used as direct aggregates which make their separation hard after treatment. Another major challenge in this respect is the economic challenge. Synthesis of adsorbing biomaterials requires almost negligible cost in comparison to that of conventional treatment technologies. In spite of the great cost-effectiveness of low-cost adsorbent biomaterials, overall expenditure may rise due to several technical hurdles associated with these materials e.g., process maintenance cost, process parameters optimization costs, etc. Membrane fouling in the case of filtration membranes is another cause of concern as it reduces the overall process efficiency. To make such processes more reliable and eco-friendly, requires great expenses. This may hike the process expenditure for wastewater treatment. Therefore, further research in this concern needed to be carried out to provide better solutions to such problems through several process improvements.

5.3 Future perspectives

Low-cost adsorbent biomaterials, because of have great adsorption potential, can extensively be utilized for adsorbing numerous organic as well as inorganic pollutants from different wastewaters. However, many of the naturally obtained materials discussed above can effectively treat wastewater; yet, several improvements or modifications in their characteristic properties particularly textural characteristics of biomaterials can enhance their adsorption capacity. Such modifications can be incorporated by physical or chemical improvement methods including sonification and freeze-drying as the physical one while introducing suitable functional groups and other chemicals as the chemical methods of their enhancement. Chemical treatment alters most of the constituent properties of the biomaterials and enhances their sorption activity effectively. For future perspectives, many of the economically and environmentally accepted methods are required to be further researched. Considering the available literature investigation, biomaterials are believed to adsorb all kinds of wastewater pollutants and improve water quality by regulating the wastewater characteristics like pH, chemical oxygen demand (COD), biological oxygen demand (BOD), turbidity, total suspended and dissolved solids in water along with oil and grease content. A few of such studies are compiled in Table 4 to show the effectiveness of biomaterials as biosorbents.

TABLE 4 Effectiveness of biomaterials in regulating wastewater characteristic.

Biomaterials	Effluent treated	BOD removal efficiency (%)	COD removal efficiency (%)	Turbidity removal efficiency (%)	Total solids removal efficiency		pH removal efficiency (%)	Oil and grease removal efficiency (%)	Treatment time	References
					TDS removal efficiency (%)	TSS removal efficiency (%)				
Banana peels	Domestic/synthetic greywater	89%	84%	90%	-	88%	-	-	1 h	113
Orange peels	Domestic/synthetic greywater	93%	87%	86%	-	87%	-	-		
Sapodilla		70%	84%	83%	-	75%	-	-		
Wood coal	Textile industry effluent	25%	30%	62.84%	-	73.74%	-	-	12h	162
Multimedia filter of sugarcane bagasse, sand, grass mulch and activated carbon	Domestic wastewater	80%	61%	74%	8%	40%	4%	-	Not specified	145
Bagasse	Dairy wastewater	76%	-	-	-	-	-	-	200min	163
Rice husk activated carbon	Sewage	-	92.37%	-	-	-	-	-	Not specified	164
Date palm seeds	Pharmaceutical industry wastewater	87%	100%	95%	70%	90%	-	-	60min	165
Coconut shell activated carbon and laterite in fix bed stationary phase (in 2:1 ratio)	Dairy wastewater	70.16%	52.1%	52.33%	-	47.53%	-	-	6min	166

Filtration media of laterite soil, orange peel, rice husk groundnut shells	Dairy wastewater	95%	94%	-	60%		19%	98%	Not specified	167
Raw date seeds	Oil produced water	96.6%	43.4%	-	66.4%	15%	-	99.9%	Not specified	168
Coconut shells		95.8%	37.4%	-	61.1%	5.3%	-	99.8%		
Drumstick peel powder	Synthetic wastewater	88.57%.	69.56%	96.15%		96.66%			3 h	169
Bagasse fly ash		85.71%	65.94%	95.93%		95%				

Additionally, these biosorbents gained enormous attention in the removal of dyes and heavy metals which are the prime cause of concern in the current time as these are the highly poisonous and dangerous pollutants found in wastewaters and impact severely to all living beings. In aspects of nanotechnology, there are many biomaterials that can be used for the production of carbon nanotubes. For instance, graphite flakes, known as one of the major constituents of carbon nanotubes, can be synthesized from waste coconut shells and are identified to show great adsorption capacity toward heavy metals such as lead (Pb^{2+}).[170] Likewise, the separation problem of graphene and its derivatives after wastewater treatment can be overcome by incorporating graphene-based nanomaterials called nanocomposites. These nanocomposites are thought to be highly effective in the removal of heavy metal ions. Antimony (Sb^{2+}), mercury (Hg^{2+}), cobalt (Co^{2+}), cadmium (Cd^{2+}), and lead (Pb^{2+}) are the probable heavy metal ions showing complete adsorption after modifying graphene with nanocomposites. These ions are shown to have variable adsorption capacity at different pH values.[171] Graphene-based nanocomposites are also found important in future perspectives due to their reusability which reduces the approximate treatment cost of the whole process. Adewuyi[172] reported cost estimation of a few commonly utilized low-cost adsorbent biomaterials illustrating much economical than the conventionally available treatment technology. The estimated cost associated with biochar as biosorbent would be 2.65 US \$ per kg, Chitosan-based bio adsorbent would be 8–10US \$ per kg, Kaolinite clay would be 0.005–0.46 US \$ per kg, and bagasse fly ash would be 0.02US \$ per kg. With few improvements, discussed biomaterials can alleviate the worldwide dilemma of wastewater treatment. It is obvious to say that naturally occurring biomaterials having great sorption potential can provide wastewater treatment options alternative to conventional methods in the near future also.

6 Conclusion

In conclusion, different waste biomaterials of minimum or no utilization can be considered as proficient agents in expelling out the toxic organic as well as inorganic pollutants from industrial and other effluents. Both customized and nanostructured biomaterials exhibit remarkable adsorbing capability, and therefore, can effectively replace conventionally utilized sorbent materials. Wide varieties of efficient natural sorbent materials are considered effective, environment-friendly and sustainable approaches for extracting wastewater contaminants. Such materials may be derived from plants, animals, and other waste materials directly or indirectly such as agricultural wastes, fruit and vegetable wastes, activated carbon synthesized from natural materials and much more. Moreover, graphene can be seen as one of the major supporting wastewater-bioadsorbent materials. Several graphene-based composites including graphene oxides, have emerged as a dye as well as heavy metal removing technology. Some metal oxides are also observed to contribute the wastewater remediation. Biomaterials like waste fruit peels are highly recommended adsorbents as they require very less treatment time and have excellent pollutant removal efficiency. Besides naturally occurring sorbent biomaterials, numerous nanomaterials like nanotubes, nanocomposites, nanozymes and much more profoundly recommended technological products that are exclusively designed for the expulsion of quite lethal heavy metal pollutants.

References

1. Tortajada C, Biswas AK. Achieving universal access to clean water and sanitation in an era of water scarcity: strengthening contributions from academia. *Curr Opin Environ Sustain*. 2018;34:21–25. https://doi.org/10.1016/j.cosust.2018.08.001.
2. Ilyas M, Ahmad W, Khan H, Yousaf S, Yasir M, Khan A. Environmental and health impacts of industrial wastewater effluents in Pakistan: a review. *Rev Environ Health*. 2019;34(2):171–186. https://doi.org/10.1515/reveh-2018-0078.
3. Zhongming Z, Linong L, Wangqiang Z, Wei L. *Half of Global Wastewater Treated, Rates in Developing Countries Still Lagging*. Earth System Science Data; 2021.
4. Ranade VV, Bhandari VM. *Industrial Wastewater Treatment, Recycling and Reuse*. Butterworth-Heinemann; 2014.
5. Anastasi EM, Matthews B, Gundogdu A, et al. Prevalence and persistence of *Escherichia coli* strains with uropathogenic virulence characteristics in sewage treatment plants. *Appl Environ Microbiol*. 2010;76(17):5882–5886. https://doi.org/10.1128/AEM.00141-10.
6. Wang H, Zhang Q. Research advances in identifying sulfate contamination sources of water environment by using stable isotopes. *Int J Environ Res Public Health*. 2019;16(11). https://doi.org/10.3390/ijerph16111914.
7. Chahal C, van den Akker B, Young F, Franco C, Blackbeard J, Monis P. *Pathogen and Particle Associations in Wastewater: Significance and Implications for Treatment and Disinfection Processes*. vol. 97. Elsevier; 2016. https://doi.org/10.1016/bs.aambs.2016.08.001.
8. Kamali M, Persson KM, Costa ME, Capela I. Sustainability criteria for assessing nanotechnology applicability in industrial wastewater treatment: current status and future outlook. *Environ Int*. 2019;125(January):261–276. https://doi.org/10.1016/j.envint.2019.01.055.
9. Shahid M, Abbas A, Cheema AI, et al. Plant-microbe interactions in wastewater-irrigated soils. In: *Plant Ecophysiology and Adaptation Under Climate Change: Mechanisms and Perspectives II*. Singapore: Springer; 2020:673–699.

10. Samer M. Biological and chemical wastewater treatment processes. In: *Wastewater Treatment Engineering.* IntechOpen; 2015:1–50. https://doi.org/10.5772/61250.

11. Al Khusaibi TM, Dumaran JJ, Devi MG, Rao LN, Feroz S. Treatment of dairy wastewater using orange and banana peels. *J Chem Pharm Res.* 2015;7(4):1385–1391.

12. Srivatsav P, Bhargav BS, Shanmugasundaram V, Arun J, Gopinath KP, Bhatnagar A. Biochar as an eco-friendly and economical adsorbent for the removal of colorants (dyes) from aqueous environment: a review. *Water.* 2020;12(12):1–28. https://doi.org/10.3390/w12123561.

13. Lingamdinne LP, Choi J-S, Angaru GKR, et al. Magnetic-watermelon rinds biochar for uranium-contaminated water treatment using an electromagnetic semi-batch column with removal mechanistic investigations. *Chemosphere.* 2022;286:131776. https://doi.org/10.1016/j.chemosphere.2021.131776.

14. Karri RR, Sahu JN, Meikap BC. Improving efficacy of Cr (VI) adsorption process on sustainable adsorbent derived from waste biomass (sugarcane bagasse) with help of ant colony optimization. *Ind Crop Prod.* 2020;143:111927. https://doi.org/10.1016/j.indcrop.2019.111927.

15. Sahu JN, Karri RR, Jayakumar NS. Improvement in phenol adsorption capacity on eco-friendly biosorbent derived from waste palm-oil shells using optimized parametric modelling of isotherms and kinetics by differential evolution. *Ind Crop Prod.* 2021;164:113333. https://doi.org/10.1016/j.indcrop.2021.113333.

16. Louis MR, Sorokhaibam LG, Bhandari VM, Bundale S. Multifunctional activated carbon with antimicrobial property derived from Delonix regia biomaterial for treatment of wastewater. *J Environ Chem Eng.* 2018;6(1):169–181. https://doi.org/10.1016/j.jece.2017.11.056.

17. Dotto GL, McKay G. Current scenario and challenges in adsorption for water treatment. *J Environ Chem Eng.* 2020;8(4):103988. https://doi.org/10.1016/j.jece.2020.103988.

18. Hassan M, Naidu R, Du J, Liu Y, Qi F. Critical review of magnetic biosorbents: their preparation, application, and regeneration for wastewater treatment. *Sci Total Environ.* 2020;702:134893. https://doi.org/10.1016/j.scitotenv.2019.134893.

19. Bamukyaye S, Wanasolo W. Performance of egg-shell and fish-scale as adsorbent materials for chromium (VI) removal from effluents of tannery industries in Eastern Uganda. *Open Access Libr J.* 2017;4(8):1–12. https://doi.org/10.4236/oalib.1103732.

20. Lingamdinne LP, Vemula KR, Chang Y-Y, Yang J-K, Karri RR, Koduru JR. Process optimization and modeling of lead removal using iron oxide nanocomposites generated from bio-waste mass. *Chemosphere.* 2020;243:125257. https://doi.org/10.1016/j.chemosphere.2019.125257.

21. Koduru JR, Karri RR, Mubarak NM. Smart materials, magnetic graphene oxide-based nanocomposites for sustainable water purification. In: *Sustainable Polymer Composites and Nanocomposites.* Springer; 2019:759–781. https://doi.org/10.1007/978-3-030-05399-4_26.

22. Muñoz AJ, Ruiz E, Abriouel H, et al. Heavy metal tolerance of microorganisms isolated from wastewaters: identification and evaluation of its potential for biosorption. *Chem Eng J.* 2012;210:325–332. https://doi.org/10.1016/j.cej.2012.09.007.

23. Li X, Liu S, Na Z, Lu D, Liu Z. Adsorption, concentration, and recovery of aqueous heavy metal ions with the root powder of *Eichhornia crassipes.* *Ecol Eng.* 2013;60:160–166. https://doi.org/10.1016/j.ecoleng.2013.07.039.

24. Reddy DHK, Ramana DKV, Seshaiah K, Reddy AVR. Biosorption of Ni(II) from aqueous phase by Moringa oleifera bark, a low cost biosorbent. *Desalination.* 2011;268(1–3):150–157. https://doi.org/10.1016/j.desal.2010.10.011.

25. Imamoglu M, Yildiz H, Altundag H, Turhan Y. Efficient removal of Cd(II) from aqueous solution by dehydrated hazelnut husk carbon. *J Dispers Sci Technol.* 2015;36(2):284–290. https://doi.org/10.1080/01932691.2014.890109.

26. Adeleke OA, Saphira MR, Daud Z, et al. Locally derived activated carbon from domestic, agricultural and industrial wastes for the treatment of palm oil mill effluent. In: *Nanotechnology in Water and Wastewater Treatment.* Elsevier; 2019:35–62. https://doi.org/10.1016/B978-0-12-813902-8.00002-2.

27. Yang X, Wan Y, Zheng Y, et al. Surface functional groups of carbon-based adsorbents and their roles in the removal of heavy metals from aqueous solutions: a critical review. *Chem Eng J.* 2019;366:608–621. https://doi.org/10.1016/j.cej.2019.02.119.

28. Dehghani MH, Karri RR, Yeganeh ZT, et al. Statistical modelling of endocrine disrupting compounds adsorption onto activated carbon prepared from wood using CCD-RSM and DE hybrid evolutionary optimization framework: comparison of linear vs non-linear isotherm and kinetic parameters. *J Mol Liq.* 2020;302:112526. https://doi.org/10.1016/j.molliq.2020.112526.

29. Reza MS, Yun CS, Afroze S, et al. Preparation of activated carbon from biomass and its' applications in water and gas purification, a review. *Arab J Basic Appl Sci.* 2020;27(1):208–238. https://doi.org/10.1080/25765299.2020.1766799.

30. Cheah W-K, Ooi C-H, Yeoh F-Y. Rice husk and rice husk ash reutilization into nanoporous materials for adsorptive biomedical applications: a review. *Open Mater Sci.* 2016;3(1):27–38. https://doi.org/10.1515/mesbi-2016-0004.

31. Khan FSA, Mubarak NM, Khalid M, et al. A comprehensive review on micropollutants removal using carbon nanotubes-based adsorbents and membranes. *J Environ Chem Eng.* 2021;9(6):106647. https://doi.org/10.1016/j.jece.2021.106647.

32. Magendran SS, Khan FSA, Mubarak NM, et al. Synthesis of organic phase change materials by using carbon nanotubes as filler material. *Nano-Struct Nano-Objects.* 2019;19:100361. https://doi.org/10.1016/j.nanoso.2019.100361.

33. Xu J, Lv H, Yang ST, Luo J. Preparation of graphene adsorbents and their applications in water purification. *Rev Inorg Chem.* 2013;33(2–3):139–160. https://doi.org/10.1515/revic-2013-0007.

34. Ali I, Basheer AA, Mbianda XY, et al. Graphene based adsorbents for remediation of noxious pollutants from wastewater. *Environ Int.* 2019;127(January):160–180. https://doi.org/10.1016/j.envint.2019.03.029.

35. Narayana PL, Lingamdinne LP, Karri RR, et al. Predictive capability evaluation and optimization of Pb(II) removal by reduced graphene oxide-based inverse spinel nickel ferrite nanocomposite. *Environ Res.* 2022;204:112029. https://doi.org/10.1016/j.envres.2021.112029.

36. Lingamdinne LP, Koduru JR, Karri RR. A comprehensive review of applications of magnetic graphene oxide based nanocomposites for sustainable water purification. *J Environ Manag.* 2019;231:622–634. https://doi.org/10.1016/j.jenvman.2018.10.063.

37. Zhao J, Ren W, Cheng HM. Graphene sponge for efficient and repeatable adsorption and desorption of water contaminations. *J Mater Chem*. 2012;22 (38):20197–20202. https://doi.org/10.1039/c2jm34128j.

38. Fraga TJM, Carvalho MN, Ghislandi MG, Da Motta Sobrinho MA. Functionalized graphene-based materials as innovative adsorbents of organic pollutants: a concise overview. *Braz J Chem Eng*. 2019;36. https://doi.org/10.1590/0104-6632.20190361s20180283.

39. Sajid M, Nazal MK, Ihsanullah, Baig N, Osman AM. Removal of heavy metals and organic pollutants from water using dendritic polymers based adsorbents: a critical review. *Sep Purif Technol*. 2018;191:400–423. https://doi.org/10.1016/j.seppur.2017.09.011.

40. Chowdhury S, Mazumder MAJ, Al-Attas O, Husain T. Heavy metals in drinking water: occurrences, implications, and future needs in developing countries. *Sci Total Environ*. 2016;569–570:476–488. https://doi.org/10.1016/j.scitotenv.2016.06.166.

41. Huber M, Welker A, Helmreich B. Critical review of heavy metal pollution of traffic area runoff: occurrence, influencing factors, and partitioning. *Sci Total Environ*. 2016;541:895–919. https://doi.org/10.1016/j.scitotenv.2015.09.033.

42. Dehghani MH, Hassani AH, Karri RR, et al. Process optimization and enhancement of pesticide adsorption by porous adsorbents by regression analysis and parametric modelling. *Sci Rep*. 2021;11(1):11719. https://doi.org/10.1038/s41598-021-91178-3.

43. Srivastav AL, Singh PK, Srivastava V, Sharma YC. Application of a new adsorbent for fluoride removal from aqueous solutions. *J Hazard Mater*. 2013;263:342–352. https://doi.org/10.1016/j.jhazmat.2013.04.017.

44. Fu F, Wang Q. Removal of heavy metal ions from wastewaters: a review. *J Environ Manag*. 2011;92(3):407–418. https://doi.org/10.1016/j.jenvman.2010.11.011.

45. Harvey PJ, Handley HK, Taylor MP. Identification of the sources of metal (lead) contamination in drinking waters in North-Eastern Tasmania using lead isotopic compositions. *Environ Sci Pollut Res*. 2015;22(16):12276–12288. https://doi.org/10.1007/s11356-015-4349-2.

46. Białowiec A, Davies L, Albuquerque A, Randerson PF. The influence of plants on nitrogen removal from landfill leachate in discontinuous batch shallow constructed wetland with recirculating subsurface horizontal flow. *Ecol Eng*. 2012;40(2):44–52. https://doi.org/10.1016/j.ecoleng.2011.12.011.

47. Rios JF, Ye M, Wang L, Lee PZ, Davis H, Hicks R. ArcNLET: a GIS-based software to simulate groundwater nitrate load from septic systems to surface water bodies. *Comput Geosci*. 2013;52:108–1116. https://doi.org/10.1016/j.cageo.2012.10.003.

48. Mahmood SJ, Taj N, Parveen F, Usmani TH, Azmat R, Uddin F. Arsenic, fluoride and nitrate in drinking water: the problem and its possible solution. *Res J Environ Sci*. 2007;1(4):179–184.

49. WHO. *Guidelines for Drinking-Water Quality Second Addendum to Third Edition WHO Library Cataloguing-In-Publication Data*. vol. 1. World Health Organization; 2008:17–19. http://www.who.int/water_sanitation_health/dwq/secondaddendum20081119.pdf.

50. Sankhla MS, Kumar R. Fluoride contamination of water in India and its impact on public health. *ARC J Forensic Sci*. 2018;3(2):10–15. https://doi.org/10.20431/2456-0049.0302002.

51. Zhang T, Li Q, Xiao H, Mei Z, Lu H, Zhou Y. Enhanced fluoride removal from water by non-thermal plasma modified CeO_2/Mg-Fe layered double hydroxides. *Appl Clay Sci*. 2013;72:117–123. https://doi.org/10.1016/j.clay.2012.12.003.

52. Oram B. *Total Phosphorus and Phosphate Impact on Surface Waters*. Dalas, USA: Water Research Center; 2009.

53. Akpor OB, Muchie M. Environmental and public health implications of wastewater quality. *Afr J Biotechnol*. 2011;10(13):2379–2387. https://doi.org/10.4314/ajb.v10i13.

54. Bowes MJ, Jarvie HP, Halliday SJ, et al. Characterising phosphorus and nitrate inputs to a rural river using high-frequency concentration-flow relationships. *Sci Total Environ*. 2015;511:608–620. https://doi.org/10.1016/j.scitotenv.2014.12.086.

55. Pikaar I, Sharma KR, Hu S, Gernjak W, Keller J, Yuan Z. Reducing sewer corrosion through integrated urban water management. *Science*. 2014;345 (6198):812–814. https://doi.org/10.1126/science.1251418.

56. Sun H, Shi B, Yang F, Wang D. Effects of sulfate on heavy metal release from iron corrosion scales in drinking water distribution system. *Water Res*. 2017;114:69–77. https://doi.org/10.1016/j.watres.2017.02.021.

57. Zhang H, Wang X, Chen C, et al. Facile synthesis of diverse transition metal oxide nanoparticles and electrochemical properties. *Inorg Chem Front*. 2016;3(8):1048–1057. https://doi.org/10.1039/c6qi00096g.

58. Sinha S, Singh R, Chaurasia AK, Nigam S. Self-sustainable Chlorella pyrenoidosa strain NCIM 2738 based photobioreactor for removal of Direct Red-31 dye along with other industrial pollutants to improve the water-quality. *J Hazard Mater*. 2016;306:386–394. https://doi.org/10.1016/j.jhazmat.2015.12.011.

59. Yao L, Zhang L, Wang R, Chou S, Dong ZL. A new integrated approach for dye removal from wastewater by polyoxometalates functionalized membranes. *J Hazard Mater*. 2016;301:462–470. https://doi.org/10.1016/j.jhazmat.2015.09.027.

60. Vymazal J, Březinová T. The use of constructed wetlands for removal of pesticides from agricultural runoff and drainage: a review. *Environ Int*. 2015;75:11–20. https://doi.org/10.1016/j.envint.2014.10.026.

61. Pandey PK, Kass PH, Soupir ML, Biswas S, Singh VP. Analysis and application of *Bacillus subtilis* sortases to anchor recombinant proteins on the cell wall. *AMB Express*. 2014;4(51):1–16.

62. Varela AR, Manaia CM. Human health implications of clinically relevant bacteria in wastewater habitats. *Environ Sci Pollut Res*. 2013;20(6): 3550–3569. https://doi.org/10.1007/s11356-013-1594-0.

63. Osiemo MM, Ogendi GM, M'Erimba C. Microbial quality of drinking water and prevalence of water-related diseases in Marigat Urban Centre, Kenya. *Environ Health Insights*. 2019;13. 1178630219836988.

64. Huang HW, Lee CH, Yu HS. Arsenic-induced carcinogenesis and immune dysregulation. *Int J Environ Res Public Health*. 2019;16(15). https://doi.org/10.3390/ijerph16152746.

65. Khanam S. Heavy metal contamination and health hazards: a review. *Trends Biotechnol Biol Sci*. 2014;1:5–8.

66. Hossin MM, Kibria G, Mallick D, Lau T, Wu R, Nugegoda D. *Pollution Monitoring in Rivers, Estuaries and Coastal Areas of Bangladesh With Artificial Mussel (AM) Technology-Findings, Ecological Significances, Implications and Recommendations*; 2015:1–57. January.

67. Miraj SS, Rao M. Clinical toxicology of copper: source, toxidrome, mechanism of toxicity, and management. In: *Metal Toxicology Handbook*. CRC Press; 2020:199–217.

68. Sharma S, Bhattacharya A. Drinking water contamination and treatment techniques. *Appl Water Sci*. 2017;7(3):1043–1067. https://doi.org/10.1007/s13201-016-0455-7.

69. Alexander J. Selenium. In: *Handbook on the Toxicology of Metals*. vol. 1. Elsevier; 2015. https://doi.org/10.1016/B978-0-444-59453-2.00052-4.

70. Brindha K, Renganayaki SP, Elango L. Sources, toxicological effects and removal techniques of nitrates in groundwater: an overview. *Indian J Environ Prot*. 2017;37(8):667–700.

71. Ghosh A, Mukherjee K, Ghosh SK, Saha B. Sources and toxicity of fluoride in the environment. *Res Chem Intermed*. 2013;39(7):2881–2915. https://doi.org/10.1007/s11164-012-0841-1.

72. Kundu S, Vassanda Coumar M, Rajendiran S, Ajay, Subba Rao A. Phosphates from detergents and eutrophication of surface water ecosystem in India. *Curr Sci*. 2015;108(7):1320–1325. https://doi.org/10.18520/cs/v108/i7/1320-1325.

73. Issabayeva G, Hang SY, Wong MC, Aroua MK. A review on the adsorption of phenols from wastewater onto diverse groups of adsorbents. *Rev Chem Eng*. 2018;34(6):855–873. https://doi.org/10.1515/revce-2017-0007.

74. Karimifard S, Alavi Moghaddam MR. Application of response surface methodology in physicochemical removal of dyes from wastewater: a critical review. *Sci Total Environ*. 2018;640-641:772–797. https://doi.org/10.1016/j.scitotenv.2018.05.355.

75. Okoh AI, Odjadjare EE, Igbinosa EO, Osode AN. Wastewater treatment plants as a source of microbial pathogens in the receiving watershed. *Afr J Biotechnol*. 2007;6:2932–2944.

76. Baldursson S, Karanis P. Waterborne transmission of protozoan parasites: review of worldwide outbreaks—an update 2004–2010. *Water Res*. 2011;45(20):6603–6614.

77. Nur Hazirah R, Nurhaslina CR, Ku Halim KH. Enhancement of biological approach and potential of Lactobacillus delbrueckii in decolorization of textile wastewater—a review. *IOSR J Environ Sci Toxicol Food Technol*. 2014;8:6–10.

78. Yadav TC, Saxena P, Srivastava AK, et al. Potential applications of chitosan nanocomposites: recent trends and challenges. In: Shahid-ul-Islam, Butola BS, eds. *Advanced Functional Textiles and Polymers: Fabrication, Processing and Applications*. Scrivener Publishing LLC; 2019:365–403.

79. Tetteh EK, Rathilal S. Application of organic coagulants in water and wastewater treatment. In: *Organic Polymers*. IntechOpen; 2019 April 3.

80. Sen Gupta AK. *Ion Exchange in Environmental Processes: Fundamentals, Applications and Sustainable Technology*. John Wiley & Sons; 2017 September 18.

81. Chuuman T, Eguchi K, Akinaga M, et al. Inhibition of silicic acid elution during the regeneration of strong base anion exchange resin column. *Bull Chem Soc Jpn*. 2019;92(4):869–874.

82. Barlokova D, Ilavsky J, Kunstek M, Kapusta O. Microfiltration in water treatment for removal of suspended solids and natural organic matter. *IOP Conf Ser: Earth Environ Sci*. 2019;362(1):012168. IOP Publishing.

83. Jun LY, Karri RR, Mubarak NM, et al. Modelling of methylene blue adsorption using peroxidase immobilized functionalized buckypaper/polyvinyl alcohol membrane via ant colony optimization. *Environ Pollut*. 2020;259:113940. https://doi.org/10.1016/j.envpol.2020.113940.

84. Pandová I, Panda A, Valíček J, Harničárová M, Kušnerová M, Palková Z. Use of sorption of copper cations by clinoptilolite for wastewater treatment. *Int J Environ Res Public Health*. 2018;15(7). https://doi.org/10.3390/ijerph15071364.

85. Tasić ŽŽ, Bogdanović GD, Antonijević MM. Application of natural zeolite in wastewater treatment: a review. *J Min Metall Sect A*. 2019;55(1):67–79.

86. Ruthiraan M, Mubarak NM, Abdullah EC, et al. An overview of magnetic material: preparation and adsorption removal of heavy metals from wastewater. In: *Nanotechnology in the Life Sciences*. Springer; 2019:131–159. https://doi.org/10.1007/978-3-030-16439-3_8.

87. Khan FSA, Mubarak NM, Tan YH, et al. Magnetic nanoparticles incorporation into different substrates for dyes and heavy metals removal—a review. *Environ Sci Pollut Res*. 2020;27(35):43526–43541. https://doi.org/10.1007/s11356-020-10482-z.

88. Farraji H. Wastewater treatment by phytoremediation methods. In: *Wastewater Engineering: Advanced Wastewater Treatment Systems*. vol. 194. ijsrpub.com; 2014. http://www.ijsrpub.com/books.

89. Chandra R, Yadav S. Potential of *Typha angustifolia* for phytoremediation of heavy metals from aqueous solution of phenol and melanoidin. *Ecol Eng*. 2010;36(10):1277–1284. https://doi.org/10.1016/j.ecoleng.2010.06.003.

90. Liu Y, Chen J, Lu S, Yang L, Qian J, Cao S. Increased lead and cadmium tolerance of *Typha angustifolia* from Huaihe River is associated with enhanced phytochelatin synthesis and improved antioxidative capacity. *Environ Technol*. 2016;37(21):2743–2749. https://doi.org/10.1080/09593330.2016.1162848.

91. Nugroho AP, Butar ESB, Priantoro EA, Sriwuryandari L, Pratiwi ZB, Sembiring T. Phytoremediation of electroplating wastewater by vetiver grass (Chrysopogon zizanoides L.). *Sci Rep*. 2021;11(1):1–8. https://doi.org/10.1038/s41598-021-93923-0.

92. Zhu S, Huo S, Feng P. Developing designer microalgal consortia: a suitable approach to sustainable wastewater treatment. In: *Microalgae Biotechnology for Development of Biofuel and Wastewater Treatment*. Springer; 2019:569–598. https://doi.org/10.1007/978-981-13-2264-8_22.

93. Wijayanti TA, Suharso S, Ansori M. Application of modified green algae nannochloropsis sp. as adsorbent in the simultaneous adsorption of methylene blue and Cu(II) cations in solution. *Sustain Environ Res*. 2021;31(1). https://doi.org/10.1186/s42834-021-00090-y.

94. Bhushan S, Rana MS, Raychaudhuri S, Simsek H, Prajapati SK. Algae-and bacteria-driven technologies for pharmaceutical remediation in wastewater. In: *Removal of Toxic Pollutants through Microbiological and Tertiary Treatment*. Elsevier; 2020:373–408.

95. Yaashikaa PR, Kumar PS, Saravanan A, Vo DVN. Advances in biosorbents for removal of environmental pollutants: a review on pretreatment, removal mechanism and future outlook. *J Hazard Mater*. 2021;420:126596.

96. Wan Ngah WS, Hanafiah MAKM. Removal of heavy metal ions from wastewater by chemically modified plant wastes as adsorbents: a review. *Bioresour Technol*. 2008;99(10):3935–3948. https://doi.org/10.1016/j.biortech.2007.06.011.

97. Hanum F, Bani O, Wirani LI. Characterization of activated carbon from rice husk by HCl activation and its application for lead (Pb) removal in car battery wastewater. *IOP Conf Ser: Mater Sci Eng*. 2017;180(1):012151. IOP Publishing.

98. Maneechakr P, Mongkollertlop S. Investigation on adsorption behaviors of heavy metal ions (Cd2+, Cr3+, Hg2+ and Pb2+) through low-cost/active manganese dioxide-modified magnetic biochar derived from palm kernel cake residue. *J Environ Chem Eng*. 2020;8(6):104467. https://doi.org/10.1016/j.jece.2020.104467.

99. Ugya AY, Hua X, Ma J. Biosorption of Cr3+ and Pb2+ from tannery wastewater using combined fruit waste. *Appl Ecol Environ Res*. 2019;17(2):1773–1787. https://doi.org/10.15666/aeer/1702_17731787.

100. Li Y, Liu J, Yuan Q, Tang H, Yu F, Lv X. A green adsorbent derived from banana peel for highly effective removal of heavy metal ions from water. *RSC Adv*. 2016;6(51):45041–45048. https://doi.org/10.1039/c6ra07460j.

101. Yahya MD, Obayomi KS, Abdulkadir MB, Iyaka YA, Olugbenga AG. Characterization of cobalt ferrite-supported activated carbon for removal of chromium and lead ions from tannery wastewater via adsorption equilibrium. *Water Sci Eng*. 2020;13(3):202–213. https://doi.org/10.1016/j.wse.2020.09.007.

102. Liu Z, Sun Y, Xu X, Qu J, Qu B. Adsorption of Hg (II) in an aqueous solution by activated carbon prepared from rice husk using KOH activation. *ACS Omega*. 2020;5:29231–29242.

103. Feng NC, Guo XY. Characterization of adsorptive capacity and mechanisms on adsorption of copper, lead and zinc by modified orange peel. *Trans Nonferrous Met Soc China*. 2012;22(5):1224–1231. https://doi.org/10.1016/S1003-6326(11)61309-5.

104. Nahar K, Chowdhury MAK, Chowdhury MAH, Rahman A, Mohiuddin KM. Heavy metals in handloom-dyeing effluents and their biosorption by agricultural byproducts. *Environ Sci Pollut Res*. 2018;25(8):7954–7967. https://doi.org/10.1007/s11356-017-1166-9.

105. Razi MAM, Gheethi AA, Za IA. Removal of heavy metals from textile wastewater using sugarcane bagasse activated carbon. *Int J Eng Technol*. 2018;7(4.30):112. https://doi.org/10.14419/ijet.v7i4.30.22066.

106. Pant G, Singh A, Panchpuri M, et al. Enhancement of biosorption capacity of cyanobacterial strain to remediate heavy metals. *Desalin Water Treat*. 2019;165(January):244–252. https://doi.org/10.5004/dwt.2019.24509.

107. Van Thuan T, Quynh BTP, Nguyen TD, Ho VTT, Bach LG. Response surface methodology approach for optimization of Cu2+, Ni2+ and Pb2+ adsorption using KOH-activated carbon from banana peel. *Surf Interfaces*. 2017;6:209–217. https://doi.org/10.1016/j.surfin.2016.10.007.

108. Van Tran T, Bui QTP, Nguyen TD, Le NTH, Bach LG. A comparative study on the removal efficiency of metal ions (Cu2+, Ni2+, and Pb2+) using sugarcane bagasse-derived ZnCl2-activated carbon by the response surface methodology. *Adsorpt Sci Technol*. 2017;35(1–2):72–85. https://doi.org/10.1177/0263617416669152.

109. Zhang Y, Zheng R, Zhao J, Ma F, Zhang Y, Meng Q. Characterization of H3PO4-treated rice husk adsorbent and adsorption of copper(II) from aqueous solution. *Biomed Res Int*. 2014;2014(II). https://doi.org/10.1155/2014/496878.

110. Javidi A. Synthesis of activated carbon from sugarcane bagasse and application for mercury adsorption pollution. *Pollution*. 2019;5(3):585. https://doi.org/10.22059/POLL.2019.269364.540.

111. Villar da Gama BM, Elisandra do Nascimento G, Silva Sales DC, Rodríguez-Díaz JM, Bezerra de Menezes Barbosa CM, Menezes Bezerra Duarte MM. Mono and binary component adsorption of phenol and cadmium using adsorbent derived from peanut shells. *J Clean Prod*. 2018;201:219–228. https://doi.org/10.1016/j.jclepro.2018.07.291.

112. Kurniawan TA, Chan GYS, Lo WH, Babel S. Comparisons of low-cost adsorbents for treating wastewaters laden with heavy metals. *Sci Total Environ*. 2006;366(2–3):409–426. https://doi.org/10.1016/j.scitotenv.2005.10.001.

113. Baloch MYJ, Mangi SH. Treatment of synthetic greywater by using banana, orange and sapodilla peels as a low cost activated carbon. *J Mater Environ Sci*. 2019;10(10):966–986.

114. Abbas A, Abussaud BA, Ihsanullah, Al-Baghli NAH, Khraisheh M, Atieh MA. Benzene removal by iron oxide nanoparticles decorated carbon nanotubes. *J Nanomater*. 2016;2016. https://doi.org/10.1155/2016/5654129.

115. Kyzas GZ, Deliyanni EA, Bikiaris DN, Mitropoulos AC. Graphene composites as dye adsorbents: review. *Chem Eng Res Des*. 2018;129:75–88. https://doi.org/10.1016/j.cherd.2017.11.006.

116. Ghenaatian HR, Shakourian-Fard M, Kamath G. The effect of sulfur and nitrogen/sulfur co-doping in graphene surface on the adsorption of toxic heavy metals (Cd, Hg, Pb). *J Mater Sci*. 2019;54(20):13175–13189. https://doi.org/10.1007/s10853-019-03791-3.

117. Shtepliuk I, Yakimova R. Interband transitions in closed-shell vacancy containing graphene quantum dots complexed with heavy metals. *Phys Chem Chem Phys*. 2018;20(33):21528–21543. https://doi.org/10.1039/c8cp03306d.

118. Zhang X, Yi G, Zhang Z, et al. Magnetic graphene-based nanocomposites as highly efficient absorbents for Cr (VI) removal from wastewater. *Environ Sci Pollut Res*. 2021;28(12):14671–14680.

119. Khan FSA, Mubarak NM, Khalid M, et al. Magnetic nanoadsorbents' potential route for heavy metals removal—a review. *Environ Sci Pollut Res*. 2020;27(19):24342–24356. https://doi.org/10.1007/s11356-020-08711-6.

120. Khan FSA, Mubarak NM, Tan YH, et al. A comprehensive review on magnetic carbon nanotubes and carbon nanotube-based buckypaper for removal of heavy metals and dyes. *J Hazard Mater*. 2021;413:125375. https://doi.org/10.1016/j.jhazmat.2021.125375.

121. Zhuang Y, Yu F, Ma J, Chen J. Enhanced adsorption removal of antibiotics from aqueous solutions by modified alginate/graphene double network porous hydrogel. *J Colloid Interface Sci*. 2017;507:250–259. https://doi.org/10.1016/j.jcis.2017.07.033.

122. Rauf MA, Ashraf SS. Survey of recent trends in biochemically assisted degradation of dyes. *Chem Eng J*. 2012;209:520–530. https://doi.org/10.1016/j.cej.2012.08.015.

123. Singh K, Arora S. Removal of synthetic textile dyes from wastewaters: a critical review on present treatment technologies. *Crit Rev Environ Sci Technol.* 2011;41(9):807–878. https://doi.org/10.1080/10643380903218376.

124. Guo H, Jiao T, Zhang Q, Guo W, Peng Q, Yan X. Preparation of graphene oxide-based hydrogels as efficient dye adsorbents for wastewater treatment. *Nanoscale Res Lett.* 2015;10(1):1.

125. Alves DC, Healy B, Yu T, Breslin CB. Graphene-based materials immobilized within chitosan: applications as adsorbents for the removal of aquatic pollutants. *Materials.* 2021;14(13):3655.

126. Shen Y, Fang Q, Chen B. Environmental applications of three-dimensional graphene-based macrostructures: adsorption, transformation, and detection. *Environ Sci Technol.* 2015;49(1):67–84.

127. Dehghani MH, Salari M, Karri RR, Hamidi F, Bahadori R. Process modeling of municipal solid waste compost ash for reactive red 198 dye adsorption from wastewater using data driven approaches. *Sci Rep.* 2021;11(1):11613. https://doi.org/10.1038/s41598-021-90914-z.

128. Lau YJ, Karri RR, Mubarak NM, et al. Removal of dye using peroxidase-immobilized Buckypaper/polyvinyl alcohol membrane in a multi-stage filtration column via RSM and ANFIS. *Environ Sci Pollut Res.* 2020;27(32):40121–40134. https://doi.org/10.1007/s11356-020-10045-2.

129. Strokal M, Spanier JE, Kroeze C, et al. Global multi-pollutant modelling of water quality: scientific challenges and future directions. *Curr Opin Environ Sustain.* 2019;36:116–125. https://doi.org/10.1016/j.cosust.2018.11.004.

130. Gao C, Zhang W, Li H, Lang L, Xu Z. Controllable fabrication of mesoporous MgO with various morphologies and their absorption performance for toxic pollutants in water. *Cryst Growth Des.* 2008;8(10):3785–3790. https://doi.org/10.1021/cg8004147.

131. Zhou J, Tang C, Cheng B, Yu J, Jaroniec M. Rattle-type carbon-alumina core-shell spheres: synthesis and application for adsorption of organic dyes. *ACS Appl Mater Interfaces.* 2012;4(4):2174–2179. https://doi.org/10.1021/am300176k.

132. Ghasemi Z, Seif A, Ahmadi TS, Zargar B, Rashidi F, Rouzbahani GM. Thermodynamic and kinetic studies for the adsorption of Hg(II) by nano-TiO$_2$ from aqueous solution. *Adv Powder Technol.* 2012;23(2):148–156. https://doi.org/10.1016/j.apt.2011.01.004.

133. Ahmad A, Razali MH, Mamat M, Mehamod FS, Amin KA. Adsorption of methyl orange by synthesized and functionalized-CNTs with 3-aminopropyltriethoxysilane loaded TiO2 nanocomposites. *Chemosphere.* 2017;168:474–482.

134. Chaudhary S, Sharma P, Renu, Kumar R. Hydroxyapatite doped CeO2 nanoparticles: impact on biocompatibility and dye adsorption properties. *RSC Adv.* 2016;6(67):62797–62809. https://doi.org/10.1039/c6ra06933a.

135. Zhu HY, Jiang R, Fu YQ, Li RR, Yao J, Jiang ST. Novel multifunctional NiFe$_2$O$_4$/ZnO hybrids for dye removal by adsorption, photocatalysis and magnetic separation. *Appl Surf Sci.* 2016;369:1–10. https://doi.org/10.1016/j.apsusc.2016.02.025.

136. Mehmood A, Khan FSA, Mubarak NM, et al. Magnetic nanocomposites for sustainable water purification—a comprehensive review. *Environ Sci Pollut Res.* 2021;28(16):19563–19588. https://doi.org/10.1007/s11356-021-12589-3.

137. Karri RR, Shams S, Sahu JN. Overview of potential applications of nano-biotechnology in wastewater and effluent treatment. In: *Nanotechnology in Water and Wastewater Treatment: Theory and Applications.* Elsevier; 2018:87–100. https://doi.org/10.1016/B978-0-12-813902-8.00004-6 [chapter 4].

138. Anwar H, Arif I, Javeed U, Javed Y. Titanium dioxide-based nanohybrids as photocatalysts for removal and degradation of industrial contaminants. In: *Nanohybrids in Environmental & Biomedical Applications.* CRC Press; 2019 July 9:255–278.

139. Anjum M, Miandad R, Waqas M, Gehany F, Barakat MA. Remediation of wastewater using various nano-materials. *Arab J Chem.* 2019;12(8):4897–4919.

140. Hafez H. *A Study on the Use of Nano/Micro Structured Goethite and Hematite as Adsorbents for the Removal of Cr (III), Co (II), Cu (II), Ni (II), and Zn (II) Metal Ions From Aqueous Solutions.* Engg Journals Publication; 2012.

141. Hua M, Zhang S, Pan B, Zhang W, Lv L, Zhang Q. Heavy metal removal from water/wastewater by nanosized metal oxides: a review. *J Hazard Mater.* 2012;211–212:317–331. https://doi.org/10.1016/j.jhazmat.2011.10.016.

142. Tanhaei B, Ayati A, Lahtinen M, Sillanpää M. Preparation and characterization of a novel chitosan/Al$_2$O$_3$/magnetite nanoparticles composite adsorbent for kinetic, thermodynamic and isotherm studies of methyl orange adsorption. *Chem Eng J.* 2015;259:1–10. https://doi.org/10.1016/j.cej.2014.07.109.

143. Yang D, Paul B, Xu W, et al. Alumina nanofibers grafted with functional groups: a new design in efficient sorbents for removal of toxic contaminants from water. *Water Res.* 2010;44(3):741–750. https://doi.org/10.1016/j.watres.2009.10.014.

144. Xiong L, Yang Y, Mai J, et al. Adsorption behavior of methylene blue onto titanate nanotubes. *Chem Eng J.* 2010;156(2):313–320.

145. Ahmad I, Ali N, Jamal Y. Treatment of domestic wastewater by natural adsorbents using multimedia filter technology. *Int J EmergTechnol Eng Res.* 2016;4(4):164–167. www.ijeter.everscience.org.

146. Gusain R, Gupta K, Joshi P, Khatri OP. Adsorptive removal and photocatalytic degradation of organic pollutants using metal oxides and their composites: a comprehensive review. *Adv Colloid Interf Sci.* 2019;272:102009. https://doi.org/10.1016/j.cis.2019.102009.

147. Cao CY, Cui ZM, Chen CQ, Song WG, Cai W. Ceria hollow nanospheres produced by a template-free microwave-assisted hydrothermal method for heavy metal ion removal and catalysis. *J Phys Chem C.* 2010;114(21):9865–9870. https://doi.org/10.1021/jp101553x.

148. Recillas S, Colón J, Casals E, et al. Chromium VI adsorption on cerium oxide nanoparticles and morphology changes during the process. *J Hazard Mater.* 2010;184(1–3):425–431. https://doi.org/10.1016/j.jhazmat.2010.08.052.

149. Le AT, Pung SY, Sreekantan S, Matsuda A, Huynh DP. Mechanisms of removal of heavy metal ions by ZnO particles. *Heliyon.* 2019;5(4): e01440. https://doi.org/10.1016/j.heliyon.2019.e01440.

150. Salehi R, Arami M, Mahmoodi NM, Bahrami H, Khorramfar S. Novel biocompatible composite (chitosan-zinc oxide nanoparticle): preparation, characterization and dye adsorption properties. *Colloids Surf B: Biointerfaces.* 2010;80(1):86–93. https://doi.org/10.1016/j.colsurfb.2010.05.039.

151. Zahid M, Nadeem N, Tahir N, Majeed MI, Naqvi SAR, Hussain T. Hybrid nanomaterials for water purification. In: *Multifunctional Hybrid Nano-materials for Sustainable Agri-Food and Ecosystems.* Elsevier; 2020:155–188.

152. Zhou C, Lai C, Zhang C, et al. Semiconductor/boron nitride composites: synthesis, properties, and photocatalysis applications. *Appl Catal B Environ.* 2018;238:6–18.

153. Chauhan R, Kumar A, Chaudhary RP. Photocatalytic degradation of methylene blue with Fe doped ZnS nanoparticles. *Spectrochim Acta A Mol Biomol Spectrosc.* 2013;113:250–256. https://doi.org/10.1016/j.saa.2013.04.087.

154. Bai W, Cai L, Wu C, et al. Alcohothermal synthesis of flower-like ZnS nano-microstructures with high visible light photocatalytic activity. *Mater Lett.* 2014;124:177–180. https://doi.org/10.1016/j.matlet.2014.03.073.

155. Madhura L, Singh S. A review on the advancements of nanomembranes for water treatment. In: *Nanotechnology in Environmental Science.* 1–2. Wiley-VCH Verlag GmbH & Co. KGaA; 2018:391–412.

156. Paulson D, Principal Partner, Water Think Tank, and Prime Membrane Partners, LLC. *Nanofiltration: The Up-and-Coming Membrane Process*; 2015. http://www.wateronline.com/doc/nanofiltration-the-up-andcoming-membrane-process-0001.

157. Khan FSA, Mubarak NM, Khalid M, et al. Functionalized multi-walled carbon nanotubes and hydroxyapatite nanorods reinforced with polypropylene for biomedical application. *Sci Rep.* 2021;11(1):843. https://doi.org/10.1038/s41598-020-80767-3.

158. Madou MJ. *Fundamentals of Microfabrication and Nanotechnology.* Boca Raton, FL, USA: CRC Press; 2018:1–3.

159. Lu H, Wang J, Stoller M, Wang T, Bao Y, Hao H. An overview of nanomaterials for water and wastewater treatment. *Adv Mater Sci Eng.* 2016;2016:4964828.

160. Sudhakar MS, Aggarwal A, Sah MK. Engineering biomaterials for the bioremediation: advances in nanotechnological approaches for heavy metals removal from natural resources. In: *Emerging Technologies in Environmental Bioremediation.* Elsevier; 2020. https://doi.org/10.1016/b978-0-12-819860-5.00014-6.

161. Yadav DK, Srivastava S. Carbon nanotubes as adsorbent to remove heavy metal ion (Mn+7) in wastewater treatment. *Mater Today: Proc.* 2017;4(2):4089–4094. https://doi.org/10.1016/j.matpr.2017.02.312.

162. Gaikwad NB, Thakur VT, Jadhav AS, Raut PD. Studies on low-cost adsorbent biomaterial like 'coconut coir' and 'wood coal' for treatment of textile industry effluent. In: *Recent Trends in Conservation and Management of Ecosystems*; 2018:99–108. ISBN:978-81-931247-7-2.

163. Inamdar SS. BOD reduction using low cost adsorbents. *J Ind Pollut Control.* 2006;22(1):111–120.

164. Mukundan U, Ratnoji SS. COD removal from sewage by activated carbon from rice husk-an agricultural by product. *Int J Innov Res Sci Eng Technol.* 2015;4(6):5003–5007. http://eprints.manipal.edu/id/eprint/145622.

165. Karunya S, Feroz S, Al Harassi S, Sakhile K. Treatment of Oman pharmaceutical industry wastewater using low cost adsorbents. *J Multidiscip Eng Sci Technol.* 2015;2(3):339–341.

166. Mohan SS, N SK. Dairy waste water treatment using coconut shell activated carbon and laterite as low cost adsorbents. *Int Res J Eng Technol.* 2008;306(June):306–309. www.irjet.net.

167. Manasa SR, Pasha SA, Ajay KM, Ganavi KG, Jeevan MK. Purification of dairy waste water using low-cost adsorbents and study on compressive strength of concrete by purified wastewater. *Int J Res Appl Sci Eng Technol.* 2020;2321-9653. 8(9).

168. Al-Abri OH, Lakkimsetty NR, Shaik F. Pre-treatment of oil produced water using low-cost adsorbents. *Int J Mech Prod Eng Res Dev.* 2020;10(3):2435–2444. ISSN (P): 2249-6890; ISSN (E): 2249-8001.

169. Burkul RM, Ranade SV, Pangarkar BL. Comparative study of removal of malathion from waste-water by using natural adsorbent. *Int J Chem Pharm Sci.* 2015;0976-9390. 6(4).

170. Hakim YZ, Yulizar Y, Nurcahyo A, Surya M. Green synthesis of carbon nanotubes from coconut shell waste for the adsorption of Pb (II) ions. *Acta Chim Asiana.* 2018;1(1):6–10. https://doi.org/10.29303/aca.vlil.2.

171. Kim S, Park CM, Jang M, et al. Aqueous removal of inorganic and organic contaminants by graphene-based nanoadsorbents: a review. *Chemosphere.* 2018;212:1104–1124. https://doi.org/10.1016/j.chemosphere.2018.09.033.

172. Adewuyi A. Chemically modified biosorbents and their role in the removal of emerging pharmaceutical waste in the water system. *Water.* 2020;12(6):1551.

Chapter 7

Nano-sorbents: A promising alternative for the remediation of noxious pollutants

Suhas[a,*], Monika Chaudhary[a], Inderjeet Tyagi[b], Ravinder Kumar[a], Vinod Kumar[c], Shubham Chaudhary[a], and Sarita Kushwaha[a]

[a]Department of Chemistry, Gurukula Kangri (Deemed to be University), Haridwar, India, [b]Centre for DNA Taxonomy, Molecular Systematics Division, Zoological Survey of India, Ministry of Environment, Forest and Climate Change, Government of India, Kolkata, West Bengal, India, [c]Special Centre for Nano Sciences, Jawaharlal Nehru University, Delhi, India

[*]Corresponding author

1 Introduction

Removal of various hazardous pollutants in water systems has become a global issue the world is facing currently due to the increasing demand for drinking water. Booming population, urbanization, and explosive growth in the industrial sector are contaminating the water resources largely.[1–5] Therefore, the treatment of wastewater to reuse it for various applications such as irrigation, industrial demands, and domestic needs has become a priority for researchers in the last few decades.[5,6] Various approaches have been considered to be effective for the treatment of wastewater, including ion exchange, coagulation, sedimentation, membrane filtration, reverse osmosis, adsorption, and electrolysis.[7–12] Adsorption, among the above-mentioned treatment methods, is a versatile technique for removing organic and inorganic noxious pollutants from wastewater due to its simplistic design and convenient operation.[13–15] Moreover, many adsorbents utilized by different researchers are low cost and can be used without any pre-treatment, making adsorption a cost-effective process.[16–19] Besides, the cost-effectiveness and the efficient removal of toxic substances make adsorption a thriving approach in the field of remediation of noxious pollutants globally.[20–23]

The recent development of nanotechnology offers ample opportunities to develop various nanoparticle-based adsorbents to remove various pollutants from water.[10,12,24–26] The adsorption of organic and inorganic contaminants from wastewater using nanoparticles as adsorbents has been reported immensely.[26–28] A wide array of literature is available to remove various pollutants using nanoparticle-based adsorbents from aqueous solutions and is presented in Table 1. Nanoparticles offer tremendous adsorptive properties due to their large surface area, outstanding chemical reactivity, mechanical and magnetic properties, and therefore, can be used to remove various noxious pollutants.[7,55–57] This chapter aims to provide a deep insight into the applicability of various nano-adsorbents for the removal of various noxious pollutants from water bodies.

2 Nanoparticle-based adsorbents for noxious pollutants removal

Nanoparticle-based adsorbents are particles having a size between 1 and 100 nm[58] and possess significant adsorbing capacities for different types of pollutants because of their high surface area and small particle size. Due to these specific characteristics, nanoparticle-based adsorbents have more active sites to interact with different materials.[59–61] Thus, nanoparticles are emerging as a growing substitute for conventional adsorbents for removing various pollutants.[10,12,25,26,62] Different types of adsorbents based on nanoparticles such as iron, alumina, titania, silica, carbon, etc., have been successfully synthesized and utilized by the researchers[63–65] to remove different noxious pollutants from the water bodies. An overview of different material-based nano-sorbents (Fig. 1), common adsorption mechanism involved (Fig. 2), and interaction of nano-sorbents with different noxious pollutants such as metal ions, dyes, PAHs, PCBs, and pharmaceuticals (Fig. 3) can be better understood from the schematic presentations. Further, a detailed discussion about the same and their applications as adsorbents for removing different noxious pollutants was discussed in the subsequent paragraphs.

Sustainable Materials for Sensing and Remediation of Noxious Pollutants. https://doi.org/10.1016/B978-0-323-99425-5.00013-X

TABLE 1 Physicochemical properties and adsorption capacities of different nanoparticle-based adsorbents for the effective removal of noxious pollutants.

Nanoparticles	Pollutants	Size (nm)/diameter (nm)	BET surface area (m²g⁻¹)	Adsorption capacity (mg/g)	Contact time (min)	pH	Effect of ions	Shape	Pore volume (cm³ g⁻¹)	Pore diameter (nm)	Reference
Akageneite nanocrystals	Cr(VI)	3–6	330	79.66	–	5.5	–	Rod-like shape	0.35	2.5	29
Akaganeite granular	Zn(II)	–	234	13.95	–	–	–	–	0.237	3.5	30
Nanocrystalline Akaganeite	Zn(II)	–	330	27.61	1200	6.5	–	–	0.35	0.237	30
Akaganeite	As(V)	2.6	330	120	1440	7.5	As(V) removal increased in the presence of potassium cations	–	–	–	31
Akaganeite	Cd(II)	2.6	330	17.1	–	8	Adsorption of Cd²⁺ ions has been decreased in the presence of K⁺ ions.	–	–	–	32
Akaganeite	Phosphates	30–100	330	49	–	7	–	–	0.35	2.5	33
Hybrid surfactant-Akaganeite	Phosphates	70–100	230.9	451.19	–	7	–	–		3.6	33
Alumina Nanoparticles	Ni(II)	15–20	–	30	120	8	–	–	–	–	34
Amino-functionalized Magnetic Nanoparticles	Cu2+	–	–	25.77	5	6	No effect of salinity on the adsorption of Cu²⁺	–	–	–	35
Carbon nanotube supported ceria	Cr(VI)	6	–	30.2	144o	3–7.4	–	–	–	–	36
Chitosan Nanoparticle	Eosin Y	20–80	–	3.33	240	5	–	Regular spheres	–	–	37
GO and oxidized CNT modified Nanohybrid	Ciprofloxacin	–	–	512	240	6	–	–	–	–	38
CNT/TNT Nanomaterials	Cu2+	–	–	83–124	–	5	–	–	–	–	39
CNT/TNT Nanomaterials	Pb2+	–	–	192–588	–	–	–	–	–	–	39
Magnetic Nanoparticles	Cu(II)	13.2	–	17.6	2	5.1	–	Spinel	–	–	40
Gum arabic modified magnetic	Cu(II)	13–67	–	38.5	2	5.1	–	Spinel	–	–	40

Hydrous titanium dioxide	As(III)	10.8	280	31.8	–	4	–	–	Anatase	–	–	41
	As(V)			33.4	–		–	–	–	–	–	
Graphene oxide-magnetic/polyrhodanine	Phenol		30.74	191	15	7	–	–	–	0.0812	10.56	43
Graphene oxide-magnetic/polyrhodanine	β-naphthol		30.74	226.2	15	7	–	–	–	0.0812	10.56	43
Maghemite nanoparticles	Cr(VI)	10	198	17	10	2.5	–	–	–	–	–	42
Maghemite nanoparticles	Cu(II)	10	198	26.8	10	6.5	–	–	–	–	–	42
Maghemite nanoparticles	Ni(II)	10	198	23.6	10	8.5	–		–	–	–	42
MnFe$_2$O$_4$ nanoparticles	Cr(VI)	10	208	31.5	5	2	–	NH$_4^+$ ion increased the adsorption at pH greater than 6.5, whereas EDTA and SO$_4^{2-}$ inhibited the uptake of Cr(VI) over the entire pH	–	–	–	44
Multi-walled carbon nanotubes	Reactive Red M-2BE dye	–	180.9	335.7	60	–	–	–	–	0.345	–	45
Nano zero-valent iron (NZVI)	cadmium ion	20 to 200	26.3	769.2	720	–	–	–	Spherical particles	–	–	46
Purified multi-walled carbon nanotubes	Zn^{2+}	8–10	297	32.68	60	–	–	–	Cylindrical shapes	0.38	–	47
Purified single-walled carbon nanotubes	Zn^{2+}	1–2	423	43.66	60	–	–	–	Curve	0.43	–	47
Micro mesoporous nano-silica	Acid Orange II	–	555	676.7	20	3	–	–	–	0.94	1.2	48
Micro mesoporous nano-silica	Acid Fuschia	–	555	621.3	20	3	–	–	–	0.94	1.2	48
Single-walled carbon nanotubes	dissolved organic matter (DOM)	–	464	21.69	120	4	–	–	–	1.07	–	49
TiO$_2$	red 195 azo	~8.8	155	87.0	60	3	–	–	–	350 cm^3 g^{-1} total pore volume	–	50

Continued

TABLE 1 Physicochemical properties and adsorption capacities of different nanoparticle-based adsorbents for the effective removal of noxious pollutants—cont'd

Nanoparticles	Pollutants	Size (nm)/ diameter (nm)	BET surface area (m^2g^{-1})	Adsorption capacity (mg/g)	Contact time (min)	pH	Effect of ions	Shape	Pore volume ($cm^3 g^{-1}$)	Pore diameter (nm)	Reference
Zero-valent iron onto activated carbon	As(III)	–	–	18.19	720	6.5	Divalent metal ions such as Mg^{2+}, Ca^{2+} increase the adsorption of arsenate while Fe^{2+} suppress arsenite adsorption	–	–	–	51
Zero-valent iron onto activated carbon	As(V)	–	–	12.02	720	6.5		–	–	–	51
Nanocomposite modified with magnetic graphene and chitosan	2-Naphthol	–	–	169.49	45	2	Electrostatic attraction	–			52
Maghemite nanoparticles	Alizarin	–	30–60	23.2	60	11	–	–	–	20–40	53
Alumina nanoparticles	Orange G	–	128	93.3	30	2.5	SO_4^{2-}, PO_4^{3-}, and $C_2O_4^{2-}$ anions exhibit a remarkable effect the adsorption		0.32	13.6	54

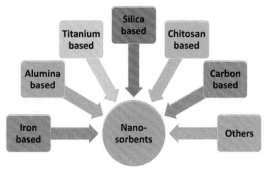

FIG. 1 Schematic overview of different material-based nano-sorbents.

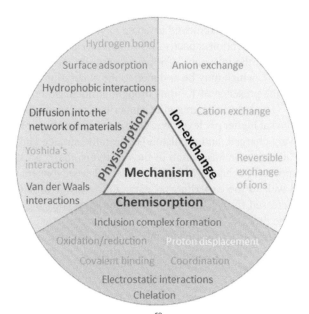

FIG. 2 Schematic presentation of most common adsorption mechanisms.[59]

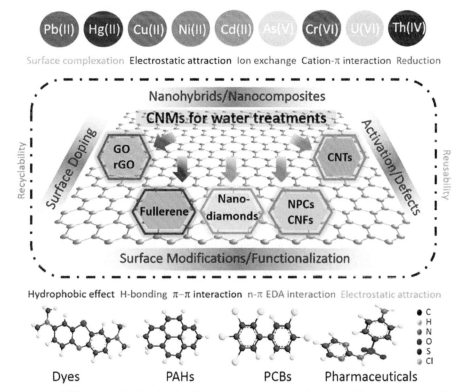

FIG. 3 Graphical presentation of different mechanisms involved between nano-sorbents and noxious pollutants such as metal ions, dyes, PAHs, PCBs, and pharmaceuticals.[59]

2.1   Iron-based nanoparticles

Iron (Fe) is one of the most abundant elements of the earth's crust and is found in the form of oxides in the atmosphere, pedosphere, biosphere, hydrosphere, and lithosphere.[66–68] Iron and iron oxide-based nanoparticles such as akaganeite, zero-valent, magnetite and maghemite nanoparticles, etc., have been extensively utilized by researchers for the treatment of different toxic substances in water bodies.[69–75] It has been found that iron-based nanoparticles have a high surface area with significant efficiency for the adsorption of pollutants. Some of the iron oxide-based nanoparticles are discussed here.

2.1.1   Akaganeite-based nanoparticles

Iron oxide-based Akaganeite-type nanoparticles have been studied extensively by Deliyanni et al.,[76] who found some interesting results from their work. They successfully removed arsenate oxyanions from water by adsorption onto synthesized akaganeite (β-FeO(OH). The maximum adsorption capacity of the material calculated using the Langmuir model was found to be 120 mg/g. Researchers also studied the influence of ionic strength on the adsorption of oxyanions and observed a decreasing trend in adsorption at pH > 7, which may be assigned to the generation of positively charged surface repelling the oxyanions. It was also seen that the presence of K^+ ions brought an increase in the removal of As (V).

In another study, Deliyanni et al.[77] used synthetic aragonite nanocrystals to adsorption Cd^{2+} in water bodies. The authors found that the Cd^{2+} is readily adsorbed at higher pH levels and adsorbed negligibly at pH below 4, since at lower pH levels, the surface of akaganeite is positively charged, and repulsion between the surface and Cd^{2+} is observed. Contrary to the previous findings of the authors,[76] decrease in adsorption of the positively charged Cd^{2+} in the presence of K^+ ions was observed, which was due to the competitive behavior of both positively charged ions for the adsorption sites. Langmuir and Freundlich models were followed well to the experimental data, and removal efficiency of 17.1 mg/g at pH of 8 was reported, which was found to be increased from 17.1 to 66.72 mg/g on increasing the temperature from 25 to 65 °C, suggesting the endothermic nature of the process.

Lazaridis et al.[78] also utilized Akaganeite to treat the Cr (VI) in water and found a maximum removal of 79.66 mg/g at pH 5.5, which dropped to 71.77 mg/g at pH 6.5. The authors demonstrated that the degree of adsorption was found to be greatly influenced by the pH of the solution.

2.1.2   Zero-valent iron nanoparticles as adsorbent

Zhu et al.[51] utilized zero-valent iron nanoparticles using activated carbon as supporting material for arsenite and arsenate removal from waters. The authors reported a maximum removal efficiency of 18.2 and 12.0 mg/g at pH 6.5 for arsenite and arsenate ions, and adsorption data were found to be well agreed with Langmuir model.[40] The authors found that the removal rate of arsenate decreased drastically when increasing the pH from 3 to 12. In contrast, the removal rate has increased from pH 3 to 6.5 and decreased readily after reaching pH 9 for arsenite. This trend was observed, because at higher pH levels, the surface of the adsorbent (ZPC-7.4) became less positive and repel the negatively charged arsenate species leading to a decrease in its adsorption, whereas for arsenite, adsorption was found to be increased due to deprotonation below pH 9. Then electrostatic repulsion is the ruling factor above pH 9, decreasing arsenite adsorption. Moreover, they also studied the effect of different anions (PO_4^{2-}, SO_4^{2-}, SiO_4^{4-}, etc.) and cations (Mg^{2+}, Ca^{2+} and Fe^{2+}) present in water on adsorption of arsenic. They found that anions present were not found to be favorable for adsorption of arsenate and arsenite due to the formation of inner surface complexes which further competed for similar adsorption sites, resulting in reduced adsorption. On the other hand, adsorption of both adsorbates was found to be enhanced by the presence of cations in water.

Kanel et al.[79] reported the removal of As (III) by zero-valent iron nanoparticles and found a maximum removal of 3.5 mg/g at pH of 7, which was well established by the Freundlich parameters and the process was found to follow pseudo-first-order kinetics. Adsorption of As (III) was found to be increased from acidic to neutral pH levels. In contrast, it was suggested to be unfavorable at alkaline pH levels due to electrostatic repulsions between the adsorbent surface and adsorbate. The impact of the adsorbent dose was also studied, and an increase in the adsorbed amount of arsenic was reported on increasing the dose from 0.05 to 1 g/L. The authors further observed that the presence of silicate and phosphate anions in water significantly decreased the arsenic adsorption, while nitrate, sulfate, and carbonate anions reduced the adsorption at higher concentration levels.

Boparai[16] successfully removed Cd^{2+} using nano zero-valent iron and reported a better fit of the Langmuir isotherm with a maximum adsorption capacity of 769.2 mg/g. The authors also investigated the effect of increasing the temperature and observed an increasing trend in the adsorption of Cd^{2+} which suggests the process to be endothermic. Pseudo-second-order equation agreed well to the kinetic data, and surface diffusion or intraparticle diffusion was found to be the rate-limiting step. Interestingly, the effect of solution pH was not explained by the authors. Thermodynamic parameters analyzed provided an indication of spontaneity and affinity of Cd^{2+} toward the adsorbent.

2.1.3 Magnetite and maghemite nanoparticles as adsorbent

The adsorptive strength of mixed magnetite and maghemite nanoparticles for As(III), As(V), and chromium removal from water bodies is also reported.[80,81] According to Chowdhury et al.[82] at an initial concentration of 1.5 mg/L and pH 2, the amount adsorbed was found to be 3.69 mg/g, 3.71 mg/g, and 2.4 mg/g for arsenic(III), arsenic(V), and chromium(VI), respectively. An adverse effect of increasing pH on the adsorption of these metal ions was observed. The authors reported that the percentage of arsenic and chromium removal decreased with the increase of phosphate concentration. The reason was suggested to be the competitive behavior for the adsorption sites due to the formation of a complex on the surface. Interestingly, authors found that anions such as chloride, sulfates, and nitrates did not significantly affect the adsorption of these metal ions.

In another work Hu et al.[83] studied the removal of Cr(VI) on maghemite nanoparticles coated with δ-FeOOH. The surface coating of maghemite was compared to uncoated maghemite was resulted, and an enhancement to the adsorptive removal of chromium hexavalent ions was reported. This was suggested to be due to the outer sphere complex formation at the mineral water surface. The authors also found that the adsorption of chromium decreased at alkaline pH levels (>8.5).

Maghemite nanoparticles[42] (surface area – 198 m^2/g) were synthesized to remove chromium, nickel, and copper ions from aqueous media. The authors reported a maximum removal of 17.0 mg/g at pH 2.5 Cr (VI), 26.8 mg/g at pH 9.5 for Ni(II), and 23.6 mg/g at pH 6.5 for Cu(II) by the maghemite nanoparticles and adsorption data fitted best to the Langmuir model. The trend for the maximum adsorption of Cr (VI) was also explained based on strong electrostatic interactions with the adsorbent surface at the investigated pH range.

2.1.4 Magnetic nano-adsorbents

Fe_3O_4 nanoparticles and Fe_3O_4 nanoparticles[40] modified with gum Arabic were synthesized and utilized as an adsorbent to treat the water containing copper metal ions. The authors observed that gum Arabic modified Fe$_3$O4 nanoparticles compared to the non-modified iron oxide nanoparticles were found to have better adsorption capacity for copper ions with a removal efficiency of 38.5 mg/g compared to 17.6 mg/g for the non-modified adsorbent. Langmuir model followed well to the experimental data with a short equilibrium time of about 2 mins, by which it can be deduced that there was no acting resistance during internal diffusion. Adsorption was found to increase on increasing the temperature, which revealed that the process is endothermic. Huang et al.[84] developed a magnetic nano-adsorbent by covalent bonding of polyacrylic acid onto the surface of iron oxide nanoparticles and further modified it with diethyl triamine to utilize it for the removal of Cu(II) and Cr(VI). They found an adsorption amount of 12.43 mg/g and 11.24 mg/g for Cu(II) and Cr(VI) ions, respectively, and suggested that the equilibrium could be achieved in a couple of minutes which was supported by the findings of Banerjee et al.[40] Ion exchange and chelation mechanisms were suggested to play a major role in the adsorption of both cations and anions.

Hao et al.[35] modified iron oxide nanoparticles with hexadiamine to prepare a magnetic nano-adsorbent and use it for the decontamination of Cu^{2+}-containing solution. The authors observed a fast achievement of the equilibrium and reported a maximum removal of 25.77 mg/g for Cu^{2+} onto the prepared adsorbent. Langmuir model was suggested to follow well to the adsorption data and reported the spontaneous and endothermic nature of the process. Hu et al.[44] also studied the adsorption of Cr(VI) on surface-modified jacobsite nanoparticles and reported 31.5 mg/g removal efficiency for Cr(VI) with a short equilibrium time. They found that ligand such as NH_4^+ would enhance the adsorption at pH levels above 6.5, whereas EDTA and SO_4^{2-} ions were found to bring a reduction in Cr(VI) adsorption over the entire studied pH range.

Liu et al.[85] utilized low-cost iron salts and humic acid followed by a co-precipitation reaction to prepare humic acid-modified iron oxide nanoparticles and tested them to remove mercury, lead, copper, and cadmium ions from aqueous media. The Langmuir adsorption capacities for Pb(II), Hg(II), Cd(II), and Cu(II) were found to be 97.7, 92.4, 50.4, and 46.3 mg/g, respectively. The authors found that coexisting ions like phosphate and calcium as well as the salinity had no considerable impact on the adsorption of metals by the prepared adsorbent, which was due to the fact that humic acid is strongly bonded with metal ions than Ca^{2+} and also has the affinity for the strong sorption of metal ions than $PO4^{3-}$.

2.1.5 Iron-based nanoparticle as reducing agent

Several adsorbents have been used to remove Cr (VI); however, the utilization of nano-adsorbents to remove Cr (VI) from water has gained quite attention in the recent works.[81,86,87] The diameter of iron-based nanoparticles plays a significant role in the removal of Cr(VI).[88] Moreover, iron nanoparticles reduced the Cr (VI) to trivalent chromium Cr (III). While comparing these materials with others, the authors suggested promising efficiency for the removal of Cr (VI) compared to iron oxide powder and fillings.

Ponder et al.[88] successfully removed chromium ions by nanoscale zero-valent iron, having a diameter of 10–30 nm synthesized by borohydride reduction of iron salt. Based on XPS and XRD analyses, the authors revealed that Cr (VI) could be reduced to Cr (III) by zero-valent modified nano-adsorbent. Niu et al.[89] used starch-stabilized zero-valent iron nanoparticles as adsorbents for removing Cr(VI) from the groundwater compared to native particles. It was observed that the removal of Cr (VI) decreased on increasing the pH and nanoparticle dose (0.40 g/L; 100% removal of 20 mg/L). Furthermore, the authors claimed starch-stabilized FeO nanoparticles as better than FeO powder and filings.

Xiao et al.[90] prepared iron nanoparticles modifying them with leaf extracts of various plants selected based on their reducing ability and utilized them to remove chromium ions. The authors observed that Cr(VI) removal was found to be consistent with the reducing ability of the employed plant extracts viz. S.jambos Alston with high reducing ability, Oolong tea and *A. moluccana* wild with moderate and weak reducing potential, respectively. The maximum chromium removal was reported for S.jambos Alston to extract modified nanoparticles, strengthening the fact as mentioned above.

2.1.6 Modified iron-based nanoparticles

Pan et al.[91] synthesized polymer-based hydrated iron oxide nanoparticles by impregnating hydrated Fe_2O_3 nanoparticles within a cation-exchange resin. The authors successfully treated the water to remove Pb, Cd, and Cu divalent ions and reported maximum adsorption of 1.6, 1.4, and 1.4 mmol/g, respectively, on the synthesized nanoparticles. The authors further stated that the working capacity of polymer-based nano-adsorbent was ~ 4 to 6 times more than that of cation-exchange resin in column experiments.

In a recent study, Parvin et al.[43] prepared two nano-adsorbents, one by coating iron nanoparticles with polyrhodanine and the other by modifying it with graphene oxide. Removal of phenol and β-naphthol was tested by the authors utilizing the prepared adsorbents. The authors observed that the prepared adsorbent exhibited great adsorption capacity for phenol (191 mg/g) and β-naphthol (226.2 mg/g) on nano-adsorbents modified with graphene oxide, which was assigned to the increased surface area of adsorbent on modification with graphene oxide. The authors revealed that the π-π bonds of adsorbates with the prepared adsorbents were responsible for the adsorption.

2.2 Alumina-based nanoparticles

Nanoparticles of alumina are another important and widely studied material because of their broad range of applications such as catalysts, electricity insulators, spacecraft accessories, and adsorption for wastewater treatment.[92] Alumina nanoparticles possesses high surface area and good adsorptive abilities. These nanoparticles are low cost and can be effectively used to treat water with different toxic substances such as phenols, dyes, metal ions, pharmaceutical compounds etc.

Sharma et al.[34] prepared alumina nanoparticles using the sol–gel process to study their potential for removing Ni(II) ions from aqueous media. The total removal efficiency of 30.8 mg/g was reported for Ni(II) by the adsorbent at 25°C, and the authors observed that it was increased from 97% to 99% when the temperature was increased from 25°C to 45°C. The experimental data were fitted well to the Langmuir and Freundlich models and governed by a first-order rate equation.

Mahmoud et al.[93] in another study, synthesized aluminum oxide nanoparticles followed by glycothermal process and further modified it with a positively charged surfactant and used it to treat the nitrobenzoic acids in water. They found that the modified adsorbent more effectively removed the studied nitrobenzoic acids than aluminum oxide nanoparticles in any analyzed condition. The authors also stated that a decrease followed adsorption of both 2-nitrobenzoic acid and 4-nitrobenzoic acid in increasing the initial concentration of adsorbates. This trend was observed due to the more dissolved molecules in the solution than the static mass of both adsorbents. Moreover, the coexisting ions such as sodium, potassium, calcium, and magnesium were not found to significantly influence the adsorption of nitrobenzoic acids on the adsorbent used.

Alumina nanoparticles utilizing the sol-gel process were prepared by Banerjee et al.[54] to treat the water containing orange G, a dye. They revealed that the pH of the solution greatly influenced the adsorption of dye. The zero point of charge reported was 7.4, and the authors observed that the adsorption was found to be decreased due to the electrostatic repulsions between the negatively charged surface and anionic groups of the adsorbate at pH above pHpzc. Langmuir model with an adsorption capacity of 93.3 mg/g was found to agree well with the adsorption data. In contrast, the kinetics of the process was well described by a pseudo-second-order equation.

In a recent study, Herrera-Barros et al.[94] prepared a biosorbent from palm oil biomass residue and Al_2O_3 nanoparticles by dispersing the biomass into the dimethyl sulfoxide (an organic solvent) to test their uptake efficiency for cadmium and nickel divalent ions from water. They found a removal efficiency of 87% and 81% at pH 6 and particle size of 0.35 mm for cadmium and nickel ions, respectively. The authors reported that the removal of metal ions followed a decreasing route on increasing the particle size of biomass from 0.35 mm to 1 mm. Moreover, the organic solvent DMSO did not play any significant role in the adsorption process.

2.3 Titanium-based nanoparticles

Many studies have been carried out to investigate the adsorption of inorganic and organic pollutants on TiO_2 nanoparticles. The TiO_2 employed in the literature can be divided into several categories: nanocrystalline TiO_2 particles, titanate nanotubes, hydrous TiO_2, granular TiO_2 and TiO_2-impregnated chitosan beads. Belessi et al.[50] worked on TiO_2 nanoparticles, which were considered to remove reactive red 195 dye. At the temperature of 30°C, the maximum adsorption capacity of ~87 mg/g was reported. In acidic conditions, high electrostatic interactions were taken place between the anionic group of the dye and the surface of the adsorbent, which was positively charged. The experimental data were found to be best fitted to Langmuir and pseudo-second-order model.

Ismaeel et al.[95] synthesized a nano-sized anatase TiO_2 and tested its efficiency for the uptake of an azo dye, Eriochrome black T. The authors found the adsorption process to be spontaneous and exothermic based on the effect of temperature and investigated thermodynamic parameters. Moreover, the pH of the solution, contact time, and adsorbent dose were found to be considerably affecting the uptake of the adsorbate.

In another study, George et al.[96] investigated the removal of arsenic, antimony, lead, and cadmium by a cationic and anionic surfactant-modified nano titania adsorbent synthesized via sol–gel process. The authors stated that the percentage removal efficiency was found to be highest for the cadmium ions (75%), and they chose it for the detailed study. Further, the effect of pH on Cd ion adsorption was also studied, and it was seen that the optimum pH value was 9.67 for all the three adsorbents prepared, viz. unmodified nano titania, cationic and anionic surfactant-modified nano titania. From the above fact, it can be described that the nature of the modifying surfactant did not influence the adsorption.

The microwave hydrothermal route was employed by Andjelkovic et al.[97] to prepare TiO_2 doped with zirconium oxide to remove arsenate and arsenite from the waters. The authors observed that alteration in pH value significantly affected the adsorption process. Freundlich isotherm in the case of arsenate and Langmuir model for arsenite was found to be a good fit to describe the process, and the pseudo-second-order model was well co-related with the experimental findings.

2.4 Silica-based nanoparticles

Zhao et al.[98] investigated the uptake of methylene blue onto vermiculite derived nano-sheets of silica followed by acid leaching in an aqueous solution. The maximum adsorbed amount for methylene blue was found to be 11.77 mg/g at an initial concentration of 28.79 mg/L at 20°C. The authors also revealed that the adsorption was not favorable at higher temperatures as the uptake was decreased by increasing the temperature from 20°C to 70°C. Langmuir and pseudo-second-order models were well suited to the experimental findings. Functionalized hollow spheres of nano-silica were synthesized by Rostamian et al.[99] to study their effect on the heavy metal (Hg^{2+}, Pb^{2+}, and Cd^{2+}) removal. They found that the adsorption of all studied heavy metals increased with an increase in their initial concentration. Redlich-Peterson isotherm in the case of Cd^{2+} ions whereas Sips isotherm model for the rest of the two metal ions were found to be best followed. The authors also reported that the prepared adsorbent had a higher adsorption potential for mercury ions than lead and cadmium.

A study to prepare the mesoporous silica nanoparticles from banana peels was carried out by Mohamad et al.[100] to test its uptake ability for phenol and a dye, methyl orange. The authors found that the maximum phenol and methyl orange removal were 91.1% and 92.6%, respectively, at 0.4 gL^{-1} of adsorbent dose. It was also observed that maximum removal was taken place at acidic pH levels for both phenol (at pH = 5) and methyl orange (at pH = 3), which was assigned to the fact that the surface of the adsorbent was positively charged and presented better electrostatic interactions with an anionic group of the adsorbates at lower pH levels.

Amino-functionalized SiO_2 nanoparticles of different sizes (10–40 nm) was prepared and used by Saleh et al.[101] to remove phenol from water bodies. They reported that the silica nanoparticles with the lowest particle size (10 nm) were presented the highest removal efficiency (35.5 mg/g) for phenol. It was seen due to the fact that the small particle size of nanoparticles possessed a high surface area, i.e., it provides more sites to adsorb the phenol. Langmuir model was found to be a good fit for the adsorption data compared to the Freundlich isotherm. A pseudo-second-order model well described kinetic data. The adsorption capacity of the prepared adsorbent for phenol followed a decreasing order with respect to their particle size, i.e., 35.2 mg/g, 31.6 mg/g, 31.07 mg/g, and 24.19 mg/g.

2.4.1 Nano-SiO₂

Lu et al.[102] also studied nano-silica and nano-kaolin adsorption potential for the atrazine dye. Based on their findings, they revealed that uptake of atrazine was found to be decreased by increasing the ionic strength of the solution for both adsorbents. Freundlich model was identified as a good fit to describe adsorption. Change in concentration and pH was found to affect the adsorption considerably. The varying concentration of nano-silica does not significantly influence the dye adsorption.

Kotsyuda et al.[103] prepared bifunctionalized new silica nanoparticles by the Stöber method using ternary alkoxysilanes systems containing 3-aminopropyl and phenyl groups and tested their adsorption potential for methylene blue and Cu^{2+}. The findings indicated that the bifunctionalized nano-silica spheres showed more than twice of adsorption capacity (99 mg/g) for methylene blue dye in comparison with the nano-silica spheres functionalized with only amino groups (48 mg/g), which may indicate better interaction of methylene blue molecules with the adsorbent functionalized with phenyl groups. The authors suggested that combining electrostatic attraction to amino groups and hydrophobic attraction to phenyl groups in the surface layer of our bifunctional samples improved the adsorption of methylene blue dye. Therefore, the sorption of the Cu(II) ions by functionalized silicas depends on the concentration of amino groups, which was suggested to depend on the isoelectric point values.

In another work, adsorption studies on the removal of dye (acid orange 8) utilizing nano-silica particles derived from sugarcane waste ash were also carried out by Rovani et al.[63] The authors found that the prepared adsorbent exhibits great adsorption potential for acid orange 8 with a removal efficiency of 230 mg/g. Liu's isotherm was found to fit better to the adsorption data. The authors suggested that the stabilizer cetyltrimethylammonium bromide (CTAB), used during the preparation of nano-silica particles, also have an important role in the adsorption of acid orange 8. The authors further suggested that owing to opposite charges of dye and stabilizer, the electrostatic attraction occurs between them, resulting in increased adsorption of the dye at the adsorbent surface.

2.5 Chitosan-based nanoparticles

Chitosan, a natural polysaccharide containing amino groups that can serve as chelation sites, can easily be combined with nanoparticles and used as adsorbents to remove different pollutants from aqueous media.[74,104,105] Chang et al.[106] developed chitosan-based Fe_2O_3 nanoparticles to recover Co(II) ions from waters. The adsorption followed the Langmuir model, and the equilibrium was achieved within one minute, which was assigned to the absence of inner diffusion resistance. The maximum amount of 27.5 mg/g for Co(II) was adsorbed at 25°C and pH of 5.5. The authors also revealed that increasing the temperature has an adverse impact on adsorption, i.e., exothermic is the preferred pathway by the process. Moreover, poor electrostatic interactions between sorbent and nanoparticles at high temperatures were also found responsible for the trend.

Chitosan nanoparticles were synthesized, followed by an ionic gelation process between chitosan and tripolyphosphate by Du et al.,[107] to remove the anionic dye, eosin Y, from the water bodies. The authors suggested that the pH of the solution greatly influences the adsorption. The authors reported a slight decrease in adsorption between pH 2 and 5, whereas the decrease happened rapidly at alkaline pH values. This trend was observed due to the interactions between negatively charged —NH₂ group of chitosan with the anionic groups of the adsorbate and amino group was tend to be decreased at higher pH levels. Langmuir model was found to be best fitted to the experimental findings with a removal efficiency of 3.333 mg/g for eosin Y.

Rebekah et al.[52] prepared magnetic graphenes and chitosan modified nano-adsorbent and utilized it to remove 2-naphthol in water bodies. The authors reported a maximum removal efficiency of 169.49 mg/g at pH 2 for 2-naphthol and revealed the favorability of the adsorption process at lower pH levels. At lower pH values, it was observed by the authors that the surface of the adsorbent is positively charged, which interacts strongly with the anionic adsorbate and results in enhanced adsorption. According to the authors, the Freundlich model followed well to the adsorption data and revealed the heterogeneity of the surface of the adsorbent.

2.6 Carbon-based nanomaterials

CNTs (Carbon nanotubes) were firstly introduced by Iijima in 1991 during the arc-evaporation synthesis of fullerene.[108] The high-specific surface area, availability of various functional groups, easily tunable surface, promising mechanical properties, and hydrophobic character make CNT's an excellent choice to use as an adsorbent for the removal of several pollutants such as metal ions, dyes, phenols, pharmaceutically active substances, etc., from aqueous solutions.[11,12,109–111]

Hu et al.[112] investigated the removal efficiency of iron oxide multi-walled CNTs grafted with cyclodextrin for 1-naphthol and lead divalent ions in liquid solution. Maximum removal of 12.29 mg/g (at pH = 5.5) and 57.49 mg/g (at pH = 6.5) for lead ions and 1-naphthol by the prepared adsorbent were reported by the authors, and the Langmuir model was also found to be well correlated with the adsorption data. A comparative assessment of the prepared adsorbent with other adsorbents, viz. MWCNT/iron oxides (9.85 mg/g), oxidized MWCNT (8.70 mg/g), and iron oxides (5.60 mg/g) was also made by the authors to establish the role of cyclodextrin modification in enhancing the removal efficiency of lead ions. The grafting of cyclodextrin into the adsorbent was found to provide more hydroxyl groups to interact with lead ions for complex formation and was resulted in enhanced adsorption. The authors also found that Fe_2O_3 content present in the adsorbent had no considerable impact on the adsorption.

In a different study, Lou et al.[49] successfully utilized SWCNTs (single-walled carbon nanotubes) to adsorb and remove the dissolved organic matter (DOM) in water. They reported that adsorption was found to be increased at low ionic strength and temperature. Maximum removal was found to be decreased from 26.10 to 20.77 mg/g on increasing the temperature from 4°C to 45°C. Authors also found that the pore diffusion was the controlling factor for the adsorption process. Machado et al.[45] synthesized MWCNTs to test their adsorptive efficiency for reactive red M-2BE and reported that 335.7 mg/g of dye was removed in 60 min on the prepared adsorbent. Experimental data were found to be in good agreement with Liu's isotherms, and the favorability of adsorption in acidic pH ranges was also observed by the authors.

In another approach, a CNT-based adsorbent was synthesized by grafting with calcium carbonate microparticles to assess the adsorption behavior of 2-naphthol.[113] The impact of coexisting adsorbates (naphthalene and 2-chlorophenol) on the degree of adsorption in real water samples was also studied by the authors. The results they obtained revealed that the adsorption of 2-naphthol was inhibited in the presence of these adsorbates due to the competitive behavior for the adsorption sites; however, 2-naphthol was found to have the maximum affinity toward the prepared adsorbent. The experimental data were found to fit with Freundlich parameters.

Yang et al.[114] prepared an adsorbent named oxidized cobalt wrapped by nitrogen-doped CNTs by single-step annealing and utilized it for the efficient removal of tetracycline, an antibiotic and rhodamine B, a dye. They reported a maximum adsorption capacity of 679.56 and 380.60 mg/g for tetracyclin and rhodamine B, respectively, onto the prepared adsorbent. The adsorption data were best fitted to the Langmuir model over Temkin and Freundlich models. A pseudo-second-order rate equation was found suitable to describe the kinetic findings. H-bonding and π-π interactions were identified as the governing factors for the adsorption of rhodamine B. In contrast, complex formation on the surface was responsible for the adsorption in the case of tetracycline.

Fares et al.[38] successfully removed ciprofloxacin from aqueous media by utilizing a nanohybrid prepared from graphene oxide crosslinked with oxidized CNTs by calcium ions. It was found that maximum removal was reported in the pH range of 6–7, which was due to the fact that adsorbate becomes neutral in this range and attached to the adsorbent surface via the strongest hydrogen bonding. Moreover, ciprofloxacin was found to bear negative and positive charges at pH > 8.9 and pH < 5.9, and this charge was responsible for weak hydrogen bonding with adsorbent surface and resulted in reduced adsorption. Four models viz. Langmuir, Freundlich, D-R, and Temkin were tested and presented a good correlation with the adsorption data. 99.2% of ciprofloxacin were eradicated in 4 h of contact time by the synthesized adsorbent.

In a recent study, different types of multi-walled CNTs with non-purified, purified, calcination, and oxidation treatment were prepared and utilized for the adsorption of benzoic acid.[115] The authors identified that the calcined CNTs exhibited the maximum adsorption capacity for benzoic acid. This was justified by the fact that the breakdown of interactions between the carbon nanotubes was enhanced on calcination and followed a good dispersion of adsorbate onto the adsorbent. The maximum adsorption capacity of 22.3 mg/g with unpurified CNTs, 23.3 mg/g with unpurified CNTs, 28.9 mg/g with oxidized CNTs, and 39.9 mg/g with calcined CNTs for benzoic acid were reported by the authors.

2.7 Other novel modified nanoparticles

A novel polymeric hybrid sorbent ZrPS-001 was developed by Zhang et al.[116] and was fabricated with impregnating $Zr(HPO_3S)_2$ (i.e., ZrPS) nanoparticles within a porous polymeric cation exchanger D-001 for increasing the sorption of heavy metal ions such as lead, cadmium, and zinc ions from aqueous solution. The exhausted ZrPS-001 beads were found to be capable of regeneration with 6 M HCl solution without any significant capacity loss.

Soltani et al.[117] utilized a novel nanocomposite modified with carboxylic acid–double hydroxide metal-organic framework to remove cadmium and lead ions from water bodies. They reported a maximum removal efficiency of 415.3 and 301.4 mg/g for Cd (II) and Pb (II), respectively, on the synthesized composite. In another study, Soliman et al.[118] prepared two novel quaternary organometal oxide nanocomposites utilizing chitosan by modifying them with CeO_2, CuO, and Fe_2O_3 (CF) and CeO_2, CuO, and Al_2O_3 (CA) for the removal of disperse red 60 dye from aqueous media.

The authors reported maximum adsorption of 100 mg/g at pH 2 and 4 for CF and CA, respectively. They also report that Monte Carlo simulation studies confirmed the strong interactions between the adsorbate and the prepared composites.

3 Future perspectives

The literature review clearly shows that the applications of nanoparticles in various fields are a fantastic and emerging area, and due to this, an approach is needed by scientists to explore more about the toxic effects of nanoparticles. In the future, nanoparticles may become a serious issue to the environment because some NPs are non-biodegradable and easily penetrated living organisms due to their small size. One of the emerging applications of nanoparticles is nano-adsorbents such as maghemite, gum arabic modified nano-adsorbents, iron nanoparticles, akaganeite, ceria nanoparticles, titanium dioxide nanoparticles, and nano-alumina particles which offer excellent removal efficiency for metals and other pollutant species. At present, NPs make a huge market value, so their production cannot be ignored. So there is a need to develop more efficient, selective, inexpensive, and eco-friendly nanoparticles for water remediation in future. Scientists developed various regeneration techniques to recover nanoparticles after adsorption, but the management of the exhausted nanoparticle is still not completely developed, and more work in this direction is expected.

4 Conclusions

This chapter focused on the adsorption capacity of different types of nanoparticles for the removal of various noxious pollutants from water and leads to the following conclusions:

- Nanoparticles these days have become economical due to low-dose requirements for removing pollutants at varied temperature and pH ranges.
- The removal time of pollutants from water by nanoparticles is quite fast compared to other adsorbents due to their small size and high surface area, so nanoparticles have gained momentum for use in water treatment technology.
- Literature shows scientists generally focussed on batch methods for removing pollutants using adsorption methodology. Some workers coupled the removal of noxious pollutants by nanoparticles using adsorption methodology with other technologies too for the complete water treatment. There is a need to elaborate their use on an industrial scale in future, and batch mode adsorption conditions should be transferred to column operations.
- Scientists need to develop some eco-friendly waste management techniques to overcome the hazardous effect of nanoparticles.

Acknowledgements

The authors are thankful for financial support to DST (Department of Science and Technology), New Delhi, India. One of the authors (Monika Chaudhary INSPIRE Fellow code IF120368) is grateful to the DST, New Delhi, India, for the award of a doctoral grant (No. DST/INSPIRE Fellowship/2012/346).

References

1. Rijsberman FR. Water scarcity: fact or fiction? *Agric Water Manag.* 2006;80(1):5–22. https://doi.org/10.1016/j.agwat.2005.07.001.
2. Nemerow NLDA. *Industrial and Hazardous Waste Treatment.* Van Nostrand Reinhold; 1991.
3. Yousefi M, Ghoochani M, Hossein MA. Health risk assessment to fluoride in drinking water of rural residents living in the Poldasht city, northwest of Iran. *Ecotoxicol Environ Saf.* 2018;148:426–430. https://doi.org/10.1016/j.ecoenv.2017.10.057.
4. Dehghani MH, Omrani GA, Karri RR. Solid waste—sources, toxicity, and their consequences to human health. In: *Soft Computing Techniques in Solid Waste and Wastewater Management.* Elsevier; 2021:205–213.
5. Karri RR, Ravindran G, Dehghani MH. Wastewater—sources, toxicity, and their consequences to human health. In: *Soft Computing Techniques in Solid Waste and Wastewater Management.* Elsevier; 2021:3–33.
6. Asano T, Burton F, Leverenz H. *Water Reuse: Issues, Technologies, and Applications.* McGraw-Hill Education; 2007.
7. Cai Z, Dwivedi AD, Lee W-N, et al. Application of nanotechnologies for removing pharmaceutically active compounds from water: development and future trends. *Environ Sci Nano.* 2018;5(1):27–47. https://doi.org/10.1039/C7EN00644F.
8. Ioannou-Ttofa L, Michael-Kordatou I, Fattas SC, et al. Treatment efficiency and economic feasibility of biological oxidation, membrane filtration and separation processes, and advanced oxidation for the purification and valorization of olive mill wastewater. *Water Res.* 2017;114:1–13. https://doi.org/10.1016/j.watres.2017.02.020.
9. Mehmood A, Khan FSA, Mubarak NM, et al. Magnetic nanocomposites for sustainable water purification—a comprehensive review. *Environ Sci Pollut Res.* 2021;28(16):19563–19588. https://doi.org/10.1007/s11356-021-12589-3.

10. Khan FSA, Mubarak NM, Khalid M, et al. Magnetic nanoadsorbents' potential route for heavy metals removal—a review. *Environ Sci Pollut Res.* 2020;27(19):24342–24356. https://doi.org/10.1007/s11356-020-08711-6.

11. Khan FSA, Mubarak NM, Tan YH, et al. A comprehensive review on magnetic carbon nanotubes and carbon nanotube-based buckypaper for removal of heavy metals and dyes. *J Hazard Mater.* 2021;413. https://doi.org/10.1016/j.jhazmat.2021.125375, 125375.

12. Khan FSA, Mubarak NM, Khalid M, et al. A comprehensive review on micropollutants removal using carbon nanotubes-based adsorbents and membranes. *J Environ Chem Eng.* 2021;9(6):106647. https://doi.org/10.1016/j.jece.2021.106647.

13. Suhas CPJM, Carrott MMLR, Singh R, Singh LP, Chaudhary M. An innovative approach to develop microporous activated carbons in oxidising atmosphere. *J Clean Prod.* 2017;156:549–555.

14. Suhas CPJM, Ribeiro Carrott MML. Lignin - from natural adsorbent to activated carbon: a review. *Bioresour Technol.* 2007;98(12):2301–2312. https://doi.org/10.1016/j.biortech.2006.08.008.

15. Suhas GVK, PJM C, Singh R, Chaudhary M, Cellulose KS. A review as natural, modified and activated carbon adsorbent. *Bioresour Technol.* 2016;216:1066–1076. https://doi.org/10.1016/j.biortech.2016.05.106.

16. Gupta VK, Carrott PJM, Ribeiro Carrott MML, Suhas. Low-cost adsorbents: growing approach to wastewater treatment—a review. *Crit Rev Environ Sci Technol.* 2009;39(10):783–842. https://doi.org/10.1080/10643380801977610.

17. Gupta VK, Ali I, Suhas MD. Equilibrium uptake and sorption dynamics for the removal of a basic dye (basic red) using low-cost adsorbents. *J Colloid Interface Sci.* 2003;265(2):257–264. https://doi.org/10.1016/S0021-9797(03)00467-3.

18. Narayana PL, Lingamdinne LP, Karri RR, et al. Predictive capability evaluation and optimization of Pb(II) removal by reduced graphene oxide-based inverse spinel nickel ferrite nanocomposite. *Environ Res.* 2022;204. https://doi.org/10.1016/j.envres.2021.112029, 112029.

19. Lingamdinne LP, Choi JS, Angaru GKR, et al. Magnetic-watermelon rinds biochar for uranium-contaminated water treatment using an electromagnetic semi-batch column with removal mechanistic investigations. *Chemosphere.* 2022;286. https://doi.org/10.1016/j.chemosphere.2021.131776, 131776.

20. Chaudhary M, Suhas SR, et al. Microporous activated carbon as adsorbent for the removal of noxious anthraquinone acid dyes: role of adsorbate functionalization. *J Environ Chem Eng.* 2021;9(5):106308. https://doi.org/10.1016/j.jece.2021.106308.

21. Nabais JMV, Gomes JA, Suhas CPJM, Laginhas C, Roman S. Phenol removal onto novel activated carbons made from lignocellulosic precursors: influence of surface properties. *J Hazard Mater.* 2009;167(1):904–910. https://doi.org/10.1016/j.jhazmat.2009.01.075.

22. Gupta VK, Suhas. Application of low-cost adsorbents for dye removal – a review. *J Environ Manage.* 2009;90(8):2313–2342. https://doi.org/10.1016/j.jenvman.2008.11.017.

23. Rathi BS, Kumar PS. Application of adsorption process for effective removal of emerging contaminants from water and wastewater. *Environ Pollut.* 2021;280. https://doi.org/10.1016/j.envpol.2021.116995, 116995.

24. Jiang Y, Liu Z, Zeng G, et al. Polyaniline-based adsorbents for removal of hexavalent chromium from aqueous solution: a mini review. *Environ Sci Pollut Res.* 2018;25(7):6158–6174. https://doi.org/10.1007/s11356-017-1188-3.

25. Mehmood A, Khan FSA, Mubarak NM, et al. Carbon and polymer-based magnetic nanocomposites for oil-spill remediation—a comprehensive review. *Environ Sci Pollut Res.* 2021;28(39):54477–54496. https://doi.org/10.1007/s11356-021-16045-0.

26. Khan FSA, Mubarak NM, Tan YH, et al. Magnetic nanoparticles incorporation into different substrates for dyes and heavy metals removal—a review. *Environ Sci Pollut Res.* 2020;27(35):43526–43541. https://doi.org/10.1007/s11356-020-10482-z.

27. Yee MJ, Mubarak NM, Abdullah EC, et al. Carbon nanomaterials based films for strain sensing application—a review. *Nano-Struct Nano-Objects.* 2019;18. https://doi.org/10.1016/j.nanoso.2019.100312, 100312.

28. Sahu JN, Zabed H, Karri RR, Shams S, Qi X. Applications of nano-biotechnology for sustainable water purification. In: *Industrial Applications of Nanomaterials*; 2019:313–340.

29. Lazaridis NK, Bakoyannakis DN, Deliyanni EA. Chromium(VI) sorptive removal from aqueous solutions by nanocrystalline akaganÄite. *Chemosphere.* 2005;58(1):65–73.

30. Deliyanni EA, Peleka EN, Matis KA. Removal of zinc ion from water by sorption onto iron-based nanoadsorbent. *J Hazard Mater.* 2007;141(1):176–184. https://doi.org/10.1016/j.jhazmat.2006.06.105.

31. Deliyanni EA, Bakoyannakis DN, Zouboulis AI, Matis KA. Sorption of As(V) ions by akaganÃ©ite-type nanocrystals. *Chemosphere.* 2003;50(1):155–163.

32. Deliyanni EA, Matis KA. Sorption of Cd ions onto akaganÃ©ite-type nanocrystals. *Sep Purif Technol.* 2005;45(2):96–102.

33. Deliyanni EA, Peleka EN, Lazaridis NK. Comparative study of phosphates removal from aqueous solutions by nanocrystalline akaganéite and hybrid surfactant-akaganéite. *Sep Purif Technol.* 2007;52(3):478–486. https://doi.org/10.1016/j.seppur.2006.05.028.

34. Sharma YC, Srivastava V, Upadhyay SN, Weng CH. Alumina nanoparticles for the removal of Ni(II) from aqueous solutions. *Ind Eng Chem Res.* 2008;47(21):8095–8100.

35. Hao YM, Man C, Hu ZB. Effective removal of Cu (II) ions from aqueous solution by amino-functionalized magnetic nanoparticles. *J Hazard Mater.* 2010;184(1–3):392–399.

36. Di Z-C, Ding J, Peng X-J, Li Y-H, Luan Z-K, Liang J. Chromium adsorption by aligned carbon nanotubes supported ceria nanoparticles. *Chemosphere.* 2006;62(5):861–865. https://doi.org/10.1016/j.chemosphere.2004.06.044.

37. Du WL, Xu ZR, Han XY, Xu YL, Miao ZG. Preparation, characterization and adsorption properties of chitosan nanoparticles for eosin Y as a model anionic dye. *J Hazard Mater.* 2008;153(1–2):152–156.

38. Fares MM, Al-Rub FAA, Mohammad AR. Ultimate eradication of the ciprofloxacin antibiotic from the ecosystem by Nanohybrid GO/O-CNTs. *ACS Omega.* 2020;5(9):4457–4468. https://doi.org/10.1021/acsomega.9b03636.

39. Doong R-A, Chiang L-F. Coupled removal of organic compounds and heavy metals by titanate/carbon nanotube composites. *Water Sci Technol J Int Assoc Water Pollut Res.* 2008;58(10):1985–1992.

40. Banerjee SS, Chen DH. Fast removal of copper ions by gum arabic modified magnetic nano-adsorbent. *J Hazard Mater.* 2007;147(3):792–799. https://doi.org/10.1016/j.jhazmat.2007.01.079.

41. Pirilä M, Martikainen M, Ainassaari K, Kuokkanen T, Keiski RL. Removal of aqueous As(III) and As(V) by hydrous titanium dioxide. *J Colloid Interface Sci.* 2011;353(1):257–262. https://doi.org/10.1016/j.jcis.2010.09.020.

42. Hu J, Chen G, Lo IM. Selective removal of heavy metals from industrial wastewater using maghemite nanoparticle: performance and mechanisms. *J Environ Eng.* 2006;132(7):709–715.

43. Parvin N, Babapoor A, Nematollahzadeh A, Mousavi SM. Removal of phenol and β-naphthol from aqueous solution by decorated graphene oxide with magnetic iron for modified polyrhodanine as nanocomposite adsorbents: kinetic, equilibrium and thermodynamic studies. *React Funct Polym.* 2020;156. https://doi.org/10.1016/j.reactfunctpolym.2020.104718, 104718.

44. Hu J, Lo IMC, Chen G. Fast removal and recovery of Cr(VI) using surface-modified jacobsite (MnFe2O4) nanoparticles. *Langmuir.* 2005;21(24): 11173–11179.

45. Machado FM, Bergmann CP, Fernandes THM, et al. Adsorption of reactive red M-2BE dye from water solutions by multi-walled carbon nanotubes and activated carbon. *J Hazard Mater.* 2011;192(3):1122–1131.

46. Boparai HK, Joseph M, O'Carroll DM. Kinetics and thermodynamics of cadmium ion removal by adsorption onto nano zerovalent iron particles. *J Hazard Mater.* 2011;186(1):458–465. https://doi.org/10.1016/j.jhazmat.2010.11.029.

47. Lu C, Chiu H. Adsorption of zinc(II) from water with purified carbon nanotubes. *Chem Eng Sci.* 2006;61(4):1138–1145. https://doi.org/10.1016/j.ces.2005.08.007.

48. Pishnamazi M, Khan A, Kurniawan TA, Sanaeepur H, Albadarin AB, Soltani R. Adsorption of dyes on multifunctionalized nano-silica KCC-1. *J Mol Liq.* 2021;338. https://doi.org/10.1016/j.molliq.2021.116573, 116573.

49. Lou JC, Jung MJ, Yang HW, Han JY, Huang WH. Removal of dissolved organic matter (DOM) from raw water by single-walled carbon nanotubes (SWCNTs). *J Environ Sci Health A Tox Hazard Subst Environ Eng.* 2011;46(12):1357–1365.

50. Belessi V, Romanos G, Boukos N, Lambropoulou D, Trapalis C. Removal of reactive red 195 from aqueous solutions by adsorption on the surface of TiO2 nanoparticles. *J Hazard Mater.* 2009;170(2–3):836–844. https://doi.org/10.1016/j.jhazmat.2009.05.045.

51. Zhu H, Jia Y, Wu X, Wang H. Removal of arsenic from water by supported nano zero-valent iron on activated carbon. *J Hazard Mater.* 2009;172 (2):1591–1596. https://doi.org/10.1016/j.jhazmat.2009.08.031.

52. Rebekah A, Bharath G, Naushad M, Viswanathan C, Ponpandian N. Magnetic graphene/chitosan nanocomposite: a promising nano-adsorbent for the removal of 2-naphthol from aqueous solution and their kinetic studies. *Int J Biol Macromol.* 2020;159:530–538. https://doi.org/10.1016/j.ijbiomac.2020.05.113.

53. Badran I, Khalaf R. Adsorptive removal of alizarin dye from wastewater using maghemite nanoadsorbents. *Sep Sci Technol.* 2020;55(14):2433–2448. https://doi.org/10.1080/01496395.2019.1634731.

54. Banerjee S, Dubey S, Gautam RK, Chattopadhyaya MC, Sharma YC. Adsorption characteristics of alumina nanoparticles for the removal of hazardous dye, Orange G from aqueous solutions. *Arab J Chem.* 2019;12(8):5339–5354. https://doi.org/10.1016/j.arabjc.2016.12.016.

55. Santhosh C, Velmurugan V, Jacob G, Jeong SK, Grace AN, Bhatnagar A. Role of nanomaterials in water treatment applications: a review. *Chem Eng J.* 2016;306:1116–1137. https://doi.org/10.1016/j.cej.2016.08.053.

56. El-sayed MEA. Nanoadsorbents for water and wastewater remediation. *Sci Total Environ.* 2020;739. https://doi.org/10.1016/j.scitotenv.2020.139903, 139903.

57. Ighalo JO, Sagboye PA, Umenweke G, et al. CuO nanoparticles (CuO NPs) for water treatment: a review of recent advances. *Environ Nanotechnol Monit Manag.* 2021;15. https://doi.org/10.1016/j.enmm.2021.100443, 100443.

58. Khan I, Saeed K, Khan I. Nanoparticles: properties, applications and toxicities. *Arab J Chem.* 2019;12(7):908–931. https://doi.org/10.1016/j.arabjc.2017.05.011.

59. Gusain R, Kumar N, Ray SS. Recent advances in carbon nanomaterial-based adsorbents for water purification. *Coord Chem Rev.* 2020;405:213111. https://doi.org/10.1016/j.ccr.2019.213111.

60. Khan FSA, Mubarak NM, Yie HT, Nizamuddin S. Magnetic nanoparticles incorporation into different substrates for dyes and heavy metals removal—a review. *Environ Sci Pollut Res Int.* 2020;27(35):43526–43541.

61. Khan FSA, Mubarak NM, Khalid M, et al. Magnetic nanoadsorbents' potential route for heavy metals removal—a review. *Environ Sci Pollut Res.* 2020;27(19):24342–24356.

62. Bui TX, Choi H. Adsorptive removal of selected pharmaceuticals by mesoporous silica SBA-15. *J Hazard Mater.* 2009;168(2):602–608. https://doi.org/10.1016/j.jhazmat.2009.02.072.

63. Rovani S, Santos JJ, Corio P, Fungaro DA. Highly pure silica nanoparticles with high adsorption capacity obtained from sugarcane waste ash. *ACS Omega.* 2018;3(3):2618–2627. https://doi.org/10.1021/acsomega.8b00092.

64. Gao S, Zhang W, An Z, Kong S, Chen D. Adsorption of anionic dye onto magnetic Fe$_3$O$_4$/CeO$_2$ nanocomposite: equilibrium, kinetics, and thermodynamics. *Adsorpt Sci Technol.* 2019;37(3–4):185–204. https://doi.org/10.1177/0263617418819164.

65. Mashkoor F, Nasar A, Inamuddin. Carbon nanotube-based adsorbents for the removal of dyes from waters: a review. *Environ Chem Lett.* 2020;18 (3):605–629. https://doi.org/10.1007/s10311-020-00970-6.

66. Braunschweig J, Bosch J, Meckenstock RU. Iron oxide nanoparticles in geomicrobiology: from biogeochemistry to bioremediation. *N Biotechnol.* 2013;30(6):793–802. https://doi.org/10.1016/j.nbt.2013.03.008.

67. Weber KA, Achenbach LA, Coates JD. Microorganisms pumping iron: anaerobic microbial iron oxidation and reduction. *Nat Rev Microbiol.* 2006;4 (10):752–764. https://doi.org/10.1038/nrmicro1490.

68. Waychunas GA, Kim CS, Banfield JF. Nanoparticulate iron oxide minerals in soils and sediments: unique properties and contaminant scavenging mechanisms. *J Nanopart Res.* 2005;7(4):409–433. https://doi.org/10.1007/s11051-005-6931-x.

69. Dragar Č, Kralj S, Kocbek P. Bioevaluation methods for iron-oxide-based magnetic nanoparticles. *Int J Pharm.* 2021;597. https://doi.org/10.1016/j.ijpharm.2021.120348, 120348.

70. Rehman AU, Nazir S, Irshad R, et al. Toxicity of heavy metals in plants and animals and their uptake by magnetic iron oxide nanoparticles. *J Mol Liq.* 2021;321. https://doi.org/10.1016/j.molliq.2020.114455, 114455.

71. Nizamuddin S, Siddiqui MTH, Mubarak NM, et al. Chapter 17: Iron oxide nanomaterials for the removal of heavy metals and dyes from wastewater. In: Thomas S, Pasquini D, Leu S-Y, Gopakumar DA, eds. *Nanoscale Materials in Water Purification.* Elsevier; 2019:447–472.

72. Deliyanni EA, Kyzas GZ, Matis KA. 14 - Inorganic nanoadsorbent: akaganéite in wastewater treatment. In: Kyzas GZ, Mitropoulos AC, eds. *Composite Nanoadsorbents.* Elsevier; 2019:337–358.

73. Wang K-S, Lin C-L, Wei M-C, et al. Effects of dissolved oxygen on dye removal by zero-valent iron. *J Hazard Mater.* 2010;182(1):886–895. https://doi.org/10.1016/j.jhazmat.2010.07.002.

74. Dehghani MH, Karri RR, Alimohammadi M, Nazmara S, Zarei A, Saeedi Z. Insights into endocrine-disrupting bisphenol-a adsorption from pharmaceutical effluent by chitosan immobilized nanoscale zero-valent iron nanoparticles. *J Mol Liq.* 2020;31. https://doi.org/10.1016/j.molliq.2020.113317, 1113317.

75. Sahu N, Rawat S, Singh J, et al. Process optimization and modeling of methylene blue adsorption using zero-valent iron nanoparticles synthesized from sweet lime pulp. *Appl Sci (Switz).* 2019;9(23). https://doi.org/10.3390/app9235112, 5112.

76. Deliyanni EA, Bakoyannakis DN, Zouboulis AI, Matis KA. Sorption of As(V) ions by akaganéite-type nanocrystals. *Chemosphere.* 2003;50(1):155–163. https://doi.org/10.1016/s0045-6535(02)00351-x.

77. Deliyanni EA, Matis KA. Sorption of cd ions onto akaganéite-type nanocrystals. *Sep Purif Technol.* 2005;45(2):96–102. https://doi.org/10.1016/j.seppur.2005.02.012.

78. Lazaridis NK, Bakoyannakis DN, Deliyanni EA. Chromium(VI) sorptive removal from aqueous solutions by nanocrystalline akaganèite. *Chemosphere.* 2005;58(1):65–73. https://doi.org/10.1016/j.chemosphere.2004.09.007.

79. Kanel SR, Manning B, Charlet L, Choi H. Removal of arsenic(III) from groundwater by nanoscale zero-valent iron. *Environ Sci Tech.* 2005;39(5):1291–1298.

80. Lingamdinne LP, Choi JS, Choi YL, et al. Process modeling and optimization of an iron oxide immobilized graphene oxide gadolinium nanocomposite for arsenic adsorption. *J Mol Liq.* 2020;299. https://doi.org/10.1016/j.molliq.2019.112261, 112261.

81. Karri RR, Sahu JN, Meikap BC. Improving efficacy of Cr (VI) adsorption process on sustainable adsorbent derived from waste biomass (sugarcane bagasse) with help of ant colony optimization. *Ind Crop Prod.* 2020;143. https://doi.org/10.1016/j.indcrop.2019.111927, 111927.

82. Chowdhury SR, Yanful EK. Arsenic and chromium removal by mixed magnetite-maghemite nanoparticles and the effect of phosphate on removal. *J Environ Manage.* 2010;91(11):2238–2247. https://doi.org/10.1016/j.jenvman.2010.06.003.

83. Hu J, Lo IMC, Chen G. Performance and mechanism of chromate (VI) adsorption by Î´-FeOOH-coated maghemite (Î³-Fe2O3) nanoparticles. *Sep Purif Technol.* 2007;58(1):76–82.

84. Huang S-H, Chen D-H. Rapid removal of heavy metal cations and anions from aqueous solutions by an amino-functionalized magnetic nanoadsorbent. *J Hazard Mater.* 2009;163(1):174–179. https://doi.org/10.1016/j.jhazmat.2008.06.075.

85. Liu JF, Zhao ZS, Jiang GB. Coating Fe3O4 magnetic nanoparticles with humic acid for high efficient removal of heavy metals in water. *Environ Sci Tech.* 2008;42(18):6949–6954.

86. Mitra T, Bar N, Das SK. Rice husk: green adsorbent for Pb (II) and Cr (VI) removal from aqueous solution—column study and GA–NN modeling. *SN Appl Sci.* 2019;1(5):486.

87. Alabi O, Olanrewaju AA, Afolabi TJ. Process optimization of adsorption of Cr (VI) on adsorbent prepared from *Bauhinia rufescens* pod by box-Behnken design. *Sep Sci Technol.* 2020;55:47–60.

88. Ponder SM, Darab JG, Mallouk TE. Remediation of Cr(VI) and Pb(II) aqueous solutions using supported, nanoscale zero-valent iron. *Environ Sci Tech.* 2000;34(12):2564–2569.

89. Niu S-F, Liu Y, Xu X-H, Lou Z-H. Removal of hexavalent chromium from aqueous solution by iron nanoparticles. *J Zhejiang Univ Sci B.* 2005;6(10):1022–1027. https://doi.org/10.1631/jzus.2005.B1022.

90. Xiao Z, Yuan M, Yang B, Liu Z, Huang J, Sun D. Plant-mediated synthesis of highly active iron nanoparticles for Cr (VI) removal: investigation of the leading biomolecules. *Chemosphere.* 2016;150:357–364. https://doi.org/10.1016/j.chemosphere.2016.02.056.

91. Pan B, Qiu H, Pan B, et al. Highly efficient removal of heavy metals by polymer-supported nanosized hydrated Fe(III) oxides: behavior and XPS study. *Water Res.* 2010;44(3):815–824. https://doi.org/10.1016/j.watres.2009.10.027.

92. Banerjee S, Gautam RK, Jaiswal A, Chattopadhyaya MC, Sharma YC. Rapid scavenging of methylene blue dye from a liquid phase by adsorption on alumina nanoparticles. *RSC Adv.* 2015;5(19):14425–14440.

93. Mahmoud ME, Abdou AEH, Shehata AK, Header HM, Hamed EA. Surface functionalized γ-alumina nanoparticles with N-cetyl-N,N,N-trimethyl ammonium bromide for adsorptive interaction with 2-nitrobenzoic and 4-nitrobenzoic acids. *J Mol Liq.* 2017;242:1248–1262. https://doi.org/10.1016/j.molliq.2017.07.102.

94. Herrera-Barros A, Tejada-Tovar C, Villabona-Ortíz A, González-Delgado AD, Benitez-Monroy J. Cd (II) and Ni (II) uptake by novel biosorbent prepared from oil palm residual biomass and Al2O3 nanoparticles. *Sustain Chem Pharm.* 2020;15. https://doi.org/10.1016/j.scp.2020.100216, 100216.

95. Ismaeel IN, Wahab HS. Adsorption of eriochrom black T azo dye onto nanosized anatase TiO_2. *Am J Environ Eng Sci*. 2015;2(6):86–92.

96. George R, Bahadur N, Singh N, Singh R, Verma A, Shukla AK. Environmentally benign TiO_2 nanomaterials for removal of heavy metal ions with interfering ions present in tap water. *Mater Today: Proc*. 2016;3(2):162–166. https://doi.org/10.1016/j.matpr.2016.01.051.

97. Andjelkovic I, Jovic B, Jovic M, et al. Microwave-hydrothermal method for the synthesis of composite materials for removal of arsenic from water. *Environ Sci Pollut Res*. 2016;23(1):469–476. https://doi.org/10.1007/s11356-015-5283-z.

98. Zhao M, Tang Z, Liu P. Removal of methylene blue from aqueous solution with silica nano-sheets derived from vermiculite. *J Hazard Mater*. 2008;158(1):43–51.

99. Rostamian R, Najafi M, Rafati AA. Synthesis and characterization of thiol-functionalized silica nano hollow sphere as a novel adsorbent for removal of poisonous heavy metal ions from water: kinetics, isotherms and error analysis. *Chem Eng J*. 2011;171(3):1004–1011. https://doi.org/10.1016/j.cej.2011.04.051.

100. Mohamad DF, Osman NS, Nazri MKHM, et al. Synthesis of mesoporous silica nanoparticle from banana peel ash for removal of phenol and methyl orange in aqueous solution. *Mater Today: Proc*. 2019;19:1119–1125. https://doi.org/10.1016/j.matpr.2019.11.004.

101. Saleh S, Younis A, Ali R, Elkady E. Phenol removal from aqueous solution using amino modified silica nanoparticles. *Korean J Chem Eng*. 2019;36(4):529–539. https://doi.org/10.1007/s11814-018-0217-3.

102. Li Y, Lu JJ, Shi BY, Wu YY. Sorption of atrazine onto nano-SiO_2 and nano-kaolin particles. *Huanjing Kexue/Environ Sci*. 2008;29(6):1687–1692.

103. Kotsyuda SS, Tomina VV, Zub YL, et al. Bifunctional silica nanospheres with 3-aminopropyl and phenyl groups. Synthesis approach and prospects of their applications. *Appl Surf Sci*. 2017;420:782–791. https://doi.org/10.1016/j.apsusc.2017.05.150.

104. Xiong Chang X, Mujawar Mubarak N, Ali Mazari S, et al. A review on the properties and applications of chitosan, cellulose and deep eutectic solvent in green chemistry. *J Ind Eng Chem*. 2021;104:362–380. https://doi.org/10.1016/j.jiec.2021.08.033.

105. Rasoulzadeh H, Dehghani MH, Mohammadi AS, et al. Parametric modelling of Pb (II) adsorption onto chitosan-coated Fe_3O_4 particles through RSM and DE hybrid evolutionary optimization framework. *J Mol Liq*. 2019;297. https://doi.org/10.1016/j.molliq.2019.111893, 111893.

106. Chang YC, Chang SW, Chen DH. Magnetic chitosan nanoparticles: studies on chitosan binding and adsorption of Co(II) ions. *React Funct Polym*. 2006;66(3):335–341. https://doi.org/10.1016/j.reactfunctpolym.2005.08.006.

107. Du WL, Xu ZR, Han XY, Xu YL, Miao ZG. Preparation, characterization and adsorption properties of chitosan nanoparticles for eosin Y as a model anionic dye. *J Hazard Mater*. 2008;153(1):152–156. https://doi.org/10.1016/j.jhazmat.2007.08.040.

108. Iijima S. Helical microtubules of graphitic carbon. *Nature*. 1991;354(6348):56–58. https://doi.org/10.1038/354056a0.

109. Khan FSA, Mubarak NM, Khalid M, et al. Functionalized multi-walled carbon nanotubes and hydroxyapatite nanorods reinforced with polypropylene for biomedical application. *Sci Rep*. 2021;11(1):843. https://doi.org/10.1038/s41598-020-80767-3.

110. Khan FSA, Mubarak NM, Khalid M, et al. Comprehensive review on carbon nanotubes embedded in different metal and polymer matrix: fabrications and applications. *Crit Rev Solid State Mater Sci*. 2021. https://doi.org/10.1080/10408436.2021.1935713.

111. Magendran SS, Khan FSA, Mubarak NM, et al. Synthesis of organic phase change materials by using carbon nanotubes as filler material. *Nano-Struct Nano-Objects*. 2019;19. https://doi.org/10.1016/j.nanoso.2019.100361, 100361.

112. Hu J, Shao D, Chen C, et al. Plasma-induced grafting of cyclodextrin onto multiwall carbon nanotube/iron oxides for adsorbent application. *J Phys Chem B*. 2010;114(20):6779–6785.

113. Xu L, Li J, Zhang M. Adsorption characteristics of a novel carbon-nanotube-based composite adsorbent toward organic pollutants. *Ind Eng Chem Res*. 2015;54(8):2379–2384. https://doi.org/10.1021/ie5041379.

114. Yang G, Li Y, Yang S, et al. Surface oxidized nano-cobalt wrapped by nitrogen-doped carbon nanotubes for efficient purification of organic wastewater. *Sep Purif Technol*. 2021;259. https://doi.org/10.1016/j.seppur.2020.118098, 118098.

115. De Luca P, Siciliano C, Macario A, Nagy JB. The role of carbon nanotube pretreatments in the adsorption of benzoic acid. *Materials (Basel)*. 2021;14(9):2118. https://doi.org/10.3390/ma14092118.

116. Zhang Q, Pan B, Zhang W, Jia K. Selective sorption of lead, cadmium and zinc ions by a polymeric cation exchanger containing nano-Zr(HPO3S)2. *Environ Sci Tech*. 2008;42(11):4140–4145.

117. Soltani R, Pelalak R, Pishnamazi M, et al. A novel and facile green synthesis method to prepare LDH/MOF nanocomposite for removal of Cd(II) and Pb(II). *Sci Rep*. 2021;11(1):1609. https://doi.org/10.1038/s41598-021-81095-w.

118. Soliman NK, Moustafa AF, El-Mageed HRA, et al. Experimentally and theoretically approaches for disperse red 60 dye adsorption on novel quaternary nanocomposites. *Sci Rep*. 2021;11(1):10000. https://doi.org/10.1038/s41598-021-89351-9.

Chapter 8

Industrial wastes as novel adsorbents for the removal of toxic impurities from wastewater

Prerona Roy and Md. Ahmaruzzaman[*]

Department of Chemistry, National Institute of Technology, Silchar, Assam, India
[*]*Corresponding author.*

1 Introduction

Water contamination has become a great threat to our natural environment over the past few decades. This has led to a drastic reduction in freshwater availability for human consumption in the 21st century. Rapid industrialization has catalyzed water pollution to a great extent lately. Many inorganic and organic pollutants[1] are found to contaminate water which include dyes, heavy metal ions, phenolic compounds, and other contaminants such as pharmaceutical wastes and radioactive wastes. These pollutants are released from several sources such as industries, mining, textile, tanneries, pigment and paint manufacturing, photographic and printing industries, herbicides, pesticides, fertilizers, and many more.[2–4] These contaminants are found to severely damage many human organs such as kidneys, lungs, central nervous system, liver, pancreas; besides, they also affect the aquatic life residing inside the waterbodies.[5,6] Therefore, treatment of water has become an alarming need.

Many methods are already in use for conventional treatment of wastewater which include precipitation,[7] coagulation,[8] reverse osmosis,[9] flocculation,[10] radiation therapy,[11] photocatalysis,[12] biological processes,[13] and adsorption.[14–16] Among these, adsorption is preferred to the rest of them as it is relatively low-cost, eco-friendly, highly efficient,[17] easily available, and operatable due to simple instrumentation.

Conventionally used adsorbents[18] are relatively less cost-effective; hence, innovation is brought about by the use of modified industrial wastes as efficient adsorbents.[19,20] Industrial waste products are not only low cost but also available in abundance.[21,22] This makes the adsorption process a lot more affordable, saving the cost of processing of conventionally used adsorbents such as activated carbon,[23,24] metal-organic framework,[25,26] biochar, hydrogels, and many more. Recently these industrial wastes have been modified in various ways, such as incorporating various nanocomposites,[27–30] grafting, and acid/base treatment, which results in enhancement in the adsorbent properties such as having greater pore size,[31–34] higher surface area, and greater porosity. This chapter gives a summary of the different efficient, novel adsorbents produced from a variety of industrial effluents used for wastewater treatment along with their limitations and challenges.

2 Different types of industrial wastes

Industrial effluents are high potential candidates for the formation of novel low-cost adsorbents that can be utilized in wastewater treatment. These are conventionally produced as by-products of various industrial processes and need some processing for enhancement of their adsorption capacities. They are low cost as they are easily available in bulk quantities. Different categories of industrial effluents that can be utilized as efficient adsorbents are fly ash (FA), lignin (LG), blast furnace slurry and sludge, red mud (RM), waste hydroxides such as ferric hydroxide, and other miscellaneous waste products such as Areca waste, residual slurry from waste biogas, residue from sea nodule, coffee husks, waste from tea and olive-oil factory, waste from battery industry, and grape stalk. The utilization of these industrial effluents as novel adsorbents for the removal of toxic impurities from wastewater is discussed below.

2.1 Fly ash (FA)

Thermal power plants heat coal at 2000°C to produce electricity and release a solid waste called FA. Annually, approximately FA weighing up to 600 million tonnes are produced by different powerplants globally.[35] The structure of the FA produced depends on its precursor coal burned. Its basic substituents are silica (Si), hematite (α-Fe_2O_3), calcium (Ca), alumina (Al_2O_3), titanium oxides (TiO_2), potassium (K), sodium (Na), and magnesium (Mg). Some metals are also responsible for its toxicity such as As, Hg, Co, Cd, Pb, Cr, Cu, and a few other heavy metals.[4] FA is a pulverized material mainly composed of fine particles that rise alongside the flue gases in the blast furnace. These particles are amorphous, porous, and spherical in nature.[36] FA can be produced from mainly burning four types of coal, which include anthracite, lignite, bituminous, and sub-bituminous[37] Among these, the anthracite coal is the purest category of coal used, while the others have varying concentrations of silica, magnetite, MgO, and CaO. The size of the particles of bituminous coal FA is smaller than 0.075 nm, while that of sub-bituminous FA is larger and rougher than that. FA is greyish white in color and its blackish hue is owed to its carbon content. The specific gravity of FA varies from 2.1 to 3.0, while its surface area may vary between a wide range of 170–1000 m^2/kg.

The FA demand of 39,548.1 million US dollars in the year 2015 is expected to rise to 64,761.9 million US dollars by the year 2022.[38] The rise has estimated a 7.3% value from the year 2016 to the upcoming year 2022 as depicted by Fig. 1.

FA acts an efficient and inexpensive adsorbent for the removal of various toxic impurities from water. This is especially favored by its charge, morphology, and its surface structure. The efficiency of FA-based adsorbents can be improved by altering the surface charge and also by using larger surface area of the adsorbent. Applications of FA in wastewater treatment via removal of toxic impurities is discussed below.[39]

2.1.1 Removal of heavy metal ions

Heavy metal ions such as Hg^{2+}, Cd^{2+}, Pb^{2+}, Zn^{2+}, Cu^{2+}, Cr^{6+}, Mn^{2+}, Ni^{2+}, As^{5+}, and many more contaminate water-bodies.[40,41] These are potent attackers to different human organs such as central nervous system, kidneys, lungs, heart, stomach, and small and large intestines on consumption of the contaminated water.[42] These also affect the aquatic life present inside the waterbodies severely. Hence, the removal of such metal ions from the waterbodies via wastewater treatment has become the need of the hour. Many adsorbents have been used for the removal of such metal ions, among which currently modified FA-based adsorbents have showed commendable efficiency and are pocket friendly. Examples of a few such adsorbents are elaborated below.

The key to low-cost production starts with using cost-effective synthetic reagents and less instrumentation in the synthetic process. Also, the production of hazard-free products counts as they cause least collateral expenses due to minimum environmental impacts. On this ground, Mondal et al.[43] synthesized efficient adsorbent with minimum expense from high Ca FA materials activated with alkali which removed Zn^{2+} ions from aqueous solution. They reported that such adsorbents are absolutely low-cost, hazard-free and showed an outstanding removal efficiency of 100% in the initial 5 min, and an adsorption capacity of 200 mg/g. Also, FA/magnetite materials[44] have been used to remove Cu^{2+} ions from wastewater. The novel adsorbent has been characterized and studied via XRD, EDAX (energy-dispersive X-ray analysis), FTIR,

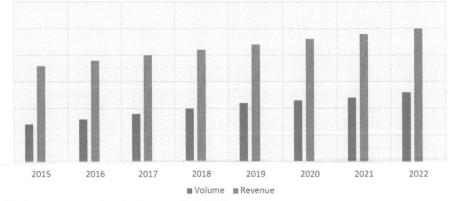

FIG. 1 Worldwide FA Growth and Annual Market Revenue (2015–2022).

SEM (scanning electron microscopy), BET, and VSM (value stream mapping) methods. Batch experiments concluded that the adsorbent showed a maximum adsorption of 17.39 mg/g. The sorption process was of pseudo-second-order kinetics and best fitted more than one adsorption isotherm. It was found to be relatively more efficient and low cost as compared to raw FA.

Zeolites are appreciably good adsorbents often used for wastewater treatment. A group of researchers recently showed efficient removal of Pb^{2+} ions from wastewater by K-F and Li-ABW (Barrer and White) zeolites along with phillipsites. These adsorbents were synthesized by hydrothermal treatment of circulated-fluid-bed FA-based geopolymers.[45] The removal process followed the pseudo-second-order kinetics and Langmuir adsorption isotherm model, and also displayed a remarkable adsorption capability of 252.70, 239.50, and 160.70 mg/g for phillipsite, K-F and Li-ABW zeolites respectively.

In the above studies, we find the removal of heavy metals from a single component adsorption system, but real wastewater is a multi-component adsorption system. It contains more than one heavy metal ion as impurity, and thus, further investigation based on multi-component wastewater system was done by various scientists to find efficient low-cost, novel adsorbents for heavy metal ion removal, taking a more practical application-based approach.

In this attempt, a comparative study has been conducted between pure zeolite and its novel carbon composite for the removal of heavy metal ions Zn^{2+} and Pb^{2+}. Both of the adsorbents were FA-derived.[46] They were characterized via SEM-EDS (scanning electron microscopy-energy-dispersive spectroscopy), XRD, particle size, analysis of elemental composition, and N-adsorption/desorption. The experiments were carried out at pH = 5.0. The authors reported that the pure zeolite proved to be a better adsorbent compared to its carbon composite, and the values of maximum adsorption capacities were Pb^{2+} = 314 and 575 mg/g and Zn^{2+} = 600 and 656 mg/g for the zeolite-C composite and the pure zeolite, respectively.

In 2020, another group of researchers[47] synthesized modified FA by treating FA with the surfactants sodium dodecyl sulfate and 2-mercaptobenzothiazole followed by 1 (M) NaOH. This adsorbent removed Hg^{2+} and Cd^{2+} ions from contaminated water. Fig. 2 shows the FESEM (field emission scanning electron microscope) images of the FA0 (untreated FA), FAN (NaOH-treated FA), FAMBT (2-mercaptobenzothiazole-treated FAN), and FASDS (sodium dodecyl sulfate modified FAN) particles indicating their size parameters pre and post functionalization. These sorbents displayed Langmuir adsorption isotherm for Cd^{2+} ions, while the Freundlich isotherm for the Hg^{2+} ion removal. The authors reported an adsorption capacity of 1.91 and 12.4 mg/g for Hg^{2+} and Cd^{2+} ions, respectively.

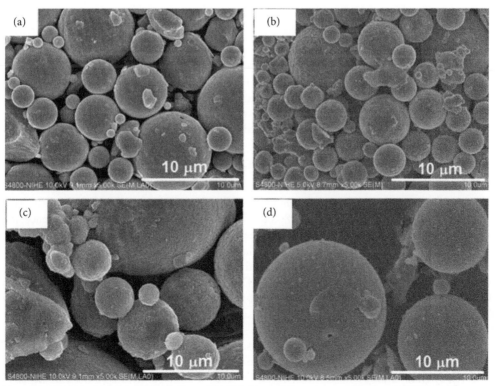

FIG. 2 FESEM images of FA0 (A), FAN (B), FAMBT (C), and FASDS (D) particles at the magnification of 5.000 times.[47]

TABLE 1 Adsorption of various heavy metal ions on FA and modified FA-based adsorbents.

Adsorbent	Metals	Reaction temperature (°C)	Adsorption capacity (mg/g)	Reference
FA	Pb^{2+}	25	444.7	[48]
Bagasse FA	Pb^{2+}	30–50	285–566	[49]
Acid-treated FA (AFA)	Pb^{2+}	25	437	[48]
Rice husk FA	Zn^{2+}	–	14.30	[50]
Bagasse FA	Zn^{2+}	30	13.21	[51]
Fe-encapsulated FA	Zn^{2+}	30–60	7.5–15.5	[52]
AFA	Cu^{2+}	25	198.5	[48]
Washed FA	Cu^{2+}	25	205.8	[48]
CFA	Cu^{2+}	30–60	178.5–249.1	[53]
FA	Cd^{2+}	–	207.3	[48]
AFA	Cd^{2+}	25	180.4	[48]
Washed FA	Cd^{2+}	25	195.2	[48]
Al-encapsulated FA	Cr^{6+}	30–60	1.67	[54]
Fe-encapsulated FA	Cr^{6+}	30–60	1.82	[54]
FA	Cr^{6+}	30–60	1.38	[54]
FA coal-char	As^{5+}	25	0.02–34.5	[55]
FA	As^{5+}	20	7.7–27.8	[56]
Al-encapsulated FA	Hg^{2+}	30–60	13.4	[54]
FA	Hg^{2+}	30–60	11.0	[54]
FA-C	Hg^{2+}	5–21	0.63–0.73	[57]

Table 1 depicts the experimental results of adsorption of various metals on FA and modified FA-based adsorbents. From the above table, we can conclude that adsorption capacity of metal (Fe and Al) impregnated FA, AFA and washed FA was reported to be higher than that of untreated FA for a given metal ion. Also, FA showed the maximum sorption capacity for Pb^{2+} among all other mentioned heavy metal ions, which was reported to be 444.7 mg/g at 25°C. The lowest sorption by FA was reported by Pattanayak et al.[55], where As^{5+} was adsorbed on the surface of FA coal-char with a sorption capacity that varied within the range of 0.02–34.5 mg/g.

2.1.2 Removal of dyes

There are about 10,000 dyes which are used in different industries, especially textile industries currently. The global annual discharge of dyes from different textile industries is 280,000 t.[39] Dyes are found to be carcinogenic, hence, they are not only detrimental to human health, but also have serious environmental hazards such as abolition of various aquatic ecosystems and deterioration in groundwater supplies. Thus, researchers have also been working on methods to free wastewater of toxic dyes via a number of cost-effective methods. A few FA-derived adsorbents used for dye removal are discussed below.

A recent study showed that hazardous dye direct blue 78 (DB78) was removed from wastewater by using raw CFA (coal fly ash) as adsorbent.[58] The batch experiments showed that an outstanding maximum removal efficiency of 98.4% was attained at 20 mg/g initial dye concentration and 4 g/L FA dosage. The sorption process was reported to better fit the Langmuir isotherm model than the Freundlich model.

In general, more experiments have been conducted by researchers by modifying FA, for an enhanced efficiency of the adsorption process. A novel ferro Schiff's base biocomposite was synthesized[59] by composting of ferro chitosan (CS) and FA particles followed by cross-linking with glyoxal (Gly) leading to formation of Schiff's base. This CS-Gly/FA/magnetite composite removed anionic reactive orange 16 dye from textile wastewater. It showed a saturation adsorption capability of

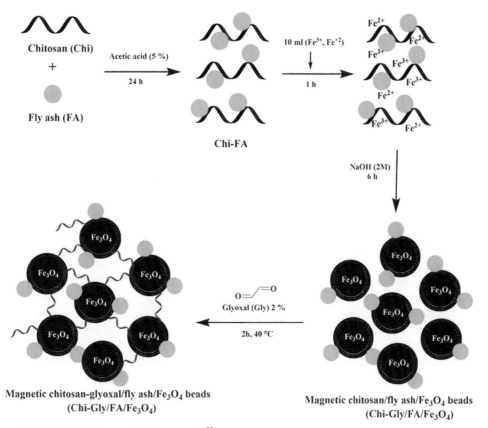

FIG. 3 Steps of synthesis of CS-Gly/FA/magnetite biocomposite.[59]

112.5 mg/g and removal efficiency of 96.20% at pH = 4.0. Fig. 3 shows the schematic representation of the steps of synthesis of the novel adsorbent. Abd Malek et al.[60] also studied the removal of the same dye via another novel adsorbent synthesized by blending magnetic biocomposite chitosan (CS)-polyvinyl alcohol (PVA) and FA microparticles. The resulting m-CS-PVA/FA removed the anionic dye with a removal efficiency of 90.3% and an adsorption capacity having a maximum value of 123.8 mg/g.

For a more practical application-based approach of real wastewater treatment, a multi-component sorption system study was conducted by Hussain et al.[61] by modifying CFA (MCFA) with NaOH and HCl solution to form adsorbents which removed four toxic dyes from industrial wastewater, namely, reactive turquoise blue KN-G dyes, direct fast scarlet 4BS, acid navy blue R, and direct sky blue 5B. These adsorbents showed up to a maximum of 96.03% removal efficiency, which were reported to follow both Freundlich and Langmuir adsorption models along with pseudo-second-order kinetics. Here we can see that the basic synthetic precursor and the reagents are considerably low cost, making it a cost-effective process.

2.1.3 Removal of phenolic compounds

Phenols and their derivatives are considered as one of the most abundant and significant organic impurity contaminating water. Although mostly phenols are biodegradable, yet, they are considered as hazardous, because they are detrimental even at low concentrations.[62,63] They are mainly released into water from pesticides, paints, petrochemical industries, polymeric resins, and coal conversions. They add an obnoxious odor to fresh water and also alter the taste a bit. Removal of phenol by FA and modified FA adsorbents is discussed below.

Khanna and Malhotra[64] were the first one to report FA as an adsorbent for the removal of phenol from water. They conducted a comparative study between FA and activated carbon and reported that the adsorption capacity of FA was approximately 3.6 times less than that of activated carbon, although the former was much more inexpensive. Hence, modification was further required to improve the adsorption efficiency of FA toward phenolic compounds.

Other researches explained that para-substituted phenol, which has less steric hindrance, is more easily adsorbed than other substituted phenol. The order of adsorption of different phenolic compounds on FA is related to electron-withdrawing properties of a functional group (m-nitrophenol > o-nitrophenol > phenol > m-cresol > o-cresol).[65,66]

In 2015, Chaudhary et al.[67] synthesized Al and Fe metal-impregnated FA adsorbents for the removal of phenols from wastewater. It was reported that Al impregnated FA showed a higher removal efficiency (85.6%) compared to the Fe-impregnated FA (82.1%) for phenol. Both the adsorbents followed pseudo-second-order kinetics and best fit the Redlich-Peterson plot.

Another study showed that Oyehan et al.[68] modified mesoporous acidified FA (AFA) by coating it with layers of poly diallyl dimethyl ammonium chloride (PDDA). The synthesized adsorbents efficiently removed phenols from water. The adsorption process followed pseudo-second-order kinetics, and the maximum efficiency of 21.420 mg/g according to the Scatchard isotherm model. This new adsorbent was found to be four times more efficient that the general AFA adsorbent.

2.1.4 Removal of other contaminants

There are a variety of other contaminants contributing to the pollution of water which include pesticides, inorganic ions such as phosphates and fluorides, other organic contaminants such as quinoline and pharmaceutical contaminants like ibuprofen (IBU), caffeine (CFN), and bisphenol A (BPA). The utility of FA and modified FA-based adsorbents in the removal of each type of impurity is discussed below.

Recently, in 2020, a high-performance, novel adsorbent was developed by impregnating La on CFA-blast furnace cement composite.[69] These adsorbents were found to have excellent pH resistance and a high ionic strength. They exhibited chemisorption via pseudo-second-order reaction. The Langmuir adsorption model showed a maximum monolayer adsorption capacity of 24.9 mg/g. Similarly, batch experiments were conducted with FA adsorbents for the removal of fluoride ions from wastewater. It was reported that the sorption process followed pseudo-second order and Elovich kinetic models, and showed a removal efficiency of 67.20%.[70]

Nowadays, pesticides and herbicides are a non-negotiable part of the agricultural industry. They have many applications, such as controlling of growth of certain unwanted plants and pests. One such herbicide is diuron, which is used for controlling the growth of grass weed and certain algal species. But, diuron is harmful to human beings as it causes nose, eyes, throat, and skin irritation. Besides it is also fatal and carcinogenic for many animals such as rats, and also harms the aquatic life.[71] So, for the removal of this harmful chemical from water,[72] industrial and agricultural waste products such as bagasse FA and the ash of rice husk were used. Techniques such as the ANN (artificial neutral network) and RSM (response surface methodology) were applied, and optimum conditions were created to obtain an outstanding maximum diuron removal efficiency of 99.78%.

In addition to this, Huang et al.[73] used FA for the removal of toxic quinoline from water. A maximum adsorption capability of 7.9 mg/g was reported from the performed batch experiments. The sorption mechanism was best explained by the Langmuir adsorption isotherm model and was reported to obey the pseudo-second-order kinetic model.

Another recent study showed that FA-derived zeolites[74] were modified using β-cyclodextrin for removing IBU, CFN, and BPA from wastewater. The batch experiments were performed to determine the intrinsic molecular interaction between the adsorbent and adsorbate molecules. The authors reported that a maximum adsorption capacities of 11.8, 32.7, and 31.3 mg/g were observed for CFN, BPA, and IBU, respectively. The sorption process followed pseudo-second-order kinetics and best fitted Langmuir and Temkin isotherm models.

2.2 Lignin (LG)

Lignin (LG) is produced from black liquor, which is an industrial effluent composed of lignocellulosic biomass. It is mainly released by pulp producing and paper industries. LG seconds the list of the most abundantly found natural polymer following cellulose. It contains various proportions of a versatile range of lignocellulosic biomass.[75]

LG is also available naturally, and its abundance in hardwood > softwood > grass.[76] But its main sources still remain the industry where it is produced in the form of a by-product of various biorefinery processes.[77] It has been predicted that due to the production increase in biofuel, the yield of lignin as industrial waste might reach up to 225 million tonnes annually, by 2030.[78]

LG has an aromatic, cross-linked, 3D, polyphenol structure consisting of numerous functionalities, such as $-OH^-$, $-COOH$, $-OMe^-$, $-CHO$, and $C_6H_5O^{-}$.[79] Presence of a wide range of anionic functionalities makes it a potential candidate for the adsorption of various toxic impurities from wastewater. Fig. 4 shows the structure of LG made of fragments of sinapyl alcohol, coniferyl alcohol, and paracoumaryl alcohol.[80,81]

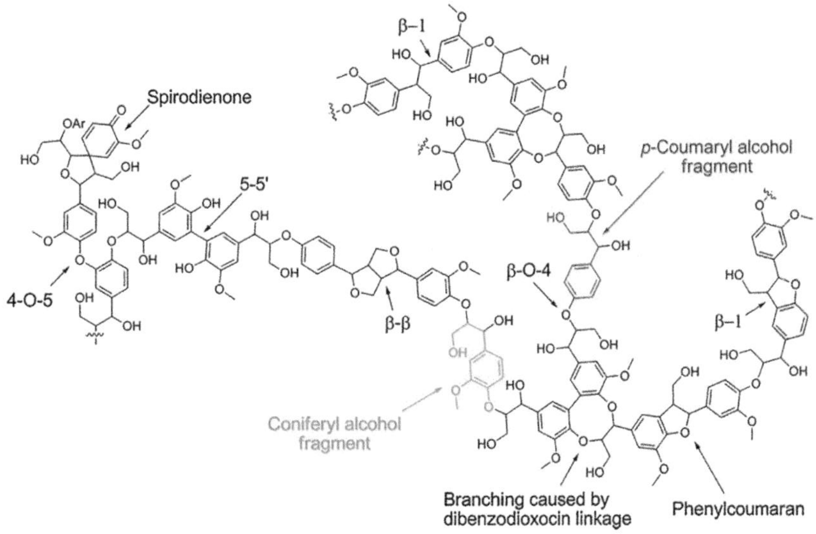

FIG. 4 Schematic representation of the structure of LG.[80]

2.2.1 Removal of heavy metals

LG has also been a potential adsorbent for heavy metals from water due to its versatility in functional groups and resin-like structure. Such adsorbents are also efficient, low cost, and has a wide applicability.[81,82] The type of heavy metal ion to be removed determines the removal mechanism. For instance, during the removal of Hg^{2+} and Cr^{6+}, the main processes dominating the adsorption mechanism are reduction and complexation, while, for Pb^{2+} and Cd^{2+} removal the dominant processes are precipitation and ion exchange.[82]

The affinity of black liquor extracted lignin toward sorption of different metal ions was investigated and the increasing order of affinity was reported as $Ni^{2+} < Zn^{2+} < Cd^{2+} < Cu^{2+} < Pb^{2+}$. It was reported that the LG surface mainly contains two categories of acidic sites, one phenolic and the other carboxylic. Among this, the phenolic type acidic site has a higher affinity for heavy metal ions as compared to the carboxylic type. The main mechanism of the sorption process is based on electrostatic attraction between the phenolic and carboxylic acid sites with the metal ions.[83]

Recently, a novel defective adsorbent was synthesized by Chen et al.[84] by treating "LG-derived Ca/Fe/Al-trimetallic layered double hydroxide (LDH) composites (LDH@LDC)" with hydrogen plasma. This novel adsorbent was reported to remove radioactive U^{6+} and Cr^{6+} ions more efficiently from water than its non-defective version. The increase in adsorption capacity was observed up to more than 50 mg/g for the toxic heavy metal ions. The significant increase in removal efficiency was owed to the ion-exchange mechanism between the LDH layer and the ions.

Lignosulfonate derived from LG is one of the most widely used compounds which act as an excellent base for preparation of mesoporous adsorbent materials. Wang et al.[85] synthesized a LG-based, green, multifunctional composite adsorbent, from Na lignosulfonate, citric acid, and acrylic acid for the removal of Pb^{2+} and Cu^{2+} ions from wastewater. Characterization of the synthesized adsorbents was carried out via SEM, XPS (X-ray photoelectron spectroscopy), TGA (thermogravimetric analysis), and FTIR. The processes dominating the mechanism of adsorption were reported to be electrostatic attraction and complexation. A maximum sorption capacities of 276 mg/g and 323 mg/g were reported for Cu^{2+} and Pb^{2+}, respectively.

2.2.2 Removal of dyes

LG-based adsorbents have been found to form effective adsorbents for dye removal from wastewater by various researchers. Many cost-effective, recyclable, highly efficient adsorbents have been synthesized and tested for various dye removals. Some of such studies are discussed below.

Regeneration studies have been conducted by many researchers to prove the cost-effectiveness of the new-generation adsorbents applied for water remediation. Such instances can be found in the case of a novel high-performance LG-based nanocomposite[86] synthesized via coprecipitation of Fe and Mn compounds with LG in NaOH solution. It was reported to efficiently remove a maximum of 252.05 mg/g of methylene blue (MB) from textile wastewater via a combination of electrostatic interaction, H bonding, and complexation in a pseudo-second-order reaction following the Langmuir adsorption isotherm. It retained 81.20% removal rate after five cycles. Similarly, another study on removal of MB was conducted by Wu et al.,[87] where a unique aerogel was synthesized from graphene oxide (GO) and alkali LG. This showed a maximum capacity of 1185.98 mg/g, and was also cost effective as it retained 1083.84 mg/g adsorption even after three cycles.

In addition to this, for a multi-component system study, a highly efficient, eco-friendly, and novel composite[88] was prepared by coking Ni/Al_2O_3 while catalytic refining of volatile particles from *co*-pyrolysis of polyethylene and LG. It was characterized and its sorption studies showed that it adsorbed a maximum of 95.71 mg/g methyl orange (MO) and 290.76 mg/g rhodamine B (RhB) from industrial wastewater. Further studies on large-scale production of such highly efficient adsorbents can be further studied for real wastewater treatment.

2.2.3 Removal of other contaminants

Among other contaminants efficiently removed by LG-based adsorbents, the most widely renowned ones are the inorganic ions such as phosphate, organic contaminants such as toluene, and pharmaceutical wastes such as antibiotics including ciprofloxacin and ofloxacin.

Phosphate and nitrogen compounds have been one of the major contaminants of water which leads to eutrophication of waterbodies; these severely affect the groundwater index too. Many groups of researchers have been studying the cost-effective eco-friendly removal of these inorganic contaminants. Li et al.[89] synthesized magnetically modified LG-based nanoparticles (NPs) M/ALFe to remove phosphate ions from industrial wastewater. The removal efficiency of this ferro adsorbent was reported to be 62% which persisted even after six cycles. The regeneration was done via magnetic influence. While another group of researchers[90] utilized LG-based bio-charcoal functionalized by MgO for phosphate removal. MgO NPs were homogeneously loaded on the surface of the bio-charcoal. The maximum adsorption capacity was 906.28 mg/g, the initial removal efficiency was 99.76% which decreased to 72.01% after six cycles of regeneration.

Toluene removal was studied by Tahari et al.[91] where they used LG/montmorillonite hydrogels in the presence of alkali, organosolv LG, and kraft LG. Remarkable removal efficiencies of 90.14, 86,081, and 85.09% were reported for alkali, organosolv, and kraft LG, respectively. The removal efficiency was seen to considerably improve by the use of alkali LG. The sorption process best fitted Freundlich adsorption model; hence it was reported to be thermodynamically spontaneous and followed pseudo-second-order kinetic model.

Among a large variety of pharmaceutical wastes, the most commonly studied ones are antibiotics such as ciprofloxacin and ofloxacin which when disposed to clean waterbodies, contaminates them by their noxious properties. Recently, a study was conducted to understand the activity-surface relation of a grafting modified adsorbent and the design for proper exploitation of the efficient adsorbent. Here, a sequence of LG-based adsorbents shaped like actinia were synthesized using LG to be core material, and for the tentacles, grafted polyacrylic acid was used.[92] These novel adsorbents efficiently removed ciprofloxacin and ofloxacin from industrial wastewater. The main interactions dominating the sorption process were reported to be hydrogen bonding and electrostatic interactions, where the former dominated the latter. The saturated adsorption capabilities were attained at a pH value of 6.0, and were reported to be 0.965 and 0.835 mmol/g for ciprofloxacin and ofloxacin, respectively.

2.3 Red mud (RM)

RM is one of the abundantly available industrial wastes available from the alumina production. It is basically a commercially valuable bauxite residue. For 1 t of alumina production, 1–1.5 t of RM is produced. Due to the increased demand of alumina worldwide, in 2020, the annual alumina production went up to 133 million t, hence, the consequent RM production reached an amount of 175 million t. Storing such a huge amount of RM can be environmentally hazardous owing to its alkalinity, and the most efficient way to utilize it is by using it in production of adsorbents for environmental remediation.

RM mainly constitutes of particles of oxides and hydroxides of Fe, Ti, Ca, Al; namely, boehmite, hematite, goethite, silica, and sodalite are some of the main constituent oxides present in it.[93] It is also found to contain small amounts of radionuclides and heavy metals like ^{40}K, ^{230}Th, ^{226}Ra, Co, Ni, Cr, V, and Cd.[94] These large variety of compounds impart a highly reactive surface to the adsorbents produced from RM, and therefore, further investigations have been done on how to exploit this industrial waste to produce potential, efficient materials which would contribute to water treatment.

2.3.1 Removal of heavy metals

RM has been widely used as a potential efficient adsorbent in the removal of toxic heavy metals from wastewater. It has been reported to remove Cd^{2+}, As^+, As^{3+}, Cu^{2+}, Cr^{5+}, Zn^2, Pb^{2+}, and Ni^{2+} from aqueous media. Raw RM has been modified in several ways to form novel, low-cost, and efficient adsorbents; few of such examples are discussed below.

We already know the harmful effects the toxic heavy metal ions have directly on the human body. Hence, direct consumption of such contaminated water can lead to serious health hazards. A comparative study of removal of Mn and As from wastewater was done using RM and pyrolusite as the adsorbents.[95] It was reported that both the heavy metals were completely removed from the solution using RM exhibiting a 100% removal efficiency in 24 h, while in case of pyrolusite the removal efficiency was considerably less, around 97 and 30% for As and Mn, respectively, within the same contact time.

Activated RM has been found to effectively remove many other heavy metals from water, such as Cr^{2+}, Cd^{2+}, Zn^{2+}, and Pb^{2+}.[96] Activated RM is a high-capacity adsorbent created by the treatment of RM by hydrogen peroxide followed by calcination at 773 K. It was reported that the sorption capacity for Cr^{2+} was 1.3 times greater than that for Pb^{2+} ions. The presence of other metals such as Na^+ and other surfactants reduces the adsorption capacity of these two ions considerably. The adsorption capacities for Zn^{2+} and Cd^{2+} were found to be the highest at pH = 4. A similar study was conducted recently by activating RM using Mg^{2+} and Ca^{2+} ions, followed by calcination.[97] This new adsorbent removed Pb^{2+} and Zn^{2+} ions from water. The maximum adsorption capabilities were reported to be 75.58 and 218.82 mg/g for Zn^{2+} and Pb^{2+} ions, respectively.

2.3.2 Removal of dyes

Numerous studies have shown that modifying RM can create green, eco-friendly, and low-cost adsorbents for the treatment of textile wastewater. A few are discussed below.

Recently, a novel adsorbent was synthesized from RM by magnetically modifying it to form a zeolite.[98] The resulting adsorbent was studied and characterized and found to remove MB from wastewater at a rate of 95.4% with a saturation capacity of 47.9 mg/g. This adsorbent could be recovered easily in the presence of an external magnet and retained 33.2 mg/g adsorption capacity after three cycles. Other experiments performed on the removal of MB by pure RM have shown considerably lower saturation capacity values.[99] The sorption process was reported to be highly pH-dependent, and the maximum adsorption capacity increased with the decrease in adsorbent dosage. Activated RM,[100] when applied for the removal of MB, yielded better results than pure RM, although it was not as efficient as the above-mentioned new-generation magnetically modified RM adsorbent. Activated RM showed an 80% removal percentage within a contact time of 75 min. The process was reported to be exothermic and best fitted by the Langmuir adsorption isotherm.

Congo red (CR) is another hazardous dye which when present in waterbodies severely affects aquatic life and its surrounding human life malignantly. To remove CR from industrial wastewater, Wang et al.[101] synthesized RM-derived α-Fe_2O_3 beads or microspheres compiling into nanosheets. These were synthesized via hydrothermal treatment in the presence of CTAB and urea. The sorption process had a maximum capacity of 342.57 mg/g, following the Langmuir adsorption isotherm, intraparticle diffusion, and the pseudo-second-order kinetic model.

A competitive study was conducted by Qian and Zhou,[102] where they synthesized RM-derived LDH along with layered metal oxides (LDO) and compared its adsorption capacities with pure RM and roasted RM in the removal of reactive brilliant blue (RBB) and reactive yellow KE-4R dye. Results showed the following order of increasing adsorption capacities for the adsorption of the two dyes in the four adsorbents used: Roasted RM < RM < LDH < LDO. The maximum capacity attained was 91.1 and 97.2 for KE-4R and RBB, respectively. This showed that modified RM creates more novel and efficient adsorbents for water remediation.

2.3.3 Removal of phenolic compounds

Phenolic species in water are released due to petrochemical operations from polymer and pharmaceutical industries. They are noxious and carcinogenic in nature, primarily due to their prolonged stability in water. RM and modified RM have been utilized for the removal of phenolic compounds from water, as discussed below.

Hydrogen peroxide-treated RM was used to remove phenol (P), 2-chlorophenol (2CP), 4-chlorophenol (4CP), 2,4-dichlorophenol (DCP).[103] Intraparticle diffusion dominated the sorption mechanism. 94–97% removal was reported for DCP and 4-CP, while, 50–80% removal was observed for 2-CP and P. The column studies showed the following removal order as P < 2-CP < 4-CP < DCP.

Recently, batch experiments were performed for the removal of phenol by RM from industrial wastewater.[104] The pH was maintained from 2.0 to 12.0; the contact time was kept in a range of 2–10 h; adsorbate dosage was 40–200 mg/L; and the adsorbent concentration was 200–700 mg/L. 87.5% removal results were obtained. The sorption process best fitted the Freundlich adsorption isotherm, followed pseudo-second-order reaction kinetics, displayed a maximum Langmuir adsorption capacity of 49.3 mg/g.

Different studies were carried out comparing the adsorption performance of HCl-modified RM and pure RM. One of the studies reported that HCl-modified RM did not show much of an improvement in adsorption performance in the removal of phenol and its derivatives, as it was theorized that HCl accelerated the desorption of the adsorbed phenol moieties easily, reducing the efficiency of the sorption process.[105] While another contradicting study showed that phenol removal by HCl activated RM was 70% and that of untreated RM was 33%.[106] Further studies are still going on in this field, for a more concrete pattern of phenol sorption.

2.3.4   Removal of other contaminants

RM has been reported to remove inorganic ions such as nitrates, phosphates, and fluorides from wastewater. Modified RM also removes pesticides and specific pharmaceutical wastes from the aqueous solution. A few such instances are discussed herewith.

Phosphate removal has been observed by both acid-activated and pure RM.[107] Acid-activated RM was found to have higher sorption capacity as compared to pure RM. Recently, comparison of phosphate removal with bauxite residues both RM and brown mud was studied.[108] It was reported that brown mud almost had 2.5 times greater adsorption capacity than RM. Also, the removal efficiency for RM was 88.6%, while that of brown mud was 95.5%. The adsorption process was reported to be highly time-dependent.

Similar studies were carried out in the case of nitrate removal, where comparison was done between HCl activated and neutral RM.[109] It was observed that the former (5.86 mmol/g) has almost five times higher removal capacity than the latter (1.86 mmol/g) at pH = 7. Removal of fluoride ions was reported via adsorption by novel nanocomposite-doped beads[110] that were synthesized from RM-derived SiO_2 NPs. Maximum adsorption capacity reported was 83.84 mg/g.

Due to the recent on-going covid-19 pandemic, the demand of psychiatric pharmaceuticals such as paroxetine, diazepam, carbamazepine, lorazepam, and fluoxetine has spiked globally. Magnetite RM-NPs[111] have proven to be excellent adsorbents for the removal of such pharmaceuticals from industrial wastewater. Carbamazepine has been removed by RM-NPs at pH = 6.5–7, with a maximum capacity of 90.5 mg/g. These novel, new-generation adsorbents are also quite cost-effective, as they can be easily regenerated under magnetic influence and reused even after five adsorption-desorption cycles. Similarly, the magnetite RM-NPs are also reported to be efficient adsorbents for organophosphorus pesticide removal from aqueous media.[112] The dominating sorption mechanism was reported to be film-diffusion mechanism.

2.4   Blast furnace sludge, slag and dust

The electroplating industry releases a dried unusable produce as a combination of metal ion precipitation in wastewater along with $Ca(OH)_2$. It consists of salts and insoluble hydroxides of metals. Steel plants release by-products of different industrial processes, among which one very useful waste product is the blast furnace slag. These are released in the granular form, and are often used as a filler for slag cement production. They are mainly composed of a mixture of silicates especially calcium-alumino and alumino-silicates. Dust released from blast furnace are also released by steel industries which can also be used for water remediation. 2–4 t of by-products and wastes were released per tonne production of steel. Sludge and dust produced are 28 kg and slag produced is 340–421 kg per tonne of the industrial waste produced.[113] Therefore, these waste products are abundantly available, and hence, they are found to behave as excellent low-cost, eco-friendly adsorbents for removing toxic impurities from contaminated water.

2.4.1   Removal of heavy metals

Slag, sludge, and dust released from steel plant blast furnace have proven to be effective adsorbents for the removal of heavy metal ions from industrial wastewater.

In the early 90s, a group of researchers studied the decontamination of both single and multi-component adsorption systems for water remediation. They studied the effectiveness of sludge as adsorbent for removal of Cu^{2+} by studying flue dust sorption, and reported that sludge is a potentially effective adsorbent for Cu^{2+} removal.[114] They also studied Pb^{2+}, Cu^{2+}, Cd^{2+}, Zn^{2+}, and Cr^{3+} ions from water on sludge. All the metals were efficiently removed. The authors reported that the lead removal was temperature-dependent and hematite/C ratio dependent. It occurred better at higher temperature and higher hematite content of the sludge sample.[115,116] The same group of researchers also studied removal of only Pb^{2+} ions from water in another experiment. They reported a maximum adsorption capacity of 80 mg/g, physical nature of the sorption process, and the main mechanism to be ion exchange.[117]

Pb^{2+} removal has also been studied by de Jesús Soria-Aguilar[118] by using blast furnace dust as the adsorbent. Dust was collected from a steel plant, and it was characterized and its chemical composition was analyzed. The presence of CaO, MgO, FeO, and metallic Fe was detected. The removal experiments showed that the presence of the above four components enhanced the adsorption capacity toward the metal ions. Bimolecular or two-degree ion exchange was observed between the metal ion and Fe and Ca which dominated the mechanism.

Landfill leachate and storm water both consist of ample amount of toxic heavy metals like Cu, Cr, Pb, Ni, and Zn. Blast furnace slag and pine bark both are reported to remove these ions from wastewater.[119] It was reported that, at lower adsorbate concentration, pine bark behaved as a better adsorbent, while, at relatively higher adsorbate concentrations, blast furnace slag removed the ions to a much larger extent. Also, the adsorption process by slag was found to be much faster than that by pine bark.

Bhatnagar et al. states that "A comparative study of the adsorbents prepared from several industrial wastes viz. carbon slurry and steel plant wastes viz. blast furnace (B.F.) slag, dust, and sludge for the removal of Pb^{2+} has been carried out.[120] The adsorption of Pb^{2+} on different adsorbents has been found in the order: B.F. sludge > B.F. dust > B.F. slag > carbonaceous adsorbent. The least adsorption of Pb^{2+} on carbonaceous adsorbent even having high porosity and consequently greater surface area as compared to other three adsorbents indicates that surface area and porosity are not important factors for Pb^{2+} removal from aqueous solutions." This also proves that these by-products form steel industry can form effective adsorbents for removal of noxious metal ions.

2.4.2 Removal of dyes

Blast furnace effluents are also found to remove water-contaminating dyes efficiently. Literature survey proved that dye removal is better performed by blast furnace slag than that in case of blast furnace dust and sludge as discussed below.

In 2015, research was conducted to utilize novel, low-cost adsorbents to treat industrial wastewater, and also to reuse the same.[121] Thus, blast furnace dust and sludge were used for the removal of RR19 dye and treatment of an oil emulsion. The adsorbents were characterized using various conventional methods, and then from the adsorption experimental results, it was concluded that these adsorbents were better at removal of the dye from the contaminated water than the treatment of an oil emulsion. For oil emulsion treatment, removal efficiency was poor, <10%, while for dye removal, the removal efficiency obtained by use of blast furnace sludge was reported to be a good 87%.

A year earlier in 2014, blast furnace sludge was used as an adsorbent for MB and the sorption performance was monitored by Malina and Rađenović.[122] Batch experiments were performed and the results showed that the saturation adsorption capacity spiked from 10.3 to 70.6 mg/g when adsorbent dosage was increased by 140 mg/L. Different investigations proved that the morphology of blast furnace sludge suggests that it is a mesoporous adsorbent material. The authors also studied the structure of the adsorbent before and after the dye adsorption, and the resulting SEM micrographs are given in Fig. 5.

Blast furnace slag has been utilized numerous times for the removal various cationic, anionic, and acidic dyes. The adsorption behavior has shown that it adsorbs acidic MO (methyl orange) dye better than cationic MB[123] at optimum conditions, where vice-versa was observed as the resistance time was decreased. Also, the chemistry of the adsorbent affected its sorption performance. Greater alkalinity of the blast furnace sludge resulted in lower removal rate for acidic MO dye at lower concentrations, and the same for cationic MB dye but when concentrations are higher.

Recently, a novel adsorbent was prepared by Wagner et al.[124] by combining granulated slag and Portland cement. The flat prisms formed from these two materials were stored in various solutions, such as aqueous solution and solution of

FIG. 5 Blast furnace sludge SEM micrographs (A) before MB adsorption; (B) after MB adsorption.[122]

FIG. 6 Hydrotalcite nanosheets SEM micrographs.[124]

sodium sulfate. The flat prisms stored in sodium sulfates solution resulted in formation of new-generation hydrotalcite nanosheets (Fig. 6). These nanosheets developed hardened binder surface, which was rich in Mg derived from the blast furnace slag. These novel hydrotalcite nanosheets were found to be good adsorbents for removal of azo dye Congo red. They displayed an adsorption capacity of 105 mg/g.

2.4.3 Removal of phenolic compounds

Phenol and its derivatives can be efficiently removed by activated or modified blast furnace slag, sludge, and dust. These wastes are often compared with activated charcoal (AC) adsorbent, as they can be a good substitute for the same.[125] Comparison between these adsorbents has been done with other carbonaceous adsorbents released by fertilizer industries. These carbonaceous adsorbents were found to be better adsorbents as compared to blast furnace dust, sludge and slag in removal of bromophenols from industrial wastewater. The author also reported that the removal efficiency of these adsorbents derived from carbonaceous industrial wastes was almost 45% of that of original AC, yet a lot more cost-effective, hence, these can be an average replacement for standard AC adsorbents for the removal of phenol and its derivatives from contaminated water.

Gupta et al.[126] used blast furnace slag-derived activated carbon as a novel adsorbent for the removal of toxic 2-aminophenol from wastewater. The sorption process was found to be endothermic in nature and it followed both the Langmuir and Freundlich adsorption models. Kinetics studies were conducted based on various parameters, and column adsorption experiments were performed in which the obtained breakthrough curve was studied for identification of ideal reaction conditions and a correlation pattern. A maximum column capacity of 312 mg/g was obtained. This adsorption performance was compared with that of fertilizer-derived activated carbon. The results suggested a better potential of activated carbon than activated slag.

Treated blast furnace slag, sludge, dust along with fertilizer-derived carbon slurry were used to remove P, 2-CP, 4-CP, and DCP.[127] And the yielded results showed that adsorption performance of the carbon slurry was better than modified blast furnace slag, sludge, and dust owing to their greater porosity due to higher carbon content. The order of adsorption was found to be DCP > 4-CP > 2-CP > P.

Das and Patnaik[128] carried out batch experiments for phenol removal by flue dust (FD) from blast furnace and alloy plant-derived slag (APS) from a prepared synthetic solution of phenol. They reported the removal efficiencies for APS and FD to be 75 and 90%, respectively. The sorption mechanism was described as external mass transfer before intraparticle diffusion mechanism. The sorption reaction was first order and followed Freundlich, Langmuir, and BDST (bed-depth service time) model for column adsorption.

2.4.4 Removal of other contaminants

Among other contaminants adsorbed by blast furnace slag, sludge, and dust, the majority is taken up by pharmaceuticals such as IBU, diclofenac (DFC), and carbamazepine (CBZ), a few inorganic ions such as phosphate and nitrate and neutral compound like ammonia. A few instances are discussed as follows.

Recently, eight natural, low-cost, and abundant materials were used as adsorbents[129] for the removal of pharmaceuticals such as IBU, DFC, and CBZ. DFC was removed at a higher extent compared to CBZ and IBU due to its anionic and hydrophobic nature. The adsorption performance order of the materials can be given as sand < brickbats < natural pyrite < blast furnace slag < LECA (light-weight, expanded clay aggregate) < waste AAC (autoclave aerated concrete) blocks < natural zeolite < wood charcoal. These adsorbents were also found to remove phosphate ions, ammonia, and nitrate ions with a saturation capacity of 90, 81, and 61.8% respectively.

Blast furnace slag has been utilized as an adsorbent for the removal of phosphate ions from wastewater.[130] The adsorbent requires a higher Ca concentration for the precipitation of phosphate. Factors affecting the sorption process such as contact time, pH, adsorbent dosage, and particle size were studied. The alkalinity of the adsorbent affected the sorption performance, and with the increase in alkalinity, the adsorption performance deteriorated. The results were found to be satisfactory, as the used adsorbent was efficient, natural, and low cost. Phosphate removal was also studied using a novel adsorbent derived from blast furnace slag[131] which had much higher adsorption capacity than the usual slag. "An ultra-high pressure water jet mill cavitation disintegrator and a controlled vacuum freeze dryer were used to disintegrate amorphous BFS. Specific surface areas of both BFS and disintegrated slag (BFS-D) were measured using the surface area analysis method. BFS-D was obtained with an average particle size of 198 nm and with 27 times bigger free specific surface area than that of original BFS. The BFS-D was tested as an adsorbent of phosphate from aqueous solutions. Adsorption data were analyzed using the Freundlich and Langmuir adsorption isotherms." The novel adsorbent exhibited 30.49 mg/g saturation adsorption capacity, which is a 126.7% increment from the phosphate removal by original blast furnace slag.

Other experiments have also been performed where along with phosphate ion, the removal of ammonium ion has been practiced using a composite adsorbent[132] from blast furnace slag and calcinated natural clay. The removal of phosphate ion was seen to be improved by 8 mg/g for the novel composite as compared to natural clay, while that of ammonium ion was reported to deteriorate by 8 mg/g. The dominating mechanism for ammonium removal was reported to be ion exchange, while that of phosphate removal was surface precipitation.

3 Desorption studies

Desorption and regeneration studies help in understanding the recyclability of the metal laden adsorbent and its reusability potential. It also gives information for a better economic analysis for the reduction of the total production, operational, and processing cost of the adsorbent. Elaborated studies have been conducted on the desorption and reusability of industrial waste-derived novel adsorbents and commendable results have been reported. Sahoo et al.[133] performed desorption studies of the metals Pb, Cd, Zn, Fe, Al, Ni, and Mn from alkali treated modified FA. The desorption studies were carried out in both neutral (pH = 6.5) as well as acidic (pH = 2.0) media. Thirty grams of modified FA was recovered. This recovered adsorbent was treated with 1 L of solution by shaking at different time intervals for a span of 7 h, and was finally filtered, collected, and analyzed. It was reported that acidic media facilitated significantly better desorption than neutral media. The results showed that at a time span of 3 h of the desorption experiment, in neutral media, a maximum of 8.20% of desorption was observed, while in the acidic media a 98.00% desorption capacity was observed. This difference was attributed to the fact that in acidic media, the modified FA was prone to protonation, and hence, released the heavy metals ions. Therefore, the protons replaced the heavy metal ions attached to the adsorbent, facilitating greater desorption. Fig. 7 depicts the desorption studies carried out by the group.

Another study showed that novel carboxymethyl LG NPs[134] were utilized as adsorbents for removal of Pb^{2+} from aqueous solution. For desorption, the metal-loaded adsorbent was treated with 0.1 (M) HNO_3 solution. The authors reported

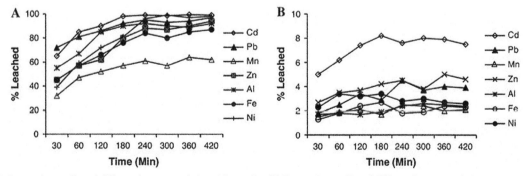

FIG. 7 (A) Desorption studies of different heavy metals in acidic media. (B) Desorption studies of different heavy metals in neutral media.[133]

that after five adsorption-desorption cycles, the adsorption capacity was found to decrease by only 15.0%, while after 10 cycles the capacity was found to decrease by 27.0%. Similar study was conducted with a RM adsorbent which has adsorbed fluoride ions.[106] On treatment of the ion laden adsorbent with HCl having 0.2 (M) concentration, a desorption efficiency of 87–46% was reported on four consecutive adsorption-desorption cycles.

Often pure industrial wastes are also used for metal adsorption from water. Their desorption studies make an excellent economic analysis and reduce the operational cost of the wastewater treatment significantly. Naiya et al.[135] utilized oxygen B.F. sludge for removal of Pb^{2+} ions from water. Batch experiments were performed to conduct the desorption studies. The metal laden sludge was treated with 0.1 (M) solutions of HNO_3 and HCl, which facilitated a desorption efficiency of 26% and 24%, respectively, in 1 h timespan. When the concentration of the acid solutions was reduced to 0.5 (M), the recovery peaked up to a remarkable 98.4%. The adsorption capacity was retained almost 90% after two consecutive cycles.

4 Future prospects

The utilization of industrial wastes or by-products as low-cost, eco-friendly, efficient, and novel adsorbents is a very environment-friendly and economic approach toward water remediation. Yet, there are a few limitations which can be worked upon and overcome for a more enhanced remediation system.

(1) Majority of the studies did not correlate the characterization with the sorption mechanism, and removal performance of the adsorbent was used.

(2) Adsorptive removal of heavy metals, dyes, and phenolic compounds is found to be highly pH-dependent generally. Very few cases were found where the pH dependency was reported along with other experimental factors such as temperature, contact time, and adsorbent dosage.

(3) Appropriate waste selection for production of novel adsorbents to achieve the maximum adsorption performance is not discussed enough.

(4) Majority of the experiments are yet to be carried out in large scale for real wastewater treatment; therefore, practicality and its challenges need to be addressed more.

(5) Other practical challenges such as leaching caused by industrial wastes in aqueous environment is an inhibition for accuracy of the obtained results of the adsorptive removal experiments. Consequently, the results obtained will be erroneous if this factor is not taken into account. Very few number researchers have considered this parameter while experimentation. Further studies on this topic are an urgent necessity.

(6) The adsorption performance of the industrial wastes used are strongly origin dependent. Such dependency and the corresponding correlation and comparison with the same adsorbent from a different origin are not discussed in majority of the cases.

(7) Using these adsorbents in large scale would also need the place to collect and store a bulk amount of such industrial wastes before using them without causing any environmental hazards. Methods of safe storage need to be studied and worked upon.

(8) Regeneration and recycling of low-cost adsorbents from industrial wastes need to be worked on more. Alternatively, if an adsorbent has a lower recovery or regeneration capacity, then disposal of the laden industrial waste is a huge issue which is harmful for the natural environment. This also affects the economic feasibility of the deemed low-cost adsorbents; hence, studies on this field need to be done for reduction of any environmental hazard.

5 Conclusion

Industrial wastes have been successfully used as novel adsorbents for water remediation. They have removed heavy metal ions, dyes, phenol and its derivatives, pharmaceutical wastes, inorganic ions, and a few other contaminants polluting water-bodies. The industrial wastes discussed in this chapter are FA, LG, RM, blast furnace slag, sludge, and FD. The adsorption processes generally are reported to best fit Freundlich and Langmuir adsorption isotherm models. Although a few cases were reported where Temkin, Redlich-Peterson, and BET models have been identified. Parameters affecting the sorption process such as temperature, pH, initial concentration of adsorbate, adsorbent dosage and in some cases chemical parameters such as composition of adsorbents and chemical environment have been studied. Characterization of the novel adsorbents via SEM, TEM, XRF, XRD, TGA, rheology, and other methods have been done. Many comparative studies have been conducted between industrial waste-derived adsorbent and other adsorbents, where both pros and cons of the eco-friendly, low-cost approach have been reported. FA has been found to be the most widely used raw material for novel adsorbents as well as an adsorbent itself for removal of various impurities from wastewater. Also, in case of most of the industrial wastes employed as adsorbents, it has been observed that removal of dyes and heavy metal ions have taken up the majority of research work. More research needs to be done in case of removal of pesticides, pharmaceutical wastes, and other hazardous organic and inorganic compounds, as they are also very detrimental for human health, aquatic life, and the surrounding natural environment. If these low-cost adsorbents can be prepared in large scale with minimum environmental hazards and economic burdens, then this would contribute hugely to environmental research and also would serve as an excellent alternative for costlier adsorbents such as activated carbon, charcoal, hydrogels, and many more leading to a significant improvement in the global economy.

References

1. Anjum M, Miandad R, Waqas M, Gehany F, Barakat MA. Remediation of wastewater using various nano-materials. *Arab J Chem.* 2019;12(8): 4897–4919.
2. Duan C, Ma T, Wang J, Zhou Y. Removal of heavy metals from aqueous solution using carbon-based adsorbents: a review. *J Water Process Eng.* 2020;37, 101339.
3. Jawed A, Saxena V, Pandey LM. Engineered nanomaterials and their surface functionalization for the removal of heavy metals: a review. *J Water Process Eng.* 2020;33, 101009.
4. Yu G, Lu Y, Guo J, et al. Carbon nanotubes, graphene, and their derivatives for heavy metal removal. *Adv Compos Hybrid Mater.* 2018;1(1):56–78.
5. Dehghani MH, Omrani GA, Karri RR. Solid waste—sources, toxicity, and their consequences to human health. In: *Soft Computing Techniques in Solid Waste and Wastewater Management.* Elsevier; 2021:205–213.
6. Karri RR, Ravindran G, Dehghani MH. Wastewater—sources, toxicity, and their consequences to human health. In: *Soft Computing Techniques in Solid Waste and Wastewater Management.* Elsevier; 2021:3–33.
7. Aigbe UO, Onyancha RB, Ukhurebor KE, Obodo KO. Correction: removal of fluoride ions using a polypyrrole magnetic nanocomposite influenced by a rotating magnetic field. *RSC Adv.* 2020;10(7):3883.
8. Aigbe UO, Ho WH, Maity A, Khenfouch M, Srinivasu V. Removal of hexavalent chromium from wastewater using PPy/Fe$_3$O$_4$ magnetic nanocomposite influenced by rotating magnetic field from two pole three-phase induction motor. *J Phys Conf Ser.* 2018;984(1), 012008. IOP Publishing.
9. Khairnar MR, Jain VM, Wadgave U, Dhole RI, Patil SJ, Chopade SR. Effect of different reverse osmosis water filters on fluoride content of drinking water. *J Indian Assoc Public Health Dent.* 2018;16(2):165.
10. Ukhurebor KE, Aigbe UO, Onyancha RB, et al. Effect of hexavalent chromium on the environment and removal techniques: a review. *J Environ Manag.* 2021;280, 111809.
11. Carvalho FGD, Fucio SBPD, Pascon FM, Kantovitz KR, Correr-Sobrinho L, Puppin-Rontani RM. Effect of gamma irradiation on fluoride release and antibacterial activity of resin dental materials. *Braz Dent J.* 2009;20:122–126.
12. Li X, Wu X, Liu S, Li Y, Fan J, Lv K. Effects of fluorine on photocatalysis. *Chin J Catal.* 2020;41(10):1451–1467.
13. Mekonen A, Kumar P, Kumar A. Integrated biological and physiochemical treatment process for nitrate and fluoride removal. *Water Res.* 2001; 35(13):3127–3136.
14. Aigbe UO, Osibote OA. A review of hexavalent chromium removal from aqueous solutions by sorption technique using nanomaterials. *J Environ Chem Eng.* 2020;8, 104503.
15. Mittal J, Ahmad R, Mittal A. Kahwa tea (*Camellia sinensis*) carbon—a novel and green low-cost adsorbent for the sequestration of titan yellow dye from its aqueous solutions. *Desalin Water Treat.* 2021;227:404–411.
16. Tyagi I, Gupta VK, Sadegh H, Ghoshekandi RS, Makhlouf AH. Nanoparticles as adsorbent; a positive approach for removal of noxious metal ions: a review. *Sci Technol Dev.* 2017;34(3):195–214.
17. Anastopoulos I, Mittal A, Usman M, et al. A review on halloysite-based adsorbents to remove pollutants in water and wastewater. *J Mol Liq.* 2018;269:855–868.
18. Anastopoulos I, Pashalidis I, Orfanos AG, et al. Removal of caffeine, nicotine and amoxicillin from (waste) waters by various adsorbents. A review. *J Environ Manag.* 2020;261, 110236.
19. Mittal J. Recent progress in the synthesis of layered double hydroxides and their application for the adsorptive removal of dyes: a review. *J Environ Manag.* 2021;295, 113017.
20. Sharma SK, Mittal J. Hen feather, a remarkable adsorbent for dye removal. In: *Green Chemistry for Dyes Removal from Wastewater.* Scrivener Publishing LLC USA; 2015.
21. Haddad B, Mittal A, Mittal J, et al. Synthesis and characterization of egg shell (ES) and egg shell with membrane (ESM) modified by ionic liquids. *Chem Data Collect.* 2021;33, 100717.
22. Mittal A, Teotia M, Soni RK, Mittal J. Applications of egg shell and egg shell membrane as adsorbents: a review. *J Mol Liq.* 2016;223:376–387.
23. Arora C, Kumar P, Soni S, Mittal J, Mittal A, Singh B. Efficient removal of malachite green dye from aqueous solution using Curcuma caesia based activated carbon. *Desalin Water Treat.* 2020;195:341–352.
24. Patel A, Soni S, Mittal J, Mittal A, Arora C. Sequestration of crystal violet from aqueous solution using ash of black turmeric rhizome. *Desalin Water Treat.* 2021;220:342–352.
25. Soni S, Bajpai PK, Bharti D, Mittal J, Arora C. Removal of crystal violet from aqueous solution using iron based metal organic framework. *Desalin Water Treat.* 2020;205:386–399.

26. Soni S, Bajpai PK, Mittal J, Arora C. Utilisation of cobalt doped Iron based MOF for enhanced removal and recovery of methylene blue dye from waste water. *J Mol Liq.* 2020;314, 113642.

27. Gupta VK, Agarwal S, Ahmad R, Mirza A, Mittal J. Sequestration of toxic Congo red dye from aqueous solution using ecofriendly guar gum/activated carbon nanocomposite. *Int J Biol Macromol.* 2020;158:1310–1318.

28. Mittal A, Ahmad R, Hasan I. Iron oxide-impregnated dextrin nanocomposite: synthesis and its application for the biosorption of Cr (VI) ions from aqueous solution. *Desalin Water Treat.* 2016;57(32):15133–15145.

29. Mittal J, Ahmad R, Ejaz MO, Mariyam A, Mittal A. A novel, eco-friendly bio-nanocomposite (Alg-Cst/Kal) for the adsorptive removal of crystal violet dye from its aqueous solutions. *Int J Phytoremediation.* 2021;1-12:1522–6514.

30. Saharan P, Kumar V, Mittal J, Sharma V, Sharma AK. Efficient ultrasonic assisted adsorption of organic pollutants employing bimetallic-carbon nanocomposites. *Sep Sci Technol.* 2021;56:1–14.

31. Mariyam A, Mittal J, Sakina F, Baker RT, Sharma AK. Adsorption behaviour of Chrysoidine R dye on a metal/halide-free variant of ordered mesoporous carbon. *Desalin Water Treat.* 2021;223:425–433.

32. Mariyam A, Mittal J, Sakina F, Baker RT, Sharma AK. Fixed-bed adsorption of the dye Chrysoidine R on ordered mesoporous carbon. *Desalin Water Treat.* 2021;229:395–402.

33. Mariyam A, Mittal J, Sakina F, Baker RT, Sharma AK, Mittal A. Efficient batch and fixed-bed sequestration of a basic dye using a novel variant of ordered mesoporous carbon as adsorbent. *Arab J Chem.* 2021;14(6), 103186.

34. Mittal J, Mariyam A, Sakina F, Baker RT, Sharma AK, Mittal A. Batch and bulk adsorptive removal of anionic dye using metal/halide-free ordered mesoporous carbon as adsorbent. *J Clean Prod.* 2021;321, 129060.

35. Ahmaruzzaman M. A review on the utilization of fly ash. *Prog Energy Combust Sci.* 2010;36(3):327–363.

36. Angaru GKR, Choi YL, Lingamdinne LP, et al. Portable SA/CMC entrapped bimetallic magnetic fly ash zeolite spheres for heavy metals contaminated industrial effluents treatment via batch and column studies. *Sci Reports.* 2022;12(1):1–17.

37. Singh LP, Ali D, Tyagi I, Sharma U, Singh R, Hou P. Durability studies of nano-engineered fly ash concrete. *Constr Build Mater.* 2019;194:205–215.

38. Surabhi S. Fly ash in India: generation vis-a-vis utilization and global perspective. *Int J Appl Chem.* 2017;13(1):29–52.

39. Visa M, Andronic L, Duta A. Fly ash-TiO$_2$ nanocomposite material for multi-pollutants wastewater treatment. *J Environ Manag.* 2015;150:336–343.

40. Gupta VK, Moradi O, Tyagi I, et al. Study on the removal of heavy metal ions from industry waste by carbon nanotubes: effect of the surface modification: a review. *Crit Rev Environ Sci Technol.* 2016;46(2):93–118.

41. Khan FSA, Mubarak NM, Tan YH, et al. Magnetic nanoparticles incorporation into different substrates for dyes and heavy metals removal—a review. *Environ Sci Poll Res.* 2020;27(35):43526–43541.

42. Perumal S, Atchudan R, Edison TNJI, Babu RS, Karpagavinayagam P, Vedhi C. A short review on recent advances of hydrogel-based adsorbents for heavy metal ions. *Metals.* 2021;11(6):864.

43. Mondal SK, Welz A, Rownaghi A, et al. Investigating the microstructure of high-calcium fly ash-based alkali-activated material for aqueous Zn sorption. *Environ Res.* 2021;198, 110484.

44. Harja M, Buema G, Lupu N, Chiriac H, Herea DD, Ciobanu G. Fly ash coated with magnetic materials: improved adsorbent for Cu (II) removal from wastewater. *Materials.* 2020;14(1):63.

45. He P, Zhang Y, Zhang X, Chen H. Diverse zeolites derived from a circulating fluidized bed fly ash based geopolymer for the adsorption of lead ions from wastewater. *J Clean Prod.* 2021;312, 127769.

46. Panek R, Medykowska M, Wiśniewska M, Szewczuk-Karpisz K, Jędruchniewicz K, Franus M. Simultaneous removal of Pb^{2+} and Zn^{2+} heavy metals using fly ash Na-X zeolite and its carbon Na-X (C) composite. *Materials.* 2021;14(11):2832.

47. Nguyen TC, Tran TDM, Dao VB, Vu QT, Nguyen TD, Thai H. Using modified fly ash for removal of heavy metal ions from aqueous solution. *J Chem.* 2020;2020, 8428473.

48. Apak R, Tütem E, Hügül M, Hizal J. Heavy metal cation retention by unconventional sorbents (red muds and fly ashes). *Water Res.* 1998;32(2):430–440.

49. Gupta VK, Mohan D, Sharma S. Removal of lead from wastewater using bagasse fly ash—a sugar industry waste material. *Sep Sci Technol.* 1998;33(9):1331–1343.

50. Bhattacharya AK, Mandal SN, Das SK. Adsorption of Zn (II) from aqueous solution by using different adsorbents. *Chem Eng J.* 2006;123(1–2):43–51.

51. Gupta VK, Sharma S. Removal of zinc from aqueous solutions using bagasse fly ash—a low cost adsorbent. *Ind Eng Chem Res.* 2003;42(25):6619–6624.

52. Banerjee SS, Jayaram RV, Joshi MV. Removal of nickel (II) and zinc (II) from wastewater using fly ash and impregnated fly ash. *Sep Sci Technol.* 2003;38(5):1015–1032.

53. Hsu TC, Yu CC, Yeh CM. Adsorption of Cu^{2+} from water using raw and modified coal fly ashes. *Fuel.* 2008;87(7):1355–1359.

54. Banerjee SS, Joshi MV, Jayaram RV. Removal of Cr (VI) and Hg (II) from aqueous solutions using fly ash and impregnated fly ash. *Sep Sci Technol.* 2005;39(7):1611–1629.

55. Pattanayak J, Mondal K, Mathew S, Lalvani SB. A parametric evaluation of the removal of As (V) and As (III) by carbon-based adsorbents. *Carbon.* 2000;38(4):589 596.

56. Diamadopoulos E, Ioannidis S, Sakellaropoulos GP. As (V) removal from aqueous solutions by fly ash. *Water Res.* 1993;27(12):1773–1777.

57. Kapoor A, Viraraghavan T. Adsorption of mercury from wastewater by fly ash. *Adsorpt Sci Technol.* 1992;9(3):130–147.

58. Etaba A, Bassyouni M, Saleh M. Removal of hazardous organic pollutants using fly ash. *Environ Ecol Res.* 2021;9(4):196–203.

59. Abd Malek NN, Jawad AH, Abdulhameed AS, Ismail K, Hameed BH. New magnetic Schiff's base-chitosan-glyoxal/fly ash/Fe$_3$O$_4$ biocomposite for the removal of anionic azo dye: an optimized process. *Int J Biol Macromol.* 2020;146:530–539.

60. Abd Malek NN, Jawad AH, Ismail K, Razuan R, Al Othman ZA. Fly ash modified magnetic chitosan-polyvinyl alcohol blend for reactive orange 16 dye removal: adsorption parametric optimization. *Int J Biol Macromol.* 2021;189:464–476.

61. Hussain Z, Chang N, Sun J, et al. Modification of coal fly ash and its use as low-cost adsorbent for the removal of directive, acid and reactive dyes. *J Hazard Mater.* 2022;422, 126778.

62. Akbal F, Onar AN. Photocatalytic degradation of phenol. *Environ Monit Assess.* 2003;83(3):295–302.

63. Calace N, Nardi E, Petronio BM, Pietroletti M. Adsorption of phenols by papermill sludges. *Environ Pollut.* 2002;118(3):315–319.

64. Khanna P, Malhotra SK. Kinetics and mechanism of phenol adsorption on fly ash. *Indian J Environ Health.* 1977;19(3):224–237.

65. Singh BK, Misra NM, Rawat NS. Sorption characteristics of phenols on fly ash and impregnated fly ash. *Indian J Environ Health.* 1994;36(1):1–7.

66. Singh BK, Nayak PS. Sorption equilibrium studies of toxic nitro-substituted phenols on fly ash. *Adsorpt Sci Technol.* 2004;22(4):295–309.

67. Chaudhary N, Balomajumder C, Agrawal B, Jagati VS. Removal of phenol using fly ash and impregnated fly ash: an approach to equilibrium, kinetic, and thermodynamic study. *Sep Sci Technol.* 2015;50(5):690–699.

68. Oyehan TA, Olabemiwo FA, Tawabini BS, Saleh TA. The capacity of mesoporous fly ash grafted with ultrathin film of polydiallyldimethyl ammonium for enhanced removal of phenol from aqueous solutions. *J Clean Prod.* 2020;263, 121280.

69. Asaoka S, Kawakami K, Saito H, Ichinari T, Nohara H, Oikawa T. Adsorption of phosphate onto lanthanum-doped coal fly ash—blast furnace cement composite. *J Hazard Mater.* 2021;406:124780.

70. Chandraker N, Jyoti G, Thakur RS, Chaudhari PK. Removal of fluoride using fly ash adsorbent. *IOP Conf Ser: Earth Environ Sci.* 2020;597(1), 012009. IOP Publishing.

71. da Silva Simões M, Bracht L, Parizotto AV, Comar JF, Peralta RM, Bracht A. The metabolic effects of diuron in the rat liver. *Environ Toxicol Pharmacol.* 2017;54:53–61.

72. Deokar SK, Gokhale NA, Mandavgane SA. A comparative study and combined application of RSM and ANN in adsorptive removal of diuron using biomass ashes. *Int J Chem React Eng.* 2021;19:1221–1230.

73. Huang L, Cao C, Xu D, Guo Q, Tan F. Intrinsic adsorption properties of raw coal fly ash for quinoline from aqueous solution: kinetic and equilibrium studies. *SN Appl Sci.* 2019;1(9):1–8.

74. Bandura L, Białoszewska M, Malinowski S, Franus W. Adsorptive performance of fly ash-derived zeolite modified by β-cyclodextrin for ibuprofen, bisphenol A and caffeine removal from aqueous solutions-equilibrium and kinetic study. *Appl Surf Sci.* 2021;562, 150160.

75. Norgren M, Edlund H. Lignin: recent advances and emerging applications. *Curr Opin Colloid Interface Sci.* 2014;19(5):409–416.

76. Calvo-Flores FG, Dobado JA. Lignin as renewable raw material. *ChemSusChem.* 2010;3(11):1227–1235.

77. Wang J, Wang S. Preparation, modification and environmental application of biochar: a review. *J Clean Prod.* 2019;227:1002–1022.

78. EPA. *Inventory of U.S. Greenhouse Gas Emmissions and Sinks.* US Environmental Protection Energy; 2018.

79. Sarkanen KV, Ludwig CH, eds. *Liguins. Occurrence, Formation, Structure, and Reactions.* John Wiley & Sons; 1971.

80. Zakzeski J, Bruijnincx PC, Jongerius AL, Weckhuysen BM. The catalytic valorization of lignin for the production of renewable chemicals. *Chem Rev.* 2010;110(6):3552–3599.

81. Ge Y, Li Z. Application of lignin and its derivatives in adsorption of heavy metal ions in water: a review. *ACS Sustain Chem Eng.* 2018;6(5): 7181–7192.

82. Shakoor MB, Ali S, Rizwan M, et al. A review of biochar-based sorbents for separation of heavy metals from water. *Int J Phytoremediation.* 2020; 22(2):111–126.

83. Guo X, Zhang S, Shan XQ. Adsorption of metal ions on lignin. *J Hazard Mater.* 2008;151(1):134–142.

84. Chen H, Gong Z, Zhuo Z, et al. Tunning the defects in lignin-derived-carbon and trimetallic layered double hydroxides composites (LDH@ LDC) for efficient removal of U (VI) and Cr (VI) in aquatic environment. *Chem Eng J.* 2021;428, 132113.

85. Wang X, Li X, Peng L, et al. Effective removal of heavy metals from water using porous lignin-based adsorbents. *Chemosphere.* 2021;279, 130504.

86. Yu H, Yang J, Shi P, Li M, Bian J. Synthesis of a lignin-Fe/Mn binary oxide blend nanocomposite and its adsorption capacity for methylene blue. *ACS Omega.* 2021;6(26):16837–16846.

87. Wu Z, Huang W, Shan X, Li Z. Preparation of a porous graphene oxide/alkali lignin aerogel composite and its adsorption properties for methylene blue. *Int J Biol Macromol.* 2020;143:325–333.

88. Zhao J, Wang Z, Shen D, Wu C, Luo K, Gu S. Coked Ni/Al$_2$O$_3$ from the catalytic reforming of volatiles from co-pyrolysis of lignin and polyethylene: preparation, identification and application as a potential adsorbent. *Catal Sci Technol.* 2021;11:4162–4171.

89. Li T, Lü S, Wang Z, Huang M, Yan J, Liu M. Lignin-based nanoparticles for recovery and separation of phosphate and reused as renewable magnetic fertilizers. *Sci Total Environ.* 2021;765, 142745.

90. Jiao GJ, Ma J, Li Y, et al. Enhanced adsorption activity for phosphate removal by functional lignin-derived carbon-based adsorbent: optimization, performance and evaluation. *Sci Total Environ.* 2021;761, 143217.

91. Tahari N, de Hoyos-Martinez PL, Abderrabba M, Ayadi S, Labidi J. Lignin-montmorillonite hydrogels as toluene adsorbent. *Colloids Surf A Physicochem Eng Asp.* 2020;602, 125108.

92. Gao B, Chang Q, Cai J, Xi Z, Li A, Yang H. Removal of fluoroquinolone antibiotics using actinia-shaped lignin-based adsorbents: role of the length and distribution of branched-chains. *J Hazard Mater.* 2021;403, 123603.

93. Antunes MLP, Couperthwaite SJ, da Conceicao FT, et al. Red mud from Brazil: thermal behavior and physical properties. *Ind Eng Chem Res.* 2012;51(2):775–779.

94. Mišík M, Burke IT, Reismüller M, et al. Red mud a by-product of aluminum production contains soluble vanadium that causes genotoxic and cytotoxic effects in higher plants. *Sci Total Environ*. 2014;493:883–890.

95. Pietrelli L, Ippolito NM, Ferro S, Dovì VG, Vocciante M. Removal of Mn and As from drinking water by red mud and pyrolusite. *J Environ Manag*. 2019;237:526–533.

96. Gupta VK, Gupta M, Sharma S. Process development for the removal of lead and chromium from aqueous solutions using red mud—an aluminium industry waste. *Water Res*. 2001;35(5):1125–1134.

97. Mi H, Yi L, Wu Q, Xia J, Zhang B. Preparation and optimization of a low-cost adsorbent for heavy metal ions from red mud using fraction factorial design and Box-Behnken response methodology. *Colloids Surf A Physicochem Eng Asp*. 2021;627, 127198.

98. Cheng Y, Xu L, Jiang Z, et al. Feasible low-cost conversion of red mud into magnetically separated and recycled hybrid $SrFe_{12}O_{19}$@ NaP1 zeolite as a novel wastewater adsorbent. *Chem Eng J*. 2021;417, 128090.

99. Martins YJC, Almeida ACM, Viegas BM, do Nascimento RA, Ribeiro NDP. Use of red mud from amazon region as an adsorbent for the removal of methylene blue: process optimization, isotherm and kinetic studies. *Int J Environ Sci Technol*. 2020;17(10):4133–4148.

100. Thakare SR, Thakare J, Kosankar PT, Pal MR. A chief, industrial waste, activated red mud for subtraction of methylene blue dye from environment. *Mater Today Proc*. 2020;29:822–827.

101. Wang J, Sun P, Xue H, Chen J, Zhang H, Zhu W. Red mud derived facile hydrothermal synthesis of hierarchical porous α-Fe_2O_3 microspheres as efficient adsorbents for removal of Congo red. *J Phys Chem Solids*. 2020;140, 109379.

102. Qian Y, Zhou S. Adsorption of two kinds of common dyes from aqueous solution by layered metal oxides based on red mud. *Chin J Environ Eng*. 2017;11(3):1402–1408.

103. Gupta VK, Ali I, Saini VK. Removal of chlorophenols from wastewater using red mud: an aluminum industry waste. *Environ Sci Technol*. 2004;38 (14):4012–4018.

104. Mandal A, Dey BB, Das SK. Thermodynamics, kinetics, and isotherms for phenol removal from wastewater using red mud. *Water Pract Technol*. 2020;15(3):705–722.

105. Tor A, Cengeloglu Y, Ersoz M. Increasing the phenol adsorption capacity of neutralized red mud by application of acid activation procedure. *Desalination*. 2009;242(1–3):19–28.

106. Tor A, Danaoglu N, Arslan G, Cengeloglu Y. Removal of fluoride from water by using granular red mud: batch and column studies. *J Hazard Mater*. 2009;164(1):271–278.

107. Liu CJ, Li YZ, Luan ZK, Chen ZY, Zhang ZG, Jia ZP. Adsorption removal of phosphate from aqueous solution by active red mud. *J Environ Sci*. 2007;19(10):1166–1170.

108. Park JH, Wang JJ, Seo DC. Sorption characteristics of phosphate by bauxite residue in aqueous solution. *Colloids Surf A Physicochem Eng Asp*. 2021;618, 126465.

109. Cengeloglu Y, Tor A, Ersoz M, Arslan G. Removal of nitrate from aqueous solution by using red mud. *Sep Purif Technol*. 2006;51(3):374–378.

110. Biftu WK, Mekala S, Ravindhranath K. De-fluoridation of polluted water using aluminium alginate beads doped with green synthesized 'nano SiO^{2+} nano CeO_2-ZrO_2', as an effective adsorbent. *ChemistrySelect*. 2020;5(47):15061–15074.

111. Aydın S, Bedük F, Ulvi A, Aydın ME. Simple and effective removal of psychiatric pharmaceuticals from wastewater treatment plant effluents by magnetite red mud nanoparticles. *Sci Total Environ*. 2021;784, 147174.

112. Aydin S. Removal of organophosphorus pesticides from aqueous solution by magnetic Fe_3O_4/red mud-nanoparticles. *Water Environ Res*. 2016; 88(12):2275–2284.

113. Das B, Prakash S, Reddy PSR, Misra VN. An overview of utilization of slag and sludge from steel industries. *Resour Conserv Recycl*. 2007;50(1): 40–57.

114. Lopez FA, Perez C, Lopez-Delgado A. *J Mater Sci Lett*. 1996;15:1310.

115. Lopez-Delgado, A., Perez, C., Lopez, F.A.., 1998. Conference: 8th, National Congress on the Science and Technology of Metallurgy. pp 164-168.

116. Lopez-Delgado A, Perez C, Lopez FA. Sorption of heavy metals on blast furnace sludge. *Water Res*. 1998;32(4):989.

117. Lopez FA, Carlos Sainz P, Alonso E, Manuel J. *J Chem Technol Biotechnol*. 1995;62(2):200.

118. de Jesús Soria-Aguilar M, Martínez-Luévanos A, Sánchez-Castillo MA, Carrillo-Pedroza FR, Toro N, Narváez-García VM. Removal of Pb (II) from aqueous solutions by using steelmaking industry wastes: effect of blast furnace dust's chemical composition. *Arab J Chem*. 2021;14(4), 103061.

119. Nehrenheim E, Gustafsson JP. Kinetic sorption modelling of Cu, Ni, Zn, Pb and Cr ions to pine bark and blast furnace slag by using batch experiments. *Bioresour Technol*. 2008;99(6):1571–1577.

120. Bhatnagar A, Jain AK, Minocha AK, Singh S. Removal of lead ions from aqueous solutions by different types of industrial waste materials: equilibrium and kinetic studies. *Sep Sci Technol*. 2006;41(9):1881–1892.

121. Dos Santos SV, Amorim CC, Andrade LN, et al. Steel wastes as versatile materials for treatment of biorefractory wastewaters. *Environ Sci Pollut Res*. 2015;22(2):882–893.

122. Malina J, Rađenović A. Kinetic aspects of methylene blue adsorption on blast furnace sludge. *Chem Biochem Eng Q*. 2014;28(4):491–498.

123. Yasipourtehrani S, Strezov V, Kan T, Evans T. Investigation of dye removal capability of blast furnace slag in wastewater treatment. *Sustainability*. 2021;13(4):1970.

124. Wagner M, Eicheler C, Helmreich B, Hilbig H, Heinz D. Removal of Congo red from aqueous solutions at hardened cement paste surfaces. *Front Mater*. 2020;7:357.

125. Bhatnagar A. Removal of bromophenols from water using industrial wastes as low cost adsorbents. *J Hazard Mater*. 2007;139(1):93–102.

126. Gupta VK, Mohan D, Suhas, Singh KP. Removal of 2-aminophenol using novel adsorbents. *Ind Eng Chem Res*. 2006;45(3):1113–1122.

127. Jain AK, Gupta VK, Jain S, Suhas. Removal of chlorophenols using industrial wastes. *Environ Sci Technol.* 2004;38(4):1195–1200.

128. Das CP, Patnaik LN. Removal of phenol by industrial solid waste. *Pract Period Hazard Toxic Radioact Waste Manage.* 2005;9(2):135–140.

129. Karthik RM, Philip L. Sorption of pharmaceutical compounds and nutrients by various porous low cost adsorbents. *J Environ Chem Eng.* 2021;9(1), 104916.

130. Yasipourtehrani S, Strezov V, Evans T. Investigation of phosphate removal capability of blast furnace slag in wastewater treatment. *Sci Rep.* 2019; 9(1):1–9.

131. Kostura B, Dvorsky R, Kukutschová JANA, Študentová SONA, Bednář JIRI, Mančík P. Preparation of sorbent with a high active sorption surface based on blast furnace slag for phosphate removal from wastewater. *Environ Prot Eng.* 2017;43(1). https://doi.org/10.37190/epe170113.

132. Samarina T, Takaluoma E. Simultaneous removal of nutrients by geopolymers made from industrial by-products. In: *Proceedings of the 5th World Congress on New Technologies (NewTech'19).* Lisbone: AVESTIA; 2019.

133. Sahoo PK, Tripathy S, Panigrahi MK, Equeenuddin SM. Evaluation of the use of an alkali modified fly ash as a potential adsorbent for the removal of metals from acid mine drainage. *Appl Water Sci.* 2013;3(3):567–576.

134. Liu C, Li Y, Hou Y. Preparation of a novel lignin nanosphere adsorbent for enhancing adsorption of lead. *Molecules.* 2019;24(15):2704.

135. Naiya TK, Bhattacharya AK, Das SK. Clarified sludge (basic oxygen furnace sludge)—an adsorbent for removal of Pb(II) from aqueous solutions–kinetics, thermodynamics and desorption studies. *J Hazard Mater.* 2009;170(1):252–262.

Chapter 9

Novel hydrochar as low-cost alternative adsorbent for the removal of noxious impurities from water

Suhas[a,*], Monika Chaudhary[a], Inderjeet Tyagi[b], Shubham Chaudhary[a], Sarita Kushwaha[a], and Ankur Kumar[c]

[a]*Department of Chemistry, Gurukula Kangri (Deemed to be University), Haridwar, India,* [b]*Centre for DNA Taxonomy, Molecular Systematics Division, Zoological Survey of India, Ministry of Environment, Forest and Climate Change, Government of India, Kolkata, West Bengal, India,* [c]*Central Instrumentation Laboratory, NIFTEM, Sonipat, India*

*Corresponding author.

1 Introduction

The removal of noxious impurities from water has become the main research topic at the present time owing to the different toxicological problems caused by them, which affect the environment and human health.[1–3] So far, activated carbon has been extensively utilized as a potential material for the treatment of various pollutants in waters.[4–13] Although activated carbon is widely used, its application is sometimes restricted due to high temperature, time taking, processing requirement, and requires a large amount of materials.[14,15] Therefore, from an economical point of view in the present time, it is essential to develop low-cost adsorbent material by applying some alternative methodology.

In recent times, hydrothermal treatment (HTT) of lignocellulosic biomass has encountered significant attention owing to its capability to develop hydrochar with fascinating characteristics.[16,17] HTT is an effective way for the carbonization of biomass to produce novel adsorbents without employing toxic chemicals.[18] Moreover, the HTT process has several advantages, such as no pre-drying of the biomass is required as water is used as the reaction medium.[19] No additional chemical besides water is required during the process, resulting in minimized corrosive problems to the reactor.[20] The lower energy amount required for the production of hydrochar in comparison to pyrolysis is also an advantage, and the prepared materials can be successfully employed as carbon precursors.[21]

During hydrothermal carbonization, dry or wet biomass is directly heated in a sealed reactor/autoclave at lower temperatures of 150-250°C under autogenous pressure.[22] The main interesting product, "hydrochar," is solid coal like product and possesses excellent features such as being rich in carbon content, and oxygenated functional groups.[23] Further, it was highly brittle and hydrophobic in nature, therefore, it can be easily separated from the liquid after the process.[24] Due to its various properties, it can be utilized in numerous applications such as biofuels, energy storage, CO_2 sequestration, and as an adsorbent for the water treatment.[16,25,26] Many researchers employed hydrochars prepared from different biomass such as bamboo,[27] walnut shell,[28] glucose and cellulose,[29] and peanut hull[30] as an adsorbent for the removal of different pollutants in water. Among the above-mentioned applications of hydrochar, the use of hydrochar as an efficient adsorbent has been preferred in this chapter, and the experimental findings were discussed on the removal of dye basic blue 3 (BB-3) and 4-nitrophenol (4-NP) from water bodies.

The escalating use of dyes in the industrial sector for coloring the different products generated a huge amount of colored water, which is a serious concern due to their toxicological, mutagenic and carcinogenic impact on the human and aquatic life.[31] Cationic dyes, which are predominantly utilized for the coloring of fibers, are termed as more toxic than other classes of dyes. Therefore, the removal of basic dyes (BB-3 in the current work) from the aqueous system is of great significance and needs to be studied. Moreover, water streams containing phenolic compounds and their derivatives are also a threat due to their persistent and carcinogenic nature and are termed as toxicological for mankind.[9] Hence, removing these compounds (4-NP in the current work) from water is essential for studies too.

Sustainable Materials for Sensing and Remediation of Noxious Pollutants. https://doi.org/10.1016/B978-0-323-99425-5.00018-9

Tectona grandis is a tall tree from Southeast Asia and is well known for its durable and precious wood.[32] The seed of this plant is a waste product and will be used as a precursor for the development of hydrochars using the novel hydrothermal carbonization technology at a specific temperature, pressure, and time using an autoclave/reactor.

Furthermore, the work is aimed to measure the efficiency of the hydrochar materials to remediate the noxious organic impurities, i.e., BB-3 and 4-NP, from aqueous solutions. Further, the effect of different parameters, i.e., contact time and initial concentration, has also been studied. The Langmuir, Freundlich, Temkin, and Dubinin-Radushkevich (D-R) adsorption isotherms have been used to test the equilibrium adsorption and kinetics data. Moreover, the adsorption comparison of H-TGs was also made with raw *T. grandis* seed (R-TGs) at 25°C.

2 Materials and methods

2.1 Precursor

T. grandis seeds were obtained from a local tree. The sample acquired was washed many times with the distilled water and oven-dried at 100°C for 24 h. The material obtained was further grounded well in a grinder to acquire the fine particles. The dried material was further sieved to get a homogenous size of the seeds. Finally, the sample was kept in a closed box at 25°C for further usage.

2.2 Hydrothermal treatment

HTT using *T. grandis* seeds was performed in 600 mL Parr stainless-steel reactor model no. 4760 with 4838 reactor controllers (Parr instrument company, USA) equipped with a pressure gauge. For this, 10 g of feedstock and 100 mL of distilled water were added into the reactor and heated at 180°C. It takes about 45 min to reach 180°C and was held at this temperature for 30 min under autogenous pressure. Furthermore, after cooling down the reactor, the hydrochar was separated by filtration and dried in an oven at 105°C for 24 h. Finally, the prepared hydrochar (H-TGs) was kept in sealed containers in desiccators for further use.

2.3 Adsorption study

The adsorption study has been investigated utilizing a batch technique that is easy and simple to study. For this, a fixed amount of the R-TGs and H-TGs (0.01 g) was added in a stoppered glass test tube containing 10 mL of varying concentrations of BB-3 and 4-NP. Further, it was kept in thermostat cum shaking assembly and was shaken well continuously at a fixed temperature until the equilibrium was achieved. After this, the remaining concentrations of BB-3 and 4-NP in an aqueous solution after adsorption were analyzed using an Agilent Carry 60 UV-visible spectrophotometer at the maximum wavelength (λ_{max}) of the BB-3 and 4-NP. Moreover, kinetic studies were carried out two different adsorbate concentrations. The amount of BB-3 and 4-NP adsorbed by R-TGs and H-TGs was obtained using the following expression:

$$\%\text{Adsorption} = \frac{C_i - C_e}{C_i} \times 100 \tag{1}$$

where C_0 (mol L^{-1}) initial concentration, C_e (mol L^{-1}) equilibrium concentrations in the solution, V is the volume of solution (L), W is the mass (g) of the R-TGs, or H-TGs used, and q_e (mol g^{-1}) is the amount of dye/phenol adsorbed.

2.4 Characterization of hydrochar

The analysis of moisture and ash content was carried out on the basis of standard methods mentioned elsewhere (ASTM D4442-07 and ASTM E1755-01), while the elemental analysis of the developed hydrochar from *T. grandis* seeds was carried out using Vario, Micro CHNS Analyzer. Further, the surface functionalization and mineralogical studies were carried out on Fourier transform infrared spectroscopy (FTIR) (Perkin Elmer Spectrum Two spectrometer) using the KBr disc method and Rigaku X-ray diffractometer using Cu-Kα radiation, respectively. The surface morphology of the hydrochar sample developed from *T. grandis* seeds was observed by a Tescan Mira 3 field-emission scanning electron microscope.

3   Results and discussion

3.1   Characterization of hydrochar

The elemental analysis of H-TGs at 180°C revealed that the developed material was rich in carbon and the CHNS analysis shows that it has carbon 51.160, hydrogen 9.637, nitrogen 1.620, and sulfur 0.055%. In addition to this, the moisture and ash content were found to be 8.70 and 3.268%, respectively, and the remaining was oxygen.

The structural parameters and features of the hydrochar obtained from the XRD analysis and the curve acquired are shown in Fig. 1A. The XRD pattern of the H-TGs shows two characteristic peaks at $2\theta \approx 16°$ and $22°$ corresponding

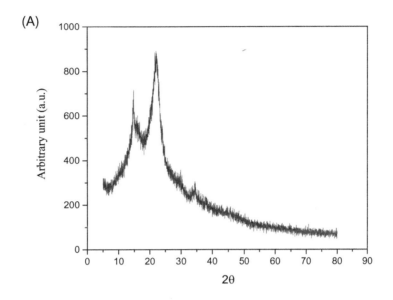

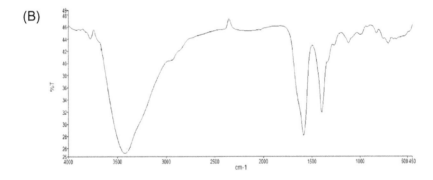

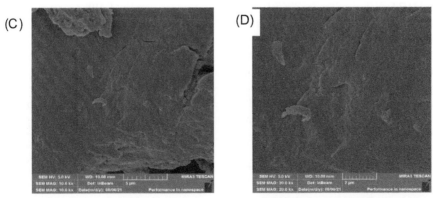

FIG. 1   (A) XRD patterns of hydrochar developed from *Tectona grandis* seeds at 180°C. (B) FTIR spectra of hydrochar developed from *T. grandis seeds* at 180°C. (C) SEM image of hydrochar developed from *T. grandis seeds* at 10.0 kx. (D) SEM image of hydrochar developed from *T. grandis seeds* at 20.0 kx (A).

to the planes (101) and (002).[33] It confirmed the fact that the sample was not completely converted into amorphous carbon at this temperature.[34]

The surface chemistry of the H-TGs was determined using FTIR, and the spectrograph obtained is shown in Fig. 1B. The broadband of hydrochar approximately at $3400 \, cm^{-1}$ was observed, and it reflects the hydroxyl stretching vibration.[35] The band at $2924 \, cm^{-1}$ corresponds to the C—H stretching vibration.[28,36] The peaks observed around $1400–1600 \, cm^{-1}$ specify aromatic C=C stretching, whereas the bands at 1114 and $1056 \, cm^{-1}$ are due to C—O—C and C—O stretching. A band in the region of $690–710 \, cm^{-1}$ is due to C—H bending.

SEM-EDX technique was used in order to examine the morphology of the H-TGs. The SEM images for hydrochar obtained from *T. grandis* seed at 180°C at different magnifications are shown in Fig. 1C and D. It can be seen from the images that the surface of the hydrochar is not smooth and non-porous in nature. The EDX result shows that hydrochar developed from *T. grandis* seeds has a high amount of Ca (2.64%) and P (0.32%) and also contain Co (0.16%), S (0.15%), Na (0.11%), K (0.09%), and Mg (0.08%); however, Cu and Zn are negligible.[37]

3.2 Effect of contact time and initial concentration

To develop a cost-effective process, it is needed to optimize the effect of contact time in the adsorption process.[38] The effect of contact time for the adsorption of BB-3 on the H-TGs was studied at 25°C. The result obtained shows that adsorption of BB-3 on H-TGs was fast during the initial stage reaching nearly 90% in 60 min. However, as expected, usually with the adsorption, it increased in this case too at a very slow speed and the equilibrium was achieved in nearly 240 min; therefore, the contact time was kept at 300 min for all the experiments. The reason behind this is that during the initial stage, large numbers of vacant sites are freely accessible, but sometimes, it is very difficult to occupy the residual vacant site owing to the repulsion between the adsorbate molecule on the solid and bulk phases.[39]

The effect of initial concentration of BB-3 on the adsorption performance on the hydrochar was investigated at different concentrations (3×10^{-5} and 4×10^{-5} M). Notably, it was observed that on increasing the initial concentration of BB-3, the amount adsorbed increased as the resistance to the mass transfer was reduced, and the interaction between the adsorbate and adsorbent improved.[40]

3.3 Adsorption isotherms

To interpret the interaction between the adsorbed amount of adsorbate per unit mass of the adsorbent, isotherm studies of BB-3 on hydrochar were carried out at equilibrium concentration (at 25°C) and are shown in Fig. 2A. Furthermore, to know the efficiency of the H-TGs, a comparative study on the R-TGs was also carried out, and the same is shown in Fig. 2A. The equilibrium adsorption capacities for BB-3 on R-TGs and H-TGs at 25°C were found to be 0.036 and $0.043 \, mmol \, g^{-1}$, respectively, and the findings obtained were compiled in Table 1. This difference clearly shows that hydrochar is superior to the raw samples in adsorbing BB-3 and can be a low-cost alternative for the removal of noxious dyes from water.

Moreover, the effect of temperature on the adsorption of BB3 on the hydrochar was also studied in the temperature range of 25–45°C, and the plots are given in Fig. 2B. It can be analyzed from Fig. 2B that with an increase in temperatures, the amount adsorbed was found to increase from 0.043 to $0.055 \, mmol \, g^{-1}$ which is an indication for the endothermic nature of the process in the case of BB-3 adsorption on hydrochar.

In addition to this, a comparison of BB-3 dye with phenol was also made, and 4-NP was selected to serve the purpose. Adsorption studies were carried out on both the samples, and interestingly negligible adsorption of 4-NP was found in both the cases (R-TGs and H-TGs) as compared to the BB-3, and therefore, no graphs (4-NP) are given. This phenomenon can be explained based on the surface charge of adsorbents (pzc of R-TGs and H-TGs are 6.43 and 6.37, respectively). As the surface of both adsorbents are negatively charged, the electrostatic repulsions dominate between 4-NP and the adsorbents resulting in a significant decrease of the removal of 4-NP. As the adsorption of 4-NP was found to be negligible in the case of both the adsorbents, therefore further findings on the adsorption of 4-NP are not discussed.

The equilibrium adsorption data obtained in this chapter were analyzed by applying four models viz. Langmuir, Freundlich, Dubinin-Radushkevich (D-R), and Temkin, and the results are shown in Table 1. The Langmuir isotherm model[41] was used to determine the maximum adsorption corresponding to the total monolayer coverage on the surface of the adsorbent and is given as follows:

$$\frac{1}{q_e} = \frac{1}{q_{max}} + \frac{1}{q_{max} b C_e} \tag{2}$$

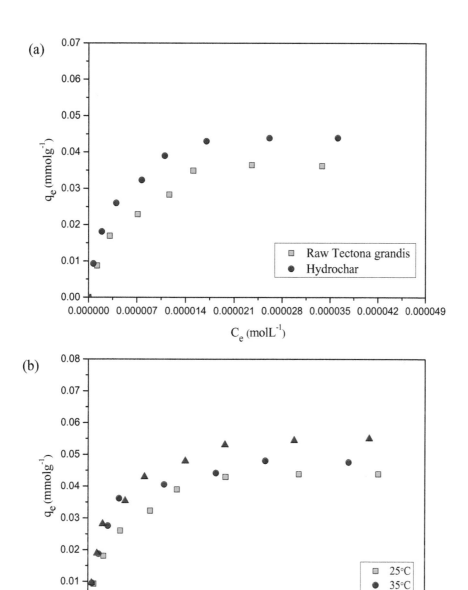

FIG. 2   (A) Adsorption isotherms for the removal of BB-3 on R-TGs and H-TGs at 25°C. (B) Adsorption isotherms for the removal of BB-3 on H-TGs at different temperatures.

where q_e (mmol g^{-1}) is the amount of the adsorbate adsorbed at equilibrium concentration; q_{max} (mmol g^{-1}) is the theoretical monolayer adsorption capacity; b (L mol^{-1}) is the Langmuir constant; and C_e (mol L^{-1}) is the equilibrium concentration of BB-3.

The plot was prepared between $\frac{1}{q_e}$ and $\frac{1}{C_e}$ to describe the removal of BB-3 on the H-TGs and is shown in Fig. 3A. The values of Langmuir parameters q_{max} and b obtained from the graph for BB-3 are provided in Table 1. The calculated value for Langmuir monolayer capacity (q_{max}) for BB-3 on H-TGs at 25°C was found to be 0.044 mmol g^{-1} which is in close vicinity of the experimental findings (0.043 mmol g^{-1}). The adsorption equilibrium constant (b) obtained from Langmuir isotherm shows the affinity of BB-3 for H-TGs, and the b value was found to be 3.86×10^5 L mmol^{-1} at 25°C. The values of q_{max}, as well as b obtained from the Langmuir isotherm model, increased on increasing the temperature indicating the endothermic nature of the process. Furthermore, the applicability of the adsorption of BB-3 on the hydrochar can be predicted with the R_L, a separation factor which is given as follows:

$$R_L = 1/(1 + bC_0) \tag{3}$$

TABLE 1 Experimental adsorption values, Langmuir, Freundlich, Temkin, and D-R isotherm parameters for the removal of basic blue 3 on the raw *Tectona grandis* (R-TGs) and hydrochar (H-TGs).

Parameters	Basic blue 3 on raw *T. grandis* (R-TGs)	Basic blue 3 on hydrochar		
temperature (°C)	25	25	35	45
Experimental				
q_{exp} (mg g^{-1})	13.1	15.7	17.3	19.8
q_{exp} (mmol g^{-1})	0.036	0.043	0.048	0.055
Langmuir isotherm parameters				
q_{max} (mmol g^{-1})	0.038	0.044	0.052	0.054
q_{max} (mg g^{-1})	12.6	15.8	18.7	19.7
b (L mmol^{-1})	2.50×10^5	3.86×10^5	4.56×10^5	4.99×10^5
R^2	0.997	0.998	0.997	0.996
Freundlich isotherm parameters				
K_F (mmol g^{-1})	11.2	12.8	17.7	21.8
N	1.91	1.98	1.96	1.93
R^2	0.977	0.985	0.935	0.951
Temkin isotherm parameters				
b (KJ mol^{-1})	0.782	0.732	0.829	0.889
A_T (L mg^{-1})	6.28	11.4	23.2	39.4
R^2	0.994	0.988	0.990	0.985
D-R isotherm parameters				
E (KJ mol^{-1})	1.8	2.3	2.7	2.9
q_m (mg g^{-1})	11.3	13.6	15.2	16.5
R^2	0.965	0.932	0.956	0.967

where C_0 (mol L^{-1}) and b (L mol^{-1}) are the initial concentration and Langmuir adsorption constant, respectively, was also calculated, and the R_L was found to be <1, indicating favorable adsorption of BB-3 on the H-TGs.

Freundlich isotherm[42] can be represented linearly as follows:

$$\log q_e = \log K_F + \frac{1}{n} \log C_e \tag{4}$$

where q_e is the amount of the adsorbate adsorbed at equilibrium concentration C_e; K_F represents the adsorption capacity; whereas n can be correlated to the adsorption intensity.

A plot between $\log q_e$ vs. $\log C_e$ was made and is shown in Fig. 3B for the adsorption of BB3 on the hydrochar. K_F and n were obtained from the intercept and slope, respectively, and the values are presented in Table 1.

Temkin model dealing with the interactions between the BB-3 and H-TGs was also taken into consideration and can be written as follows:

$$q_e = B_T \ln K_T + B_T \ln C_e \tag{5}$$

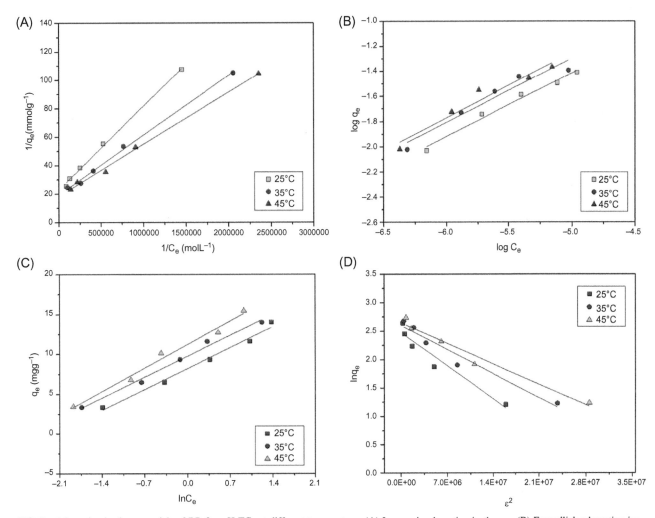

FIG. 3 Adsorption isotherm models of BB-3 on H-TGs at different temperatures (A) Langmuir adsorption isotherms; (B) Freundlich adsorption isotherms; (C) Temkin adsorption isotherms; (D) Dubinin-Radushkevich (D-R) adsorption isotherms.

where B_T is the heat of adsorption, T is the temperature in Kelvin, and K_T is Temkin binding constant at equilibrium in $L\,mg^{-1}$. The plot of q_e vs. $\ln C_e$ is illustrated in Fig. 3C.

The corresponding parameters, along with correlation coefficients, were obtained from the slope and the intercept of the plot and are given in Table 1.

Moreover, to predict the nature of the adsorption process, the Dubinin-Radushkevich (D-R) isotherm model was employed.

The linear form of the D-R model can be expressed as follows:

$$\ln q_e = \ln q_m - \beta_{DR}\varepsilon^2 \tag{6}$$

$$\varepsilon = RT \ln\left(1 + 1/C_e\right) \tag{7}$$

$$E = 1/\sqrt{(2\beta_{DR})} \tag{8}$$

where ε, q_m ($mg\,g^{-1}$), and β_{DR} ($mol^2\,kJ^{-2}$) are Polanyi potential, D-R monolayer adsorption capacity, and D-R constant, respectively. The mean free energy E ($kJ\,mol^{-1}$) can be calculated using Eq. (8) and utilized to describe the nature of the adsorption process: chemisorption or physisorption. A plot of $\ln q_e$ vs. ε^2 (Fig. 3D) was drawn, and the corresponding parameters, calculated from the slope and the intercept of the drawn plot, are depicted in Table 1. It can be observed from Table 1 that in the case of BB3 adsorption on the hydrochar, the experimental adsorption capacity (q_{exp}) obtained was found to be in close agreement with theoretical monolayer adsorption capacity (q_{max}) obtained from Langmuir as compared to other models. In addition to this on comparing the R^2 values (Table 1), the Langmuir model was found to fit better to the experimental data as compared to the other studied models.

3.4 Thermodynamics of adsorption

The thermodynamic properties associated with the adsorption process can be determined by calculating $\Delta G°$, $\Delta H°$, and $\Delta S°$, using the following equations:

$$\Delta G° = -RT \ln(b) \tag{9}$$

$$\ln b = -\frac{\Delta H°}{RT} + \frac{\Delta S°}{R} \tag{10}$$

where R is the gas constant $(8.314\,\mathrm{J\,mol^{-1}K^{-1}})$, T is the temperature (°C), and b is the thermodynamic equilibrium constant.

The slope and intercept of the plot were used for calculating the values of $\Delta H°$ and $\Delta S°$ (Fig. 4A). Results obtained in terms of thermodynamic parameters are presented in Table 2. In the study, $\Delta G°$ value obtained for basic blue 3 was found to be 31.9–34.7 kJ mol^{-1}. The negative value of $\Delta G°$ suggests that the process is favorable for the adsorption of BB-3 on H-TGs. The value of $\Delta H°$ for BB-3 was found to be 10.1 kJ mol^{-1}, i.e., the process is endothermic in nature. Besides this, the positive $\Delta S°$ value indicates the enhancement in randomness at the solid/solution interface during the sorption.[43]

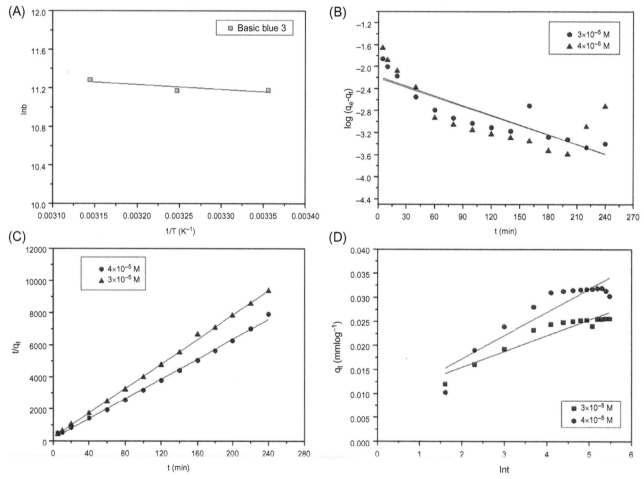

FIG. 4 (A) Van't Hoff plots for the removal of BB-3 on H-TGs, and kinetic plots for the removal of BB-3 on H-TGs at different initial concentrations: 3×10^{-5} M and 4×10^{-5} M at T: 25°C; (B) pseudo-first order; (C) pseudo-second order; (D) Elovich kinetic plot.

TABLE 2 Thermodynamic parameters for the adsorption of basic blue 3 on hydrochar.

Temperature	$-\Delta G^\circ$ (kJ mol^{-1})	ΔS° (J mol^{-1} K^{-1})	ΔH° (kJ mol^{-1})
25	31.9	140.9	10.1
35	33.4		
45	34.7		

3.5   Kinetic studies

The experimental data acquired from the adsorption study on the hydrochar were analyzed by applying the pseudo-first-order, pseudo-second-order and Elovich kinetic models in order to know the mechanisms and dynamics of the adsorption process.

Pseudo-first-order kinetic equation has been used widely and is expressed by the following equation[44]:

$$\log (q_e - q_t) = \log q_e - \frac{k_1}{2.303} t \tag{11}$$

where q_e and q_t are the amount of BB3 adsorbed (mmol g^{-1}) on the *hydrochar* at equilibrium and at time t (min), respectively, and k_1 is the Lagergren adsorption rate constant (min^{-1}). The values of k_1 and q_e were obtained from the slope and intercept by the plot made between $\log (q_e - q_t)$ vs. t. The plot obtained for BB3 on the hydrochar is presented in Fig. 4B and the values are provided in Table 3, and the calculated value ($q_{e(\text{cal})}$) for BB3 is less than the experimental values ($q_{e(\text{exp})}$).

Pseudo-second-order kinetic model[45] was also used and is expressed by the following equation:

$$\frac{t}{q_t} = \frac{1}{k_2 q_e^2} + \frac{1}{q_e} t \tag{12}$$

where q_e and q_t are the amount of BB3 adsorbed (mmol g^{-1}) on the hydrochar at equilibrium and time t, respectively, and the parameter k_2 (g mmol^{-1} min^{-1}) is the rate constant. The parameters of q_e and k_2 were calculated from the slope and intercept of the plot between t/q_t vs. t (Fig. 4C) and the results are given in Table 3.

The Elovich kinetic model was also used to simulate the adsorption data and can be represented as follows:

$$q_t = 1/\beta \ln (\alpha\beta) + 1/\beta \ln (t) \tag{13}$$

where α (mmol g^{-1} min^{-1}) is the initial adsorption rate, β (g mmol^{-1}) is the constant related to the surface coverage or desorption rate constant, and q_t (mmol g^{-1}) is the amount of adsorbed BB3 at time t (min). The plot between $\ln t$ and q_t was drawn and is shown in Fig. 4D. The parameters obtained from the plot are given in Table 3.

Further, to determine the better applicability of the analyzed models to the adsorption data, R^2 values for all the applied models were also determined and are shown in Table 3. Table 3 illustrates that the R^2 for the pseudo-second order is greater than the pseudo-first order and Elovich kinetic models. Besides, the $q_{e(\text{cal})}$ values of pseudo-second order are quite close to the $q_{e(\text{exp})}$, showing the applicability of the model as compared to the pseudo-first order and Elovich kinetic models.

4   Conclusions

This chapter presented the applicability of the hydrochar (H-TGs) prepared at 180°C via HTT using *T. grandis* seeds. In this chapter, it is shown that the prepared hydrochar can be utilized as a low-cost alternative adsorbent for the adsorptive removal of the noxious pollutant (dye), and more studies (at higher temperature/pressure or on other noxious impurities) may be performed to find newer aspects. The results obtained showed that the adsorption of BB3 was found to be 13.1 (0.036 mmol g^{-1}) and 15.7 mg g^{-1} (0.043 mmol g^{-1}) on raw *T. grandis* and the prepared hydrochar, respectively, and can be a low-cost alternative for the removal of noxious dyes from water. The adsorption of 4-NP was found to be negligible on both adsorbents, i.e., raw *T. grandis* and hydrochar owing to the negatively charged surface of the adsorbents. Four isotherm models viz. Langmuir, Freundlich, Temkin, and D-R were analyzed to test the adsorption data, and the Langmuir isotherm model was found to well-fit the process. The adsorption of BB3 on the hydrochar was endothermic in nature. The negative values of ΔG° suggested that the process is favorable for the adsorption. Three kinetic models, viz. pseudo-first order, pseudo-second order and Elovich models, were applied on the adsorption data, and among these the adsorption kinetic data fitted well to the pseudo-second-order model for the adsorption of BB-3 on the hydrochar developed from *T. grandis* seeds.

TABLE 3 Kinetic parameters for the removal of basic blue 3 on hydrochar.

| Adsorbent | Dyes | C_0 (mol L^{-1}) | Experimental | Pseudo-first order | | | Pseudo-second order | | | Elovich | | |
			$q_{e(exp)}$ (mmol g^{-1})	$q_{e(cal)}$ (mmol g^{-1})	K_1 (min^{-1})	R^2	$q_{e(cal)}$ (mmol g^{-1})	K_2 (g mmol^{-1} min^{-1})	R^2	α (mmol g^{-1} min^{-1})	β (g mmol^{-1})	R^2
Hydrochar	BB-3	3×10^{-5}	0.0256	0.986	5.02	0.800	0.0262	6.67	0.998	0.0496	304.2	0.897
		4×10^{-5}	0.0322	0.987	5.08	0.549	0.0323	6.08	0.997	0.0229	206.1	0.846

Acknowledgments

Authors are thankful to DST, New Delhi, India, for the financial support under Water Technology Initiative (Project No: DST/TMD/EWO/WTI/2K19/EWFH/2019/90).

References

1. Couto CF, Lange LC, Amaral MC. Occurrence, fate and removal of pharmaceutically active compounds (PhACs) in water and wastewater treatment plants—a review. *J Water Process Eng.* 2019;32:100927.
2. Dehghani MH, Omrani GA, Karri RR. Solid waste—sources, toxicity, and their consequences to human health. In: *Soft Computing Techniques in Solid Waste and Wastewater Management.* Elsevier; 2021:205–213.
3. Karri RR, Ravindran G, Dehghani MH. Wastewater—sources, toxicity, and their consequences to human health. In: *Soft Computing Techniques in Solid Waste and Wastewater Management.* Elsevier; 2021:3–33.
4. Gupta VK, Suhas. Application of low-cost adsorbents for dye removal—a review. *J Environ Manag.* 2009;90(8):2313–2342. https://doi.org/10.1016/j.jenvman.2008.11.017.
5. Gupta VK, Carrott PJM, Ribeiro Carrott MML, Suhas. Low-cost adsorbents: growing approach to wastewater treatment—a review. *Crit Rev Environ Sci Technol.* 2009;39(10):783–842. https://doi.org/10.1080/10643380801977610.
6. Suhas, Gupta VK, Carrott PJM, Singh R, Chaudhary M, Kushwaha S. Cellulose: a review as natural, modified and activated carbon adsorbent. *Bioresour Technol.* 2016;216:1066–1076. https://doi.org/10.1016/j.biortech.2016.05.106.
7. Chaudhary M, Suhas, Singh R, et al. Microporous activated carbon as adsorbent for the removal of noxious anthraquinone acid dyes: role of adsorbate functionalization. *J Environ Chem Eng.* 2021;9(5):106308. https://doi.org/10.1016/j.jece.2021.106308.
8. Suhas, Carrott PJM, Ribeiro Carrott MML. Lignin—from natural adsorbent to activated carbon: a review. *Bioresour Technol.* 2007;98(12):2301–2312. https://doi.org/10.1016/j.biortech.2006.08.008.
9. Nabais JMV, Gomes JA, Suhas, Carrott PJM, Laginhas C, Roman S. Phenol removal onto novel activated carbons made from lignocellulosic precursors: influence of surface properties. *J Hazard Mater.* 2009;167(1–3):904–910. https://doi.org/10.1016/j.jhazmat.2009.01.075.
10. Mostafa M, Sarma S, Yousef A. Removal of organic pollutants from aqueous solution: part 1, adsorption of phenol by activated carbon. *Indian J Chem.* 1989;28A:946–948.
11. Santoso E, Ediati R, Kusumawati Y, Bahruji H, Sulistiono DO, Prasetyoko D. Review on recent advances of carbon based adsorbent for methylene blue removal from waste water. *Mater Today Chem.* 2020;16:100233. https://doi.org/10.1016/j.mtchem.2019.100233.
12. Karri RR, Sahu JN. Process optimization and adsorption modeling using activated carbon derived from palm oil kernel shell for Zn (II) disposal from the aqueous environment using differential evolution embedded neural network. *J Mol Liq.* 2018;265:592–602. https://doi.org/10.1016/j.molliq.2018.06.040.
13. Dehghani MH, Karri RR, Yeganeh ZT, et al. Statistical modelling of endocrine disrupting compounds adsorption onto activated carbon prepared from wood using CCD-RSM and DE hybrid evolutionary optimization framework: comparison of linear vs non-linear isotherm and kinetic parameters. *J Mol Liq.* 2020;302:112526. https://doi.org/10.1016/j.molliq.2020.112526.
14. Elaigwu SE, Rocher V, Kyriakou G, Greenway GM. Removal of Pb2+ and Cd2+ from aqueous solution using chars from pyrolysis and microwave-assisted hydrothermal carbonization of *Prosopis africana* shell. *J Ind Eng Chem.* 2014;20(5):3467–3473. https://doi.org/10.1016/j.jiec.2013.12.036.
15. Alatalo S-M, Repo E, Mäkilä E, Salonen J, Vakkilainen E, Sillanpää M. Adsorption behavior of hydrothermally treated municipal sludge & pulp and paper industry sludge. *Bioresour Technol.* 2013;147:71–76. https://doi.org/10.1016/j.biortech.2013.08.034.
16. Jain A, Balasubramanian R, Srinivasan MP. Hydrothermal conversion of biomass waste to activated carbon with high porosity: a review. *Chem Eng J.* 2016;283:789–805. https://doi.org/10.1016/j.cej.2015.08.014.
17. Hairuddin MN, Mubarak NM, Khalid M, Abdullah EC, Walvekar R, Karri RR. Magnetic palm kernel biochar potential route for phenol removal from wastewater. *Environ Sci Pollut Res.* 2019;26(34):35183–35197. https://doi.org/10.1007/s11356-019-06524-w.
18. Wei L, Sevilla M, Fuertes AB, Mokaya R, Yushin G. Hydrothermal carbonization of abundant renewable natural organic chemicals for high-performance supercapacitor electrodes. *Adv Energy Mater.* 2011;1(3):356–361. https://doi.org/10.1002/aenm.201100019.
19. Heilmann SM, Jader LR, Sadowsky MJ, Schendel FJ, von Keitz MG, Valentas KJ. Hydrothermal carbonization of distiller's grains. *Biomass Bioenergy.* 2011;35(7):2526–2533. https://doi.org/10.1016/j.biombioe.2011.02.022.
20. Ruiz HA, Rodríguez-Jasso RM, Fernandes BD, Vicente AA, Teixeira JA. Hydrothermal processing, as an alternative for upgrading agriculture residues and marine biomass according to the biorefinery concept: a review. *Renew Sust Energ Rev.* 2013;21:35–51. https://doi.org/10.1016/j.rser.2012.11.069.
21. Falco C, Perez Caballero F, Babonneau F, et al. Hydrothermal carbon from biomass: structural differences between hydrothermal and pyrolyzed carbons via ^{13}C solid state NMR. *Langmuir.* 2011;27(23):14460–14471. https://doi.org/10.1021/la202361p.
22. Libra JA, Ro KS, Kammann C, et al. Hydrothermal carbonization of biomass residuals: a comparative review of the chemistry, processes and applications of wet and dry pyrolysis. *Biofuels.* 2011;2(1):71–106. https://doi.org/10.4155/bfs.10.81.
23. Chowdhury SR, Yanful EK. Arsenic and chromium removal by mixed magnetite-maghemite nanoparticles and the effect of phosphate on removal. *J Environ Manag.* 2010;91(11):2238–2247. https://doi.org/10.1016/j.jenvman.2010.06.003.
24. Hoekman SK, Broch A, Robbins C. Hydrothermal carbonization (HTC) of lignocellulosic biomass. *Energy Fuel.* 2011;25(4):1802–1810. https://doi.org/10.1021/ef101745n.

25. Titirici M-M, Thomas A, Antonietti M. Back in the black: hydrothermal carbonization of plant material as an efficient chemical process to treat the CO_2 problem? *New J Chem*. 2007;31(6):787–789. https://doi.org/10.1039/B616045J.

26. Béguin F, Presser V, Balducci A, Frackowiak E. Carbons and electrolytes for advanced supercapacitors. *Adv Mater*. 2014;26(14):2219–2251. https://doi.org/10.1002/adma.201304137.

27. Li Y, Meas A, Shan S, Yang R, Gai X. Production and optimization of bamboo hydrochars for adsorption of Congo red and 2-naphthol. *Bioresour Technol*. 2016;207:379–386. https://doi.org/10.1016/j.biortech.2016.02.012.

28. Román S, Nabais JMV, Laginhas C, Ledesma B, González JF. Hydrothermal carbonization as an effective way of densifying the energy content of biomass. *Fuel Process Technol*. 2012;103:78–83. https://doi.org/10.1016/j.fuproc.2011.11.009.

29. Lu X, Berge ND. Influence of feedstock chemical composition on product formation and characteristics derived from the hydrothermal carbonization of mixed feedstocks. *Bioresour Technol*. 2014;166:120–131. https://doi.org/10.1016/j.biortech.2014.05.015.

30. Pellera F-M, Giannis A, Kalderis D, et al. Adsorption of Cu(II) ions from aqueous solutions on biochars prepared from agricultural by-products. *J Environ Manag*. 2012;96(1):35–42. https://doi.org/10.1016/j.jenvman.2011.10.010.

31. Jalil AA, Triwahyono S, Adam SH, et al. Adsorption of methyl orange from aqueous solution onto calcined Lapindo volcanic mud. *J Hazard Mater*. 2010;181(1):755–762. https://doi.org/10.1016/j.jhazmat.2010.05.078.

32. Palanisamy K, Hegde M, Yi J-S. Teak (*Tectona grandis* Linn. f.): a renowned commercial timber species. *J For Environ Sci*. 2009;25.

33. Sevilla M, Fuertes AB. The production of carbon materials by hydrothermal carbonization of cellulose. *Carbon*. 2009;47(9):2281–2289. https://doi.org/10.1016/j.carbon.2009.04.026.

34. Melo CA, Junior FHS, Bisinoti MC, Moreira AB, Ferreira OP. Transforming sugarcane bagasse and vinasse wastes into hydrochar in the presence of phosphoric acid: an evaluation of nutrient contents and structural properties. *Waste Biomass Valoriz*. 2017;8(4):1139–1151. https://doi.org/10.1007/s12649-016-9664-4.

35. Suhas, Carrott PJM, Carrott MMLR, Singh R, Singh LP, Chaudhary M. An innovative approach to develop microporous activated carbons in oxidising atmosphere. *J Clean Prod*. 2017;156:549–555.

36. Kang S, Li X, Fan J, Chang J. Characterization of hydrochars produced by hydrothermal carbonization of lignin, cellulose, d-xylose, and wood meal. *Ind Eng Chem Res*. 2012;51(26):9023–9031. https://doi.org/10.1021/ie300565d.

37. Apaydın-Varol E, Pütün AE. Preparation and characterization of pyrolytic chars from different biomass samples. *J Anal Appl Pyrolysis*. 2012;98:29–36. https://doi.org/10.1016/j.jaap.2012.07.001.

38. Ghaedi M, Nasab AG, Khodadoust S, Rajabi M, Azizian S. Application of activated carbon as adsorbents for efficient removal of methylene blue: kinetics and equilibrium study. *J Ind Eng Chem*. 2014;20(4):2317–2324. https://doi.org/10.1016/j.jiec.2013.10.007.

39. Ai L, Zhang C, Liao F, et al. Removal of methylene blue from aqueous solution with magnetite loaded multi-wall carbon nanotube: kinetic, isotherm and mechanism analysis. *J Hazard Mater*. 2011;198:282–290. https://doi.org/10.1016/j.jhazmat.2011.10.041.

40. Dursun G, Çiçek H, Dursun AY. Adsorption of phenol from aqueous solution by using carbonised beet pulp. *J Hazard Mater*. 2005;125(1):175–182. https://doi.org/10.1016/j.jhazmat.2005.05.023.

41. Langmuir I, et al. *J Am Chem Soc*. 1918;40(9):1361–1403. https://doi.org/10.1021/ja02242a004.

42. Freundlich H. Over the adsorption in solution. *J Phys Chem*. 1906;57(385471):1100–1107.

43. Deniz F, Karaman S. Removal of basic red 46 dye from aqueous solution by pine tree leaves. *Chem Eng J*. 2011;170(1):67–74.

44. Lagergren SK. About the theory of so-called adsorption of soluble substances. *Sven Vetenskapsakad Handingarl*. 1898;24:1–39.

45. Ho Y-S, McKay G. Sorption of dye from aqueous solution by peat. *Chem Eng J*. 1998;70(2):115–124.

Chapter 10

Agricultural and agro-wastes as sorbents for remediation of noxious pollutants from water and wastewater

Tarun Kumar Kumawat[a,*], Vishnu Sharma[a], Varsha Kumawat[b], Anjali Pandit[a], and Manish Biyani[c]

[a]*Department of Biotechnology, Biyani Girls College, University of Rajasthan, Jaipur, Rajasthan, India,* [b]*Biyani Institute of Pharmaceutical Sciences, Rajasthan University for Health Sciences, Jaipur, Rajasthan, India,* [c]*Department of Bioscience and Biotechnology, Japan Advanced Institute of Science and Technology, Nomi, Ishikawa, Japan*

[*]*Corresponding author.*

Abbreviations

AWs	agro-wastes
EDS	energy-dispersive X-ray spectroscopy
FTIR	Fourier transform infrared
HMs	heavy metals
MNRE	Ministry of New and Renewable Energy
MWCNTs	multi-walled carbon nanotubes
PAHs	polycyclic aromatic hydrocarbons
SEM	scanning electron microscopy

1 Background

Water is a vital component of life on Earth and plays an important part in both the economic and social development of the globe.[1] Water contamination is a major global problem due to industrialization, population growth, and climate change.[2] Wastewater discharged by industries containing pollutants such as heavy metals, dyes, and other hazardous compounds causes many problems for the environment and people (Fig. 1).[3–6] Waterborne diseases kill 10–20 million people annually, while nonfatal illnesses harm over 200 million. Diarrhea induced by polluted water kills 5000–6000 children each day.[7]

Globally, agricultural waste is increasing due to population growth and technological advancement. Agriculture wastes are mostly made up of crops (such as stubble, fruits, and vegetables), animal waste (such as manure, waste feed, and animal carcasses), and by-products of food processing that have little or no economic value to the farmer.[8] The incineration of agricultural wastes (agro-wastes) results in landfilled ash that, in most instances, creates an environmental hazard due to disposal.[9] Thus, methods to their use, reuse, and processing are required to remove noxious pollutants from water and decrease pollution.[10] Increasing levels of agricultural waste, water pollution, and energy scarcity have prompted us to combine the process of water remediation with agricultural waste and renewable energy production for the sake of society's sustainable development.[11]

Water purification has been achieved through physical, chemical, and biological techniques.[2] Agricultural wastes are leftovers from the agricultural output, including fruit and vegetable harvesting and processing, and grape, banana, olive, and milk processing. These wastes may be a treasure when turned into useful applications.[12] Many years have passed since agricultural solid wastes were employed to remove metal ions and colors from aqueous phase effluent.[13] Biosorption is the passive and metabolically independent absorption of pollutants by biomass. Biosorption performance is influenced by biosorbent selection, modification technique, process parameters, and biosorption processes.[14] This chapter discusses many methods for agro-waste adsorbent as corrective strategies for decreasing the toxicity of organic and inorganic noxious pollutants so that people may drink clean water.

FIG. 1   Major water pollutants.

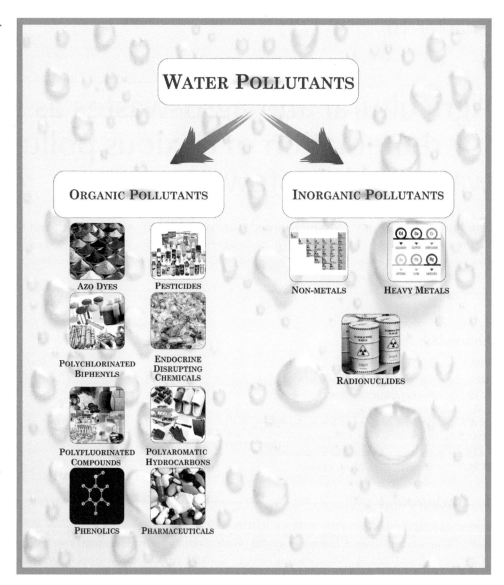

2   Agricultural or agro-waste

The agricultural industry has unavoidably attracted attention due to increasing concern about population growth, climate change, and resource depletion. Technological advancements have resulted in a rapid increase in agricultural output and a massive generation of agro-waste.[15,16] Agricultural wastes are residues from the cultivation, harvesting, and processing of main agricultural products such as fruit and vegetables, animals, milk-based products, and harvests.[17] Their composition changes based on the system and agricultural operations. The kind of farming impacts the amount of waste produced.[18] Agricultural wastes are plentiful, a nuisance to the surrounding region, and some are renowned for their foul odor. Decomposing agricultural wastes can alter soil pH.[19]

Today, the amassing of unregulated agricultural, industrial, and other solid wastes contributes to a rise in environmental concern, particularly in emerging nations.[20] Intensive farming production has increased animal waste, field crop leftovers, and agriculture industry by-products. Agricultural by-products are projected to surge if developing countries keep expanding farming techniques. Agricultural development is often accompanied by wastes from poor agricultural practices and pesticide misuse.[18] Agro-waste comprises organic food waste, farm waste, livestock feed, gardening, fisheries, and other sources (Fig. 2). According to the Ministry of New and Renewable Energy (MNRE) data in 2009, agriculture produces approximately 140 billion metric tons of organic matter per year globally, and India generates roughly 500 metric

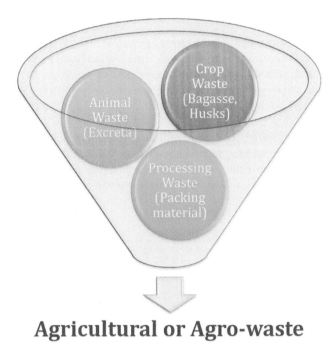

Agricultural or Agro-waste

FIG. 2 Agriculture waste.

tons of agro-waste annually.[21,22] There is, nevertheless, an urgent need to decrease agro-waste and utilize it in a constructive and organized manner, perhaps through biological composting, power generation, livestock production, and other alternatives.[23] Using anaerobic and aerobic methods, waste may be transformed into organic manures, compost or disposed of in landfills with minimal environmental impact.[24] At the moment, agricultural waste products are being touted as cost-effective and environmentally beneficial.[25]

3 Composition of agricultural wastes

The agro-wastes are produced from cultivation activities, livestock production, and plant wastes (Fig. 3).

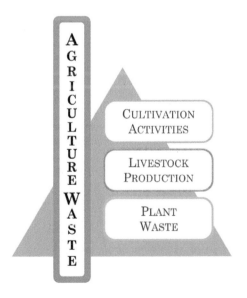

FIG. 3 Areas of agricultural waste.

3.1 Cultivation activities

The major cause of agricultural waste creation is cultivation operations. Every phase of farming activities, from land clearance through harvest, leads to the creation of agricultural waste. Agricultural expansion is often followed by wastes from improper intensive agricultural methods and chemical misuse, both of which have a major effect on rural settings and the broader environment globally.[18] Agro-waste materials come from a variety of places. Pesticides, such as herbicides and insecticides, are these contributors. If pesticide use is fully eliminated, it is projected that world food output will drop by roughly 42%.[26]

Cultivation activities include wheat, cotton, paddy, rice hulls, and jute stalks. Straw is a wheat harvest by-product that is utilized as cattle feed, carpets, dried flowers, hats, and a range of other handmade goods. Rice hulls are commonly used as fuel. Cotton canes are leftovers from the cotton crop. Power generation plants, hardwood industries, particleboard manufacturing, and compost all employ canes. They are utilized for mulching, composting, timber, and animal feed, bedding, and shelter.[17]

3.2 Waste from livestock production

Solid waste including manure and organic matters in the slaughterhouse; wastewater, like urine, water from animal washing and cleaning; pollutants in the air, such as hydrogen sulfide and methane; and odors are all examples of waste from livestock activities.[17] Livestock waste contributes significantly to greenhouse gas emissions that causes pollution, diseases, and stinks. Agriculture and animal by-products create 40% of worldwide methane, with waste disposal accounting for the remaining 18%.[27] It is a rich source of energy and nutrients that may be recovered for agricultural use. Effective management of animal manure as biofuel, biocompost, and vermicompost production may significantly improve crop productivity and sustainability.[28]

3.3 Plant waste

Plant trash comprises banana stalks and leaves, coconut garbage, sugarcane waste, horticulture waste, and other agro-industrial wastes.[29,30] Banana leaves and stalks are used in paint and waterproofing. Straw, coconut husks, coir fiber, coconut shell, and other by-products of coconuts can be obtained. Boards, fibers, roofs, pavements, and mats are all made from them.[17] Ruined, unwanted, and damaged vegetables or fruits, as well as roots, stems, and leaves, and perished plants, are all examples of horticulture waste. Such wastes are usually composted or utilized as cattle feed.[31]

4 Agricultural waste adsorbents

Agricultural, domestic, and industrial by-products are acknowledged as sustainable wastewater treatment solutions. They remove contaminants from wastewater and contribute to waste reduction, recovery, and reuse.[32] Adsorbents are made from agricultural wastes and residues, such as rice husks, starch-based waste products, and saw dust[33] and cashew-almond shells.[34] Also, residues like potato wastes, maize wastes, tea wastes, etc., can be effectively used to remove toxic elements from water.[35,36] Rice residual wastes are a common type of agricultural trash produced in significant quantities, especially in developing countries. Rice bran, rice straw, and husk are examples of such wastes.[37] Heavy metals have been demonstrated to be efficiently removed from aqueous solutions using these waste products.[38] Biosorbents produced from agricultural waste are similarly low cost and have high adsorption properties. Wastes account for a substantial percentage of global agricultural production. Various estimates estimate that these wastes account for almost 30% of worldwide agricultural output. Agricultural waste may be both beneficial and detrimental to the environment.[39]

4.1 Bagasse-based adsorbents

An agricultural waste, bagasse is from the sugar industry regarded as a low-cost adsorbent.[40] This is a naturally occurring lignocellulosic substance derived from sugarcane that may be used to absorb textile dyes, heavy metals, etc.[41,42] Activated carbon from bagasse was studied to effectively remove Cd and Zn, while the adsorption of Cd (II) was slightly more than the later and this was reported to be a result of high temperature.[43]

4.2 Adsorbents made from potato waste

Agricultural waste materials are getting a lot of attention these days as adsorbents for removing dyes, metals from wastewater because of their cheap cost and wide availability. The potato plant, *Solanum tuberosum* (potato), is a Solanaceae herbaceous perennial. The tuber is the only edible component of the plant; the remainder of the plant is discarded as agricultural waste after harvesting.[44] Potato peel waste is utilized as a low-cost adsorbent, and its surface area is very low compared to other adsorbents since it is not activated chemically.[45] Potato peels were utilized to make charcoal, and the material's Cu (II) adsorptive competence in industrial effluent samples was determined.[46]

4.3 Adsorbents from maize waste

Scientists and environmentalists must manage natural resources and waste streams to produce a safe and healthy environment.[47] The world's most commonly cultivated grain, maize, is used to clean water. Corn residues from the starch and glucose manufacturing industries were utilized to make activated carbons.[48] Chemical alterations would improve their inherent ion-exchange performance and contribute positively to the by-products, resulting in increased removal efficiency.[49,50]

4.4 Adsorbents based on tea waste

Tea residue is called bio-waste and is another organic waste. It has a high amount of aromatic, hydroxyl, carboxylate, and phenolic groups in its leaves, allowing a longer ion replacement period, increasing its pollutant uptake ability.[51,52] Before treating it with Trichoderma reesei cellulases *Bacillus* sp., alkali, magnetite, and other methods have been used to enhance tea litter pollution absorption.[53–55]

4.5 Adsorbents from peanut waste

Activated carbon and ashes from peanut plants have been shown to remove contaminants such as heavy metals, dyes, and other chemicals. Several research studies indicate that peanut husks may be used as a low-cost, biosorbent for heavy metal ions.[56,57]

5 Characterization of agro-waste adsorbent

It is critical to use sustainable methods that use waste resources to create useful goods to develop economically viable technologies.[58] Solutes may be removed from aqueous solutions via adsorption. Various biosorbents may be used in wastewater treatment systems.[59] SEM, EDS, XRD, and FTIR spectroscopy were utilized to evaluate the agro-based adsorbent.[60,61] SEM is used to determine the adsorbent's morphology, EDS for elemental analysis, and XRD for crystallinity. FTIR spectroscopy was utilized to examine the adsorbent's functional groups.[62] Agricultural waste adsorbents have a coarse-grained solid surface with a limited pore volume. FTIR showed hydrophobic methyl, hyddeterminesroxylic acid, and carboxylic groups.[63]

6 Process of agro-waste adsorption

The adsorption process involves concentrating adsorbate at the interface of two layers, either in the gas or liquid phase, onto the surface of the adsorbent.[13] For wastewater treatment operations, an adsorption system that deals with liquid/solid interfaces are called adsorption system.[64] Tannin-based adsorbents naturally adsorb heavy metals, dyes, surfactants, and medicinal compounds from contaminated water, making them ideal for water treatment.[59] The absorption process can be physisorption or chemisorption process.[65]

The association between the solid surface and the attached molecules in the physical adsorption process is physical since it is based on weak van der Waal contact force, making the process permanent. This physisorption process is considered a poor system, requiring a minimal energy barrier of 5 to 40 kJ/mol.[66,67] This adsorption method also reduces adsorption entropy and free energy, suggesting exothermic physisorption.[68]

Chemisorption occurs when adsorbed particles and the solid adsorbent surface establish chemical crosslinks by transfer of electrons or by pairing. Chemisorption necessitates large activation energies, ranging from 40 to 800 kJ/mol, and always takes place as a monolayer.[69] In contrast to physisorption, the bound material on the adsorbent surface is harder to eliminate by chemisorption because greater contact forces are involved.[70] It is worth noting that, depending on the wastewater treatment system, both adsorption processes may occur at the same time or alternately.

7 Adsorption of inorganic contaminants from water using agro-waste

Inorganic pollutants polluting surface water sources are a major problem in more industrialized countries; therefore, it is essential to eliminate them using eco-friendly techniques.[71] Due to the massive quantity of agricultural waste produced globally, a pragmatic change in waste material management and wastewater treatment technologies is now underway.[72] Agro-waste-derived products as low-cost adsorbent materials have become more important in removing inorganic pollutants from contaminated waterways.[73] Agricultural waste is porous, loose, and includes functional groups like carboxyl and hydroxyl.[74]

7.1 Heavy metal pollution removal from wastewater

Intensified population expansion, industrialization, urbanization and rural activities are depleting water supplies.[75] Anthropogenic pollutions have grown rapidly over the past several decades, reaching worrisome levels in terms of detrimental impacts on living creatures.[76] Heavy metal contamination of surface and groundwater was caused by fertilizer, mining activities, fuel production, metal plating facilities, paper, pesticide, tanneries, electroplating, battery, aerospace and atomic energy, photography, metal surface treatment, and leather industries.[77] Globally, researchers are working to preserve water quality and prevent contamination. Heavy metal removal from wastewater has been extensively studied.[78–80] Electrochemical precipitations, membrane separation, ion exchange, and adsorption are examples of such methods.[38] Adsorption has been identified as a viable technology due to its simplicity, practicality, and economic viability. Adsorbents having a high metal-binding capacity that is inexpensive are extensively utilized for metal removal (Fig. 4).[81] Activated carbon is often employed as an adsorbent treatment of wastewater containing heavy metals. Surprisingly, activated carbon remains a costly material.[82] The adsorbents from palm kernel shell, coconut shell, and bamboo stem were used to treat paint effluent for Cu, Ni, and Pb.[83] Almond shells, discarded potato, and Petha wastes are revalued and managed sustainably for chromium removal.[84]

Many research utilized adsorption to remove heavy metals from agricultural wastes. Singhal et al.[11] examined tea trash as a bioadsorbent for copper removal from water (91.98%). Sugarcane bagasse, orange peel, passion fruit waste, pineapple peel, and activated carbon were used as adsorbents to remediate nickel-containing effluents.[85] Rwiza et al.[86] removed Pb^{2+} from an aqueous solution using agro-wastes (corn and rice husks). The synthesized biosorbents demonstrated a removal efficiency of >90% for Pb^{2+}. With the use of different horticulture wastes such as mango (*Mangifera indica*) waste in both its forms such as seed and shell waste and papaya waste (*Carica papaya*), Sharma and Ayub[87] were able to extract hexavalent chromium from tannery wastewater.

Farasati et al.[82] examined the potential of adsorbing cadmium (Cd) using cheap agro-based adsorbents (*Phragmites australis* and sugarcane straw). Industrial wastewaters are discharged directly into the river without being treated in any way. Elevated levels of some heavy metals, such as Ni^{+2}, Cu^{+2}, Zn^{+2}, Co^{+2}, and Cd^{+2}, may lead to various health issues, including liver and kidney illness.[88] Feng et al.[89] used two agricultural wastes to make adsorbents for removing ammonium and phosphate from swine effluent (Chinese medicinal herbal residue and *Pleurotus ostreatus*). Sugarcane bagasse may also eliminate Zr(IV) (Table 1).[90]

7.2 Radioactive contaminant removal from wastewater

Recent public and scientific concern over radionuclide processing has been prompted by radionuclide's rapid release into the environment, as well as its mobility. Numerous instances of groundwater contamination with various radioactive wastes have been reported.[97] These radioactive contaminations were removed from biosorption technique using agricultural waste as adsorbents. Potato peels, a common agricultural waste, were utilized to investigate the effectiveness of adsorption of long-lived radioisotopes ^{133}Ba and ^{134}Cs. At optimal conditions, the biosorbent could efficiently absorb ∼99% of ^{133}Ba and ∼48% of ^{134}Cs.[98] Kapashi et al.[99] found that alkaline-modified *Aloe vera* waste biosorbents had the greatest removal efficiency of thorium and barium (170 and 107.5 mg g^{-1}, respectively) from aqueous. The biosorption of uranium utilizing *A. vera* wastes was studied, and it was discovered that the biosorbent treated with alkaline reagent had a high sorption capacity for uranium, with q^{max} values of 370.4 mg g^{-1}.

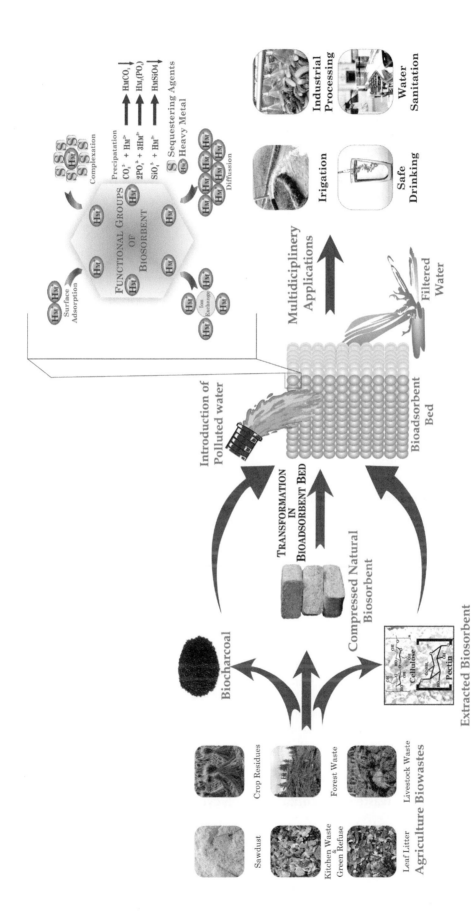

FIG. 4 Metal pollution removal from water utilizing agro–waste–based adsorbents.

TABLE 1 Adsorption of heavy metal pollutants from agro-wastes.

Heavy metal pollutant	Name of agro-waste (Adsorbent)	Adsorption capacity	Removal efficiency	References
Cr(IV)	Sugarcane bagasse	–	92.0%	91
Cr(IV)	Modified wheat bran (M-WB)	5.28 mg	90.0%	92
Pb(II)	Oil palm empty fruit bunch, *Ceiba pentandra* and cellulose	49 mg g^{-1}	99.4%	93
Cd (II) Ni (II)	Oil palm bagasse	–	87% 81%	94
Ni^{2+}	Sugarcane bagasse	–	96.77%.	95
Pb^{2+} Cd^{2+}	Coffee husk	116.3 mg g^{-1} 139.5 mg g^{-1}	89.6% 81.5%	96

8 Adsorption of organic contaminants from water using agro-waste

Synthetic organic contaminants and emerging pollutants pollute aquatic systems, posing major public and environmental health concerns.[100] Organic pollution is caused by sewage, urban runoff, municipal wastewater, agricultural, and industrial effluents.[101] Most persistent organic contaminants are by-products of everyday human activity.[102] The removal of such pollutants is difficult in poor nations owing to the high expense of sophisticated water treatment techniques.[100] The idea of utilizing agricultural biomass in the next decades has grown toward a future of green, sustainable, and renewable goods. Biomass material is a significant source of alternative material for biocomposite product manufacturing.[103] The potential of lignocellulosic materials to preserve the environment by adsorbing various organic contaminants from aqueous solution to purify wastewater has been extensively researched.[104]

8.1 Dye pollution removal from wastewater

The rapid industrialization and uncontrolled growth of dye-using businesses, combined with inadequate wastewater treatment, have exacerbated water contamination, particularly from textile industries.[105] Environmental degradation occurs as a result of the presence of hazardous dyes in industrial effluent.[106] The detrimental consequences of dyes in natural water systems are a global issue. While many dye removal methods exist, adsorption is a potential technique for removing dyes from wastewater.[107,108] Chemical dyes may be removed from wastewater by soaking them in cheap solid sorbents.[109] Meili et al.[110] removed methylene blue from an aqueous medium (>90%) using low-cost biosorbents produced from soursop and sugarcane. Jain et al.[111] investigated the adsorption of Acid Blue 25 using tea waste residue in a batch and continuous operation and determined that the maximum dye absorption was 127.14 mg g^{-1}.

Cuong Nguyen et al.[112] investigated the adsorption of methyl orange (MO) dye using agro-waste and invasive plant biochars. Ahmad et al.[113] used potassium hydroxide impregnation and carbon (IV) oxide gasification techniques to create pomegranate fruit peel-activated carbon (PFPAC), which was then used to remediate an aqueous solution of methylene blue (removal effectiveness 83%). Potato (*S. tuberosum* L.) and pumpkin (*Cucurbita pepo* L.) shells were used to remove methylene blue dye from aqueous solutions. The adsorption capacity of pumpkin seed shells was 161.4 mg g^{-1}, followed by potato shells at 277.8 mg g^{-1}.[114] Batch adsorption experiments using biochar produced from agricultural wastes (coconut shell, groundnut shell, and rice husk) by Praveen et al.[115] have demonstrated a cost-effective method for eliminating Basic Red 09 from wastewaters (Table 2).

8.2 Pollutant removal from wastewater containing polycyclic aromatic hydrocarbons (PAHs)

Polycyclic aromatic hydrocarbon (PAH)-based water pollution has lately received a lot of attention due to its ubiquitous occurrence in the environment and detrimental effects on human health. Polluted water may be difficult to clean because of the presence of PAHs. Naphthalene (NP), the most basic and common PAH, has long been removed from wastewater using the adsorption technique.[122,123] In batch biosorption studies, Chen et al.[124] examined the capacity of plant residues, wood chips, ryegrass root, orange peels, bamboo leaves, and pine needles to adsorb PAHs. Adsorption of phenanthrene onto

TABLE 2 Adsorption of dye pollutants from agro-wastes.

Name of dye pollutant	Name of agro-waste (Adsorbent)	Adsorption capacity	Removal efficiency	References
Indigo carmine	*Grewia venusta* peel	188.68 mg g^{-1}	99%	116
Oranges olimax TGL	Malt bagasse	232 mg g^{-1}	96%	117
Methylene blue	*Ficus palmate* leaves	6.89 mg g^{-1}	98%	118
Methylene blue	*Cymbopogon citrates* leaves	15.97 mg g^{-1}	63.87%	119
Brilliant green	Rice husk Barley husk	– –	76% 65%	120
Malachite green	*Cornulaca monacantha* root	44.84 mg g^{-1}	96.98%	121

activated carbons derived from *Vitis vinifera* leaf litter was studied by Awe et al.[125] The rising awareness of oil spills in waterways has prompted the quest for low-cost alternative remediation methods. Oluwatoyin and Olalekan[126] used activated agro-wastes (rice husks, banana peels, and groundnut husks) to extract crude oil from an aqueous solution. The average effectiveness of oil removal by adsorbents was increasing: peels of bananas > husks of groundnuts > husks of rice (Table 3).

8.3 Removal of pharmaceutical pollutants from wastewater

Pollution has been an issue for decades. Artificial pollutants damage natural areas and impair water quality.[58] Pharmaceuticals are used in huge amounts on a worldwide scale on an annual basis.[130] By and large, the water body is an accessible route for wastewater disposal since it contains a variety of pollutants of increasing concern, such as medicines and personal care items.[131] A critical need exists to reduce the pollution of water resources and wastewaters by antibiotics and pharmaceuticals, which are mostly emitted from healthcare effluents and discharged into the environment.[132] Conventional wastewater treatment methods such as physicochemical treatment, bioremediation, filtering, and sedimentation are incapable of removing medicines.[133] Agro-based adsorbents are used to remove contaminants, particularly pharmaceutical-based products, from the environment. El-Azazy et al.[134] examined eco-structured and efficient sarafloxacin (SARA) removal (82.39%) from wastewater using agro-wastes (AWs) (Table 4).

8.4 Removal of pesticide pollutants from wastewater

Water demand is rising rapidly, but pollution is growing.[141] Agriculture has been polluted due to modern farming methods.[142] Pesticide use on a worldwide scale has risen by more than 1.5 times in the past three decades.[143] Water pollution is caused by agricultural runoff and industrial discharge of pesticides into the surface and groundwater.[144] The removal of residual pesticide from wastewater would be accomplished via the use of agriculturally derived absorbents.

TABLE 3 Adsorption of PAHs pollutants from agro-wastes.

Name of PAHs pollutant	Name of agro-waste (Adsorbent)	Adsorption capacity	Removal efficiency	References
Naphthalene	Sugarcane bagasse	192.30 mg g^{-1}	99%	127
Phenanthrene Pyrene	Rice husks	50.4 mg g^{-1} 104.5 mg g^{-1}	– –	128
Naphthalene	AC from palm shell AC from anthracite modified with nitric acid	44.87 mg g^{-1} 90.91 mg g^{-1}	100% 91.97%	129

TABLE 4 Adsorption of pharmaceutical pollutants from agro-wastes.

Name of pharmaceutical pollutant	Name of agro-waste (Adsorbent)	Adsorption capacity (mg g^{-1})	References
Tetracycline	Corn stover	132.9	135
Doxycycline	Pomegranate peel	11.50	
Doxycycline	Rice straw biochar	432.9	
Doxycycline	Spent black tea leaves	48.80	136
Ciprofloxacin		131.58	
Ibuprofen	Bean husk activated carbon	50	137
Lumefantrine (LUMF)	Banana stalk (BSAC)	102.1	138
Tetracycline hydrochloride (TC)	Corn stover	132.9	132
Ciprofloxacin (CIP)	Functionalized banana stalk	49.7	139
Balofloxacin (BALX)	Spent coffee grounds	142.55	140

TABLE 5 Adsorption of pesticide pollutants from agro-wastes.

Name of pesticide pollutant	Name of agro-waste (Adsorbent)	Adsorption capacity (mg g^{-1})	References
Profenofos	Roasted date pits powder	14.49	146
Bendiocarb	Tangerine seed-activated carbon (TSAC)	7.97	147
Metolcarb		9.11	
Isoprocarb		13.95	
Pirimicarb		39.37	
Carbaryl		44.64	
Methiocarb		93.46	
Carbaryl	Banana pith	45.9	148
Hexachlorocyclohexane	Coconut shell AC	244.78	143
2,4-Dichlorophenoxyacetic acid	Date stones AC	238.10	
Ethylparaben	Biochar from oil palm fiber	349.65	
Atrazine	Biochar from banana peel	14	

Eucalyptus, barley, citrus, mustard, rice husk, sugarcane bagasse, peanut husk and sawdust are some adsorbents used for pollutant adsorption (Table 5).[145]

9 Challenges and perspectives for future research

Numerous nations have struggled with the severe issue of water over the past few decades. Due to anthropogenic activities, the accessible freshwater is severely polluted. Pollution arises from industries such as leather tanning, metal, fertilizer, and photography polluting water supplies. A wide variety of water contaminants are now being removed using low-cost, environmentally friendly techniques. Adsorption efficiency, low cost, and accessibility of many agricultural wastes have gained interest in cleaning hazardous pollutants.

Multiple adsorbents have been noted in literary works; however, such experiments associated with functional adsorbent materials are restricted to particular adsorbent sorts of operating variables. More research should be done before the large-scale commercial water treatment implementation. Relevant adsorption features should be optimized for designing and development. Even though the adsorption mechanism can accumulate a range of contaminants, including medical products, heavy metals, dyes, and agrochemicals, there is limited research on concurrent pollutants uptake. As a result, efficient multipurpose adsorbents with maximum productivity for removing many contaminants are needed. It is hard to ascertain an analysis for real-world applications since few studies on real-world wastewater treatment have been published, necessitating additional study using real-world circumstances and effluent discharged.

The techniques for disposing of these agricultural leftovers have sparked widespread environmental concerns due to their potential impact on the ecosystem. The impacts of residual or leftover materials on the environment and public health, and their segregation and toxic effects should be thoroughly investigated, with substantial recovery strategies established. Adsorbent recovery is critical, and researchers are continually exploring strategies to conserve both adsorbent and adsorbate in in-suite operations. The renewal approach for the used adsorbent is also an important phase that needs improvement to enhance the process's long-term viability. Biomass adsorbents produced from agro-waste offer several advantages over commercial activated carbon.

10 Conclusions

All living things need water as a basic resource. Contamination of aquatic systems by toxic organic and inorganic pollutants is a serious problem that most countries are now dealing with. Water bodies are polluted by industrial, agricultural, and residential waste. The environment and humans are harmed by heavy metals, dyes, and other harmful substances contained in industry wastewater. Adsorption is widely acknowledged as the most effective method for removing contaminants from aqueous systems, and selecting and using the correct adsorbents is critical to its long-term success. This method is usually thought of as an alternative to standard water/wastewater treatment. Finding a cost-effective adsorbent might be a substantial impediment to adsorption's widespread application. The agricultural industry generates a significant amount of trash, which is often discarded. The adoption of sustainable practices that make use of waste resources to create value goods is required for the development of commercially viable technology. Due to their particular features, agricultural waste residues have been widely explored as adsorbents. This chapter of the book describes agricultural waste-based adsorbents for wastewater sequestration that may be employed in environmental remediation techniques.

Acknowledgments

The authors wish to express their appreciation and obligation to the Chairman, Biyani Group of Colleges, Jaipur, India.

References

1. Srivastav AL, Ranjan M. Inorganic water pollutants. In: Devi P, Singh P, Kansal SK, eds. *Inorganic Pollutants in Water*. Elsevier; 2020:1–15.
2. Singh NB, B.H. Susan MA, Guin M. Applications of green synthesized nanomaterials in water remediation. *Curr Pharm Biotechnol*. 2021;22 (6):733–761.
3. Mohamed Khalith SB, Rishabb Anirud R, Ramalingam R, et al. Synthesis and characterization of magnetite carbon nanocomposite from agro waste as chromium adsorbent for effluent treatment. *Environ Res*. 2021;202, 111669.
4. Singh M, Pandey S, Kumar A, Pandey KD. Microbial biofilms for the remediation of contaminated water. In: Kumar A, Singh VK, Singh P, Mishra VK, eds. *Microbe Mediated Remediation of Environmental Contaminants*. Elsevier; 2021:255–269.
5. Karri RR, Gobinath R, Dehghani MH. Wastewater—sources, toxicity, and their consequences to human health. In: *Soft Computing Techniques in Solid Waste and Wastewater Management*. Elsevier; 2021:3–33.
6. Dehghani MH, Ghasem AO, Rama Rao K. Solid waste—Sources, toxicity, and their consequences to human health. In: *Soft Computing Techniques in Solid Waste and Wastewater Management*. Elsevier; 2021:205–213.
7. Amin MT, Alazba AA, Manzoor U. A review of removal of pollutants from water/wastewater using different types of nanomaterials. *Adv Mater Sci Eng*. 2014;2014:1–24.
8. Ohanaka AUC, Ukonu EC, Etuk IF, Charles OI, Ifeanyi O. Evaluation of the physic-chemical properties of agro-wastes derived activated charcoal as a potential feed additive in poultry production. *Int J Agric Res Dev*. 2021;24(1).
9. Nguyen H, Jamali Moghadam M, Moayedi H. Agricultural wastes preparation, management, and applications in civil engineering: a review. *J Mater Cycles Waste Manag*. 2019;21(5):1039–1051.
10. Rao P, Rathod V. Valorization of food and agricultural waste: a step towards greener future. *Chem Rec*. 2019;19(9):1858–1871.
11. Singhal S, Agarwal S, Sharma R, Arora S, Singhal N, Kumar A. Integrated waste-metal-bioenergy: an operative approach for water remediation waste to bio-energy. *Int J Pharm Sci Rev Res*. 2016;37(2):258–263.
12. El-Ramady H, El-Henawy A, Amer M, et al. Agricultural waste and its nano-management: mini review. *Egypt J Soil Sci*. 2020;60(4):349–364.
13. Afroze S, Sen TK. A review on heavy metal ions and dye adsorption from water by agricultural solid waste adsorbents. *Water Air Soil Pollut*. 2018;229(7):225.
14. Hiew BYZ, Lee LY, Thangalazhy-Gopakumar S, Gan S. Biosorption. In: Show PL, Ooi CW, Ling TC, eds. *Bioprocess Engineering*. CRC Press; 2019:143–164.
15. Vaish B, Srivastava V, Kumar Singh P, Singh P, Pratap SR. Energy and nutrient recovery from agro-wastes: rethinking their potential possibilities. *Environ Eng Res*. 2019;25(5):623–637.

16. Khounani Z, Hosseinzadeh-Bandbafha H, Nazemi F, et al. Exergy analysis of a whole-crop safflower biorefinery: a step towards reducing agricultural wastes in a sustainable manner. *J Environ Manage.* 2021;279, 111822.

17. Agrawa P, Srivastava AK. Agricultural-domestic waste management in India. *Int J Mod Agric.* 2021;10(2):2145–2154.

18. Obi F, Ugwuishiwu B, Nwakaire J. Agricultural waste concept, generation, utilization and management. *Niger J Technol.* 2016;35(4):957.

19. Bello OS, Owojuyigbe ES, Babatunde MA, Folaranmi FE. Sustainable conversion of agro-wastes into useful adsorbents. *Appl Water Sci.* 2017;7 (7):3561–3571.

20. Verma D, Sanal I. Agro wastes/natural fibers reinforcement in concrete and their applications. In: Kharissova O, Martínez L, Kharisov B, eds. *Handbook of Nanomaterials and Nanocomposites for Energy and Environmental Applications.* Springer; 2021:1953–1974.

21. Sindhu N, Shehrawat PS, Singh B. Agricultural waste utilization for healthy environment and sustainable lifestyle. *Ann Agric Bio Res.* 2015;20 (1):110–114.

22. Pratap Singh D, Prabha R. Bioconversion of agricultural wastes into high value biocompost: a route to livelihood generation for farmers. *Adv Recycl Waste Manag.* 2018;2(3).

23. Lim SF, Matu SU. Utilization of agro-wastes to produce biofertilizer. *Int J Energy Environ Eng.* 2015;6(1):31–35.

24. Adani F, Tambone F, Gotti A. Biostabilization of municipal solid waste. *Waste Manag.* 2004;24(8):775–783.

25. Kumar B, Kumar U. Removal of malachite green and crystal violet dyes from aqueous solution with bio-materials: a review. *Glob J Res Eng.* 2014;14 (4):50–60.

26. Adejumo IO, Adebiyi OA. Agricultural solid wastes: causes, effects, and effective management. In: Saleh HM, ed. *Strategies of Sustainable Solid Waste Management.* IntechOpen; 2021.

27. Sorathiya LM, Fulsoundar AB, Tyagi KK, Patel MD, Singh RR. Eco-friendly and modern methods of livestock waste recycling for enhancing farm profitability. *Int J Recycl Org Waste Agric.* 2014;3(1):50.

28. Vijay VK. Biogas enrichment and bottling technology for vehicular use. *Biogas Forum.* 2011;1(1):12–15.

29. Karri RR, Sahu JN, Meikap BC. Improving efficacy of Cr (VI) adsorption process on sustainable adsorbent derived from waste biomass (sugarcane bagasse) with help of ant colony optimization. *Ind Crops Prod.* 2020;143:111927.

30. Lingamdinne LP, Vemula KR, Chang YY, Yang JK, Karri RR, Koduru JR. Process optimization and modeling of lead removal using iron oxide nanocomposites generated from bio-waste mass. *Chemosphere.* 2020;243:125257.

31. Cao JP, Xiao XB, Zhang SY, et al. Preparation and characterization of bio-oils from internally circulating fluidized-bed pyrolyses of municipal, livestock, and wood waste. *Bioresour Technol.* 2011;102(2):2009–2015.

32. De Gisi S, Lofrano G, Grassi M, Notarnicola M. Characteristics and adsorption capacities of low-cost sorbents for wastewater treatment: a review. *Sustain Mater Technol.* 2016;9:10–40.

33. Malik PK. Use of activated carbons prepared from sawdust and rice-husk for adsorption of acid dyes: a case study of acid yellow 36. *Dyes Pigments.* 2003;56(3):239–249.

34. Flores-Cano JV, Sanchez-Polo M, Messoud J, Velo-Gala I, Ocampo-Perez R, Rivera-Utrilla J. Overall adsorption rate of metronidazole, dimetridazole and diatrizoate on activated carbons prepared from coffee residues and almond shells. *J Environ Manage.* 2016;169:116–125.

35. Velazquez-Jimenez LH, Pavlick A, Rangel-Mendez JR. Chemical characterization of raw and treated agave bagasse and its potential as adsorbent of metal cations from water. *Ind Crops Prod.* 2013;43:200–206.

36. Joseph L, Jun BM, Flora JRV, Park CM, Yoon Y. Removal of heavy metals from water sources in the developing world using low-cost materials: a review. *Chemosphere.* 2019;229:142–159.

37. Hairuddin MN, Mubarak NM, Khalid M, Abdullhah EC, Walvekar R, Karri RR. Magnetic palm kernel biochar potential route for phenol removal from wastewater. *Environ Sci Pollut Res.* 2019;26(34):35183–35197.

38. Singha B, Das SK. Adsorptive removal of Cu(II) from aqueous solution and industrial effluent using natural/agricultural wastes. *Colloids Surf B Biointerfaces.* 2013;107:97–106.

39. Tsade H, Ananda Murthy HC, Muniswam D. Bio-sorbents from agricultural wastes for eradication of heavy metals: a review. *J Mater Environ Sci.* 2020;11(10):1719–1735.

40. Sharma DC, Forster CF. A preliminary examination into the adsorption of hexavalent chromium using low-cost adsorbents. *Bioresour Technol.* 1994;47(3):257–264.

41. Nguyen TAH, Ngo HH, Guo WS, et al. Applicability of agricultural waste and by-products for adsorptive removal of heavy metals from wastewater. *Bioresour Technol.* 2013;148:574–585.

42. Tony MA. An industrial ecology approach: green cellulose-based bio-adsorbent from sugar industry residue for treating textile industry wastewater effluent. *Int J Environ Anal Chem.* 2021;101(2):167–183.

43. Mohan D, Singh KP. Single- and multi-component adsorption of cadmium and zinc using activated carbon derived from bagasse—an agricultural waste. *Water Res.* 2002;36(9):2304–2318.

44. Gupta N, Kushwaha AK, Chattopadhyaya MC. Application of potato (*Solanum tuberosum*) plant wastes for the removal of methylene blue and malachite green dye from aqueous solution. *Arab J Chem.* 2016;9:S707–S716.

45. Palabıyık BB, Selcuk H, Oktem YA. Cadmium removal using potato peels as adsorbent: kinetic studies. *Desalin Water Treat.* 2019;172:148–157.

46. Aman T, Kazi AA, Sabri MU, Bano Q. Potato peels as solid waste for the removal of heavy metal copper(II) from waste water/industrial effluent. *Colloids Surf B Biointerfaces.* 2008;63(1).116–121.

47. Othmani A, Magdouli S, Senthil Kumar P, Kapoor A, Chellam PV, Gökkuş Ö. Agricultural waste materials for adsorptive removal of phenols, chromium (VI) and cadmium (II) from wastewater: a review. *Environ Res.* 2022;204, 111916.

48. Abdel-Ghani NT, El-Chaghaby GA, Zahran EM. Pentachlorophenol (PCP) adsorption from aqueous solution by activated carbons prepared from corn wastes. *Int J Environ Sci Technol.* 2015;12(1):211–222.

49. Randall JM, Hautala E, McDonald G. Binding of heavy metal ions by formaldehyde-polymerized peanut skins. *J Appl Polym Sci.* 1978;22(2): 379–387.

50. Kausar A, MacKinnon G, Alharthi A, Hargreaves J, Bhatti HN, Iqbal M. A green approach for the removal of Sr(II) from aqueous media: kinetics, isotherms and thermodynamic studies. *J Mol Liq.* 2018;257:164–172.

51. Hussain S, Jabeen S, Khalid AN, et al. Underexplored regions of Pakistan yield five new species of *Leucoagaricus*. *Mycologia.* 2018;110(2): 387–400.

52. Anastopoulos I, Pashalidis I, Hosseini-Bandegharaei A, et al. Agricultural biomass/waste as adsorbents for toxic metal decontamination of aqueous solutions. *J Mol Liq.* 2019;295, 111684.

53. Weng CH, Lin YT, Hong DY, Sharma YC, Chen SC, Tripathi K. Effective removal of copper ions from aqueous solution using base treated black tea waste. *Ecol Eng.* 2014;67:127–133.

54. Gupta A, Balomajumder C. Simultaneous removal of Cr(VI) and phenol from binary solution using *Bacillus* sp. immobilized onto tea waste biomass. *J Water Process Eng.* 2015;6:1–10.

55. Wen T, Wang J, Li X, et al. Production of a generic magnetic Fe3O4 nanoparticles decorated tea waste composites for highly efficient sorption of Cu(II) and Zn(II). *J Environ Chem Eng.* 2017;5(4):3656–3666.

56. Witek-Krowiak A, Szafran RG, Modelski S. Biosorption of heavy metals from aqueous solutions onto peanut shell as a low-cost biosorbent. *Desalination.* 2011;265(1–3):126–134.

57. Tahir N, Bhatti HN, Iqbal M, Noreen S. Biopolymers composites with peanut hull waste biomass and application for crystal violet adsorption. *Int J Biol Macromol.* 2017;94:210–220.

58. Das S, Goud VV. Characterization of a low-cost adsorbent derived from agro-waste for ranitidine removal. *Mater Sci Energy Technol.* 2020;3: 879–888.

59. Bacelo HAM, Santos SCR, Botelho CMS. Tannin-based biosorbents for environmental applications – a review. *Chem Eng J.* 2016;303:575–587.

60. Dutta R, Nagarjuna TV, Mandavgane SA, Ekhe JD. Ultrafast removal of cationic dye using agrowaste-derived mesoporous adsorbent. *Ind Eng Chem Res.* 2014;53(48):18558–18567.

61. Poudel BR, Aryal RL, Bhattarai S, et al. Agro-waste derived biomass impregnated with TiO2 as a potential adsorbent for removal of as(iii) from water. *Catalysts.* 2020;10(10):1125.

62. Alves CCO, Franca AS, Oliveira LS. Evaluation of an adsorbent based on agricultural waste (corn cobs) for removal of tyrosine and phenylalanine from aqueous solutions. *Biomed Res Int.* 2013;2013:1–8.

63. Sepulveda LA, Cuevas FA, Contreras EG. Valorization of agricultural wastes as dye adsorbents: characterization and adsorption isotherms. *Environ Technol.* 2015;36(15):1913–1923.

64. Du Z, Deng S, Liu D, et al. Efficient adsorption of PFOS and F53B from chrome plating wastewater and their subsequent degradation in the regeneration process. *Chem Eng J.* 2016;290:405–413.

65. Sharipova AA, Aidarova SB, Bekturganova NE, et al. Triclosan as model system for the adsorption on recycled adsorbent materials. *Colloids Surf A Physicochem Eng Asp.* 2016;505:193–196.

66. Kim YS, Kim DH, Yang JS, Baek K. Adsorption characteristics of as(III) and as(V) on alum sludge from water purification facilities. *Sep Sci Technol.* 2012;47(14–15):2211–2217.

67. Parker HL, Budarin VL, Clark JH, Hunt AJ. Use of starbon for the adsorption and desorption of phenols. *ACS Sustain Chem Eng.* 2013;1(10): 1311–1318.

68. Wang L, Shi C, Wang L, Pan L, Zhang X, Zou J-J. Rational design, synthesis, adsorption principles and applications of metal oxide adsorbents: a review. *Nanoscale.* 2020;12(8):4790–4815.

69. Baraka A. Adsorptive removal of tartrazine and methylene blue from wastewater using melamine-formaldehyde-tartaric acid resin (and a discussion about pseudo second order model). *Desalin Water Treat.* 2012;44(1–3):128–141.

70. Paşka OM, Păcurariu C, Muntean SG. Kinetic and thermodynamic studies on methylene blue biosorption using corn-husk. *RSC Adv.* 2014;4 (107):62621–62630.

71. Sarker TC, Azam SMGG, El-Gawad AMA, Gaglione SA, Bonanomi G. Sugarcane bagasse: a potential low-cost biosorbent for the removal of hazardous materials. *Clean Technol Environ Policy.* 2017;19(10):2343–2362.

72. Hossain N, Bhuiyan MA, Pramanik BK, Nizamuddin S, Griffin G. Waste materials for wastewater treatment and waste adsorbents for biofuel and cement supplement applications: a critical review. *J Clean Prod.* 2020;255, 120261.

73. Alghamdi AA. An investigation on the use of date palm fibers and coir pith as adsorbents for Pb(II) ions from its aqueous solution. *Desalin Water Treat.* 2016;57(26):12216–12226.

74. Dai Y, Sun Q, Wang W, et al. Utilizations of agricultural waste as adsorbent for the removal of contaminants: a review. *Chemosphere.* 2018;211: 235–253.

75. Kumar PS, Gayathri R, Rathi BS. A review on adsorptive separation of toxic metals from aquatic system using biochar produced from agro-waste. *Chemosphere.* 2021;285, 131438.

76. Bandela NN, Babrekar MG, Jogdand OK, Kaushik G. Removal of copper from aqueous solution using local agricultural wastes as low cost adsorbent. *J Mater Environ Sci.* 2016;7(6):1972–1978.

77. Saravanan A, Kumar PS, Yashwanthraj M. Sequestration of toxic Cr(VI) ions from industrial wastewater using waste biomass: a review. *Desalin Water Treat.* 2017;68:245–266.

78. Rahman DZ, Vijayaraghavan J, Thivya J. A comprehensive review on zinc(II) sequestration from wastewater using various natural/modified low-cost agro-waste sorbents. *Biomass Convers Biorefin*. 2021.

79. Khan FSA, Mubarak NM, Tan YH, et al. Magnetic nanoparticles incorporation into different substrates for dyes and heavy metals removal—a review. *Environ Sci Pollut Res*. 2020;27(35):43526–43541.

80. Khan FSA, Mubarak NM, Khalid M, et al. Magnetic nanoadsorbents' potential route for heavy metals removal—a review. *Environ Sci Pollut Res*. 2020;27(19):24342–24356.

81. Arthi D, Michael Ahitha Jose J, Edinsha Gladis EH, Shajin Shinu PM, Joseph J. Removal of heavy metal ions from water using adsorbents from agro waste materials. *Mater Today Proc*. 2021;45:1794–1798.

82. Farasati M, Haghighi S, Boroun S. Cd removal from aqueous solution using agricultural wastes. *Desalin Water Treat*. 2016;57(24):11162–11172.

83. Olatunji MA, Okafor JO, Kovo AS. Development of nano-composite adsorbents for the removal of heavy metals from industrial effluents via batch study. *J Appl Chem Sci Int*. 2018;9(2):131–144.

84. Sharma PK, Ayub S, Shukla BK. Cost and feasibility analysis of chromium removal from water using agro and horticultural wastes as adsorbents. In: Ashish DK, de Brito J, Sharma SK, eds. *3rd International Conference on Innovative Technologies for Clean and Sustainable Development. ITCSD 2020. RILEM Bookseries*; 2021:449–463.

85. Dotto GL, Meili L, de Souza Abud AK, Tanabe EH, Bertuol DA, Foletto EL. Comparison between Brazilian agro-wastes and activated carbon as adsorbents to remove Ni(II) from aqueous solutions. *Water Sci Technol*. 2016;73(11):2713–2721.

86. Rwiza MJ, Oh S-Y, Kim K-W, Kim SD. Comparative sorption isotherms and removal studies for Pb(II) by physical and thermochemical modification of low-cost agro-wastes from Tanzania. *Chemosphere*. 2018;195:135–145.

87. Sharma PK, Ayub S. Economic feasibility study of horticultural wastes for chromium adsorption from tannery wastewater. *IWRA*. 2019;8(2):51–58.

88. Dwidar E. Removal of heavy metal ions from some wastewater by using different agricultural wastes. *Ann Agric Sci Moshtohor*. 2018;56(1). 272–242.

89. Feng C, Zhang S, Wang Y, et al. Synchronous removal of ammonium and phosphate from swine wastewater by two agricultural waste based adsorbents: performance and mechanisms. *Bioresour Technol*. 2020;307, 123231.

90. Kausar A, Bhatti HN, MacKinnon G. Re-use of agricultural wastes for the removal and recovery of Zr(IV) from aqueous solutions. *J Taiwan Inst Chem Eng*. 2016;59:330–340.

91. Garg UK, Kaur MP, Garg VK, Sud D. Removal of hexavalent chromium from aqueous solution by agricultural waste biomass. *J Hazard Mater*. 2007;140(1–2):60–68.

92. Kaya K, Pehlivan E, Schmidt C, Bahadir M. Use of modified wheat bran for the removal of chromium(VI) from aqueous solutions. *Food Chem*. 2014;158:112–117.

93. Daneshfozoun S, Abdullah MA, Abdullah B. Preparation and characterization of magnetic biosorbent based on oil palm empty fruit bunch fibers, cellulose and *Ceiba pentandra* for heavy metal ions removal. *Ind Crops Prod*. 2017;105:93–103.

94. Herrera-Barros A, Tejada-Tovar C, Villabona-Ortíz A, González-Delgado AD, Benitez-Monroy J. Cd (II) and Ni (II) uptake by novel biosorbent prepared from oil palm residual biomass and Al_2O_3 nanoparticles. *Sustain Chem Pharm*. 2020;15, 100216.

95. Aldalbahi A, El-Naggar M, Khattab T, et al. Development of green and sustainable cellulose acetate/graphene oxide nanocomposite films as efficient adsorbents for wastewater treatment. *Polymers (Basel)*. 2020;12(11):2501.

96. thi Quyen V, Pham TH, Kim J, et al. Biosorbent derived from coffee husk for efficient removal of toxic heavy metals from wastewater. *Chemosphere*. 2021;284:131312.

97. Vandana UK, Gulzar ABM, Laskar IH, Meitei LR, Mazumder PB. Role of microbes in bioremediation of radioactive waste. In: Panpatte DG, Jhala YK, eds. *Microbial Rejuvenation of Polluted Environment. Microorganisms for Sustainability*. Springer; 2021:329–352.

98. Naskar N, Banerjee K. Development of sustainable extraction method for long-lived radioisotopes, [133]Ba and [134]Cs using a potential bio-sorbent. *J Radioanal Nucl Chem*. 2020;325(2):587–593.

99. Kapashi E, Kapnisti M, Dafnomili A, Noli F. Aloe Vera as an effective biosorbent for the removal of thorium and barium from aqueous solutions. *J Radioanal Nucl Chem*. 2019;321(1):217–226.

100. Chaukura N, Gwenzi W, Tavengwa N, Manyuchi MM. Biosorbents for the removal of synthetic organics and emerging pollutants: opportunities and challenges for developing countries. *Environ Dev*. 2016;19:84–89.

101. Adelodun B, Ajibade FO, Ogunshina MS, Choi KS. Dosage and settling time course optimization of *Moringa oleifera* in municipal wastewater treatment using response surface methodology. *Desalin Water Treat*. 2019;167:45–56.

102. Katsoyiannis A, Samara C. Persistent organic pollutants (POPs) in the sewage treatment plant of Thessaloniki, Northern Greece: occurrence and removal. *Water Res*. 2004;38(11):2685–2698.

103. Ali HR, Hassaan MA. Applications of bio-waste materials as green synthesis of nanoparticles and water purification. *Adv Mater*. 2017;6(5):85.

104. Liu J, Li E, You X, Hu C, Huang Q. Adsorption of methylene blue on an agro-waste oiltea shell with and without fungal treatment. *Sci Rep*. 2016;6 (1):38450.

105. Selvam K. Recent trends in agro-waste based activated carbons for the removal of emerging textile pollutants. *Int J Environ Anal Chem*. 2021;1-17.

106. Baloo L, Isa MH, Bin SN, et al. Adsorptive removal of methylene blue and acid orange 10 dyes from aqueous solutions using oil palm wastes-derived activated carbons. *Alex Eng J*. 2021;60(6):5611–5629.

107. Adelodun B, Ajibade FO, Abdulkadir TS, Bakare HO, Choi KS. SWOT analysis of agro-waste based adsorbents for persistent dye pollutants removal from wastewaters. In: Kumar V, Singh J, Kumar P, eds. *Environmental Degradation: Causes and Remediation Strategies*. Agro Environ Media - Agriculture and Ennvironmental Science Academy; 2020:88–103.

108. Mashkoor F, Nasar A. Environmental application of agro-waste derived materials for the treatment of dye-polluted water: a review. *Curr Anal Chem*. 2021;17(7):904–916.

109. Jasper EE, Ajibola VO, Onwuka JC. Nonlinear regression analysis of the sorption of crystal violet and methylene blue from aqueous solutions onto an agro-waste derived activated carbon. *Appl Water Sci.* 2020;10(6):132.

110. Meili L, Lins PVS, Costa MT, et al. Adsorption of methylene blue on agroindustrial wastes: experimental investigation and phenomenological modelling. *Prog Biophys Mol Biol.* 2019;141:60–71.

111. Jain SN, Tamboli SR, Sutar DS, et al. Batch and continuous studies for adsorption of anionic dye onto waste tea residue: kinetic, equilibrium, breakthrough and reusability studies. *J Clean Prod.* 2020;252, 119778.

112. Cuong Nguyen X, Thanh Huyen Nguyen T, Hong Chuong Nguyen T, et al. Sustainable carbonaceous biochar adsorbents derived from agro-wastes and invasive plants for cation dye adsorption from water. *Chemosphere.* 2021;282:131009.

113. Ahmad MA, Eusoff MA, Oladoye PO, Adegoke KA, Bello OS. Optimization and batch studies on adsorption of methylene blue dye using pomegranate fruit peel based adsorbent. *Chem Data Collect.* 2021;32, 100676.

114. Cuce H, Temel FA. Reuse of agro-wastes to treat wastewater containing dyestuff: sorption process with potato and pumpkin seed wastes. *Int J Glob Warm.* 2021;24(1):14.

115. Praveen S, Gokulan R, Pushpa TB, Jegan J. Techno-economic feasibility of biochar as biosorbent for basic dye sequestration. *J Indian Chem Soc.* 2021;98(8), 100107.

116. Agbahoungbata MY, Fatombi JK, Ayedoun MA, et al. Removal of reactive dyes from their aqueous solutions using *Moringa oleifera* seeds and *Grewia venusta* peel. *Desalin Water Treat.* 2016;57(47):22609–22617.

117. Fontana KB, Chaves ES, Sanchez JDS, Watanabe ERLR, Pietrobelli JMTA, Lenzi GG. Textile dye removal from aqueous solutions by malt bagasse: isotherm, kinetic and thermodynamic studies. *Ecotoxicol Environ Saf.* 2016;124:329–336.

118. Fiaz R, Hafeez M, Mahmood R. *Ficcus palmata* leaves as a low-cost biosorbent for methylene blue: thermodynamic and kinetic studies. *Water Environ Res.* 2019;91(8):689–699.

119. Ahmad MA, Ahmed NB, Adegoke KA, Bello OS. Sorption studies of methyl red dye removal using lemon grass (*Cymbopogon citratus*). *Chem Data Collect.* 2019;22, 100249.

120. Garamon SE. Sequestration of hazardous Brilliant Green dye from aqueous solution using low-cost agro-wastes: activated carbonprepared from rice and barley husks. *Res Crops.* 2019;20(4):886–891.

121. Mishra S, Cheng L, Maiti A. The utilization of agro-biomass/byproducts for effective bio-removal of dyes from dyeing wastewater: a comprehensive review. *J Environ Chem Eng.* 2021;9(1), 104901.

122. Alshabib M. Removal of naphthalene from wastewaters by adsorption: a review of recent studies. *Int J Environ Sci Technol.* 2021;19:4555–4586.

123. Parthipan P, Cheng L, Rajasekar A, Angaiah S. Microbial surfactants are next-generation biomolecules for sustainable remediation of polyaromatic hydrocarbons. In: Sarma H, Prasad MNV, eds. *Biosurfactants for a Sustainable Future*; 2021.

124. Chen B, Yuan M, Liu H. Removal of polycyclic aromatic hydrocarbons from aqueous solution using plant residue materials as a biosorbent. *J Hazard Mater.* 2011;188(1–3):436–442.

125. Awe AA, Opeolu BO, Fatoki OS, Ayanda OS, Jackson VA, Snyman R. Preparation and characterisation of activated carbon from *Vitisvinifera* leaf litter and its adsorption performance for aqueous phenanthrene. *Appl Biol Chem.* 2020;63(1):12.

126. Oluwatoyin AO, Olalekan AA. Adsorption of crude oil spill from aqueous solution using agro-wastes as adsorbents. *J Sci Res Rep.* 2021;27–52.

127. Eslami A, Borghei SM, Rashidi A, Takdastan A. Preparation of activated carbon dots from sugarcane bagasse for naphthalene removal from aqueous solutions. *Sep Sci Technol.* 2018;53(16):2536–2549.

128. Chaukura N, Masilompane TM, Gwenzi W, Mishra AK. Biochar-based adsorbents for the removal of organic pollutants from aqueous systems. In: Mishra AK, Hussain CM, Mishra SB, eds. *Emerging Carbon-Based Nanocomposites for Environmental Applications*. Wiley; 2020:147–174.

129. Alshabib M. Removal of naphthalene from wastewaters by adsorption: a review of recent studies. *Int J Environ Sci Technol.* 2021;19:4555–4586.

130. Chaukura N, Mahamadi C, Muzawazi E, Sveera T. Occurrence and attenuation of antibiotics in water using biomass-derived materials. In: Kumar V, Prasad R, Kumar M, eds. *Rhizobiont in Bioremediation of Hazardous Waste*. Springer; 2021:511–530.

131. Inyinbor AA, Bello OS, Dada OA, Oreofe TA. Emerging water pollutants and wastewater treatments. In: Das R, ed. *Two-Dimensional (2D) Nanomaterials in Separation Science*. Springer; 2021:13–42.

132. Haghighat GA, Saghi MH, Anastopoulos I, et al. Aminated graphitic carbon derived from corn Stover biomass as adsorbent against antibiotic tetracycline: optimizing the physicochemical parameters. *J Mol Liq.* 2020;313, 113523.

133. Sires I, Brillas E. Remediation of water pollution caused by pharmaceutical residues based on electrochemical separation and degradation technologies: a review. *Environ Int.* 2012;40:212–229.

134. El-Azazy M, El-Shafie AS, Elgendy A, Issa AA, Al-Meer S, Al-Saad KA. A comparison between different agro-wastes and carbon nanotubes for removal of sarafloxacin from wastewater: kinetics and equilibrium studies. *Molecules.* 2020;25(22):5429.

135. Aniagor CO, Igwegbe CA, Ighalo JO, Oba SN. Adsorption of doxycycline from aqueous media: a review. *J Mol Liq.* 2021;334, 116124.

136. Zeng Z, Tan X, Liu Y, et al. Comprehensive adsorption studies of doxycycline and ciprofloxacin antibiotics by biochars prepared at different temperatures. *Front Chem.* 2018;6.

137. Bello OS, Alao OC, Alagbada TC, Olatunde AM. Biosorption of ibuprofen using functionalized bean husks. *Sustain Chem Pharm.* 2019;13, 100151.

138. Agboola OS, Akanji SB, Bello OS. Functionalized banana stalk for lumefantrine drug removal. *Phys Chem Res.* 2021;9(3):483–507.

139. Agboola OS, Bello OS. Enhanced adsorption of ciprofloxacin from aqueous solutions using functionalized banana stalk. *Biomass Convers Biorefin.* 2020. https://doi.org/10.1007/s13399-020-01038-9.

140. El-Azazy M, El-Shafie AS, Morsy H. Biochar of spent coffee grounds as per se and impregnated with TiO_2: promising waste-derived adsorbents for balofloxacin. *Molecules.* 2021;26(8):2295.

141. Guo Q, Zang Z, Ma J, Li J, Zhou T, Han R. Adsorption of copper ions from solution using xanthate wheat straw. *Water Sci Technol.* 2020;82 (10):2029–2038.

142. Singh H, Sharma A, Bhardwaj SK, Arya SK, Bhardwaj N, Khatri M. Recent advances in the applications of nano-agrochemicals for sustainable agricultural development. *Environ Sci Process Impacts.* 2021;23(2):213–239.

143. Ponnuchamy M, Kapoor A, Senthil Kumar P, et al. Sustainable adsorbents for the removal of pesticides from water: a review. *Environ Chem Lett.* 2021;19(3):2425–2463.

144. Mandal A, Singh N. Optimization of atrazine and imidacloprid removal from water using biochars: designing single or multi-staged batch adsorption systems. *Int J Hyg Environ Health.* 2017;220(3):637–645.

145. Khalid Q, Khan A, Bhatti HN, et al. Cellulosic biomass biocomposites with polyaniline, polypyrrole and sodium alginate: insecticide adsorption-desorption, equilibrium and kinetics studies. *Arab J Chem.* 2021;14(7), 103227.

146. Hassan SS, Al-Ghouti MA, Abu-Dieyeh M, McKay G. Novel bioadsorbents based on date pits for organophosphorus pesticide remediation from water. *J Environ Chem Eng.* 2020;8(1), 103593.

147. Wang Y, Wang S, Xie T, Cao J. Activated carbon derived from waste tangerine seed for the high-performance adsorption of carbamate pesticides from water and plant. *Bioresour Technol.* 2020;316, 123929.

148. Sathishkumar M, Choi JG, Ku CS, Vijayaraghavan K, Binupriya AR, Yun SE. Carbaryl sorption by porogen-treated banana pith carbon. *Adsorpt Sci Technol.* 2008;26(9):679–686.

Chapter 11

Carbon-based adsorbents for remediation of noxious pollutants from water and wastewater

R. Suresh[*] and Saravanan Rajendran
Department of Mechanical Engineering, Faculty of Engineering, University of Tarapaca, Arica, Chile
[*]*Corresponding author.*

1 Introduction

1.1 Need for water treatment

Water is one of the vital components for living organisms in the biosphere, and it is generally distributed in the hydrosphere (ocean), cryosphere (iceberg), lithosphere, and atmosphere.[1] About 70% of our Earth is covered by water as oceans, glaciers, rivers, lakes, and ponds. But, only ∼2.5% is available as fresh water from the Earth. Among these, 68.5% of fresh water is stagnant as ice sheets and glaciers in polar regions. Nearly, 31% of fresh water is stored in the form of surface and underground water. The least amount (0.1%) of fresh water (as vapor, liquid, and solid) is aggregated in the atmosphere.[1] Overall, nearly 1% of the total water is freshwater that is accessible for human consumption. Further, human population explosion and economic development lead to water demand significantly.[2] Besides, ecological pollution progressively happens due to the discharge of toxic metal ions (Hg, Pb, Cd, Cr, etc.), metalloids (As, Sb, etc.),[3,4] organic (textile dyes, health care products, pesticides, etc.),[5] and organometallic (roxarsone, tributyltin, etc.)[6,7] contaminants from various manufacturing industries. Human activities such as urbanization and mining are also causing environmental pollution considerably.[8,9] Water pollution is becoming a serious issue by creating a pure water shortage. According to the United Nations World Water Development report, by 2050, approximately six billion people will be affected due to scarcity of pure water.[2]

To tackle the pure water crisis, cheap and efficient water treatment needs to be developed. So far, many water purification methods like flocculation, chemical oxidation, precipitation, reverse osmosis, membrane filtration, ion exchange, biological process, phytoremediation, adsorption, advanced oxidation process, and electrochemical techniques are used for wastewater purification purposes.[10,11] Nonetheless, every wastewater purification methods have merits and demerits.[12] Among other methods, the adsorption method is widely practiced for abatement of harmful inorganic and organic contaminants from polluted water because of the following merits[13]: It is (i) cheaper, (ii) easy to perform faster, (iii) environmentally benign due to the use of waste materials as adsorbents, and (iv) recycling capability of adsorbents without much loss in adsorption capacity.

1.2 Adsorption method

Adsorption is an exothermic process that involves capturing substances (adsorbate) from a liquid or gas medium by a solid surface (adsorbent). In the batch adsorption method,[14] an appropriate quantity of adsorbent powder is added to polluted water, and the resulting suspension is shaken to reach adsorption equilibrium.[15] Pollutants may attach to the surface of the adsorbent through physical (physisorption) and/or chemical (chemisorption) interactions. Physisorption involves weak physical forces like π-π interactions, hydrogen bonding, van der Waals force, etc. At the same time, chemisorption governs by the formation of chemical bonds between functional groups of adsorbent and adsorbate molecules.[15] The specific surface area and quantity of the adsorbent, nature of the pollutant, contact time, temperature, solution pH, and initial concentration of adsorbate are the major deciding factors in the adsorption efficiency.[15] Contact systems such as batch,

FIG. 1 Various types of adsorbents used in the removal of contaminants from water.

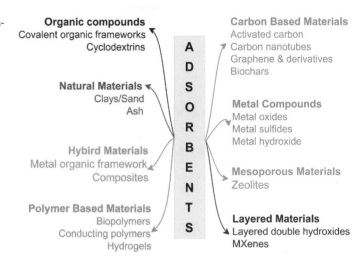

fixed-bed, fluidized and pulsed bed methods were adopted to perform adsorption experiments in the lab and practical applications.[16]

Noticeably, the selection of proper adsorbents for wastewater treatment is an essential requirement. To attain, an effective adsorptive removal process, adsorbents should have large active surface sites, exceptional stability, low cost, and eco-friendly material. Different kinds of adsorbents, including metal oxides, metal chalcogenides, zeolites, layered double hydroxides, polymers and carbon-based materials, have been explored as adsorbents for the purification of polluted water.[17] Fig. 1 displays different kinds of adsorbents and examples for removing aquatic pollutants from water. Among others, carbon-based materials have gained high popularity as adsorbents in wastewater treatment due to their unique properties, discussed in the next section.

2 Carbon-based materials

Carbon (symbol—C; atomic number—6) is one of the important known elements in the Earth. Carbon atoms will form sp, sp^2, and sp^3 hybridization with each other and/or other elements such as hydrogen, oxygen, etc.[18] Specifically, sp^2 hybridization is responsible for the formation of unique carbon allotropes, including fullerene, carbon nanotubes, and graphene. Carbon materials are generally found in different forms, such as graphite, charcoal, diamond, fullerenes, carbon dots, carbon nanofibers, carbon spheres, carbon nanotubes, activated carbon, porous carbon biochar, graphene, and their derivatives (Fig. 2).

Carbon-based materials have applications in various fields including food safety,[19] sensors,[20] medical,[21] surface coatings,[22] electronics,[23] energy,[24,25] and wastewater treatment.[26]

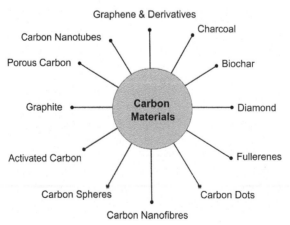

FIG. 2 Different kinds of carbon materials.

In this chapter, the application of some common carbon-based adsorbents, particularly activated carbon, porous carbon, carbon nanotubes (CNTs), carbon spheres, coal, graphite and its derivatives, and biochar for removal of organic and inorganic pollutants from aqueous solutions has been described. Due to the following reasons, carbon materials have attracted attention in the adsorption process.

(i) Abundant and renewable source

Carbon materials like coal are naturally abundant materials.[27] For instance, naturally available Wanli and Meng Dong coal was used for Congo dye adsorption studies.[28] Further, biomass, a renewable resource, is used to prepare different carbon materials. For example, porous carbon and biochar were prepared from rice stalk[29] and rice husk,[30] respectively. Corn stover was used to prepare reduced graphene oxide (rGO) nanomaterial.[31] Industrial wastes can also be used as source materials for carbon adsorbents.[32] Discarded waste cigarette butts were used for the preparation of porous carbon adsorbents.[33] More examples of carbon sources used to synthesize carbon nanomaterials are listed in Table 1.

(ii) Morphology

Generally, less particle size and unique porous morphology provide more active sites for the adsorption of pollutants. It is possible to tune the size and morphology of carbon-based adsorbents and thus possible to improve their adsorption capacity easily.[43]

(iii) Physical properties

Carbon materials have excellent mechanical strength and low density.[44] It is also possible to prepare highly hydrophobic and hydrophilic carbon materials and thus be able to efficiently adsorb hydrophobic and hydrophilic adsorbate molecules, respectively.[45]

(iv) Chemical properties

Carbon materials with diverse functional groups were also reported.[46] Depending on the solution pH, a charge of surface groups of carbon nanomaterials can be manipulated. The $pH > pH_{pzc}$ condition provides the negative charge to the adsorbent, while if $pH < pH_{pzc}$, the adsorbent acquires a positive surface charge. Hence, oppositely charged pollutants could be adsorbing efficiently and selectively from water.

(v) Stability

The thermal and chemical stabilities of carbon materials are also high. For instance, biochar is stable at over $700\,°C$.[30] Higher calcination temperature will lead to generating graphitic biochar with huge surface area and carbon defects.[44]

TABLE 1 Some carbon materials along with their carbon precursors.

Carbon nanomaterial	Carbon source	Synthesis method	References
Activated carbons	Tea waste	Chemical treatment followed carbonization	34
Porous carbon	Cotton cloth	Carbonization	35
Carbon nanoparticles	Candle soot	Chemical treatment	36
Carbon black	Waste tire	Thermal decomposition	37
Activated carbon	Palm oil kernel shell	Physical activation followed chemical modification	38
Biochar	Fargesia leaves	Carbonization	39
Biochar	Date and Delonix regia seeds	Pyrolysis	40
Porous carbons	Bamboo shoot	Hydrothermal treatment followed activation	41
Graphene	Black sesame	Pyrolysis followed activation	27
Carbon spheres	Larch sawdust	Ultrasonic-assisted method	42

3 Carbon-based materials as adsorbents

3.1 Activated carbon

Activated carbon has a random and imperfect structure with high porosity and visible cracks and crevices. Synthesis of activated carbon involves pretreatment of carbon source material, carbonization (calcination under inert atmosphere), and activation.[47] Medhat et al.[48] have used corn cob for the preparation of activated carbon powder. The dried corn powder was calcined at 700°C for 2h in a muffle furnace under an N_2 atmosphere. The carbon powder was activated using KOH or ammonium sulfate. Suhas et al.[49] have synthesized activated carbon with relatively high surface area (569 $m^2 g^{-1}$) by hydrothermal method. The carbon source used in this research was *Phyllanthus emblica* fruit stone. Generally, the as-synthesized activated carbon contains surface groups such as hydroxyl (—OH), carboxyl (—COOH), carbonyl (C=O), phenol (C_6H_5OH), lactones (—(C=O)—O—) etc., which are responsible for adsorption of aquatic contaminants. Importantly, carbon sources have a significant impact on the properties of activated carbon. Commonly, activated carbon is prepared from high carbon content with a low concentration of inorganic substances like coal, biowastes (leaves, rice husks, coconut shells, waste paper, etc.) and non-degradable wastes (plastics, tires, etc.).[50]

(i) Pure activated carbon

Owing to availability and excellent adsorption capacity, pure activated carbon was much studied as an adsorbent in wastewater treatment. For example, Shokry et al.[51] have prepared activated carbon nanoparticles (average diameter=38 nm) with a pore volume of 0.183 $cm^3 g^{-1}$ using raw Maghara coal (volatile matter=50.6% and ash content=4.12%), an eco-friendly, abundant natural material and an aqueous solution of NaOH (50%) as an activating agent for removal of methylene blue from water. NaOH activates the carbon via multiple chemical reactions (Eqs. 1–4).[51]

$$4NaOH + C \rightarrow Na_2CO_3 + Na_2O + 2H_2 \tag{1}$$

$$Na_2CO_3 + 2C \rightarrow 2Na + 3CO \tag{2}$$

$$Na_2O + C \rightarrow C—O—Na + Na \tag{3}$$

$$C—O—Na + H_2O \rightarrow C—O—H + NaOH \tag{4}$$

From the batch adsorption experimental results, it was found that a monolayer methylene blue adsorption (28.09 $mg g^{-1}$) on activated carbon occurred. The enhanced adsorption capability of Maghara coal-derived activated carbon was due to surface hydroxyl (—OH) groups that form ionic interactions with charged methylene blue (MB^+) molecules (Eq. 5). Moreover, spent activated carbon could be effectively regenerated using HCl aqueous solution.

$$\equiv C—O^- + MB^+ \rightarrow \equiv C—O^- \, MB^+ \tag{5}$$

Studies have also been performed to explore the utilization of wastes for the generation of activated carbons. For example, the tannery sludge biomass was converted to mesoporous activated carbons via carbonization (500°C/3 h, inert atmosphere) followed by a physical activation route.[52] The surface area and average pore diameter of the synthesized activated carbon are 187.21 $m^2 g^{-1}$ and 3.24 nm, respectively. This activated carbon was effectively utilized for the adsorption of malachite green (231.34 $mg g^{-1}$), 2,4-dichlorophenoxyacetic acid (20.09 $mg g^{-1}$), and Cr(VI) ions ($Cr_2O_7^{2-}$, 86.26 $mg g^{-1}$) from their respective aqueous solutions. The palm shell (agricultural waste) derived activated carbon (surface area=506.84 $m^2 g^{-1}$) with the aid of oleic acid activation followed by 50 kHz ultrasonic irradiation was used for acenaphthene (poly aromatic hydrocarbons) adsorption from wastewater.[53] Under the optimized condition, this porous activated carbon showed a higher adsorption capacity of 52.75 $mg g^{-1}$, while normal palm shell-derived activated carbon (without activation) showed only 15 $mg g^{-1}$ toward acenaphthene adsorption. Thermal treatment was successfully applied to regenerate spent activated carbon. Chaudhary et al.[54] have investigated the role of functional groups of a few selected anthraquinone acid dyes on the adsorption ability of activated carbons. These researchers have found that acid blue 129 dye molecules containing methyl group show enhanced adsorption than other dyes (acid blue 25 and acid blue 40). They also found that tautomerism is responsible for the adsorption of acid blue 40 on the synthesized activated carbons.

(ii) Functionalized activated carbon

The surface functionalization of activated carbon has remarkably improved the adsorption capacities. The incorporation of oxygen, nitrogen, sulfur, and phosphorus containing functional groups maximizes active sites in activated carbon.[55] In this regard, activated carbon functionalized with ethylenediaminetriacetic acid was reported for the adsorption of Nd(III) ions from wastewater.[56] Based on the Langmuir adsorption model, the highest adsorption ability was observed for Nd(III) ions with a good recovery rate. In another study, activated carbon was functionalized by 8-hydroxyquinoline.[57] This adsorbent was used for the extraction of Cd(II), Ni(II), Mn(II), Zn(II), and Pb(II) ions from contaminated groundwater by the solid phase extraction method. The 8-hydroxyquinoline functionalized sample showed more than 50% enhanced adsorption capacities than the pure adsorbent. The order of adsorption capacity (mmol g^{-1}) is Mn(II) = 0.393 > Ni(II) = 0.345 > Zn(II) = 0.314 > Cd(II) = 0.170 > Pb(II) = 0.092.

(iii) Activated carbon-based composites

Activated carbon was mixed with definite proportions of other adsorbent materials for the betterment of its adsorption performance. Biopolymers, metal oxides, clay, and other nano carbons were employed as counterparts in activated carbon composites. For example, coffee waste-derived activated carbon containing $Zn(OH)_2$ nanoparticles were efficiently utilized for adsorptive removal of malachite green from water.[58] Adsorption ability of activated carbon/$Zn(OH)_2$ composite was enhanced with in increase of the initial dye concentration, pH (7.0), and temperature (318 K). The maximum adsorption capacity was 303.03 mg g^{-1}. Activated carbon/graphene oxide binary composite adsorbent was also reported for the removal of methylene blue from the solution.[59] When compared to individual components, activated carbon/graphene oxide nanocomposite showed an excellent adsorption capacity of 1000 mg g^{-1}. This is due to the strong electrostatic force of attraction between the charged groups of adsorbent and methylene blue and pores in activated carbon. Moreover, this composite adsorbent is highly selective toward the adsorption of cationic dyes. On the other hand, an activated carbon/Fe_3O_4/CuO ternary composite powder was also prepared and used for the removal of fuchsin acid ($C_{20}H_{17}N_3Na_2O_9S_3$), methyl green ($C_{27}H_{35}BrClN_3$ $ZnCl_2$), murexide ($C_8H_8N_6O_6$), methyl orange ($C_{14}H_{14}N_3NaO_3S$), and bromophenol blue ($C_{19}H_{10}Br_4O_5S$) in water.[60] Among others, this ternary composite exhibited the highest adsorption capacity for bromophenol blue (88.60 mg g^{-1}; 97%) under the optimum conditions (initial dye concentration = 20 mg L^{-1}; pH = 9, adsorbent dose = 0.06 g L^{-1}).

3.2 Porous carbon

Carbon with a huge surface area and hierarchical porosity ($d_{macropores} > 50$ nm, $d_{mesopores} < 50$ nm and > 2 nm, and $d_{micropores} < 2$ nm) is called porous carbon. They have tunable specific surface and pore structure and good stability. Like activated carbon, porous carbon nanoparticles are also synthesized by the carbonization of carbon source materials.[61] For instance, the as-synthesized metal azolate framework-6 was thermally decomposed at 800°C for 6 h under a nitrogen atmosphere.[62] The obtained carbon product was only washed with a 2 M HCl solution to achieve porous carbon. In some other studies,[33,63] the activation step was also additionally adopted after the carbonization of precursor powder. Many carbon sources were used to prepare porous carbon with high porosity and surface area. Table 2 lists some porous carbons with their sources and surface characteristics.

(i) Pure porous carbon

Pure porous carbon powder, synthesized by calcination of cellulose, was used for adsorptive removal of chrysoidine (598.8 mg g^{-1}), methyl orange (381.7 mg g^{-1}), crystal violet (310.6 mg g^{-1}), and rhodamine B (226.2 mg g^{-1}) dyes.[67] Porous carbon prepared at higher temperatures has a huge surface area and becomes a more effective adsorbent in dye adsorption. Noticeably, depending on the structural features of dye molecules, porous carbon adsorbent showed adsorption capacities. Considerably, this adsorbent was regenerated by soaking in methanol at 60°C.

(ii) Defects rich porous carbon

Defects in carbon structures significantly enhance its adsorption capability. For instance, defect-rich porous carbon adsorbent[33] synthesized from cigarette butt showed high bisphenol A adsorption capacity (865 mg g^{-1}) and rapid adsorption rate (186.9 mg g^{-1} min^{-1}).

(iii) Hierarchical structure of porous carbon

Hierarchical porous carbon was synthesized and employed as an adsorbent for Cu(II) ions removal from water.[68] The adsorption ability of this adsorbent was intensely affected by the meso/micropore ratio. Porous carbon with 2330 m^2 g^{-1} (surface area) and 81% mesopore showed a greater adsorption capacity of 265 mg g^{-1} toward Cu(II) ions removal in water. Hierarchical porous carbon powder was also synthesized from coal gangue (solid wastes from mining industries), and this adsorbent showed exceptional adsorption capacity for rhodamine B (3086.42 mg g^{-1}) and Cr(VI) species (320.51 mg g^{-1}).[69]

TABLE 2 List of porous carbon samples and their sources with surface features.

Source of porous carbon	Synthesis method	Synthesis condition	Surface area ($m^2\,g^{-1}$)	Pore volume ($cm^3\,g^{-1}$)	Pore size (nm)	References
The fluffy yellow-brown substance of discarded cigarette butts	Hydrothermal followed KOH activation	800°C	1838.0	0.95	2.08	33
The zinc-based metal organic framework of MOF-5	Direct carbonization	950°C	1512.0	0.94	4.0	61
Wheat straw char Rice stalk	Pyrolysis followed microwave activation	Chemical modification with 10% H_2O_2	433.0 273.9	0.29 0.18	1.32 1.36	29
Waste cellulose fibers	Spray drying followed by heat treatment	800°C	1105.3	0.49	0.9	64
Asphalt	Carbonization	800°C	827.3	0.43	21.02Å	65
Shells of bamboo shoot	Hydrothermal followed KOH activation	800°C	3250.0	1.85	5.8	41
Polyaniline	Carbonization followed KOH activation and HCl washing	800°C	2752.0	1.90	–	66

(iv) Doping in porous carbon

Reports also reveal that the adsorption efficiency of porous carbon could also be improved by heteroatoms (N, S, P, etc.) doping process. For instance, the N (3.79–5.24%) doped porous carbons with a large surface area (803–1002 $m^2\,g^{-1}$) have been prepared by solvent-free pyrolysis route[70] and utilized as adsorbents for the removal of ibuprofen from aqueous solution. It was found that both meso/microporous structure and high nitrogen content are responsible for higher ibuprofen adsorption capacity (113 mg g^{-1}) and shorter equilibrium time (60 min). A dual doping strategy was also reported for enhancing adsorption performance of porous carbon adsorbents. For example, N- and S-doped porous carbons exhibit enhanced adsorption capacities (295.8 mg g^{-1} for bisphenol F and 308.7 mg g^{-1} for bisphenol S) within 30 min of contact time.[71] The notable performance is mainly due to N/S doping, large surface area, and unique hierarchical structure.

(v) Composites of porous carbon

To improve adsorption capacity further, porous carbon particles were composited with metal oxide and polymers, etc. For example, porous carbon/ZnO was synthesized using metal organic framework-74 (Zn) as a precursor.[72] Porous carbon/ZnO composites exhibited rapid rhodamine B adsorption capability due to their high surface area (782.971 $m^2\,g^{-1}$) and pore volume (0.698 $m^3\,g^{-1}$). Particularly, porous carbon/ZnO synthesized at 1000°C efficiently adsorbs nearly complete (40 mg L^{-1}) rhodamine B molecules from water within 30 min. A ternary composite, porous carbon microspheres/chitosan/$Ti_3C_2T_x$ with huge surface area (>1800 $m^2\,g^{-1}$) was used as adsorbent for the removal of crystal violet from solution.[73] The highest crystal violet adsorption capacity is 2750 mg g^{-1}. The adsorption pathway is attributed to the occurrence of π-π, hydrogen bonding, and electrostatic interactions between ternary composite adsorbent and crystal violet molecules.

3.3 Carbon nanotubes

Carbon nanotubes (CNTs), a hollow cylindrical form of carbon allotrope, were first reported by Iijima in 1991.[74] CNTs have a one-dimensional structure of rolled-up graphene layers with diameters ranging from one to a few tens of nanometers. The cylindrical carbon structure was composed of carbon atoms covalently linked to each other with a hexagonal structure.[75] CNTs have unique optical properties, high conductivity and surface area, and less weight features which can also be tuned by the functionalization process.[76] Generally, they have strong hydrophobicity, pore structure, and multiple surface groups. Based on the number of layers, CNTs are usually classified as single-walled and multiwalled carbon nanotubes. Single-walled carbon nanotubes (SWCNTs) have a single layer of graphene, while multiwalled carbon nanotubes (MWCNTs) have many concentrically arranged walls of graphene sheets with a distance of 0.34 nm.[77] Apart from the number of layers, some other parameters also substantially differentiate SWCNTs from MWCNTs.[77] In the case of SWCNT, large-scale synthesis with high purity is difficult to achieve. Also, the catalyst is essential for its synthesis. However, high defects can be created within SWCNTs by functionalization, while highly pure MWCNTs can be prepared on a large scale even without using catalysts. However, defects will be less for MWCNTs in comparison with SWCNTs.

(i) Forms and nature of CNTs

CNT strip (length = 30 cm and diameter = 3 mm) with mesopores and the hydrophobic surface was used for removing o-cresol from water.[78] Compared to pure CNT powder, CNT strip has less surface area and lower o-cresol adsorption capacity. However, strips can be recovered more easily than powder form and utilize 10 successive processes. Further, the strength of the strips is also high under compression.

(ii) Functionalized carbon nanotubes

Like activated carbon, functionalized CNTs also performed well in the adsorption process. To remove uranyl ions (UO_2^{2+}) from water (pH = 4; time = 120 min), poly(amidoxime) functionalized CNTs were used as adsorbent.[79] Amidoxime functional group is a ligand that shows a strong affinity with uranyl ions. Functionalized CNTs showed excellent uranyl ion adsorption (adsorption capacity = $\sim$247 mg g^{-1}) due to its abundant amidoxime group and porosity (surface area = 58.169 m^2 g^{-1}). Based on the Fourier transform infrared and X-ray photoelectron spectroscopic studies, two ways of coordination between amidoxime groups and uranyl ions were proposed (Fig. 3A). In another investigation, poly(acrylic acid) and polyacrylamide functionalized CNTs were fabricated by oxidative polymerization method and applied as sorbent for phenol removal in water.[80] This adsorbent exhibited great performance for four cycles which is due to π-π interactions, hydrogen bonding, and electrostatic interactions between phenol rings and functionalized CNTs (Fig. 3B).

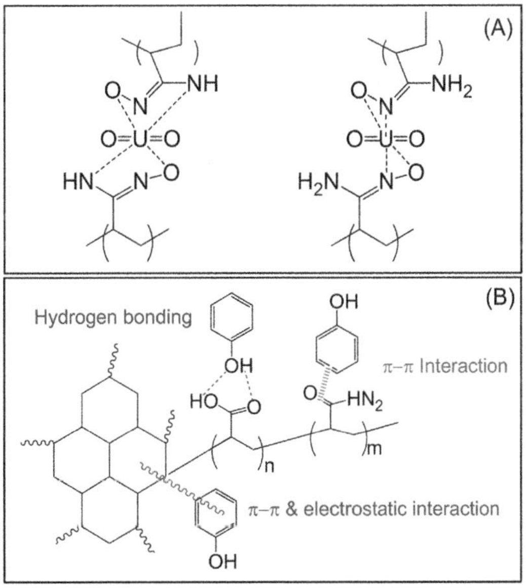

FIG. 3 Proposed interaction of pollutants with adsorbents (A) adsorption mode of uranyl ions on poly(amidoxime) functionalized CNTs[79] and (B) phenol adsorption on polyacrylic acid and polyacrylamide functionalized CNTs.[80] *(Figures were redrawn with modifications from the references.)*

MWCNTs functionalized with tetra-n-butyl-ammonium bromide showed high adsorption performance toward Ni(II) ions.[81] After 68 min of contact time, the highest Ni(II) adsorption capacity of 115.8 mg g^{-1} (removal rate = 93%) was obtained. Coulombic force of attraction is the key reason for the adsorption of Ni(II) ions on functionalized MWCNTs.

(iii) CNT-based composites

Numerous CNT-based binary and ternary composites were reported as adsorbents.[82–84] For example, SWCNT/Fe$_3$O$_4$ composite adsorbent was fabricated for the elimination of xylene (C$_8$H$_{10}$) from aqueous solutions.[85] Under the optimized conditions (pH = 8, dosage = 2000 mg L^{-1}, time period = 20 min), 99.2% (q_e = 50 mg g^{-1}) xylene (initial concentration = 100 mg L^{-1}) was captured by SWCNT/Fe$_3$O$_4$ composite adsorbent. In this process, contact time showed maximum effect among other influencing factors. Guo et al.[86] have determined the bisphenol A adsorption property of MWCNT/Fe$_3$O$_4$/MnO$_2$ ternary composite in water. The q_{max} and contact time were 132.9 mg g^{-1} and 150 min, respectively. The external magnet was used for the separation of this ternary composite conveniently and quickly from the reaction aliquot.

3.4 Carbon spheres

Carbon spheres are also important carbon materials that have high encapsulation ability, surface functionality, high surface area, and high stability. The application of carbon spheres in the adsorptive removal of aquatic contaminants is quite high.[87]

(i) Pure carbon spheres

A hollow carbon sphere with a discrete cage-like structure was synthesized by the nitric acid oxidation method.[88] The prepared hollow carbon sphere powder was used to adsorb Pb^{2+} ions from water, and the observed absorption capacity is 280.79 mg g^{-1}. Furthermore, the carbon sphere exhibited great adsorption capacity even after 5 recycles.

(ii) Doped carbon spheres

The mild hydrothermal approach was adapted to synthesize N-doped carbons spheres.[89] N-doped carbons spheres exhibit the highest adsorption capacity of 181.82 mg g^{-1} toward Cr(VI) ions. In the case of real electroplating wastewater, this adsorbent (4 g L^{-1}) decreases Cr concentration from 652.96 to 113.42 mg L^{-1} by means of effective adsorption. Cr(VI) ions adsorption on N-doped carbons spheres was ascribed by the following way[89]: (a) The presence of functional groups such as C=O, C=C, and C—OH donates (Lewis base) electrons to Cr(VI) from Cr(III) ions (Eq. 6).

$$3R - CH_2OH + 2HCrO_4^- + 8H^+ \rightarrow 3R - CHO + 2Cr^{3+} + 8H_2O \qquad (6)$$

(b) Also, O-containing (—COOH) and N-containing functional (—NH$_2$) groups protonated in acidic solution and that can attract Cr(VI) anions by ionic interaction (Eqs. 7, 8).

$$-NH_3^+ + HCrO_4^- \rightarrow -NH_3^+...HCrO_4^- \qquad (7)$$

$$-OH_2^+ + HCrO_4^- \rightarrow -OH_2^+...HCrO_4^- \qquad (8)$$

A multi-elemental doping process was also implemented in carbon spheres. For instance, S, N, O-co-doped porous carbon spheres (surface area = 2617 m^2 g^{-1}) were synthesized from starch and used for adsorptive removal of Cr(VI) ions from wastewater.[90] The abundant reductive surface functional groups such as C—SH, —NH—, C—H, and C—OH of carbon spheres considerably enriched the Cr(VI) adsorption and subsequently led to the reduction of Cr(VI) ions to Cr(III) ions. About 97% of Cr(VI) ions removal efficiency (adsorption capacity = 434 mg g^{-1}) was achieved by this adsorbent.

(iii) Carbon sphere-based composites

Like other carbon materials, carbon spheres were also composited with other nanomaterials. For example, hollow carbon sphere/polyaniline composite adsorbents with different mass ratios were reported for the removal of Cr(VI) ions.[91] Like doped carbon spheres, hollow carbon sphere/polyaniline composite adsorbent also adsorbs Cr(VI) ions and then reduces them as Cr(III) ions. The maximum adsorption capacity of this composite was 250.0 mg g^{-1}. Cr(VI) ions adsorption on carbon sphere/polyaniline occurs (acidic condition) via electrostatic attraction with protonated imino groups of polyaniline and amine group in polyaniline reduced Cr(VI) as Cr(III), followed by Cr(III) chelated on imino groups.

3.5 Coal

Coal is naturally formed by the degradation of plants beneath the Earth and is composed mainly of carbon along with hydrogen, sulfur, oxygen, and nitrogen. Coal is also evaluated as an adsorbent for the removal of harmful contaminants in water.[92]

(i) Nature of coals

The nature of coals determines their adsorption ability. For example, three types of coals, synthesized with different temperatures (coalification degrees), were applied as adsorbents for removal of phenol (C_6H_6O, $200\,mg\,L^{-1}$) and Cr(VI) ions ($5\,mg\,L^{-1}$) from simulated wastewater.[93] These coal samples have C—H and O-containing functional groups (hydroxyl, carbonyl, and ether). Compared to raw coal samples, the adsorption capacity of synthesized coal samples toward phenol and Cr(VI) ions is high. Brown coal and its underground coal gasification coke exhibited maximum removal rates toward phenol (72.44%) and Cr(VI) ions (21.77%).

(ii) Effect of bubbles

Micro (diameter $= 2$ to $600\,\mu m$) and nanobubbles (diameter $= 50$ to $200\,nm$) influence the adsorption property of low-rank coals (less density with high surface area and humic acid) such as Meng Dong and Wanli coal toward Congo red dye adsorption.[28] It was found that bubbles, generated by hydrodynamic and acoustic cavitation methods, accelerated the Congo red adsorption capacity of low-rank coals. With micro and nanobubbles, the adsorption rate was two times greater than that without bubbles for Meng Dong and Wanli coal. This is because of the reduction of Congo red by hydroxyl radicals, generated by bubbles and betterment in adsorption on low-rank coals.

(iii) Coal-based composites

Polyurethane foam modified with coal was constructed and used for the removal of brilliant green from water.[94] About 99.40% ($q_{max} = 134.95\,mg\,g^{-1}$) brilliant green removal efficiency was obtained when the coal content and contact time were 4% and 200 min, respectively. Further, this adsorbent exhibited excellent regeneration capability. The adsorption between coal modified polyurethane foam and brilliant green dye molecules was attributed to electrostatic interactions, hydrogen bonding formation, and π-π electron-donor-acceptor attractions. A low-grade coal (low carbon and high silicate content)/chitosan composite was prepared and utilized as an adsorbent for removal of diethyl phthalate.[95] Coal/chitosan with weight ratio of 9:1 showed 91.1% ($q_{max} = 42.67\,mg\,g^{-1}$) diethyl phthalate adsorption within 4 h (pH $= 5.8$; adsorbent dose $= 4\,mg\,mL^{-1}$). The adsorption is attributed to ionic attraction and hydrogen bonding between the diethyl phthalate and the surface of adsorbent particles.

3.6 Graphite and its derivatives

Graphite is a highly conductive form of carbon allotrope. It is purely composed of carbon atoms which are bonded with three other carbon atoms and forms hexagonal sheets via covalent bonds (sp^2 hybridization).[18] These carbon sheets are bounded by weak forces.

(i) Graphite-based adsorbents

Exfoliated graphite (surface area $= 40.95\,m^2\,g^{-1}$), prepared by Hummer's method, followed by exfoliation using a microwave system, was used for the adsorption of Congo red.[96] The obtained maximum Congo red adsorption capacity is $80.775\,mg\,g^{-1}$ (contact time $= 150\,min$). Apart from pure graphite, graphite-based composite was also developed for adsorption of aquatic pollutants. As an example, an expanded graphite/MgO composite with a surface area of $26.92\,m^2\,g^{-1}$ was prepared and used to remove phosphate ions from water.[97] The adsorption capacity of expanded graphite/MgO (dosage $= 1\,g\,L^{-1}$) toward phosphate ions is $491.6\,mg\,g^{-1}$ (pH $= 7$; time $= 90\,min$). Moreover, coexisting ions such as sulfate, chloride, and nitrate did not affect the phosphate adsorption capacity of expanded graphite/MgO composite. The expanded graphite/Fe_3O_4 composite adsorbent (surface area $= 34.8\,m^2\,g^{-1}$; magnetic saturation $= 17.5\,emu\,g^{-1}$) was also reported for adsorptive removal of methylene blue dye from water.[98] The reported methylene blue adsorption capacity is $78.06\,mg\,g^{-1}$.

(ii) Graphene-based adsorbents

Graphene has a hexagonal, close-packed honeycomb structure composed of sp^2 hybridized carbon atoms (Fig. 4A). It possesses a huge surface area, high conductivity, strength, and flexibility. Recently, industrial graphite waste was converted

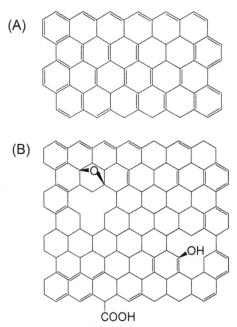

FIG. 4   Structure of (A) graphene and (B) reduced graphene oxide.[26]

as exfoliated graphene and applied as an adsorbent for the removal of royal blue, turquoise blue, black supra, navy blue, and deep red dyes in water.[99] The adsorption capacity of exfoliated graphene depends on the structure of dye molecules. The order of adsorption capacity is deep red ($q_{max} = 74.074$ mg g^{-1}) > turquoise blue ($q_{max} = 68.027$ mg g^{-1}) > black supra ($q_{max} = 60.975$ mg g^{-1}) > navy blue ($q_{max} = 54.054$ mg g^{-1}) > royal blue ($q_{max} = 36.231$ mg g^{-1}).

However, graphene-based adsorbents suffer in practical application due to the following reasons: (a) Strong interplanar interactions will lead to agglomeration of graphene nanolayers and may restack to form graphite and (b) graphene may not be simply detached from treated water. These drawbacks can be eliminated by creating composites with other adsorbent materials.[100] Functionalized graphene and graphene-based composites including graphene-polymer, graphene-metal compounds, graphene-other carbon materials and graphene-metal nanoparticles were widely applied in wastewater treatment for the removal of harmful inorganic and organic pollutants.[26,101–103]

(iii)   Reduced graphene oxide

The rGO is a reduced form of graphene oxide (Fig. 4B). It shows lower conductivity than graphene. Like graphene, rGO-based adsorbents are also widely studied as adsorbents in the waste water purification process.[104] For instance, corn stover, a renewable agro-waste, was used to prepare porous rGO functionalized with nitrogen-containing functional groups (surface area = 493.5 m^2 g^{-1}) for the removal of tetracycline from solution.[31] When the initial concentrations of tetracycline and adsorbent dosage are 48.74 mg L^{-1} and 0.98 g L^{-1}, respectively, the adsorption capacity was found to be 132.9 mg g^{-1} (contact time = 51.6 min and pH = 7.4). Porous rGO aerogel was synthesized for adsorption of antimony (Sb) ions from an aqueous solution.[105] This adsorbent shows an adsorption capacity of 168.59 and 206.72 mg g^{-1} toward Sb(III) and Sb(V) ions, respectively. The adsorption of Sb(V) on the rGO surface is higher than Sb(III) ions. Moreover, this adsorbent exhibited good selectivity among competing ions, excellent practicability, and reusability.

In order to adsorb phenolic pollutants, hydrophobic N-doped rGO adsorbent was reported by Zhao et al.[106] The obtained q_{max} for phenol and p-nitrophenol were 155.82 and 80.60 mg g^{-1}, respectively (pH = 6). The enhanced adsorption property is due to π-π and hydrophobic attraction between adsorbent and pollutant molecules. Importantly, >80% adsorptive removal efficiency was maintained over five cycles of reuse.

Attention to the development of rGO-based binary and ternary composites is also significant.[107,108] For example, rGO/Fe$_2$O$_3$ composite exhibited 89.6% acid blue 113 adsorption removal efficiency within 440 min of contact time.[109] The rGO-porphyrin/Ag ternary composite was reported with a maximum methylene blue adsorption capacity of 130.37 mg g^{-1}.[110]

3.7 Biochar

Biochar, a stable form of carbon, is a charcoal-like substance that is black, highly porous,[111] and weightless powder with high large surface area. Nearly 70% of its composition is carbon, while the rest of the composition comprises hydrogen (H), oxygen (O), and nitrogen (N) atoms. Waste biomass, including agricultural residues and industrial wastes, is used to prepare biochar via pyrolysis and gasification under an inert atmosphere. The structure of biochar is mainly dependent on its source materials. The possible biochar structures synthesized using oil palm residues,[112] pine residues,[113] and sugarcane bagasse[114] are shown in Fig. 5. Biochar structure is also dependent on the reaction temperature of the same source material.[115] Due to the availability of sources, functional groups and vast surface area, biochar was also extensively investigated as an adsorbent in water remediation.

(i) Nature of biochar

Biochar with abundant redox and inert functional moieties was achieved using rice husk, wheat straw, and corncob as carbon sources. These biochar samples have amorphous character and defects in their structures. The order of specific surface area of biochar samples is rice husk-derived biochar ($255.78 \, m^2 \, g^{-1}$) > wheat straw-derived biochar ($24.5 \, m^2 \, g^{-1}$) > corn cob-derived biochar ($8.99 \, m^2 \, g^{-1}$). By using batch experiments, the prepared biochar samples were used to remove Pb(II) and Cd(II) ions from water.[30] Under the optimal conditions, Pb(II) adsorption efficiencies of biochar derived from rice husk, wheat straw, and corncob are 96.41%, 95.38%, and 96.92%, respectively. In the case of the Cd(II) ions removal process, the adsorption efficiencies of biochar derived from rice husk, wheat straw, and corncob are 94.73%, 93.68%, and 95.78%, respectively.

Cork is a carbon-rich component of the bark of *Quercus suber* L.[32] It is discharged as a waste powder (>50,000 t per annum) from the cork industry.[116] For recycling purposes, cork-derived biochar with a hollow polyhedral cell structure was prepared by the slow pyrolysis method.[117] The biochar produced at higher temperature (750 °C) and pyrolysis time (0.5 h) gain a high degree of aromaticity and surface area ($392.5 \, m^2 \, g^{-1}$). This biochar was found to be a good adsorbent for the removal of Cu(II) ions from simulated solutions. After 4 h of contact time, the maximum Cu(II) ion adsorption capacity is $18.5 \, mg \, g^{-1}$.

(ii) Modified biochar

Organosulfur modified biochar adsorbent was prepared from sugarcane bagasse and used for removing Cd(II), Ni(II), Pb(II), and Cr(III) ions from aqueous solutions.[114] The highest adsorption capacity was obtained for modified biochar prepared at 700 °C. The order of adsorption capacity is as followed: Pb(II) > Cr(III) > Cd(II) ≅ Ni(II). Metal ion adsorption mechanisms involve physisorption, surface complexation, precipitation, and ion exchange. Coating okra (*Abelmoschus esculentus* L.) mucilage modified biochar was reported for adsorption of methylene blue from an aqueous solution.[118]

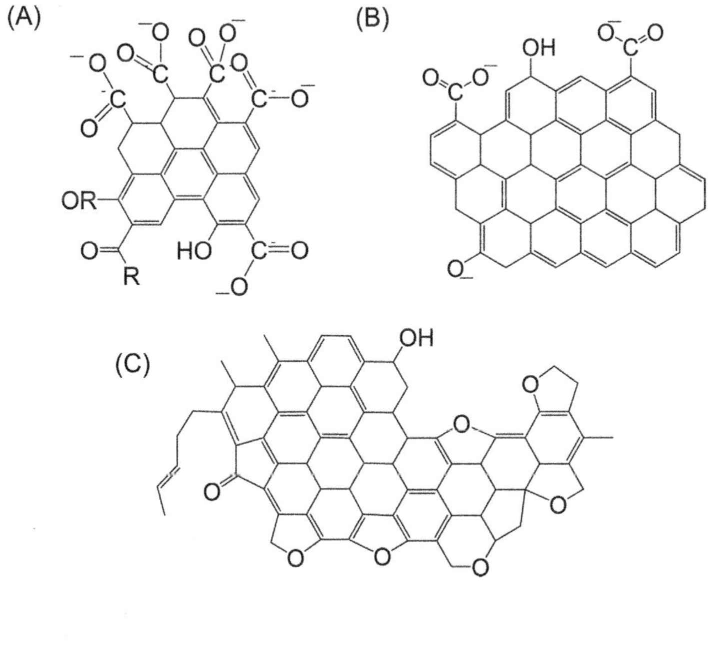

FIG. 5 Structure of biochar prepared from different sources: (A) oil palm residues,[112] (B) pine residues,[113] and (C) sugarcane bagasse.[114]

At pH $= 8.1$ modified biochar (dose $= 1\,g\,L^{-1}$) shows methylene blue (concentration $= 100\,mg\,L^{-1}$) adsorption capacity of $78.13\,mg\,g^{-1}$ which is higher than that of unmodified biochar ($27.47\,mg\,g^{-1}$). Interestingly, methylene blue adsorption follows chemisorption in modified biochar, while physisorption dominates in unmodified biochar.

(iii) Surface nature of biochar

Hydrophobicity of biochar plays a major role in the absorption of nutrients (total nitrogen and phosphorus, ammonia-nitrogen), organic matter, and metals ions (Na, Al, Mg, Fe, Mn, Cu, Zn, Ni, Cd, Pb, and Cr) in water.[45] Hydrophilic biochar leaches greater loads of nutrients and organic matter than hydrophobic biochar. However, hydrophilic biochar leached lower metal loads.

(iv) Biochar-based composites

To enhance adoption rate and capacity, selectivity, pure biochar samples synthesized from various sources are also composited with other nanomaterials. Some examples are listed in Table 3.

4 Current research trend

Enormous studies on carbon-based adsorbents in water treatment were reported recently. Based on the survey of recent research articles, current research trends on carbon-based adsorbents could be identified. (a) Keeping in mind production cost and sustainability, researchers are targeting household and industrial wastes and biomass as carbon precursors for the

TABLE 3 List of biochar-based composites and their pollutant adsorption performances.

Biochar-based adsorbent	Source of biochar	Pollutant	Experiment (time)	Adsorbent dosage	Adsorption capacity (mg g^{-1})	References
Biochar/polysaccharide/reduced iron oxide	Jackfruit peel	Phosphate (PO_4^{3-}) Nitrate (NO_3^-)	Batch isotherm	0.1 g/100 mL	7.94 5.26	14
Biochar/MnO$_2$	Lignin	Methylene blue	Batch adsorption (24 h)	20 mg	248.96	119
Functionalization biochar/Fe$_3$O$_4$	Date leaves and stalks	Pb(II) Cd(II)	Batch experiment	20 mg	61.25 53.75	120
Biochar/CoFe$_2$O$_4$	Banana pseudostem	Amoxicillin	Batch experiment	50 mg	99.99	115
Biochar/Fe	Bamboo Bagasse Tyre	Pb Cu	Fixed-bed adsorption	Soil +10% Biochar or Biochar/Fe (10%)	0.15 (75.75%; Tyre derived biochar/Fe (10%)) 0.12 (59.44%; 10% Bamboo derived biochar)	121
Biochar/iron oxide/attapulgite	*Camellia oleifera* shells	Cr(VI) ions	Batch adsorption experiments (24 h)	0.1 g	49.99	122
Biochar/Mg/Al-layered double hydroxides	Almond shell	Phosphate (PO_4^{3-}) Nitrate (NO_3^-)	Batch adsorption (12 h)	0.05 g	6.10 (47.5%) 0.93 (27.7%)	123
Watermelon rinds biochar/Fe$_3$O$_4$	Watermelon rinds	U(VI) ions	Batch adsorption	0.2 g L^{-1}	323.56	124

synthesis of carbon-based adsorbents.[125] (b) For reuse purposes, a focus on magnetic carbon-based adsorbent materials was given.[115,120] (c) New types of carbon materials were also reported as adsorbents for abatement of water contaminants.[126,127] (d) Interest in carbon-based aerogel/hydrogel, membrane, strips as adsorbent was also given.[128] (e) Research on selective removal of pollutants by modified carbon adsorbents was also reported.[129] (f) Heteroatoms, rich carbons, and composites of carbon-carbon materials were adopted to obtain a high adsorption rate, thereby producing clean water efficiently.[130,131]

5 Summary and perspectives

Carbon-based materials were developed as adsorbents and studied in the removal of organic and inorganic aquatic pollutants from water. The basics of the adsorption method and properties of carbon materials were outlined. The adsorption properties of activated carbon, porous carbon, CNTs, carbon spheres, coal, graphite derivatives, and biochar in aquatic pollutants elimination were described. To achieve high adsorption capacity/efficiency, researchers have followed different improvement strategies in pure carbon materials. The following conclusive points were extracted from this chapter: (i) pure carbon materials have considerable interest in the removal of different kinds of pollutants due to the abundance of carbon precursors, low cost for production, huge surface area, unique morphology, and porosity. (ii) Adsorption capacity of pure carbon-based adsorbents depends on the active surface area, defects, hydrophobicity/hydrophilicity, surface functional groups, and nature of carbon source and pollutants. (iii) Adsorption ability of carbon materials was further improved by surface functionalization, doping process, and construction of composite with other suitable nanomaterials. (iv) Adsorbent dosage, pH of the solution, reaction temperature, nature and quantity of adsorbents, and initial concentration of pollutants also determine the adsorption capacity of carbon-based adsorbents. (v) Adsorption of pollutants on carbon-based adsorbents happens via π-π attraction, electrostatic interactions, hydrophobic, and hydrogen bonding. Besides, the following perspectives are given: (a) For commercialization, shortcomings in large-scale preparation of adsorbents, elimination of pollutants in real wastewaters need to be eliminated. (b) Economic analysis should also be published along with the synthesis of adsorbents. (c) Most of the reports concentrated only on lab-scale adsorption experiments. Hence, pilot scale studies should also need to be considered. (d) More focus on stability and biocompatibility of adsorbents is needed.

References

1. Stephens GL, Slingo JM, Rignot E, et al. Earth's water reservoirs in a changing climate. *Proc R Soc A*. 2020;476:20190458. https://doi.org/10.1098/rspa.2019.0458.
2. Boretti A, Rosa L. Reassessing the projections of the world water development report. *npj Clean Water*. 2019;2:15. https://doi.org/10.1038/s41545-019-0039-9.
3. Ledezma CZ, Bolagay DN, Figueroa F, et al. Heavy metal water pollution: a fresh look about hazards, novel and conventional remediation methods. *Environ Technol Innov*. 2021;22:101504. https://doi.org/10.1016/j.eti.2021.101504.
4. Briffa J, Sinagra E, Blundell R. Heavy metal pollution in the environment and their toxicological effects on humans. *Heliyon*. 2020;6:e04691. https://doi.org/10.1016/j.heliyon.2020.e04691.
5. Wen Y, Schoups G, Giesen NVD. Organic pollution of rivers: combined threats of urbanization, livestock farming and global climate change. *Sci Rep*. 2017;7:43289. https://doi.org/10.1038/srep43289.
6. Singh AK, Kumar A, Bilal M, Chandra R. Organometallic pollutants of paper mill wastewater and their toxicity assessment on Stinging catfish and sludge worm. *Environ Technol Innov*. 2021;24:101831. https://doi.org/10.1016/j.eti.2021.101831.
7. Tang L, Liu YL, Qin G, Lin Q, Zhang YH. Effects of tributyltin on gonad and brood pouch development of male pregnant lined seahorse *(Hippocampus erectus)* at environmentally relevant concentrations. *J Hazard Mater*. 2021;408:124854. https://doi.org/10.1016/j.jhazmat.2020.124854.
8. Dehghani MH, Omrani GA, Karri RR. Solid waste—sources, toxicity, and their consequences to human health. In: *Soft Computing Techniques in Solid Waste and Wastewater Management*. Elsevier; 2021:205–213. https://doi.org/10.1016/B978-0-12-824463-0.00013-6.
9. Karri RR, Ravindran G, Dehghani MH. Wastewater—sources, toxicity, and their consequences to human health. In: *Soft Computing Techniques in Solid Waste and Wastewater Management*. Elsevier; 2021:3–33. https://doi.org/10.1016/B978-0-12-824463-0.00001-X.
10. Shahid MK, Kashif A, Fuwad A, Choi Y. Current advances in treatment technologies for removal of emerging contaminants from water—a critical review. *Coord Chem Rev*. 2021;442:213993. https://doi.org/10.1016/j.ccr.2021.213993.
11. Rajasulochana P, Preethy V. Comparison on efficiency of various techniques in treatment of waste and sewage water—a comprehensive review. *Resour-Effic Technol*. 2016;2:175–184. https://doi.org/10.1016/j.reffit.2016.09.004.
12. Crini G, Lichtfouse E. Advantages and disadvantages of techniques used for wastewater treatment. *Environ Chem Lett*. 2019;17:145–155. https://doi.org/10.1007/s10311-018-0785-9.
13. Rashid R, Shafiq I, Akhter P, Iqbal MJ, Hussain M. A state-of-the-art review on wastewater treatment techniques: the effectiveness of adsorption method. *Environ Sci Pollut Res*. 2021;28:9050–9066. https://doi.org/10.1007/s11356-021-12395-x.

14. Nayak A, Bhushan B, Gupta V, Kotnala S. Fabrication of microwave assisted biogenic magnetite-biochar nanocomposite: a green adsorbent from jackfruit peel for removal and recovery of nutrients in water sample. *J Ind Eng Chem*. 2021;100:134–148. https://doi.org/10.1016/j.jiec.2021.05.028.

15. Senthil Rathi B, Senthil Kumar P. Application of adsorption process for effective removal of emerging contaminants from water and wastewater. *Environ Pollut*. 2021;280:116995. https://doi.org/10.1016/j.envpol.2021.116995.

16. Patel H. Fixed-bed column adsorption study: a comprehensive review. *Appl Water Sci*. 2019;9:45. https://doi.org/10.1007/s13201-019-0927-7.

17. Kumari P, Weqar MA, Siddiqi A. Usage of nanoparticles as adsorbents for waste water treatment: an emerging trend. *Sustain Mater Technol*. 2019;22:e00128. https://doi.org/10.1016/j.susmat.2019.e00128.

18. Ying P, Liang T, Du Y, Zhang J, Zeng X, Zhong Z. Thermal transport in planar sp^2-hybridized carbon allotropes: a comparative study of biphenylene network, pentaheptite and graphene. *Int J Heat Mass Transf*. 2022;183:122060. https://doi.org/10.1016/j.ijheatmasstransfer.2021.122060.

19. Shi X, Wei W, Fu Z, et al. Review on carbon dots in food safety applications. *Talanta*. 2019;194:809–821. https://doi.org/10.1016/j.talanta.2018.11.005.

20. Joshi P, Mishra R, Narayan RJ. Biosensing applications of carbon-based materials. *Curr Opin Biomed Eng*. 2021;18. https://doi.org/10.1016/j.cobme.2021.100274, 100274.

21. Riley PR, Narayan RJ. Recent advances in carbon nanomaterials for biomedical applications: a review. *Curr Opin Biomed Eng*. 2021;17:100262. https://doi.org/10.1016/j.cobme.2021.100262.

22. Fu Q, Zhang P, Zhuang L, et al. Micro/nano multiscale reinforcing strategies toward extreme high-temperature applications: take carbon/carbon composites and their coatings as the examples. *J Mater Sci Technol*. 2022;96:31–68. https://doi.org/10.1016/j.jmst.2021.03.076.

23. Liu L, Shen Z, Zhang X, Ma H. Highly conductive graphene/carbon black screen printing inks for flexible electronics. *J Colloid Interface Sci*. 2021;582:12–21. https://doi.org/10.1016/j.jcis.2020.07.106.

24. Prasankumar T, Jose S, Ajayan PM, Ashokkumar M. Functional carbons for energy applications. *Mater Res Bull*. 2021;142:111425. https://doi.org/10.1016/j.materresbull.2021.111425.

25. Gabris MA, Ping J. Carbon nanomaterial-based nanogenerators for harvesting energy from environment. *Nano Energy*. 2021;90:106494. https://doi.org/10.1016/j.nanoen.2021.106494.

26. Ali I, Basheer AA, Mbianda XY, et al. Graphene based adsorbents for remediation of noxious pollutants from wastewater. *Environ Int*. 2019;127:160–180. https://doi.org/10.1016/j.envint.2019.03.029.

27. Li K, Liu Q, Rimmer SM, Huggett WW, Zhang S. Investigation of the carbon structure of naturally graphitized coals from Central Hunan, China, by density-gradient centrifugation, X-ray diffraction, and high-resolution transmission electron microscopy. *Int J Coal Geol*. 2020;232:103628. https://doi.org/10.1016/j.coal.2020.103628.

28. He Q, Cui R, Miao Z, et al. Improved removal of Congo Red from wastewater by low-rank coal using micro and nanobubbles. *Fuel*. 2021;291:120090. https://doi.org/10.1016/j.fuel.2020.120090.

29. Li Y, Liu Y, Yang W, Liu L, Pan J. Adsorption of elemental mercury in flue gas using biomass porous carbons modified by microwave/hydrogen peroxide. *Fuel*. 2021;291:120152. https://doi.org/10.1016/j.fuel.2021.120152.

30. Amen R, Yaseen M, Mukhtar A, et al. Lead and cadmium removal from wastewater using eco-friendly biochar adsorbent derived from rice husk, wheat straw, and corncob. *Clean Eng Technol*. 2020;1:100006. https://doi.org/10.1016/j.clet.2020.100006.

31. Haghighat GA, Saghi MH, Anastopoulos I, et al. Aminated graphitic carbon derived from corn Stover biomass as adsorbent against antibiotic tetracycline: optimizing the physicochemical parameters. *J Mol Liq*. 2020;313:113523. https://doi.org/10.1016/j.molliq.2020.113523.

32. Demertzi M, Dias AC, Matos A, Arroja LM. Evaluation of different end-of life management alternatives for used natural cork stoppers through life cycle assessment. *Waste Manag*. 2015;46:668–680. https://doi.org/10.1016/j.wasman.2015.09.026.

33. Shao P, Pei J, Tang H, et al. Defect-rich porous carbon with anti-interference capability for adsorption of bisphenol A via long-range hydrophobic interaction synergized with short-range dispersion force. *J Hazard Mater*. 2021;403:123705. https://doi.org/10.1016/j.jhazmat.2020.123705.

34. Akbayrak S, Özçifçi Z, Tabak A. Activated carbon derived from tea waste: a promising supporting material for metal nanoparticles used as catalysts in hydrolysis of ammonia borane. *Biomass Bioenergy*. 2020;138:105589. https://doi.org/10.1016/j.biombioe.2020.105589.

35. Jiang S, Shao H, Cao G, et al. Waste cotton fabric derived porous carbon containing Fe_3O_4/NiS nanoparticles for electrocatalytic oxygen evolution. *J Mater Sci Technol*. 2020;59:92–99. https://doi.org/10.1016/j.jmst.2020.04.055.

36. Pandey R, Siddiqui S, Saurabh S, et al. Waste candle soot derived carbon nanoparticles: a competent alternative for the management of *Helicoverpa armigera*. *Chemosphere*. 2021;264:128537. https://doi.org/10.1016/j.chemosphere.2020.128537.

37. Gomez-Hernandez R, Panecatl-Bernal Y, Mendez-Rojas MA. High yield and simple one-step production of carbon black nanoparticles from waste tires. *Heliyon*. 2019;5:e02139. https://doi.org/10.1016/j.heliyon.2019.e02139.

38. Karri RR, Sahu JN. Process optimization and adsorption modeling using activated carbon derived from palm oil kernel shell for Zn (II) disposal from the aqueous environment using differential evolution embedded neural network. *J Mol Liq*. 2018;265:592–602. https://doi.org/10.1016/j.molliq.2018.06.040.

39. Ma C, Huang H, Gao X, et al. Honeycomb tubular biochar from Fargesia leaves as an effective adsorbent for tetracyclines pollutants. *J Taiwan Inst Chem Eng*. 2018;91:299–308. https://doi.org/10.1016/j.jtice.2018.05.032.

40. Pal DB, Singh A, Jha JM, et al. Low-cost biochar adsorbents prepared from date and Delonix regia seeds for heavy metal sorption. *Bioresour Technol*. 2021;339:125606. https://doi.org/10.1016/j.biortech.2021.125606.

41. Liang Y, Huang G, Zhang Q, Yang Y, Zhou J, Cai J. Hierarchical porous carbons from biowaste: hydrothermal carbonization and high-performance for rhodamine B adsorptive removal. *J Mol Liq*. 2021;330:115580. https://doi.org/10.1016/j.molliq.2021.115580.

42. Zhao X, Li W, Kong F, et al. Carbon spheres derived from biomass residue via ultrasonic spray pyrolysis for supercapacitors. *Mater Chem Phys*. 2018;219:461–467. https://doi.org/10.1016/j.matchemphys.2018.08.055.

43. Zhao X, Chen H, Kong F, et al. Fabrication, characteristics and applications of carbon materials with different morphologies and porous structures produced from wood liquefaction: a review. *Chem Eng J*. 2019;364:226–243. https://doi.org/10.1016/j.cej.2019.01.159.

44. Xu Z, He M, Xu X, Cao X, Tsang DCW. Impacts of different activation processes on the carbon stability of biochar for oxidation resistance. *Bioresour Technol*. 2021;338:125555. https://doi.org/10.1016/j.biortech.2021.125555.

45. Hong N, Cheng Q, Goonetilleke A, Bandala ER, Liu A. Assessing the effect of surface hydrophobicity/hydrophilicity on pollutant leaching potential of biochar in water treatment. *J Ind Eng Chem*. 2020;89:222–232. https://doi.org/10.1016/j.jiec.2020.05.017.

46. Lee KM, Wong CPP, Tan TL, Lai CW. Functionalized carbon nanotubes for adsorptive removal of water pollutants. *Mater Sci Eng B*. 2018;236–237:61–69. https://doi.org/10.1016/j.mseb.2018.12.004.

47. Suhas, Carrott PJM, Carrott MMLR, Singh R, Singh LP, Chaudhary M. An innovative approach to develop microporous activated carbons in oxidising atmosphere. *J Clean Prod*. 2017;156:549–555. https://doi.org/10.1016/j.jclepro.2017.04.078.

48. Medhat A, El-Maghrabi HH, Abdelghany A, et al. Efficiently activated carbons from corn cob for methylene blue adsorption. *Appl Surf Sci Adv*. 2021;3:100037. https://doi.org/10.1016/j.apsadv.2020.100037.

49. Suhas, Gupta VK, Singh LP, Chaudhary M, Kushwaha S. A novel approach to develop activated carbon by an ingenious hydrothermal treatment methodology using Phyllanthus emblica fruit stone. *J Clean Prod*. 2021;288:125643. https://doi.org/10.1016/j.jclepro.2020.125643.

50. Peralta DRL, Brito ED, Cortes AA, et al. Advances in activated carbon modification, surface heteroatom configuration, reactor strategies, and regeneration methods for enhanced wastewater treatment. *J Environ Chem Eng*. 2021;9:105626. https://doi.org/10.1016/j.jece.2021.105626.

51. Shokry H, Elkady M, Hamad H. Nano activated carbon from industrial mine coal as adsorbents for removal of dye from simulated textile wastewater: operational parameters and mechanism study. *J Mater Res Technol*. 2019;8:4477–4488. https://doi.org/10.1016/j.jmrt.2019.07.061.

52. Ramya V, Murugan D, Lajapathirai C, Saravanan P, Sivasamy A. Removal of toxic pollutants using tannery sludge derived mesoporous activated carbon: experimental and modelling studies. *J Environ Chem Eng*. 2019;7:102798. https://doi.org/10.1016/j.jece.2018.11.043.

53. Aravind Kumar J, Senthil Kumar P, Krithiga T, et al. Acenaphthene adsorption onto ultrasonic assisted fatty acid mediated porous activated carbon-characterization, isotherm and kinetic studies. *Chemosphere*. 2021;284:131249. https://doi.org/10.1016/j.chemosphere.2021.131249.

54. Chaudhary M, Suhas R, Singh I, Tyagi J, Ahmed S, Chaudhary SK. Microporous activated carbon as adsorbent for the removal of noxious anthraquinone acid dyes: role of adsorbate functionalization. *J Environ Chem Eng*. 2021;9:106308. https://doi.org/10.1016/j.jece.2021.106308.

55. Yang X, Wan Y, Zheng Y, et al. Surface functional groups of carbon-based adsorbents and their roles in the removal of heavy metals from aqueous solutions: a critical review. *Chem Eng J*. 2019;366:608–621. https://doi.org/10.1016/j.cej.2019.02.119.

56. Babu CM, Binnemans K, Roosen J. Ethylenediaminetriacetic acid-functionalized activated carbon for the adsorption of rare earths from aqueous solutions. *Ind Eng Chem Res*. 2018;57:1487–1497. https://doi.org/10.1021/acs.iecr.7b04274.

57. AlQadhi NF, AlSuhaimi AO. Chemically functionalized activated carbon with 8-hydroxyquinoline using aryldiazonium salts/diazotization route: green chemistry synthesis for oxins-carbon chelators. *Arab J Chem*. 2020;13:1386–1396. https://doi.org/10.1016/j.arabjc.2017.11.010.

58. Altıntıg E, Yenigun M, Sarı A, Altundag H, Tuzen M, Saleh TA. Facile synthesis of zinc oxide nanoparticles loaded activated carbon as an eco-friendly adsorbent for ultra-removal of malachite green from water. *Environ Technol Innov*. 2021;21:101305. https://doi.org/10.1016/j.eti.2020.101305.

59. Bhattacharyya A, Ghorai S, Rana D, et al. Design of an efficient and selective adsorbent of cationic dye through activated carbon—graphene oxide nanocomposite: study on mechanism and synergy. *Mater Chem Phys*. 2021;260:124090. https://doi.org/10.1016/j.matchemphys.2020.124090.

60. Alorabi AQ, Hassan MS, Azizi M. Fe$_3$O$_4$-CuO-activated carbon composite as an efficient adsorbent for bromophenol blue dye removal from aqueous solutions. *Arab J Chem*. 2020;13:8080–8091. https://doi.org/10.1016/j.arabjc.2020.09.039.

61. Teng W, Bai N, Chen Z, Shi J, Fan J, Zhang WX. Hierarchically porous carbon derived from metal-organic frameworks for separation of aromatic pollutants. *Chem Eng J*. 2018;346:388–396. https://doi.org/10.1016/j.cej.2018.04.051.

62. Bhadra BN, Jhung SH. A remarkable adsorbent for removal of contaminants of emerging concern from water: porous carbon derived from metal azolate framework-6. *J Hazard Mater*. 2017;340:179–188. https://doi.org/10.1016/j.jhazmat.2017.07.011.

63. Li T, Ma R, Xu X, Sun S, Lin J. Microwave-induced preparation of porous graphene nanosheets derived from biomass for supercapacitors. *Microporous Mesoporous Mater*. 2021;324:111277. https://doi.org/10.1016/j.micromeso.2021.111277.

64. Sun B, Yuan Y, Li H, et al. Waste-cellulose-derived porous carbon adsorbents for methyl orange removal. *Chem Eng J*. 2019;371:55–63. https://doi.org/10.1016/j.cej.2019.04.031.

65. Siddiqui MN, Ali I, Asim M, Chanbasha B. Quick removal of nickel metal ions in water using asphalt-based porous carbon. *J Mol Liq*. 2020;308:113078. https://doi.org/10.1016/j.molliq.2020.113078.

66. Khan NA, An HJ, Yoo DK, Jhung SH. Polyaniline-derived porous carbons: remarkable adsorbent for removal of various hazardous organics from both aqueous and non-aqueous media. *J Hazard Mater*. 2018;360:163–171. https://doi.org/10.1016/j.jhazmat.2018.08.001.

67. Hao Y, Wang Z, Wang Z, He Y. Preparation of hierarchically porous carbon from cellulose as highly efficient adsorbent for the removal of organic dyes from aqueous solutions. *Ecotoxicol Environ Saf*. 2019;168:298–303. https://doi.org/10.1016/j.ecoenv.2018.10.076.

68. Cuong DV, Liu NL, Nguyen VA, Hou CH. Meso/micropore-controlled hierarchical porous carbon derived from activated biochar as a high-performance adsorbent for copper removal. *Sci Total Environ*. 2019;692:844–853. https://doi.org/10.1016/j.scitotenv.2019.07.125.

69. Wu J, Yan X, Li L, et al. High-efficiency adsorption of Cr(VI) and RhB by hierarchical porous carbon prepared from coal gangue. *Chemosphere*. 2021;275:130008. https://doi.org/10.1016/j.chemosphere.2021.130008.

70. Lei X, Huang L, Liu K, Ouyang L, Shuai Q, Hu S. Facile one-pot synthesis of hierarchical N-doped porous carbon for efficient ibuprofen removal. *J Colloid Interface Sci*. 2021;604:823–831. https://doi.org/10.1016/j.jcis.2021.07.055.

71. Wang T, Xue L, Zheng L, et al. Biomass-derived N/S dual-doped hierarchically porous carbon material as effective adsorbent for the removal of bisphenol F and bisphenol S. *J Hazard Mater*. 2021;416:126126. https://doi.org/10.1016/j.jhazmat.2021.126126.

72. Wang L, Tang P, Liu J, et al. Multifunctional ZnO-porous carbon composites derived from MOF-74(Zn) with ultrafast pollutant adsorption capacity and supercapacitance properties. *J Colloid Interface Sci*. 2019;554:260–268. https://doi.org/10.1016/j.jcis.2019.07.015.

73. Wu Z, Deng W, Tang S, Hitzky ER, Luo J, Wang X. Pod-inspired MXene/porous carbon microspheres with ultrahigh adsorption capacity towards crystal violet. *Chem Eng J*. 2021;426:130776. https://doi.org/10.1016/j.cej.2021.130776.

74. Iijima S. Helical microtubules of graphitic carbon. *Nature*. 1991;354:56–58. https://doi.org/10.1038/354056a0.

75. Hussein F, Abdulrazzak F, Alkaim A. Synthesis, characterization and general properties of carbon nanotubes. In: *Nanomaterials: Biomedical, Environmental, and Engineering Application*. Wiley & Sons; 2018:1–59.

76. Rathinavel S, Priyadharshini K, Panda D. A review on carbon nanotube: an overview of synthesis, properties, functionalization, characterization, and the application. *Mater Sci Eng B*. 2021;268:115095. https://doi.org/10.1016/j.mseb.2021.115095.

77. Onyancha RB, Aigbe UO, Ukhurebor KE, Muchiri PW. Facile synthesis and applications of carbon nanotubes in heavy-metal remediation and biomedical fields: a comprehensive review. *J Mol Struct*. 2021;1238:130462. https://doi.org/10.1016/j.molstruc.2021.130462.

78. Yin Z, Duoni, Chen H, et al. Resilient, mesoporous carbon nanotube-based strips as adsorbents of dilute organics in water. *Carbon*. 2018;132:329–334. https://doi.org/10.1016/j.carbon.2018.02.074.

79. Zhuang S, Wang J. Poly amidoxime functionalized carbon nanotube as an efficient adsorbent for removal of uranium from aqueous solution. *J Mol Liq*. 2020;319:114288. https://doi.org/10.1016/j.molliq.2020.114288.

80. Saleh TA, Elsharif AM, Asiri S, Mohammed ARI, Dafalla H. Synthesis of carbon nanotubes grafted with copolymer of acrylic acid and acrylamide for phenol removal. *Environ Nanotechnol Monit Manag*. 2020;14:100302. https://doi.org/10.1016/j.enmm.2020.100302.

81. Rahmati N, Rahimnejad M, Pourali M, Muallah SK. Effective removal of nickel ions from aqueous solution using multi-wall carbon nanotube functionalized by glycerol-based deep eutectic solvent. *Colloids Interface Sci Commun*. 2021;40:100347. https://doi.org/10.1016/j.colcom.2020.100347.

82. Khan FSA, Mubarak NM, Khalid M, et al. A comprehensive review on micropollutants removal using carbon nanotubes-based adsorbents and membranes. *J Environ Chem Eng*. 2021;9:106647. https://doi.org/10.1016/j.jece.2021.106647.

83. Khan FSA, Mubarak NM, Tan YH, et al. A comprehensive review on magnetic carbon nanotubes and carbon nanotube-based buckypaper for removal of heavy metals and dyes. *J Hazard Mater*. 2021;413:125375. https://doi.org/10.1016/j.jhazmat.2021.125375.

84. Khan FSA, Mubarak NM, Khalid M, et al. Comprehensive review on carbon nanotubes embedded in different metal and polymer matrix: fabrications and applications. *Crit Rev Solid State Mater Sci*. 2021. https://doi.org/10.1080/10408436.2021.1935713.

85. Pourzamani H, Parastar S, Hashemi M. The elimination of xylene from aqueous solutions using single wall carbon nanotube and magnetic nanoparticle hybrid adsorbent. *Process Saf Environ Prot*. 2017;109:688–696. https://doi.org/10.1016/j.psep.2017.05.010.

86. Guo X, Huang Y, Yu W, Yu X, Han X, Zhai H. Multi-walled carbon nanotubes modified with iron oxide and manganese dioxide (MWCNTs-Fe$_3$O$_4$–MnO$_2$) as a novel adsorbent for the determination of BPA. *Microchem J*. 2020;157:104867. https://doi.org/10.1016/j.microc.2020.104867.

87. Sivaranjanee R, Senthil Kumar P. A review on cleaner approach for effective separation of toxic pollutants from wastewater using carbon Sphere's as adsorbent: preparation, activation and applications. *J Clean Prod*. 2021;291:125911. https://doi.org/10.1016/j.jclepro.2021.125911.

88. Zhang X, Wang W, Luo S, Lin Q. Preparation of discrete cage-like oxidized hollow carbon spheres with vertically aligned graphene-like nanosheet surface for high performance Pb^{2+} absorption. *J Colloid Interface Sci*. 2019;553:484–493. https://doi.org/10.1016/j.jcis.2019.06.050.

89. Wei J, Cai W. One-step hydrothermal preparation of N-doped carbon spheres from peanut hull for efficient removal of Cr(VI). *J Environ Chem Eng*. 2020;8:104449. https://doi.org/10.1016/j.jece.2020.104449.

90. Li H, Li N, Zuo P, Qu S, Shen W. Efficient adsorption-reduction synergistic effects of sulfur, nitrogen and oxygen heteroatom co-doped porous carbon spheres for chromium (VI) removal. *Colloids Surf A Physicochem Eng Asp*. 2021;618:126502. https://doi.org/10.1016/j.colsurfa.2021.126502.

91. Wang F, Zhang Y, Fang Q, Li Z, Lai Y, Yang H. Prepared PANI@nano hollow carbon sphere adsorbents with lappaceum shell like structure for high efficiency removal of hexavalent chromium. *Chemosphere*. 2021;263:128109. https://doi.org/10.1016/j.chemosphere.2020.128109.

92. Simate GS, Maledi N, Ochieng A, Ndlovu S, Zhang J, Walubita LF. Coal-based adsorbents for water and wastewater treatment. *J Environ Chem Eng*. 2016;4:2291–2312. https://doi.org/10.1016/j.jece.2016.03.051.

93. Xu B, Yang M, Xing B, et al. Removal of pollutants from aqueous solutions by coals and residual cokes obtained from simulated underground coal gasification experiments. *Fuel*. 2021;292:120292. https://doi.org/10.1016/j.fuel.2021.120292.

94. Kong L, Qiu F, Zhao Z, et al. Removal of brilliant green from aqueous solutions based on polyurethane foam adsorbent modified with coal. *J Clean Prod*. 2016;137:51–59. https://doi.org/10.1016/j.jclepro.2016.07.067.

95. Shaida MA, Dutta RK, Sen AK. Removal of diethyl phthalate via adsorption on mineral rich waste coal modified with chitosan. *J Mol Liq*. 2018;261:271–282. https://doi.org/10.1016/j.molliq.2018.04.031.

96. Pham TV, Tran TV, Nguyen TD, et al. Adsorption behavior of Congo red dye from aqueous solutions onto exfoliated graphite as an adsorbent: kinetic and isotherm studies. *Mater Today Proc*. 2019;18:4449–4457. https://doi.org/10.1016/j.matpr.2019.07.414.

97. Shirazinezhad M, Faghihinezhad M, Baghdadi M, Ghanbari M. Phosphate removal from municipal effluent by a porous MgO-expanded graphite composite as a novel adsorbent: evaluation of seawater as a natural source of magnesium ions. *J Water Process Eng*. 2021;43:102232. https://doi.org/10.1016/j.jwpe.2021.102232.

98. Wu KH, Huang WC, Hung WC, Tsai CW. Modified expanded graphite/Fe₃O₄ composite as an adsorbent of methylene blue: adsorption kinetics and isotherms. *Mater Sci Eng B*. 2021;266:115068. https://doi.org/10.1016/j.mseb.2021.115068.

99. Ambika S, Srilekha V. Eco-safe chemicothermal conversion of industrial graphite waste to exfoliated graphene and evaluation as engineered adsorbent to remove toxic textile dyes. *Environ Adv*. 2021;4:100072. https://doi.org/10.1016/j.envadv.2021.100072.

100. Choudhury S, Balasubramanian R. Recent advances in the use of graphene-family nanoadsorbents for removal of toxic pollutants from wastewater. *Adv Colloid Interf Sci*. 2014;204:35–56. https://doi.org/10.1016/j.cis.2013.12.005.

101. Lingamdinne LP, Koduru JR, Karri RR. A comprehensive review of applications of magnetic graphene oxide based nanocomposites for sustainable water purification. *J Environ Manag*. 2019;231:622–634. https://doi.org/10.1016/j.jenvman.2018.10.063.

102. Koduru JR, Karri RR, Mubarak NM. Smart materials, magnetic graphene oxide-based nanocomposites for sustainable water purification. In: *Sustainable Polymer Composites and Nanocomposites*; 2019:759–781. https://doi.org/10.1007/978-3-030-05399-4_26.

103. Lingamdinne LP, Choi JS, Choi YL, et al. Process modeling and optimization of an iron oxide immobilized graphene oxide gadolinium nanocomposite for arsenic adsorption. *J Mol Liq*. 2020;299:112261. https://doi.org/10.1016/j.molliq.2019.112261.

104. Peng W, Li H, Liu Y, Song S. A review on heavy metal ions adsorption from water by graphene oxide and its composites. *J Mol Liq*. 2017;230:496–504. https://doi.org/10.1016/j.molliq.2017.01.064.

105. Nundy S, Ghosh A, Nath R, Paul A, Tahir AA, Mallick TK. Reduced graphene oxide (rGO) aerogel: efficient adsorbent for the elimination of antimony (III) and (V) from wastewater. *J Hazard Mater*. 2021;420:126554. https://doi.org/10.1016/j.jhazmat.2021.126554.

106. Zhao R, Li Y, Ji J, et al. Efficient removal of phenol and p-nitrophenol using nitrogen-doped reduced graphene oxide. *Colloids Surf A Physicochem Eng Asp*. 2021;611:125866. https://doi.org/10.1016/j.colsurfa.2020.125866.

107. Narayana PL, Lingamdinne LP, Karri RR, et al. Predictive capability evaluation and optimization of Pb(II) removal by reduced graphene oxide-based inverse spinel nickel ferrite nanocomposite. *Environ Res*. 2022;204:112029. https://doi.org/10.1016/j.envres.2021.112029.

108. Lingamdinne LP, Koduru JR, Chang YY, Karri RR. Process optimization and adsorption modeling of Pb(II) on nickel ferrite-reduced graphene oxide nano-composite. *J Mol Liq*. 2018;250:202–211. https://doi.org/10.1016/j.molliq.2017.11.174.

109. Suresh R, Udayabhaskar R, Sandoval C, et al. Effect of reduced graphene oxide on the structural, optical, adsorption and photocatalytic properties of iron oxide nanoparticles. *New J Chem*. 2018;42:8485–8493. https://doi.org/10.1039/c8nj00321a.

110. Zheng ALT, Phromsatit T, Boonyuen S, Andou Y. Synthesis of silver nanoparticles/porphyrin/reduced graphene oxide hydrogel as dye adsorbent for wastewater treatment. *FlatChem*. 2020;23:100174. https://doi.org/10.1016/j.flatc.2020.100174.

111. Gan F, Cheng B, Jin Z, et al. Hierarchical porous biochar from plant-based biomass through selectively removing lignin carbon from biochar for enhanced removal of toluene. *Chemosphere*. 2021;279:130514. https://doi.org/10.1016/j.chemosphere.2021.130514.

112. Purwanto H, Rozhan AN, Salleh HM. Innovative process to enrich carbon content of EFB-derived biochar as an alternative energy source in iron making. *Adv Mater Sci Eng*. 2018;4067237. https://doi.org/10.1155/2018/4067237 [7 pages].

113. Zheng Y, Wang J, Li D, et al. Insight into the KOH/KMnO₄ activation mechanism of oxygen-enriched hierarchical porous biochar derived from biomass waste by in-situ pyrolysis for methylene blue enhanced adsorption. *J Anal Appl Pyrolysis*. 2021;158:105269. https://doi.org/10.1016/j.jaap.2021.105269.

114. Macedo JCA, Gontijo ESJ, Herrera SG, et al. Organosulphur-modified biochar: an effective green adsorbent for removing metal species in aquatic systems. *Surf Interfaces*. 2021;22:100822. https://doi.org/10.1016/j.surfin.2020.100822.

115. Chakhtouna H, Benzeid H, Zari N, Qaiss AEK, Bouhfid R. Functional CoFe₂O₄-modified biochar derived from *banana pseudostem* as an efficient adsorbent for the removal of amoxicillin from water. *Sep Purif Technol*. 2021;266:118592. https://doi.org/10.1016/j.seppur.2021.118592.

116. Gama NV, Soares B, Freire CS, et al. Rigid polyurethane foams derived from cork liquefied at atmospheric pressure. *Polym Int*. 2015;64:250–257. https://doi.org/10.1002/pi.4783.

117. Wang Q, Lai Z, Mu J, Chu D, Zang X. Converting industrial waste cork to biochar as Cu (II) adsorbent via slow pyrolysis. *Waste Manag*. 2020;105:102–109. https://doi.org/10.1016/j.wasman.2020.01.041.

118. Nath H, Saikia A, Goutam PJ, Saikia BK, Saikia N. Removal of methylene blue from water using okra (Abelmoschus esculentus L.) mucilage modified biochar. *Bioresour Technol Rep*. 2021;14:100689. https://doi.org/10.1016/j.biteb.2021.100689.

119. Liu XJ, Li MF, Singh SK. Manganese-modified lignin biochar as adsorbent for removal of methylene blue. *J Mater Res Technol*. 2021;12:1434–1445. https://doi.org/10.1016/j.jmrt.2021.03.076.

120. Zahedifar M, Seyedi N, Shafiei S, Basij M. Surface-modified magnetic biochar: highly efficient adsorbents for removal of Pb(II) and Cd(II). *Mater Chem Phys*. 2021;271:124860. https://doi.org/10.1016/j.matchemphys.2021.124860.

121. Ramola S, Rawat N, Shankhwar AK, Srivastava RK. Fixed bed adsorption of Pb and Cu by iron modified bamboo, bagasse and tyre biochar. *Sustain Chem Pharm*. 2021;22:100486. https://doi.org/10.1016/j.scp.2021.100486.

122. Sun S, Zeng X, Gao Y, et al. Iron oxide loaded biochar/attapulgite composites derived camellia oleifera shells as a novel bio-adsorbent for highly efficient removal of Cr(VI). *J Clean Prod*. 2021;317:128412. https://doi.org/10.1016/j.jclepro.2021.128412.

123. Li S, Ma X, Ma Z, et al. Mg/Al-layered double hydroxide modified biochar for simultaneous removal phosphate and nitrate from aqueous solution. *Environ Technol Innov*. 2021;23:101771. https://doi.org/10.1016/j.eti.2021.101771.

124. Lingamdinne LP, Choi JS, Angaru GKR, et al. Magnetic-watermelon rinds biochar for uranium-contaminated water treatment using an electromagnetic semi-batch column with removal mechanistic investigations. *Chemosphere*. 2022;286:131776. https://doi.org/10.1016/j.chemosphere.2021.131776.

125. Yadav SK, Dhakate SR, Singh BP. Carbon nanotube incorporated eucalyptus derived activated carbon-based novel adsorbent for efficient removal of methylene blue and eosin yellow dyes. *Bioresour Technol*. 2022;344:126231. https://doi.org/10.1016/j.biortech.2021.126231.

126. Wang Y, Cai M, Chen T, et al. Oxide of porous graphitized carbon as recoverable functional adsorbent that removes toxic metals from water. *J Colloid Interface Sci*. 2022;606:983–993. https://doi.org/10.1016/j.jcis.2021.08.082.

127. Pan C, Wang C, Fang Y, Zhu Y, Deng H, Guo Y. Graphdiyne: an emerging two-dimensional (2D) carbon material for environmental remediation. *Environ Sci Nano*. 2021;8:1863–1885. https://doi.org/10.1039/D1EN00231G.

128. Han M, Xu B, Zhang M, et al. Preparation of biologically reduced graphene oxide-based aerogel and its application in dye adsorption. *Sci Total Environ*. 2021;783:147028. https://doi.org/10.1016/j.scitotenv.2021.147028.

129. Andrade M, Parnell AJ, Bernardo G, Mendes A. Propane selective carbon adsorbents from phenolic resin precursor. *Microporous Mesoporous Mater*. 2021;320:111071. https://doi.org/10.1016/j.micromeso.2021.111071.

130. Wei C, Xiang C, Ren E, et al. Synthesis of 3D lotus biochar/reduced graphene oxide aerogel as a green adsorbent for Cr(VI). *Mater Chem Phys*. 2020;253:123271. https://doi.org/10.1016/j.matchemphys.2020.123271.

131. Liu Y, Peng X, Hu Z, Yu M, Fu J, Huang Y. Fabrication of a novel nitrogen-containing porous carbon adsorbent for protein-bound uremic toxins removal. *Mater Sci Eng C*. 2021;121:111879. https://doi.org/10.1016/j.msec.2021.111879.

Chapter 12

Luminescent metal-organic frameworks for sensing of toxic organic pollutants in water and real samples

Luis D. Rosales-Vázquez[a], Alejandro Dorazco-González[a,*], and Víctor Sánchez-Mendieta[b]

[a]Institute of Chemistry, National Autonomous University of Mexico, Mexico City, Mexico, [b]CCIQS—Joint Center for Research in Sustainable Chemistry UAEM-UNAM, Toluca, Estado de México, Mexico

*Corresponding author: A. D.-G.

1 Introduction

In the coming years, food production and supply are considered two of the most important and challenging issues, mainly, due to the ever-increasing problem of maintaining the production rate as the world population increases.[1] Since decades ago, the growing of crops, fruits, and vegetables, principally, have been challenging to achieve without the application of chemicals, for instance, pesticides and herbicides. Despite emerging techniques to produce more efficiently and trying to avoid the treatment with those hazardous chemicals, to this day, most of the world's production of vegetable-origin food stocks makes use of them.[2] Consequently, it will surely take several years more before an effective and massive food production process could eliminate the usage of pesticides and herbicides. Due to the well-known health implications that these noxious chemicals provoke in humans and animals, it is relevant to have analytical tools that detect toxic organic pollutants in our food and the water used for irrigation in their production.[3] Moreover, it has been acknowledged that only around 1% of the applied pesticides reach the aimed application.[4] The remaining 99% of the used pesticides and herbicides tend to bio-accumulate and reach water sources, triggering irreversible damage and extending to humans as serious health-related problems, usually via the food chain.[4–8]

Currently, few analytical techniques can detect pesticides and herbicides, with good sensitivity and detection limit, in diverse media. However, methods like gas chromatography, surface-enhanced Raman spectroscopy, mass spectrometry, among others, are challenging to perform in situ since they usually require special sample preparation and expensive equipment.[9–11] Here is where luminescence analytical tools come as very efficient and simpler detection techniques of organic compounds, particularly those, that can be harmful to humans and the environment by their chemical nature. In recent years, photoluminescence has been demonstrated to be an efficient, highly sensitive, and selective method for the sensing of widespread types of substances ranging from metal ions, inorganic and organic anions, solvents, and explosives, to more complex organic molecules.[12]

Luminescent-sensing techniques require probe compounds that could possess luminescence properties. Among the divergent chemical structures found in luminescent compounds, metal-organic frameworks (MOFs) have lately delivered outstanding results as fast, highly selective, sensitive, and sometimes, recyclable luminescent probes of a great variety of analytes.[13] Nowadays, crystalline CPs and MOFs are used as fine luminescent chemosensors due to two main advantages. Firstly, their relatively facile structural changes can be performed based on coordination chemistry principles, which can tune their luminescent properties. Secondly, the fact that there are increasingly examples of these coordination arrays that are becoming more stable in water, is a desirable condition for practical applications as luminescent sensors, particularly, on real samples. Many MOFs having luminescent properties are assembled using mixed ligands, among them, the aromatic carboxylates and N-donor π-conjugated ligands are the most used. Thus, the luminescence process in these systems can occur mainly from processes such as ligand-centered emission (LC), metal-centered emission (MC), ligand-to-metal charge transfer (LMCT), metal-to-ligand charge transfer (MLCT), and ligand-to-ligand charge transfer (LLCT) mechanisms. The most frequent emissions reported for Zn(II)- and Cd(II)-MOFs are related to LC, LMCT, and LLCT processes.[14,15]

Sustainable Materials for Sensing and Remediation of Noxious Pollutants. https://doi.org/10.1016/B978-0-323-99425-5.00005-0

Hence, this chapter emphasizes contemporary developments of the application and efficiency of luminescent Zn(II)- and Cd(II)-MOFs as low-cost fluorescent sensors to detect pesticides and herbicides in water and real samples, such as fruits and vegetables. Thus, aiming these functional hybrid materials could efficiently detect and ultimately help control these dangerous pollutants, diminishing their undesired consequences on the environment and human health.

2 General features for the design of LMOF in the detection of pesticides and herbicides

The general strategy for the development of Zn(II) and Cd(II)-LMOFs with application in optical detection of common pesticides and herbicides has to include light-emitting materials with the following features.

(1) Chemical stability in complex samples and real-world samples (e.g., irrigation water, lake water, fruits, vegetables, and physiological samples); (2) stable photoluminescence units; and (3) complementary interaction sites with pesticides capable of inducing a selective binding process and analytical response. These general characteristics are shown in Fig. 1.[16]

2.1 Water stability in Zn(II)-, Cd(II)-LMOFs

Zn(II)- and Cd(II)-LMOFs have been extensively used as chemosensors for selective quantification of inorganic anions (halides, oxyanion),[17,18] metal ions,[19–23] explosives based on nitroaromatic derivatives,[24–26] volatile organic compounds,[27,28] amino acids,[29–31] pollutant gases,[32] and antibiotics.[33–35] The literature features relative few examples of chemosensors based on Zn(II)- and Cd(II)-LMOFs for the detection of common pesticides (Fig. 2) such as parathion,[36] azinphos-methyl,[37,38] 2,6-dichloro-4nitroaniline,[39] parathion-methyl,[40,41] acephate,[42] and chlorpyrifos[43] in real-world samples (e.g., irrigation water, lake water, tap water, fruits, vegetables, and living cells).

Despite considerable progress, some analytical parameters, such as an efficient luminescent response and sensitivity in complex samples, remain a central challenge in these systems.

Water stability of LMOFs is a crucial property and requisite when considering these compounds for luminescent sensing applications in real-world samples.

In general, a water-stable MOF is a compound that can be handled in aqueous media without any change in its structure. The water stability can be probed experimentally by exposing the LMOFs in contact with water and subsequent comparison of the structural and spectroscopic properties before and after water contact through spectroscopic tools (e.g., fluorescence spectroscopy, Raman, IR spectroscopy, X-ray photoelectron spectroscopy, and powder X-ray diffraction).[44,45]

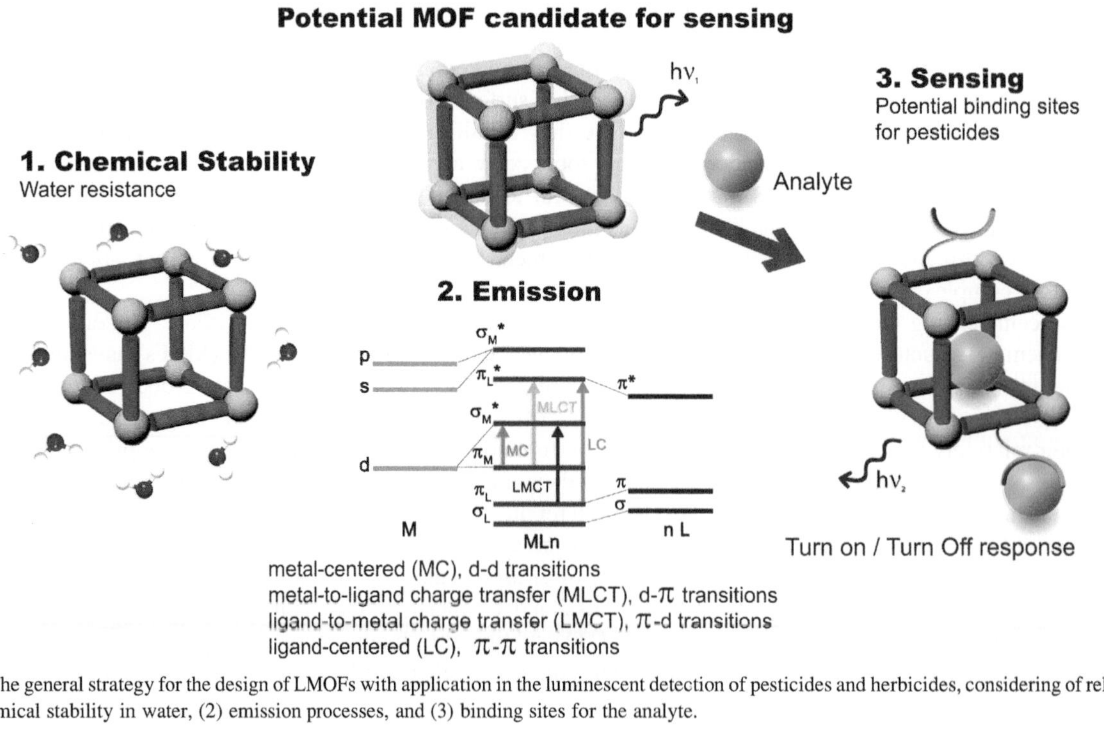

Potential MOF candidate for sensing

1. Chemical Stability
Water resistance

$h\nu_1$

3. Sensing
Potential binding sites
for pesticides

Analyte

2. Emission

metal-centered (MC), d-d transitions
metal-to-ligand charge transfer (MLCT), d-π transitions
ligand-to-metal charge transfer (LMCT), π-d transitions
ligand-centered (LC), π-π transitions

$h\nu_2$

Turn on / Turn Off response

FIG. 1 The general strategy for the design of LMOFs with application in the luminescent detection of pesticides and herbicides, considering of relevance of (1) chemical stability in water, (2) emission processes, and (3) binding sites for the analyte.

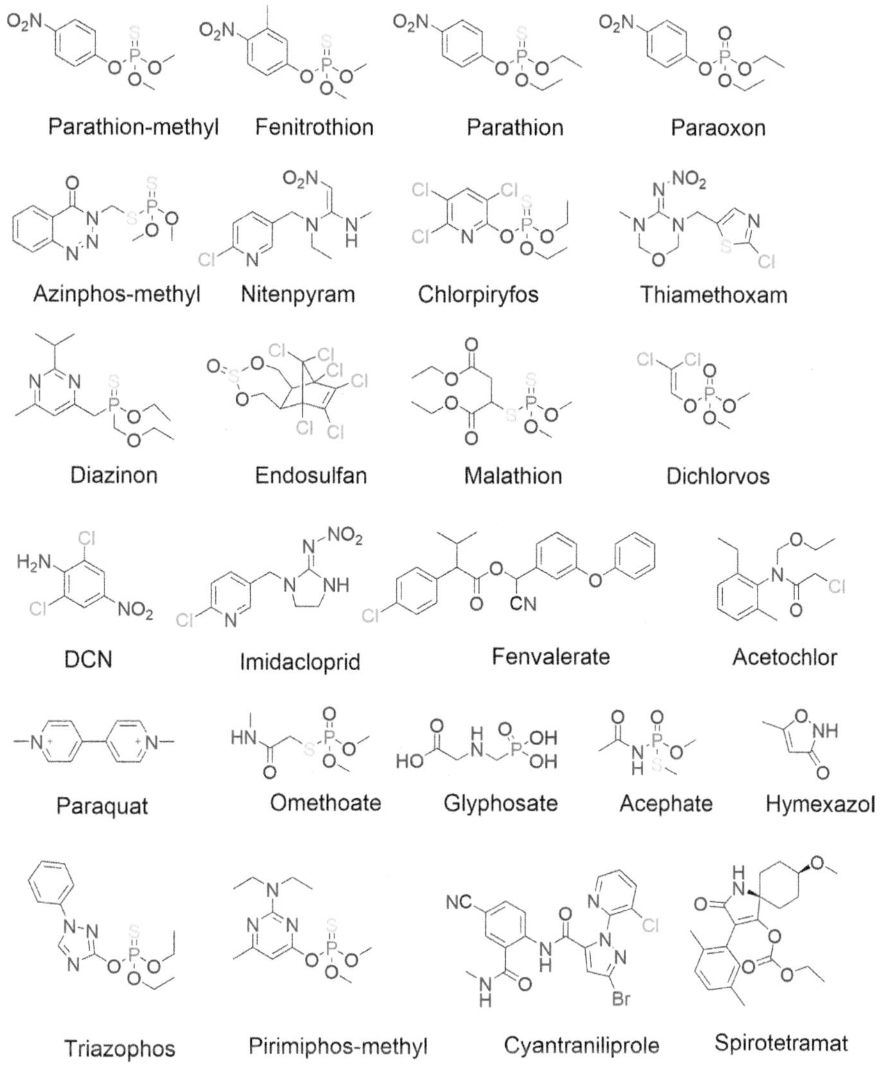

FIG. 2 Structure of common herbicides and pesticides.

In structural and thermodynamic terms, water-stable LMOFs comprehend a suitable combination of inert organic linkers and hydrostable metal clusters, which makes energetically unfavorable the following hydrolysis reaction.

$$\Delta G_{hyd} = \Delta G_{prod}(LMOF + nH_2O) + \Delta G_{react}(LMOF + nH_2O)$$

where ΔG_{prod} is the Gibbs free energy of the LMOF, formed between the LMOF and the water after the hydrolysis process takes place, and ΔG_{react} is the Gibbs free energy of the LMOF and the water molecules before the hydrolysis occurs.

In these compounds, the strength of metal-ligand coordination bonds can be a strong indicator of their hydrolytic stability. The major drawback of synthetic LMOFs is that their metal-ligand bond strength is lower than natural zeolite frameworks based on aluminosilicate minerals.[45]

Hence, the water stability of the MOFs requires sufficient metal-ligand binding Gibbs free energy to overcome the very high hydration energies of divalent Zn and Cd ions ($-\Delta G^0 = 2040$ and -1840 kJmol^{-1} for Zn(II) and Cd(II), respectively).[46]

2.2 Emission

The origin of the emission of LMOFs based on d^{10} transition metals that include π-conjugated carboxylates ligands and N-donor ligands stems from photophysical mechanisms such as follows: (1) ligand-centered photoemission (LC),

(2) metal-centered photoemission (MC), (3) ligand-to-metal charge transfers (LMCTs), (4) metal-to-ligand charge transfers (MLCTs), and (5) ligand-to-ligand charged transfers (LLCTs). The electronic transitions involved in the photoluminescence mechanisms of Zn(II)- and Cd(II)-based LMOFs are depicted in Fig. 1. To understand the general concepts of these photoluminescence phenomena, the readers are referred to particular reviews on this issue.[13,15,16]

Among these emission processes, the most commonly reported for LMOFs containing divalent Zn and Cd ions are ligand-centered photoemission and charge transfers.[16,45] Metal-centered photoemission is exhibited particularly when MOFs have metal clusters with high nuclearity as nodes. Ligand-centered photoemission and ligand charge transfers in Zn(II)- and Cd(II)-LMOFs are determined basically by the symmetry of the single ground of the ligands and their vibrational excited states, which can induce electronic transitions of the kind n-π^* and π-π^*.[14] This emission originates from the lowest excited singlet state to the singlet ground state of the ligand and corresponds to a short lifetime's emission ($\sim 10^{-9}$ s).[15]

On the other hand, the emissions based on ligand-to-metal charge transfers are displayed in a large number of Zn(II)- and Cd(II)-LMOFs. It involves electronic transitions from a π-conjugated ligand localized orbital to a metal-centered orbital. This mechanism is primarily in structures bearing benzene-multicarboxylate and aromatic N-donors ligands.[14] Typically, this kind of emitting materials show visible strong emission from blue to green, in the range of 390–520 nm.[13–16,47]

A frequent emission mechanism for the sensing of pesticides and herbicides based on LMOFs includes: (1) Photoinduced electron transfer (PET), (2) Charge transfers (CT), and (3) Förster resonance energy transfer (FRET).[22]

2.3 Selectivity

The exposition of commercial pesticides is associated with impaired health (neurotoxic effect, particularly Parkinson's disease, and risk of cancer)[48] and severe adverse environmental effects such as water and soil contamination,[4] which force the creation of efficient chemosensors. While the need for practical chemosensors for pesticides and herbicides is evident, to date, very few luminescent MOFs have been reported capable of operating in real samples and cellular environments.

For LMOFs with a stable emission source, the change in luminescence induced by the interaction of a pesticide or herbicide is enough for the accurate quantitative analysis. However, the selectivity is still not comparable to the high-performance liquid chromatography techniques, electrochemical methodologies, and high-resolution mass spectrometry.[45] Table 1 lists some outstanding examples of Zn(II) and Cd(II) LMOFs for the luminescent detection of pesticides in real samples with their analytical and structural features.

Despite notable advances in the development of sensitive chemosensors for pesticides based on LMOFs, its application in real samples remains largely unexplored. The most outstanding examples consist of nanomaterials, 3D LMOFs, such as $[Zn_4(TCPP)_2(TCPB)_2]$ (1),[36] $[Cd_3(PDA)(tz)_3Cl(H_2O)_4]$ (3),[38] and $[Zn_3(DDB)(DPE)]$ (4)[39] (see Table 1 for abbreviations).

In general, LMOF-based chemosensors have large possibilities for chemical and structural modulation of interaction sites and the emission properties, in addition to pore-volume modulation, which makes these LMOFs very versatile systems for the efficient detection of pesticides.[2]

Thus, Cd(II)-LMOFs chemosensors (2) and (3) are noteworthy examples where the selectivity toward azinphos-methyl, over several organophosphate derivatives, is attributed to various convergent supramolecular interactions, such as hydrogen bond interactions, π stacking, and coordination bonds (Cd $\cdots$ S=P) between the metal center and the azinphos-methyl.

A selective chemosensor for parathion over several organophosphate derivatives (e.g., azinphos-methyl, chlorpyrifos, etc.) is the LMOF (1) $[Zn_4(TCPP)_2(TCPB)_2]$ where the preference toward parathion is attributed to strong electron transfer from the electron-donating ligands to the electron-withdrawing nitroaromatic moieties of parathion.[36,40] Meanwhile, other more elaborated systems, such as $Ru(bpy)_3^{2+}$-ZIF-90-MnO_2 (5), AuNCs@ZIF-8 (6), and CBZ-BOD@ZIF-8 (7), utilize the inhibition of specific enzymes to accomplish the selectivity toward organophosphate pesticides.[40,42,43]

3 Zn(II) and Cd(II) LMOFs as fluorescent chemosensor for pesticides in real samples

Organophosphate pesticides (OPPs) (Fig. 2) are chemicals that attack the nervous system and permanently stop the acetylcholinesterase-enzyme activity.[36] The MOF $[Cd_{2.5}(PDA)(tz)_3]$ (2) (PDA = 1,4-phenylenediacetate and tz = 1,2,4-triazolate)[38] (Table 1) was assembled through a solvothermal methodology. The 3D MOF structure consisted of a cage connected 3D structure having three different coordination geometries in the Cd(II) centers. This MOF exhibited emission at 290 nm, upon excitation at 225 nm, in an aqueous medium. Thus, this hybrid material was

TABLE 1 Structural features and analytical conditions of selected Zn(II)- and Cd(II)-LMOF-based sensors for pesticides in real sample.

	Formula	Structure	Synthesis method	Fluorimetric response	Pesticide/ herbicide	Real sample	λ_{em}	Limit of detection (M)	Refs.
1	[Zn$_4$(TCP)$_2$(TCPB)$_2$]	[a]	Solvothermal	Turn off	Parathion	– Irrigation water	380	6.69×10^{-9}	36
2	[Cd$_{2.5}$(PDA)(tz)$_3$]	3D	Solvothermal	Turn off	Azinphos-methyl	– Extract from apples	290	5×10^{-5}	38
3	[Cd$_3$(PDA)(tz)$_3$Cl (H$_2$O)$_4$]	3D	Slow evaporation	Turn off	Azinphos-methyl	– Extract from apples and tomatoes	290	2.5×10^{-8}	37
4	[Zn$_3$(DDB)(DPE)]	3D	Solvothermal	Turn off	DCN	– Extract from carrots, nec-tarines, and grapes	450	2.7×10^{-7}	39
5	Nanocomposite Ru(bpy)$_3^{2+}$-ZIF-90-MnO$_2$	3D	Solvothermal	Turn off	Parathion-methyl	– Tap and lake water – Extract from cabbages	600	1.40×10^{-10}	40
6	Nanocomposite AuNCs@ZIF-8	3D	Solvothermal	Turn on	Acephate	– Tap and lake water – Extract from lettuce	600	3.66×10^{-9}	42
7	CBZ-BOD@ZIF-8	3D	Solvothermal	Turn on	Chlorpyrifos	Living cells	624	1.6×10^{-7}	43

Ligands: *TCP*, tetrakis (4-carboxyphenyl)-porphyrin; *TCPB*, 1,2,4,5-tetrakis (4-carboxyphenyl) benzene; *PDA*, 1,4-phenylenediacetate; *tz*, 1,2,4-triazolate; *DDB*, 3,5-di(2′,4′-dicarboxyphenyl)benzoic acid; *DPE*, 1,2-di(4-pyridyl)ethylene; *DCN*, 2,6-dichloro-4-nitroaniline; *bpy*, 2,2′-bipyridine; *ZIF*, zeolitic imidazolate framework; *NC*, nanocluster; *CBZ-BOD*, (*E*)-4-(3-(2-(9-ethyl-9H-carbazol-2-yl)vinyl)-5,5-difluoro-1,7,9-trimethyl-5H-5,6-dipyrrolo[1,2-c:2′,1′-f][1,3,2]diazaborinin-10-yl)phenyl-benzoate.
[a]Not specified.

capable to detect azinphos-methyl pesticide in water through luminescence quenching; remarkably, this was performed with a limit of detection of 16 ppb in the presence of several other pesticides, such as parathion, chlorpyrifos, diazinon, endosulfan, malathion, and dichlorvos (see Fig. 2). As discussed in the corresponding article, $\pi \cdots \pi$ interactions, along with the coordination vacancy generated in Cd1 (distorted-trigonal bipyramidal coordination geometry), are both accountable for the azinphos-methyl sensing (Fig. 3).[38] Azinphos-methyl is a usual component in the formulation of commercial pesticides. It is widely used to treat insect pests in fruits (e.g., apples, grapes), vegetables, nuts, and crop fields. Another structurally similar compound 3D MOF $[Cd_3(PDA)(tz)_3Cl(H_2O)_4] \cdot 3H_2O$ (**3**)[37] (Table 1), also exhibited a luminescence quenching response for the azinphos-methyl; apparently, the interaction of this Cd(II)-MOF with the pesticide is restricted to the $\pi \cdots \pi$ interactions between the aromatic ligands (PDA and tz ligands) and the azinphos-methyl molecule (Fig. 3). In this case, the limit of detection of azinphos-methyl in water was 8 ppb. Moreover, this water-stable MOF has been capable of detecting azinphos-methyl in apples and tomato extracts. According to the Food and Agriculture Organization (FAO), the maximum residue limits of azinphos-methyl in tomatoes is 3.15×10^{-6} M[37]; therefore, the low detection limits obtained in these previously mentioned studies make these Cd(II)-LMOF sensors susceptible and selective to comply with those limits.

Since these two last Cd(II)-LMOFs systems have equivalent emission signal, which is provoked by the PDA ligand, and both are capable of azinphos-methyl detection, the luminescence quenching mechanism can be ascribed to a resonance energy transfer process promoted by the absorption of the excitation light from the MOF to the azinphos-methyl molecule.[37,38]

Organochlorine pesticides (OCPs) are another important branch of harmful organic compounds pollutants. Particularly, 2,6-dichloro-4-nitroaniline (DCN) is a very popular pesticide, which is employed in crops to combat infections such as cotton rotten bell and wheat powdery mildew. DCN possesses high toxicity, slow-degradation rate, and it is insoluble, which may cause severe damage to living organisms. It can get into human skin and lungs.[6,49] A water-stable trinuclear MOF $[Zn_3(DDB)(DPE)] \cdot H_2O$ (**4**) (H5DDB = 3,5-di(2′,4′-dicarboxylphenyl) benzoic acid and DPE = 1,2-di(4-pyridyl)ethylene)[39] (Table 1) has efficiently achieved the sensing of DCN in carrot, grape, and nectarine extracts, with nanomolar sensing limits.

The secondary building units in this MOF are made of three Zn(II) ions connected by one μ_3-OH$^-$ and four carboxylates to form the trinuclear cluster: $[Zn_3(OH)(COO)_4]$. Moreover, two of these trinuclear units are connected by two carboxylate groups to generate an hexanuclear cluster $[Zn_6(OH)_2(COO)_{10}]$, yielding thus a 4,12-c net (Fig. 4).[39]

Fluorescence quenching coming from DCN was the only effect revealed when the experiments were performed in the presence of eight other OCPs. An increase to 93.5% in the quenching efficiency of the DCN sensing was achieved, reaching a limit of detection of 2.7×10^{-7} M in water. These outcomes prompted the sensing of DCN using this Zn(II)-MOF in carrot, grape, and nectarine extracts. The fluorescence quenching mechanism occurring was supported by theoretical studies, which resulted in a proposal of a PET mechanism triggered by the nitro moiety in the DCN molecule.

In an interesting and novel approach, the luminescent recognition of organic pesticides has taken a step further by not just focusing solely on the pesticide by itself. Instead, those strategies rely directly on the mechanism of how the corresponding enzyme is affected by the pesticide. Consequently, an all-around sensor based on (Ru(bpy)$_3$-ZIF-90-MnO$_2$) has been synthesized[40] (**5**), Table 1. This MOF design exploits all the advantages of its components in a synergic way. First, it is assembled in a robust, and facile to prepare, ZIF-90,[40] which is used as a platform, where the complex $[Ru(bpy)_3]^{2+}$ is encapsulated within the ZIF architecture. The $[Ru(bpy)_3]^{2+}$ molecules exhibit modest luminescent properties.

However, by being attached to the MOF, the fluorescent behavior is improved by limiting the aggregation-induced quenching of the fluorophore and increasing its photostability. The complementary fragment (MnO$_2$) is incorporated in the form of nanosheets to the composite, developing a dual role in the sensing probe, acting as a quencher and, at the same time, bringing a colorimetric response in the detection of OPPs. The nanocomposite (Ru(bpy)$_3$-ZIF-90-MnO$_2$) is a stable brownish probe in water solution, and its activity is centralized in the inhibition of the acetylcholinesterase (AChE) by OPPs. As depicted in Fig. 5, the process of determination of the OPPs occurs in the presence of composite (**5**) following the enzymatic reaction of the acetylthiocholine (ATCh) to thiocholine (TCh) by the AChE. The product, TCh, simultaneously reacts with the MnO$_2$ molecules of the composite, reducing the MnO$_2$ nanosheets to Mn(II) ions, "turning-on" the fluorescent feature of (Ru(bpy)$_3$-ZIF-90), and changing the brown color into a colorless solution. The introduction of an organophosphorus pesticide to the system parathion-methyl blocks the production of TCh by the AChE, leaving the (Ru(bpy)$_3$-ZIF-90-MnO$_2$) probe unaltered. Consequently, the detection of parathion-methyl could be traced by naked eye or through fluorescence with a limit of detection equals to 1.40×10^{-10} M. The selectivity for parathion-methyl was examined in the competition of several coexisting electrolytes and biological species in water such as Ca^{2+}, K$^+$, Mg^{2+}, Na$^+$,

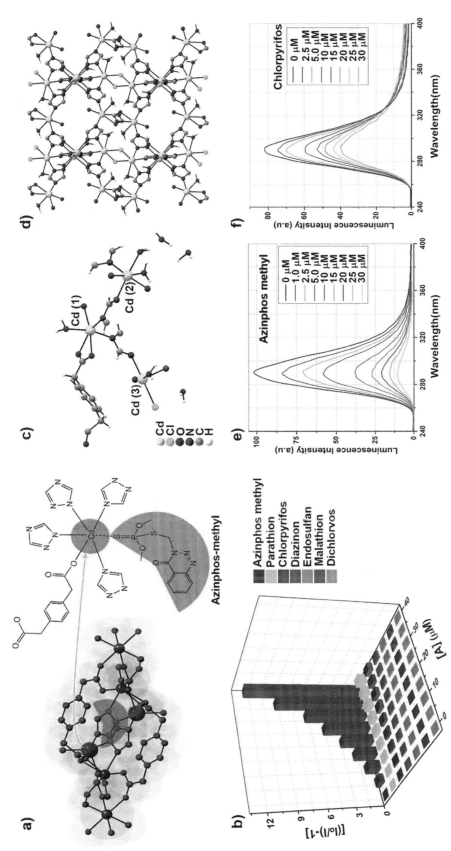

FIG. 3 (A) Scheme showing the penta-coordinated Cd(II) ions and the suggested interaction of azinphos-methyl, through the sulfur atom, within the coordination cage. (B) Changes in emission spectra of [Cd₂.₅(PDA)(tz)₃]⁻ (2) dispersed in water upon increasing amounts of azinphos-methyl. (C) Perspective view of the asymmetric unit and (D) two-dimensional arrangement in the crystal of [Cd₃(PDA)(tz)₃Cl (H₂O)₄]·3H₂O (3). (E) Changes in the emission spectra of (3) in water as concentration of azinphos-methyl and (F) chlorpyrifos increase ($\lambda_{ex} = 225$ nm). (Adapted with permission from (B) Singha DK, Majee P, Mondal SK, Mahata P. Highly selective aqueous phase detection of azinphos-methyl pesticide in ppb level using a cage-connected 3D MOF. ChemistrySelect. 2017;2(20):5760–5768. doi:10.1002/slct.201700963, ©2020, Wiley Online Lib. (F) Singha DK, Majee P, Mandal S, Mondal SK, Mahata P. Detection of pesticides in aqueous medium and in fruit extracts using a three-dimensional metal-organic framework: experimental and computational study. Inorg Chem. 2018;57(19):12155–12165. https://doi.org/10.1021/acs.inorgchem.8b01767, ©2018, ACS.)

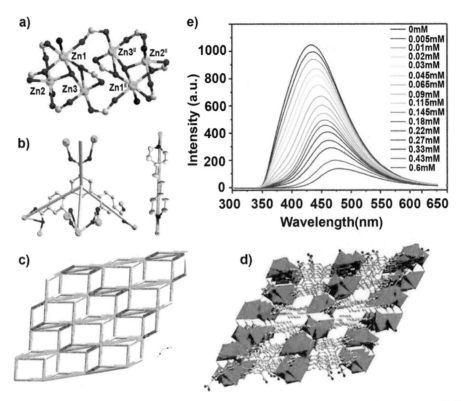

FIG. 4 (A) Secondary building unit of sensor (**4**), [Zn₃(DDB)(DPE)]·H₂ (B) Coordination modes of DPE and DDB ligands. (C) Simplified topological scheme of the array along the *b* axis. (D) Projection view of the 3D open framework along the *b* axis. (E) Fluorescence spectra of the MOF at several concentrations of DCN in water, including carrot extract (100 μL). *(Adapted with permission from Wang XQ, Feng DD, Tang J, Zhao YD, Li J, Yang J, Kim CK, Su F. A water-stable zinc(II)-organic framework as a multiresponsive luminescent sensor for toxic heavy metal cations, oxyanions and organochlorine pesticides in aqueous solution. Dalton Trans. 2019;48(44):16776–16785. https://doi.org/10.1039/c9dt03195b, ©2019, The Royal Society of Chemistry.)*

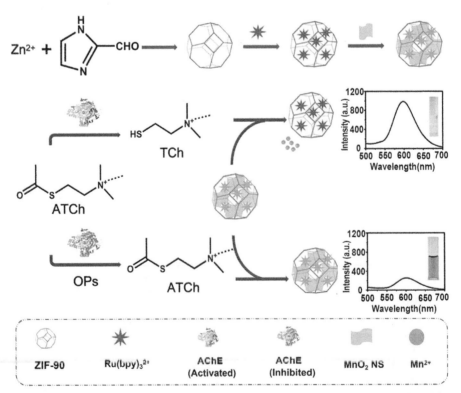

FIG. 5 Synthetic path of the (Ru(bpy)₃-ZIF-90-MnO₂) nanocomposite (**5**) and its schematic detection mechanism of OPPs. *(Adapted with the permission from Li J, Weng Y, Shen C, Luo J, Yu D, Cao Z. Sensitive fluorescence and visual detection of organophosphorus pesticides with a Ru(bpy)32 +-ZIF-90-MnO2 sensing platform. Anal Methods. 2021;13(26):2981–2988. https://doi.org/10.1039/d1ay00841b, ©2021, The Royal Chemical Society.)*

Zn^{2+}, Cl^-, NO_3^-, SO_4^{2-}, glycine, L-cysteine, bovine serum albumin (BSA), catalase, glucose oxidase, and L-glutamic acid, with negligible interference. More importantly, other pesticides were also tested: acetochlor, fenvalerate, imidacloprid, nitenpyram, and quaternary ammonium compounds as paraquat with the same good results. Nevertheless, the organophosphorus-based pesticides like chlorpyrifos, dichlorvos, malathion, omethoate, triazophos, and paraoxon also showed an inhibition effect in the AChE, interfering with the specific detection of parathion-methyl, as can be obtained an almost identical response for paraoxon in comparison with parathion-methyl. The sensing platform has been applied in real samples such as lake water, tap water and cabbage extract with successful results.

In a remarkable analogy, the election of ZIF-8 in the AuNCs@ZIF-8 sensor (**6**)[42] obeys the same intention of ZIF-90 for (Ru(bpy)$_3$-ZIF-90-MnO$_2$), to serve as matrix support for the guest species and enhance its luminescent properties. The emission of AuNCs (Au-nanoclusters) is linked to the restriction of intramolecular movement. Hence, their disposition inside the ZIF conveys increased fluorescence as the intramolecular movement was limited.[42]

The luminescent-sensing process addressed by the probe can be divided into three consecutive steps as illustrated in Fig. 6. (1) The enzymatic hydrolyzation of the acetylcholine by the AChE-producing choline. (2) Choline in the presence of O_2 is catalyzed by the choline oxidase generating H_2O_2. (3) The H_2O_2 molecules rip the structure of the ZIF-8, liberating the AuNCs and quenching the fluorescence. When an OPPs participate in the first step of the reaction cascade, inhibition in the AChE arises, causing a decrease in the production of H_2O_2. By this means, the presence of OPPs avoids the destruction of the AuNCs@ZIF-8 probe, allowing fluorescence emission.

Additionally, the authors have also considered including colorimetric techniques besides the fluorimetric determination through the addition of 3′,5,5′-tetramethylbenzidine (TMB).[42] Thus, once the release of the AuNCs from the MOFs has been ensured in the absence of OPPs, the TMB molecules become prone to be oxidized, transforming the colorless solution into a blue one (oxTMB). The oxidation proceeds as the Au atoms in AuNCs (which tend to exist in both of the oxidation states, Au(0) and Au(I)) decompose the O—O bond in H_2O_2, forming hydroxyl radicals (•OH) that are responsible for reacting with TMB, turning the solution blue.

The study contemplated numerous pesticides to be analyzed, including OPPs (acephate, fenitrothion, glyphosate, malathion, parathion, and pirimiphos-methyl) and non-OPPs (cyantraniliprole, hymexazol thiamethoxam, tebuconazole, and spirotetramat). As expected, the non-OPPs caused no signal in the detection process, but the OPPs did it with a similar intensity. A detection limit was given for acephate as a representative pesticide (3.66×10^{-9} M). Notably, further studies were accomplished by implementing the compounds AuNCs@ZIF-8 and TMB in the paper stripes tests.

Taking advantage of the colorimetric signals, the paper stripes were evaluated as a portable kit in a quantitative measurement of OPPs in lake water, tap water, and lettuce extract with the utilization of photography and a downloaded application of a smartphone concluding in a good and replicable detection performance by the sensor.[42] The use of colorimetric methods has been contemplated in other two enzymatic pesticide MOFs sensors (MIP-CoZn-ZIF[50] and AChE@ZIF-8[51]), although those probes exhibited no luminescent properties.

Remarkably, the use of ZIF-8 as a sensing platform has also been reached in the indirect detection of chlorpyrifos by CBZ-BOD@ZIF-8 due to its robustness, practicality, and biocompatibility.[43] Still, the enzyme target for the analysis was carboxylesterase (CES1) as an alternative for AChE. Likewise, CES1 gets inhibited in the presence of OPPs. The background strategy in the synthetic design of CBZ-BOD was to construct a selective and dual fluorogenic and colorimetric molecule capable of achieving a specific binding to CES1, which can be encapsulated inside the LMOF.[43] The recognition process starts with the incubation of CES1 and the red emission probe (CBZ-BOD@ZIF-8) in the interior of the micropores of the ZIF architecture. Then, fluorescence intensity changes as the hydrolysis of the carboxylic group in CBZ-BOD evolves into a non-emitting molecule (BOD-OH) triggered by CES1. The pesticide determination proceeds as chlorpyrifos blocks the CES1, impeding the catalytic reaction over CBZ-BOD@ZIF-8, leaving the characteristic fluorescence and red colored probe unaltered (Fig. 7).[43] Significantly, all those steps in the study had taken place via intracellular, with minor or no interference of many biological species including: α-CT, trypsin, bovine serum albumin, CES2, lysing enzymes, lipase, proteinase K, pepsase, D-beta-hydroxybutyrate (BHb), lysozyme, and AchE.

Furthermore, dimethoate and fenitrothion were also compared against chlorpyrifos showing a more remarkable turn-on behavior toward chlorpyrifos, estimating its detection limit as 1.6×10^{-7} M.

The authors also investigated kinetic parameters, and monitored the real-time activity of CES1 coexisting with chlorpyrifos in living cells, obtaining encouraging results headed for more complex and sophisticated designs in LMOFs as a sensing platform in their application toward toxic pollutants.

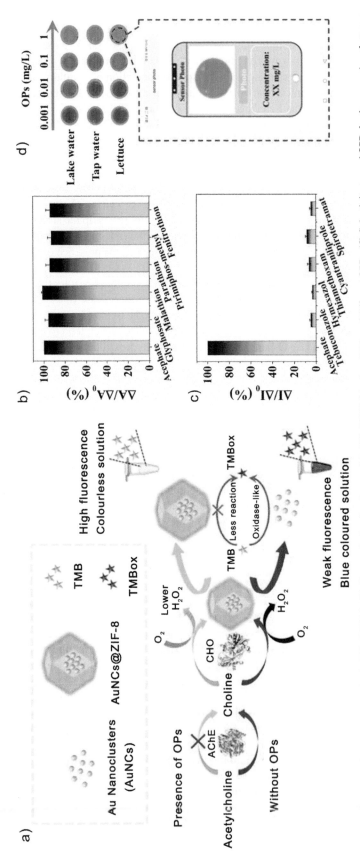

FIG. 6 (A) OPPs detection mechanism by AuNCs@ZIF-8. (B) Changes in the absorbance of AuNCs@ZIF-8 by the addition of different OPPs. (C) Selectivity toward OPPs by the sensor (acephate as an example), (D) Chromogenic sensing in real samples and the use of a smartphone to analyze the determination of OPPs in the problem solution. *(Adapted with permission from Cai Y, Zhu H, Zhou W, Qiu Z, Che, C, Qileng A, Li K, Liu Y. Capsulation of AuNCs with AIE effect into metal-organic framework for the marriage of a fluorescence and colorimetric biosensor to detect organophosphorus pesticides. Anal Chem. 2021;93(19):7275–7282. https://doi.org/10.1021/acs.analchem.1c00616, ©2021, ACS.)*

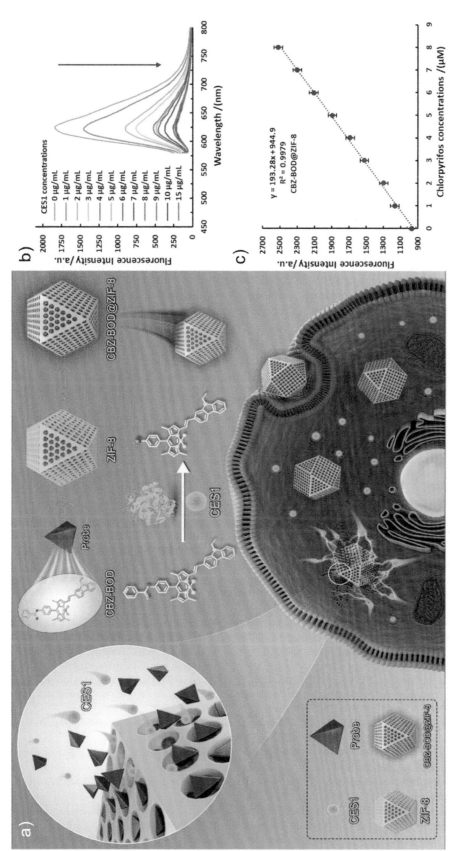

FIG. 7 (A) Designing of the nanocomposite CBZ-BOD@ZIF-8 (**6**) and its applications in living cell sensing. (B) Fluorimetric titration experiment of CBZ-BOD@ZIF-8 (10 μM) among the presence of different concentrations of CES1. (C) Fluorescence changes of CBZ-BOD@ZIF-8 in the presence of increasing chlorpyrifos concentrations. (*Adapted with the permission from Shen B, Ma C, Ji Y, Dai J Li B, Zhang X, Huang H. Detection of carboxylesterase 1 and chlorpyrifos with ZIF-8 metal-organic frameworks using a red emission BODIPY-based probe. ACS Appl Mater Interfaces. 2021;13(7):8718–8726. https://doi.org/10.1021/acsami.0c19811, ©2021, ACS.*)

4 Conclusions

The current global challenges caused by the indiscriminate use of pesticides are evident and urgently need to be addressed. There is an obvious need to develop functional chemosensors with synergy between their structural and luminescent properties to develop sensitive and selective luminescent probes to detect these toxic compounds.

Among optical sensory materials, luminescent MOFs containing d^{10} transition metals (mainly Zn(II) and Cd(II) ion) are at the forepart as chemosensors owing to their simple preparation. Their highly modifiable structures, which by simply chemical functionalization of their components can generate an extended range of photoluminescent mechanisms and structural properties effective in sensing/recognition of pesticides through supramolecular interactions. Under this approach, the selective and quantitative detection of target pesticides is achieved through the MOF-pesticide interaction process that can be transduced to a change in its emission properties. Thus, more elaborated systems can be achieved by using water-stable LMOFs as a cornerstone platform in which other molecules can be placed toward a specific recognition of notorious pollutants or by and indirect detection considering other relevant biological targets like enzymes.

In summary, the suitable design of chemosensor based on Zn(II)- and Cd(II)-LMOFs can generate new analytical methodologies toward efficient quantitative detection of industrially scale produced agro molecules such as pesticides. However, the luminescent MOF-based sensors for pesticides have the following challenges: (1) Improve water stability because water is the main component in environmental and biological media. High water stability can endow sensors with high recyclability and render them inexpensive. (2) The detection limit for the targeted pesticides needs to be further improved or well matched with the range of complex samples. (3) The sensors should operate in real samples, requiring the sensors to be studied under real conditions at various concentrations of both analytes and common interfering species.

References

1. Lustig WP, Mukherjee S, Rudd ND, Desai AV, Li J, Ghosh SK. Metal-organic frameworks: functional luminescent and photonic materials for sensing applications. *Chem Soc Rev*. 2017;46(11):3242–3285. https://doi.org/10.1039/c6cs00930a.
2. Vikrant K, Tsang DCW, Raza N, Giri BS, Kukkar D, Kim KH. Potential utility of metal-organic framework-based platform for sensing pesticides. *ACS Appl Mater Interfaces*. 2018;10(10):8797–8817. https://doi.org/10.1021/acsami.8b00664.
3. Karri RR, Ravindran G, Dehghani MH. Wastewater—sources, toxicity, and their consequences to human health. In: *Soft Computing Techniques in Solid Waste and Wastewater Management*. Elsevier; 2021:3–33.
4. Zulkifli SN, Rahim HA, Lau WJ. Detection of contaminants in water supply: a review on state-of-the-art monitoring technologies and their applications. *Sens Actuators B Chem*. 2018;255:2657–2689. https://doi.org/10.1016/j.snb.2017.09.078.
5. Vasylevskyi SI, Bassani DM, Fromm KM. Anion-induced structural diversity of Zn and cd coordination polymers based on bis-9,10-(pyridine-4-yl)-anthracene, their luminescent properties, and highly efficient sensing of nitro derivatives and herbicides. *Inorg Chem*. 2019;58(9):5646–5653. https://doi.org/10.1021/acs.inorgchem.8b03628.
6. Di L, Xia Z, Li J, et al. Selective sensing and visualization of pesticides by ABW-type metal-organic framework based luminescent sensors. *RSC Adv*. 2019;9(66):38469–38476. https://doi.org/10.1039/c9ra08940c.
7. Yu CX, Hu FL, Song JG, et al. Ma LF ultrathin two-dimensional metal-organic framework nanosheets decorated with tetra-pyridyl calix[4]arene: design, synthesis and application in pesticide detection. *Sens Actuators B Chem*. 2020;310(January):127819. https://doi.org/10.1016/j.snb.2020.127819.
8. Zhao Y, Xu X, Qiu L, Kang X, Wen L, Zhang B. Metal-organic frameworks constructed from a new thiophene-functionalized dicarboxylate: luminescence sensing and pesticide removal. *ACS Appl Mater Interfaces*. 2017;9(17):15164–15175. https://doi.org/10.1021/acsami.6b11797.
9. Saito-Shida S, Nemoto S, Akiyama H. Multiresidue method for determining multiclass acidic pesticides in agricultural foods by liquid chromatography-tandem mass spectrometry. *Anal Methods*. 2021;13(7):894–902. https://doi.org/10.1039/d0ay02101f.
10. Vargas-Pérez M, Domínguez I, González FJE, Frenich AG. Application of full scan gas chromatography high resolution mass spectrometry data to quantify targeted-pesticide residues and to screen for additional substances of concern in fresh-food commodities. *J Chromatogr A*. 2020;1622:461118. https://doi.org/10.1016/j.chroma.2020.461118.
11. Beneito-Cambra M, Gilbert-López B, Moreno-González D, et al. Ambient (desorption/ionization) mass spectrometry methods for pesticide testing in food: a review. *Anal Methods*. 2020;12(40):4831–4852. https://doi.org/10.1039/d0ay01474e.
12. Li HY, Zhao SN, Zang SQ, Li J. Functional metal-organic frameworks as effective sensors of gases and volatile compounds. *Chem Soc Rev*. 2020;49(17):6364–6401. https://doi.org/10.1039/c9cs00778d.
13. Dong J, Zhao D, Lu Y, Sun WY. Photoluminescent metal-organic frameworks and their application for sensing biomolecules. *J Mater Chem A*. 2019;7(40):22744–22767. https://doi.org/10.1039/c9ta07022b.
14. Parmar B, Bisht KK, Rachuri Y, Suresh E. Zn(II)/Cd(II) based mixed ligand coordination polymers as fluorosensors for aqueous phase detection of hazardous pollutants. *Inorg Chem Front*. 2020;7(5):1082–1107. https://doi.org/10.1039/c9qi01549c.
15. Allendorf MD, Bauer CA, Bhakta RK, Houk RJT. Luminescent metal-organic frameworks. *Chem Soc Rev*. 2009;38(5):1330–1352. https://doi.org/10.1039/b802352m.

16. Heine J, Müller-Buschbaum K. Engineering metal-based luminescence in coordination polymers and metal-organic frameworks. *Chem Soc Rev.* 2013;42(24):9232–9242. https://doi.org/10.1039/c3cs60232j.

17. Karmakar A, Samanta P, Dutta S, Ghosh SK. Fluorescent "turn-on" sensing based on metal–organic frameworks (MOFs). *Chem Asian J.* 2019;14 (24):4506–4519. https://doi.org/10.1002/asia.201901168.

18. Yang L, Song Y, Wang L. Multi-emission metal-organic framework composites for multicomponent ratiometric fluorescence sensing: recent developments and future challenges. *J Mater Chem B.* 2020;8(16):3292–3315. https://doi.org/10.1039/c9tb01931f.

19. Guo XY, Dong ZP, Zhao F, Liu ZL, Wang YQ. Zinc(II)-organic framework as a multi-responsive photoluminescence sensor for efficient and recyclable detection of pesticide 2,6-dichloro-4-nitroaniline, Fe(III) and Cr(VI). *New J Chem.* 2019;43(5):2353–2361. https://doi.org/10.1039/c8nj05647a.

20. Pamei M, Puzari A. Luminescent transition metal–organic frameworks: an emerging sensor for detecting biologically essential metal ions. *Nano-Struct Nano-Objects.* 2019;19:100364. https://doi.org/10.1016/j.nanoso.2019.100364.

21. Li Y, Ma D, Chen C, et al. A hydrostable and bromine-functionalized manganese-organic framework with luminescence sensing of Hg^{2+} and antiferromagnetic properties. *J Solid State Chem.* 2019;269(July 2018):257–263. https://doi.org/10.1016/j.jssc.2018.09.034.

22. Kumar P, Deep A, Kim K-H. Metal organic frameworks for sensing applications. *TrAC Trends Anal Chem.* 2015;73:39–53. https://doi.org/10.1016/j.trac.2015.04.009.

23. Gu TY, Dai M, Young DJ, Ren ZG, Lang JP. Luminescent Zn(II) coordination polymers for highly selective sensing of Cr(III) and Cr(VI) in water. *Inorg Chem.* 2017;56(8):4668–4678. https://doi.org/10.1021/acs.inorgchem.7b00311.

24. Zhang L, Kang Z, Xin X, Sun D. Metal–organic frameworks based luminescent materials for nitroaromatics sensing. *CrstEngComm.* 2016;18(2):193–206. https://doi.org/10.1039/C5CE01917F.

25. Qin JH, Huang YD, Shi MY, et al. Aqueous-phase detection of antibiotics and nitroaromatic explosives by an alkali-resistant Zn-MOF directed by an ionic liquid. *RSC Adv.* 2020;10(3):1439–1446. https://doi.org/10.1039/c9ra08733h.

26. Dalapati R, Biswas S. Post-synthetic modification of a metal-organic framework with fluorescent-tag for dual naked-eye sensing in aqueous medium. *Sens Actuators B Chem.* 2017;239:759–767. https://doi.org/10.1016/j.snb.2016.08.045.

27. Jackson SL, Rananaware A, Rix C, Bhosale SV, Latham K. Highly fluorescent metal-organic framework for the sensing of volatile organic compounds. *Cryst Growth Des.* 2016;16(6):3067–3071. https://doi.org/10.1021/acs.cgd.6b00428.

28. Zhang J, Zhang X, Chen J, et al. Highly selective luminescent sensing of xylene isomers by a water stable Zn-organic framework. *Inorg Chem Commun.* 2016;69:1–3. https://doi.org/10.1016/j.inoche.2016.04.013.

29. Liu G, Li Y, Chi J, et al. Various Cd(II) coordination polymers induced by carboxylates: multi-functional detection of Fe3+, anions, aspartic acids and bovine serum albumin. *Dalton Trans.* 2020;49(3):737–749. https://doi.org/10.1039/c9dt04103f.

30. Chandrasekhar P, Mukhopadhyay A, Savitha G, Moorthy JN. Remarkably selective and enantiodifferentiating sensing of histidine by a fluorescent homochiral Zn-MOF based on pyrene-tetralactic acid. *Chem Sci.* 2016;7(5):3085–3091. https://doi.org/10.1039/c5sc03839a.

31. Qiu L-G, Li Z-Q, Wu Y, Wang W, Xu T, Jiang X. Facile synthesis of nanocrystals of a microporous metal–organic framework by an ultrasonic method and selective sensing of organoamines. *Chem Commun.* 2008;31:3642. https://doi.org/10.1039/b804126a.

32. Wu S, Min H, Shi W, Cheng P. Multicenter metal–organic framework-based ratiometric fluorescent sensors. *Adv Mater.* 2020;32(3):1–14. https://doi.org/10.1002/adma.201805871.

33. Li C, Yang W, Zhang X, et al. A 3D hierarchical dual-metal-organic framework heterostructure up-regulating the pre-concentration effect for ultrasensitive fluorescence detection of tetracycline antibiotics. *J Mater Chem C.* 2020;8(6):2054–2064. https://doi.org/10.1039/c9tc05941e.

34. Yao XQ, Xiao GB, Xie H, et al. Solvent-induced structural diversity of two luminescent metal-organic frameworks as dual-functional sensor for the detection of nitroaromatic compounds and highly selective detection of ofloxacin antibiotics. *CrstEngComm.* 2019;21(15):2559–2570. https://doi.org/10.1039/C8CE02122H.

35. Ying YM, Tao CL, Yu M, et al. In situ encapsulation of pyridine-substituted tetraphenylethene cations in metal-organic framework for the detection of antibiotics in aqueous medium. *J Mater Chem C.* 2019;7(27):8383–8388. https://doi.org/10.1039/c9tc02229e.

36. Wang L, He K, Quan H, Wang X, Wang Q, Xu X. A luminescent method for detection of parathion based on zinc incorporated metal-organic framework. *Microchem J.* 2020;153:104441. https://doi.org/10.1016/j.microc.2019.104441.

37. Singha DK, Majee P, Mandal S, Mondal SK, Mahata P. Detection of pesticides in aqueous medium and in fruit extracts using a three-dimensional metal-organic framework: experimental and computational study. *Inorg Chem.* 2018;57(19):12155–12165. https://doi.org/10.1021/acs.inorgchem.8b01767.

38. Singha DK, Majee P, Mondal SK, Mahata P. Highly selective aqueous phase detection of azinphos-methyl pesticide in ppb level using a cage-connected 3D MOF. *ChemistrySelect.* 2017;2(20):5760–5768. https://doi.org/10.1002/slct.201700963.

39. Wang XQ, Feng DD, Tang J, et al. A water-stable zinc(II)-organic framework as a multiresponsive luminescent sensor for toxic heavy metal cations, oxyanions and organochlorine pesticides in aqueous solution. *Dalton Trans.* 2019;48(44):16776–16785. https://doi.org/10.1039/c9dt03195b.

40. Li J, Weng Y, Shen C, Luo J, Yu D, Cao Z. Sensitive fluorescence and visual detection of organophosphorus pesticides with a $Ru(bpy)_3^{2+}$-ZIF-90-MnO_2 sensing platform. *Anal Methods.* 2021;13(26):2981–2988. https://doi.org/10.1039/d1ay00841b.

41. Xu X, Guo Y, Wang X, et al. Sensitive detection of pesticides by a highly luminescent metal-organic framework. *Sens Actuators B Chem.* 2018;260:339–345. https://doi.org/10.1016/j.snb.2018.01.075.

42. Cai Y, Zhu H, Zhou W, et al. Capsulation of AuNCs with AIE effect into metal-organic framework for the marriage of a fluorescence and colorimetric biosensor to detect organophosphorus pesticides. *Anal Chem.* 2021;93(19):7275–7282. https://doi.org/10.1021/acs.analchem.1c00616.

43. Shen B, Ma C, Ji Y, et al. Detection of carboxylesterase 1 and chlorpyrifos with ZIF-8 metal-organic frameworks using a red emission BODIPY-based probe. *ACS Appl Mater Interfaces.* 2021;13(7):8718–8726. https://doi.org/10.1021/acsami.0c19811.

44. Burtch NC, Jasuja H, Walton KS. Water stability and adsorption in metal-organic frameworks. *Chem Rev.* 2014;114(20):10575–10612. https://doi.org/10.1021/cr5002589.

45. Rosales-Vázquez LD, Dorazco-González A, Sánchez-Mendieta V. Efficient chemosensors for toxic pollutants based on photoluminescent Zn(II) and Cd(II) metal-organic networks. *Dalton Trans.* 2021;50(13):4470–4485. https://doi.org/10.1039/d0dt04403b.

46. Kepp KP. Free energies of hydration for metal ions from heats of vaporization. *J Phys Chem A.* 2019;123(30):6536–6546. https://doi.org/10.1021/acs.jpca.9b05140.

47. Rosales-Vázquez LD, Valdes-García J, Bazany-Rodríguez IJ, et al. A sensitive photoluminescent chemosensor for cyanide in water based on a zinc coordination polymer bearing ditert-butyl-bipyridine. *Dalton Trans.* 2019;48(33):12407–12420. https://doi.org/10.1039/c9dt01861a.

48. Sidhu GK, Singh S, Kumar V, Dhanjal DS, Datta S, Singh J. Toxicity, monitoring and biodegradation of organophosphate pesticides: a review. *Crit Rev Environ Sci Technol.* 2019;49(13):1135–1187. https://doi.org/10.1080/10643389.2019.1565554.

49. Kumar T, Venkateswarulu M, Das B, Halder A, Koner RR. Zn(ii)-based coordination polymer: an emissive signaling platform for the recognition of an explosive and a pesticide in an aqueous system. *Dalton Trans.* 2019;48(33):12382–12385. https://doi.org/10.1039/c9dt02224d.

50. Amirzehni M, Hassanzadeh J, Vahid B. Surface imprinted CoZn-bimetalic MOFs as selective colorimetric probe: application for detection of dimethoate. *Sens Actuators B Chem.* 2020;325(July):128768. https://doi.org/10.1016/j.snb.2020.128768.

51. Kukkar P, Kukkar D, Younis SA, et al. Colorimetric biosensing of organophosphate pesticides using enzymatic nanoreactor built on zeolitic imdiazolate-8. *Microchem J.* 2021;166(December 2020):106242. https://doi.org/10.1016/j.microc.2021.106242.

Chapter 13

Metal-organic frameworks for remediation of noxious pollutants

Jafar Abdi[a,*], Seyyed Hamid Esmaeili-Faraj[a], Golshan Mazloom[b], and Tahereh Pirhoushyaran[c]

[a]*Faculty of Chemical and Materials Engineering, Shahrood University of Technology, Shahrood, Iran,* [b]*Department of Chemical Engineering, Faculty of Engineering, University of Mazandaran, Babolsar, Iran,* [c]*Department of Chemical Engineering, Dezful Branch, Islamic Azad University, Dezful, Iran*
[*]*Corresponding author.*

1 Introduction

Recent statistics have shown that nearly one-third of the world's population suffers from access to clean water sources, and this increases daily.[1] In addition to climate change and lack of rain, the existence of harmful compounds in wastewater discharged into the environment without adequate pretreatment leads to severe contamination. According to the database reported by the European Chemicals Agency, plenty of noxious pollutants (more than 106,000 compounds) threaten human health and the aquatic ecosystem.[2–4] Among these noxious pollutants, heavy metals, dyes, pharmaceuticals and personal care products (PPCPs), pesticides and herbicides, industrial preservatives and additives comprise a huge group of highly toxic chemical substances which have attracted great attention by the scientific and legislative communities.[5,6] The development of urban systems and progress in industries are expected to enhance noxious pollutants in water bodies. Therefore, researches, selection, and implementation of effective methods for water purification and wastewater treatment will become a vital issue. Among new materials employed to remove water pollutions, porous materials confer a sufficient candidate to replace the most common and traditional cases. Activated carbons, Mobil composition of matter (MCM), and zeolites have already been utilized in pollutant removal processes as a series of porous materials. However, these traditional materials possess restrictions for different operational applications. Metal-organic frameworks (MOFs) as a new class of porous compounds have emerged in the last two decades. The MOF structure comprises two parts of organic (as linker) and inorganic (metal ions) connected through coordination bonds. In recent years, MOFs have enticed many researchers for remedying hazardous pollutants from water and wastewater due to their amazing properties, including tunable porous structures, high surface area (up to $7000 \, m^2/g$), and notable chemical, thermal and mechanical stability. These specific properties make MOFs notable materials with high potential for use in various fields, such as separation, gas storage/adsorption, biomedical sciences, drug delivery, photocatalyst, sensing, batteries, etc.[7]

Moreover, MOFs can be employed for designing different water purification systems due to their potential in treating numerous types of pollutants. In this chapter, the origin of noxious pollutants and recent advances and developments of MOFs in the removal of hazardous water contaminants, such as heavy metal ions, dyes, PPCPs, and pesticides and herbicides, will be introduced. The properties and synthesis methods of MOF materials are investigated and compared with those of other conventionally used materials. In addition, the removal mechanism of the mentioned noxious pollutants by different methods, including adsorption, photocatalytic degradation, and membrane purification, is also illustrated. Furthermore, the remaining future challenges regarding the promising potential for noxious pollutant remediation using MOFs are assessed.

2 Metal-organic frameworks

MOFs as a subclass of porous coordination polymers (PCPs) have emerged in the field of coordination chemistry. MOFs are a kind of hybrid materials constructed by a combination of organic linkers (e.g., carboxylates, azoles, phosphonate, sulfonates, and heterocyclic compounds) and inorganic clusters (metal ion nodes), also named secondary unit buildings (SBUs).[7] Using different techniques to connect these two main components lead to preparing plenty of MOFs (up to 20,000 types) with unique features. However, different parameters affect the synthesis procedure, such as temperature, pressure, solvent, reaction time, modulators, and metal ions salts. In this regard, different methods have been introduced and employed to prepare MOF crystals with specific properties, including solvothermal/hydrothermal,[8,9] electrochemical,[10,11] mechanochemical,[12,13] diffusion,[14,15] solvent evaporation, microwave-assisted fabrication,[16,17] and heating and ultrasound methods.[18,19] Fig. 1 represents different the synthesis approaches for the construction of MOFs. To get more perception, an intuitive comparison between these approaches attributed to the advantages and disadvantages of each technique are listed in Table 1. Nevertheless, the approaches mentioned above possess some restrictions regarding green synthesis and environmentally friendly subjects. These limitations can be overcome using six basic principles which can be employed in the synthesis procedure to obtain better quality MOFs: (1) implementing biocompatible precursors, (2) reducing energy consumption, (3) increasing the use of green solvents under suitable reaction conditions, (4) development of solvent-free

FIG. 1 Overview of different synthesis approaches for constructing MOFs. *(Reproduced with permission from Tchinsa A, et al. Removal of organic pollutants from aqueous solution using metal organic frameworks (MOFs)-based adsorbents: a review.* Chemosphere. *2021; 131393.)*

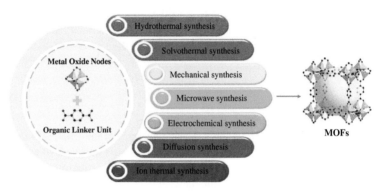

TABLE 1 Advantages and disadvantages of various methods for synthesizing MOFs.

Technique	Advantages	Disadvantages
Electrochemical	<ul><li>High yielding</li><li>Usable in low temperatures</li><li>High kinetic reaction</li></ul>	<ul><li>Accumulation of particles</li><li>Depends on temperature</li><li>The electrode defects</li></ul>
Mechanochemical	<ul><li>No need for solvent</li><li>Environmentally friendly</li><li>Applicable in atmospheric condition</li><li>High kinetic reaction</li></ul>	<ul><li>Need external forces</li><li>Power consumption</li></ul>
Microwave	<ul><li>Reaction temperature can be controlled</li><li>MOFs can be shaped</li><li>Particle size can be regulated</li><li>Fast process</li></ul>	<ul><li>Power consumption</li><li>Need microwave device</li></ul>
Solvent evaporation	<ul><li>Nucleation in low temperature</li></ul>	<ul><li>Need saturated solution</li><li>Reactants solubility needs high temperature</li></ul>
Solvothermal and hydrothermal	<ul><li>Usable for different MOFs</li><li>High efficiency</li></ul>	<ul><li>Low kinetic reaction</li><li>Need high pressure and temperature</li><li>Need autoclave</li><li>Need organic and toxic solvent</li></ul>
Sonochemistry	<ul><li>Applicable in atmospheric condition</li><li>High kinetic reaction</li><li>Environmentally friendly</li><li>Easy and simple procedure</li><li>Usable for industrial scale</li><li>Reactants solubility is high</li></ul>	<ul><li>Power consumption</li><li>Need ultrasonic device</li></ul>

technique, (5) using continuous processes to prepare MOFs, and (6) using theoretical calculations to design unique MOFs with improved performance.

The various MOFs are classified based on the metal agents and organic ligands. Among them, Zeolitic imidazolate frameworks (ZIFs),[20,21] materials institute Lavoisier (MIL-n type),[22] Zr-based MOFs (UiO),[23,24] HKUST-1, IRMOFs, are the most important groups. Researchers conducted a techno-economic analysis to evaluate the cost of large-scale synthesis of MOFs. They identified that the production prices for representative MOFs (i.e., MOF-5, HKUST-1, Ni_2(dobdc), Mg_2(dobdc), where dobdc = 1,4-dioxido-2,5-benzene-dicarboxylate) could be controlled to 8–10 \$/kg with an optimized process and alternative synthesis methods like liquid-assisted grinding (LAG) and aqueous synthesis. It was also recognized that a fully continuous synthesis operation incorporating chemical recycling designs has the potential for further cost opportunities.[7]

3 Removal of noxious pollutants using MOFs

Different techniques have been developed to effectively remove different pollutants such as adsorption, photocatalytic degradation, filtration, biodegradation, coagulation, and advanced oxidation processes. Among them, adsorption is one of the most common methods due to the simple design, mild operation conditions, high efficiency, and low energy consumption. Adsorption happens during the uptake of ions, atoms, molecules, and particles denoted as adsorbates over the surface of the adsorbent. Interactions between contaminants and the adsorbents can occur via diverse mechanisms, including electrostatic, acid-base, π-π and hydrophobic interactions and hydrogen bonding.

(1) Electrostatic interaction: This mechanism is the most commonly proposed mechanism for adsorption of different pollutants from water. This interaction occurs between charged ions of adsorbate and oppositely charges on the adsorbent surface. Net charges of the MOFs encounter over the surface through grafting with foreign species. In addition, the pH of the aqueous solution can lead to protonation or deprotonation, which is in favor of electrostatic interaction.

(2) Acid-base interaction: This mechanism plays a critical role in the adsorption process, which significantly improves the adsorptive performance of different MOFs regarding various pollutants. For example, various researches revealed more adsorption capacity of functionalized MOFs with basic (—NH_2) or acidic groups (—SO_3H) regarding the pristine MOFs.

(3) π-π interaction: Likewise, to the electrostatic interaction where occurs between opposite charges, π-π stacking involves the interaction between an electron-rich π system with another anionic, cationic, neutral molecule or another π system.

(4) Hydrophobic interaction: This mechanism is observed where hydrophobic molecules accumulate in the aqueous solution. Hydrophobic molecules are nonpolar molecules with a long carbon chain and low solubility in water. Some of the MOFs have a hydrophobic feature with a high tendency to remove pollutants such as oil droplets from an aqueous solution.

(5) Hydrogen bonding: This mechanism involves the interaction between a hydrogen atom and highly electronegative O, N, or F atoms which can significantly enhance the adsorptive performance of MOFs.

Based on the different mechanisms, it can be concluded that the adsorption capability of MOFs is highly affected by different structural characteristics such as surface area, porosity, active sites, and functional groups. Despite the many advantages of MOFs as adsorbents, the use of powdered materials for commercial purposes has been limited: because powdery forms are difficult to be separated. To exploit the characteristic features of the MOFs and overcome the limitations of powder materials in commercial applications, generally, MOFs can be deposited over a polymeric matrix to form a membrane. In this way, some concerns relating to powder regeneration and process safety can be overcome. Nowadays, membrane separation technology has been significantly investigated in water treatment processes due to its high efficiency, selective performance, strong compatibility, and easy maintenance.

The adsorption process suffers from mass transfer limitation from one phase to another.[25] In addition, the contaminant concentrations are decreased by adsorption rather than elimination. In contrast, photocatalysis is a more precise approach, since the organic pollutants are completely converted to H_2O and CO_2 under facile conditions without the production of secondary pollutants.[26] During a photocatalysis process, the organic linkers of MOFs absorb the light leading to generating some free electrons, which transfer from the valence band (VB) to the conduction band (CB). However, other mechanisms are also used to interpret photocatalyst degradation: charge transfer from metal to ligand, ligand to ligand, and metal to metal.[27] Finally, these electrons can reach the metal-oxo clusters surface, be trapped by oxygen molecules, and °O_2^- superoxide radicals. In addition, the holes (h^+) on the HOMO produced by light are other oxidizing agents which can directly participate in the oxidizing reaction. Also, (h^+) can react with H_2O to form °OH.[28] Therefore, three kinds of active species are produced under light: $°O_2^-$, h^+, and $°OH$ that can decompose the pollutants. Therefore for designing a suitable MOF for photocatalysis reactions, some critical issues should be considered: Firstly, absorption of light irradiation.[29] For this purpose, suitable metals and linkers that can heavily absorb light could be employed. For example, porphyrins as light-harvesting materials have been utilized for a MOF photocatalyst with high activity. In addition, the range and intensity of light absorbed can be improved by the amino-substituted organic linker. Secondly, photos' recombination of electrons and holes should be suppressed. It can be tuned by functionalized MOF construction; thirdly, the suitable porosity structure for rapid diffusion.[30]

In the following section, the application of these common techniques is investigated using MOFs systems for decontamination of different noxious pollutants, including heavy metals, dyes, PPCPS, pesticides, and herbicides.

3.1 Heavy metals

Heavy metals (HMs) in wastewater have been expanded with the development of industries, such as batteries, papers, rayon, mining, metal purifying, and petrochemicals. The heavy metals are released into the environment through wastewater and threaten human health and the environment. The foremost prevalent heavy metals in water resources include arsenic (As), mercury (Hg), lead (Pb), copper (Cu), zinc (Zn), nickel (Ni), chromium (Cr), and cadmium (Cd), which are harmful due to the non-biodegradability and carcinogenicity.[6,31] Qasem et al.[32] have classified some heavy metals in terms of emission source, their effects on health, and their permissible amount in drinking water. In addition to the metals mentioned above, some other metals, such as calcium (Ca), cobalt (Co), silver (Ag), antimony (Sb), manganese (Mn), iron (Fe), boron (B), and molybdenum (Mo) may be present in water resources and therefore must be eliminated. From 2011 to 2019, the number of articles that have examined the applications of various types of MOFs has been increased from 250 to 2250, while the number of articles published in heavy metal adsorption only grew from 2 to 62 papers. This shows that the portion of HMs adsorption is less than 3% among the applications of MOFs.[33] In the last decade, studies on the removal of heavy metals have focused on several techniques such as electrocoagulation (EC), adsorption, magnetic field usage, advanced oxidation processes (AOPs), membranes, etc.[6,31] In these studies, the advantages and disadvantages of different methods have been investigated.[32]

3.1.1 Adsorption

The performance of the adsorption process depends on the physical and chemical properties of the adsorbent and the heavy metals. In addition, operating parameters such as temperature, pressure, adsorption time, and adsorbent amount affect the removal efficiency. Because heavy metal ions are adsorbed on the adsorbent surface, the adsorbent's surface is also important. The adsorption method can be selected due to the lower operating costs, high adsorption capacity, practical and recyclability of the adsorbents.[34] Different types of adsorbents have been used for wastewater treatment, which has been introduced and categorized by Qasim et al., as follows[32]:

- Carbon-based adsorbents such as activated carbon (AC), carbon nanotubes (CNTs), graphene (G), and graphene oxide (GO).
- Chitosan-based adsorbents such as chitosan/clinoptilolite and chitosan/GO (CSGO).
- Mineral adsorbents based on zeolite, silica, and clay.
- Magnetic adsorbents containing magnetic nanoparticles, such as Fe_3O_4.
- Bio-sorbents such as activated charcoal, coconut husk, etc.
- Metal-organic framework (MOF) adsorbents.

Unlike other adsorbents introduced, MOFs have both the advantages of organic and inorganic adsorbents due to their constituent compounds.[35] These highly porous nanomaterials can capture heavy metal ions through physical or chemical adsorption mechanisms. Fig. 2A illustrates different possible mechanisms for the adsorption of heavy metals by MOFs from aqueous media. Physical adsorption is based on electrostatic interaction, diffusion effect, and van der Waals force, and chemical adsorption of HMs on MOFs is classified as coordination interaction, chemical bonding, acid-base interactions.[35] Chemical adsorption plays the main role in the adsorption mechanism of HMs on MOFs.[35]

Bedia et al.[2] categorized MOF systems often applied for adsorbing HMs based on the type of MOF or the HMs. The MOF systems, including virgin, modified, and magnetic MOFs, have been made and implemented for removing HM cations because of the structural properties of MOFs. For the adsorption of divalent heavy metal ions, MOF adsorbents are more efficient than other types of adsorbents.[35] In addition, modified MOFs have been used to improve the adsorption process of heavy metal ions. Post-synthesis techniques, including hybridization or coating, are used to modify virgin MOF

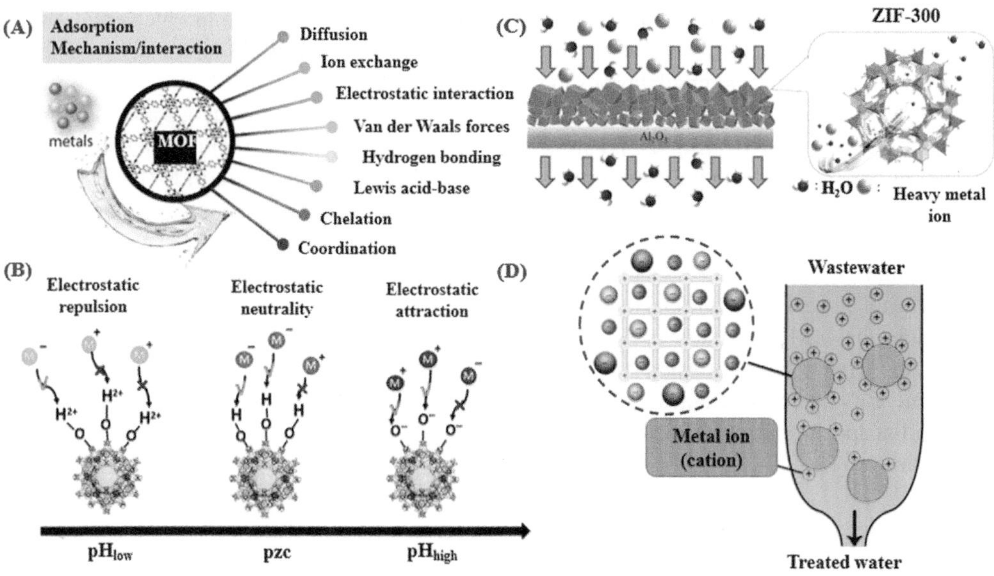

FIG. 2 Schematic illustration of different methods for removal of heavy metal ions: (A) adsorption mechanism, the (B) effect of pH value in adsorption mechanism, (C) MOF-based MMM for ion filtration, and (D) mechanism of ion exchange process comprising MOF particles. *Reproduced with permission from (A) Manousi N, et al. Extraction of metal ions with metal–organic frameworks. Molecules 2019;24(24):4605; (B) Ramanayaka S, et al. Performance of metal–organic frameworks for the adsorptive removal of potentially toxic elements in a water syste a critical review. RSC Adv. 2019;9(59):34359–34376; (C) Yuan J, et al. Fabrication of ZIF-300 membrane and its application for efficient removal of heavy metal ions from wastewater. J Membr Sci. 2019;572:20–27; (D) Qasem, NA, Mohammed RH, Lawal DU. Removal of heavy metal ions from wastewater: a comprehensive and critical review. npj Clean Water 2021;4(1):1–15.)*

types. Thiol, amino, azine, and quinine functional groups are the most common functional groups in modified MOFs.[35] Among the modified MOFs, S-MOFs are used more than other types because the sulfur group is softer and more polarized than other functional groups and, therefore, tends to absorb heavy metal ions. In addition, S-MOFs will receive more attention in the future due to their multi-functionalization. Li et al. specifically focused on S-MOFs. They have collected various synthesis methods and applications of S-MOFs as adsorbents for heavy metals, especially mercury.[36] The collection and recovery of MOF adsorbents is an important factor that has led researchers to use magnetic particles such as iron oxide nanoparticles in the MOF structure. Moreover, some researches performed on the ability of MOFs/magnetic nanocomposites to adsorb heavy metals yielded satisfactory results.[37]

Most heavy metals in the environment are divalent metals. However, there are some common multivalent metals in the environment, such as arsenic (As(III) and As(V)), chromium (Cr(III) and Cr(VI)), and manganese (Mn(VII)). The toxicity of Cr(III) is less than that of Cr(VI); however, removing this ion from water has attracted the researcher's attention.[38] For example, UiO-66 and UiO-66-NHC(S) were used to adsorb Cr(III), and the results showed the adsorption capacity of 67.3 and 117 mg/g, respectively.[39] Recently, Wen et al.[35] have reported a list of different MOFs utilized for the adsorption of high toxic Cr(VI). In addition, they assessed different adsorption capacities for arsenic adsorption by different types of MOFs, such as UiO, MILs, and ZIFs possessing the highest adsorption capacity.[35]

3.1.1.1 Effect of pH value

Due to the importance of pH, its effect on the adsorption process should be considered. Zhang et al.[40] studied the adsorption of Pb^{2+} and Cd^{2+} ions on MOF-5 and HS-mSi@MOF-5 at different pH values. They reported that the adsorption capacity for the two studied ions slowly increases with pH increment so that the highest value is obtained at pH 7. Due to the presence of positive hydrogen ions in the acidic pH range, there will be competition between this ion and heavy metal ions for adsorption on the MOF surface. As the pH increases, the concentration of hydrogen ions decreases, and therefore the adsorption capacity of heavy metals increases.[40] The effect of pH value in the adsorption mechanism is schematically represented in Fig. 2B.

3.1.2 Mixed matrix membranes

One of the applications of MOFs in separating metals from aqueous solutions is to combine them in the membrane structure as mixed matrix membranes (MMMs). MMMs comprise some heterogeneous inorganic compounds uniformly dispersed in the polymeric matrix and usually improve membrane performance. Among various techniques employed to prepare the MMMs, using MOFs as fillers in electrospun nanofibers has been considered by researchers.[41,42] The electrospinning method can produce high-mechanical strength porous membranes at the lowest cost. MMMs systems bring the highly efficient process for the purification of aqueous solutions due to their high filtration performance and membrane renewability and the hydraulic stability of the MOFs (Fig. 2C). Efome et al.[43] utilized the electrospinning method for preparing MMM containing iron and zinc-based MOFs in polyacrylonitrile (PAN) and polyvinylidene fluoride (PVF) polymers. They concluded that the MOF-based MMM had a high adsorption capacity for lead and mercury ions in an aqueous solution. The best performance was obtained for iron-based MOFs/PAN polymer membrane with the flux of $348 \, L/m^2$ h and the permeance of $870 \, L/m^2$ h bar. Although S-MOFs have shown excellent performance in adsorbing mercury ions, the use of such MOFs in membrane matrixes is not currently operational due to their low stability in the liquid phase and their recovery challenges.[44] Gao et al.[44] fabricated the MMM containing S-MOF (UiO-66-NHC(S)NHMe@NWF-g-MAH) using in situ synthesis method to enhance the mercury adsorption and stability of MOFs. The results in different aqueous environments indicated high selectivity for mercury and excellent adsorption rate. It is better to use thinner membrane layers to accelerate the separation and increase the specific adsorption capacity. In addition, the MMM could be recovered quickly, and its performance did not decrease.

3.1.3 Ion exchange

The ion exchange method is a reversible chemical reaction-based separation technique in which undesirable heavy metal ions are replaced by harmless and environmentally friendly ions on the adsorbent surface. Accordingly, the heavy metal in the wastewater is attached to a fixed solid particle and replaces the cation on the surface of the solid particle, as shown in Fig. 2D. The solids used in this process can be natural compounds such as zeolites or synthesized organic compounds such as resins. The ion exchange method can remove all types of divalent or polyvalent heavy metals from wastewater. The mechanism of ion exchange for the removal of heavy metals is expressed as the following reaction[32]:

$$M^-EC^+ + HMC^+ \leftrightarrow M^-HMC^+ + EC^+ \tag{1}$$

where M^-EC^+ is an ion exchanger with a fixed solid particle of M^- and the exchange cation of EC^+ and HMC^+ is a heavy metal cation. Sodium and hydrogen ions (Na^+ and H^+) are often used as exchange cations. Also, zeolite has a high potential for use as a fixed solid particle. MOFs have emerged as good options for use as ion exchangers in recent years. Some of the MOFs used as ion exchangers are AMOF-1, ZIF-8, and ZIF-67.[32,45,46] Studies on the use of MOFs for ion exchange methods are much smaller than other methods, and further research on their stability and recovery is required. Zheng et al.[47] used Cd-MOF-74 to investigate selectivity in removing Cu^{2+} from a mixture of seven other metal ions containing Mg^{2+}, Co^{2+}, Zn^{2+}, Fe^{2+}, Ni^{2+}, Na^+, and K^+. Their experiments were accomplished in pH 6.7, and the result showed $189.5 \, mg/g$ as removal capacity. Also, Li et al.[46] used ZIF-67 to remove Cr(III) from aqueous solutions. Their experiments were carried out at room temperature, and various pH values from 3 to 11, and the maximum removal capacity was attained at $13.34 \, mg/g$.

In recent years, publications and citations to MOF-based works have grown exponentially, with more than 27,000 different MOFs synthesized by researchers, many of which have been used to remove heavy metals from aqueous solutions. Comparing MOF adsorbents performance for removal of heavy metals is presented in Table 2.

3.2 Dyes

As one of the most important causes of water pollution with the natural or artificial origin, dyes have always been considered by many studies to remediate due to lack of water resources in the world.[1,3] Moreover, since many dyes are toxic and have adverse results on human health and aquatic life, various methods have been used to remove them till now, such as adsorption, coagulation, advanced chemical oxidation, membranes, and microbial degradation.[6,59,60] These technologies have their advantages and disadvantages. For example, the use of Fenton's reagent is, although on the one hand, a suitable chemical treatment process, on the other hand, generates sludge or employing the photochemical process leads to the formation of byproducts even if it removes the sludge.[61] Among the removal methods classified as physical, chemical, and biological, those based on MOFs system are discussed.

3.2.1 Adsorption

MOFs as new adsorbents have exhibited great potential to adsorb and remove dyes from wastewater in recent years. The easily adjustable pore size of MOFs and highly active sites, excellent mechanical and thermal stability, and the variable functional group make them talented in adsorption of dye molecules.[62] Generally speaking, the adsorption mechanism of

TABLE 2 Application of MOFs for removal of heavy metals using different methods.

Mechanism	MOFs	Heavy metals	Removal efficiency	Adsorption capacity/permeance	Operation parameters	Refs.
Adsorption	MOF-808-EDTA	22 HMs (e.g., Ce^{3+}, Gd^{3+}, Zr^{4+}, Fe^{3+}, Mn^{2+}, Hg^{2+}, Cd^{2+}, Pb^{2+}, Sn^{2+}, etc.)	>99%	528 mg/g for Cd^{2+} 313 mg/g for Pb^{2+} 592 mg/g for Hg^{2+}	pH=2, Eq. time=24 h, $T=25°C$, $C_0=10$ ppm	48
	Zr-DMBD	Hg^{2+}	99.64%	171.5 mg/g	pH=6, Eq. time=6 h, $T=25°C$, $C_0=500$ mg/L	49
	UiO-66-NHC(S) NHMe	Cd^{2+} Cr^{3+} Pb^{2+} Hg^{2+}	37% 70% 90% 99%	49 mg/g 117 mg/g 232 mg/g 769 mg/g	Eq. time=5–240 min, $T=55°C$, $C_0=10$–200 mg/L	39
	MIL-100(Fe)	As^{5+}	86%	110 mg/g	pH=7, Eq. time=1 h, $T=25°C$, $C_0=10$ ppm	50
	SH-Fe_3O_4/ $Cu_3(BTC)_2$	Pb^{2+}	>95%	198 mg/g	pH=6, Eq. time=15 min, $T=150°C$, $C_0=0.01$ mg/L	51
Mixed matrix membrane	UiO-66-NHC(S) NHMe (MOF/PP)	Hg^{2+}	98%	–	pH=2.6, Eq. time=2 h, $T=25°C$	44
	Nanofibrous MOF membranes	Pb^{2+}	>90%	870 L/m² h bar	pH<5, $T=25°C$, $P=0.4$ bar, $C_0=20$ ppm, contact time=3 h	43
	ZIF-300 MOF membrane	Cu^{2+}	99.21%	39.2 L/m² h bar	$T=25°C$, $P=1$ bar, $C_0=40$ mg/L	52
	PC-HMO MMs	Cd^{2+} Cu^{2+}	98%	–	pH=2–8, Eq. time=1 min, $T=40°C$, $C_0=40$ mg/L	53
	PSf/GO/DMF	As^{5+}	83.65%	–	pH=3–11, Eq. time=1 min, $T=25°C$, $C_0=0.3$ mg/L, $P=4$ bar	54
Ion exchange	ZIF-8	Cu^{2+}	>95%	800 mg/g	pH=3–6, Eq. time=30 min, $T=20$–50°C, $C_0=150$ mg/L	45
	TMU-16-NH_2	Cd^{2+}	98.91%	126.6	pH=6, Eq. time=30 min, $T=25$–45°C, $C_0=50$ mg/L	55
	UiO-66	As(V)	–	303.34 mg/g	pH=2, Eq. time=48 h, $T=25°C$, $C_0=10$–200 ppm	56
	MOR-1-HA	Cr (VI)	99%	242–280 mg/g	pH=3, Eq. time=3 min, $T=25°C$, $S_{BET}=833$ m²/g, $C_0=21.2$ ppm	57
	NH_2-MIL-53(Al)	Hg^{2+}	>96%	153.85 mg/g	pH=4–10, Eq. time=60 min, $T=25$–45°C, $C_0=50$–300 mg/L	58

dye removal is mainly due to the electrostatic attraction, π-π interaction, hydrogen bonding, hydrophobic interactions, acid-base interaction, and physical adsorption.[61] Actually, from one viewpoint, dyes are classified into three groups: cationic, anionic, and natural dyes.[63] MOFs can have positive charges to uptake the opposite charges in anionic dyes, and in cationic dyes, they possess negative charges. Therefore, electrostatic attraction can be formed by the MOF's net charge and opposite charges of adsorbates, indicating the most significant adsorption mechanism. Of course, an important parameter named pH can affect the net surface charge in the solution. For instance, a cationic dye like methylene blue (MB) can be adsorbed by a MOF with negative surface charges like UiO-66 at pH = 7.5[64] or Zn-MOF.[65] Another interaction that is not as strong as electrostatic interaction but can be considered as the second important mechanism of adsorption is the π-π interaction formed between two aromatics rings. As an example, for eliminating methyl orange (MO) and MB with UiO-66, the benzene rings in both of them are connected via a π-π interaction.[66] Hydrogen bonding is less reported as a mechanism of dye adsorption. Still, acid red 1 (AR1) can be eliminated by the formation of this bond between —COOH in MIL-101(Cr) MOF and —OH groups in dye molecule.[67] The remaining mechanisms are less noticeable for dye removal. The schematic of possible interactions between MOFs and dye molecules is represented in Fig. 3.

The effective factors on dye adsorption onto the MOFs include solution pH, initial dye concentration, temperature, adsorption time, the existence of ions, adsorbent dosage and mixing speed. For more details, the effect of each parameter is explained in order of importance as below[61]:

(1) Solution pH: One of the significant factors that affect adsorption capacity is pH. The number of electrostatic charges on the surface of ionized dye molecules is influenced by pH, which determines adsorption rate. The best pH values for the adsorption of cationic and anionic dyes can be reached at higher and lower than pH at the charge of zero (pH^0), respectively. For instance, Tehrani et al.[68] showed that the optimum pH for the remediation of MB by MIL-68(Al) is

FIG. 3 Possible interactions between MOFs and dye molecules as adsorbate. *(Reproduced with permission from Dhaka, S., et al. Metal–organic frameworks (MOFs) for the removal of emerging contaminants from aquatic environments. Coord Chem Rev 2019;380:330–352.)*

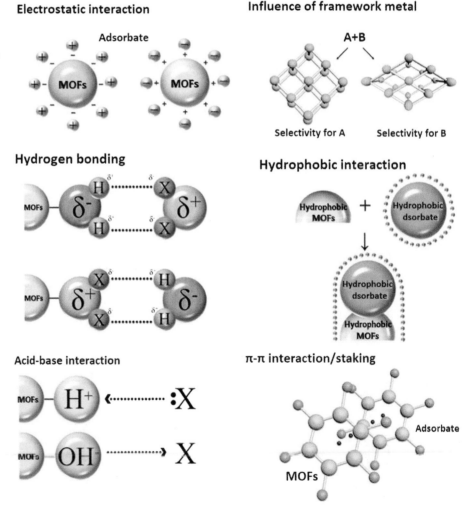

7.9 and this value for the adsorption of MO by MOF-235 is 4.[69] Actually, at pH < pH 0, the MOF's surface gets positive charges due to the presence of H^+ ions or the reaction of these ions with the surface groups of hydroxyl. In contrast, for cationic dyes in higher pH due to the elimination of H^+ ions, the positive charges of dye are attached to the negative sites of MOF.

(2) Initial dye concentration: Another important parameter in water decontamination is the initial concentration of dyes since it intensifies the mass transfer of dye molecules from the bulk to the adsorbent surface. Usually, the rate of adsorption efficiency decreases when the initial dye concentration increases because of the saturation of the binding sites on the MOF surface. On the other hand, a high concentration of dye can increase the adsorption capacity until the point in which the concentration is equal to the binding sites.

(3) Temperature: Depending on whether the dye adsorption on MOFs is exothermic or endothermic, the increase of temperature causes to decrease or increase of adsorption efficiency, respectively. This can be deduced that the temperature elevation enhances dye molecule movements and functional adsorptive sites in an endothermic reaction. In contrast, in an exothermic reaction, this increase leads to a decrease in the binding forces between the adsorbate molecules and the active sites on the adsorbent surface.

(4) The amount of adsorbent: It is obvious that by enhancing the adsorbent dosage, more amount of dye is eliminated, and naturally, it is a result of an increase in the available surface area of the adsorbent. But there is a point that the level of adsorption is reduced by approaching the equilibrium point. Since the use of a high amount of adsorbent is not economical and can lead to the accumulation of sludge after the process, it is important to determine a point at which maximum adsorption occurs with the consumption of the least amount of adsorbent.

(5) Adsorption time: Since the contact time is one of the serious design parameters in the real water treatment applications, it is favorable to find the minimum equilibrium time to get high efficiency. Besides, the equilibrium time affects the restoration of the adsorbent. It means a short equilibrium time results in a short restoration time, and MOFs possess this outstanding attribute. For instance, $H_6P_2W_{18}O_6$/MOF-5 nanohybrid could be capable of removing MB in 10 min.

(6) Competing ions: The presence of competing ions such as cations and anions in the sewer can affect the dye adsorption onto MOFs. Negative anions may compete with anionic dyes, thereby decreasing the adsorption efficiency, and the same is true for positive cation and cationic dyes. The reason can be justified with the statement that electrostatic attraction between the adsorbent and the adsorbate may decline in higher ion concentrations. Of course, there are some exceptions. Some divalent cations and anions can enhance the adsorption, and contrast divalent co-ions may reduce the adsorption rate. Thus, it is essential to check and test the effect of diverse ionic concentrations in water treatment by MOFs.

(7) Mixing speed: The high speed of mixing can increase dye adsorption since the thickness of the external boundary layer and, therefore, the resistance to mass transfer of dye molecules to the MOFs is reduced by agitation with high speed. Some researches show that at first, the rate of uptake increases as the agitation speed increases but up to a certain amount of speed. For more transparency, the mixing speed higher than 400 rpm is ineffective in removing Remazol Deep Black RGB on MIL-101, which can be the cause of adsorbent saturation.

3.2.2 Photocatalytic degradation

Due to the lack of former catalytic and photocatalytic reactions such as Fenton and Fe-based catalyst (for example, non-selectivity, non-recoverability, and generation of lots of sludge or agglomeration problem), researchers have synthesized MOF-based catalysts for dye removal from water.[61] For more explanation, it can be said that electron and hole pair is created on the surface of the catalyst in the presence of light. After that, electrons stay in the conduction band, and holes are located in the valance band. The evoked electron reacts with O_2 and leads to the formation of reactive superoxide (O_2^-) radicals. Positively charged holes destroy hydroxyl ions or water molecules by generating highly reactive hydroxyl radicals.[70] For example, Fe_3O_4@MIL-100(Fe) composite, a magnetic MOF catalyst, has been used for MB degradation, and the results showed 100% elimination of dye in the presence of H_2O_2.[71] The additives such as hydrogen peroxide and peroxymonosulfate act as electron receivers in the photocatalytic process to solve a problem referred to as recombination of photogenerated carriers. They do this by the generation of active sulfate or hydroxyl radicals. Also, superior MOFs act just under UV light, but recent advances in the construction of new MOFs have led to the solution of this problem, and some MOFs like MIL-88A and NTU-9 act well for eliminating MB or Rhodamine B (RhB) under the visible light.

3.2.3 MOF-based membrane

MOFs can be used in the form of membranes for the adsorption of dye contaminants. The MOFs membranes have been synthesized since the usual application of other adsorbents was powder or pellet, thereby leading to low performance. Also, the pressure of liquids moving through the system is reduced, and the probability of contamination by small particles highly exists. Therefore, the idea of using membranes seems appealing. However, common materials used in membrane adsorbents are polymers, but they still suffer from some problems such as poor chemical and thermal stability.[72] Therefore, advanced membranes based on MOFs have entered the arena and have shown high capacity in the adsorption of a wide variety of dye pollutants, having some individual characteristics, including high tunable porosities, large surface areas, numerous channels, high selectivity, and the ability to apply in large-scale processes. MOF-based membranes have been employed in various applications, such as ultrafiltration (UF), forward-osmosis (FO), nanofiltration (NF), and reverse-osmosis (RO) processes.[73] NF membranes like Polyamide/UiO-66 have shown appropriate results in the separation of some dyes, including MO, rose bengal (RB), thiazole yellow G (TY), crystal violet (CV), and safranin O from various solvents. Also, as a UF membrane, MIL-53(Fe)/polyvinylidene fluoride has an impressive effect on MB removal. ZIF-8 and ZIF-L membranes have excellent performance for removing RB, and also HKUST-1/reduced GO membranes can eliminate MB and CR dyes significantly.[74] Some other examples for MOFs application for removing dyes are listed in Table 3.

3.3 Pharmaceuticals and personal care products

Pharmaceuticals and personal care products (PPCPs) which can induce physiological effects on human beings, are a prominent group of contaminants.[95] More than 4000 commercial PPCPs are currently employed for different purposes such as antibacterial, anticonvulsant, antibiotics, artificial sweeteners, cholesterol-lowering, hormones, anti-inflammatory, drug preservatives, an asthma drug, beta-blocker, antipsychotic, sunscreen lotion, analgesic, anti-inflammatory, anticancer, and antimalarial. However, some fractions of PPCPs are metabolized in the living body. Still, they can continuously enter the surface and groundwater resources through different pathways, including industries, hospitals, runoff from farms, and aquaculture applications.[96,97] Surveying the literature showed that MOFs have been mainly used in the PPCPs removal based on the two common technologies: (i) as porous adsorbent and (ii) as (photo)-catalyze degradation via an oxidation process.

3.3.1 Adsorption

Adsorption of different types of PPCPs using MOFs is widely investigated. Wang et al.[98] have studied the adsorption capacity of MIL-101(Fe), MIL-53(Fe), and MIL-100(Fe) for the adsorption of tetracycline, the most common antibiotic for treating bacterial infections. The maximum tetracycline adsorption capacities of 43.4, 12.5, and 51.5 mg/g were reported for MIL-101(Fe), MIL-53(Fe), and MIL-100(Fe), respectively. Simultaneous adsorption of oxytetracycline and tetracycline was studied by Li et al.[99] using ZIF-8. Both tetracycline and oxytetracycline were efficiently adsorbed over ZIF-8. The observed excellent performance of ZIF-8 was attributed to the imidazolate rings, which induced π-π interaction with antibiotics. Dehghan et al.[100] have synthesized ZIF-67 with different morphologies under various conditions, such as cobalt sources, i.e., acetate, chloride, and nitrate. ZIF-67 prepared with acetate salt performed the highest removal of tetracycline with an adsorption capacity of 447 mg/g. The effects of different experimental conditions were studied using response surface methodology. The optimum conditions were obtained: contact time of 26.8 min, adsorbent dosage of 0.63 g/L, pH of 5.9, and an antibiotic concentration of 74.6 mg/L. Moreover, the proposed kinetic model showed that the process was controlled by chemisorption. The water stability of different MOFs, MOF-5, MOF-505, HKUST-1, MIL-100, UMCM-150, and MOF-177 was studied by Cychosz et al.[101] It was shown that the stability of MOFs is related to the metal clusters presented in the framework. MIL-100(Cr) was found to be more stable with excellent adsorption capacity for furosemide and sulfasalazine, which are usually used to treat edema and inflammatory bowel diseases. Cychosz et al.[101] have reported the adsorption capacity of 11.8 and 6.2 mg/g obtained over MIL-100(Cr) for initial concentrations of 7.5 and 1.4 mg/L for furosemide and sulfasalazine, respectively. The microporous MIL-53 represents breathing behavior relating to the quest molecule.[102] Gao et al.[103] have investigated the effects of the metal ions in MIL-53(Fe), MIL-53(Cr), and MIL-53(Al) on the breathing behavior and adsorption of sulfamethoxazole (SMZ). All of the synthesized MIL-53 series exhibited narrow pore sizes. The solution observed a phase transition to larger pore sizes for MIL-53 with Cr and Al metal ions. At the same time, Fe-type retained its small pore sizes. Interestingly, MIL-53(Al) and MIL-53(Cr) with larger pore sizes presented an excellent performance for SMZ adsorption. On the other hand, MIL-53(Fe) with a narrow pore size did not adsorb large molecules of SMZ effectively. Lin et al.[104] have evaluated the ability of Zr-based MOF (MOF-808, MOF-802, and UiO-66) for adsorptive removal of several pharmaceuticals, including naproxen, salicylic acid, ibuprofen, indomethacin, ketoprofen and furosemide, and acetophenone. UiO-66 indicated the highest performance. The excellent performance of UiO-66 was attributed to the high tendency of incomplete-coordinated Zr for interaction with anionic pharmaceuticals. In addition, π-π interaction between benzene rings of pharmaceuticals and MOFs

TABLE 3 The summary of different studies employed MOFs to remove dyes using various methods.

Adsorption				
MOF	Dye molecule	Adsorption capacity (mg/g)	Operation conditions	Refs.
Ni-MOF	CR	276.7	–	75
Zn-MOF		132.2	–	
Co-MOF		4885.20	–	76
$[Ni_2F_2(4,4bipy)_2(H_2O)_2]$ $(VO_3)_2 \cdot 8H_2O$		242.1	pH$=$4, $T=$318K	77
Fe-MIL-88NH$_2$		167.547	$T=$323K	78
UiO-66	MB	69.8	–	64
Fe$_3$O$_4$@MIL-100(Fe)		221	–	79
BUT-29		1119	–	80
Co-Fe-LDH@UiO-66-NH$_2$		555.62	pH$=$8, $T=$338K	81
UiO-66	MO	454	pH$=$4, $T=$298K	82
ZIF-8@SiO$_2$@MnFe$_2$O$_4$		78.12	pH$=$5, $T=$293K	83
TMU-1		100%	–	84
Ce(III)-doped UiO-67		401.2	–	63

Photocatalytic degradation				
Catalyst	Dye molecule	Degradation (%)	Operating condition	Refs.
MIL-100(Fe)	MO	64	UV light, $t=$420min	85
ZIF-8	MB	82.3	UV light, $t=$120min	86
Bi$_2$WO$_6$/UiO-66	RhB	100	UV light, $t=$180min	87
MIL-53(Fe)	MB(H$_2$O$_2$)	99	Vis light, t$=$20min	88
NTU-9	RhB	100	Vis light, $t=$80min	89
MIL-53(Fe)	AO7(persulfate)	100	Vis light, $t=$90min	90

MOF-based membrane				
Membrane	Dye molecule	Permeance L/(m^2 h bar)/ retention%	Operating condition	Refs.
Interfacial synthesis of ZIF-8 membranes	RB	90L/m^2 h bar 85%	Dead-end $C_0=$17.5μM $P=$2bar	91
67-MIL-polyvinylidene fluoride	MB	~250L/m^2 h 60−>95%	Dead-end $C_0=$20mg/L $\Delta P=$2bar	92
Polydopamine-modified reduced RO/ MOFs Nanocomposite	MB	185L/m^2 h 99.8%	Vacuum suction $C_0=$40mg/L $P=$0.09MPa	93
Cu(terephthalate)@ GO/polyethersulfone	MB CR MO	~150L/m^2 h ~20−90%	Dead-end $C_0=$100mg/L $\Delta P=$200KPa	94

ligand is another alternative involved for adsorption. The pore size of MOFs is an important factor for the adsorption of large-sized PPCPs. Creating large pores from defects and the hierarchical porous structure can enhance the PPCPs adsorption and is highly investigated by different researchers. The adsorptive performance of meso-MIL-53(Al) synthesized using Pluronic F127 was evaluated in respect to micro-MIL-53(Al) in the removal of triclosan, a popular additive in toothpaste and soap detergents.[105] The meso-MIL-53(Al) indicated the higher capacity for adsorption via the improving mass transfer caused by larger pore sizes. Also, the adsorption over meso-MIL-53(Al) was 4.4 times faster than micro-MIL-53(Al). Jung et al.[106] have studied the adsorption of p-arsanilic acid using ZIF-8, meso-ZIF-8, MIL-53(Cr), MIL-101(Cr), zeolite, and activated carbon. Meso-ZIF-8 efficiently adsorbed this organoarsenic pollutant with an adsorption capacity of 791.1 mg/g, which is significantly higher than 293.2 mg/g obtained over activated carbon. Li et al.[107] have introduced defect sites to the UiO-66 structure by controlling the amount of benzoic acid (modulator). The effects of defect creation on the roxarsone adsorption were discussed. Obtained results indicated that porosity increased with increasing defect sites, and the adsorption capacity followed the same trend of increasing defect sites number, pore volume, and surface area.

Another approach for improving the adsorption performance of MOFs is installing the functional groups. Functionality can adjust the environment of pores using two methods:

(1) Coordination to metal clusters [including urea, aminomethane sulfonic acid (AMSA), ethylenediamine (ED), diethanolamine (DEA), melamine, triethanolamine (TEA)].
(2) Coordination to linkers (e.g., —SO_3H, furan-Br, —NH_2, furan).

Interaction between acid and base sites and the tendency to form hydrogen bonds was promoted by introducing different functional groups. Hasan et al.[108] have investigated the effects of different functional groups in the MIL-101(Cr) structure for adsorptive removal of naproxen and clofibric acid. Functionalizing of coordinatively unsaturated sites (CUSs) with —SO_3H and —NH_2 by aminomethane sulfonic acid ethylenediamine, respectively, was conducted over parent MIL-101(Cr). However, the porosity of functionalized MIL-101(Cr) decreased significantly compared to the parent sample. Meanwhile, the performance of the basic MIL-101(Cr)-NH_2 was better than those of MIL-101(Cr) and the acidic MIL-101(Cr)-SO_3H. The interaction between basic structures of MIL-101(Cr)-NH_2 with acidic COOH group of clofibric acid and naproxen is responsible for the excellent operation. Hasan et al.[109] have reported the effects of functionalizing linkers using UiO-66-NH_2, UiO-66, and UiO-66-SO_3H on the adsorptive removal of diclofenac sodium. The adsorption capacity increased with the order of UiO-66-NH_2 < UiO-66 < UiO-66-SO_3H. The excellent performance was provided by acidic UiO-66-SO_3H due to the acid-base attraction. The performance of MIL-68(In)-NH_2 in the adsorption of p-Arsanilic was compared with pristine MIL-68(In) by Lv et al.[110] The adsorption capacities of 340.1 and 401.6 mg/g were obtained for MIL-68(In) and MIL-68(In)-NH_2, respectively. The excellent performance of functionalized In-based MOF is attributed to hydrogen bonding and π-π interaction. Seo et al.[111] have synthesized CUS-functionalized MIL-101(Cr) using urea and melamine and evaluated their ability for the adsorption of different antibiotics, including dimetridazole, tinidazole, and metronidazole. After urea functionalizing, the amount of NIABs adsorbed was significantly increased compared to MIL-101(Cr)-melamine and pristine MIL-101(Cr). The high tendency between the —NO_2 group of antibiotics and —NH_2 group of MOF to form hydrogen bonding causes the excellent performance of MIL-101(Cr)-urea. The functionalization of MIL-101(Cr) with —OH (using EA), —OH_2 (using DEA), —NO_2 and —NH_2 groups was performed to uptake different PPCPs, such as naproxen, oxybenzone, and ibuprofen.[112] The three selected PPCPs have different functional groups, including ether, alcohols, and ketones, effectively interacting with adsorbents. The adsorption capacity for all three selected PPCPs followed the trend of MIL-101(Cr)-OH > MIL-101(Cr)-$(OH)_2$ > MIL-101(Cr)-NH_2 > MIL-101(Cr) > MIL-101(Cr)-NO_2. The functional groups (—NH_2, —$(OH)_2$, and —OH) with hydrogen donor feature interact with polar functional groups of PPCPs as hydrogen acceptors.

3.3.2 Photocatalytic degradation

In addition to using MOFs as adsorbents in the removal of PPCPs, many studies have used them as photocatalysts for degrading different PPCPs. M@MIL-100(Fe) (M = Au, Pt, and Pd) was employed by Liang et al.[113] to photodegradation of theophylline, bisphenol A, and ibuprofen under visible light. This paper was the first report related to the application of MOF in photocatalysis of PPCPs and confirmed that MOF functionalization with metal nanoparticles could effectively enhance the activity of MOF. Pd@MIL-100(Fe) doped with 1% Pd displayed the highest efficiency. The incorporation of Pd nanoparticles suppressed the recombination of electrons and holes generated by photos, leading to improved photocatalytic activity. The Pd-$H_3PW_{12}O_4$@MIL-100(Fe) nanocomposite was synthesized by Liang et al.[114] and employed for photodegradation of theophylline and ibuprofen. The encapsulated $H_3PW_{12}O_4$ is a UV-switchable reducing agent. In addition, this component can stabilize the decoration of Pd nanoparticles in the MIL framework. Generated electrons and holes can be captured by H_2O_2 and H_2O, respectively, to produce OH•.

On the other hand, the transformation of H_2O to produce OH• can be catalyzed with Fe^{3+}-O clusters presented in the MIL structure. Therefore, the formation of OH• has promoted with Pd-$H_3PW_{12}O_4$@MIL-100(Fe) catalyst and considerably accelerated the PPCP removal. The removal of tetracycline with photocatalysis was evaluated by Wang et al.[115] using In_2S_3@MIL-125 core-shell composite. Based on the obtained results, the composite activity was much higher than the pristine In_2S_3 and MIL-125. It may be due to the synergic effects of In_2S_3 and MIL-125, structure with open pores, effective electron transfer between Ti^{3+}-Ti^{4+}, and effective transfer of electrons-holes pairs generated by photos. Dong et al.[116] have investigated the performance of In_2S_3/UiO-66 that Zr/In molar ratio equals 0.37/1 for photodegradation of tetracycline. The large pore diameter of UiO-66 and many functional groups over the tetracycline can lead to increased contact between the catalyst and pollutant. On the other hand, excellent photocatalysis can be assigned to the more formation of active $°O_2^-$ and h^+ species owning to the expanding light absorption and suppressing recombination of electron-holes pairs. Three MOFs, MIL-101(Fe), MIL-53(Fe), and MIL-100(Fe), were evaluated for degradation of tetracycline under visible light.[98] 96.6% of tetracycline was decomposed over MIL-101(Fe) after 180 min, much higher than 54.7% obtained with MIL-100(Fe) and 40.6% with MIL-53(Fe). Because the MIL-101(Fe) had a strong capacity for visible light absorption, confirmed by UV-Vis analysis. The photocatalysis removal of amoxicillin using MIL-68(In)-NH_2/GO hybrid has been reported by Yang et al.[117] Ninety-three percent decomposition of amoxicillin and 80% removal of total organic carbon (TOC) were achieved under visible light after 120 and 210 min, respectively. The excellent performance of the composite was attributed to GO, which can serve as an efficient electron transporter and suppresses the recombination of electron-hole pairs. In addition, visible light absorption can be improved by GO introduction.

3.3.3 Membrane technology

The use of adsorptive membrane technology by MOF-containing membranes appears promising for wastewater treatment. However, rarely studies have explored the MOF-containing membranes for adsorption of PPCPs from wastewater. Basu et al.[118] have investigated the performance of a thin-film nanocomposite (TFN) membrane with ZIF-8 to remove acetaminophen. Two types of TFN were prepared: a) support made of polysulfone (PSF) with ZIF-8 containing polyamide (PA) and b) deposition of ZIF-8 and PA in separation layers over PSF support. The obtained results indicated that adding ZIF-8 to PA resulted in membrane defects, so increasing solvent permeance was observed accompanied by deteriorating retention of acetaminophen. At the same time, the layer-by-layer membrane showed a defect-free structure which improves acetaminophen separation. Zhao et al.[119] have employed three MOFs, MIL-53(Al), ZIF-8, and UiO-66-NH_2, to prepare PA nanocomposite. The cross-linking grade of the membrane decreased by MOF incorporation leading to enhancement of water permeability up to 30%, with negligible effects of their rejection performance. MIL-53(Al) was strongly bound with PA. At the same time, weaker interaction occurred between PA and other MOFs. This difference explained the observed variable perm-selectivities of three membranes for phenacetin, nalidixic, atenolol, and sulpiride. Yadav et al.[120] have synthesized polyvinylidene fluoride-*co*-hexafluoro propylene (PVDF-*co*-HFP) membrane containing carbon nanotube (CNT) functionalized by ZIF-8. The authors investigated the membrane performance regarding the flux and removal efficiency of tetracycline, doxycycline, norfloxacin, and ciprofloxacin. Porosity and roughness of the membrane surface increased by impregnation of ZIF-8 and CNT. The best performance with the removal of >99.4% for all investigated antibiotics was achieved over 18% PVDF-*co*-HFP/0.5% CNT@ZIF-8. An ultrafiltration (UF) membrane containing MIL-101(Cr) hybrid system was reported for adsorptive removal of ibuprofen, 17α-ethinyl estradiol, and a mixture of humic acid and tannic acid by Kim et al.[121] The removal efficiency and flux of the hybrid system were compared with commercial activated carbon UF. The higher removal efficiency was achieved using a MOF-containing membrane which can be ascribed to its greater pore volume.

3.4 Pesticides and herbicides

The increasing usage of pesticides and herbicides has become a global concern. They are highly toxic and stable and do not decompose biologically in nature.[122,123] Many organic materials have been employed as pesticides and herbicides such as atrazine, diazinon, organophosphates pesticides (OPPs), and 2,4-dichlorophenoxyacetic acid (2,4-D). Elimination of pesticides and herbicides in wastewater using MOFs has been rarely investigated. Jhung et al.[124] have studied an adsorption removal of 2,4-D over MIL-53(Cr). They obtained the adsorption capacity of 556 mg/g, greater than those obtained for activated carbon (286 mg/g) and USY zeolite (256 mg/g) after 1 h. The adsorption mechanism was proposed to proceed via electrostatic interaction between 2,4-D anions and the positively charged surface of MIL-53(Cr). The adsorption of diazinon by MIL-101(Cr) was studied by Mirsoleimani-Azizi et al.[125] in a continuous fixed bed reactor. Interestingly, the capacity of 260 mg/g was achieved only after 3 min. The effects of different operating variables such as pH, diazinon concentration, feed flow rate, and bed height were examined on the adsorptive removal. Akpinar et al.[126] have compared the performance of two Zr_6-based MOFs, NU-1000 and NU-1008, to eliminate atrazine. Both MOFs have possessed a similar surface area of 1400 m^2/g and mesoporous structure. NU-1000 with pyrene linkers exhibited high adsorption capacity. It was suggested the mesoporous structure of NU-1000 improved mass transfer of atrazine. In addition, pyrene-based linkers presented sufficient active sites for the π-π interaction leading to 93% removal in only 5 min. While NU-1008 without pyrene and thereby lack of active sites to form π-π interaction adsorbed less than 20% of atrazine, despite having the same surface area and pore size. ZIF-8/magnetic-CNT was evaluated for simultaneous adsorption of eight OPPs from tap water and soil samples.[127] The authors demonstrated effectively and rapidly removal of almost all the pesticides. Owning to high activity and facile separation magnetically, ZIF-8/magnetic-CNT was suggested as a promising adsorbent for pesticides removal.

There are very few papers dealing with the decomposition of these agrochemicals by photocatalysis. Mohaghegh et al.[128] reported the first study of atrazine photodegradation.[128] They synthesized a complex composite containing $Ag_3PO_4/BiPO_4$ heterojunction assembled on the copper terephthalate (Cu(TPA) surface with high activity under visible light and UV. The excellent activity was assigned to the heterojunction of coupling copper terephthalate, which facilitates effective electron transfer and effective separation of electron-hole pairs. Xue et al.[129] have evaluated the BiOBr/UiO-66 composite for atrazine degradation with visible light. They studied the effects of different environmental parameters on photocatalysis performance, including pH and the presence of different organic and inorganic ions. The obtained results revealed that decomposition of atrazine decreased with increasing the pH solution. The presence of inorganic cations had negligible effects on atrazine degradation.

In contrast, the organic anions HCO_3^-, SO_4^{2-}, and Cl^- severely inhibited its decomposition. In addition, a reaction mechanism was proposed based on the three parts: dechlorination, dealkylation, and breakage of CN bond, with the participation of $°O_2^-$ and h^+ as active species. Oladipo et al.[130] have investigated photodegradation of chlorpyrifos and methyl malathion by $AgIO_3$/MIL-53(Fe) under solar light. About 90% of decomposition of chlorpyrifos was achieved after 60 min irradiation in tap and distilled water, respectively. The high surface area of the composite, large content of hydroxyl groups over the surface, and also effective suppressing of recombination of electrons-holes pairs were the key factors for enhanced activity. In another work, Oladipo[131] has studied the photodegradation of 2,4-D using WO_3/MIL-53(Fe) under solar light. The synergic effects between MIL-53(Fe) and WO_3 resulted in higher photocatalytic activity than WO_3 and MIL-53(Fe). Also, the composite can retain its initial activity after five cycles, indicating its reasonable stability. Table 4 summarizes the list of different studies that employed MOFs to remove PPCPs and pesticides using various methods.

TABLE 4 The summary of different studies employed MOFs to remove PPCPs and pesticides using various methods.

MOFs	Pollutant	Adsorptive removal q_{max} (mg/g)/removal (%)	Operating condition	Refs.
micro-ZIF-8	p-Arsanilic acid	729.9	$T=25°C$, pH=1–11 $C_0=3$–350 mg/L	19
meso-ZIF-8		791.1		
micro-MIL-53(Al)	Triclosan	447	$T=25°C$, pH=6.5 $C_0=10$–60 mg/L	18
meso-MIL-53(Al)		488		
MIL-101(Cr)	Naproxen	131	$T=25°C$, pH=2–12 $C_0=10$–18 mg/L	20
MIL-101(Cr)-NH_2		154		
MIL-101(Cr)-SO_3H		93		
MIL-101(Cr)	Clofibric acid	315	$T=25°C$, pH=2–12 $C_0=10$–18 mg/L	20
MIL-101(Cr)-NH_2		347		
MIL-101(Cr)-SO_3H		105		

TABLE 4 The summary of different studies employed MOFs to remove PPCPs and pesticides using various methods—cont'd

Adsorptive removal				
MOFs	**Pollutant**	**q_{max} (mg/g)/removal (%)**	**Operating condition**	**Refs.**
UiO-66	Diclofenac sodium	189	$T=25°C$, pH$=5.4$ $C_0=30$ mg/L	21
UiO-66-NH$_2$		106		
UiO-66-SO$_3$H		363		
NU-1000	Atrazine	93%	$T=25°C$, $C_0=10$ mg/L	36
NU-1008		20%		
ZIF-8/magnetic-CNT	Triazophos	3.12	$T=25°C$, pH$=2–10$ $C_0=0.2–2$ mg/L	37
	Diazinon	2.59		
	Phosalone	3.80		
	Profenofos	3.89		
	Methidathion	2.34		
	Ethoprop	2.18		
	Sulfotep	2.84		
	Isazofos	3.0	3.0	3.0

Membrane adsorption				
Membrane	**Pollutant**	**Permeance L/(m^2 h bar)/ retention%**	**Operating condition**	**Refs.**
PSF/PA-ZIF-8 layer-by-layer	Acetaminophen	5.5–22.5 L/(m^2 h bar) 55%	$T=25°C$, pH$=8$, $P=4$ bar, $C_0=0.1$ mg/L	25
PSF/PA-UiO-66-NH$_2$	Sulpiride	7.2 L/(m^2 h bar) 90%	$T=20°C$, pH$=7$, $P=10$ bar, $C_0=50$ µg/L	28

Photocatalytic degradation				
Catalyst	**Pollutant**	**Degradation (%)**	**Operating condition**	**Refs.**
In$_2$S$_3$@MIL-125	Tetracycline	63.3	$T=25°C$, pH$=5.9$, $C_0=46$ mg/L, visible light	31
Pd@MIL-100(Fe)	Theophylline	45.2	$T=25°C$, pH$=2–6$, $C_0=20$ mg/L, 40 µL H$_2$O$_2$, visible light	29
	Ibuprofen	69.2		
	Bisphenol A	20.5		
Pd-H$_3$PW$_{12}$O$_4$@MIL-100 (Fe)	Theophylline	99	$T=30°C$, pH$=2–6$, $C_0=20$ mg/L, 40 µL H$_2$O$_2$, visible light	30
	Ibuprofen	98		
BiOBr/UiO-66	Atrazine	88	$T=25°C$, pH$=3.1–9.4$, $C_0=5$ mg/L, visible light	39
MIL-53(Fe)	2,4 D	58	$T=25°C$, pH$=7$, $C_0=45$ mg/L, sun light	132
WO$_3$/MIL-53(Fe)		100		

4 Industrial limitations of MOFs

Despite the many advantages of MOFs as wastewater treatment, their industrial applications are limited due to some drawbacks such as thermal and chemical instability, production with a lack of sustainability and scalability. In addition, producing large quantities of MOFs at a reasonable price for commercial purposes still requires more effort. On the other hand, MOFs are usually synthesized in powder form, which limits their use in commercial purposes:

- MOFs in powder forms are difficult to be recycled.
- Compression of powdery MOFs in columns leads to poor mass transfer and high-pressure drop.
- Nano-sized MOF powders can be easily agglomerated and then cause the blockage of the water treatment instruments, resulting in secondary contamination and reduced process efficiency.
- Powdered MOFs can be easily wasted during practical applications leading to increasing the operating cost of the process.
- Synthesis of MOFs with controllable size and shape and high efficiency is a complex process.

Finally, most MOFs and their composites are synthesized through environmentally harmful methods using organic solvent and high temperature.

5 Conclusions

MOFs possess excellent and specific advantages due to their high surface area, tunable porosity, easy modification with functional groups and integration with different practical systems, which give them a great potential for eliminating noxious pollutants from water and wastewater. The large number of researches published during the last decades confirms this potential. However, there are still some challenges to revealing the comprehensive potential of these amazing materials. First, the removal performance of noxious pollutants should be sufficiently enhanced. Second, the recyclability and shelf life of the utilized MOFs must be carefully assessed to implement these materials truly. Third, environmentally friendly approaches should be clearly established for synthesizing MOFs. And finally, the economic viability and the performance of the techniques containing MOFs must be examined in water purification systems with different hazardous pollutants and real wastewaters.

Acknowledgments

The authors thank the Shahrood University of Technology for their supports.

References

1. Wan L, Wang H. Control of urban river water pollution is studied based on SMS. *Environ Technol Innov*. 2021;22:101468.
2. Bedia J, et al. Metal–organic frameworks for water purification. In: Bonelli B, et al., eds. *Nanomaterials for the Detection and Removal of Wastewater Pollutants*. Elsevier; 2020:241–283. [chapter 9].
3. Karri RR, Ravindran G, Dehghani MH. Wastewater—sources, toxicity, and their consequences to human health. In: *Soft Computing Techniques in Solid Waste and Wastewater Management*. Elsevier; 2021:3–33.
4. Dehghani MH, Omrani GA, Karri RR. Solid waste—sources, toxicity, and their consequences to human health. In: *Soft Computing Techniques in Solid Waste and Wastewater Management*. Elsevier; 2021:205–213.
5. Lee B, et al. Submerged arc plasma system combined with ozone oxidation for the treatment of wastewater containing non-degradable organic compounds. *Front Environ Sci Eng*. 2021;15(5):1–9.
6. Khan FSA, et al. Magnetic nanoparticles incorporation into different substrates for dyes and heavy metals removal—a review. *Environ Sci Pollut Res*. 2020;27(35):43526–43541.
7. Abdi J, et al. State of the art on the ultrasonic-assisted removal of environmental pollutants using metal-organic frameworks. *J Hazard Mater*. 2022;424:127558.
8. Shen L, et al. Highly dispersed palladium nanoparticles anchored on UiO-66 (NH 2) metal-organic framework as a reusable and dual functional visible-light-driven photocatalyst. *Nanoscale*. 2013;5(19):9374–9382.
9. Zhang Y, et al. Electrocatalytically active cobalt-based metal–organic framework with incorporated macroporous carbon composite for electrochemical applications. *J Mater Chem A*. 2015;3(2):732–738.
10. Van Assche TR, et al. Electrochemical synthesis of thin HKUST-1 layers on copper mesh. *Microporous Mesoporous Mater*. 2012;158:209–213.
11. Campagnol N, et al. Luminescent terbium-containing metal–organic framework films: new approaches for the electrochemical synthesis and application as detectors for explosives. *Chem Commun*. 2014;50(83):12545–12547.
12. James SL, et al. Mechanochemistry: opportunities for new and cleaner synthesis. *Chem Soc Rev*. 2012;41(1):413–447.

13. Masoomi MY, Morsali A, Junk PC. Rapid mechanochemical synthesis of two new Cd (II)-based metal–organic frameworks with high removal efficiency of Congo red. *CrystEngComm.* 2015;17(3):686–692.

14. Wang D, et al. A 3D porous metal–organic framework constructed of 1D zigzag and helical chains exhibiting selective anion exchange. *CrystEngComm.* 2010;12(4):1041–1043.

15. Arcís-Castillo Z, et al. [Fe (TPT) 2/3 {MI (CN) 2} 2]· nSolv (MI = Ag, Au): new bimetallic porous coordination polymers with spin-crossover properties. *Chem Eur J.* 2013;19(21):6851–6861.

16. Phang WJ, et al. pH-dependent proton conducting behavior in a metal–organic framework material. *Angew Chem.* 2014;126(32):8523–8527.

17. Sabouni R, Kazemian H, Rohani S. Microwave synthesis of the CPM-5 metal organic framework. *Chem Eng Technol.* 2012;35(6):1085–1092.

18. Jin L-N, Liu Q, Sun W-Y. An introduction to synthesis and application of nanoscale metal–carboxylate coordination polymers. *CrystEngComm.* 2014;16(19):3816–3828.

19. Morsali A, et al. Ultrasonic irradiation assisted syntheses of one-dimensional di (azido)-dipyridylamine Cu (II) coordination polymer nanoparticles. *Ultrason Sonochem.* 2015;23:208–211.

20. Abdi J, et al. Synthesis of metal-organic framework hybrid nanocomposites based on GO and CNT with high adsorption capacity for dye removal. *Chem Eng J.* 2017;326:1145–1158.

21. Abdi J, et al. Synthesis of amine-modified zeolitic imidazolate framework-8, ultrasound-assisted dye removal and modeling. *Ultrason Sonochem.* 2017;39:550–564.

22. Mahmoodi NM, et al. Metal-organic framework (MIL-100 (Fe)): synthesis, detailed photocatalytic dye degradation ability in colored textile wastewater and recycling. *Mater Res Bull.* 2018;100:357–366.

23. Abdi J, et al. Synthesis of porous TiO2/ZrO2 photocatalyst derived from zirconium metal organic framework for degradation of organic pollutants under visible light irradiation. *J Environ Chem Eng.* 2019;7(3):103096.

24. Abdi J, Banisharif F, Khataee A. Amine-functionalized Zr-MOF/CNTs nanocomposite as an efficient and reusable photocatalyst for removing organic contaminants. *J Mol Liq.* 2021;334:116129.

25. Choi K-J, Kim S-G, Kim S-H. Removal of antibiotics by coagulation and granular activated carbon filtration. *J Hazard Mater.* 2008;151(1):38–43.

26. Wen M, et al. Metal–organic framework-based nanomaterials for adsorption and photocatalytic degradation of gaseous pollutants: recent progress and challenges. *Environ Sci Nano.* 2019;6(4):1006–1025.

27. Wen M, et al. Design of single-site photocatalysts by using metal–organic frameworks as a matrix. *Chem Asian J.* 2018;13(14):1767–1779.

28. Zhang A-Y, et al. Degradation of refractory pollutants under solar light irradiation by a robust and self-protected ZnO/CdS/TiO2 hybrid photocatalyst. *Water Res.* 2016;92:78–86.

29. Qiu J, et al. Modified metal-organic frameworks as photocatalysts. *Appl Catal B Environ.* 2018;231:317–342.

30. Yao P, et al. Enhanced visible-light photocatalytic activity to volatile organic compounds degradation and deactivation resistance mechanism of titania confined inside a metal-organic framework. *J Colloid Interface Sci.* 2018;522:174–182.

31. Khan FSA, et al. Magnetic nanoadsorbents' potential route for heavy metals removal—a review. *Environ Sci Pollut Res.* 2020;27(19):24342–24356.

32. Qasem NAA, Mohammed RH, Lawal DU. Removal of heavy metal ions from wastewater: a comprehensive and critical review. *npj Clean Water.* 2021;4(1):36.

33. Chen Y, Bai X, Ye Z. Recent progress in heavy metal ion decontamination based on metal–organic frameworks. *Nanomaterials.* 2020;10(8):1481.

34. Ruthiraan M, et al. An overview of magnetic material: preparation and adsorption removal of heavy metals from wastewater. In: *Magnetic Nanostructures.* Cha Springer; 2019:131–159.

35. Wen J, Fang Y, Zeng G. Progress and prospect of adsorptive removal of heavy metal ions from aqueous solution using metal–organic frameworks: a review of studies from the last decade. *Chemosphere.* 2018;201:627–643.

36. Li X, et al. Sulfur-functionalized metal-organic frameworks: synthesis and applications as advanced adsorbents. *Coord Chem Rev.* 2020;408:213191.

37. Fang Y, et al. From nZVI to SNCs: development of a better material for pollutant removal in water. *Environ Sci Pollut Res.* 2018;25(7):6175–6195.

38. Karri RR, Sahu JN, Meikap BC. Improving efficacy of Cr (VI) adsorption process on sustainable adsorbent derived from waste biomass (sugarcane bagasse) with help of ant colony optimization. *Ind Crop Prod.* 2020;143:111927.

39. Saleem H, Rafique U, Davies RP. Investigations on post-synthetically modified UiO-66-NH2 for the adsorptive removal of heavy metal ions from aqueous solution. *Microporous Mesoporous Mater.* 2016;221:238–244.

40. Zhang J, et al. Exploring a thiol-functionalized MOF for elimination of lead and cadmium from aqueous solution. *J Mol Liq.* 2016;221:43–50.

41. Lau YJ, et al. Removal of dye using peroxidase-immobilized Buckypaper/polyvinyl alcohol membrane in a multi-stage filtration column via RSM and ANFIS. *Environ Sci Pollut Res.* 2020;27(32):40121–40134.

42. Jun LY, et al. Modeling and optimization by particle swarm embedded neural network for adsorption of methylene blue by jicama peroxidase immobilized on buckypaper/polyvinyl alcohol membrane. *Environ Res.* 2020;183:109158.

43. Efome JE, et al. Metal–organic frameworks supported on nanofibers to remove heavy metals. *J Mater Chem A.* 2018;6(10):4550–4555.

44. Gao J, et al. Rapid removal of mercury from water by novel MOF/PP hybrid membrane. *Nanomaterials.* 2021;11(10):2488.

45. Zhang Y, et al. Unveiling the adsorption mechanism of zeolitic imidazolate framework-8 with high efficiency for removal of copper ions from aqueous solutions. *Dalton Trans.* 2016;45(32):12653–12660.

46. Li X, et al. Mechanistic insight into the interaction and adsorption of Cr(VI) with zeolitic imidazolate framework-67 microcrystals from aqueous solution. *Chem Eng J.* 2015;274:238–246.

47. Zheng T-T, et al. A luminescent metal organic framework with high sensitivity for detecting and removing copper ions from simulated biological fluids. *Dalton Trans.* 2017;46(8):2456–2461.

48. Peng Y, et al. A versatile MOF-based trap for heavy metal ion capture and dispersion. *Nat Commun*. 2018;9(1):187.

49. Ding L, et al. Thiol-functionalized Zr-based metal–organic framework for capture of Hg(II) through a proton exchange reaction. *ACS Sustain Chem Eng*. 2018;6(7):8494–8502.

50. Cai J, et al. Selective adsorption of arsenate and the reversible structure transformation of the mesoporous metal–organic framework MIL-100(Fe). *Phys Chem Chem Phys*. 2016;18(16):10864–10867.

51. Wang Y, et al. Preparation of magnetic metal organic frameworks adsorbent modified with mercapto groups for the extraction and analysis of lead in food samples by flame atomic absorption spectrometry. *Food Chem*. 2015;181:191–197.

52. Yuan J, et al. Fabrication of ZIF-300 membrane and its application for efficient removal of heavy metal ions from wastewater. *J Membr Sci*. 2019;572:20–27.

53. Delavar M, Bakeri G, Hosseini M. Fabrication of polycarbonate mixed matrix membranes containing hydrous manganese oxide and alumina nano-particles for heavy metal decontamination: characterization and comparative study. *Chem Eng Res Des*. 2017;120:240–253.

54. Rezaee R, et al. Fabrication and characterization of a polysulfone-graphene oxide nanocomposite membrane for arsenate rejection from water. *J Environ Health Sci Eng*. 2015;13(1):61.

55. Roushani M, Saedi Z, Baghelani YM. Removal of cadmium ions from aqueous solutions using TMU-16-NH2 metal organic framework. *Environ Nanotechnol Monit Manag*. 2017;7:89–96.

56. Wang C, et al. Superior removal of arsenic from water with zirconium metal-organic framework UiO-66. *Sci Rep*. 2015;5(1):16613.

57. Rapti S, et al. Rapid, green and inexpensive synthesis of high quality UiO-66 amino-functionalized materials with exceptional capability for removal of hexavalent chromium from industrial waste. *Inorg Chem Front*. 2016;3(5):635–644.

58. Zhang L, et al. NH2-MIL-53(Al) metal–organic framework as the smart platform for simultaneous high-performance detection and removal of Hg2+. *Inorg Chem*. 2019;58(19):12573–12581.

59. Wong S, et al. Effective removal of anionic textile dyes using adsorbent synthesized from coffee waste. *Sci Rep*. 2020;10(1):1–13.

60. Khan FSA, et al. A comprehensive review on magnetic carbon nanotubes and carbon nanotube-based buckypaper for removal of heavy metals and dyes. *J Hazard Mater*. 2021;413:125375.

61. Uddin MJ, Ampiaw RE, Lee W. Adsorptive removal of dyes from wastewater using a metal-organic framework: a review. *Chemosphere*. 2021;131314.

62. Foo ML, Matsuda R, Kitagawa S. Functional hybrid porous coordination polymers. *Chem Mater*. 2014;26(1):310–322.

63. Yang J-M, et al. Rapid adsorptive removal of cationic and anionic dyes from aqueous solution by a Ce (III)-doped Zr-based metal–organic framework. *Microporous Mesoporous Mater*. 2020;292:109764.

64. Molavi H, et al. Selective dye adsorption by highly water stable metal-organic framework: long term stability analysis in aqueous media. *Appl Surf Sci*. 2018;445:424–436.

65. Zhang J, Li F, Sun Q. Rapid and selective adsorption of cationic dyes by a unique metal-organic framework with decorated pore surface. *Appl Surf Sci*. 2018;440:1219–1226.

66. Chen D, Feng P-f, Wei F-h. Preparation of Fe (III)-MOFs by microwave-assisted ball for efficiently removing organic dyes in aqueous solutions under natural light. *Chem Eng Process Process Intensif*. 2019;135:63–67.

67. Wang K, et al. Rational construction of defects in a metal–organic framework for highly efficient adsorption and separation of dyes. *Chem Eng J*. 2016;289:486–493.

68. Tehrani MS, Zare-Dorabei R. Highly efficient simultaneous ultrasonic-assisted adsorption of methylene blue and rhodamine B onto metal organic framework MIL-68 (Al): central composite design optimization. *RSC Adv*. 2016;6(33):27416–27425.

69. Haque E, Jun JW, Jhung SH. Adsorptive removal of methyl orange and methylene blue from aqueous solution with a metal-organic framework material, iron terephthalate (MOF-235). *J Hazard Mater*. 2011;185(1):507–511.

70. Du J, et al. Mesoporous sulfur-modified iron oxide as an effective Fenton-like catalyst for degradation of bisphenol A. *Appl Catal B Environ*. 2016;184:132–141.

71. Li W, et al. Magnetic porous Fe3O4/carbon octahedra derived from iron-based metal-organic framework as heterogeneous Fenton-like catalyst. *Appl Surf Sci*. 2018;436:252–262.

72. Ting H, et al. High-permeance metal–organic framework-based membrane adsorber for the removal of dye molecules in aqueous phase. *Environ Sci Nano*. 2017;4(11):2205–2214.

73. Jun B-M, et al. Adsorption of selected dyes on Ti3C2Tx MXene and Al-based metal-organic framework. *Ceram Int*. 2020;46(3):2960–2968.

74. Jun B-M, et al. Applications of metal-organic framework based membranes in water purification: a review. *Sep Purif Technol*. 2020;247:116947.

75. Yang M, Bai Q. Flower-like hierarchical Ni-Zn MOF microspheres: efficient adsorbents for dye removal. *Colloids Surf A Physicochem Eng Asp*. 2019;582:123795.

76. Chen N, et al. Effect of structures on the adsorption performance of cobalt metal organic framework obtained by microwave-assisted ball milling. *Chem Phys Lett*. 2018;705:23–30.

77. Zolgharnein J, et al. Application of a new metal-organic framework of [Ni2F2 (4, 4′-bipy) 2 (H2O) 2](VO3) 2.8 H2O as an efficient adsorbent for removal of Congo red dye using experimental design optimization. *Environ Res*. 2020;182:109054.

78. Fu J, et al. A review on anammox process for the treatment of antibiotic-containing wastewater: linking effects with corresponding mechanisms. *Front Environ Sci Eng*. 2021;15(1):1–15.

79. Aslam S, et al. In situ one-step synthesis of Fe3O4@ MIL-100 (Fe) core-shells for adsorption of methylene blue from water. *J Colloid Interface Sci*. 2017;505:186–195.

80. Yang Q, et al. An anionic In (III)-based metal-organic framework with Lewis basic sites for the selective adsorption and separation of organic cationic dyes. *Chin Chem Lett*. 2019;30(1):234–238.

81. Khajeh M, et al. Co-Fe-layered double hydroxide decorated amino-functionalized zirconium terephthalate metal-organic framework for removal of organic dyes from water samples. *Spectrochim Acta A Mol Biomol Spectrosc*. 2020;234:118270.

82. Ahmadijokani F, et al. Superior chemical stability of UiO-66 metal-organic frameworks (MOFs) for selective dye adsorption. *Chem Eng J*. 2020;399:125346.

83. Abdi J, et al. Synthesis of magnetic metal-organic framework nanocomposite (ZIF-8@ SiO2@ MnFe2O4) as a novel adsorbent for selective dye removal from multicomponent systems. *Microporous Mesoporous Mater*. 2019;273:177–188.

84. Hu M-L, Hashemi L, Morsali A. Pore size and interactions effect on removal of dyes with two Lead (II) metal-organic frameworks. *Mater Lett*. 2016;175:1–4.

85. Guesh K, et al. Sustainable preparation of MIL-100 (Fe) and its photocatalytic behavior in the degradation of methyl orange in water. *Cryst Growth Des*. 2017;17(4):1806–1813.

86. Jing H-P, et al. Photocatalytic degradation of methylene blue in ZIF-8. *RSC Adv*. 2014;4(97):54454–54462.

87. Sha Z, et al. Bismuth tungstate incorporated zirconium metal–organic framework composite with enhanced visible-light photocatalytic performance. *RSC Adv*. 2014;4(110):64977–64984.

88. Du J-J, et al. New photocatalysts based on MIL-53 metal–organic frameworks for the decolorization of methylene blue dye. *J Hazard Mater*. 2011;190(1-3):945–951.

89. Gao J, et al. A p-type Ti (IV)-based metal–organic framework with visible-light photo-response. *Chem Commun*. 2014;50(29):3786–3788.

90. Gao Y, et al. Accelerated photocatalytic degradation of organic pollutant over metal-organic framework MIL-53 (Fe) under visible LED light mediated by persulfate. *Appl Catal B Environ*. 2017;202:165–174.

91. Li Y, et al. Interfacial synthesis of ZIF-8 membranes with improved nanofiltration performance. *J Membr Sci*. 2017;523:561–566.

92. Ren Y, et al. MIL-PVDF blend ultrafiltration membranes with ultrahigh MOF loading for simultaneous adsorption and catalytic oxidation of methylene blue. *J Hazard Mater*. 2019;365:312–321.

93. Liu Y, et al. A polydopamine-modified reduced graphene oxide (RGO)/MOFs nanocomposite with fast rejection capacity for organic dye. *Chem Eng J*. 2019;359:47–57.

94. Makhetha T, Moutloali R. Antifouling properties of Cu (tpa)@ GO/PES composite membranes and selective dye rejection. *J Membr Sci*. 2018;554:195–210.

95. Ebele AJ, Abdallah MA-E, Harrad S. Pharmaceuticals and personal care products (PPCPs) in the freshwater aquatic environment. *Emerg Contam*. 2017;3(1):1–16.

96. Boxall AB, et al. Pharmaceuticals and personal care products in the environment: what are the big questions? *Environ Health Perspect*. 2012;120 (9):1221–1229.

97. Price OR, et al. Improving emissions estimates of home and personal care products ingredients for use in EU risk assessments. *Integr Environ Assess Manag*. 2010;6(4):677–684.

98. Wang D, et al. Simultaneously efficient adsorption and photocatalytic degradation of tetracycline by Fe-based MOFs. *J Colloid Interface Sci*. 2018;519:273–284.

99. Li N, et al. Simultaneous removal of tetracycline and oxytetracycline antibiotics from wastewater using a ZIF-8 metal organic-framework. *J Hazard Mater*. 2019;366:563–572.

100. Dehghan A, et al. Tetracycline removal from aqueous solutions using zeolitic imidazolate frameworks with different morphologies: a mathematical modeling. *Chemosphere*. 2019;217:250–260.

101. Cychosz KA, Matzger AJ. Water stability of microporous coordination polymers and the adsorption of pharmaceuticals from water. *Langmuir*. 2010;26(22):17198–17202.

102. Serre C, et al. Very large breathing effect in the first nanoporous chromium (III)-based solids: MIL-53 or CrIII (OH)⊙{O2C− C6H4− CO2}⊙{HO2C− C6H4 − CO2H} x⊙ H2O y. *J Am Chem Soc*. 2002;124(45):13519–13526.

103. Gao Y, et al. Understanding the adsorption of sulfonamide antibiotics on MIL-53s: metal dependence of breathing effect and adsorptive performance in aqueous solution. *J Colloid Interface Sci*. 2019;535:159–168.

104. Lin S, Zhao Y, Yun Y-S. Highly effective removal of nonsteroidal anti-inflammatory pharmaceuticals from water by Zr (IV)-based metal–organic framework: adsorption performance and mechanisms. *ACS Appl Mater Interfaces*. 2018;10(33):28076–28085.

105. Dou R, et al. High efficiency removal of triclosan by structure-directing agent modified mesoporous MIL-53 (Al). *Environ Sci Pollut Res*. 2017;24 (9):8778–8789.

106. Jung BK, et al. Adsorptive removal of p-arsanilic acid from water using mesoporous zeolitic imidazolate framework-8. *Chem Eng J*. 2015;267:9–15.

107. Li B, et al. Defect creation in metal-organic frameworks for rapid and controllable decontamination of roxarsone from aqueous solution. *J Hazard Mater*. 2016;302:57–64.

108. Hasan Z, Choi E-J, Jhung SH. Adsorption of naproxen and clofibric acid over a metal–organic framework MIL-101 functionalized with acidic and basic groups. *Chem Eng J*. 2013;219:537–544.

109. Hasan Z, Khan NA, Jhung SH. Adsorptive removal of diclofenac sodium from water with Zr-based metal–organic frameworks. *Chem Eng J*. 2016;284:1406–1413.

110. Lv Y, et al. Removal of p-arsanilic acid by an amino-functionalized indium-based metal–organic framework: adsorption behavior and synergetic mechanism. *Chem Eng J*. 2018;339:359–368.

111. Seo PW, Khan NA, Jhung SH. Removal of nitroimidazole antibiotics from water by adsorption over metal–organic frameworks modified with urea or melamine. *Chem Eng J*. 2017;315:92–100.

112. Seo PW, et al. Adsorptive removal of pharmaceuticals and personal care products from water with functionalized metal-organic frameworks: remarkable adsorbents with hydrogen-bonding abilities. *Sci Rep*. 2016;6(1):1–11.

113. Liang R, et al. A simple strategy for fabrication of Pd@ MIL-100 (Fe) nanocomposite as a visible-light-driven photocatalyst for the treatment of pharmaceuticals and personal care products (PPCPs). *Appl Catal B Environ*. 2015;176:240–248.

114. Liang R, et al. Facile in situ growth of highly dispersed palladium on phosphotungstic-acid-encapsulated MIL-100 (Fe) for the degradation of pharmaceuticals and personal care products under visible light. *Nano Res*. 2018;11(2):1109–1123.

115. Wang H, et al. In situ synthesis of In2S3@ MIL-125 (Ti) core–shell microparticle for the removal of tetracycline from wastewater by integrated adsorption and visible-light-driven photocatalysis. *Appl Catal B Environ*. 2016;186:19–29.

116. Dong W, et al. Facile synthesis of In2S3/UiO-66 composite with enhanced adsorption performance and photocatalytic activity for the removal of tetracycline under visible light irradiation. *J Colloid Interface Sci*. 2019;535:444–457.

117. Yang C, et al. A novel visible-light-driven In-based MOF/graphene oxide composite photocatalyst with enhanced photocatalytic activity toward the degradation of amoxicillin. *Appl Catal B Environ*. 2017;200:673–680.

118. Basu S, Balakrishnan M. Polyamide thin film composite membranes containing ZIF-8 for the separation of pharmaceutical compounds from aqueous streams. *Sep Purif Technol*. 2017;179:118–125.

119. Zhao Y-Y, et al. Impacts of metal–organic frameworks on structure and performance of polyamide thin-film nanocomposite membranes. *ACS Appl Mater Interfaces*. 2019;11(14):13724–13734.

120. Yadav A, et al. CNT functionalized ZIF-8 impregnated PVDF-co-HFP mixed matrix membranes for antibiotics removal from pharmaceutical industry wastewater by vacuum membrane distillation. *J Environ Chem Eng*. 2021;106560.

121. Kim S, et al. A metal organic framework-ultrafiltration hybrid system for removing selected pharmaceuticals and natural organic matter. *Chem Eng J*. 2020;382:122920.

122. Dehghani MH, et al. Optimizing the removal of organophosphorus pesticide malathion from water using multi-walled carbon nanotubes. *Chem Eng J*. 2017;310:22–32.

123. Dehghani MH, et al. Process optimization and enhancement of pesticide adsorption by porous adsorbents by regression analysis and parametric modelling. *Sci Rep*. 2021;11(1):11719.

124. Jung BK, Hasan Z, Jhung SH. Adsorptive removal of 2, 4-dichlorophenoxyacetic acid (2, 4-D) from water with a metal–organic framework. *Chem Eng J*. 2013;234:99–105.

125. Mirsoleimani-Azizi SM, et al. Diazinon removal from aqueous media by mesoporous MIL-101 (Cr) in a continuous fixed-bed system. *J Environ Chem Eng*. 2018;6(4):4653–4664.

126. Akpinar I, et al. Exploiting π–π interactions to design an efficient sorbent for atrazine removal from water. *ACS Appl Mater Interfaces*. 2019;11(6):6097–6103.

127. Liu G, et al. Adsorption and removal of organophosphorus pesticides from environmental water and soil samples by using magnetic multi-walled carbon nanotubes@ organic framework ZIF-8. *J Mater Sci*. 2018;53(15):10772–10783.

128. Mohaghegh N, et al. Comparative studies on Ag3PO4/BiPO4–metal-organic framework–graphene-based nanocomposites for photocatalysis application. *Appl Surf Sci*. 2015;351:216–224.

129. Xue Y, et al. Efficient degradation of atrazine by BiOBr/UiO-66 composite photocatalyst under visible light irradiation: environmental factors, mechanisms and degradation pathways. *Chemosphere*. 2018;203:497–505.

130. Oladipo AA, Vaziri R, Abureesh MA. Highly robust AgIO3/MIL-53 (Fe) nanohybrid composites for degradation of organophosphorus pesticides in single and binary systems: application of artificial neural networks modelling. *J Taiwan Inst Chem Eng*. 2018;83:133–142.

131. Oladipo AA. MIL-53 (Fe)-based photo-sensitive composite for degradation of organochlorinated herbicide and enhanced reduction of Cr (VI). *Process Saf Environ Prot*. 2018;116:413–423.

132. Manousi N, et al. Extraction of metal ions with metal–organic frameworks. *Molecules*. 2019;24(24):4605.

Chapter 14

Adsorptive removal and concentration of rare-earth elements from aquatic media using various materials: A review

Alexandr Burakov[a], Inderjeet Tyagi[b], Rama Rao Karri[c], Irina Burakova[a], Anastasia Memetova[a], Vladimir Bogoslovskiy[d], Gulnara Shigabaeva[e], and Evgeny Galunin[e,*]

[a]*Department of Technology and Methods of Nanoproducts Manufacturing, Tambov State Technical University, Tambov, Russian Federation,* [b]*Centre for DNA Taxonomy, Molecular Systematics Division, Zoological Survey of India, Ministry of Environment, Forest and Climate Change, Government of India, Kolkata, West Bengal, India,* [c]*Petroleum and Chemical Engineering, Faculty of Engineering, Universiti Teknologi Brunei, Bandar Seri Begawan, Brunei Darussalam,* [d]*Research School of Chemistry & Applied Biomedical Sciences, Tomsk Polytechnic University, Tomsk, Russian Federation,* [e]*Department of Organic and Ecological Chemistry, University of Tyumen, Tyumen, Russian Federation*
*Corresponding author.

1 Introduction

Developing an efficient system for the safe disposal and control of radioactive waste (RW) appears to be the most important issue for the successful functioning of nuclear power engineering and industry. A multi-barrier system can provide underground storage and isolation of RW to protect RW disposal repositories (RWDR), which includes artificial engineered barriers and the surrounding natural geological environment—rocks that meet certain requirements for the placement of underground disposals.

One of the priority directions for increasing the efficiency of RW underground storage is the study of changes in the primary properties of reinforced concrete containers located in deep RW repositories under the influence of an aggressive aqueous medium that destroys the original structure of concrete with the formation of new mineral components, and the accumulation of these new formations in the pores and capillaries of the primary structure. Deep RWDRs are conventionally located in special soil zones representing a combination of clay layers characterized by extremely low water permeability. However, the voids and pores of clay rocks are filled with free and bound water. Thus, despite the rather low culvert capacity of clays, their pores contain a certain amount of moisture which will inevitably interact with the engineered barriers of the RWDRs. As a result of this interaction, the penetration of water into the RW repository is inevitable. It should be noted that the corrosion of concrete (especially concrete, from which RW containers are made) is extremely slow. In the case of water penetration into the RWDR, the chemical composition of the water cannot be predicted with certainty. However, the most dangerous case is acidic aqueous systems. Under the influence of acidic solutions, significant degradation of the concrete structure of the engineered barrier occurs, which is associated, first of all, with the predominant transition of calcium into the solution. The loss of calcium is accompanied by the mechanical destruction of concrete with the formation of a network of cracks, the disintegration of the monolithic cement mass, and the fallout of filler particles from the samples. Consequently, in an acidic aqueous medium, simultaneously with the loss of calcium, there is a decrease in the concrete strength properties in protective containers located in the deep RWDRs. The acidic medium of aqueous underground solutions has a negative impact on the safety of the concrete engineered barriers and the conditions for the migration of radionuclides.

In the present work, the authors studied the extraction of lanthanide group rare-earth elements (REEs), which represent a chemical analogue of the family of radioactive actinides, from model aqueous solutions, as well as sulfuric-chloride solutions, including those simulating leaching solutions, particularly proceeding from uranium ores.

Sustainable Materials for Sensing and Remediation of Noxious Pollutants. https://doi.org/10.1016/B978-0-323-99425-5.00011-6

Studies on the adsorption of REEs, which can be used to solve issues associated with their production technology, ecology, environmental protection, prevention, and treatment of diseases, etc., are of special interest. For instance, the purification of wastewater from REE ions is relevant both to exclude pollution of water bodies and to utilize scarce raw materials.

The ion exchange sorption method is one of the widespread REE extraction techniques. Adsorption seems to be the most attractive option due to the non-toxicity of materials and reagents used, reuse possibility, an abundance of various sorption materials, and is intensively used for the REE extraction and concentration.

Unfortunately, in the existing literature, the data previously reported on the removal of REEs from aquatic media using novel sorbent materials (especially those based on carbon nanostructures developed) are not completely systematized, and the prospects of using such materials are not properly outlined. In this regard, the present review tends to overcome this problem by providing summarized updated information on the existing methods, parameters, and conditions of the REE sorption, with an emphasis on employing carbon nanomaterials such as carbon nanotubes, graphene, graphene oxide, etc., and on estimating the prospects for their use in this direction.

2 Rare-earth elements and the ways of their removal from aquatic media

The group of REEs encompasses 17 elements, including scandium, yttrium, lanthanum, cerium, samarium, praseodymium, neodymium, promethium, europium, gadolinium, terbium, dysprosium, holmium, erbium, ytterbium, thulium, and lutetium.[1] The family of lanthanides (4f-elements) is a chemical analog of the family of actinides (5f-elements). The proximity to the lanthanides plays an important role in the geochemistry of uranium and thorium. In terms of ionic radii, U^{4+} is closer to the group of "heavy" lanthanides ("yttrium group"), whereas Th^{4+} belongs to the "light" group ("cerium group"). Due to the similarity of the actinides and lanthanides, all the rare-earth minerals contain variable amounts of uranium and thorium in the form of an isomorphic impurity. At the same time, U prevails in "yttrium" minerals, and Th in "cerium" ones.[2]

The REEs find their wide application in different areas such as chemical engineering, medicine, electronics, computer manufacturing, metallurgy, and nuclear energy. A variety of processed raw materials and the complexity of production processes determine a large yield of technological water and wastewater and a high degree of their contamination with toxic REE ions. In this regard, their effective extraction from contaminated aquatic media is an important problem requiring appropriate attention (Fig. 1).

There are many ways to extract the REEs, such as solvent extraction, filtration, precipitation, etc.; however, they do not seem economically viable. In this regard, adsorption has attracted wide consideration due to its simpleness, high effectiveness, and low cost of materials (adsorbents) used. This report will provide an overview of the latest literature (mainly for 2015–20) on the REE removal from aquatic media using various adsorbents and consider the prospects for employing nanomaterials such as carbon nanotubes (CNTs) and graphene. The REE adsorption process will be discussed, taking into account the model approach under equilibrium conditions (kinetics, isotherms), thermodynamics, and other factors (e.g., aqueous solution pH, adsorbent dose, contact time, and temperature).

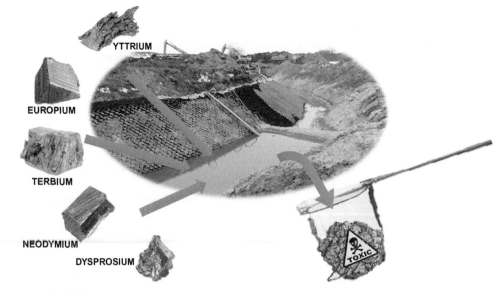

FIG. 1 Toxic paragenesis of REEs in aquatic media.

3 Relevance and expediency of the adsorptive removal and concentration of REEs

Despite their title, the REEs are quite common in the Earth's crust; however, capital costs for their extraction are relatively high (about $ 40,000 per ton—according to Roskill Information Services). The United States, China, and Australia are the leading REE manufacturers.[3,4] Due to their unique physical-chemical characteristics, the REEs are used in different fields: in the production of high-energy permanent magnets, modern structural materials, optics and glass, radio electronics, nuclear technology, mechanical engineering, chemical, petrochemical, and glass industries. The automotive industry consumes thousands of tons of REE every year, and modern military technologies cannot do without them. Many green innovations also rely on these elements, including wind turbines, energy-efficient light bulbs, and hybrid car engines.[4,5]

Although primary deposits containing bastnaesite [$(La,Ce,Y)CO_3F$], monazite [$(Ce,La,Nd,Th)PO_4$], and xenotime [YPO_4] are the main REE sources, they can also be found in abundance in secondary resources such as mine byproducts, household and industrial wastes.

It should be noted that nuclear waste contains different toxic contaminants, and radioactive elements including trivalent actinides.[6] Thus, its disposal is of great significance from an environmental point of view. The trivalent actinides are highly radiotoxic and therefore difficult to use under laboratory conditions. In this regard, the REEs (especially lanthanides like europium and neodymium) can be employed as homologs due to the simplicity in their physical-chemical characteristics.[7]

Recently, due to the constantly growing demand for high-purity REEs and their compounds, the production of pure REEs has attracted special attention.[8,9] There are many methods for separating and concentrating metals, such as chemical precipitation, ion exchange, liquid-liquid extraction, and adsorption.[10] Among them, adsorption has been found as the most promising method due to the ease of execution, high efficiency, and availability of materials (adsorbents).[11-27]

There exist various adsorbents (both native and modified) that are used to remove the REEs from aqueous solutions, such as silica gel particles modified with diglycolamic acid groups,[12] magnetic chitosan nanoparticles functionalized with cysteine,[14] modified red clays,[15] granular hybrid,[28] carbonized nanocarbon shells of polydopamine,[29] bottom sediments of water bodies,[30] biopolymer microcapsules containing the extractant di-(2-ethylhexyl) phosphoric acid/trioctylphosphine oxide,[31] and biosorbents.[32]

Besides, there is currently a tendency to use nanomaterials as adsorbents (including the purification of wastewater from the REEs) in "single" and "hybrid" systems.[33] Some of the most "popular" and perspective nanomaterials are CNTs, graphene, and graphene oxide (GO), and they possess unique physical-chemical and mechanical properties: they are distinguished by high strength, conductivity, and thermal stability.[34-36] Furthermore, these materials' great specific surface area is associated with various types of intermolecular interactions, thereby promoting their efficient use in many adsorption systems. The only obstacle to their full employment as adsorbents is their high price. Nevertheless, with the transition to their commercial, large-scale production and with the further development of nanotechnology, the costs of such nanomaterials will go down, and they will become more and more affordable.

4 Modeling the equilibrium state of adsorption systems

To understand the REE adsorption from aquatic media and to elucidate all the occurring mechanisms, it is very important to simulate the equilibrium state of adsorption systems using the equilibrium, kinetic, desorption and thermodynamic approaches (Table 1).[38]

Adsorption isotherm equations (models) are widely employed to receive information on the qualitative and quantitative adsorbent-adsorbate interactions.[39] In most studies, it was observed that the Langmuir and Freundlich models are better suitable compared to many isotherm models with two, three, four, and five parameters.[40] The Langmuir isotherm presumes the same affinity of all binding adsorbent surface sites for the adsorbate, together with forming a monolayer of adsorbed molecules (adsorbent saturation) on the surface. The Freundlich isotherm explains adsorption on heterogeneous surfaces, the sites of which possess different energies (affinities); using this model, it is assumed that the adsorbent is not saturated.

For isotherms and kinetic studies: q_e (mg/g) and C_e (mg/L) are the amount of the metal adsorbed under equilibrium and the equilibrium metal concentrations in the liquid phase, respectively, T is the absolute temperature (K), and R is the gas constant (8.314 J/mol K).

Isotherms: Langmuir parameters—q_m (mg/g, maximum adsorption capacity), b_L (L/mol, associated with the adsorption energy and equilibrium constant). Freundlich parameters—K_F (adsorption capacity constant) and n (associated with the intensity of adsorption or the degree of adsorption favorableness).

Kinetics: The amount of the metal adsorbed at any time, q_t (mg/g, amount adsorbed over time t, min). Pseudo-first-order—k_1 (min^{-1}, rate constant) and pseudo-second-order—k_2 (min^{-1}, rate constant).

TABLE 1 Isotherm and kinetic models, thermodynamic equations.[37]

Model	Equation	Graph	Parameters
Non-linear Langmuir sorption	$q_e = q_m \frac{b_L C_e}{1 + b_L C_e}$	–	q_m & b_L
Linear Langmuir sorption-1	$\frac{C_e}{q_e} = \frac{1}{q_m} C_e + \frac{1}{b_L q_m}$	$\frac{C_e}{q_e}$ vs C_e	
Linear Langmuir sorption-2	$\frac{1}{q_e} = \left(\frac{1}{b_L q_m}\right)\frac{1}{C_e} + \frac{1}{q_m}$	$\frac{1}{q_e}$ vs $\frac{1}{C_e}$	
Linear Langmuir sorption-3	$q_e = q_m - \left(\frac{1}{b_L}\right)\frac{q_e}{C_e}$	q_e vs $\frac{q_e}{C_e}$	
Linear Langmuir sorption-4	$\frac{q_e}{C_e} = b_L q_m - b_L q_e$	$\frac{q_e}{C_e}$ vs q_e	
Non-linear Freundlich sorption	$q_e = K_F C_e^{\frac{1}{n}}$	–	K_F, n
Linear Freundlich sorption	$\ln q_e = \ln K_F + \frac{1}{n} \ln C_e$	$\ln q_e$ vs $\ln C_e$	
Non-linear pseudo-first-order kinetics	$q_t = q_e(1 - \exp^{-k_1 t})$	–	q_t, k_1
Linear pseudo-first-order kinetics	$\ln(q_e - q_t) = \ln q_e - k_1$	$\ln(q_e - q_t)$ vs t	
Non-linear pseudo-second-order kinetics	$q_t = \frac{k_2 q_e^2 t}{1 + k_2 q_e t}$	–	q_t, k_2
Linear pseudo-second-order kinetics	$\frac{t}{q_t} = \frac{1}{k_2 q_e^2} + \frac{1}{q_e} t$	$\frac{t}{q_t}$ vs t	
Gibbs thermodynamics	$\Delta G^0 = -RT \ln b_L$ $\Delta G^0 = -RT \ln\left(\frac{q_e}{C_e}\right)$ $\Delta G^0 = -RT \ln K$ $\Delta G^0 = -RT \ln K_0$	–	Free energy change
Van't Hoff thermodynamics	$\ln(b_L) = -\frac{\Delta H^0}{RT} + \frac{\Delta S^0}{R}$ $\ln\left(\frac{q_e}{C_e}\right) = -\frac{\Delta H^0}{RT} + \frac{\Delta S^0}{R}$ $\ln(K) = -\frac{\Delta H^0}{RT} + \frac{\Delta S^0}{R}$ $\ln(K_0) = -\frac{\Delta H^0}{RT} + \frac{\Delta S^0}{R}$	$\ln(b_L) =$ vs $\frac{1}{T}$ $\ln\left(\frac{q_e}{C_e}\right) =$ vs $\frac{1}{T}$ $\ln(K) =$ vs $\frac{1}{T}$ $\ln(K_0) =$ vs $\frac{1}{T}$	Entropy change Enthalpy change
Clapeyron thermodynamics	$\Delta H^0 = \frac{RT_1 T_2}{T_1 - T_2}\left(\frac{\ln C_{e1}}{\ln C_{e2}}\right)$	–	Enthalpy change

A kinetic study is a useful tool in determining optimum adsorption conditions.[41] The kinetic modeling elucidates adsorption mechanisms and stages limiting the rate of the process, such as mass transfer or chemical reactions.[41,42] The most generic kinetic models are pseudo-first and second-order models. It should be noted that they encompass all adsorption stages (i.e., film diffusion, adsorption, and intraparticle diffusion). However, they cannot qualitatively determine adsorption mechanisms.[43] Therefore, diffusion models such as the Weber-Morris (intraparticle diffusion)[44] and Boyd[45] models are used to interpret kinetic data.

The desorption study is required to assess the adsorbent regenerability and provide information on the adsorption mechanism. Distilled water usually desorbs the REEs not strongly attached to the adsorbent surface (i.e., they are physically adsorbed) and, therefore, it is easily removed into the liquid phase.[46] As a rule, at low pH values, the REEs represent ions with a positive (+ve) charge, whereas, at higher pH values, they begin to precipitate as hydroxides (depending on the REE nature, metal-$(OH)_3$ bonds are formed).[7] Since the REEs possess a +ve charge at low pH values, the presence of HNO_3 and HCl increases the proton concentration. Hydrogen ions can easily extract REE cations, thereby leading to the desorption of the adsorbed metals via the ion exchange mechanism.[47–50] EDTA is a commonly used chelator able to selectively extract the REE adsorbed on the adsorbent surface through complexation.[51,52]

The thermodynamic parameters represent the changes in free energy (ΔG^0), enthalpy (ΔH°), and entropy (ΔS°), and provide additional information about internal energy changes associated with the adsorption process.[53]

5 REE adsorption on various materials

The most suitable models of adsorption isotherms and kinetics and maximum adsorption capacities for a specific REE ion regarding various adsorbents are presented in Table 2.

6 Nd^{3+}, Dy^{3+}, and Ce^{3+} adsorption

The adsorption capacity of granular-grafted hydrogel composites, grapefruit peel, modified biohydrogels, and plant and animal origin materials toward Ce^{3+} was studied by various research teams.[54–56] Batch tests were carried out, and q_m was achieved at pH 4.0–7.0,[28] 5.0[54,55], and 6.0.[55,56] Adsorption equilibrium was reached within 40,[28] 60,[54,56] 80,[56] 240,[55] and 360[55] min. Kinetic tests indicate that the intraparticle and film diffusion affects the adsorption process.[56] In the case of the granular-grafted hydrogel composite, the average desorption efficiency was ~95%, even after five cycles, thereby indicating a good reusing possibility for this material.[28]

Oxidized multiwalled carbon nanotubes (MWCNTs) were used to extract Dy^{3+} from aquatic media.[57] The q_{max} was found at pH of 5.0–6.0. The average ΔG^0 was found to be 15.08 kJ/mol, demonstrating that the Dy^{3+} adsorption proceeds through the ion exchange chemical reaction mechanism. Desorption experiments were carried out by varying the solution pH, and at 1.5, it was found that the ion extraction from the material is about 70%, whereas, at pH of 5.0, desorption is not observed.

Magnetic nanohydroxyapatite was used as a new adsorbent for the removal of Nd^{3+}.[73] The ion exchange and chemical adsorption became the main mechanisms of the process and optimum adsorption at the pH of 5.0. The materials were regenerated with HCl and NaOH solutions. It was found that these solutions provide the highest desorption percentage (98%) at a concentration of 0.5 mol/L. After three adsorption-desorption cycles, no significant changes in the adsorption of Nd^{3+} took place.

The Nd^{3+} adsorption on a silicon dioxide/urea-formaldehyde (SiO_2/UF) composite material impregnated with an organophosphorus extractant.[67] The equilibrium in the system was reached within 120 min, and it was also found that the adsorption increases when increasing the pH from 1.0 to 6.0, and an increase in the temperature from 25 to 50 °C leads to an improvement in the removal capacity regarding this pollutant.

Calcium alginate (ALG) and hybrid ALG polyglutamic acid gels (ALG-PGA) were sufficiently investigated for their ability to remove Nd^{3+}.[68] The q_{max} of ALG and ALG-PGA of 194.73 and 238 mg/g, respectively, indicates that the ALG-PGA has a better efficiency. Desorption studies conducted using 0.01–2 mol/L HCl, and according to the results obtained, it turned out that the concentration of 0.1 mol/L of this acid is enough to desorb 99% of Nd^{3+} for 20 min. After eight adsorption/desorption cycles, no surface damage to the ALG-PGA gel was observed, thereby confirming its excellent reusability. Alginate tetramethyl orthosilicate (TMOS) and alginate-silicon dioxide (silica) M600 microspheres[78] were used to remove Nd^{3+} from aqueous solutions. Maximum adsorption was observed at pH of 5.0–6.0, and the sequence of the materials regarding their Nd^{3+} adsorption capacity was as follows: alginate-silica M600 (wet) > alginate-silicon dioxide M600 (dry) > alginate-TMOS. It is noted that in this adsorbent, a higher weight ratio is achieved by using the silica matrix material. Based on the alginate material proportion, it was found that the relative amount of adsorbed Nd^{3+} is higher for alginate-TMOS, probably due to its higher specific surface area (216.1 m^2/g) compared to the surface of alginate-silicon dioxide M600 (4.6 m^2/g). Besides, compared to the wet microspheres, the air-dried alginate-silica M600 microspheres possessed a lower q_{max}, due to the compression of the micropores during the drying process, which leads to a decrease in porosity and, therefore, adsorption itself.

7 La^{3+} adsorption

Granados-Correa et al.[66] investigated adsorption of La^{3+} by hydroxyapatite and found that the adsorption is fast, and equilibrium is reached within 20–30 min. Isotherm studies showed that the adsorption is cooperative, and the adsorbent can be reused. Biosorbents of animal (fish scales, crab shell, shrimp shell, eggshells) and vegetable (orange peel, corn grains, neem sawdust, pineapple greens (crown)) origin were also used to extract La^{3+}.[67,68] The q_{max} was determined under the optimum conditions: pH of 6.0, the temperature of 50 °C, and contact time of 4 h (animal materials) and 3 h (plant materials). FT-IR analysis showed that functional groups such as amines, amide, and alkynes were involved in the adsorption process. To remove La^{3+} from the aquatic medium, grapefruit peel was also used.[54] In this process, pH played an important role, and optimum adsorption was achieved at a pH of 5.0. The FT-IR spectra revealed that the removal process involves the carboxyl and hydroxyl functional groups.

TABLE 2 Isotherm and kinetic models of the REE adsorption on various materials.

Adsorbent	REE	Isotherm model	Kinetic model	Maximum adsorption capacity	Refs.
EDTA-β-cyclodextrin	Ce^{3+}	L	Ps2	50.16	5
Granular-grafted hydrogel composites			Ps2	169.49–243.9	28
Grapefruit peel			Ps2	159.3	54
Orange peel			–	71.4	55
Fish scales			–	200	55
Shrimp shell		F	–	1000	55
Eggshell		L	–	166.6	55
Corn kernels			–	250	55
Pineapple greens (crown)			–	142.8	55
Crab shell			–	90.9	55
Neem sawdust			–	200	55
Sporopollenin-modified biohydrogel			Ps2	333.3	56
Xylan-modified biohydrogel		F	Ps2	200	56
Oxidized MWCNTs	Dy^{3+}	F	Ps2	78.12	57
EDTA-β-cyclodextrin	Eu^{3+}	L	Ps2	55.62	5
Activated carbon			E	86	58
Malt roots			Ps2	156	58
Fresh cactus fibers			Ps1	0.16	59
Modified cactus fibers (MnO_2 coated)			Ps1	0.46	59
Modified cactus fibers (phosphorylated)			Ps1	0.045	59
Chitosan nanoparticles			Ps2	114.9	60
Crab shells			Ps2	3.238	60
Sulfonated GO			Ps2	125	61
GO			Ps2	142.8	61
Ethylenediamine propyl salicylaldimine-functionalized SBA-15 mesoporous silica			Ps2	15.6	62
N-Propyl salicylaldimine-functionalized SBA-15 mesoporous silica		L, F	Ps2	5.1	62
Fe_3O_4 @ cyclodextrin magnetic composite (pH of 3.5)		L	Ps2	$5.03*10^{-5}$ (mol/g)	63
Fe_3O_4@ cyclodextrin magnetic composite (pH of 5.0)		L	Ps2	$8.35*10^{-5}$ (mol/g)	63
SiO_2/UF composite material		–	–	0.23	64
SiO_2		–	–	0.12	64
Organophosphate extractant-impregnated SiO_2/UF		–	Ps2	3.1	64
Bones powder		L	–	12.7	65
Hydroxyapatite		F	Ps2	0.25	66

TABLE 2 Isotherm and kinetic models of the REE adsorption on various materials—cont'd

Adsorbent	REE	Isotherm model	Kinetic model	Maximum adsorption capacity	Refs.
Hydroxyapatite	La^{3+}	F	Ps2	0.94	66
Crab shell			–	90.9	67
Shrimp shell			–	200	67
Eggshell			–	100	67
Corn kernels			–	76.9	67
Pineapple greens (crown)			–	100	67
Orange peel			–	125	67
Neem sawdust			Ps1	166.6	68
Fish scales		L	Ps1	250	68
Grapefruit peel			Ps2	171.2	54
Pleurotus ostreatus mushrooms			–	54.54	69
Oxidized MWCNTs		F	Ps2	99.01	57
Stichococcus bacillaris		L	Ps2	51.02	70
Desmodesmus multivariabilis			Ps2	100	70
Chlorella vulgaris			Ps2	74.6	70
Scenedes musacuminatus			Ps2	111.1	70
Chloroidium saccharophilum			Ps2	129.87	70
Chlamydomonas reinhardtii			Ps2	142.86	70
EDTA-β-cyclodextrin			Ps2	47.78	5
Granular-grafted hydrogel composite			Ps2	256.41–333.33	28
Bones powder			–	8.7	65
Surface-modified mesoporous nanoparticles (one-step method) (MNSP-N-2)	Gd^{3+}	L	Ps2	85.38	71
Surface-modified mesoporous nanoparticles (two-step method) (MNSP-N-1)			Ps2	56.22	71
SiO_2	Nd^{3+}	–	–	0.1	64
Calcium alginate		L	Ps2	194.73	72
Calcium alginate-polyglutamic acid hybrid gels		L	Ps2	238	72
SiO_2/UF composite material		–	–	0.18	64
Organophosphate extractant-impregnated SiO_2/UF		–	Ps2	2.8	64
Bones powder		L	–	10.9	65
Magnetic nanohydroxyapatite		L	Ps2	323	73
Magnetic nanohydroxyapatite	Sm^{3+}	L	Ps2	370	73
Activated biochar from cactus fiber (pH 3.0)			Ps1	90	74
Activated biochar from cactus fiber (pH 6.5)			Ps1	350	74

Continued

TABLE 2 Isotherm and kinetic models of the REE adsorption on various materials—cont'd

Adsorbent	REE	Isotherm model	Kinetic model	Maximum adsorption capacity	Refs.
Sargassum wightii (brown algae) in native form	Pr^{3+}	L, F, S	Ps1	131.4	75
Polysulfone-immobilized *Sargassum wightii*			Ps2	111.2	75
Turbinaria conoides (brown algae) in native form			Ps1	146.4	75
Polysulfone-immobilized *Turbinaria conoides*			Ps2	119.5	75
Orange peel		L	Ps2	58.8	56
Crab shell		L	Ps1	66.6	56
Lysine-modified mesoporous material (Fmoc-SBA-15)	Sc^{3+}	F	Ps2	30.51	76
NaOH-modified *Pleurotus ostreatus* fungi	Y^{3+}	L	Ps2	45.45	77

L, Langmuir; *F*, Freundlich; *RP*, Redlich-Peterson; *S*, Sips; *J*, Jovanovic; *T*, Thoth; *K*, Khan; *RPZ*, Radke-Prausnitz; *FS*, Fritz-Schlander; *H*, Hill; *KC*, Coble-Corrigan; *B*, Baudu; *Ps1*, Pseudo-first-order; *Ps2*, Pseudo-second-order kinetic model; *E*, Elovich.

The La^{3+} extraction with freshwater algae (*Desmodesmus multivariabilis, Chloroidium saccharophilum, Stichococcus bacillaris, Scenedesmus acuminatus, Chlorella vulgaris*, and *Chlamydomonas reinhardtii*) was studied.[70] It was determined that *Desmodesmus multivariabilis* possesses a high q_{max} and a high affinity for the adsorbate. In the case of this biomaterial, under the action of 0.1 mol/L HNO_3, 99.63% of La^{3+} was desorbed, whereas, in the case of *Stichococcus bacillaris*, the desorption degree was minimum.

Oxidized MWCNTs were considered to extract La^{3+} from aqueous solutions.[57] The q_{max} was found at pH 5.0–6.0, and the optimum adsorbent dose was 1 g/L. Desorption studies were carried out at varying pH values, and the maximum desorption degree ($\sim 65\%$), was obtained at a pH of 1.5; however, no metal recovery from the adsorbent was observed at pH $= 5.0$.

Batch adsorption tests were described to study the La^{3+} extraction on granular-grafted hydrogel composites.[28] The q_{max} was observed at the ratio of 20:1 and pH at 3.0–7.0. Kinetic studies demonstrated that 40 min is sufficient to achieve equilibrium. The average desorption efficiency was $\sim 95\%$ up to five cycles using 0.5 mol/L HCl as eluent. Moreover, the authors noted that the adsorption capacity increases after the second and third cycles, which indicates that new active adsorption sites can be formed during the regeneration. After the third cycle, the adsorption capacity decreases slightly.

Pleurotus ostreatus fungi was also used as an adsorbent to remove La^{3+} from aquatic media.[69] The optimum conditions observed in this process at a temperature of 40 °C were as follows: adsorbent dose of 0.5 g/L and pH of 6.8. At 0.1 mol/L HNO_3, the desorption rate found for La^{3+} was 96.89%.

8 Gd^{3+}, Sm^{3+}, Y^{3+}, Pr^{3+}, and Sc^{3+} adsorption

Magnetic nanohydroxyapatite was studied for the removal of Sm^{3+} from aqueous solutions.[73] It was found that this metal is optimally extracted at pH 5.5, the equilibrium in the system is achieved within 150 min, and the average ΔG^0 is 13.6 kJ/mol, thereby indicating the chemical nature of the adsorption. In addition, it was noted that an increase in the temperature from 20 to 50 °C leads to an increase in the Sm^{3+} removal degree.[73] Activated biochar obtained from cactus fibers was used as an adsorbent for Sm^{3+}.[79] The optimum recovery was found at a pH of 7.0. FT-IR spectra showed the formation of inner-sphere complexes with Sm^{3+} on the biochar surface.[74]

The adsorption of Gd^{3+} using mesoporous silica-based hybrid materials was studied by Zheng et al.[71] It was found that the adsorption occurs quickly within the first 60 min, and the equilibrium is achieved in 2–4 h. The material was synthesized using modified hybrid material, MSNP-N-2 (single-step method), and showed the highest q_{max} among the adsorbents tested. Adsorbents showed good efficiency up to five adsorption/desorption cycles. This made it possible to assume that MSNP-N-2 could become a promising candidate due to its high chemical stability.

Several researchers studied the Pr^{3+} removal from aqueous solutions by green algae,[79] brown algae (in native and poly-sulfone matrix-immobilized forms),[75] and other biological wastes (orange peel and crab shell).[56] The FT-IR spectra revealed the presence of alcohol and ketone functional groups in the orange peel and crab shell, respectively, which indi-cates convincing evidence of the participation of these groups in the Pr^{3+} adsorption.[56] As for the contact time, it was found that to achieve equilibrium, 240 min is sufficient in the case of the immobilized algae and 90 min in the case of the native algae.[75] The desorption percentage was 99.1% using 0.1 mol/L HCl for all the algae (native and immobilized), thereby demonstrating good recyclability of these adsorbents. However, upon exposure to the acid, biomass losses were observed for the native algae, whereas the immobilized algae remained stable even after 10 cycles (weight losses—3.7%).[75] The other biomaterials (orange peel and crab shell) were successfully reused up to seven cycles.[56]

Ma et al. synthesized a new lysine-functionalized mesoporous material (Fmoc-SBA-15), and its adsorption capacity was tested for Sc^{3+}.[76] The following characteristics were determined: BET-specific surface area—223 m^2/g, pore diameter—58.2 Å, and total pore volume—0.35 cm^3/g. It was also revealed that the Sc^{3+} adsorption increases during the first 10 min until equilibrium is reached.

Hussien et al. investigated NaOH-modified *Pleurotus ostreatus* fungi for the Y^{3+} adsorption.[77] The optimal process parameters were found at a temperature of 50 °C, pH of 7.0, contact time of 30 min, and stirring speed of 175 rpm, which resulted in high desorption of 94.89%.

9 Eu^{3+} adsorption

The Eu^{3+} removal using hydroxyapatite was investigated by Granados-Correa et al.[66] The process equilibrium was reached within 20–30 min, and it was found that the intraparticle diffusion limits the adsorption. The kinetic and isothermal data determined that the Eu^{3+} adsorption is cooperative, and the adsorbent can be reused.

Used malt roots found their application as a promising adsorbent for removing Eu^{3+}.[58] The q_{max} was observed at pH of 4.5; however, this process was not investigated at pH > 4.5 to avoid precipitation of the metal as $Eu(OH)_3$. It was found that Eu^{3+} is extracted from aqueous solutions rather quickly, and 60 min time is enough to reach equilibrium. The value of the average ΔG^0, determined according to the DRIM, was in the range of 8–16 kJ/mol, thereby indicating the chemical nature of the process (i.e., chemisorption).

Anagnostopoulos and Symeopoulos studied the ability of untreated and treated cactus fibers (MnO_2-coated and phos-phorylated) to adsorb Eu^{3+} from aquatic media.[59] The untreated material showed q_{max} at a pH of 4.0, whereas the modified ones possessed the best capacity at a pH of 6.0. The highest q_{max} resulted for the MnO_2 coated cactus fibers, followed by the untreated and phosphorylated ones. The average ΔG^0 estimated by the DRIM was 11.5, 14.7, and 17.7 kJ/mol, respectively.

Chitosan nanoparticles and crab shell particles were also considered as adsorbents for extracting Eu^{3+} from water.[60] The adsorption proceeded rapidly within the first 15 min, and the equilibrium reached within an hour. Kinetic studies indicated that the intraparticle diffusion is not the only stage limiting the rate of the process. It was also found that pH affects the q_{max} of the materials, and the optimum adsorption occurs at pH 3.0. Moreover, it turned out that the q_{max} of the chitosan nano-particles is much higher than that of the crab shell particles, which indicates the suitability and superiority of the former for the adsorption process.

Yao et al. investigated sulfonated and native GO for adsorption of Eu^{3+}.[61] The results are provided in Table 3. The native GO demonstrated a better adsorption capacity due to the presence of a larger number of oxygen functional groups. The isoelectric point was determined to be 2.06 (for the native GO) and 1.87 (for the sulfonated GO). It was found that the Eu^{3+} adsorption on the native and sulfonated GO's is explained by the formation of two inner-sphere surface complexes, $SO Eu^{2+}$ и $(SO)_2 Eu(OH)^{2-}$.

Using Fe_3O_4 and a magnetic Fe_3O_4/cyclodextrin composite to remove Eu^{3+} was investigated by Guo et al.[63] The com-posite showed a higher adsorption capacity than Fe_3O_4. At low pH values, the adsorption mechanism represented intra-sphere surface complexation, whereas, at higher pH values, it combined the precipitation and intrasphere complexation. Based on the kinetic data, it was found that 180 min is sufficient to reach equilibrium.

Dolatyari et al. investigated that the mesoporous silica SBA-15 functionalized with *N*-propyl salicylaldimine (SBA/SA) and ethylenediamine propyl salicylaldimine (SBA/EnSA) was employed as adsorbents regarding Eu^{3+}.[62] Batch experi-ments were carried out under equilibrium conditions, and the results showed that optimum adsorption occurs at pH of 4.0. It was established that an increase in the ionic strength of an aqueous solution (0–1 mol/L KNO_3) does not affect the Eu^{3+} adsorption capacity. Considering the above and the strong influence of pH on the efficiency of the adsorption process, it was assumed that the dominant mechanism of the process represents chemical inner-sphere complexation. The removal efficiency was found to be significant up to nine cycles using 0.1 mol/L HNO_3.

TABLE 3 Thermodynamic parameters of the REE adsorption on various materials.

Adsorbent	REE	T	ΔG^0 (kJ/mol)	ΔH^0 (kJ/mol)	ΔS^0 (kJ/mol)	Refs.
Sporopollenin-modified biohydrogel	Ce^{3+}	293	−0.1	17.48	0.060	56
		303	−0.7			
		313	−1.3			
Xylan-modified biohydrogel		293	−0.93	18.72	0.067	56
		303	−1.58			
		313	−2.25			
Grapefruit peel		293	−19.32	39.09	0.199	54
		303	−21.32			
		313	−23.31			
		323	−25.3			
Unprocessed cactus fibers	Eu^{3+}	283	−28.5	39.7	0.237	59
		398	−31.1			
		313	−32			
		323	−33			
		333	−34.4			
Modified cactus fibers (phosphorylated)		283	−31.8	80.9	0.397	59
		398	−38.2			
		313	−40.5			
		323	−42.5			
		333	−44.5			
Modified cactus fibers (MnO_2-coated)		283	−22.7	73.6	0.342	59
		398	−29.5			
		313	−32.3			
		323	−35.4			
		333	−38.4			
Mesoporous silica SBA-15, functionalized N-propylsalicylaldimine		298	−6.5	40.4	0.157	62
Ethylenediamine propyl salicylaldimine-functionalized mesoporous silica SBA-15		298	−8	34.5	0.143	62
Sulfonated GO		293	−23.34	3.733	0.096	61
		303	−25.29			
		313	−26.26			
GO		293	−27.53	33.21	0.200	61
		303	−27.66			
		313	−29.82			
SiO_2/UF, impregnated with organophosphate extractant		298	−1.22	46.89	0.162	64
		303	−2.49			
		313	−3.93			
		323	−4.9			
Hydroxyapatite		293	−15.2	34.1	0.17	66
		303	−16.9			
		313	−18.5			
		323	−20.2			
Hydroxyapatite	La^{3+}	293	−16.2	5.9	0.08	66
		303	−17.0			
		313	−17.8			
		323	−18.5			
Fish scales		293	−24.17	13.04	0.038	68
		303	−24.55			

TABLE 3 Thermodynamic parameters of the REE adsorption on various materials—cont'd

Adsorbent	REE	T	ΔG^0 (kJ/mol)	ΔH^0 (kJ/mol)	ΔS^0 (kJ/mol)	Refs.
		313	−24.93			
		323	−25.31			
Neem sawdust		293	−25.47	13.75	0.040	68
		303	−25.87			
		313	−26.27			
		323	−26.67			
Grapefruit peel		293	−19.88	36.57	0.193	54
		303	−21.81			
		313	−23.74			
		323	−25.66			
Pleurotus ostreatus fruits		293	−11.56	6.65	0.039	69
		303	−12.0			
		313	−12.44			
		323	−12.75			
		333	−13.14			
Organophosphate extractant-impregnated SiO_2/UF	Nd^{3+}	298	−1.14	38.27	0.141	64
		303	−1.47			
		313	−2.45			
		323	−4.45			
Magnetic nanohydroxyapatite		293.15	−21.8	87.7	0.374	73
		298.15	−23.6			
		303.15	−25.5			
		323.15	−29.2			
			−33.0			
Magnetic nanohydroxyapatite	Sm^{3+}	293.15	−20.1	81.7	0.343	73
		298.15	−21.8			
		303.15	−23.6			
		323.15	−27.0			
			−30.5			
NaOH-modified *Pleurotus ostreatus*	Y^{3+}	293	−11.94	7.18	0.040	79
		303	−12.35			
		313	−12.75			
		323	−13.07			
		333	−13.56			

A silicon dioxide/urea-formaldehyde (SiO_2/UF) composite material impregnated with an organophosphorus extractant was used to adsorb Eu^{3+} from aqueous solutions.[64] It was checked that the impregnation improves the adsorption capacity. The maximum adsorption was found at a pH of 6, the equilibrium in the system was achieved in 120 min, and an increase in the temperature from 25 to 50 °C caused an increase in the efficiency of the Eu^{3+} extraction. It was also found that the adsorption is controlled by intraparticle diffusion.

10 Multi-component adsorption

The bone powder was evaluated for its ability to adsorb La^{3+}, Eu^{3+}, and Nd^{3+} from the liquid phase.[65] The order of the REE adsorption (determined by the maximum adsorption capacity value) was as follows: $Eu^{3+} > Nd^{3+} > La^{3+}$.

Zhao et al. investigated that the removal of Eu^{3+}, Ce^{3+}, and La^{3+} was carried out using EDTA-β-cyclodextrin.[5] Optimum adsorption was observed at pH of 4.0, and equilibrium in the system was achieved within 240 min. From isotherms, it was found that in this case, the adsorption proceeds on a homogeneous surface. In a single adsorption system, this process occurs in the following sequence: $Eu^{3+} > Ce^{3+} > La^{3+}$, whereas, in a multi-component system, the adsorption order is already different: $Eu^{3+} \gg Ce^{3+} > La^{3+}$, which indicates selectivity concerning Eu^{3+}. It was also determined that after cycles, the regeneration efficiency remains at 91.3%, 93.7%, and 88.6% for Eu^{3+}, Ce^{3+}, and La^{3+}, respectively, using 1 M HNO_3.

Modified (as H-MPRH and Na-MPRNa) polymer resins (hydrogels) and a silica composite (SC) were developed to adsorb Ce^{3+}, Nd^{3+}, Eu^{3+}, and La^{3+}.[76] The adsorption was found to be higher on the SC than on the other materials. This may be due to the presence of silica which provides large cavities and facilitates the REE movement. In addition, silica is also a source of active functional groups. It was found that the adsorption proceeds with an increase on all the adsorbents until equilibrium is reached after 3 h. The metals were extracted in the following sequence: $La^{3+} > Ce^{3+} > Nd^{3+} > Eu^{3+}$, corresponding to the chemical activity decrease in the order from La^{3+} to Eu^{3+}.

11 Carbon nanomaterials as REE adsorbents

Increasingly, carbon nanomaterials, such as graphene, CNT, and GO, are employed to remove the REEs when treating contaminated aquatic media due to the uniqueness of their physical-chemical properties: large specific surface area, high strength, stability, electrical conductivity, etc. Moreover, the high specific surface area of these nanomaterials is associated with intermolecular interactions, which allows them to be effectively used in various adsorption systems.[80]

12 REE adsorption on CNTs

Studies of the REE adsorption on CNTs-based composites were mainly carried out in ultrapure (Milli-Q),[81] in which an aqueous 0.5 mol/L HCl solution was used. As for the contact time between the nanocomposite and the adsorbate, it varied over a wide range but did not exceed 96 h; in most cases, it lasted 2–4 h.[82–86] The temperature varied in the range of 20–65 °C, although most of the described studies were carried out at 30 °C.[82–84] In many works, oxidized MWCNTs were chosen since they are more efficient and cheaper than single-walled CNTs (SWCNTs). Besides, it was shown that the functionalization of the CNTS makes it possible to increase their efficiency, adsorption properties, and imparts the other properties, such as magnetic, by modifying, for instance, magnetite (Fe_3O_4).[86,87] The literature shows that the adsorbent weight (mass)-to-solution volume ratio is ranging from 600 to 100,000 mg/L for the REE extraction. However, generally, the values of 600, 1000, and 5000 mg/L were used. For instance, in Refs. 87–89, the authors report the lowest adsorbent mass-to-solution volume value (600 mg/L), which was used to extract Eu^{3+} from multi-component solutions at pH 5.0–6.0. To date, the highest maximum adsorption capacity for the REEs (Gd^{3+}, 121.51 mg/g) has been achieved on a magnetically retrievable imprinted chitosan/CNTs composite.[86]

Studies on the REEs in single-component systems included Sc^{3+} and Eu^{3+}, and in multi-component systems, Ce^{3+}, Sm^{3+}, La^{3+}, Dy^{3+}, Tb^{3+}, Lu^{3+}, and Gd^{3+}; Y^{3+} was tested in both types of systems. La^{3+} and Eu^{3+} are the most studied ions. In multi-component systems, sorption tests were limited to a maximum of three ions.[83,90] A wide concentration range was considered, from 30[90] to 1,000,000 μg/L.[83] A concentration of 10,000 μg/L was used in multi-component solutions for various REEs, such as La^{3+} and Dy^{3+} or Ce^{3+} and Sm^{3+}.[86] It was determined that the maximum values of the REE adsorption on the CNTs-based composites strongly depend on the selected working pH, which affects the surface charge, and consequently, the adsorption of metal ions on the CNTs. In general, an increase in pH leads to an increase in metal adsorption; this is because, at pH above pHPZC (zero charge point), positively charged metal ions can be adsorbed on negatively charged oxidized CNTs.[91] Thus, the pH range of 5.0–7.0 was tested to determine the optimum pH and/or working pH. pH of 5.0 is the most used working pH[82,84,87,89,90]; in addition, pH of 1.5 is the lowest pH value,[85,90] whereas pH of 8.0 is the highest working pH.[82,89] However, at least two literature sources state that an adsorption percentage of approximately 100% was achieved at a pH of 5.0[82] and 5.5.[87]

13 REE adsorption on graphene nanomaterials

Several papers describe the use of graphene nanomaterials (i.e., GO, graphene nanoplatelets, etc.) for the REE extraction from aquatic media in both batch and column experiments. Almost all studies were carried out in Milli-Q water, with the exception of work in which an aqueous 0.01 mol/L $HClO_4$ solution was used.[92] In addition to the native GO, studies were carried out with the GO functionalized to increase its adsorption efficiency using Fe_3O_4[93] and polyaniline (PANI).[92] In the literature, one can find the adsorbent weight: solution volume ratios of 40–5000 m/L and the most common ratio is 1000 mg/L. In the studies by Chen et al.,[94,95] the lowest ratio, 40 mg/L, were used to extract Gd^{3+} and Y^3, respectively, at pH of 5.9 ± 0.1.

Most of the studies were carried out in single-component systems, and only in the few papers, multi-component systems were considered.[96,97] Eu^{3+} is the most studied REE ion, whereas Ce^{2+} was studied by Fakhri et al.[98] and Farzin et al.,[99] Gd^{3+} and Y^{3+} by Chen et al.,[95] and Sc^{3+} by Kilian et al.[85] Ashour et al.[96] used a four-component quaternary system based on La^{3+}, Nd^{3+}, Gd^{3+}, and Y^{3+}. Su et al.[97] used a mixture of 15 REEs. Different REE concentration ranges were considered, from 10^{61} to $300{,}000^{83}$ μg/L, and the most common range being 10,000–100,000 μg/L.[92–96,98,100,101] Lower Eu^{3+} and Ce^{3+} concentrations (10 and 50 μg/L) were used by Farzin et al.[99] and Xie et al.,[102] respectively. However, the concentration of each metal out of 15 ones in a multi-component system was taken equal to 10 μg/L by Su et al.[97]

As noted above, the REE adsorption strongly depends on the pH. Some authors conducted experiments at pH 2.0–11.0 to find the optimum pH value. Thus, it turned out that pH 6.0 is the most used working pH value,[94–96,98,101] whereas pH 2.0 and 7.0 are the lowest[83,101] and highest[93,101,103] respectively. To date, the highest REE capacity (Eu^{3+}, 250.74 mg/g) has been achieved on the PANI/GO composite.[92]

The adsorption turns out to be a promising, efficient, and economical method for extracting the REEs from aquatic media. Langmuir and pseudo-second-order models were found to correlate well with the experimental data. Further research should be focused on the use of adsorbents for REE extraction from real wastewater. Moreover, batch experiments are increasingly required to be accompanied by dynamic (column) tests to better understand the mechanisms of adsorbent-REE interactions.

Carbon nanomaterials (CNTs, graphene, GO) are increasingly used to extract metals, including the REEs, due to their unique physical-chemical properties such as a large surface area and many functional groups that promote the adsorption process. Besides, it is advisable to use directional functionalization of the carbon nanostructures, which makes it possible to improve the adsorption efficiency or impart additional operational properties to the materials used, such as magnetic ones, which facilitate the adsorbent removal from the solution by applying an external magnetic field.

14 Conclusions

The relevance and scientific novelty of the present review study is due to the low degree of elaboration and the lack of systematics in the existing literature regarding the removal of REEs from aquatic media using nanostructured graphene materials developed. Therefore, novel methods directed at synthesizing such materials and studying their physical-chemical and functional characteristics arouse the undoubted scientific interest of the world community since they would provide prospects of a significant improvement in the quality of fine sorption purification of aquatic media, and also would contribute to the development of efficient engineered barriers for deep disposal of radioactive waste.

It should be noted that an important direction of such research is the establishment of the regularities of the effect of the physical-chemical and structural parameters of the developed nanostructured materials on the selective sorption of target contaminants (lanthanide group metals, which are chemical analogs of radioactive actinides). Furthermore, it is necessary to comprehensively study the physical mechanisms of the process, which, in turn, will make it possible to efficiently simulate the REE extraction from aquatic media using nanostructured graphene sorbents.

References

1. Moldoveanu GA, Papangelaki VG. Recovery of rare earth elements adsorbed on clay minerals: I. Desorption mechanism. *Hydrometallurgy*. 2012;117–118:71–78. https://doi.org/10.1016/j.hydromet.2012.10.011.

2. Fisher A, Kara D. Determination of rare earth elements in natural water samples—a review of sample separation, preconcentration and direct methodologies. *Anal Chim Acta*. 2016;935:1–29. https://doi.org/10.1016/j.aca.2016.05.052.

3. Mineral Commodity Summaries. *US Geological Survey*; 2014.

4. Yulusov PS, Surkova TY, Amanzholova LU, Barmenshinova MB. On sorption of the rare-earth elements. *J Chem Technol Metall*. 2018;53(1):79–82.

5. Zhao F, Repo E, Meng Y, Wang X, Yin D, Sillanpaa M. An EDTA-β-cyclodextrin material for the adsorption of rare earth elements and its application in preconcentration of rare earth elements in seawater. *J Colloid Interface Sci*. 2016;465:215–224. https://doi.org/10.1016/j.jcis.2015.11.069.

6. Kedari C, Das S, Ghosh S. Biosorption of long lived radionuclides using immobilized cells of *Saccharomyces cerevisiae*. *World J Microbiol Biotechnol*. 2001;17(8):789–793. https://doi.org/10.1023/A:1013547307770.

7. Anagnostopoulos VA, Symeopoulos BD. Significance of age, temperature, and aeration of yeast cell culture for the biosorption of europium from aquatic systems. *Desalin Water Treat*. 2014;57(9):3957–3963. https://doi.org/10.1080/19443994.2014.987177.

8. Li K, Gao Q, Yadavalli G, et al. Selective adsorption of Gd^{3+} on a magnetically retrievable imprinted chitosan/carbon nanotube composite with high capacity. *ACS Appl Mater Interfaces*. 2015;7(38):21047–21055. https://doi.org/10.1021/acsami.5b07560.

9. Tian M, Song N, Wang D, et al. Applications of the binary mixture of sec-octylphenoxyacetic acid and 8-hydroxyquinoline to the extraction of rare earth elements. *Hydrometallurgy*. 2012;111–112:109–113. https://doi.org/10.1016/j.hydromet.2011.11.002.

10. Zhu Y, Zheng Y, Wang A. A simple approach to fabricate granular adsorbent for adsorption of rare elements. *Int J Biol Macromol*. 2015;72:410–420. https://doi.org/10.1016/j.ijbiomac.2014.08.039.

11. Ogata T, Narita H, Tanaka M. Adsorption behavior of rare earth elements on silica gel modified with diglycol amic acid. *Hydrometallurgy*. 2015;152:178–182. https://doi.org/10.1016/j.hydromet.2015.01.005.

12. Ogata T, Narita H, Tanaka M, Hoshino M, Kon Y, Watanabe Y. Selective recovery of heavy rare earth elements from apatite with an adsorbent bearing immobilized tridentate amido ligands. *Sep Purif Technol*. 2016;159:157–160. https://doi.org/10.1016/j.seppur.2016.01.008.

13. Galhoum AA, Mafhouz MG, Abdel-Rehem ST, et al. Cysteine-functionalized chitosan magnetic nano-based particles for the recovery of light and heavy rare earth metals: uptake kinetics and sorption isotherms. *Nanomaterials (Basel, Switzerland)*. 2015;5(1):154–179. https://doi.org/10.3390/nano5010154.

14. Gładysz-Płaska A, Majdan M, Grabias E. Adsorption of La, Eu and Lu on raw and modified red clay. *J Radioanal Nucl Chem*. 2014;301(1):33–40. https://doi.org/10.1007/s10967-014-3111-4.

15. Kala R, Biju VM, Rao TP. Synthesis, characterization, and analytical applications of erbium(III) ion imprinted polymer particles prepared via γ-irradiation with different functional and crosslinking monomers. *Anal Chim Acta*. 2005;549(1–2):51–58. https://doi.org/10.1021/ie049313j.

16. Kamio E, Matsumoto M, Valenzuela F, Kondo K. Sorption behavior of Ga(III) and In(III) into a microcapsule containing long-chain alkylphosphonic acid monoester. *Ind Eng Chem Res*. 2005;44(7):2266–2272. https://doi.org/10.1016/j.aca.2005.06.024.

17. Daniel S, Babu P, Rao T. Preconcentrative separation of palladium(II) using palladium(II) ion-imprinted polymer particles formed with different quinoline derivatives and evaluation of binding parameters based on adsorption isotherm models. *Talanta*. 2005;65(2):441–452. https://doi.org/10.1016/j.talanta.2004.06.024.

18. Li C, Zhuang Z, Huang F, Wu Z, Hong Y, Lin Z. Recycling rare earth elements from industrial wastewater with flowerlike nano-Mg(OH)$_2$. *ACS Appl Mater Interfaces*. 2013;5(19):9719–9725. https://doi.org/10.1021/am4027967.

19. Diniz V, Volesky V. Biosorption of La, Eu and Yb using Sargassum biomass. *Water Res*. 2005;39(1):239–247. https://doi.org/10.1016/j.watres.2004.09.009.

20. Palmieri MC, Volesky B, Garcia O. Biosorption of lanthanum using *Sargassum fluitans* in batch system. *Hydrometallurgy*. 2002;67(1–3):31–36. https://doi.org/10.1016/S0304-386X(02)00133-0.

21. Kazy SK, Das SK, Sar P. Lanthanum biosorption by a Pseudomonas sp.: equilibrium studies and chemical characterization. *J Ind Microbiol Biotechnol*. 2006;33(9):773–783. https://doi.org/10.1007/s10295-006-0108-1.

22. Xu S, Zhang S, Chen K, Han J, Liu H, Wu K. Biosorption of La3 +and Ce3 + by Agrobacterium sp. HN1. *J Rare Earths*. 2011;29(3):265–270. https://doi.org/10.1016/S1002-0721(10)60443-7.

23. Awwad N, Gad H, Ahmad M, Aly H. Sorption of lanthanum and erbium from aqueous solution by activated carbon prepared from rice husk. *Colloids Surf B: Biointerfaces*. 2010;81(2):593–599. https://doi.org/10.1016/j.colsurfb.2010.08.002.

24. Vlachou A, Symeopoulos BD, Koutinas AA. A comparative study of neodymium sorption by yeast cells. *Radiochim Acta*. 2009;97(8):437–441. https://doi.org/10.1524/ract.2009.1632.

25. Palmieri MC, GarciaO MP. Neodymium biosorption from acidic solutions in batch system. *Process Biochem*. 2000;36(5):441–444. https://doi.org/10.1016/S0032-9592(00)00236-3.

26. Xiong C, Chen X, Yao C. Enhanced adsorption behavior of Nd(III) onto D113-III resin from aqueous solution. *J Rare Earths*. 2011;29(10):979–985. https://doi.org/10.1016/S1002-0721(10)60582-0.

27. Anastopoulos I, Kyzas GZ. *Citrus Residues as Super-Adsorbents. Citrus Fruits: Production, Consumption and Health Benefits*. USA: Nova Science Publishers; 2016:119–134.

28. Zhu Y, Zheng Y, Wang A. Preparation of granular hydrogel composite by the redox couple for efficient and fast adsorption of La(III) and Ce(III). *J Environ Chem Eng*. 2015;3(2):1416–1425. https://doi.org/10.1016/j.jece.2014.11.028.

29. Sun X, Luo H, Mahurin SM, Liu R, Hou X, Dai S. Adsorption of rare earth ions using carbonized polydopamine nano carbon shells. *J Rare Earths*. 2016;34(1):77–82. https://doi.org/10.1016/S1002-0721 (14) 60582-2.

30. Liatsou I, Efstathiou M, Pashalidis I. Adsorption of trivalent lanthanides by marine sediments. *J Radioanal Nucl Chem*. 2014;304(1):41–45. https://doi.org/10.1007/s10967-014-3448-8.

31. Delrish E, Khanchi A, Outokesh M, Tayyebi A, Tahvildari K. Study on the adsorption of samarium and gadolinium ions by a biopolymer microcapsules containing DEHPA/TOPO extract. *J Appl Chem Res*. 2014;8(2):61–69.

32. Das N, Das D. Recovery of rare earth metals through biosorption: an overview. *J Rare Earths*. 2013;31(10):933–943. https://doi.org/10.1016/S1002-0721(13)60009-5.

33. Postnov VN, Rodinkov OV, Moskvin LN, Novikov AG, Bugaichenko AS, Krokhina OA. From carbon nanostructures to high-performance sorbents for chromatographic separation and preconcentration. *Russ Chem Rev*. 2016;85(2):115–138. https://doi.org/10.1070/RCR4551.

34. Burakov A, Romantsova I, Kucherova A, Tkachev A. Removal of heavy-metal ions from aqueous solutions using activated carbons: effect of adsorbent surface modification with carbon nanotubes. *Adsorpt Sci Technol*. 2014;32(9):737–747. https://doi.org/10.1260/0263-6174.32.9.737.

35. Dervishi E, Li Z, Xu Y, et al. Carbon nanotubes: synthesis, properties, and applications. *Part Sci Technol*. 2009;27(2):107–125. https://doi.org/10.1080/02726350902775962.

36. Melezhyk AV, Kotov VA, Tkachev AG. Optical properties and aggregation of graphene nanoplatelets. *J Nanosci Nanotechnol*. 2016;16(1):1067–1075. https://doi.org/10.1166/jnn.2016.10496.

37. Karri RR, Sahu JN, Jayakumar NS. Optimal isotherm parameters for phenol adsorption from aqueous solutions onto coconut shell based activated carbon: error analysis of linear and non-linear methods. *J Taiwan Inst Chem Eng*. 2017;80:472–487. https://doi.org/10.1016/j.jtice.2017.08.004.

38. Anastopoulos I, Bhatnagar A, Lima EC. Adsorption of rare earth metals: a review of recent literature. *J Mol Liq*. 2016;221:954–962. https://doi.org/10.1016/j.molliq.2016.06.076.

39. Bharathi KS, Ramesh ST. Removal of dyes using agricultural waste as low-cost adsorbents: a review. *Appl Water Sci*. 2013;3(4):773–790. https://doi.org/10.1007/s13201-013-0117-y.

40. Rangabhashiyam S, Anu N, Nandagopal MG, Selvaraju N. Relevance of isotherm models in biosorption of pollutants by agricultural byproducts. *J Environ Chem Eng*. 2014;2(1):398–414. https://doi.org/10.1016/j.jece.2014.01.014.

41. Febrianto J, Kosasih AN, Sunarso J, Ju Y-H, Indraswati N, Ismadji S. Equilibrium and kinetic studies in adsorption of heavy metals using biosorbent: a summary of recent studies. *J Hazard Mater*. 2009;162(2–3):616–645. https://doi.org/10.1016/j.jhazmat.2008.06.042.

42. Park D, Yun Y-S, Yun PJM. The past, present, and future trends of biosorption. *Biotechnol Bioprocess Eng*. 2010;15(1):86–102. https://doi.org/10.1007/s12257-009-0199-4.

43. Crini G, Badot P-M. Application of chitosan, a natural aminopolysaccharide, for dye removal from aqueous solutions by adsorption processes using batch studies: a review of recent literature. *Prog Polym Sci*. 2008;33(4):399–447. https://doi.org/10.1016/j.progpolymsci.2007.11.001.

44. Weber WJ, Morris JC. Kinetics of adsorption on carbon from solution. *J Sanit Eng Div*. 1963;89(2):31–59. https://doi.org/10.1061/JSEDAI.0000430.

45. Boyd GE, Adamson AW, Myers LS. The exchange adsorption of ions from aqueous solutions by organic zeolites. II kinetics. *J Am Chem Soc*. 1947;69(11):2836–2848. https://doi.org/10.1021/ja01203a066.

46. Anastopoulos I, Massas I, Ehaliotis C. Composting improves biosorption of Pb^{2+} and Ni^{2+} by renewable lignocellulosic materials. Characteristics and mechanisms involved. *Chem Eng J*. 2013;231:245–254. https://doi.org/10.1016/j.cej.2013.07.028.

47. Akhtar N, Iqbal M, Zafar SI, Iqbal J. Biosorption characteristics of unicellular green alga *Chlorella sorokiniana* immobilized in loofa sponge for removal of Cr(III). *J Environ Sci*. 2008;20(2):231–239. https://doi.org/10.1016/S1001-0742(08)60036-4.

48. Ngah WSW, Hanafiah MAKM. Biosorption of copper ions from dilute aqueous solutions on base treatedrubber (*Hevea brasiliensis*) leaves powder: kinetics, isotherm, and biosorption mechanisms. *J Environ Sci*. 2008;20(10):1168–1176. https://doi.org/10.1016/S1001-0742(08)62205-6.

49. Ofomaja AE, Ho Y-S. Effect of temperatures and pH on methyl violet biosorption by Mansonia wood sawdust. *Bioresour Technol*. 2008;99(13):5411–5417. https://doi.org/10.1016/j.biortech.2007.11.018.

50. Zhang Y, Liu W, Xu M, Zheng F, Zhao M. Study of the mechanisms of Cu^{2+} biosorption by ethanol/caustic-pretreated baker's yeast biomass. *J Hazard Mater*. 2010;178(1–3):1085–1093. https://doi.org/10.1016/j.jhazmat.2010.02.051.

51. Fang L, Zhou C, Cai P, et al. Binding characteristics of copper and cadmium by cyanobacterium Spirulina platensis. *J Hazard Mater*. 2011;190(1–3):810–815. https://doi.org/10.1016/j.jhazmat.2011.03.122.

52. Li Y, Yue Q, Gao B. Adsorption kinetics and desorption of Cu(II) and Zn(II) from aqueous solution onto humic acid. *J Hazard Mater*. 2010;178(1–3):455–461. https://doi.org/10.1016/j.jhazmat.2010.01.103.

53. Ahmed MJ. Application of agricultural based activated carbons by microwave and conventional activations for basic dye adsorption: review. *J Environ Chem Eng*. 2016;4(1):89–99. https://doi.org/10.1016/j.jece.2015.10.027.

54. Torab-Mostaedi M, Asadollahzadeh M, Hemmati A, Khosravi A. Biosorption of lanthanum and cerium from aqueous solutions by grapefruit peel: equilibrium, kinetic and thermodynamic studies. *Res Chem Intermed*. 2013;41(2):559–573. https://doi.org/10.1007/s11164-013-1210-4.

55. Varshini CJS, Das D, Das N. Recovery of cerium (III) from electronic industry effluent using novel biohydrogel: batch and column studies. *Pharm Lett*. 2015;7(6):166–179.

56. Varshini CJS, Das D, Das N. Optimization of parameters for praseodymium(III) biosorption onto biowaste materials using response surface methodology: equilibrium, kinetic and regeneration studies. *Ecol Eng*. 2015;81:321–327. https://doi.org/10.1016/j.ecoleng.2015.04.072.

57. Koochaki-Mohammadpour SMA, Torab-Mostaedi M, Talebizadeh-Rafsanjani A, Naderi-Behdani F. Adsorption isotherm, kinetic, thermodynamic, and desorption studies of lanthanum and dysprosium on oxidized multiwalled carbon nanotubes. *J Dispers Sci Technol*. 2014;35(2):244–254. https://doi.org/10.1080/01932691.2013.785361.

58. Anagnostopoulos VA. Symeopoulos BD sorption of europium by malt spent rootlets, a low cost biosorbent: effect of pH, kinetics and equilibrium studies. *J Radioanal Nucl Chem*. 2012;295(1):7–13. https://doi.org/10.1007/s10967-012-1956-y 0.

59. Prodromou M, Pashalidis I. Europium adsorption by non-treated and chemically modifiedopuntia ficus indicacactus fibres in aqueous solutions. *Desalin Water Treat*. 2015;57(11):5079–5088. https://doi.org/10.1080/19443994.2014.1002431.

60. Cadogan EI, Lee C-H, Popuri SR, Lin H-Y. Efficiencies of chitosan nanoparticles and crab shell particles in europium uptake from aqueous solutions through biosorption: synthesis and characterization. *Int Biodeterior Biodegradation*. 2014;95:232–240. https://doi.org/10.1016/j.ibiod.2014.06.003.

61. Yao T, Xiao Y, Wu X, Guo C, Zhao Y, Chen X. Retraction notice to "Adsorption of Eu (III) on sulfonated graphene oxide: combined macroscopic and modeling techniques". *J Mol Liq*. 2020;308:112620. https://doi.org/10.1016/j.molliq.2020.112620 [J Mol Liq 215, March 2016, 443–448].

62. Dolatyari L, Yaftian MR, Rostamnia S. Adsorption characteristics of Eu(III) and Th(IV) ions onto modified mesoporous silica SBA-15 materials. *J Taiwan Inst Chem Eng*. 2016;60:174–184. https://doi.org/10.1016/j.jtice.2015.11.004.

63. Guo Z, Li Y, Pan S, Xu J. Fabrication of Fe_3O_4@cyclodextrin magnetic composite for the high-efficient removal of Eu(III). *J Mol Liq*. 2015;206:272–277. https://doi.org/10.1016/j.molliq.2015.02.034.

64. Naser A, El-deen GS, Bhran AA, Metwally S, El-Kamash A. Elaboration of impregnated composite for sorption of europium and neodymium ions from aqueous solutions. *J Ind Eng Chem*. 2015;32:264–272. https://doi.org/10.1016/j.jiec.2015.08.024.

65. Butnariu M, Negrea P, Lupa L, et al. Remediation of rare earth element pollutants by sorption process using organic natural sorbents. *Int J Environ Res Public Health*. 2015;12(9):11278–11287. https://doi.org/10.3390/ijerph120911278.

66. Granados-Correa F, Vilchis-Granados J, Jiménez-Reyes M, Quiroz-Granados L. Adsorption behaviour of La(III) and Eu(III) ions from aqueous solutions by hydroxyapatite: kinetic, isotherm, and thermodynamic studies. *J Chem.* 2013;2013:1–9. https://doi.org/10.1155/2013/751696.

67. Varshini C, Das N. Relevant approach to assess the performance of biowaste materials for the recovery of lanthanum (III) from aqueous medium. *Res J Pharm, Biol Chem Sci.* 2014;5(6):88–94.

68. Das D, Varshini CJS, Das N. Recovery of lanthanum(III) from aqueous solution using biosorbents of plant and animal origin: batch and column studies. *Miner Eng.* 2014;69:40–56. https://doi.org/10.1016/j.mineng.2014.06.013.

69. Hussien S. Biosorption lanthanum pleurotus ostreatus basidiocarp. *Int J Biomed Res.* 2014;2:26–36.

70. Birungi ZS, Chirwa EMN. The kinetics of uptake and recovery of lanthanum using freshwater algae as biosorbents: comparative analysis. *Bioresour Technol.* 2014;160:43–51. https://doi.org/10.1016/j.biortech.2014.01.033.

71. Zheng X, Wang C, Dai J, Shi W, Yan Y. Design of mesoporous silica hybrid materials as sorbents for the selective recovery of rare earth metals. *J Mater Chem A.* 2015;3(19):10327–10335. https://doi.org/10.1039/C4TA06860B.

72. Wang F, Zhao J, Wei X, et al. Adsorption of rare earths (III) by calcium alginate-poly glutamic acid hybrid gels. *J Chem Technol Biotechnol.* 2013;89(7):969–977. https://doi.org/10.1002/jctb.4186.

73. Gok C. Neodymium and samarium recovery by magnetic nano-hydroxyapatite. *J Radioanal Nucl Chem.* 2014;301(3):641–651. https://doi.org/10.1007/s10967-014-3193-z.

74. Hadjittofi L, Charalambous S, Pashalidis I. Removal of trivalent samarium from aqueous solutions by activated biochar derived from cactus fibres. *J Rare Earths.* 2016;34(1):99–104. https://doi.org/10.1016/S1002-0721(14)60584-6.

75. Vijayaraghavan K, Jegan J. Entrapment of brown seaweeds (Turbinaria conoides and Sargassum wightii) in polysulfone matrices for the removal of praseodymium ions from aqueous solutions. *J Rare Earths.* 2015;33(11):1196–1203. https://doi.org/10.1016/S1002-0721(14)60546-9.

76. Ma J, Wang Z, Shi Y, Li Q. Synthesis and characterization of lysine-modified SBA-15 and its selective adsorption of scandium from a solution of rare earth elements. *RSC Adv.* 2014;4(78):41597–41604. https://doi.org/10.1039/C4RA07571D.

77. Hussien SS, Desouky OA. Biosorption studies on yttrium using low cost pretreated biomass of Pleurotus ostreatus. In: *Proceedings of the 4th International Conference on Radiation Sciences and Application. Taba: Egypt*; 2014:139–150.

78. Roosen J, Pype J, Binnemans K, Mullens S. Shaping of alginate–silica hybrid materials into microspheres through vibrating-nozzle technology and their use for the recovery of neodymium from aqueous solutions. *Ind Eng Chem Res.* 2015;54(51):12836–12846. https://doi.org/10.1021/acs.iecr.5b03494.

79. Vijayaraghavan K. Biosorption of lanthanide (praseodymium) using *Ulva lactuca*: mechanistic study and application of two, three, four and five parameter isotherm models. *J Environ Biotechnol Res.* 2015;1(1):10–17.

80. Pastrana-Martínez LM, Morales-Torres S, Gomes HT, Silva AMT. Nanotubos e grafeno: os primos mais jovens na família do carbono! *Química.* 2013;128:21–27.

81. Yadav KK, Dasgupta K, Singh DK, Anitha M, Varshney L, Singh H. Solvent impregnated carbon nanotube embedded polymeric composite beads: an environment benign approach for the separation of rare earths. *Sep Purif Technol.* 2015;143:115–124. https://doi.org/10.1016/j.seppur.2015.01.032.

82. Behdani FN, Rafsanjani AT, Torab-Mostaedi M, Mohammadpour SMAK. Adsorption ability of oxidized multiwalled carbon nanotubes towards aqueous Ce(III) and Sm(III). *Korean J Chem Eng.* 2012;30(2):448–455. https://doi.org/10.1007/s11814-012-0126-9.

83. Cardoso CED, Almeida JC, Lopes C, Trindade T, Vale C, Pereira E. Recovery of rare earth elements by carbon-based nanomaterials—a review. *Nanomaterials (Basel, Switzerland).* 2019;9(6):814. https://doi.org/10.3390/nano9060814.

84. Chen C, Li X, Zhao D, Tan X, Wang X. Adsorption kinetic, thermodynamic and desorption studies of Th(IV) on oxidized multi-wall carbon nanotubes. *Colloids Surf A Physicochem Eng Asp.* 2007;302(1–3):449–454. https://doi.org/10.1016/J.COLSURFA.2007.03.007.

85. Kilian K, Pyrzyńska K, Pęgier M. Comparative study of Sc(III) sorption onto carbon-based materials. *Solvent Extr Ion Exch.* 2017;35(6):450–459. https://doi.org/10.1080/07366299.2017.1354580.

86. da Silva Alves DC, Healy B, de Almeida Pinto LA, Sant'Anna Cadaval Jr TR, Breslin CB. Recent developments in chitosan-based adsorbents for the removal of pollutats from aqueous environments. *Molecules.* 2021;26(3):594. https://doi.org/10.3390/molecules26030594.

87. Chen CL, Wang XK, Nagatsu M. Europium adsorption on multiwall carbon nanotube/iron oxide magnetic composite in the presence of polyacrylic acid. *Environ Sci Technol.* 2009;43(7):2362–2367. https://doi.org/10.1021/es803018a.

88. Chen C, Hu J, Xu D, Tan X, Meng Y, Wang X. Surface complexation modeling of Sr(II) and Eu(III) adsorption onto oxidized multiwall carbon nanotubes. *J Colloid Interface Sci.* 2008;323(1):33–41. https://doi.org/10.1016/j.jcis.2008.04.046.

89. Fan QH, Shao DD, Hu J, Chen CL, Wu WS, Wang XK. Adsorption of humic acid and Eu(III) to multi-walled carbon nanotubes: effect of pH, ionic strength and counterion effect. *Radiochim Acta.* 2009;97(3):141–148. https://doi.org/10.1524/ract.2009.1586.

90. Tong S, Zhao S, Zhou W, Li R, Jia Q. Modification of multi-walled carbon nanotubes with tannic acid for the adsorption of La, Tb and Lu ions. *Microchim Acta.* 2011;174(3–4):257–264. https://doi.org/10.1007/s00604-011-0622-3.

91. Pyrzynska K, Kubiak A, Wysocka I. Application of solid phase extraction procedures for rare earth elements determination in environmental samples. *Talanta.* 2016;154:15–22. https://doi.org/10.1016/j.talanta.2016.03.022.

92. Sun Y, Shao D, Chen C, Yang S, Wang X. Highly efficient enrichment of radionuclides on graphene oxide-supported polyaniline. *Environ Sci Technol.* 2013;47(17):9904–9910. https://doi.org/10.1021/es401174n.

93. Li D, Zhang B, Xuan F. The sorption of Eu(III) from aqueous solutions by magnetic graphene oxides: a combined experimental and modeling studies. *J Mol Liq.* 2015;211:203–209. https://doi.org/10.1016/j.molliq.2015.07.012.

94. Chen W, Wang L, Zhuo M, Liu Y, Wang Y, Li Y. Facile and highly efficient removal of trace Gd(III) by adsorption of colloidal graphene oxide suspensions sealed in dialysis bag. *J Hazard Mater.* 2014;279:546–553. https://doi.org/10.1016/j.jhazmat.2014.06.075.

95. Chen W, Wang L, Zhuo M, et al. Reusable colloidal graphene oxide suspensions combined with dialysis bags for recovery of trace Y(iii) from aqueous solutions. *RSC Adv.* 2014;4(102):58778–58787. https://doi.org/10.1039/C4RA09175B.

96. Ashour RM, Abdelhamid HN, Abdel-Magied AF, et al. Rare earth ions adsorption onto graphene oxide nanosheets. *Solvent Extr Ion Exch.* 2017;35(2):91–103. https://doi.org/10.1080/07366299.2017.1287509.

97. Su S, Chen B, He M, Hun B, Xiao Z. Determination of trace/ultratrace rare earth elements in environmental samples by ICP-MS after magnetic solid phase extraction with Fe3O4@SiO2@polyaniline–graphene oxide composite. *Talanta.* 2014;119:458–466. https://doi.org/10.1016/j.talanta.2013.11.027.

98. Fakhri H, Mahjoub AR, Aghayan H. Effective removal of methylene blue and cerium by a novel pair set of heteropoly acids based functionalized graphene oxide: adsorption and photocatalytic study. *Chem Eng Res Des.* 2017;120:303–315. https://doi.org/10.1016/j.cherd.2017.02.030.

99. Farzin L, Shamsipur M, Shanehsaz M, Sheibani S. A new approach to extraction and preconcentration of Ce(III) from aqueous solutions using magnetic reduced graphene oxide decorated with thioglycolic-acid-capped CdTe QDs. *Int J Environ Anal Chem.* 2017;97(9):854–867. https://doi.org/10.1080/03067319.2017.1364376.

100. Arunraj B, Talasila S, Rajesh V, Rajesh N. Removal of Europium from aqueous solution using *Saccharomyces cerevisiae* immobilized in glutaraldehyde cross-linked chitosan. *Sep Sci Technol.* 2018;54(10):1620–1631. https://doi.org/10.1080/01496395.2018.1556303.

101. Sun Y, Wang Q, Chen C, Tan X, Wang X. Interaction between Eu(III) and graphene oxide nanosheets investigated by batch and extended X-ray absorption fine structure spectroscopy and by modeling techniques. *Environ Sci Technol.* 2012;46(11):6027. https://doi.org/10.1021/es300720f.

102. Xie Y, Helvenston EM, Shuller-Nickles LC, Powell BA. Surface complexation modeling of Eu(III) and U(VI) interactions with graphene oxide. *Environ Sci Technol.* 2016;50(4):1821–1827. https://doi.org/10.1021/acs.est.5b05307.

103. Li C, Huang Y, Lin Z. Fabrication of titanium phosphate@graphene oxide nanocomposite and its super performance on Eu3+ recycling. *J Mater Chem A.* 2014;2(36):14979–14985. https://doi.org/10.1039/C4TA02983F.

Chapter 15

Ion-selective membranes as potentiometric sensors for noxious ions

Bhavana Sethi[a,*] and Saurabh Ahalawat[b]

[a]*Academy of Business and Engineering Sciences, Ghaziabad, UP, India,* [b]*Central Research and Development, Ultratech Cement LTD, Khor Neemuch, MP, India*

[*]*Corresponding author.*

1 Introduction

Environmental impact and health hazards due to noxious ions in various mediums are of great concern to modern society. These ions have toxic properties, leading to unfavorable effects on the ecosystem and the health of humans even if present in minute quantities.[1] Among various noxious ions, the metal ions (especially the heavy metal ions) are even more potential pollutants because these are non-biodegradable, resulting in bioaccumulation and biomagnification, thus causing a large environmental impact.[2,3] Technological advancements have accelerated the emissions of these anthropogenic noxious ion pollutants. Human activities have led the way for metal ions into the environment, which has influenced and changed the natural cycles. The industrial revolution added multiple dimensions to the usage of metals which affected the environment to a greater extent. The combustion of fossil fuels also added an enormous amount of heavy metals into the environment.

Metal ions (more specifically, heavy metals) show a complexation and binding tendency with biological matter ligands because they generally have oxygen, sulfur, and nitrogen donor atoms in them.[4]This binding may result in changes in the protein structure at the molecular level, inhibition of enzymatic activity, or breaking bonds. Some heavy metal ions also result in mutations, abnormal fetal growth, or may be carcinogenic.[5] Thus, the determination of noxious ions is important and is a demanding subject for analytical chemists with regard to the concentration ranges set by standards and guidelines for reasons of toxicity.

In recent years, the analytical field has expanded as tremendous know-how on the pollutants needs to be obtainable in context with environmental welfare. These requirements have resulted in the development of different and useful analytical tools. Various instrumental techniques viz. spectral methods, separatory methods, and electroanalytical methods are available for analytical chemists. Although these methods are sensitive and precise but require chemical manipulation of the sample and are expensive techniques, and thus unsuitable for routine and "online" analysis. Moreover, these techniques may not be used as effectively in colored and turbid solutions, which may lead to erroneous results. Presently, another area of analytical chemistry, the chemical sensors, is greatly explored in the determination of ions as it is advantageous over other methods of analysis in several ways.[6,7] Ion-selective electrodes (ISEs) find application in most of the most frequent routine analysis, and they have the longest history. The most attractive features of this technique are as follows:

- the samples can be analyzed at greater speed,
- equipment is portable,
- non-destruction of the sample,
- online analysis,
- affordability, and
- range of measurement is large.

Moreover, the ISEs can be fabricated in the laboratory easily and may become commercially available soon after their establishment. Therefore, the usefulness and accessibility of ISEs have succeeded them than other wet analytical methods. The use of ISEs has found applications everywhere in the determination of ions, from laboratory to field, viz. analysis of

Sustainable Materials for Sensing and Remediation of Noxious Pollutants. https://doi.org/10.1016/B978-0-323-99425-5.00020-7

blood sugar by glucometer in clinical labs to portable water quality monitoring kit for the analysis of environmental samples. Thus, the sensing of noxious ions by ISEs using electroactive sensing materials has always been a topic of vast research and development.

2 Chemical sensors

Nature has provided an excellent sensing system to living beings. The nose, tongue, eye, ear, and skin are the five natural sensors for sensing smell, taste, sight, sound, and feel, respectively. Analytical chemists are continuously working to develop sensors for ionic species, drugs, metals, organics, inorganics, etc. The discovery of a pH electrode is one such example, wherein H^+ concentration is measured in solution irrespective of the surrounding environment without any interference.

Chemical sensors are the sensors that measure chemical substances by chemical or physical responses. Further, these devices are connected with transducers so as to observe the response visually. These electrochemical transducers can be categorized into four types mainly potentiometric, conductometric, voltametric, and field-effect transistors type.

The potentiometric sensor (primarily focused in this chapter) measures the equilibrium potential of an indicator electrode (which is prepared by careful choice of electroactive material), selecting other electrodes as the reference electrode (like the calomel electrode). The whole assembly is constructed to operate effectively at no current. Depending on the electrode material chosen, the electrode will provide good selectivity to one particular ion and will have minimal interference from other ions.

3 Chronological developments

Potentiometric sensors involve the generation of electric potential on a membrane, solid or liquid when placed between two solutions of different concentrations of an appropriate electrolyte. The history dates back to 1791 when Galvani observed a bioelectric phenomenon while dissecting a frog. However, the study of the potential difference across semipermeable membranes by Ostwald in 1890 was the first major contribution in this field.[8] Cremer, in 1906, discovered the tendency of some glasses for selective hydrogen ion responses. Initially, the potential response was interpreted in terms of Donnan equilibrium, but in the 1930s, Nicolsky suggested that the electrode response is dependent on the active sites of the glass capable of ion exchange, and the concept enabled the selectivity coefficient of an electrode to be calculated.[9] Studies have shown the potentialities of solid silver halide membrane-based electrodes[10] and initiated the use of materials other than the glass in the construction of electrodes, while silver iodide-based electrodes were reported in 1961.[11] An electrode membrane was prepared by embedding silver iodide in paraffin which was selective for iodide ions. Subsequently, Pungor et al. developed a whole range of electrodes of heterogeneous membranes consisting of an active material supported in an inert matrix of silicone rubber. The commercial development of these electrodes began when calcium and fluoride electrodes were developed.[12] The selective fluoride electrode, based on lanthanum fluoride doped with europium fluoride, is the second-best electrode developed after the glass electrode. Advancements in the field of medical sciences and physiology also contributed to the development of ISEs. Liquid membrane electrodes were introduced, and these electrodes were responsive for both organic and inorganic ions.[13] Simon et al. has contributed to a great extent in designing highly selective ionophore-based liquid membrane sensors.[14] The different important procedures and techniques used for PVC membranes were thoroughly discussed by researchers.[15] Therefore, the PVC matrix-based ISEs have attracted much attention and have been used in the construction of chemical sensors for a variety of noxious ions.

4 Potentiometric sensors: Components and design

4.1 Sensing elements

This component holds an important place in the design of components of any ISE. The selectivity imparted by these sensing elements permits the sensor to be selective to a specific ion or a group of ions, thereby avoiding interferences from other ions.

4.2 Membrane composition

The carefully selected membrane composition will decide how well an ion-selective layer will be fabricated. The first step involves the inclusion of ionophore molecules inside the bulk of a polymeric membrane. The most commonly used polymeric matrix in the membranes of ISEs is the plasticized poly (vinyl chloride). The adhesion properties of the obtained sensing layers can be improved by the application of polysiloxane or polyurethane having intrinsic elastomeric properties. Apart from selecting appropriate ionophores, the lipophilic ionic sites are also required in the membranes to further improve the evaluation factors like useful range of concentration, theoretical Nernstian slope, and also to obtain stable sensor signals. These ionic sites will additionally enhance the selectivity of the potentiometric sensors.

4.3 Transducers/detecting system

It is used for the amplification of the primary signal to the level which is usable. The chemical information is directed to the receptor unit, which translates it into a kind of energy acceptable by the transducer, and then finally, the signal is produced. Generally, electrochemical transducers have been employed in sensors as they are easy to construct and affordable.

4.4 Analyte recognition

The complexation reaction between the guest ions and the selected ionophore forms the basis of analyte recognition. Different types of attractive forces present operating between ion and dipole, dipole and dipole, hydrogen bond interactions or seldom covalent bonds are the driving force for the reversible process. Two important structural parameters of the receptor, viz. complementary cavity shape and a suitable array of the binding sites (functional groups), are responsible for the key lock configuration between the guest ion and ionophore. In the design of the chemical sensor, one of the surfaces of the membrane (containing the receptor molecule of proper choice) is in contact with the analyte solution in question. The complexation reactions occur in the membrane, resulting in the generation of potential because the process of ion exchange happens between the test solution and the membrane phase, which results in the signal generation. The potential difference developed across the membrane depends on the concentration of the analyte. The resultant selectivity of the membrane for a particular ion is governed by the selectivity of the ion-ionophore complex formation and by the partition coefficients of different ions between the aqueous solution and the membrane phase, which is relatively less polar in nature. Fig. 1 shows the design process of a chemical sensor.

5 Membrane materials

The polymeric membrane used in ISEs consists of the following components.

5.1 Ionophore

Ionophore is the primary component of polymeric membranes. The process which is sensed by the membrane of ISE is the complexation between the electroactive material and target ion. The complexation is considered ideally between ionophore and target ion and not with the opposite ions.

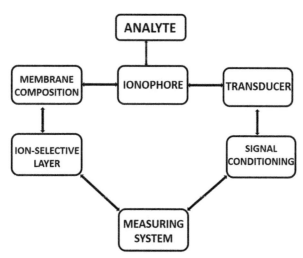

FIG. 1 Steps in the design of a chemical sensor.

For effective complexation, the ionophore should fulfil these conditions:

a. The physical compatibility of the ionophore with the matrix should be high.
b. The solubility product should be low.
c. It should have some electrical conductivity.
d. There should be a striking balance between two important energies, viz. ion-hydration energy and the free energy of ion and ligand interaction; these result in rapid ion exchange at the membrane sample interface.
e. Retention of ionophores within the membrane should be high, which could be achieved by the presence of numerous lipophilic groups other than the binding sites.

A wide range of substances has been tried as ionophores by different researchers. These include different classes of calixarenes, crown ethers, dendrimers, cryptands, chelating ligands, macrocyclic compounds, ion exchangers, etc. Some of the commercially available ionophores are given in Fig. 2.

5.2 Polymeric matrix

The important requirement for the membrane to function efficiently is that it has to be elastic and should be both physically and mechanically stable. These requirements are fulfilled by a polymeric matrix which provides an inert base to the membrane. Polymer matrix used in membrane preparation should be inert toward chemicals, water resistance, tough, flexible, resistant to cracks, should not be porous, and should retain their size in sample solutions. Various classes of compounds, viz. polyamides, polystyrenes, polyurethanes, methacrylates, etc., have been explored as polymeric matrix materials.[16–18] However, the most commonly used polymer in the membrane preparation is polyvinyl chloride (PVC). It is because it

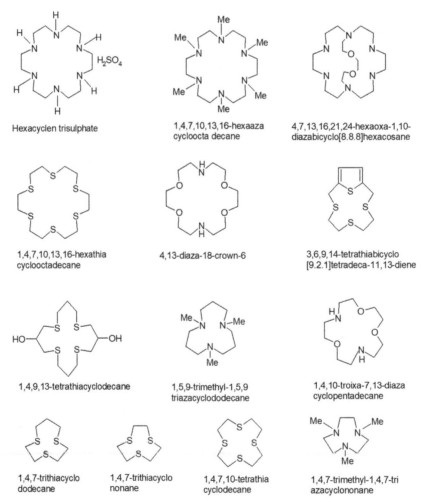

FIG. 2 Some commercially available ionophores are used in ISEs.

is cost-effective, has good mechanical properties, is chemically inert, and offers durability to mechanical and pressure damage.[19] Another interesting feature of the PVC matrix is its high compatibility with the electroactive materials, which leads to its reduced oozing out from the membrane, further leading to increased electrode life. A comparative analysis of different polymeric matrices with PVC, taking into account their response characteristics, was provided.[20] The presence of dissociated fixed exchange sites present in PVC was found to be a factor responsible for the response and selectivity of the membrane.[21,22] Different pros and cons associated with the use of PVC in the membranes were also studied.[23]

5.3 Plasticizer

The plasticizer employed in membranes should have low vapor pressure, high lipid affinity, and high relative molecular mass, should exhibit the least tendency to ooze out from the polymer matrix, and should be capable of dissolving the other components of the membrane.[24,25] It has been observed that the addition of plasticizers also improves the important performance parameters of the membrane, which include the lowering of detection limit, stabilization of the readings, and increase in the shelf life of the ISEs.[26] A number of organic solvents like adipates, phthalates, terephthalates, sebacates, octyl ethers, and benzyl acetate have been explored as a plasticizer in the membrane preparation of ISEs.

5.4 Lipophilic additive

The lipophilic ionic additive is a salt of lipophilic anion/cation and a counter ion which are non-exchangeable and exchangeable, respectively. The addition of this component to the membrane provides perm selectivity to the ion-selective membrane, thus enhancing the sensing selectivity and reducing the bulk membrane impedance.[27,28] The presence of this component in membrane electrodes reduces the resistance and also enhances the sensitivity of membrane electrodes.[29,30] Presently, different derivatives of tetraphenylborate and certain tetraalkylammonium salts are employed as anionic additives and cationic additives, respectively, in the membrane preparation.

6 Theory and methodology

The utility of ISEs depends on the membrane potential determination. The value of membrane potential can be obtained from the *emf* values of a complete electrochemical cell, which consists of the two solutions 1 and 2, which are separated by a membrane and have two reference electrodes.[31] A schematic representation of a cell employing ISE is given in Fig. 3.

6.1 Potential of an ion-exchange membrane

A semipermeable ion-exchange membrane is one, which allows the counter ion A to permeate through and restricts the co-ion, Y. When the two solutions are separated, the membrane potential is developed because of the diffusion of counter ions from higher to lower concentrations. This potential, E_m, is given as follows:

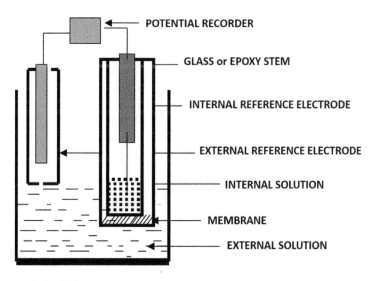

FIG. 3 Schematic cell diagram employing ISE.

$$E_m = \frac{2.303RT}{Z_A F} \left[\log \frac{a_A}{a_B} - (z_Y - z_A) \int_1^2 t_Y^- d\log a^\pm \right] \tag{1}$$

where t_Y^- represents the transport number of co-ions, $a^\pm$ = mean ionic activity of the electrolyte, and a_A and a_B are the activities of the counter ions in the two solutions.

In the above expression, the right-hand side consists of two terms. If the membrane is ideally perm-selective membrane ($t_Y^- = 0$), then Eq. (1) reduces to

$$E_m = \frac{2.303RT}{Z_A F} \left[\log \frac{a_A}{a_B} \right] \tag{2}$$

Eq. (2) is the so-called Nernst equation and gives the value of Donnan potential for a membrane that is ideally considered as a perm-selective membrane. From the above equation, it is clear that the potential developed is directly proportional to the activity of the cation in the test solution. At 298 K, the value of 2.303 $RT/Z_A F$ becomes $0.059/Z_A$ volts. The plots between cell potential and log activity are termed Nernst plots, and the slope of these plots are called the *Nernstian slope*.[32,33]

6.2 Selectivity of electrodes

The ISEs developed for a particular ion do not only respond to that ion but also other ions as well, although its response is higher for the ion for which it is designed. The ion of interest for which the electrode is developed is conventionally called the primary ion, and all other ions that interfere in its analysis are known as interfering ions. If the concentration of the interfering ions is comparable to or larger concerning the primary ion, then the membrane potential developed at the membrane may be due to both primary as well as interfering ions. The extent of selectivity of the electrode for primary ion A over the interfering ion, B, is expressed by the potentiometric selectivity coefficient ($K_{A,B}^{Pot}$), which is given by the following equation:

$$E = E^0 \frac{2.303RT}{Z_A F} \left[\log a_A + K_{A,B}^{Pot}(a_B)^{Z_A/Z_B} \right] \tag{3}$$

This equation is called the Nicolsky equation, where Z_A, Z_B and a_A, a_B are the charges and activities of ions A and B, respectively. Numerous methods for the experimental determination of selectivity coefficient have been reported.[34] Broadly, they are classified as separate solution methods (SSMs) and mixed solution methods (MSMs). The above methods can be performed by following different procedures. One such procedure of the mixed solution method is the *fixed interference method (FIM)*. The IUPAC has recommended it as a procedure for the determination of selectivity coefficients. The methodology adopted behind this procedure is that it considers the similar working conditions at the membrane and solution interface as were present while analyzing the sample.

6.3 Significance of selectivity coefficient values

The selectivity coefficient values (given by the Nicolsky equation) give the selectivity of the electrode. The selectivity coefficient should have been a constant value, but it depends on the conditions at the time of the experiment, like the concentration of ions as well as the method of determination. The values of selectivity coefficients are lower if determined at a lower concentration of interfering ions and vice versa. Different methods give different values of selectivity coefficients, as the conditions prevailing at the membrane-solution interface are not the same.[35,36]

7 Experimental aspects

7.1 Pre-starting procedure

Pre-conditioning of electrodes is a procedure followed to get the most stable results. Prior to its use, both the ISE and the reference electrodes are dipped in the standard solution for at least 10 min. If, after that, stable voltage readings are obtained, then the system is considered to be working aptly. Most accurate results are obtained by repeating the wetting and preconditioning techniques until three consecutive readings are within 1 mV of each other. However, if the electrode is being used for the first time or is used after prolonged storage, it may require further soaking overnight to get stable voltage readings. The preconditioning step should be followed by washing the electrodes well with distilled water and drying them with soft tissue.

7.2   Methodology of measurements

7.2.1   Direct potentiometry

Direct potentiometry is the most common method of using ISEs because of its simplicity. In this method, firstly, the system is calibrated, which is followed by measuring the electrode response in a solution of unknown concentration from the meter. The concentration can then be directly determined from the calibration graph. Thus, this method offers the advantage that it can be used to rapidly analyze the solutions of very different concentrations at a very fast speed without the need of recalibrating the instrument repeatedly or involving any tedious calculations.

7.2.2   Incremental methods

The incremental methods are as follows:

- Standard addition
- Sample addition
- Sample subtraction

7.2.2.1   Standard addition and sample addition methods

These methods involve the voltage measurement of standard or sample solutions for standard addition and sample addition methods, respectively. The volumes of these solutions are large and measured with great accuracy. The initial voltage measurement is followed by the addition of a much smaller volume and then taking the next reading after the voltage is stabilized in the mixture.

7.2.2.2   Sample subtraction method

In this method, a small quantity of reagent is added to the standard solution of the ion. This reagent will react with the ion in stoichiometric proportions to form a complex or precipitate. This will produce a decrease in ion concentration. This method offers a big advantage to be used in the case of ions whose measurements are unstable or are highly toxic.

7.2.3   Potentiometric titrations

The above-described methods often have a disadvantage of their relative imprecise results. The ISE titrations are based on the fact that at the end point of titrations, there is often a sharp change in reactant concentration which results in a very different value of electrode potential. Potentiometric titrations are thus advantageous over other methods as they depend on the volumetric measurements and not on the measurement of the electrode potential.

All these methodologies of measurements are given in Fig. 4.

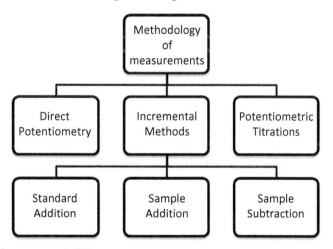

FIG. 4   Different methodologies of measurement in ISEs.

8 Advancements in the field of ISEs for certain highly noxious ions

8.1 Pb^{2+}-selective electrodes

Pb^{2+}-selective electrodes have been prepared using different types of electroactive materials. The mixture of tetraphenylborate salt of lead and polyalkoxylates could be effectively used as electroactive components of heterogeneous membranes for the preparation of Pb^{2+} electrodes. The reported ISEs were sufficiently Pb^{2+}-selective, gave near-Nernstian response between 10^{-5} and 10^{-1} M for Pb^{2+} and were highly selective for lead ions. A number of crown ethers, viz. dicyolohexano-18-orown-6, benzo-13-crown-6, and dibenzo-18-crown-6, have been used by various workers as the electroactive phase in the heterogeneous membranes to prepare lead selective electrodes. These electrodes generally show overall better performance compared to electrodes prepared by using other electroactive materials. A lead electrode has been developed using derivatives of calixarene as electroactive material in PVC matrix.[37] Another ISE was reported by using the derivative of crown ether as an ionophore. This ionophore was directly coated on a graphite electrode. It showed utility in real samples of oil and water.[38] A liquid membrane ISE containing 1,4-bis[2-(3,4,5-trimethoxyphenyl) ethenyl] benzene types of compounds has been reported for lead ions. They were reported to be highly selective for lead ions.[39] Calixarenes and their derivatives have always been an attractive class of ionophores for researchers. Working on this class of ionophore, 4-*tert*-butylcalix arene and 15-crown-5 were explored as ionophores and were used for the determination of Pb^{2+} in polluted water.[40,41] A Schiff base-based ISE for the determination of Pb(II) ions in an aqueous solution has been developed.[42] Polymeric electrodes have been prepared for lead ions using a diamine derivative as an ion carrier.[43] A lead ISE based on electrically conducting microparticles of sulfonic phenylenediamine *co*-polymer as an ionophore was developed.[44] Acridono-crown ether was also tried successfully as an ionophore by the potentiometric method.[45] Recently, a new generation of potentiometric sensors showing high sensitivity for lead ions has been reported. It is based on ion-imprinted polymeric nanoparticles.[46] A single-layer Pb^{2+} potentiometric sensor with MoS_2 nanoflakes as an ion-to-electron transducer was developed.[47]

8.2 Hg^{2+} selective electrodes

The determination of mercury is important as it is one of the highly toxic environmental pollutants. Several mercury-selective electrodes based on AgI-Ag_2S and dithia-crown ethers have been reported. In addition to these, liquid membrane electrodes using Hg(II)-chelates have also been tried for Hg^{2+} determination.[48] However, these electrodes generally have a slow response time, poor selectivity, and interference from Ag^+ ions. Cerium(IV) selenite and antimony(III) arsenate, based on inorganic ion-exchange membranes using polystyrene as a binder, has also been reported as mercury ion ISE.[49,50] These electrodes responded in a Nernstian manner with good reproducibility with high response time (30-40s). Another group of workers have used pentathia-15-crown-5 and a two-armed crown ether in PVC matrix.[51,52] An electrode utilizing diamine donor ligand was also developed by the same group of workers.[53] An electrode based on a highly chelating ligand as an ionophore has been reported to determine Hg^{2+} ions up to a 1.0 ppm level.[54] A derivative of the calixarene family has been explored as an ionophore in the determination of anionic mercury species. The researchers have employed tri dodecyl methylammonium chloride as an ion exchange for preparing this ISE.[55] Another derivative of calixarene in which the lower rim of thiacalixarene has been functionalized has been reported as a cationic receptor for mercury ions.[56] Real sample analysis for the potentiometric determination of mercury (II) ion was carried out by using a novel modified screen-printed electrode.[57] A novel ionophore was tested for the fabrication of a potentiometric sensor for Hg^{2+} metal ions.[58] Recently, the determination of Hg^{2+} in a variety of real samples was carried out by using a new ionophore based on a new Schiff base.[59]

8.3 Fe^{3+} selective electrodes

Iron plays an important role in the various biological processes in the body.[60,61] However, the excess of this element results in its accumulation in organs like the heart and liver,[62,63] which further puts these vital organs at risk for serious damage.[64] Certain iron compounds can cause carcinogenic activity. Thus, there is a demand for the determination of iron. Various ISEs have been reported. One such highly selective Fe(III) ion sensor was reported by utilizing a phenol derivative as a suitable carrier.[65] An Fe(III) selective sensor using μ-bis(tridentate) as an electroactive material was found to exhibit high sensitivity from iron ions.[66] Ferroin-TPB was also tried as an ionophore for preparing a highly selective iron sensor.[67] Iron selective electrodes based on formyl salicylic acid derivatives have been successfully tried and reported.[68] Another electrode has been developed for iron(III) which was based on crown ether.[69] A highly efficient chelating ligand has been tried and utilized for the successful development of iron sensors.[70] Recently, a new iron sensor using benzo-18-crown-6 has been successfully reported.[71] A novel potentiometric and voltammetric sensor for creating a cationic response for Fe(II) was recently introduced.[72] Another new iron ISE with membranes based on iron(III) phosphate and silver sulfide integrated into a completely new electrode body design has been developed for the determination of iron(III) cations.[73]

8.4 Cu^{2+} selective electrodes

Copper is found to be present in various environmental and biological samples, and thus its analysis is imperative. The determination of copper assumes importance given its widespread occurrence in environmental and biological samples. Thus, several potentiometric sensors have been reported for copper determination. A number of solid-state heterogeneous membrane electrodes for Cu^{2+} have been developed using $CuS-Ag_2S$ mixture, copper tungstoarsenate, Dowex 50W-X4, 13,14-benzo-1,5-tetrathiacyclopentadecane and neutral carrier, o-xylylene bis (dithiocarbamates) as electroactive materials.[74,75] These electrodes work in the concentration range of $10^{-5}-10^{-1}$ M, and the electrode based on Dowex 50W-X4 was used in determining the stability constant of copper sulphosalicylic acid chelate. In addition to solid-state electrodes, a number of liquid membrane electrodes for Cu^{2+} using copper chelates of 2-hydroxy-3-ethyl-5-methylhexanophenoneoxime, salicylanilide, 5-(octyloxymethyl) quinoline-8-ol and bis(trifluoroacetylacetone)-ethylene diamine have also been reported, which work satisfactorily in the concentration range of $10^{-5}-10^{-1}$ M. Thin-film chemical microelectrodes were developed, showing a good response for copper ions. These miniaturized electrodes exhibited Nernstian response and showed a low detection limit of 1.0×10^{-7} M for Cu^{2+}.[76] Another electrode using diamine derivative as an ionophore which was coated as a thin film on graphite has been reported.[77] The effect of co-polymers on the membranes of the Schiff Base complex was studied by researchers. The lifetime of these electrodes was found to be increased by the co-polymers.[78] RNA-based Cu(II) sensors have been reported.[79] Dithiane-based copper ISE was reported for copper ions.[80] Cerium(IV) phosphomolybdate and tetraaza macrocycles[81,82] and ethambutol, benzo tetraazacyclo tetradecane[83] have been used in preparing selective electrodes for Cu^{2+} ions. A comparative investigation on Cu^{2+} selective sensors was carried out using thiohydrazone and thiosemicarbazone, dithiosalicylic and thiosalicylic acids as chelating ionophores.[84] These sensors exhibited interference due to Hg^{2+}, Cl^-, SO_4^{2-} ions. In addition to these, phenylglyoxal-a-monoxime, 9,10-anthraquinone derivative, and various Schiff's bases[85,86] have been explored for the fabrication of copper ions in the past years. A copper (II) ISE based on a highly chelating ligand as an ionophore was reported.[87] A Cu(II) ion sensor using quinone derivative as an ionophore has been reported.[88] A group of workers modified Cu(II)-selective electrodes by introducing a solid contact layer based on graphene or graphene oxide.[89] A novel highly selective and sensitive Cu(II) ions modified carbon paste electrode (MCPE) has been recently reported.[90]

8.5 NO_3^- selective electrodes

Nitrate determination finds an important place in analytical chemistry. Nitrate converts into nitrite either through a chemical route or through the process of biological degradation. The interaction of nitrite ions with haemoglobin produces methemoglobinemia which has ill effects on humans.[91] A polystyrene-block-polybutadiene-block-polystyrene (SBS)-based nitrate selective sensors have been reported as nitrate ion sensors.[92] A group of researchers have tried derivatives of porphyrin as neutral ionophores and fabricated a nitrate ISE.[93] An anionic receptor-based sensor utilizing tris(2-aminoethyl) amino triamide derivative was reported as nitrate ISE. The sensor has a working range of 1.0×10^{-5} to 1.0×10^{-1} M.[94] Cyclic bis-thiourea derivatives have also been tried successfully in the preparation of nitrate ISE.[95] Doped polypyrroles have also been tried as an ionophore in nitrate sensors.[96] PVC-based membranes using a dendrimer molecule as an ionophore were reported.[97] Recently, a new nitrate potentiometric sensor using trioctylmethylammonium chloride (TOMACl) as an ionophore has been reported.[98] A new type of solid-state-ISEs sensitive to nitrate(V) ions using multi-walled carbon nanotubes-ionic liquid nanocomposite was prepared.[99] A group of researchers have recently developed solid-state ISEs as potentiometric ion sensors for the determination of chloride (Cl^-) or nitrate (NO_3^-) ions.[100]

8.6 Cd^{2+} selective electrodes

Cadmium is one of the important toxic metals, and the effect of its acute poisoning is manifested in a variety of symptoms. Jain and coworkers[101] reported cerium(IV) vanadate in polystyrene matrix as Cd(II) ISE. Gupta and coworkers exploited several crown compounds such as dibenzo-24-crown-8,[102] monoaza-18-crown-6,[103] dicyclohexano-18-crown-6,[104] and dicyclohexano-24-crown-8[105] in PVC matrix for construction of Cd(II)-selective electrodes, while Shamsipur et al. utilized tetrathia-12-crown-4[106] as electroactive material for determination of cadmium. Singh and coworkers also explored different macrocyclic ionophores,[107] aza containing[108] and thia containing ligands[109] for the fabrication of Cd(II) selective electrodes. In recent years, Cd(II) selective electrodes based on several neutral carriers, viz. N,N,N´,N´-tetrado-decyl-3,6-dioxaoctanedithioamide[110] and Schiff base,[111] have been reported. An ion-selective potentiometric electrode (IPE) was prepared based on the sales material (bis(salicylaldehyde) ethylenediamine) as a suitable carrier for the determination of cadmium ions.[112] Cadmium (II)-selective sensor using a novel PVC membrane containing a carbothioamide as an ionophore[113] was developed. New selective and sensitive electrochemical sensors were designed based on the deposition of a promising ion-imprinted polymer (IIP) on the surface of a glassy carbon electrode (GCE) for the detection and monitoring of Cd(II) in different real samples.[114]

9 Conclusions

The analysis of noxious ions is one of the most important aspects of analytical chemistry as these ions contribute to environmental pollution due to their toxic effects on plants, animals, and human beings. The field of chemical sensing is one of the oldest and largest among the several analytical techniques available today for the analysis of such noxious ions. The utility of ISEs for environmental, clinical, food samples, etc., has tremendously increased with the commercial availability of electrodes for cations and anions. The procedure for analysis by these electrodes is fast, involves non-destruction of samples, and can be carried with a little volume of samples. The results are accurate, and the method is quite affordable. A large number of electrodes have been reported for different ions using a variety of electroactive sensing materials. The development of a large number of electrodes for a particular ion has been due to the efforts made by the researchers in this field to improve the working concentration range, selectivity, shelf life, response time, etc. For most of the ions, even the best electrode so far developed is not the last word, and the efficacy can be further improved with continuous R&D. Future challenges in ISEs such as good electrode materials, miniaturization, high selectivity, etc., must be focused on and addressed.

10 Challenges and future outlook

Despite the diverse range of applications of chemical sensors, various aspects still remain unextended. There is a need for sensors with still lowered detection limits in the fields of environmental and industrial process monitoring as well as for analyses in clinical chemistry. The use of newer materials with molecular recognition properties, such as carbon nanotubes, micro- and nanoparticles, magnetic beads, polymers, metal complexes, etc., can thus be explored. Limited lifetime is one of the drawbacks of ion-selective sensors; thus, additional efforts should be given to the development of new immobilization procedures that increase the stability of the electroactive component in the membrane. Efforts are needed in this direction for the further development of electrochemical sensors for uncharged analytes. New approaches, viz. backside calibration potentiometry, controlled current coulometry and controlled current chronopotentiometry (pulstrodes), are other useful directions for ISEs that are recently being explored. These approaches open new attractive horizons for these selective materials and further expand the field of ISEs.

References

1. Karri RR, Ravindran G, Dehghani MH. Wastewater—sources, toxicity, and their consequences to human health. In: *Soft Computing Techniques in Solid Waste and Wastewater Management.* Elsevier; 2021:3–33.
2. Brower JB, Ryan RL, Pazirandeh M. Comparison of ion-exchange resins and biosorbents for the removal of heavy metals from plating factory wastewater. *Environ Sci Technol.* 1997;31(10):2910–2914. https://doi.org/10.1021/es970104i.
3. Ugwu EI, Karri RR, Nnaji CC, et al. Application of green nanocomposites in removal of toxic chemicals, heavy metals, radioactive materials, and pesticides from aquatic water bodies. In: *Sustainable Nanotechnology for Environmental Remediation.* Elsevier; 2022:321–346.
4. Sigg L, Werner S. *Aquatische Chemie: eine Einführung in die Chemie wässriger Lösungen und in die Chemie natürlicher Gewässer.* Verlag der Fachvereine; 1991.
5. Merian E. *Metals and Their Compounds in the Environment: Occurrence, Analysis and Biological Relevance.* New York, Basel, Cambridge: VCH. Weinheim; 1990:790–801.
6. Chandra S, Ruzicka S, Svec P, Lang H. Organotin compounds: an ionophore system for fluoride ion recognition. *Anal Chim Acta.* 2006;577(1):91–97. https://doi.org/10.1016/j.aca.2006.06.036.
7. Gospel W, Jones TA, Kleitz M, Lundstrom I, Siyama T. *Sensors: A Comprehensive Survey Chemical and Biochemical Sensors.* vol. 3. Germany: Wiley-VCH; 1991.
8. Koryta J. *Ions, Electrodes and Membranes.* Chichester, New York, Brisbane, Toronto, Singapore: John Wiley & Sons Ltd.; 1982.
9. Florido A. *Handbook of Chemistry and Physics.* Boca Raton, Florida: CRC Press Inc.; 1988.

10. Kolthoff IM, Sanders HL. Electric potentials at crystal surfaces, and at silver halide surfaces in particular. *J Am Chem Soc*. 1937;59(2):416–420. https://doi.org/10.1021/ja01281a059.

11. Pungor E, Hollos-Rokosinyi E. The use of membrane electrodes in the analysis of ionic concentrations. *Acta Chim Acad Sci Hung*. 1961;27:63–68.

12. Frant MS, Ross JW. Electrode for sensing fluoride ion activity in solution. *Science*. 1966;154(3756):1553–1555. https://doi.org/10.1126/science.154.3756.1553.

13. Higuchi T, Illian CR, Tossounian JL. Plastic electrodes specific for organic ions. *Anal Chem*. 1970;42(13):1674–1676. https://doi.org/10.1021/ac60295a045.

14. Buck RP, Lindner E. Tracing the history of selective ion sensors. *Anal Chem*. 2001;73(3):88–97. https://doi.org/10.1021/ac012390t.

15. Moody GJ, Oke RB, Thomas JDR. A calcium-sensitive electrode based on a liquid ion exchanger in a poly(vinyl chloride) matrix. *Analyst*. 1970;95:910–918. https://doi.org/10.1039/AN9709500910.

16. Tsujimura Y, Sunagawa T, Yokoyama M, Kimura K. Sodium ion-selective electrodes based on silicone-rubber membranes covalently incorporating neutral carriers. *Analyst*. 1996;121:1705–1709. https://doi.org/10.1039/AN9962101705.

17. Qin Y, Peper S, Bakker E. Plasticizer-free polymer membrane ion-selective electrodes containing a methacrylic co-polymer matrix. *Electroanalysis*. 2002;14(19):1375–1381. https://doi.org/10.1002/1521-4109(200211)14:19/20<1375: AID-ELAN1375>3.0.CO;2-8.

18. Yun SY, Hong YK, Oh BK, Cha GS, Nam H. Potentiometric properties of ion-selective electrode membranes based on segmented polyether urethane matrices. *Anal Chem*. 1997;69(5):868–873. https://doi.org/10.1021/ac9605455.

19. Khan FSA, Mubarak NM, Khalid M, et al. Comprehensive review on carbon nanotubes embedded in different metal and polymer matrix: fabrications and applications. *Crit Rev Solid State Mater Sci*. 2021;1–28.

20. Moody GJ, Saad BB, Thomas JDR. Glass transition temperatures of poly(vinyl chloride) and polyacrylate materials and calcium ion-selective electrode properties. *Analyst*. 1987;112(8):1143–1147. https://doi.org/10.1039/AN9871201143.

21. Horvai G, Graf E, Toth K, Pungor E, Buck RP. Plasticisedpoly(vinyl chloride) properties and characteristics of valinomycin electrodes. 1. High-frequency resistance and dielectric properties. *Anal Chem*. 1986;58(13):2735–2740. https://doi.org/10.1021/ac00126a034.

22. Toth K, Graf E, Horvai G, Pungor E, Buck RP. Plasticised poly(vinyl chloride) properties and characteristics of valinomycin electrodes. 2. Low-frequency, surface-rate, and Warburg impedance characteristics. *Anal Chem*. 1986;58(13):2741–2744. https://doi.org/10.1021/ac00126a035.

23. Mikhelson KN. Ion-selective electrodes in PVC matrix. *Sensors Actuat B Chem*. 1994;18(1):31–37. https://doi.org/10.1016/0925-4005(94)87051-9.

24. Arada-Perez MA, Marin LP, Quintana JC, Yazdani-Pedram M. Influence of different plasticisers on the response of chemical sensors based on polymeric membranes for nitrate ion determination. *Sensors Actuat B Chem*. 2003;89(3):262–268. https://doi.org/10.1016/S0925-4005(02)00475-6.

25. Sakaki T, Harada T, Kawahara Y, Shinkai S. On the selection of the optimal plasticiser for calix[n]arene-based ion-selective electrodes: possible correlation between the ion selectivity and the 'softness' of the plasticiser. *J Incl Phenom Macrocycl Chem*. 1994;17(4):377–392. https://doi.org/10.1007/BF00707133.

26. Craggs A, Keil L, Moody GJ, Thomas JDR. An evaluation of solvent mediators for ion-selective electrode membranes based on calcium bis(dialkylphosphate) sensors trapped in poly(vinyl chloride) matrices. *Talanta*. 1975;22(10–11):907–910. https://doi.org/10.1016/0039-9140(75)80191-3.

27. Ammann D, Pretsch E, Simon W, Lindler E, Bezegh A, Pungor E. Lipophilic salts as membrane additives and their influence on the properties of macro-and micro-electrodes based on neutral carriers. *Anal Chim Acta*. 1985;171:119–129. https://doi.org/10.1016/S0003-2670(00)84189-6.

28. Gehring P, Morf WE, Welti M, Pretsch E, Simon W. Catalysis of ion transfer by tetraphenylborates in neutral carrier-based ion-selective electrodes. *Helv Chim Acta*. 1990;73(1):203–212. https://doi.org/10.1002/hlca.19900730124.

29. Rosatzin T, Bakker E, Suzuki K, Simon W. Lipophilic and immobilised anionic additives in solvent polymeric membranes of cation-selective chemical sensors. *Anal Chim Acta*. 1993;280(2):197–208. https://doi.org/10.1016/0003-2670(93)85122-Z.

30. Wakida S, Masadome T, Imato T, et al. Additive-salt effect on low detection limit and slope sensitivity in response of potassium- and sodium-selective neutral carrier-based electrodes and their liquid-membrane based ion-sensitive field-effect transistor. *Anal Sci*. 1999;15(1):47–51. https://doi.org/10.2116/analsci.15.47.

31. Baiulescu GE, Cosofret VV. *Applications of Ion-Selective Membrane Electrodes in Organic Analysis*. Chichester: Ellis Horwood; 1977.

32. Helfferich F. *Ion-Exchange*. New York: McGraw Hill; 1962.

33. Moody GJ, Thomas JDR. Development and publication of work with selective ion-sensitive electrodes. *Talanta*. 1972;19(5):623–639. https://doi.org/10.1016/0039-9140(72)80202-9.

34. Bailey PL. *Analysis with Ion-Selective Electrodes*. 2nd ed. London: Heyden; 1980.

35. Levins RJ. Barium ion-selective Electrode based on a neutral carrier complex. *Anal Chem*. 1971;43(8):1045–1047. https://doi.org/10.1021/ac60303a008.

36. Moody GJ, Thomas JDR. *Selective Ion-Sensitive Electrodes*. Watford: Merrow Publishing Co.; 1971.

37. Bhat VS, Ijeri VS, Srivastava AK. Coated wire lead(II) selective potentiometric sensor based on 4-tert-butylcalix[6]arene. *Sensors Actuat B Chem*. 2004;99(1):98–105. https://doi.org/10.1016/j.snb.2003.11.001.

38. Ganjali MR, Hosseini M, Basiripour F, et al. Novel coated-graphite membrane sensor based on N,N′-dimethylcyanodiaza-18-crown-6 for the determination of ultra-trace amounts of lead. *Anal Chim Acta*. 2002;464(2):181–186. https://doi.org/10.1016/S0003 2670(02)00478-6.

39. Szymanska I, Ocicka K, Radecki H, et al. Methoxy-substituted derivatives of 1,4-bis(2-phenylethenyl)benzene and of 1,4-bis(2-phenylethyl)benzene as ligands in ion-selective electrodes for lead ions. *Mater Sci Eng C*. 2001;18(1-2):171–176. https://doi.org/10.1016/s0928-4931(01)00348-4.

40. Gupta VK, Mangla R, Agarwal S. Pb(II) selective potentiometric sensor based on 4-tert-butylcalix[4]arene in PVC matrix. *Electroanalysis*. 2002;14(15–16):1127–1132. https://doi.org/10.1002/1521-4109(200208)14:15/163.0.CO;2-7.

41. Srivastava SK, Gupta VK, Jain S. Determination of lead using poly(vinyl chloride) based crown ether membrane. *Analyst*. 1995;120:495–498. https://doi.org/10.1039/AN9952000495.

42. Ardakani MM, Kashani MK, Salavati-Niasari M, Ensafi AA. Lead ion-selective electrode prepared by sol-gel and PVC membrane techniques. *Sensors Actuat B Chem*. 2005;107(1):438–445. https://doi.org/10.1016/J.SNB.2004.10.036.

43. Kim H, Lee HK, Choi AY, Jeon S. Polymeric lead(II)-selective electrode based on N,N′-Bis-thiophenthiophene-2-ylmethylene-pyridine-2,6-diamine as an ion carrier. *Bull Korean Chem Soc*. 2007;28(4):538–542. https://doi.org/10.5012/bkcs.2007.28.4.538.

44. Huang MR, Ding YB, Li XG. Lead-ion potentiometric sensor based on electrically conducting microparticles of sulfonic phenylenediamine. *Analyst*. 2013;138:3820–3829. https://doi.org/10.1039/C3AN00346A.

45. Golds Á, Horváth V, Huszthy P, Tóth T. Fast potentiometric analysis of lead in aqueous medium under competitive conditions using an acridono-crown ether neutral ionophore. *Sensors*. 2018;18(5):1407–1421. https://doi.org/10.3390/s18051407.

46. Ardalani M, Shamsipur M, Seidani AB. A new generation of highly sensitive potentiometric sensors based on ion-imprinted polymeric nanoparticles/multiwall carbon nanotubes/polyaniline/graphite electrode for sub-nanomolar detection of lead(II) ions. *J Electroanal Chem*. 2020;879(9). https://doi.org/10.1016/j.jelechem.2020.114788, 114788.

47. Lan J, Tao W, Meng-Xia X, Nan Z. Single-layer Pb^{2+} potentiometric sensor with MoS$_2$ nanoflakes as ion-to-electron transducer. *Int J Electrochem Sci*. 2021;16. https://doi.org/10.20964/2021.07.36, 210751.

48. Radic Njegomir. Solid-state Electrode sensitive to mercury ions. In: *Ion Selective Electrode Review*. vol. 11. Pergamon; 1989:177–188.

49. Jain AK, Gupta VK, Singh LP. A polystyrene-based heterogeneous solid membrane of cerium(IV) selenite as sensor for Hg(II) ions. *Indian J Chem Technol*. 1995;2:189–192. http://nopr.niscair.res.in/handle/123456789/31123.

50. Jain AK, Gupta VK, Singh LP. A solid membrane sensor for Hg(II) ions. *Bull Electrochem*. 1996;12:418–420.

51. Gupta VK, Jain S, Khurana U. A PVC-based pentathia-15-crown-5 membrane potentiometric sensor for mercury(II). *Electroanalysis*. 1997;9(6):478–480. https://doi.org/10.1002/elan.1140090609.

52. Gupta VK, Chandra S, Agarwal S. Mercury selective electrochemical sensor based on a double-armed crown ether as ionophore. *Indian J Chem, Sect A*. 2003;42A(4):813–818. http://nopr.niscair.res.in/handle/123456789/18196.

53. Gupta VK, Chandra S, Lang H. A highly selective mercury electrode based on a diamine donor ligand. *Talanta*. 2005;66(3):575–580. https://doi.org/10.1016/j.talanta.2004.11.028.

54. Singh LP, Bhatnagar JM. Chelating ionophores based electrochemical sensor for Hg(II) ions. *J Appl Electrochem*. 2004;34(4):391–396. https://doi.org/10.1023/B:JACH.0000016612.99921.b6.

55. Liang R, Wang Q, Qin W. Highly sensitive potentiometric sensor for detection of mercury in Cl⁻-rich samples. *Sensors Actuat B Chem*. 2015;208:267–272. https://doi.org/10.1016/j.snb.2014.11.040.

56. Gupta VK, Sethi B, Sharma RA, Agarwal S, Bharti A. Mercury selective potentiometric sensor based on low rim functionalisedthiacalix [4]-arene as a cationic receptor. *J Mol Liq*. 2013;177:114–118. https://doi.org/10.1016/j.molliq.2012.10.008.

57. Aglan RF, Saleh HM, Mohamed GG. Potentiometric determination of mercury (II) ion in various real samples using novel modified screen-printed electrode. *Appl Water Sci*. 2018;8(5):141–152. https://doi.org/10.1007/s13201-018-0781-z.

58. Jumal J, Bohari MY, Ahmad M, Heng L. Mercury Ion-selective electrode with self-plasticizing poly(n-butyl acrylate) membrane based on 1,2-bis-(N′-benzoylthioureido)cyclohexane as ionophore. *APCBEE Proc*. 2012;3:116–123. https://doi.org/10.1016/j.apcbee.2012.06.056.

59. Alharthi S, Fallatah A, Hamed S. Design and characterization of electrochemical sensor for the determination of mercury(II) ion in real samples based upon a new schiff Base derivative as an ionophore. *Sensors*. 2021;21(9):3020. https://doi.org/10.3390/s21093020.

60. Goyer RA. Toxic effects of metals. In: Klaassen CD, ed. *Casarett&Doull's Toxicology: The Basic Science of Poisons*. 5th ed. New York City, NY: McGraw-Hill; 1996:715–716. https://accesspharmacy.mhmedical.com/content.aspx?bookid=958&sectionid=53483748.

61. Greentree WF, Hall JO. Iron toxicosis. In: Bonagura JD, ed. *Kirk's Current Therapy XII Small Animal Practice*. Philadelphia, PA: WB Saunders Co; 1995:240–242.

62. Porter JB, Garbowski M. The pathophysiology of transfusional iron overload. *Hematol Oncol Clin North Am*. 2014;28(4):683–701. https://doi.org/10.1016/j.hoc.2014.04.003.

63. Andrews NC. Disorders of iron metabolism. *N Engl J Med*. 1999;341(26):1986–1995. https://doi.org/10.1056/NEJM199912233412607.

64. Cabantchik ZI, Breuer W, Zanninelli G, Cianciulli P. LPI-labile plasma iron in iron overload. *Best Pract Res Clin Haematol*. 2005;18(2):277–287. https://doi.org/10.1016/j.beha.2004.10.003.

65. Mashhadizadeh MH, Shoaeil S, Monadi N. A novel ion-selective membrane potentiometric sensor for direct determination of Fe(III) in the presence of Fe(II). *Talanta*. 2004;64(4):1048–1052. https://doi.org/10.1081/AL-100104956.

66. Gupta VK, Jain AK, Agarwal S, Maheshwari G. An iron(III) ion-selective sensor based on a μ-bis(tridentate) ligand. *Talanta*. 2007;71(5):1964–1968. https://doi.org/10.1016/j.talanta.2006.08.038.

67. Hassan SSM, Marzouk SAM. A novel ferroin membrane sensor for potentiometric determination of iron. *Talanta*. 1994;41(6):891–899. https://doi.org/10.1016/0039-9140(94)e0042-p.

68. Saleh MB. Iron(III) ionophores based on formylsalicylic acid derivatives as sensors for ion-selective electrodes. *Analyst*. 2000;125:179–183. https://doi.org/10.1039/A905530D.

69. Ekmekci G, Uzun D, Somer G, Kalaycı S. A novel iron(III) selective membrane electrode based on benzo-18-crown-6 crown ether and its applications. *J Membr Sci*. 2007;288(1-2):36–40. https://doi.org/10.1016/j.memsci.2006.10.044.

70. Gupta VK, Sethi B, Upadhyay N, Kumar S, Singh R, Singh LP. Iron (III) selective electrode based on S-methyl N-(methylcarbamoyloxy) thioacetimidate as a sensing material. *Int J Electrochem Sci*. 2011;6(3):650–663.

71. Badakhshan S, Ahmadzadeh S, Bandpei AM, Aghasi M, Basiri A. Potentiometric sensor for iron (III) quantitative determination: experimental and computational approaches. *BMC Chem*. 2019;13(131):131–143. https://doi.org/10.1186/s13065-019-0648-x.

72. Sanjeev K, Susheel KM, Navneet K, Ravneet K. Improved performance of Schiff based ionophore modified with MWCNT for Fe(II) sensing by potentiometry and voltammetry supported with DFT studies. *RSC Adv.* 2017;7:16474–16483. https://doi.org/10.1039/C7RA00393E (Paper).

73. Paul A, Prkić A, Mitar I, et al. Potentiometric response of solid-state sensors based on ferric phosphate for iron(III) determination. *Sensors.* 2021;21 (5):1612–1625. https://doi.org/10.3390/s21051612.

74. Leest RE. Solid-state ion-selective electrodes for metal ions. *Analyst.* 1977;102:509–514. https://doi.org/10.1039/AN9770200509.

75. Palanivel A, Riyazuddin P. Silver sulphide-copper sulphide coated graphite as an electrode for selective determination of Ag(I) &Cu(II). *Indian J Chem Sect A.* 1984;23A(12):1051–1052. http://nopr.niscair.res.in/handle/123456789/49058.

76. Mourzina YG, Schubert J, Zander W, et al. Development of multisensor systems based on chalcogenide thin-film chemical sensors for the simultaneous multicomponent analysis of metal ions in complex solutions. *Electrochim Acta.* 2001;47(1–2):251–258. https://doi.org/10.1016/j.proeng.2012.09.148.

77. Ganjali MR, Poursaberi T, Babaei LH, et al. Highly selective and sensitive copper(II) membrane coated graphite electrode based on a recently synthesised Schiff's base. *Anal Chim Acta.* 2001;440(2):81–87. https://doi.org/10.2116/analsci.18.289.

78. Gupta KC, D'Arc MJ. Effect of concentration of ion exchanger, plasticiser and molecular weight of cyanocopolymers on selectivity and sensitivity of Cu(II) ion-selective electrodes. *Anal Chim Acta.* 2001;437(2):199–216. https://doi.org/10.1016/S0003-2670(01)00995-3.

79. Hassan SSM, Mahmoud WH, Othman AHM. Ribonucleic acid as a novel ionophore for potentiometric membrane sensors of some transition metal ions. *Talanta.* 1998;47(2):377–385. https://doi.org/10.1016/s0039-9140(98)00142-8.

80. Abbaspour A, Kamyabi MA. Copper (II)-selective electrode based on dithioacetal. *Anal Chim Acta.* 2002;455:225–231. https://doi.org/10.1016/S0003-2670(01)01622-1.

81. Jain AK, Singh P, Singh LP. A polystyrene-based heterogeneous ion- exchange membrane of cerium(IV) phosphomolybdate as copper(II) ion-selective electrode. *Indian J Chem Sect A.* 1994;33A(3):272–273.

82. Jain AK, Gupta VK, Sahoo BB, Singh LP. Copper(II)-selective electrodes based on macrocyclic compounds. *Anal Proc Incl Anal Commun.* 1995;32:99–101. https://doi.org/10.1039/AI9953200099.

83. Gupta VK, Prasad R, Kumar A. Preparation of ethambutol-copper(II) complex and fabrication of PVC based membrane potentiometric sensor for copper. *Talanta.* 2003;60(1):149–160. https://doi.org/10.1016/S0039-9140(03)00118-8.

84. Gismera MJ, Procopio JR, Sevilla MT, Hernandez L. Copper(II) ion-selective electrodes based on dithiosalicylic and thiosalicylic acids. *Electroanalysis.* 2003;15(2):126–132. https://doi.org/10.1016/j.jelechem.2011.11.024.

85. Jain AK, Gupta VK, Singh LP, Raisoni JR. Chelating ionophore based membrane sensors for copper(II) ions. *Talanta.* 2005;66(5):1355–1361. https://doi.org/10.1016/j.talanta.2005.02.001.

86. Gholivand MB, Nasrabadi MR, Ganjali MR, Nisar MS. Highly selective and sensitive copper membrane electrode based on a new synthesised Schiff base. *Talanta.* 2007;73(3):553–560. https://doi.org/10.1016/j.talanta.2007.04.010.

87. Gupta VK, Singh R, Upadhyay N, Kaur SP, Singh LP, Sethi B. A novel copper(II)-PVC membrane potentiometric sensor based on dimethyl 4,4′-(o-phenylene)bis(3-thioallophanate). *J Mol Liq.* 2012;174:11–16. https://doi.org/10.1016/j.molliq.2012.07.016.

88. Fraga EYZ, Mohameda ME, Alia AE, Mohamed GG. Potentiometric sensors selective for Cu(II) determination in real water samples and biological fluids based on graphene and multiwalled carbon nanotubes modified graphite electrodes. *Indian J Chem.* 2020;59A:162–173. http://nopr.niscair.res.in/handle/123456789/54009.

89. Magdalena P, Katarzyna F, Joanna S, Robert P, Beata PB. High selective potentiometric sensor for determination of nanomolar con-centration of Cu(II) using a polymeric electrode modified by a graphene/7,7,8,8-tetracyanoquinodimethane nanoparticles. *Talanta.* 2017;170:41–48. https://doi.org/10.1016/j.talanta.2017.03.068.

90. Marwa EBM, Eman YF, Mohamed HEB. Rapid potentiometric sensor for determination of Cu(II) ions in food samples. *Microchem J.* 2021;0026-265X. 164:106065. https://doi.org/10.1016/j.microc.2021.106065.

91. Eaton AD, Clesceri LS, Greenberg AE. *Standard Methods for Examination of Water and Waste Water.* Washington: American Public Health Association; 1995.

92. Goff TL, Braven J, Ebdon L, Scholefield D. High-performance nitrate-selective electrodes containing immobilized amino acid betaines as sensors. *Anal Chem.* 2002;74(11):2596–2602. https://doi.org/10.1021/ac010985i.

93. Lee HK, Song K, Seo HR, Jeon S. Nitrate-selective electrodes based on meso-tetrakis[(2-arylphenylurea)-phenyl]porphyrins as neutral lipophilic ionophores. *Talanta.* 2004;62(2):293–297. https://doi.org/10.1016/j.talanta.2003.07.016.

94. Ortuno JA, Exposito R, Pedreno CS, Albero MI, Espinosa A. A nitrate-selective electrode based on a tris(2-aminoethyl)aminetriamide derivative receptor. *Anal Chim Acta.* 2004;525(2):231–237. https://doi.org/10.1016/j.aca.2004.08.036.

95. Watts AS, Gavalas VG, Cammers A, Andrada PS, Alajarın M, Bachas LG. Nitrate-selective electrode based on a cyclic bis-thiourea ionophore. *Sensors Actuat B Chem.* 2007;121(1):200–207. https://doi.org/10.1016/j.snb.2006.09.048.

96. Rawat A, Chandra S, Sarkar A. Nitrate selective polymeric membrane electrode based on bis-thiourea ligand as carrier. *Sens Lett.* 2009;7(6):1100–1105. https://doi.org/10.1166/sl.2009.1242.

97. Gupta VK, Singh LP, Kumar S, Singh R, Chandra S, Sethi B. Anion recognition through amide-based dendritic molecule: a PVC based sensor for nitrate ions. *Talanta.* 2011;85(2):970–974. https://doi.org/10.1016/j.talanta.2011.05.014.

98. Pérez MAA, Florián KYN. A new potentiometric sensor for nitrate using diethyl phthalate (DEP) as plasticiser and triocthylmethylammonium chloride (TOMACl) as ionophore. *Rev Cuba Quím.* 2018;30(2):277–288.

99. Karolina P, Cecylia W. Comparative study of nitrate all-solid-state ion-selective electrode based on multiwalled carbon nanotubes-ionic liquid nanocomposite. *Sensors Actuat B Chem.* 2021;348. https://doi.org/10.1016/j.snb.2021.130720, 130720.

100. Tsuchiya K, Akatsuka T, Abe Y, Komaba S. Design of all-solid-state chloride and nitrate ion-selective electrodes using anion insertion materials of electrodeposited poly(allylamine)-MnO$_2$ composite. *Electrochim Acta*. 2021;389. https://doi.org/10.1016/j.electacta.2021.138749, 138749.

101. Jain AK, Singh LP. A new polystyrene-based heterogeneous membrane of cerium(IV) vanadate as cadmium(II) ion-selective electrode. *Indian J Chem*. 1994;33A(12):1122–1123. http://nopr.niscair.res.in/handle/123456789/41158.

102. Gupta VK, Kumar P. Cadmium (II)-selective sensors based on dibenzo-24- crown-8 in PVC matrix. *Anal Chim Acta*. 1999;389:205–212. https://doi.org/10.1016/S0003-2670(99)00154-3.

103. Gupta VK, Kumar P, Mangla R. PVC based monoaza-18-crown-6 membrane potentiometric sensor for cadmium. *Electroanalysis*. 2000;12(9):752–756.

104. Gupta VK, Chandra S, Mangla R. Dicyclohexano-18-crown-6 as active material in PVC matrix membrane for the fabrication of cadmium selective potentiometric sensor. *Electrochim Acta*. 2002;47:1579–1586. https://doi.org/10.1002/1521-4109(200006)12:103.0.CO;2-V.

105. Gupta VK, Jain AK, Kumar P. PVC based membranes of dicyclohexano24-crown-8 as Cd(II)-selective sensor. *Electrochim Acta*. 2006;52(2):736–741. https://doi.org/10.1016/S0003-2670(99)00154-3.

106. Shamsipur M, Mashhadizadeh MH. Cadmium ion-selective electrode based on tetrathia-12-crown-4. *Talanta*. 2001;53(5):1065–1071. https://doi.org/10.1016/s0039-9140(00)00602-0.

107. Panwar A, Baniwal S, Sharma CL, Singh AK. A polystyrene-based membrane electrode for cadmium(II) ions. *Fresenius J Anal Chem*. 2000;368(8):768–772. https://doi.org/10.1007/s002160000601.

108. Singh AK, Saxena P, Singh R. New cadmium(II)-selective electrode based on a tetraazacyclohexadeca macrocyclic ionophore. *Anal Sci*. 2005;21(2):179–181. https://doi.org/10.2116/analsci.21.179.

109. Singh AK, Mehtab S, Singh UP, Aggarwal V. Comparative studies of tridentate sulfur and nitrogen-containing ligands as ionophores for construction of cadmium ion-selective membrane sensors. *Electroanalysis*. 2007;19(11):1213–1221. https://doi.org/10.1002/elan.200703846.

110. In AC, Bakker E, Pretsch E. Potentiometric Cd^{2+}-selective electrode with a detection limit in the low ppt range. *Anal Chim Acta*. 2001;440(2):71–79. https://doi.org/10.1016/S0003-2670(01)01052-2.

111. Mashhadizadeh MH, Sheikhshoaie I, Saeid-Nia S. Asymmetrical Schiff bases as carriers in PVC membrane electrodes for cadmium (II) ions. *Electroanalysis*. 2005;17(8):648–654. https://doi.org/10.1002/elan.200403134.

112. Ghazizadeh M, Asadollahzadeh H. A rapid cadmium determination based on ion-selective membrane potentiometric sensor by bis (salicylaldehyde) ethylenediamine as carrier. *Anal Methods Environ Chem J*. 2021;4(02):25–33. https://doi.org/10.24200/amecj.v4.i02.136.

113. Oguz Ö, Ömer I, Meliha BG, Caglar B. Cadmium(II)-selective potentiometric sensor based on synthesised (*E*)-2-benzylidenehydrazinecarbothioamide for the determination of Cd^{2+} in different environmental samples. *Int J Environ Anal Chem*. 2020;1–16. https://doi.org/10.1080/03067319.2020.1817427.

114. Abdallah AB, Mohamed RE, Molouk AFS, Tamer AA, El-Shafei AA, Magdi EK. Selective and sensitive electrochemical sensors based on an ion imprinting polymer and graphene oxide for the detection of ultra-trace Cd(II) in biological samples. *RSC Adv*. 2021;11:30771–30780. https://doi.org/10.1039/D1RA05489A [Paper].

Chapter 16

Potential of *Cassia fistula* pod-based absorbent in remediating water pollutants: An analytical study

Rakesh Bhutiani[a], R.C. Tiwari[b], Parul Chauhan[b], Faheem Ahamad[c,*], Ved Bhushan Sharma[b], Inderjeet Tyagi[d], and Pooja Singh[b]

[a]*Limnology and Ecological Modelling Lab, Department of Zoology and Environmental Science, Gurukula Kangri Vishwavidyalaya, Haridwar, UK, India,* [b]*Department of Agadtantra, Uttarakhand Ayurvedic University, Haridwar, India,* [c]*Department of Environmental Science, Keral Verma Subharti College of Science (KVSCOS), Swami Vivekanand Subharti University, Meerut, UP, India,* [d]*Centre for DNA Taxonomy, Molecular Systematics Division, Zoological Survey of India, Ministry of Environment, Forest and Climate Change, Government of India, Kolkata, West Bengal, India*
*Corresponding author.

1 Introduction

Water pollution is an issue of great concern worldwide. It can be broadly divided into three main categories: contamination by organic compounds, inorganic compounds (e.g., heavy metals), and microorganisms. In recent years, the number of research studies concerning the use of efficient processes to clean up and minimize the pollution of water bodies has been increasing.[1] The available water is under stress due to an ever-escalating demand from the agriculture, industrial, and domestic sectors.[2–5] Like all other natural resources, water is essential to the living organism. Contaminated water is world's one of the leading health dangers and remains to hover both value of life and community health.[6–8] Heavy metals and toxic compounds, discharged from industries gather in adjoining lakes and rivers, are lethal to oceanic fish and crustacean.[9–12] Worldwide approximately 1.2 billion people at present are living in natural water-scarce areas, and this figure is expected to increase by about 1.8 billion by 2025.[13–15] Consumption of these pollutants by humans causes congenital disabilities, different kinds of malignancies, and features of lead poisoning, mercury poisoning, arsenic poisoning, and many others. Microbial contaminants from sewage cause infectious diseases like cholera, diarrhea, typhoid fever, viral hepatitis, amoebiasis, worm infestation, and many other diseases.[16] Suspended particles make the water turbid, which results in reduced photosynthesis due to less availability of sunlight.

As per the central pollution control board (CPCB) report, a treatment facility for approximately 30% of the total generated domestic effluent is available. A similar scenario is in the case of industrial effluent. To avoid the risk of health and for the conservation of this natural entity, wastewater treatment is necessary. Several methods are already being used to clean up the environment from these kinds of contaminants, but most are costly and far away from their optimum performance.[17] The chemical technologies generate large volumetric sludge and increase the costs of chemical and thermal methods which are both technically challenging and expensive that all of these methods can also degrade the valuable component of soils.[18,19] There are many processes described in modern science for water cleansing. Some of them are physical processes such as aeration, separation, sedimentation and refinement, chemical processes, biological processes, biologically active carbon, and the use of ultraviolet rays, reverse osmosis (RO); but none of these are satisfactory.[20,21] All of them have some disadvantages such as RO removes essential nutrients from the water, some methods are time-consuming, some are costly, and some have wastage of water. Despite the tremendous advancement of the modern water purification system, an ideal system for the purification of water is not yet available. For industrial and domestic effluents, effluent treatment plant (ETP) and sewage treatment plant (STP) are used. Still, these require a lot of energy and skilled manpower and are therefore considered costly. Adsorption is another technique used nowadays to treat wastewater, but pilot scale studies are not available till now because of the slow speed of the treatment.

Cassia fistula Linn. is a native to the Indian continent, having a height ranging from 30 to 40 ft. Flowers are yellow and have a profuse bloom from May to July, and the pod is cylindrical, almost straight, smooth, and dark brown colored.[22] It

Sustainable Materials for Sensing and Remediation of Noxious Pollutants. https://doi.org/10.1016/B978-0-323-99425-5.00001-3

possesses therapeutic potential in diseases like all skin diseases, cardiac problems, gout, fever, blood disorders, diabetes mellitus, herpes, etc. Mainly its fruit pulp is used for medicinal purposes. Its pods and fruit pulp contain many alkaloids like tannin, anthraquinone, cetyl alcohol, fistulic acid, rhein, methionine, argentine, leucine, etc. Tannin has antioxidant and antimicrobial properties; it accelerates blood clotting, reduces BP, decreases serum lipid level, and modulates the immune response. Anthraquinone is water-soluble, laxative, antifungal, antiviral, antimicrobial, diuretic, and anti-inflammatory. Cetyl alcohol is a lubricant and has hydrating properties. Fistulic acid has mild laxative, antiparasitic, and antifertility properties. Glycoside is water-soluble, antibacterial, and converts toxic material to nontoxic material.[23-29]

There is a need to develop a methodology for wastewater treatment that will require low amount of energy and manpower and possess a high speed of water treatment. In the current study, an effort has been made to advance low-cost methodology by combining sand filtration and biosorption techniques.

2 Materials and methods

2.1 Authentication of plant material

The plant materials for the authentication were a genuine sample of *C. fistula* Linn. with leaves and flowers. The sample was collected from Uttarakhand Ayurved University (UAU), Rishikul Campus, Haridwar. The authentication of the sample was carried out by Botanical Survey of India (BSI), Dehradun [Tech. /Herb (Ident.)/2018-2019/261].

2.2 Collection of *Aragvadha* pods and preparation of the ash

Pods of ripened *Aragvadha* were collected from UAU, Rishikul Campus, Haridwar. For the preparation of pod ash, it was kept for 7 days inside the sand, dried in sunlight, and burnt in *Rasashala* using an open method in the *Rasashastra* department of the Rishikul campus. The obtained ash was ground and washed three times with water. The sand and gravel were purchased from a building material shop from the market. The material was first washed three times with tap water and then three times with distilled water. After washing and drying the material, the material was oven-dried in the oven at 100°C. The schematic diagram of ash preparation is given in Fig. 1.

FIG. 1 Process of ash preparation of *C. fistula* Linn.

2.3 Design of filtering apparatus

The reactor design prepared for the present study is given in Fig. 2. The present filtering apparatus is a combination of slow sand filtration technology and biosorption technology which is made up of four layers as presented below:

(1) First layer: The first or upper layer was prepared using previously washed, dried, and sieved sand. The width of this layer was fixed at 10 cm.
(2) Second layer: The second layer was prepared using previously prepared biosorbents (using pod of *C. fistula* Linn.). The width of this layer was fixed at 10 cm.
(3) Third layer: The third layer was prepared using previously washed and dried gravel. The width of this layer was fixed at 10 cm.
(4) Fourth layer: The fourth or bottom layer was made up of cotton. The width of this layer was fixed at 5 cm.

All the four layers were separated from each other with the help of markeen cloth (a very light cloth).

2.4 Examination of water

Four water samples were collected from four different sites of Haridwar. For study purpose, the collected water sample were coded as follows:

SS 1—Ganga water.
SS 2—Hand pump water.
SS 3—Industrial effluent.
SS 4—Tap water.

The samples were collected and analyzed following the standard methods described in APHA, Trivedi and Goel, and Khanna and Bhutiani.[30–32] The parameters taken in the study were temperature, pH, color, TDS, DO, BOD, COD, alkalinity, acidity, hardness, chloride, *Escherichia coli, and* heavy metals (Hg, Pb, Zn, Cd, and Cu) estimation. For microbial analysis, samples were collected in sterile bottles. The efficiency of reactors was calculated using the following equation:

$$\text{Efficiency of the reactor} = \left(\frac{C_i - C_f}{C_i}\right) \times 100$$

where C_i is the concentration before treatment and C_f is the concentration after treatment.

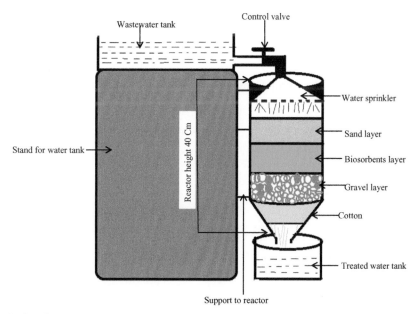

FIG. 2 System for water treatment.

3 Results and discussion

3.1 Physicochemical study of *C. fistula* Linn.

The plant *C. fistula* Linn. and its physicochemical tests are described in Ayurvedic Pharmacopeia of India (API). The physical property of the *C. fistula* Linn. pod ash sample is given in Table 1.

3.1.1 Moisture content

Moisture content can affect the shelf life of any sample. The moisture content of the *C. fistula* Linn. pod ash was found 10.14%. The standard value of moisture content for *C. fistula* Linn. is not available in API.

3.1.2 Ash value

The ash value is the indicator of inorganic matter in the plant. The higher ash value suggests heat-stable or inorganic constituents present in the sample. The weight of total ash was found to be 48.17% which is beyond the standard value of API (not more than 6% for *C. fistula* Linn.). The value of acid-insoluble ash was 2.62% which is beyond the standard value of API (not to exceed 1% for *C. fistula* Linn.). The value of water doable ash was 2.34%. The standard value of water doable ash for *C. fistula* Linn. is not available in API.

3.1.3 Extractive value

An extractive value of the medicinal plant is related to its phytochemical constituent. The water-soluble extractive value of ash was 2.56%, which is said not to be more than 46% in the case of *Aragvadha*.

3.2 Analytical study of water

3.2.1 Examination of water on Ayurvedic parameters

Water examination on *Ayurvedic* parameters was performed in which *gandha, varna, pichhilta, phenodgam, raji,* and *aakriti* were examined. Ganga water was found turbid, while industrial water was found yellowish-green in color, pungent in the smell with *phenodgam*. The other two samples were found with a standard range of parameters (tap water and hand pump water).

3.2.2 Shalidhanya pariksha

It was carried out on all water samples. None of the tested samples contained *kotha*, but *varnavikriti* was present in industrial water. The presence of *varnavikriti* in industrial water shows the polluted nature of that sample.

TABLE 1 Showing the physical properties of *C. fistula* Linn. (*Aragvadha*) pod ash.

SN	Test parameters	Results	Method reference
1.	Description	A black colored powder	Visual
2.	Color	Black	Visual
3.	Loss on drying (% w/w)[a]	10.14	API
4.	Total ash (% w/w)[a]	48.17	API
5.	Acid-insoluble ash (% w/w)[a]	2.62	API
6.	Water-soluble ash (% w/w)[a]	2.34	API
7.	Water-soluble extractive (% w/w)[a]	2.56	API

[a]API, Ayurvedic Pharmacopeia of India.

3.3 Examination of water on modern parameters

The parameters analyzed during the study period were physicochemical parameters such as total dissolved solids (TDS), pH, acidity, alkalinity, hardness, chloride, dissolved oxygen (DO), biochemical oxygen demand (BOD), chemical oxygen demand (COD); heavy metals such as mercury (Hg), lead (Pb), zinc (Zn), cadmium (Cd), copper (Cu); and microbiological parameters such as *E. coli*. Water was filtered thrice from the filtration reactor. Physicochemical parameters were analyzed before and after each filtration, while heavy metals and *E. coli* were analyzed before and after final filtration. The results of physicochemical parameters are presented in Table 2 and Figs. 3–6. In contrast, the heavy metals and microbiological results are given in Table 3 and results of ANOVA are shown in Table 4.

TDS is the indicator of the amount of dissolved solids in water. Reduction in TDS ranged from 21.1% to 45.6% in all water samples. Maximum reduction was observed in tap water. *Aragvadha* pod ash contains glycosides (anthraquinone),

TABLE 2 Showing physicochemical parameters of samples before and after filtration using *C. fistula* Linn. (*Aragvadha*) pod ash.

Sample		TDS (mg/L)	pH	Acidity (mg/L)	Alkalinity (mg/L)	Hardness (mg/L)	Chloride (mg/L)	DO (mg/L)	BOD (mg/L)	COD (mg/L)
SS 1	Raw water	650	7.6	40	330	66	18.2	8.1	5.6	85
	1st filtration	440	9.3	28	687	14	15.5	8.7	4.1	74
	2nd filtration	265	8.3	22	450	20	12.4	8.9	3.6	52
	3rd filtration	185	7.9	15	386	16	9.8	9.2	2.9	45
SS 2	Raw water	685	6.5	320	114	400	248.5	4.5	5.6	14.5
	1st filtration	435	7.6	146	366	320	298.5	4.1	6.1	11.6
	2nd filtration	256	7.5	192	324	240	240	3.4	6.5	9.7
	3rd filtration	188	7.4	200	306	200	212.5	2.8	6.7	6.3
SS 3	Raw water	1258	6.3	410	150	280	65.7	0.9	76	655
	1st filtration	755	8.9	100	450	170	55.3	2.5	45	425
	2nd filtration	412	8.2	187.5	400	90	42.7	4.1	22	295
	3rd filtration	325	7.4	230	350	70	38.9	5.5	15	220
SS 4	Raw water	789	7.7	134	400	180	142	6.2	3.2	56
	1st filtration	512	8.5	34	650	130	156.5	6.9	2.7	39
	2nd filtration	395	8.2	38	610	124	106.5	7.6	2.1	28
	3rd filtration	215	8.1	45	584	80	104	7.9	1.4	22

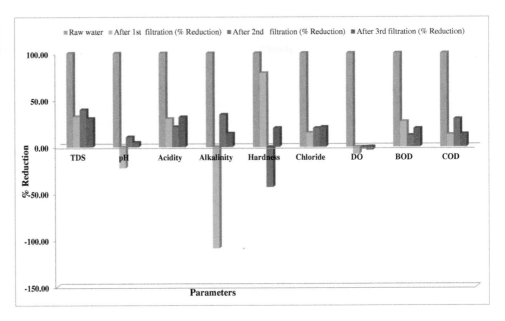

FIG. 3 Representing the % reduction in physicochemical parameters of raw and treated water in case of SS 1.

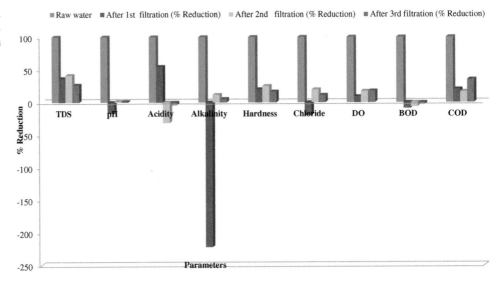

FIG. 4 Showing the % reduction in physicochemical parameters of raw and treated water in case of SS 2.

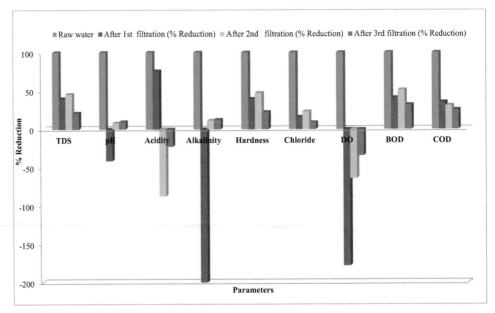

FIG. 5 Showing the % reduction in physicochemical parameters of raw and treated water in case of SS 3.

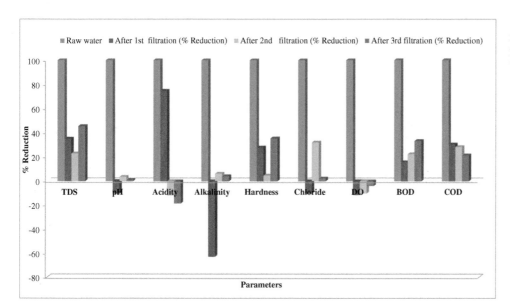

FIG. 6 Showing the % reduction in physicochemical parameters of raw and treated water in case of SS 4.

converting toxic material into nontoxic material. This might help in the deduction of TDS. The statistic result shows that *P*-value is less than 0.05, which depict a significant difference among TDS values in all groups. Rak and Ismail observed the 56.4% reduction in solids after treating *C. alata* leaves based on coagulant.[33] pH is the extent of the acid or alkali in a liquid or aqueous solution of a solid material. The ideal pH of water is 6.5–8.5. In all four types of water samples, pH tremendously increased in starting and after three times filtration, this went in the normal range. This increase in pH might be due to the alkaline nature of *Aragvadha* ash. For example, raw Ganga water pH was 7.63, after first-time filtration, this was 9.3, after second time 8.3, and lastly, this went on a range of 7.9. The statistic result shows that *P*-value is greater than 0.05, so there is no significant difference found among pH values in all groups. Díaz et al. also observed no significant difference between the average pH.[22] The pH range also affects the reduction of other metallic ions and other ions from the water.[34,35] Therefore, pH balancing is the necessary step of water treatment.

Acidity is the sum of all titratable acid present in the water sample. Increased acidity of water causes cancer, cardiac, and kidney diseases. In all four types of samples, acidity was decreased due to the richness of alkaloids in *Aragvadha* ash, which are little alkaline in nature. For example, in industrial effluent, firstly acidity level was 250, and after 3-time filtration, this was 62.50. The statistic result shows that *P*-value is less than 0.05, so there is a significant difference found among acidity values in all groups. Alkalinity is the degree of water's ability to neutralize acidity. The ideal alkalinity range for drinkable water is 20–200 mg/L. Its increased value causes skin irritation and metabolic alkalosis. After the treatment, the increase in alkalinity of all the samples may be due to the presence of alkaloids in an absorbent, which also lowers the acidity of the treated water. The statistic result shows *P*-value is less than 0.05, so there is a significant difference among alkalinity values in all four groups.

Hardness is considered the capacity of water to reduce and destroy the lather of soap. Hardness in water is present due to the natural accumulation of salts from contact with soil and other geological formations, or it may enter the water from direct pollution caused by industrial effluents. Hardness is temporary if it is associated with carbonates and bicarbonates and permanent if associated with sulfate and chlorides. The ideal hardness of drinkable water is less than 60 mg/L; if its high concentration is present in water, it deposits in the pipe and indicates an increased level of metals in water. In our study, hardness is magically decreased in all samples. For example, hardness of raw tap water was reported 180, and after three times filtration, this was 80. The statistic result shows that *P*-value is less than 0.05, so there is a significant difference among hardness values in all four groups. The standard value of chloride is 250 mg/L. Its higher concentration causes high blood pressure in humans and damage to plants. In our study, value of chloride was decreased in all samples. The statistic result shows *P*-value is less than 0.05, so there is a significant difference among chloride values in all groups. Chloride reduction may be due to hydrogen interaction (hydroxyl group) in coagulation or bridging process.[5]

DO represent the amount of oxygen in the water. Both increased and decreased values of DO cause harmful effects on aquatic life. An ideal range of DO is 5–15 mg/L. DO was increased in all samples because of the continuous contact of water with air. In raw industrial effluent, this was 0.90, and at the end of an experiment, this was 5.5. The statistic result shows that *P*-value is less than 0.05, so there is a significant difference among values of DO in all groups. BOD is the amount of

TABLE 3 Showing concentration of metal and microbes in samples before and after filtration using *C. fistula* Linn. (*Aragvadha*) pod ash.

Metals/ sample	SS 1			SS 2			SS 3			SS 4		
	Before filtration	After filtration	% Reduction	Before filtration	After filtration	% Reduction	Before filtration	After filtration	% Reduction	Before filtration	After filtration	% Reduction
Mercury (ppm)	BDL	BDL	–	BDL	BDL	–	BDL	BDL	–	BDL	BDL	–
Lead (ppm)	0.93	0.046	95	BDL	BDL	–	0.09	BDL	100	BDL	BDL	–
Zinc (ppm)	0.10	BDL	100	0.27	BDL	100	13.79	4.73	70	BDL	BDL	–
Cadmium (ppm)	BDL	BDL	100	BDL	BDL	–	BDL	BDL	–	BDL	BDL	–
Copper (ppm)	0.24	BDL	100	0.24	BDL	100	0.35	BDL	100	BDL	BDL	–
E. coli (per 100mL)	5	3	40	BDL	BDL	–	BDL	BDL	–	BDL	BDL	–

TABLE 4 Showing values of ANOVA.

Sample	TDS		pH		Acidity		Alkalinity		Hardness		Chloride		DO		BOD		COD	
	P-value	F-value	P-value	F-value	P-value	F-value	P-value	F-value	P-value	F-value	P-value	F-value	P-value	F-value	P-value	F-value	P-value	F-value
SS 1	0.971	0.438	1.605	0.240	5.248	0.015	4.046	0.033	9.617	0.002	81.801	0.000	17.017	0.000	6.894	0.006	14.383	0.000
SS 2																		
SS 3																		
SS 4																		

$P > 0.05$ (95% confidence level).
SS 1: Ganga water. SS 2: Hand pump water. SS 3: Industrial water. SS 4: Tap water.

dissolved oxygen used in the biological process of metabolizing organic matter in water by microorganisms. It gives an approximate index of organic pollution. Microorganisms utilize animate material as the birthplace of carbon and consume DO present in water for the oxidation process. The ideal BOD of water is 0. In the present study, BOD was decreased in all samples because some alkaloids like fistulic acid, anthraquinone possess antimicrobial activity. The statistic result shows that P-value is less than 0.05, so there is a significant difference among values of BOD in all groups. COD is the oxygen required by the organic substance of water to oxidize them by a strong chemical oxidant.

The increased value of COD indicates the presence of an organic pollutant in water. The ideal COD of water is 0. In the present study, COD was decreased in all samples due to the antimicrobial action of some alkaloids. The statistic result shows that P-value is less than 0.05, so there is a significant difference found among values of COD in all groups. Reduction in COD can also be enhanced using *C. fistula* plant parts extract with synthetic coagulants such as ferric chloride.[22,36] Reduction in most of the physicochemical parameters is due to coagulation, adsorption, and neutralization.[37] It may be due to water-soluble positive globular protein that absorbs or neutralize the negatively charged pollutant.[22,38]

3.4 Heavy metal's study (Hg, Pb, Zn, Cd, and Cu)

In SS 1, Hg was absent. Removal of all the studied metals was 100% except Pb (95%). In SS 2, Hg, Pb, and Cd were missing, and the removal of the rest of the studied metal was 100%. In SS 3, Hg and Cd were missing, and the removal of the rest of the studied metal was 100% except Zn (70%). In SS 4, all the studied metal was found absent. Heavy metal's removal may be due to the chemisorption process between ions of metals and functional group (hydroxyl, aliphatic, carboxyl, amide, and carbonyl group) present in biosorbent[10,21,39] and may be due to glycosides and galactomannans present in *Aragvadha* ash which bound with metals and form inorganic substance.[4,5] Imran et al. observed 98% removal of Cd using the adsorbent prepared from *C. fistula* leaves. Effective heavy metal removal may also be due to the large surface area, which provides more active sites on which metal ions get attached and also due to the porous nature of absorbent.[5,40,41] Rak and Ismail observed an increase in iron and manganese concentrations after the treatment with *C. alata* leaves based coagulant.[33]

3.5 Microbial study (*E. coli*)

E. coli causes many diseases in humans like bloody diarrhea, UTI, food poisoning, etc. In the present study, *E. coli* was found in below detectable limit (BDL) in all types of water except Ganga water. In Ganga water, *the E. coli* number was observed as 5/100 mL of water. In treated water, this was observed as 3/100 mL; this may be due to the antimicrobial action of alkaloids of *Aragvadha*. As per standard, the normal value of *E. coli* is Nil/100 mL of water.

4 Conclusion

From the observations of the present study, it can be concluded that pod ash of *C. fistula* is a suitable adsorbent for removing water pollutants as it showed positive results against almost every parameter taken under the study. Batch experiments proved optimum conditions for maximum removal efficiency for TDS (45%) and hardness (35.5%). The metal reduction was found as Ganga water (95%–100%), hand pump water (100%) and in case of industrial effluent (70%–100%). The optimum temperature recorded was 27°C, which is usually the ambient mean temperature throughout the year in a tropical country like India.

In *Ayurvedic* literature, *Aragvadha* is told as a *Krimighna*, *Vishaghana*, and *jwaraghana dravya*. Removal activity of *C. fistula* against *E. coli*, BOD, and COD strengthens the claim of *Krimighna* and *Jwaraghna* effects of *Aragvadha*. At the same time, removal activity against metals, hardness, and TDS proves *Vishaghana* effect of *C. fistula*. This study also demonstrates *Acharya Sushruta's* quote about *Aragvadha's* water purifying properties. Based on the present study's findings, it can be concluded that *C. fistula* Linn. can be used as potential adsorbents for wastewater treatment, although further studies to explore the potential are required.

References

1. Coelho LM, Rezende HC, Coelho LM, de Sousa PA, Melo DF, Coelho NM. Bioremediation of polluted waters using microorganisms. In: *Advances in Bioremediation of Wastewater and Polluted Soil*. vol. 10. InTech Open; 2015:60770.
2. De Gisi S, Lofrano G, Grassi M, Notarnicola M. Characteristics and adsorption capacities of low-cost sorbents for wastewater treatment: a review. *Sustain Mater Technol*. 2016;9:10–40.

3. Edebali S, Oztekin Y, Arslan G. Metallic engineered nanomaterial for industrial use. In: *Handbook of Nanomaterials for Industrial Applications.* Elsevier; 2018:67–73.

4. Amari A, Alalwan B, Eldirderi MM, Mnif W, Rebah FB. Cactus material-based adsorbents for the removal of heavy metals and dyes: a review. *Mater Res Express.* 2019;7(1), 012002.

5. Dao MT, Nguyen VC, Tran TN, et al. Pilot-scale study of real domestic textile wastewater treatment using *Cassia fistula* seed-derived coagulant. *J Chem.* 2021;2021.

6. Ahmad I, Akhtar MJ, Jadoon IB, Imran M, Ali S. Equilibrium modeling of cadmium biosorption from aqueous solution by compost. *Environ Sci Pollut Res.* 2017;24(6):5277–5284.

7. Hemavathy RR, Kumar PS, Suganya S, Swetha V, Varjani SJ. Modelling on the removal of toxic metal ions from aquatic system by different surface modified *Cassia fistula* seeds. *Bioresour Technol.* 2019;281:1–9.

8. Karri RR, Ravindran G, Dehghani MH. Wastewater—sources, toxicity, and their consequences to human health. In: *Soft Computing Techniques in Solid Waste and Wastewater Management.* Elsevier; 2021:3–33.

9. Wołowiec M, Komorowska-Kaufman M, Pruss A, Rzepa G, Bajda T. Removal of heavy metals and metalloids from water using drinking water treatment residuals as adsorbents: a review. *Minerals.* 2019;9(8):487.

10. Ali H, Khan E, Ilahi I. Environmental chemistry and ecotoxicology of hazardous heavy metals: environmental persistence, toxicity, and bioaccumulation. *J Chem.* 2019;2019.

11. Dehghani MH, Omrani GA, Karri RR. Solid waste—sources, toxicity, and their consequences to human health. In: *Soft Computing Techniques in Solid Waste and Wastewater Management.* Elsevier; 2021:205–213.

12. Ugwu EI, Karri RR, Nnaji CC, et al. Application of green nanocomposites in removal of toxic chemicals, heavy metals, radioactive materials, and pesticides from aquatic water bodies. In: *Sustainable Nanotechnology for Environmental Remediation.* Elsevier; 2022:321–346.

13. Ferro G, Fiorentino A, Alferez MC, Polo-López MI, Rizzo L, Fernandez-Ibanez P. Urban wastewater disinfection for agricultural reuse: effect of solar driven AOPs in the inactivation of a multidrug resistant *E. coli* strain. *Appl Catal B Environ.* 2015;178:65–73.

14. Dehghani R, Yunesian M, Sahraian MA, Gilasi HR, Moghaddam VK. The evaluation of multiple sclerosis dispersal in Iran and its association with urbanization, life style and industry. *Iran J Public Health.* 2015;44(6):830.

15. Bhutiani R, Ahamad F. Efficiency assessment of sand intermittent filtration technology for waste water treatment. *Int J Adv Res Sci Eng.* 2018;7 (03):503–512.

16. Dehghani R, Miranzadeh MB, Tehrani AM, Akbari H, Iranshahi L, Zeraatkar A. Evaluation of raw wastewater characteristic and effluent quality in Kashan Wastewater Treatment Plant. *Membr Water Treat.* 2018;9(4):273–278.

17. Tangahu BV, Sheikh Abdullah SR, Basri H, Idris M, Anuar N, Mukhlisin M. A review on heavy metals (As, Pb, and Hg) uptake by plants through phytoremediation. *Int J Chem Eng.* 2011;2011.

18. Rakhshaee R, Giahi M, Pourahmad A. Studying effect of cell wall's carboxyl–carboxylate ratio change of Lemna minor to remove heavy metals from aqueous solution. *J Hazard Mater.* 2009;163(1):165–173.

19. Negri MC, Hinchman RR, Gatliff EG. *Phytoremediation: Using Green Plants to Clean Up Contaminate Soil, Groundwater, and Wastewater.* IL (United States): Argonne National Lab; 1996.

20. Bhatnagar A, Sillanpää M, Witek-Krowiak A. Agricultural waste peels as versatile biomass for water purification—a review. *Chem Eng J.* 2015;270:244–271.

21. Imran M, Suddique M, Shah GM, et al. Kinetic and equilibrium studies for cadmium biosorption from contaminated water using *Cassia fistula* biomass. *Int J Environ Sci Technol.* 2019;16(7):3099–3108.

22. Díaz JJ, Ramos LJ, Barreto JD. Efficiency of *Cassia fistula* seed as a natural coagulant in raw water treatment from Sinú River, Colombia. *Indian J Sci Technol.* 2018;11(11).

23. Hemavathy RV, Saravanan A, Kumar PS, Vo DV, Karishma S, Jeevanantham S. Adsorptive removal of Pb (II) ions onto surface modified adsorbents derived from *Cassia fistula* seeds: optimization and modelling study. *Chemosphere.* 2021;, 131276.

24. Sushruta A. In: Shastri AD, ed. *Sushruta Samhita. Part—1.* Varanasi: Chaukhambha Sanskrit Sansthan; 2010. Sutra sthana, Dravyasangrahniya adhayaya.

25. Agnivesha A. *Charak Samhita.* Commented by Pandit Kashinath Shastri, Varanasi: Chaukhambha Bharti Academy; 2015. Part 1&2, Siddhi sthana, Uttarbasti siddhi adhayaya.

26. Harita A. *Harita Samhita.* Translated and edited by Vaidya Jaymin Pandey, 1st ed. Varanasi: Chaukhambha Visvabharati; 2010. Tritiya sthana, Jwara chikitsa.

27. Vagabhatta A. In: Gupta KA, ed. *Astanga Hridyam.* Varanasi: Chaukhambha Prakashan; 2015. Sutra sthana, Shodhanadi gana sangraha.

28. Vagabhatta A. *Astanga Samgraha. Volume 1.* translated by prof. K. R. Srikantha Murthy, Varanasi: Chaukhambha Orientalia; 2015. Sutra sthana, Vividha dravya ganasangraha adhyaya, Dravdravya vijnaniya adhyaya.

29. Yogaratnakar A. In: Laxmipatishastri S, ed. *Yogaratnakar.* Varanasi: Chaukambha sanskrita sansthana; 2005. purvardha, ritucharya vidhi.

30. Federation WE, APH Association. *Standard Methods for the Examination of Water and Wastewater.* Washington, DC, USA: American Public Health Association (APHA); 2012.

31. Trivedy RK, Goel PK. *Chemical and Biological Methods for Water Pollution Studies.* Environmental Publications; 1984.

32. Khanna DR, Bhutiani R. *Laboratory Manual of Water and Wastewater Analysis.* Delhi: Daya Publishing House; 2008.

33. Rak AE, Ismail AA. *Cassia alata* as a potential coagulant in water treatment. *Res J Recent Sci.* 2012;1(2):28–33.

34. Peavy HS, Rowe DR, Tchobanoglous G. *Environmental Engineering.* New York: McGraw-Hill; 1985.

35. Dao MT, Le HA, Nguyen TK, Nguyen VC. Effectiveness on color and COD of textile wastewater removing by biological material obtained from *Cassia fistula* seed. *J Viet Environ.* 2016;8(2):121–128.

36. Nacheva PM, Bustillos LT, Camperos ER, Armenta SL, Vigueros LC. Characterization and coagulation-flocculation treatability of Mexico City wastewater applying ferric chloride and polymers. *Water Sci Technol.* 1996;34(3–4):235–247.

37. Rodiño-Arguello JP, Feria-Diaz JJ, Paternina-Uribe RD, Marrugo-Negrete JL. Sinú River raw water treatment by natural coagulants. *Rev Fac Ing Univ Antioq.* 2015;76:90–98.

38. Yin CY. Emerging usage of plant-based coagulants for water and wastewater treatment. *Process Biochem.* 2010;45(9):1437–1444.

39. Lupea M, Bulgariu L, Macoveanu M. Biosorption of Cd (II) from aqueous solution on marine green algae biomass. *Environ Eng Manag J.* 2012;11(3).

40. Grieco G, Merlini A, Porta M, et al. The dissemination of geoscience education through geoparks and geosites: the SOLE (Social Open Learning Environment) Erasmus+ project. In: *World Multidisciplinary Earth Sciences Symposium*; 2015.

41. Naushad M, Ahamad T, Al-Maswari BM, Alqadami AA, Alshehri SM. Nickel ferrite bearing nitrogen-doped mesoporous carbon as efficient adsorbent for the removal of highly toxic metal ion from aqueous medium. *Chem Eng J.* 2017;330:1351–1360.

Chapter 17

Cesium lead bromide (CsPbBr$_3$) perovskite nanocrystals for sensing applications

Ananthakumar Soosaimanickam*, Pedro J. Rodríguez-Cantó, Juan P. Martínez-Pastor, and Rafael Abargues

UMDO, Institute of Materials Science (ICMUV), University of Valencia, Valencia, Spain

*Corresponding author.

1   Introduction—All-inorganic metal halide perovskite nanocrystals

Recently emerged metal halide perovskite compounds are showing outstanding performance for the photovoltaic applications. Because of this, nanostructured perovskite compounds also gain interest to investigate them for the future generation optoelectronic devices. Metal halide perovskite nanocrystals are seeming to be potential candidates for numerous optoelectronic applications, and their amazing structural and optical properties are currently examined by different research groups.[1–4] The field has been diversified into different categories such as synthetic aspects, surface chemistry, functionalization, and fabricating potential devices. Compared with organic-inorganic perovskites, the applications of pure inorganic lead halide perovskite nanocrystals (LHP NCs) are highly interested because of their exemplary optical properties. High photoluminescent quantum yield (PLQY), tunable photoluminescence (PL) spectra, large absorption coefficient, extreme defect tolerance, etc., are the salient properties of the LHP NCs.[5–7] Lead halide perovskite nanocrystals possess outstanding structural and optical properties which are quite useful for the light-emitting diodes (LEDs), photovoltaics, photodetectors, sensors, lasers, photocatalysis, etc.[8–10] Out of others, potential use of LHP NCs for the sensing applications is investigated in the recent years and different kinds of promising directions are explored.[11,12] Cesium lead halide perovskite NCs are having the general formula CsPbX$_3$ (where X = Cl, Br, and I) and the optical spectra of these NCs is modified through altering the halide composition. Because of their unique surface, the ligands and atoms present in the surface of the LHP NCs are quite active with foreign elements, additives, and compounds.[13–16] Also, even a small amount of water or organic ligands are sufficient to convert the LHP NCs to different phases and morphologies. Moreover, exposure of gaseous molecules and organic polar solvents are found to be altering the surface texture of LHP NCs which essentially make them potential candidates for the sensing applications.

For the sensing of explosives, trace amount of detection is important and this can be achieved by coupling sensitive molecules on the perovskite surface. Furthermore, these NCs are also sensitive with the polarity of the solvents used for the purification and this inherently affect the optical properties of the prepared NCs.[17] Improving the quality of the LHP NCs after purification is also achieved through the incorporation of wide variety of compounds such as ionic salts, organic compounds, and metal ions.[18–20]

As established in the literature, different kinds of ligands are found to be useful in converting LHP NCs into various interesting morphologies such as nanoplatelets, nanowires, nanorods, nanocubes, etc. All these morphologies are quite sensitive with the halide compounds, ionic salts, and therefore undoubtedly LHP NCs can be used for the sensing applications. Furthermore, because of the ligand-induced surface defects, the molecular interaction between the incoming molecules and the defect sites present in the surface of the LHP NCs is interesting for further analysis. Also, owing to the toxicity associated with Pb^{2+}, integrating perovskite nanostructures into the metal oxide templates is also found to be useful for such purposes. Because of these advantages, LHP NCs are widely analyzed for the sensing applications in the recent years.[21–23,11,12] The response and recovery time of the perovskite NCs is important when we use for sensing purpose. Furthermore, properties such as sensitivity, limit of detection, resolution, dynamic range, selectivity, reversibility, linearity, and hysteresis are the important factors that govern the performance of a sensor.[24] In this regard, the potential use of LHP NCs for sensing applications recently received intensive interest and various kinds of their sensing properties are analyzed.

Potential use of LHP NCs for the sensing applications is a promising field of interest, for example, in detecting the explosives. This is also elaborated toward electrochemical sensors, detection of metal ions, and so on. Because of their

Sustainable Materials for Sensing and Remediation of Noxious Pollutants. https://doi.org/10.1016/B978-0-323-99425-5.00010-4

excellent PLQY values, it is imperative to study on the physicochemical properties of the LHP NCs. In all these above-mentioned properties, ligands such as oleylamine (OAm) and oleic acid (OA) are significantly influencing on the LHP NCs.[17] Thus, it is essential to understand the current trend of LHP NCs for the sensing applications so that considerable improvement can be achieved in the near future. Although different kinds of perovskites, namely hybrid halide perovskites,[25] all-inorganic LHPs, lead-free halide perovskites,[26,27] double perovskites,[28] are showing promising directions for sensing applications, due to their impressive optical properties, all-inorganic LHP NCs attracted much attention in the recent years. Out of other halide perovskite NCs, cesium lead bromide ($CsPbBr_3$) nanocrystals are studied much for the sensing applications owing to their extraordinary optical properties and high stability over other halide composition. The PL properties of these nanocrystals are impressive and reached up to near-unity through post-treatment strategies. In this regard, this chapter is aiming to summarize the existing scenario of the $CsPbBr_3$ NCs for the sensing applications and future perspectives of the LHP NCs for the same.

2 Why all-inorganic metal halide perovskite nanocrystals are preferred for sensing?

As mentioned in the introduction section, the sensing properties of the LHP NCs are quite interesting especially due to their structural features. The structure is actually built with octahedra building blocks and the surface atoms are showing weak binding to the ligand molecules attachment owing to ionicity. Because of the ionic nature, the surface ligands in the LHP NCs are not showing strong binding behavior and so leaving easily. Furthermore, the amine ligands, which form alkylammonium ions in the reaction, are substituting in the place of the Cs^+ ions and modifying the functional properties.[29–32,15] This substitution which drastically changes the structural features of LHP NCs is making important consequences on sensing properties. Also, since LHP NCs are quite sensitive with moisture, it is important to organize the selective molecules for the sensing applications. This might be correlated with the formation of different phases in the presence of capping ligands with different ratios.[33] Because of their higher sensitivity with gases, moisture, and other chemical compounds, the structural integration of LHP NCs is greatly disturbed with the interaction and hence can be applied for sensing purpose. The self-assembling tendency, PL variation with respect to halide vacancies, and different phase formations in LHP NCs are given in Fig. 1.

The sensing mechanism of the LHP NCs and the analyte is usually described through quenching process. This quenching is attributed with different types of the charge transfer processes such as: (a) Forster resonance energy transfer (FRET), (b) intramolecular charge transfer (ICT), (c) photoinduced electron transfer, and (d) exciplex/excimer complex.[34] In the case of photoinduced electron transfer, electrons from the highest occupied molecular orbital (HOMO) are transferred into lowest unoccupied molecular orbital (LUMO) and then transferred into analyte's LUMO which results in PL quenching. In LHP NCs, it is proposed that the halide vacancies are taking part in the sensing mechanism and with the exposure with the analyte compound, the density of halide vacancies is found to decrease and so high amount of photoexcited charges are available for electrical transport.[35,36] Inner filter effect (IFE) is an emerging sensing mechanism of perovskite NCs in which absorption band of an absorber is overlapping with the excitation and/or emission band of a donor. Combination of IFE and FRET was also postulated in the case of detection of metal ion, for example, Co^{2+}.[37] Although several mechanisms are postulated, a clear understanding in this area is further required. The schematic diagram of the quenching mechanism due to the electron transfer is given in Fig. 2. It is important to note that halide vacancies in the LHP NCs significantly affect the PL properties.[32] For the analysis of quality of sensors, two factors namely response and recovery times are quite important and these two factors are varied with respect to several factors. The response time is defined as the time taken by the electrical current to attain 90% of its maximum value and the recovery time is defined as the time taken by current to reach the 10% of its maximum value.[38] Nature of analyte, chemical composition of the LHP NCs and analyte compound/molecule, interaction of these two compounds/molecules under different solvent atmospheres are critically influencing the sensing ability of the LHP NCs and therefore much attention is required in choosing the appropriate compounds and molecules in order to couple with the perovskite NCs surface.

3 Current trends in sensing applications of cesium lead bromide ($CsPbBr_3$) perovskite nanocrystals

Although all-inorganic lead halide perovskite nanocrystals are generally considered suitable for sensing applications, because of the versatility and interesting PL properties, $CsPbBr_3$ NCs are widely studied for sensing purposes in several fields. These investigations on the $CsPbBr_3$ NCs are classified and discussed in the forthcoming sections.

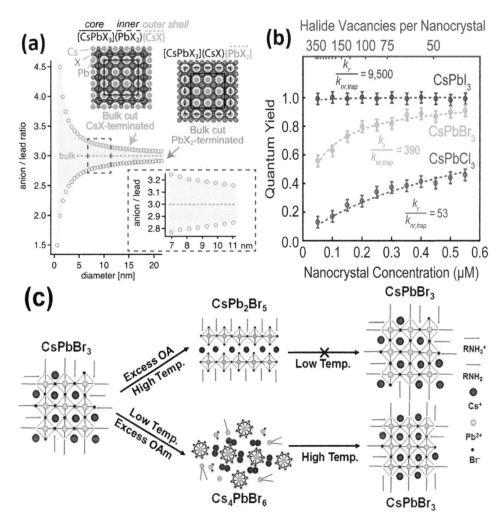

FIG. 1 (A) Surface assembly of LHP NCs. (B) Variation of PLQY with respect to the halide vacancies. (C) Formation of different phases of LHP NCs under different conditions. *(Panel (A): Reproduced with permission from Bodnarchuk MI, Boehme SC, Brinck S, Bernasconi C, Shynkarenko Y, Krieg F, Widmer R, Aeschlimann, Gunther D, Kovalenko MV, Infante I. Rationalizing and controlling the surface structure and electronic passivation of cesium lead halide nanocrystals. ACS Energy Lett 2019;4(1):63–74. https://doi.org/10.1021/acsenergylett.8b01669 (https://pubs.acs.org/doi/10.1021/ acsenergylett.8b01669) Copyright@2019 American Chemical Society. Further permissions related to the material excerpted should be directed to the ACS. Panel (B): Reproduced with permission from Nenon DP, Pressler K, Kang J, Koscher BA, Olshansky JH, Osowiecki WT, Koc MA, Wang L-W, Alivisatos AP. Design principles for trap-free CsPbX3 nanocrystals: enumerating and eliminating surface halide vacancies with softer Lewis bases. J Am Chem Soc 2018;140(50):17760–17772. https://doi.org/10.1021/jacs.8b11035 Copyright@2018 American Chemical Society. Panel (C): Reproduced from Ding H, Jiang H, Wang X. How organic ligands affect the phase transition and fluorescent stability of perovskite nanocrystals. J Mater Chem C 2020;8:8999–9004. https://doi.org/10.1039/D0TC01028F with permission from the Royal Society of Chemistry.)*

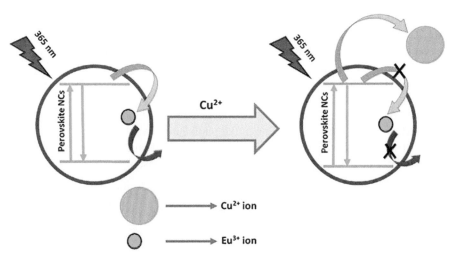

FIG. 2 Schematic diagram of the fluorescence quenching mechanism in halide perovskite NCs due to electron transfer. *(Adapted from Ding N, Zhou D, Pan G, Xu W, Chen X, Li D, Zhang X, Zhu J, Ji Y, Song H. Europium-doped lead-free Cs₃Bi₂Br₉ perovskite quantum dots and ultrasensitive Cu²⁺ detection. ACS Sustain Chem Eng 2019;7(9):8397–8404. https://doi.org/10.1021/acssuschemeng.9b00038.)*

3.1   In the detection of metal ions

Incorporation of CsPbBr$_3$ QDs with poly-methyl methacrylate (PMMA) polymer is found to be useful for the multipurpose applications like detection of metal ions, pH, and to sense biomolecules. Wang et al. demonstrated this concept using cyclam-functionalized CsPbBr$_3$ QDs/PMMA composite and used for the sensing of Cu^{2+} ions in aqueous medium.[39] For the increased concentration of Cu^{2+} ions, the PL intensity is found to be increased and for the optimized concentration, the detection limit is downed up to 10^{-15} M. The special quenching behavior of Cu^{2+} ions with respect to CsPbBr$_3$ QDs is interesting and unique compared with other divalent ions. It is observed that the PL quenching is due to the electronic configuration of Cu^{2+} ions which make a steric shielding with respect to its concentration.[40] The extreme sensitivity of CsPbBr$_3$ QDs with water is also used to detect the metal ion Hg^{2+}. In this case, the CsPbBr$_3$ QDs dissolved in CCl$_4$ is interacted with aqueous solution of Hg^{2+} ions.[41] These investigations clearly indicate the potential use of CsPbBr$_3$ NCs for the metal ion detection.

3.2   In the detection of gases

Because of the sensitivity with the water, the LHP NCs can be dispersed in non-aqueous solvents and employed for the sensing applications. To find out the toxic gas such as H$_2$S, Chen et al. fabricated a fluorescence sensor using CsPbBr$_3$ QDs and used it in the rat brain.[42] Firstly, H$_2$S is allowed to phosphoric acid solution and then to CsPbBr$_3$ QDs (9–15 nm) in n-hexane. It is evident that the fluorescence intensity of the CsPbBr$_3$ QDs is decreased due to the interaction of H$_2$S, whereas no change in the spectrum is observed for other tested compounds. Interestingly, it is proposed that the OAm and OA ligands on the surface help to reach H$_2$S into the surface Pb^{2+} that results in the formation of PbS nanoparticles. This decomposition of structure finally leads to the quenching in the PL spectrum. Compared with the as-synthesized NCs, use of nanocomposites of LHP NCs is found to be more beneficial for the sensing applications due to their prolonged optical stability. To find out the gases such as chlorine and iodine in the sewage and domestic water, composite structure of the CsPbBr$_3$ QDs/cellulose is successfully employed.[43] The size of the CsPbBr$_3$ QDs synthesized through this reaction is 12.61 ± 3.61 nm. Here, this composite in 1-ODE is turned out significant changes in the optical spectra while interacting with the chlorine and iodine. Moreover, the limit of detection of Cl$^-$ and I$^-$ is found to be 4.11 and 2.56 mM. Like in the previous case, the interaction of CsPbBr$_3$ QDs with other ions such as Na$^+$, Co^{2+}, Ca^{2+}, Mg$^+$, Al^{3+}, HCO$_3^-$, SO$_4^{2-}$, and NO$_3^-$ did not result any change in the optical spectra. These evidences are clearly showing the selectivity preference of the LHP NCs for specific candidates. The mechanism of gas-sensing properties of LHP NCs is associated with different factors including surface defects. For example, while interacting with acetone/ethanol vapors, the photoexcited electrons and holes are undergoing oxidation process which results in the decrease of concentration of electrons and holes.[44] Sensing ability of LHP NCs is strongly governed by the surface functionalization of the molecules. For example, it is observed that CsPbBr$_3$ QDs modified by tetraphenylporphyrin tetrasulfonic acid (TPPS) hardly quenched the PL spectrum of the QDs.[45] Despite lack of PL properties while exposing gases on the LHP NCs, the properties can be recovered with the suitable further treatment. It is observed that the composite of CsPbBr$_3$ QDs (size: ~12.12 nm)/boron nitride(BN) fibers is efficient in sensing the ammonia (NH$_3$) gas.[46] Boron nitride fibers are more efficient in adsorbing the LHP NCs within their hollow structure and hence the prepared composite shows long-term optical properties. Specifically, the average lifetime of the CsPbBr$_3$ QDs is increased from ~10.796 to ~18.477 ns. In this case, the loss of PL properties of the composite structure is recovered through further exposure with N$_2$. In more detail, the CsPbBr$_3$ QDs/BN fibers turned out yellow to white color when exposed with NH$_3$ and this is recovered back with the treatment of nitrogen. This appearance and disappearance of structural changes are also observed through the observation in the PL spectrum. Hence, the selectivity, reversibility, and stability of the composite are realized with the exposure of NH$_3$ and this kind of novel composite systems is useful for the sensing of different kinds of toxic gases. Similarly, composite of CsPbBr$_3$ nanofibers/polystyrene is found to be useful in the detection of NH$_3$.[47] Here, CsPbBr$_3$ NCs are synthesized using hot-injection method and converted into nanofibers through electrospinning method for the exposure of NH$_3$. With the high stability and resistive against halide exchange, this composite PL intensity has shown sensitive to the NH$_3$ even below 10 mg/L. Nitrogen is used to recover the PL properties of the CsPbBr$_3$ NCs, and in another experiment, the photocurrent value of the CsPbBr$_3$ NCs is recovered upon N$_2$ exposure.[44] Brintakis et al. synthesized CsPbBr$_3$ nanocubes (500 nm to 1 μm size) using a fast reprecipitation method on the patterned substrate.[38] While exposing ozone (O$_3$) on these nanocubes, increase in the current is observed. Furthermore, even for the lowest O$_3$ concentration, a high sensitivity is observed. The electrical response of the CsPbBr$_3$ NCs with respect to different concentrations of O$_3$ and response and recovery time curves as a function of gas concentration and schematic diagram of the interaction of O$_3$ gas on the CsPbBr$_3$ NCs surface are given in Fig. 3A–F.

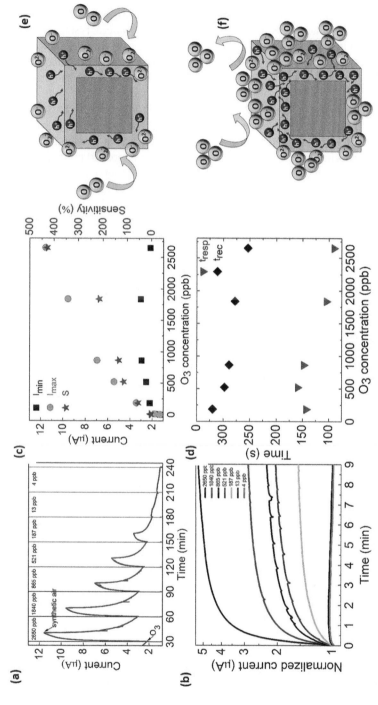

FIG. 3 Electrical response of the CsPbBr₃ NCs as sensing materials upon applying various O₃ concentrations from 2650 down to 4 ppb as a function of the O₃ exposure time (A, B). Sensitivity (S) and response (t_res) and recovery time (t_rec) as a function of gas concentration of the CsPbBr₃ NC-based sensor (C, D) and schematic representation of the interaction of O₃ with the CsPbBr₃ NCs surface (E, F). *(Reproduced from: RBrintakis K, Gagaoudakis E, Kostopoulou A, Faka V, Kiriakidis G, Stratakis E. Ligand-free all-inorganic metal halide nanocubes for fast, ultra-sensitive and self-powered ozone sensors. Nanoscale Adv 2019;1:2699–2706. https://doi.org/10.1039/C9NA00219G with permission from the Royal Society of Chemistry.)*

3.3 In the detection of chemicals and explosives

It is also possible to use the molecularly imprinted silica encapsulated with $CsPbBr_3$ QDs to sense chemical such as 2,2-dichlorovinyl dimethyl phosphate.[48] In this case, the quenching mechanism is attributed with the charge transfer between QDs and dichlorovinyl dimethyl phosphate. Although several studies are indicating that quenching mechanism is the sensing tool in analyzing the efficiency of the sensor, there is no much investigation available to prove the exact steps involved in this process. Despite of numerous kinds of materials availability for detecting the explosives, LHP NCs are showing interesting results due to their uniqueness. It is proved that $CsPbBr_2I$ microcrystals are efficient in sensing trinitrophenol (TNT), an explosive used in military.[49] In this case, it is stated that π-π stacking of the benzene ring or the interaction of hydroxyl group of the TNT with the perovskite is found to be the reason for the quenching in the PL spectrum. Depending on the number of electron-withdrawing groups attached with the benzene ring, quenching is found to be efficient.

To detect the dye such as Rohdamine-6G, composite structure of LHP NCs is used. This is because of the ultrahigh stability of composite structure of LHP NCs against water. When $CsPbBr_3$ QDs are embedded with the polystyrene matrix, this composite structure is resistant with the exposure of aqueous solution and it is useful to detect Rhodamine-6G in water.[50] The sensitivity of this composite structure in this case is found to be up to 0.01 ppm. Specifically, the PL intensity of the composite at $\sim 513\,nm$ decreases, while an increase in the intensity is observed at $\sim 560\,nm$. Because of the overlapping of the emission spectrum of the composite with the absorption spectrum of R6G in solution, FRET is realized in this case which further influence on the PL mechanism. Similar to this, $CsPbBr_3$ QDs/PMMA composite is efficient in detecting the biomolecule trypsin with the lowest detection limit $0.1\,\mu g\,mL^{-1}$.[39] Together with ZnO inverse opal photonic crystals, the $CsPbBr_3$ QDs/ZnO composite electrode is proved as useful in detecting dihydronicotinamide adenine dinucleotide (NADH). The photocurrent response of the prepared composite electrode is increased while increasing the concentration of NADH. With the large linear range (0.1–$250\,\mu M$), the prepared composite electrode has showed very good sensitivity and selectivity. Likewise, $CsPbBr_3$ NCs@$BaSO_4$ composite structure is found to be efficient for the detection of melamine.[51] Here, the composite structure is synthesized through an emulsion approach and this composite is tagged with the citrate-stabilized Au nanoparticles (dia: $20 \pm 5\,nm$) for the sensing of melamine. Here, due to the inner filter effect of Au nanoparticles, the fluorescence spectrum of the $CsPbBr_3$ NCs@$BaSO_4$ composite is quenched. In this case, the detection limit of the melamine is found to be $0.42\,nmol/L$. Investigation with the TiO_2 inverse opal photonic crystals for the detection of dopamine using $CsPbBr_{1.5}I_{1.5}$ QDs (size: $\sim 6.8\,nm$) has delivered a linear range from 0.1 to $250\,\mu M$ with the detection limit $0.012\,\mu M$.[44] Moreover, the photocurrent of the assembled sensor remained as 91% even after 2 weeks. In this case, the inverse opal structures are obtained through the polymeric template, PMMA. All these results are suggesting the promising path of LHP NCs combined with the polymeric system for the photoelectrochemical sensor applications. Since graphene is an effective material to decorate any kind of NCs, incorporation of $CsPbBr_3$ NCs into the graphene matrix is found to be useful for the detection of benzene and toluene.[52] All these results are clearly indicating the importance of nanocomposite structured LHP NCs for the sensing application.

Molecularly imprinted polymers (MIPs) also have attracted much to fabricate sensor based on perovskite nanocomposites. A nanocomposite fabricated by (3-aminopropyl) triethoxysilane (APTES)-coated $CsPbBr_3$ QDs (diameter: 8–16 nm) can be coupled with an MIP and this is used to detect the pesticide, omethate.[53] Because of the interaction with the recognition cavities of MIP@$CsPbBr_3$ QDs, omethoate showed higher sensitivity compared with other pesticide compounds. The detection limit of the MIP@$CsPbBr_3$ QDs is found to be $18.8\,ng/L$. Surprisingly, because of the closest resemblance with the structure, the selectivity factor (γ) for the diomethoate is observed as 1.16. The fabrication of $CsPbBr_3$ QDs@MIP sensor, fluorescence intensity curves of the fabricated sensor, and the chemical structures of the compounds studied in this investigation are collectively given in Fig. 4A–D. Recently, Azar et al. have incorporated $CsPbBr_3$ NCs into the polycaprolactone polymer through 3-nitrotoluene and nitromethane as template molecules and studied the effect of this nanocomposite for sensing of nitro explosive, 3-nitrotoluene.[54] In this case, the molecular imprinting process takes place through baking the mixture at 100°C for 10 min to eliminate the template molecules and solvent. The authors have observed that the imprinted polymers show 69% PL quenching which is higher than the one observed from non-imprinted polymer (53%). Also, the sensor fabricated using nitromethane as template molecule showed higher selectivity and fast detection time (below 2–3 s) owing to the formation of large number of selective binding sites.

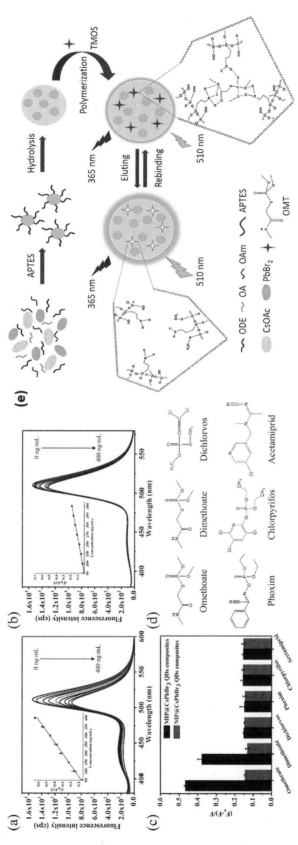

FIG. 4 Fluorescence intensity curves of (A) molecularly imprinted polymer (MIP)/CsPbBr₃ QDs composite (B) molecularly non-imprinted polymer(NIP)/CsPbBr₃ QDs composite (C) selectivity of MIP@CsPbBr₃ QDs and NIP@CsPbBr₃ QDs on omethoate over other compounds (D) chemical structures of the compound studied and (E) fabrication of MIP@CsPbBr₃ QDs sensor. *(Reproduced with permission from Huang S, Guo M, Geng Y, Wu J, Tang Y, Su C, Lin CC, Liang Y. Novel fluorescence sensor based on all-inorganic perovskite quantum dots coated with molecularly imprinted polymers for highly selective and sensitive detection of omethoate. ACS Appl Mater Interfaces 2018;10(45):39056–39063. https://doi.org/10.1021/acsami.8b14472 Copyright@2018 American Chemical Society.)*

3.4 Miscellaneous

Tan et al. have prepared highly luminescent $CsPbBr_3$ QDs with the molecularly imprinted polymer N-(benzyl)-N'-(3-(triethoxysilyl)propyl)urea (BUPTEOS), and this composite structure is used to sense the pesticide, phoxim.[55] The imprinted cavities available in the composite structure is used to accommodate the pesticide. It is found that the fluorescence intensity of the $CsPbBr_3$ QDs/BUPTEOS in this case is decreased after the addition of the pesticide. This was however recovered after the elution and the detection limit of phoxim was found to be 1.45 ng/mL. The selectivity observed in this case lies between 3.53 and 5.68. Detection of humidity using LHP NCs is another interesting application since the halide perovskite surface is quite sensitive with the moisture. When moisture is interacted with $CsPbBr_3$ perovskite surface, it undergoes phase change and this property is used to detect water content in herbal medicine.[56] The selectivity property of sensor is usually estimated through the impedance analysis measurements. With respect to the increase of relative humidity (RH), the impedance response of the $CsPbBr_3$ NCs is also increased linearly.[57] Here, the incorporation of water molecules additionally imparts a new semicircle in the impedance spectra and this also repairs the bromide vacancies (V_{Br}) in the NCs. Similar to this, when $CsPbBr_3$ NCs are used to detect the total polar materials (TPM) in the edible oils, the fluorescence intensity is found to be decreased with respect to increasing the TPM values.[58] Other aspects such as checking the food and quality of food materials are also one of the important applications of the $CsPbX_3$ nanomaterials in this area.

Because of its sensitivity, $CsPbBr_3$ QDs are used for the detection of antibiotic such as tetracycline.[34] When APTES-modified $CsPbBr_3$ QDs are typically used in this case, high selectivity and sensitivity are observed. Together with photoluminescence quenching, the interaction of tetracycline with the APTES-$CsPbBr_3$ QDs also shows reduced lifetime. Using $CsPbBr_3$ NCs, a bipolar electrode system (BPE) can be fabricated and this could be an efficient tool to sense the tetracycline.[59,60] These investigations are confirming that $CsPbX_3$ nanomaterials are the potential candidates for the sensing of organic compounds. Through the composite structure of $CsPbBr_3$/ZnO, Zhu et al. applied to sense dihydronicotinamide adenine dinucleotide (NADH) with the detection limit of 0.010 μM.[61]

Use of LHP NCs for the detection of acid number, 3-chloro 1,2-propanediol (3-MCPD) and moisture content in the food safety measurements is also demonstrated. Zhao et al. prepared $CsPbBr_{1.5}I_{1.5}$ QDs using hot-injection method, and the prepared QDs were applied for the analysis of food safety.[62] The QDs prepared using this method have oily soluble nature and cannot be used for the moisture detecting purpose. To detect the moisture content, the nanosheets of the QDs are coated with silica (SiO_2) spheres in order to protect the surface. Here, it is observed that the PL intensity of $CsPbBr_{1.5}I_{1.5}$ QDs is decreased with respect to increasing the value of acid number. Interestingly, the use of these QDs with the MCPD detection causes halide exchange due to the presence of chloride in MCPD. This halide exchange is evidenced through the PL shift and XRD measurements, which implies that sensing mechanism of LHP NCs is comprised with several factors. This kind of halide exchange is also realized when $CsPbBr_3$ NCs are used to detect Cl^- ions in sweat in the presence of hexane/water system.[63]

It is already demonstrated that rare-earth-doped LHP NCs are showing exemplary PLQY. Rare-earth dopant such as Tb^{3+} is found to be efficient in improving the sensitivity of the LHP NCs. The Tb^{3+}-doped $CsPbI_3$ glass is found to be showing higher absolute and relative temperature sensitivity in the range of 80–480 K.[64] Similar to the doped system, a field-effect transistor system with the device structure Au/Al_2O_3/$CsPbBr_3$ QDs/PMMA/pentacene/Au on polyimide substrate has shown promising outcome for the strain sensor applications.[65] A photosensor fabricated through the deposition of $CsPbBr_3$ QDs on the interdigitated electrode with the heat treatment at 300 °C has shown excellent performance.[66] Here, together with molybdenum sulfide (MoS_2) nanoflakes, the fabricated sensor shows about 20-fold higher sensitivity than the sensor without MoS_2. The important findings of the prepared nanocomposite system are given in Fig. 5. Similar to rare-earth metal doping, Mn^{2+}-doped $CsPb(Cl/Br)_3$ QDs incorporated with the glass composite showed potential material for the cryogenic temperature-sensing applications.[67]

Since different wavelength-emitting $CsPbX_3$ NCs are obtained with respect to its halide composition, it is possible to synthesize LHP NCs which cover blue to red region for the desired applications. For instance, ($CsPb(Br_{0.4}I_{0.6})_3$) QDs prepared by hot-injection method are coupled with the paraffin and this thermally stable composite is efficiently performed for the UV sensor purpose.[68] Here the previously prepared QDs are mixed with paraffin and heated at 65°C under vacuum drying. When NCs are incorporated with the polymer matrix, the thermal as well as photo stability of the prepared nanocomposites are much improved. Similar kind of UV sensor is also fabricated with polydimethylsiloxane (PDS) using organic-inorganic hybrid perovskite NCs.[69] These investigations are clearly representing the potential avenue of the LHP NCs for UV-sensing applications.

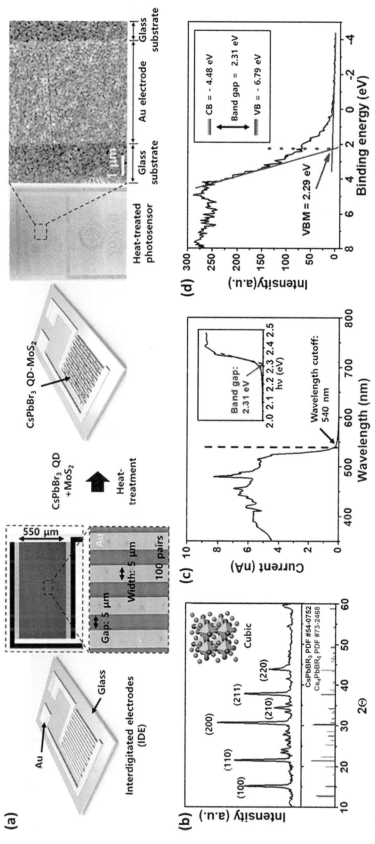

FIG. 5 (A) CsPbBr$_3$ QDs/MoS$_2$ nanoflake-based photosensor (B) XRD pattern of the prepared CsPbBr$_3$ QDs/MoS$_2$ nanoflake composite (C) photoresponse of the photosensor with respect to the incident light and (D) estimation of valence band maximum (VB) and band edge position through XPS measurement of the CsPbBr$_3$ QDs/MoS$_2$ composite. *(Reproduced with permission from Kim H-R, Bong J-H, Park J-H, Song Z, Kang M-J, Son DH, Pyun J-C. Cesium lead bromide (CsPbBr3) perovskite quantum dot-based photosensor for chemiluminescence immunoassays. ACS Appl Mater Interfaces 2021;13 (25):29392–29405. https://doi.org/10.1021/acsami.1c08128 Copyright@2021 American Chemical Society.)*

4 Conclusion

The discussion in the above sessions shows that LHP NCs have demonstrated potential materials for the fabrication of future generation sensors. The so far developed approaches and results are declaring the new directions toward in this area and hopefully LHP NCs can be employed for sensing applications of wide variety of compounds. In this sense, developing new kind of polymer matrix to stabilize the LHP NCs is helpful for achieving highly stable sensors using LHP NCs. Also, new analyte compounds should be explored to utilize the different kinds of ligands capped LHP NCs with tunable halide composition. In most of the cases, $CsPbBr_3$ NCs (or) QDs are used for the sensing applications and hence this chapter has discussed the current trends of use of $CsPbBr_3$ NCs for sensing. However, the possible use of other halides, mixed halide composition systems should be studied. Furthermore, doping binary, ternary metal ions and incorporation of alkali metal ions could make significant influence on the sensing properties since the PL properties of these metal ions incorporated LHP NCs are greatly enhanced. Also, core–shell nanostructures and hybridization with other structures such as inverse opal assembly are showing promising directions to achieve high sensitivity and selectivity in the case of LHP NCs. Furthermore, other bromide perovskite phases such as Cs_4PbBr_6 and $CsPb_2Br_5$ are also interesting compounds to fabricate as sensor, and investigations on these compounds could deliver promising results. Extending these concepts toward other potential perovskites such as double perovskites, lead-free perovskite nanostructures, and copper halide perovskite nanostructures are quite useful for the fabrication of long-term withstanding perovskite sensors. Research in these directions is expected to establish a new platform in all-inorganic perovskite nanostructure-based sensors in different sectors.

Acknowledgments

This work was supported through the NATO Science for Peace and Security Programme Project SPS (no. G5361), Retos-Colaboración 2016 Project Safetag (no. RTC-2016-5197-2) from MINECO, Programme Valorización Project Hidronio (no. INNVAL10/18/032) from Agencia Valenciana de la innovación, and Ramón y Cajal Programme (no. RYC-2015-18349) from MINECO.

References

1. Ha S-T, Su R, Zhang Q, Xiong Q. Metal halide perovskite nanomaterials: synthesis and applications. *Chem Sci.* 2017;8:2522–2536. https://doi.org/10.1039/C6SC04474C.
2. Shamsi J, Urban AS, Imran M, Trizio LD, Manna L. Metal halide perovskite nanocrystals: synthesis, post-synthesis modifications, and their optical properties. *Chem Rev.* 2019;119(5):3296–3348. https://doi.org/10.1021/acs.chemrev.8b00644.
3. Liu D, Guo Y, Que M, et al. Metal halide perovskite nanocrystals: application in high-performance photodetectors. *Mater Adv.* 2021;2:856–879. https://doi.org/10.1039/D0MA00796J.
4. Hills-Kimball K, Yang H, Cai T, Wang J, Chen O. Recent advances in ligand design and engineering in lead halide perovskite nanocrystals. *Adv Sci.* 2021;8(12):2100214. https://doi.org/10.1002/advs.202100214.
5. Xue J, Wang R, Yang Y. The surface of halide perovskites from nano to bulk. *Nat Rev Mater.* 2020;5:809–827. https://doi.org/10.1038/s41578-020-0221-1.
6. Zhang Y, Sigler TD, Thomas CJ, et al. A "tips and tricks" practical guide to the synthesis of metal halide perovskite nanocrystals. *Chem Mater.* 2020;32(12):5410–5423. https://doi.org/10.1021/acs.chemmater.0c01735.
7. Li Y, Zhang X, Huang H, Kershaw SV, Rogach AL. Advances in metal halide perovskite nanocrystals: synthetic strategies, growth mechanisms, and optoelectronic applications. *Mater Today.* 2020;32:204–221. https://doi.org/10.1016/j.mattod.2019.06.007.
8. Akkerman QA, Raino G, Kovalenko MV, Manna L. Genesis, challenges and opportunities for colloidal lead halide perovskite nanocrystals. *Nat Mater.* 2018;17:394–405. https://doi.org/10.1038/s41563-018-0018-4.
9. Fu H. Colloidal metal halide perovskite nanocrystals: a promising juggernaut in photovoltaic applications. *J Mater Chem A.* 2019;7:14357–14379. https://doi.org/10.1039/C8TA12509K.
10. Liu M, Zhang H, Gedamu D, et al. Halide perovskite nanocrystals for next-generation optoelectronics. *Small.* 2019;15(28):1900801. https://doi.org/10.1002/smll.201900801.
11. Shellaiah M, Sun KW. Review on sensing applications of perovskite nanomaterials. *Chem.* 2020;8(3):55. https://doi.org/10.3390/chemosensors8030055.
12. George JK, Halali VV, Sanjayan CG, Suvina V, Sakar M, Balakrishna RG. Perovskite nanomaterials as optical and electrochemical sensors. *Inorg Chem Front.* 2020;7:2702–2725. https://doi.org/10.1039/D0QI00306A.
13. Soosaimanickam A, Adl HA, Chirvony V, Rodriguez-Canto PJ, Martinez-Pastor JP, Abargues R. Effect of alkali metal nitrate treatment on the optical properties of $CsPbBr_3$ nanocrystal films. *Mater Lett.* 2021;305. https://doi.org/10.1016/j.matlet.2021.130835, 130835.
14. Luo B, Naghadeh SB, Zhang JZ. Lead halide perovskite nanocrystals: stability, surface passivation, and structural control. *ChemNanoMat.* 2017;3(7):456–465. https://doi.org/10.1002/cnma.201700056.
15. Yang D, Li X, Zeng H. Surface chemistry of all inorganic halide perovskite nanocrystals: passivation mechanism and stability. *Adv Mater Interfaces.* 2018;5(8):1701662. https://doi.org/10.1002/admi.201701662.

16. Dutt VGV, Akhil S, Mishra N. Surface passivation strategies for improving photoluminescence and stability of cesium lead halide perovskite nanocrystals. *ChemNanoMat.* 2020;6(12):1730–1742. https://doi.org/10.1002/cnma.202000495.

17. Ananthakumar S, Kumar JR, Babu SM. Cesium lead halide (CsPbX$_3$, X = Cl, Br,I) perovskite quantum dots-synthesis, properties, and applications: a review of their present status. *J Photonics Energy.* 2016;6(4):1–8. https://doi.org/10.1117/1.JPE.6.042001.

18. Nakahara S, Tahara H, Yumoto G, et al. Suppression of trion formation in CsPbBr$_3$ perovskite nanocrystals by postsynthetic surface modification. *J Phys Chem C.* 2018;122(38):22188–22193. https://doi.org/10.1021/acs.jpcc.8b06834.

19. Liu Y, Li F, Liu Q, Xia Z. Synergetic effect of postsynthetic water treatment on the enhanced photoluminescence and stability of CsPbX$_3$ (X = Cl, Br, I) perovskite nanocrystals. *Chem Mater.* 2018;30(19):6922–6929. https://doi.org/10.1021/acs.chemmater.8b03330.

20. Chen C, Cai Q, Luo F, et al. Sensitive fluorescent sensor for hydrogen sulfide in rat brain microdialysis via CsPbBr$_3$ quantum dots. *Anal Chem.* 2019;91(24):15915–15921. https://doi.org/10.1021/acs.analchem.9b04387.

21. Zhu Z, Sun Q, Zhang Z, et al. Metal halide perovskites: stability and sensing-ability. *J Mater Chem C.* 2018;6:10121–10137. https://doi.org/10.1039/C8TC03164A.

22. Huang Y, Lai Z, Jin J, et al. Ultrasensitive temperature sensing based on ligand-free alloyed CsPbCl$_x$Br$_{3-x}$ perovskite nanocrystals confined in hollow mesoporous silica with high density of halide vacancies. *Small.* 2021;17(46):2103425. https://doi.org/10.1002/smll.202103425.

23. Aamir M, Shahiduzzaman M, Taima T, Akhtar J, Nunzi J-M. It is an all-rounder! On the development of metal halide perovskite-based fluorescent sensors and radiation detectors. *Adv Opt Mater.* 2021;9(24):2101276. https://doi.org/10.1002/adom.202101276.

24. Senesac L, Thundat TG. Nanosensors for trace explosive detection. *Mater Today.* 2008;11(3):28–36. https://doi.org/10.1016/S1369-7021(08)70017-8.

25. Wu W, Wang X, Han X, et al. Flexible photodetector arrays based on patterned CH$_3$NH$_3$PbI$_{3-x}$Cl$_x$ perovskite film for real-time photosensing and imaging. *Adv Mater.* 2019;31(3):1805913. https://doi.org/10.1002/adma.201805913.

26. Zhang Z-X, Li C, Lu Y, et al. Sensitive deep ultraviolet photodetector and image sensor composed of inorganic lead-free Cs$_3$Cu$_2$I$_5$ perovskite with wide bandgap. *J Phys Chem Lett.* 2019;10(18):5343–5350. https://doi.org/10.1021/acs.jpclett.9b02390.

27. Ding N, Zhou D, Pan G, et al. Europium-doped lead-free Cs$_3$Bi$_2$Br$_9$ perovskite quantum dots and ultrasensitive Cu^{2+} detection. *ACS Sustain Chem Eng.* 2019;7(9):8397–8404. https://doi.org/10.1021/acssuschemeng.9b00038.

28. Weng Z, Qin J, Umar AA, et al. Lead-free Cs$_2$BiAgBr$_6$ double perovskite-based humidity sensor with superfast recovery time. *Adv Funct Mater.* 2019;29(24):1902234. https://doi.org/10.1002/adfm.201902234.

29. Grisorio R, Clemente MED, Fanizza E, et al. Exploring the surface chemistry of cesium lead halide perovskite nanocrystals. *Nanoscale.* 2019;11:986–999. https://doi.org/10.1039/C8NR08011A.

30. Bodnarchuk MI, Boehme SC, Brinck S, et al. Rationalizing and controlling the surface structure and electronic passivation of cesium lead halide nanocrystals. *ACS Energy Lett.* 2019;4(1):63–74. https://doi.org/10.1021/acsenergylett.8b01669.

31. Smock SR, Chen Y, Rossini AJ, Brutchey RL. The surface chemistry and structure of colloidal lead halide perovskite nanocrystals. *Acc Chem Res.* 2021;54(3):707–718. https://doi.org/10.1021/acs.accounts.0c00741.

32. Nenon DP, Pressler K, Kang J, et al. Design principles for trap-free CsPbX$_3$ nanocrystals: enumerating and eliminating surface halide vacancies with softer Lewis bases. *J Am Chem Soc.* 2018;140(50):17760–17772. https://doi.org/10.1021/jacs.8b11035.

33. Ding H, Jiang H, Wang X. How organic ligands affect the phase transition and fluorescent stability of perovskite nanocrystals. *J Mater Chem C.* 2020;8:8999–9004. https://doi.org/10.1039/D0TC01028F.

34. Wang T, Wei X, Zong Y, Zhang S, Guan W. An efficient and stable fluorescent sensor based on APTES-functionalized CsPbBr$_3$ perovskite quantum dots for ultrasensitive tetracycline detection in ethanol. *J Mater Chem C.* 2020;8:12196–12203. https://doi.org/10.1039/D0TC02852E.

35. Chen H, Zhang M, Bo R, et al. Superior self-powered room-temperature chemical sensing with light-activated inorganic halides perovskites. *Small.* 2018;14(7):1702571. https://doi.org/10.1002/smll.201702571.

36. Chen H, Zhang M, Fu X, et al. Light-activated CsPbBr$_2$I perovskite for room-temperature self-powered chemical sensing. *Phys Chem Chem Phys.* 2019;21:24187–24193. https://doi.org/10.1039/C9CP03059J.

37. George JK, Ramu S, Halali VV, Balakrishna GB. Inner filter effect as a boon in perovskite sensing systems to achieve higher sensitivity levels. *ACS Appl Mater Interfaces.* 2021;13(48):57264–57273. https://doi.org/10.1021/acsami.1c17061.

38. Brintakis K, Gagaoudakis E, Kostopoulou A, et al. Ligand-free all-inorganic metal halide nanocubes for fast, ultra-sensitive and self-powered ozone sensors. *Nanoscale Adv.* 2019;1:2699–2706. https://doi.org/10.1039/C9NA00219G.

39. Wang Y, Zhu Y, Huang J, et al. Perovskite quantum dots encapsulated in electrospun fiber membranes as multifunctional supersensitive sensors for biomolecules, metal ions and pH. *Nanoscale Horiz.* 2017;2:225–232. https://doi.org/10.1039/C7NH00057J.

40. Liu Y, Tang X, Zhu T, et al. All-inorganic CsPbBr$_3$ perovskite quantum dots as a photoluminescent probe for ultrasensitive Cu^{2+} detection. *J Mater Chem C.* 2018;6:4793–4799. https://doi.org/10.1039/C8TC00249E.

41. Wang J, Hu Y-L, Zhao R-X, et al. Liquid-liquid extraction and visual detection of Hg^{2+} in aqueous solution by luminescent CsPbBr$_3$ perovskite nanocrystals. *Microchem J.* 2021;170. https://doi.org/10.1016/j.microc.2021.106769, 106769.

42. Chen C, Cai Q, Luo F, et al. Sensitive fluorescent sensor for hydrogen sulfide in rat brain microdialysis via CsPbBr$_3$ quantum dots. *Anal Chem.* 2019;91(24):15915–15921.

43. Park B, Kang S-M, Lee G-W, Kwak CH, Rethinasabapathy M, Huh YS. Fabrication of CsPbBr$_3$ perovskite quantum dots/cellulose-based colorimetric sensor: dual-responsive on-site detection of chloride and iodide ions. *Ind Eng Chem Res.* 2020;59(2):793–801. https://doi.org/10.1021/acs.iecr.9b05946.

44. Chen X, Li D, Pan G, et al. All-inorganic perovskite quantum dot/TiO$_2$ inverse opal electrode platform: stable and efficient photoelectrochemical sensing of dopamine under visible irradiation. *Nanoscale.* 2018;10:10505–10513. https://doi.org/10.1039/C8NR02115E.

45. Wang H, Li Q, Niu X, Du J. Tetraphenylporphyrin-modified perovskite nanocrystals enable ratiometric fluorescent determination of sulfide ion in water samples. *J Mater Sci.* 2021;56:15029–15039. https://doi.org/10.1007/s10853-021-06253-x.

46. He X, Yu C, Yu M, et al. Synthesis of perovskite CsPbBr$_3$ quantum dots/porous boron nitride nanofiber composites with improved stability and their reversible optical responses to ammonia. *Inorg Chem*. 2020;59(2):1234–1241. https://doi.org/10.1021/acs.inorgchem.9b02947.

47. Park B, Kim S, Kwak CH, et al. Visual colorimetric detection of ammonia under gaseous and aqueous state: approach on cesium lead bromide perovskite-loaded porous electrospun nanofibers. *J Ind Eng Chem*. 2021;97:515–522. https://doi.org/10.1016/j.jiec.2021.03.006.

48. Huang S, Tan L, Zhang L, et al. Molecularly imprinted mesoporous silica embedded with perovskite CsPbBr$_3$ quantum dots for the fluorescence sensing of 2,2′-dichlorovinyl dimethyl phosphate. *Sens Actuators B*. 2020;325. https://doi.org/10.1016/j.snb.2020.128751, 128751.

49. Aamir M, Khan MD, Sher M, et al. A facile route to cesium lead bromoiodide perovskite microcrystals and their potential application as sensors for nitrophenol explosives. *Eur J Org Chem*. 2017;31:3755–3760. https://doi.org/10.1002/ejic.201700660.

50. Wang Y, Zhu Y, Huang J, et al. CsPbBr$_3$ perovskite quantum dots-based monolithic electrospun fiber membrane as an ultrastable and ultrasensitive fluorescent sensor in aqueous medium. *J Phys Chem Lett*. 2016;7(21):4253–4258. https://doi.org/10.1021/acs.jpclett.6b02045.

51. Li Q, Wang H, Yue X, Du J. Perovskite nanocrystals fluorescence nanosensor for ultrasensitive detection of trace melamine in dairy products by the manipulation of inner filter effect of gold nanoparticles. *Talanta*. 2020;211. https://doi.org/10.1016/j.talanta.2019.120705, 120705.

52. Casanova-Chafer J, Garcia-Aboal R, Atienzar P, Llobet E. The role of anions and cations in the gas sensing mechanisms of graphene decorated with lead halide perovskite nanocrystals. *Chem Commun*. 2020;56:8956–8959. https://doi.org/10.1039/D0CC02984J.

53. Huang S, Guo M, Geng Y, et al. Novel fluorescence sensor based on all-inorganic perovskite quantum dots coated with molecularly imprinted polymers for highly selective and sensitive detection of omethoate. *ACS Appl Mater Interfaces*. 2018;10(45):39056–39063. https://doi.org/10.1021/acsami.8b14472.

54. Aznar E, Sanchez-Alarcon I, Soosaimanickam A, et al. Molecularly imprinted nanocomposites of CsPbBr$_3$ nanocrystals: an approach towards fast and selective gas sensor of explosive taggants. *J Mater Chem C*. 2020;1–24. https://doi.org/10.1039/D1TC05169E.

55. Tan L, Guo M, Tan J, et al. Development of high-luminescence perovskite quantum dots coated with molecularly imprinted polymers for pesticide detection by slowly hydrolysing the organosilicon monomers in situ. *Sensor Actuat B Chem*. 2019;291:226–234. https://doi.org/10.1016/j.snb.2019.04.079.

56. Xiang X, Ouyang H, Fu Z. Humidity-sensitive CsPbBr$_3$ perovskite based photoluminescent sensor for detecting water content in herbal medicines. *Sensor Actuat B Chem*. 2021;346. https://doi.org/10.1016/j.snb.2021.130547, 130547.

57. Wu Z, Yang J, Sun X, et al. An excellent impedance-type humidity sensor based on halide perovskite CsPbBr$_3$ nanoparticles for human respiration monitoring. *Sensor Actuat B Chem*. 2021;337. https://doi.org/10.1016/j.snb.2021.129772, 129772.

58. Huangfu C, Feng L. High-performance fluorescent sensor based on CsPbBr$_3$ quantum dots for rapid analysis of total polar materials in edible oils. *Sensor Actuat B Chem*. 2021;344. https://doi.org/10.1016/j.snb.2021.130193, 130193.

59. Hao N, Qiu Y, Lu J, et al. Flexibly regulated electroluminescence of all-inorganic perovskite CsPbBr$_3$ quantum dots through electron bridge to across interfaces between polar and non-polar solvents. *Chin Chem Lett*. 2021;32(9):2861–2864.

60. Hao N, Lu J, Dai Z, et al. Analysis of aqueous systems using all-inorganic perovskite CsPbBr$_3$ quantum dots with stable electrochemiluminescence performance using a closed bipolar electrode. *Electrochem Commun*. 2019;108. https://doi.org/10.1016/j.elecom.2019.106559, 106559.

61. Zhu Y, Tong X, Song H, et al. CsPbBr$_3$ perovskite quantum dots/ZnO inverse opal electrodes: photoelectrochemical sensing for dihydronicotinamide adenine dinucleotide under visible irradiation. *Dalton Trans*. 2018;47:10057–10062. https://doi.org/10.1039/C8DT01790E.

62. Zhao Y, Xu Y, Shi L, Fan Y. Perovskite nanomaterial-engineered multiplex-mode fluorescence sensing of edible oil quality. *Anal Chem*. 2021;93 (31):11033–11042. https://doi.org/10.1021/acs.analchem.1c02425.

63. Li F, Feng Y, Huang Y, et al. Colorimetric sensing of chloride in sweat based on fluorescence wavelength shift via halide exchange of CsPbBr$_3$ perovskite nanocrystals. *Microchim Acta*. 2021;188(2):1–8. https://doi.org/10.1007/s00604-020-04653-5.

64. Zhang Y, Liu J, Zhang H, He Q, Liang X, Xiang W. Ultra-stable Tb^{3+}:CsPbI$_3$ nanocrystal glasses for wide-range high-sensitivity optical temperature sensing. *J Eur Ceram Soc*. 2020;40(15):6023–6030. https://doi.org/10.1016/j.jeurceramsoc.2020.07.016.

65. Li M-Z, Guo L-C, Ding G-L, et al. Inorganic perovskite quantum dot-based strain sensors for data storage and in-sensor computing. *ACS Appl Mater Interfaces*. 2021;13(26):30861–30873. https://doi.org/10.1021/acsami.1c07928.

66. Kim H-R, Bong J-H, Park J-H, et al. Cesium lead bromide (CsPbBr$_3$) perovskite quantum dot-based photosensor for chemiluminescence Immunoassays. *ACS Appl Mater Interfaces*. 2021;13(25):29392–29405. https://doi.org/10.1021/acsami.1c08128.

67. Zhuang B, Liu Y, Yuan S, Huang H, Chen J, Chen D. Glass stabilized ultra-stable dual-emitting Mn-doped cesium lead halide perovskite quantum dots for cryogenic temperature sensing. *Nanoscale*. 2019;11:15010–15016. https://doi.org/10.1039/C9NR05831A.

68. Wu H, Zhang W, Wu J, Chi Y. A visual solar UV sensor based on paraffin-perovskite quantum dot composite film. *ACS Appl Mater Interfaces*. 2019;11(18):16713–16719. https://doi.org/10.1021/acsami.9b02495.

69. Amjadi A, Hosseini MS, Ashjari T, Roghabadi FA, Ahmadi V, Jalili K. Durable perovskite UV sensor based on engineered size-tunable polydimethyl-siloxane microparticles using a facile capillary microfluidic device from a high-viscosity precursor. *ACS Omega*. 2020;5(2):1052–1061.

Chapter 18

Green synthesized silver nanoparticles for the sensing of pathogens

Juliana Botelho Moreira[a,*], Ana Luiza Machado Terra[a], Suelen Goettems Kuntzler[a], Michele Greque de Morais[a], and Jorge Alberto Vieira Costa[b]

[a]Laboratory of Microbiology and Biochemistry, College of Chemistry and Food Engineering, Federal University of Rio Grande, Rio Grande, RS, Brazil,
[b]Laboratory of Biochemical Engineering, College of Chemistry and Food Engineering, Federal University of Rio Grande, Rio Grande, RS, Brazil
*Corresponding author.

1 Introduction

Innovative technologies have been investigated to increase productivity in sustainable agricultural.[1] Microalgal biotechnology stands out in the search for developing bioproducts from innovative processes that do not harm the environment. Microalgae are unicellular or multicellular photosynthetic microorganisms that use sunlight, water, and carbon dioxide to produce biomass.[2] In the agricultural sector, microalgae are interesting because they produce biologically active components such as plant hormones and antimicrobial compounds.[1]

Biologically synthesized metallic nanoparticles have shown effective results concerning disease management and the protection of crops.[1,3] Microalgae can also synthesize silver nanoparticles,[2,4] which can be applied as a biocide and/or sensors in the control of pathogens. Microalgae technology for the synthesis of silver nanoparticles uses materials that do not cause toxicity to the product and the waste from cultivation. Microalgae growth and biosynthesis are simultaneous, and the resulting biomass can potentiate the antibacterial and antifungal properties of silver nanoparticles.[4] Thus, microalgae become a sustainable alternative for the green synthesis of silver nanoparticles that can replace or minimize the use of chemicals applied in controlling and detecting pathogens.

The physicochemical and antimicrobial properties of silver nanoparticles indicate the potential of these nanostructures for the control of pathogenic microorganisms.[5] Furthermore, the contact surface area of silver is increased when available in the nanometer range, resulting in greater efficiency.[4] The controlled release of silver ions gives the nanoparticles excellent bactericidal and fungicidal activities. These free ions act as inhibitors, interfering with the organisms' metabolic processes.[6] The antimicrobial effect of silver nanoparticles depends on the particle size, concentration, and exposure time.[4]

Silver nanoparticles have also gained attention for developing sensors for detecting pathogens. These nanostructures can modify electrodes and improve the specificity, electrocatalytic properties, stability, and conductivity.[7] The high surface permeability, greater surface/volume ratio, reactivity, and high penetrability allow for a more efficient functionalization,[8] improving the properties of biosensors, such as sensitivity, specificity, speed, and precision.[9]

Given the above, the chapter presents the green and simultaneous production of silver nanoparticles and biomass from microalgae technology. The chapter also discusses the antimicrobial properties of microalgal biomass and silver nanoparticles produced by microalgae and their ability to act as a biocide. Moreover, it also addresses the use of silver nanoparticles for pathogen sensors and the challenges regarding the nanoparticles synthesized by microalgae as an ecological and sustainable process.

2 Microalgae biomass and its antimicrobial property

Microalgae have biological properties of commercial interest, mainly due to the bioactive compounds (Fig. 1) in their composition, which can be used as biomass, extracts, or isolated form.[10,11] The production of these compounds may vary according to the taxonomic group of the microalgae, nutrients available in the culture medium, light intensity, temperature, and pH of the microalgal culture.[12] In this context, several microalgal compounds present antimicrobial properties, indicating the potential of this biomass for application as a biocide.

Sustainable Materials for Sensing and Remediation of Noxious Pollutants. https://doi.org/10.1016/B978-0-323-99425-5.00008-6

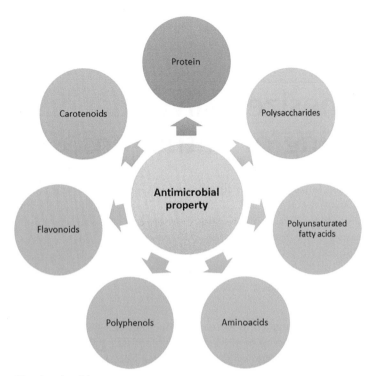

FIG. 1 Active compounds found in microalgae biomass.

Shaima et al.[13] investigated the antibacterial potential of *Chlorella sorokiniana*, *Chlorella* sp., and *Scenedesmus* sp. The microalgae were cultivated in bold basal medium at an incubation temperature of $25 \pm 2°C$ for 15 days. The biomass of *Chlorella* sp. showed higher activity against methicillin-resistant *Staphylococcus aureus* (13.8 mm), *Staphylococcus epidermidis* (11.3 mm), *S. aureus* (11 mm), *Bacillus thuringiensis* (11 mm), *Pseudomonas aeruginosa* (10.9 mm), *Escherichia coli* (9 mm), and two strains of *Bacillus subtilis* (9.1 and 9.5 mm) concerning other microalgae strains. Ferreira et al.[14] cultivated microalgae strains with swine effluents and evaluated the antifungal property of *Tetradesmus obliquus*, *Chlorella prototecoides*, *Chlorella vulgaris*, and *Synechocystis* sp. *Tetradesmus obliquus* and *C. prototecoides* achieved higher yields than *Synechocystis* sp. and *C. vulgaris*. Regarding the effect of microalgae as a biopesticide, an antifungal test was carried out with the *Fusarium oxysporum* strain. The results showed that microalgae *C. vulgaris* and *Tetradesmus obliquus* inhibited 40% of fungal growth, *C. prototecoides* showed 35% antifungal activity, and *Synechocystis* sp. had no significant effect.

Alsenani et al.[15] analyzed the antibacterial activity of compounds from 14 microalgal strains against gram-positive bacteria. The microalgal cultures were grown in a 200-L outdoors closed photobioreactor, at a temperature of 23.6°C, during the day and 13.9°C at night, with a photoperiod of 12/12 h. Among the strains studied, cultures of *Isochrysis galbana*, *Scenedesmus* sp., and *Chlorella* sp. showed satisfactory antibacterial activity. The authors found that the minimum inhibitory concentration (MIC) for *I. galbana* was 500 mg/mL, and for *Chlorella* sp. and *Scenedesmus* sp. was 1 mg/mL. The study also found that linoleic acid, oleic acid, docosahexaenoic acid, and eicosapentaenoic acid had the potential to inhibit the growth of gram-positive bacteria, such as *Listeria monocytogenes*, *S. aureus*, *B. subtilis*, *Clavibacter michiganensis*, *S. epidermidis*, and *Enterococcus faecalis*.

Maadane et al.[16] evaluated microalgal compounds' antibacterial and antifungal activities from *Nannochloropsis gaditana*, *Dunaliella salina*, *Dunaliella* sp., *Phaeodactylum tricornutum*, *Isochrysis* sp., *Navicula* sp., and *Tetraselmis* sp. The cultures were maintained in 5-L flasks and grown in sterile natural seawater, enriched with nutrients from the F/2 medium at a temperature of 25°C. The most excellent antibacterial activity was found in the extract of *Tetraselmis* sp., which presented MIC of 2.6–3.0 mg/mL against *E. coli*, *P. aeruginosa*, and *S. aureus*, and MIC >5 mg/mL against *Candida albicans*. The extracts of all microalgae studied inhibited the growth of *C. albicans*, with the highest activity obtained by *N. gaditana* with a MIC of 4.0 mg/mL. The authors also found that the antimicrobial activities observed were associated with the content of fatty acids, carotenoids, and phenolic compounds constituted in microalgal extracts.

3 Advantages of microalgae cultivation for silver nanoparticles production

Most of the conventional syntheses of silver nanoparticles consume high energy and use toxic chemicals that can cause risks to human health and environmental damage. The biological synthesis methods of silver nanoparticles are an alternative to chemical and physical processes to reduce costs and the impacts on the environment.[17,18] The diversity of species and taxonomic groups highlights photosynthetic microorganisms' potential for the production of nanoparticles.[17]

Microalgae are a source of several secondary metabolites such as pigments and proteins, which can act in synthesizing of silver nanoparticles.[19,20] Other molecules found in microalgae, such as proteins, polysaccharides, tannins, and steroids, are involved in reducing metallic salts into metallic nanoparticles.[21–23] In addition, this production method reduces downstream processing steps that reduce the cost of its synthesis and does not involve toxic solvents.[24,25] Thus, microalgae technology is an emerging, non-toxic, simple, and economical processing alternative for the green synthesis of silver nanoparticles.[2]

Microalgae can use carbon dioxide as a nutrient source and remediate industrial effluents. Besides, they improve efficiency in generating bio-based products with high-added value.[26] These microorganisms show rapid cell growth and high biomass productivity.[4,27] Microalgae can be found in fresh and saltwater environments and present the ability to adapt to adverse conditions,[28,29] reaching a higher biomass production per unit area when compared to superior plants.[1] Another advantage of using these microorganisms is the presence of greater amounts of reducing agents, which reduce metallic salts to nanoparticles without releasing harmful by-products.[29] Moreover, the antimicrobial properties of microalgae can enhance the biocidal action of silver nanoparticles.[2,30]

4 Synthesis of silver nanoparticles from microalgae

The chelation process is used by microalgae to convert toxic metals into non-toxic forms by successive chemical reactions.[4] Some genera of microalgae strains were studied for their ability to form silver nanoparticles, such as *Anabaena*,[31] *Synechocystis*,[32] *Oscillatoria*,[33] *Botryococcus*,[34] *Chlamydomonas*,[35] *Chlorella*,[29] *Scenedesmus*,[36] *Desmodesmus*,[37] and *Spirulina*.[38]

The production of silver nanoparticles by microalgae occurs through the reduction of metal ions to nanoparticles in just one step (Fig. 2), using compounds and cellular constituents with low or no toxicity (terpenoids, carbonyl groups, phenolics, flavanones, amines, amides, proteins, pigments, alkaloids, reductase enzymes, and electron transporting quinones). The reduction of the silver ion (Ag^+) to its metallic form (Ag^0) occurs through electrostatic interactions between metallic ions and microalgal compounds that capture the ions of silver salts and promote charge reduction, forming neutral atoms. These atoms collide and form a stable nucleus with a subsequent growth process, in which more atoms collide and aggregate, forming larger particles. Finally, stabilization occurs by the depletion of metallic ions in the solution or by coating the particle with compounds present in the extract.[2,4,39,40]

The size, shape, and agglomeration of nanoparticles are influenced by parameters such as pH, temperature, and reaction time. High pH increases the reducing power of the functional groups, preventing the agglomeration of the nanoparticles. The basic pH contributes to stabilizing nanoparticles by interacting with the amino groups of surface-bound proteins and their residual amino acids.[41] The reaction temperature will influence the adsorption/desorption of the surfactant, the stability of complexation, and the rate of biosynthesis, in addition to the shape and size distribution of the produced nanoparticles. The use of low temperature decreases the formation of nanoparticles and increases the time needed to complete the reduction reaction. The enhance in temperature can increase the rate of synthesis and particle size. With a higher temperature, the reaction can become more active in reducing silver ions. Therefore, high temperatures are crucial for nucleation. On the other hand, lower temperatures are needed for growth in wet chemical conditions for silver

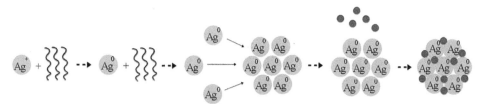

FIG. 2 Schematic representation of silver nanoparticle biosynthesis by microalgae. *(Source: Terra ALM, Kosinski RC, Moreira JB, Costa JAV, Morais MG. Microalgae biosynthesis of silver nanoparticles for application in the control of agricultural pathogens. J Environ Sci Health B 2019;54(8):709–716. https://doi.org/10.1080/03601234.2019.1631098.)*

nanoparticles synthesis. Concerning reaction time, an increase in contact time can accelerate the biosynthesis of silver nanoparticles.[42] In addition to these factors, the type and initial concentration of metal, microalgal species, the phytochemical composition of algal extracts, the solvent type are approaches that affect nanoparticle biosynthesis of silver (Table 1).

In this context, Kashyap et al.[48] evaluated different concentrations of $AgNO_3$ (0.1 mM, 0.2 mM, 0.5 mM, and 1.0 mM) for the biosynthesis of nanoparticles from an alcoholic extract prepared from biomass of *Chlorella* sp., which was screened out of *Chlorella* sp., *Lyngbya putealis*, *Oocystis* sp., and *Scenedesmus vacuolatus*. This study found that 0.5 mM $AgNO_3$ was the best concentration of metallic precursor. Furthermore, the authors observed that the biomolecules present in the extract help the synthesis of nanoparticles, reducing the metallic salt to metallic ions and acting as capping agents to stabilize the particles. The silver nanoparticles had a size of 90.6 nm.

Microalgal green synthesis of silver nanoparticles can occur through extracellular or intracellular reducing agents.[2,4] Öztürk[37] evaluated the synthesis of silver nanoparticles by the microalga *Desmodesmus* sp. intracellular and extracellular routes. The solution of 5 mM $AgNO_3$ was added to both the microalgal biomass and the supernatant of the culture. Subsequently, both conditions were incubated at 28°C for 72 h. The results showed that silver nanoparticles presented different sizes according to the via used to synthesis process. The intracellular and extracellular routes produced nanoparticles with 10 to 30 nm and 4 to 8 nm, respectively. Moraes et al.[17] studied the intracellular production of silver nanoparticles by different microalgal strains. *Coelastrum astroideum*, *Desmodesmus armatus*, *Cosmarium punctulatum*, *Klebsormidium*

TABLE 1 Green synthesis of silver nanoparticles from microalgae cultivation.

Microalgae	Microalgae cultivation conditions	Biosynthesis conditions	Morphology and size of nanoparticles	Reference
Acutodesmus dimorphus	35°C under 60 μmol m^{-2} s^{-1} and 12:12 h of light:dark period for 8 days	Dried de-oiled biomass (500 mg) mixed with distilled water, heated at 100°C for 5 min. Extract mixed with 1 mM of silver nitrate ($AgNO_3$) solution and stirred at room temperature for 24 h	Polydispersed and predominantly spherical shaped with the size range of 2–15 nm	43
Chlorella vulgaris	Erlenmeyer flasks (250 mL), 25°C, 12:12 h light/dark cycles, 123.47 ± 8.23 μmol m^{-2} s^{-1} for 21 days	Cell-free culture supernatant was incubated with 3.5 mM $AgNO_3$ at 25°C for 5 days, and protected from light	Heterogeneous morphology, predominance of particles with near-spherical shape, 9.8 ± 5.7 nm	44
Oscillatoria limnetica	Erlenmeyer flasks (500 mL), continuous illumination (57.75 M mol m^{-2} s^{-1}) at 28 ± 1°C for 22 days	0.1 M phosphate buffer at pH 7 and 1 mL of silver nitrate solutions (0.1, 0.2, 0.3, 0.4, and 0.5 mM) was added to the microalgal extract. The reaction mixture was incubated at 35 ± 1°C and 75.9 μmol m^{-2} s^{-1}	Quasi-spherical in shape with size ranging from 3.30 to 17.97 nm	45
Microchaete sp. NCCU-342	Erlenmeyer flask (500 mL) at 2000 ± 200 lx, photoperiod 12:12 h light:dark and 30 ± 1°C	Biomass of 80 μg/mL, pH 5.5, 60°C, duration of 60 min of exposure to ultraviolet light and concentration of 1 mM of $AgNO_3$	Spherical, polydispersed and in the range of 60–80 nm	46
Chaetoceros sp., *Skeletonema* sp., *Thalassiosira* sp.	Erlenmeyer flasks (500 ML) containing artificial seawater enriched with f/2-Si medium at pH 8.3, 12:12 light dark conditions, 23°C, and 100 μmol m^{-2} s^{-1}	The harvested biomass of each species was added slowly to the aqueous solution of $AgNO_3$	Homogenous uniform distribution. Sizes: 149.03 ± 3.0 nm (*Chaetoceros* sp.); 186.73 ± 4.9 nm (*Skeletonema* sp.); 239.46 ± 44.3 nm (*Thalassiosira* sp.)	47

flaccidum, Synechococcus elongatus, and *Microcystis aeruginosa* cells were harvested in the exponential growth phase and concentrated by filtration. AgNO$_3$ solution (1.0 mmol/L) was added to each microalgal strain and kept under stirring at room temperature for 20 h. The authors showed that the strains studied were capable of dispersed silver nanoparticles with spherical morphology with a size of ~2–3 nm. In another study, Terra et al.[38] presented the synthesis of silver nanoparticles simultaneously with the cultivation of microalgae *Spirulina* sp. LEB 18, combining extra and intracellular production. The microalgae were cultivated in a modified Zarrouk medium with the addition of 1 mM of silver sulfate (Ag$_2$SO$_4$). The authors obtained 70-nm polydisperse nanoparticles synthesized in the fifth cultivation period. On the 11th day of culture, the silver nanoparticles were reduced to sizes between 10 and 80 nm. There was a predominance of nanoparticles of 20 nm. Thus, the authors concluded that it is possible to synthesize silver nanoparticles in a non-toxic way simultaneously with microalgal culture and with fewer processing steps.

5 Potential of silver nanoparticles for biocide development

Nanobiocides are an emerging technology and may have an improved spectrum of characteristics compared to conventional biocides. Nanotechnology promotes controlled release and increased solubility of active compounds. There is a reduction in the degradation or oxidation of these substances, extending the product's shelf life.[49] The high surface area allows more reactive sites on the material's surface to interact with target molecules in the medium. Therefore, properties of the silver particle, such as antifungal, antibacterial, and thermal and electrical conductivity, are enhanced in the form of nanoparticles.[50]

The physicochemical and antimicrobial characteristics allow the application of silver nanoparticles in the control of pathogenic microorganisms,[5] in cultivars,[4] water disinfection,[51] and against biofouling on surfaces.[52] The control of microorganisms occurs through several biocidal mechanisms of silver nanoparticles, as the Ag$^+$ ions are released. After directly interacting with fungal spores, silver nanoparticles affect the formation of germ tubes, reducing disease progression in plants.[3,53] Another biocidal mechanism is the adhesion of silver nanoparticles to the surface of microorganisms, which can alter the exchange of nutrients, salts, and water.[54] In this context, Manukumar et al.[55] encapsulated silver nanoparticles (75 nm) synthesized from *Thymus vulgaris* and tested in vitro the antimicrobial efficiency against the *S. aureus* 090. The authors observed that the MIC to inactivate *S. aureus* 090, after 60 min of exposure to silver nanoparticles, was 1 mg/mL. Besides, morphological analyses of *S. aureus* 090 cells showed alterations in the cytoplasmic membrane after treatment with silver nanoparticles. The change in the membrane may have stimulated the release of cellular materials into the medium, such as DNA, potassium ions, among others, promoting the irreversible inactivation of the microorganism.

Dong et al.[56] immobilized silver nanoparticles on the surface of commercial polyamide reverse osmosis membrane and verified the antibacterial activity against *B. subtilis* (Gram-positive) and *E. coli* (Gram-negative). The study found that silver nanoparticles (10–20 nm) evenly distributed across the membrane surface, inactivated approximately 99% of Gram-positive and Gram-negative bacteria. The authors concluded that the mean diameter of the silver nanoparticles was responsible for the broad-spectrum antibacterial effect. The reduced diameter of the nanoparticles allowed penetration into the cytoplasmic membrane, causing the formation of reactive species and inactivation at the level of DNA in bacteria.

Despite their satisfactory biocidal activities, silver nanoparticles tend to aggregate in solution and, as a result, lose their antimicrobial efficiency.[57] For this aggregation not to occur, it is necessary to use stabilizing agents, such as surfactants,[58,59] polymers,[60,61] and reducing agents.[62,63] However, the stabilizing agent addition can modify the surface and dissolution of silver nanoparticles, altering the biocidal action and toxicity levels.[60,62,64,65] Thus, the microalgal synthesis of silver nanoparticles becomes an alternative for the stabilization of nanoparticles.[4] Biosynthesis by microalgae results in stable silver nanoparticles due to the presence of peptides that prevent aggregation.[17] Stability occurs through steric interactions where reducing molecules that act in the green synthesis remain attached to the nanoparticles, promoting their stability and dispersion in a non-toxic way.[3,66,67]

6 Silver nanoparticles in sensor engineering and pathogens control

Green technology approaches new methodologies and sustainable materials with the purpose of decreasing the damage to the ecosystem and the economic impact.[68] In this sense, nanomaterials can improve the efficiency and quality of biosensors, reducing the analysis time.[69] Thus, nanostructures have gained prominence to develop pathogen detection methods.[9] Microfluidic nano biosensors used to detect food-borne pathogens, for example, may reduce assay time, increasing sensitivity and accuracy.[70] Several nanoparticles have been investigated in microfluidic detection to functionalize, recognize, capture, and concentrate pathogens in food samples.[71]

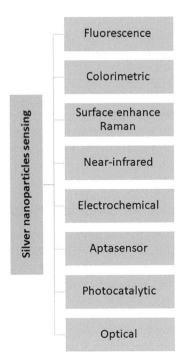

FIG. 3 Types of sensing methods with silver nanoparticles for pathogen detection.

Furthermore, for the development of optical, electrochemical and piezoelectric biosensors, and chemosensors, several nanostructures with transduction functions have been investigated (Fig. 3). Carbon nanotubes, graphene, gold nanoparticles, and silver nanoparticles were used to develop pathogen detection platforms.[69] Nanocomposites and nanoparticles have been applied to modify electrodes and increase the surface area for immobilizing elements, improving the sensor's sensitivity. Metallic nanoparticles, for example, can be used to modify electrodes and improve the specificity, electrocatalytic properties, stability, and conductivity of sensors. These nanomaterials allow an efficient functionalization with biological materials.[7]

The mechanism of optical sensors with silver nanoparticles is based on the quantification of the analyte through refractive index measurements, amount of absorbed light, fluorescent properties, or surface transduction.[68] In pathogen detection, optical sensors are used to consist the presence of a biomarker coupled with silver nanoparticles, that through transduction technology, convert the optical signal into data.[72] Electrochemical devices produced with silver nanoparticles can convert the analyte–electrode interaction by oxidation or reduction process into electrical signal.[68]

The characteristics of nanostructures, such as high surface permeability, greater surface/volume ratio, reactivity, and high penetrability allow for efficient functionalization in physical and chemical reactions.[8] High surface area to volume ratio of nanomaterials is one of the main characteristics responsible for the excellent properties of nanobiosensors, such as sensitivity, specificity, precision, and accuracy.[9] The relatively simple synthesis, combined with the particular surface area and high capacity to be functionalized with various biological fractions, makes metallic nanoparticles a good platform for detecting a range of biologically relevant analytes.[73]

Conventional methods for detecting pathogens take a long time to process. Generally, it is difficult to carry out the morphological and biochemical characterization of bacterial isolates with these detection methods.[74,75] Kumar et al.[76] studied silver ions to detect bacterial contamination in water samples. Colorimetry and electrochemical techniques were performed to verify cell membrane adhesion to silver ions. The authors demonstrated that the detection mechanism occurs in two stages in the colorimetric assay. The first stage is characterized by the sequestration of ions by pathogenic microorganisms, while the second is by the inhibition of the catalytic activity of urease. Thus, the correlation of enzymatic activity was used to determine the bacterial concentration in water. Urease activity can be observed by changes in pH after the addition of urea and phenol red as an indicator. In the electrochemical method, antibodies were used as a primary capture to detect *Salmonella enterica* serovar Typhi. The use of functionalized surface antibodies is necessary to capture silver ions and subsequent incubation for bacterial growth. Therefore, ions present on the surface of cells bound to the antibody layer are detected by an electrochemical probe and correlated with the presence of bacteria.[76]

TABLE 2 Applications of silver nanoparticles as a sensor in several fields.

Type of sensor	Silver nanoparticles concentration	Detection	Application field	Reference
Colorimetric	0.2% (w/v)	pH	Food packaging	80
Chemosensor	–	Mutagenic drug nitrofurazone	Environmental, clinical, and food sciences	81
Fluorescent	10 mM	Organophosphorus pesticides	Water pollution, food pollution, and environmental pollution	82
Electrochemical	0.2 g/mL	Chloramphenicol	Water environment	83
Near-infrared	Approximately 16% (v/v)	Ciprofloxacin and chloramphenicol	Water environment	184
Aptasensor	2% (v/v)	Zearalenone	Agricultural products	85
Electrochemical	0.017 g/mL	Superoxide anions	Human cells	86
Chemosensor	0.1698 mg/mL	2-Mercapto-5-methyl-1,3,4-thiadiazole	Food	87

–, data not available.

Moreover, the development of silver nanoparticle sensors is an alternative for identifying pathogens. They have a high surface area, high electrical conductivity, adsorption capacity for molecules, and magnetic and physicochemical properties.[77–79] Silver nanoparticles can be used in electrochemical sensors, fluorescent sensors, and biosensors for different areas (Table 2). Alex et al.[88] investigated silver nanoparticles' photocatalytic and sensing properties produced by green synthesis with silver nitrate and neem leaf extract. The biosensor property of silver nanoparticles was performed with a mancozeb fungicide solution. The results showed that the biosensor property produced a linear response with increasing fungicide concentration, obtaining a sensitivity of 39.1 nm/mM. Furthermore, photocatalytic activity occurred through the formation of reactive oxygen species. In this way, silver nanoparticles can be effective as a biosensor for application in agricultural cultivars.[88]

7 Challenges

Microalgal biomass productivity and the yield/characteristics of compounds are crucial approaches to increase the viability of using microalgae as biocides. Microalgae cultivation conditions can be manipulated to stimulate the achievement of the desired compound. However, this becomes a challenge since the variation in growing conditions can reduce biomass productivity and compound yield.[27,89]

Despite the numerous advantages of microalgae, large-scale production is still an economic challenge. Alternative sources of nutrients and the biorefinery concept can be applied to overcome this challenge.[89] Economic feasibility analyses regarding the synthesis of silver nanoparticles by microalgae must be investigated to assess the potential of these nanomaterials for application as biocide and biosensors.

Although several studies report that biosensors detect different pathogens, few biomaterials have been commercially manufactured.[9] The main challenge for applying biosensors in foods is their specificity, sensitivity, and detection time to assess the presence of food-borne pathogens in complex samples.[8] Furthermore, distinguishing between live and dead pathogens in food samples is another challenge.[9]

Despite the progress and advances in nanotechnology in detecting pathogens, challenges remain in practical applications. Most published studies do not present the cost-benefit ratio of their proposed biosensors. Thus, before biosensors based on nanomaterials reach the market, an economic evaluation must be carried out. The stability of nanoparticles in food samples should also be further investigated, as these nanomaterials can easily aggregate due to their high surface energy. Moreover, other aspects must be considered, including the affinities of the biorecognition elements during the immobilization and storage process and the miniaturization of the detector and multiplexing analysis.[9]

8 Conclusions and future perspectives

Biological approaches to the production of silver nanoparticles, such as microalgae technology, reduce the environmental impact caused by non-renewable synthesis methods. Advances in biotechnology make it possible to use microalgae to fix atmospheric carbon dioxide, carry out the treatment of effluents and, thus, generate nanoparticles with potential application for the control and detection of pathogens. In addition to synthesizing silver nanoparticles, microalgae have several anti-microbial compounds that can enhance the biocidal properties of nanoparticles. Furthermore, simultaneously with the green synthesis of these nanostructures, microalgae produce potential biomass for the development of biocides, among other diverse bioproducts. The antimicrobial potential of silver nanoparticles and microalgal biomass produced during synthesis indicates that this green technology is a promising and sustainable way for pathogen's control and should be more investigated.

Another essential property of silver nanoparticles is their ability to detect pathogenic microorganisms. The nanometric size improves the sensors' specificity, sensitivity, efficiency, and accuracy. Furthermore, the costs of the synthesis process of silver nanoparticles can be reduced when applying the concept of microalgal biorefineries. On the other hand, for these bioprocesses to be developed on a commercial/industrial scale, in-depth analyses on the economic viability of using micro-algal culture for the synthesis of silver nanoparticles and application as a biocide and biosensor must be presented. Fur-thermore, the long-term environmental impact of these nanostructures and their effects on human health also deserves further investigation in future research.

Acknowledgments

This research was developed within the scope of the Capes-PrInt Program (Process # 88887.310848/2018-00). The authors also are grateful to the Coordenação de Aperfeiçoamento de Pessoal de Nível Superior—Brazil (CAPES)—Finance Code 001.

References

1. Ranglová K, Lakatos GE, Manoel JAC, et al. Growth, biostimulant and biopesticide activity of the MACC-1 *Chlorella* strain cultivated outdoors in inorganic medium and wastewater. *Algal Res.* 2021;53:102136. https://doi.org/10.1016/j.algal.2020.102136.
2. Costa JAV, Terra ALM, Cruz ND, et al. Microalgae cultivation and industrial waste: new biotechnologies for obtaining silver nanoparticles. *Mini-Rev Org Chem.* 2019;16:1–8. https://doi.org/10.2174/1570193X15666180626141922.
3. Narware J, Yadav RN, Keswani C, Singh SP, Singh HB. Silver nanoparticle-based biopesticides for phytopathogens: scope and potential in agri-culture. In: Koul O, ed. *Nano-Biopesticides Today and Future Perspectives.* Academic Press; 2019:303–314. https://doi.org/10.1016/B978-0-12-815829-6.00013-9.
4. Terra ALM, Kosinski RC, Moreira JB, Costa JAV, Morais MG. Microalgae biosynthesis of silver nanoparticles for application in the control of agri-cultural pathogens. *J Environ Sci Health B.* 2019;54(8):709–716. https://doi.org/10.1080/03601234.2019.1631098.
5. Ugwoke E, Aisida SO, Mirbahar AA, et al. Concentration induced properties of silver nanoparticles and their antibacterial study. *Surf Interfaces.* 2020;18:100419. https://doi.org/10.1016/j.surfin.2019.100419.
6. Long Y, Hu L, Yan X, et al. Surface ligand controls silver ion release of nanosilver and its antibacterial activity against *Escherichia coli. Int J Nanomed.* 2017;12:3193–3206. https://doi.org/10.2147/IJN.S132327.
7. Vidic J, Manzano M. Electrochemical biosensors for rapid pathogen detection. *Curr Opin Electrochem.* 2021;100750. https://doi.org/10.1016/j.coelec.2021.100750.
8. Kumar H, Kuča K, Bhatia SK, et al. Applications of nanotechnology in sensor-based detection of foodborne pathogens. *Sensors.* 2020;20 (7):1966. https://doi.org/10.3390/s20071966.
9. Zhang R, Belwal T, Li L, Lin X, Xu Y, Luo Z. Nanomaterial-based biosensors for sensing key foodborne pathogens: advances from recent decades. *Compr Rev Food Sci Food Saf.* 2020;19(4):1465–1487. https://doi.org/10.1111/1541-4337.12576.
10. Vaz BS, Moreira JB, Morais MG, Costa JAV. Microalgae as a new source of bioactive compounds in food supplements. *Curr Opin Food Sci.* 2016;7:73–77. https://doi.org/10.1016/j.cofs.2015.12.006.
11. Costa JAV, Moreira JB, Fanka LS, Kosinski RC, Morais MG. Microalgal biotechnology applied in biomedicine. In: Konur O, ed. *Handbook of Algal Science, Technology and Medicine.* Elsevier; 2020:429–439. https://doi.org/10.1016/B978-0-12-818305-2.00027-9.
12. Falaise C, Francois C, Travers MA, et al. Antimicrobial compounds from eukaryotic microalgae against human pathogens and diseases in aquaculture. *Mar Drugs.* 2016;14:159. https://doi.org/10.3390/md14090159.
13. Shaima AF, NHM Y, Ibrahim N, Takriff MS, Gunasekaran D, MYY I. Unveiling antimicrobial activity of microalgae *Chlorella sorokiniana* (UKM2), *Chlorella* sp. (UKM8) and *Scenedesmus* sp. (UKM9). *Saudi J Biol Sci.* 2021. https://doi.org/10.1016/j.sjbs.2021.09.069.
14. Ferreira A, Melkonyan L, Carapinha S, et al. Biostimulant and biopesticide potential of microalgae growing in piggery wastewater. *Environ Adv.* 2021;4:100062. https://doi.org/10.1016/j.envadv.2021.100062.
15. Alsenani F, Tupally KR, Chua ET, et al. Evaluation of microalgae and cyanobacteria as potential sources of antimicrobial compounds. *Saudi Pharm J.* 2020;28(12):1834–1841. https://doi.org/10.1016/j.jsps.2020.11.010.

16. Maadane A, Merghoub N, Mernissi NE, et al. Antimicrobial activity of marine microalgae isolated from moroccan coastlines. *J Microbiol Biotechnol Food Sci.* 2017;6(6):1257–1260. https://doi.org/10.15414/jmbfs.2017.6.6.1257-1260.

17. Moraes LC, Figueiredo RC, Ribeiro-Andrade R, et al. High diversity of microalgae as a tool for the synthesis of different silver nanoparticles: a species-specific green synthesis. *Colloids Interface Sci Commun.* 2021;42:100420. https://doi.org/10.1016/j.colcom.2021.100420.

18. Zhang X-F, Liu Z-G, Shen W, Gurunathan S. Silver nanoparticles: synthesis, characterization, properties, applications and therapeutic approaches. *Int J Mol Sci.* 2016;17(9):1534. https://doi.org/10.3390/ijms17091534.

19. Khanna P, Kaur A, Goyal D. Algae-based metallic nanoparticles: synthesis, characterization and applications. *J Microbiol Methods.* 2019;163:105656. https://doi.org/10.1016/j.mimet.2019.105656.

20. El-Naggar NEA, Hussei MH, El-Sawah AA. Phycobiliprotein-mediated synthesis of biogenic silver nanoparticles, characterization, in vitro and in vivo assessment of anticancer activities. *Sci Rep.* 2018;8:8925. https://doi.org/10.1038/s41598-018-27276-6.

21. Arya A, Gupta K, Chundawat TS, Vaya D. Biogenic synthesis of copper and silver nanoparticles using green alga *Botryococcus braunii* and its antimicrobial activity. *Bioinorg Chem Appl.* 2018;2018:7879403. https://doi.org/10.1155/2018/7879403.

22. Madkour LH. Ecofriendly green biosynthesized of metallic nanoparticles: bio-reduction mechanism, characterization and pharmaceutical applications in biotechnology industry. *Glob Drug Ther.* 2021;3(1):1–11. https://doi.org/10.15761/GDT.1000144.

23. Jin J, Dupre C, Legrand J, Grizeau D. Extracellular hydrocarbon and intracellular lipid accumulation are related to nutrient-sufficient conditions in pH-controlled chemostat cultures of the microalga *Botryococcus braunii* SAG 30.81. *Algal Res.* 2016;17:244–252. https://doi.org/10.1016/j.algal.2016.05.007.

24. Vishwanath R, Negi R. Conventional and green methods of synthesis of silver nanoparticles and their antimicrobial properties. *Curr Res Green Sustain Chem.* 2021;4:100205. https://doi.org/10.1016/j.crgsc.2021.100205.

25. Hamida RS, Ali MA, Redhwan A, Bin-Meferij MM. Cyanobacteria—a promising platform in green nanotechnology: a review on nanoparticles fabrication and their prospective applications. *Int J Nanomed.* 2020;15:6033–6066. https://doi.org/10.2147/IJN.S256134.

26. Jacob JM, Ravindran R, Narayanan M, Samuel SM, Pugazhendhi A, Kumar G. Microalgae: a prospective low cost green alternative for nanoparticle synthesis. *Curr Opin Environ Sci Health.* 2021;20:100163. https://doi.org/10.1016/j.coesh.2019.12.005.

27. Costa JAV, Moraes L, Moreira JB, Rosa GM, Henrard ASA, Morais MG. Microalgae-based biorefineries as a promising approach to biofuel production. In: Tripathi B, Kumar D, eds. *Prospects and Challenges in Algal Biotechnology.* Springer; 2017:113–140. https://doi.org/10.1007/978-981-10-1950-0_4.

28. Bhuyar P, Hong DD, Mandia E, Rahim MHA, Maniam GP, Govindan N. Salinity reduction from poly-chem-industrial wastewater by using microalgae (*Chlorella* sp.) collected from coastal region of peninsular Malaysia. *J Bio Med Open Access.* 2020;1(1):105.

29. Rajkumar R, Ezhumalai G, Gnanadesigan M. A green approach for the synthesis of silver nanoparticles by *Chlorella vulgaris* and its application in photocatalytic dye degradation activity. *Environ Technol Innov.* 2021;21:101282. https://doi.org/10.1016/j.eti.2020.101282.

30. Gonçalves AL. The use of microalgae and cyanobacteria in the improvement of agricultural practices: a review on their biofertilising, biostimulating and biopesticide roles. *Appl Sci.* 2021;11(2):871. https://doi.org/10.3390/app11020871.

31. Ebrahimzadeh Z, Salehzadeh A, Naeemi AS, Jalali A. Silver nanoparticles biosynthesized by *Anabaena flos-aquae* enhance the apoptosis in breast cancer cell line. *Bull Mater Sci.* 2020;43(92). https://doi.org/10.1007/s12034-020-2064-1.

32. Fathy W, Elsayed K, Essawy E, et al. Biosynthesis of silver nanoparticles from *Synechocystis* sp. to be used as a flocculant agent with different microalgae strains. *Curr Nanomater.* 2020;5(2):175–187. https://doi.org/10.2174/2468187310999200605161200.

33. Geetha S, Vijayakumar K, Aranganayagam KR, Thiruneelakandan G. Biosynthesis, characterization of silver nanoparticles and antimicrobial screening by *Oscillatoria annae.* *AIP Conf Proc.* 2020;2270:110026. https://doi.org/10.1063/5.0024262.

34. Arévalo-Gallegos A, Garcia-Perez JS, Carrillo-Nieves D, Ramirez-Mendoza RA, Iqbal HMN, Parra-Saldíva R. *Botryococcus braunii* as a bioreactor for the production of nanoparticles with antimicrobial potentialities. *Int J Nanomed.* 2018;13:5591–5604. https://doi.org/10.2147/IJN.S174205.

35. Rahman A, Kumar S, Bafana A, Dahoumane SA, Jeffryes C. Biosynthetic conversion of Ag^+ to highly stable Ag^0 nanoparticles by wild type and cell wall deficient strains of *Chlamydomonas reinhardtii.* *Molecules.* 2019;24(1):98. https://doi.org/10.3390/molecules24010098.

36. Darwesh OM, Matter IA, Eida MF, Moawad H, Oh Y-K. Influence of nitrogen source and growth phase on extracellular biosynthesis of silver nanoparticles using cultural filtrates of *Scenedesmus obliquus.* *Appl Sci.* 2019;9(7):1465. https://doi.org/10.3390/app9071465.

37. Öztürk BY. Intracellular and extracellular green synthesis of silver nanoparticles using *Desmodesmus* sp.: their antibacterial and antifungal effects. *Caryologia.* 2019;72(1):29–43. https://doi.org/10.13128/cayologia-249.

38. Terra ALM, Cruz ND, Henrard ASA, Costa JAV, Morais MG. Simultaneous biosynthesis of silver nanoparticles with *Spirulina* sp. LEB 18 cultivation. *Ind Biotechnol.* 2019;15(4):263–267. https://doi.org/10.1089/ind.2018.0022.

39. Chugh D, Viswamalya VS, Das B. Green synthesis of silver nanoparticles with algae and the importance of capping agents in the process. *J Genet Eng Biotechnol.* 2021;19:126. https://doi.org/10.1186/s43141-021-00228-w.

40. Cui J, Shao Y, Zhang H, Zhang H, Zhu J. Development of a novel silver ions-nanosilver complementary composite as antimicrobial additive for powder coating. *Chem Eng J.* 2021;420(2):127633. https://doi.org/10.1016/j.cej.2020.127633.

41. Chaudhary R, Nawaz K, Khan AK, Hano C, Abbasi BH, Anjum S. An overview of the algae-mediated biosynthesis of nanoparticles and their biomedical applications. *Biomolecules.* 2020;10:1498. https://doi.org/10.3390/biom10111498.

42. Shalaby EA. Algae-mediated silver nanoparticles: synthesis, properties, and biological activities. In: Abd-Elsalam KA, ed. *Green Synthesis of Silver Nanomaterials.* Elsevier; 2022:525–545. https://doi.org/10.1016/B978-0-12-824508-8.00009-5.

43. Chokshi K, Pancha I, Ghosh T, et al. Green synthesis, characterization and antioxidant potential of silver nanoparticles biosynthesized from de-oiled biomass of thermotolerant oleaginous microalgae *Acutodesmus dimorphus.* *RSC Adv.* 2016;6:72269. https://doi.org/10.1039/c6ra15322d.

44. Ferreira VS, ConzFerreira ME, Lima LMTR, Frasés S, Souza W, Sant'Anna C. Green production of microalgae-based silver chloride nanoparticles with antimicrobial activity against pathogenic bacteria. *Enzyme Microb Technol.* 2017;97:114–121. https://doi.org/10.1016/j.enzmictec.2016.10.018.

45. Hamouda RA, Hussein MH, Abo-elmagd RA, Bawazir SS. Synthesis and biological characterization of silver nanoparticles derived from the cyanobacterium *Oscillatoria limnetica. Sci Rep.* 2019;9:13071. https://doi.org/10.1038/s41598-019-49444-y.

46. Husain S, Afreen S, Hemlata, Yasin D, Afzal B, Fatma T. Cyanobacteria as a bioreactor for synthesis of silver nanoparticles-an effect of different reaction conditions on the size of nanoparticles and their dye decolorization ability. *J Microbiol Methods.* 2019;162:77–82. https://doi.org/10.1016/j.mimet.2019.05.011.

47. Mishra B, Saxena A, Tiwari A. Biosynthesis of silver nanoparticles from marine diatoms *Chaetoceros* sp., *Skeletonema* sp., *Thalassiosira* sp., and their antibacterial study. *Biotechnol Rep.* 2020;28:e00571. https://doi.org/10.1016/j.btre.2020.e00571.

48. Kashyap M, Samadhiya K, Gosh A, Anand V, Shirage PM, Bala K. Screening of microalgae for biosynthesis and optimization of Ag/AgCl nano hybrids having antibacterial effect. *RSC Adv.* 2019;9:25583–25591. https://doi.org/10.1039/c9ra04451e.

49. Bapat MS, Singh H, Shukla SK, et al. Evaluating green silver nanoparticles as prospective biopesticides: an environmental standpoint. *Chemosphere.* 2022;286:131761. https://doi.org/10.1016/j.chemosphere.2021.131761.

50. Liao C, Li Y, Tjong SC. Bactericidal and cytotoxic properties of silver nanoparticles. *Int J Mol Sci.* 2019;20(2):449. https://doi.org/10.3390/ijms20020449.

51. Wei F, Zhao X, Li C, Han X. A novel strategy for water disinfection with a AgNPs/gelatin sponge filter. *Environ Sci Pollut Res.* 2018;25:19480–19487. https://doi.org/10.1007/s11356-018-2157-1.

52. Linhares AMF, Borges CP, Fonseca FV. Investigation of biocidal effect of microfiltration membranes impregnated with silver nanoparticles by sputtering technique. *Polymers.* 2020;12(8):1686. https://doi.org/10.3390/polym12081686.

53. Mansoor S, Zahoor I, Baba TR, et al. Fabrication of silver nanoparticles against fungal pathogens. *Front Nanotechnol.* 2021;3:67. https://doi.org/10.3389/fnano.2021.679358.

54. Ferdous Z, Nemmar A. Health impact of silver nanoparticles: a review of the biodistribution and toxicity following various routes of exposure. *Int J Mol Sci.* 2020;21:2375. https://doi.org/10.3390/ijms21072375.

55. Manukumar HM, Yashwanth B, Umesha S, Rao JV. Biocidal mechanism of green synthesized thyme loaded silver nanoparticles (GTAgNPs) against immune evading tricky methicillin-resistant *Staphylococcus aureus* 090 (MRSA090) at a homeostatic environment. *Arab J Chem.* 2020;13(1):1179–1197. https://doi.org/10.1016/j.arabjc.2017.09.017.

56. Dong C, Wang Z, Wu J, Wang Y, Wang J, Wang S. A green strategy to immobilize silver nanoparticles onto reverse osmosis membrane for enhanced anti-biofouling property. *Desalination.* 2017;401:32–34. https://doi.org/10.1016/j.desal.2016.06.034.

57. Singh A, Sharma B, Deswal R. Green silver nanoparticles from novel Brassicaceae cultivars with enhanced antimicrobial potential than earlier reported Brassicaceae members. *J Trace Elem Med Biol.* 2018;47:1–11. https://doi.org/10.1016/j.jtemb.2018.01.001.

58. Pisárčik M, Lukáč M, Jampílek J, et al. Silver nanoparticles stabilized with phosphorus-containing heterocyclic surfactants: synthesis, physicochemical properties, and biological activity determination. *Nanomaterials.* 2021;11(8):1883. https://doi.org/10.3390/nano11081883.

59. Bekhit M, Abuel-naga MN, Sokary R, Fahim RA, El-Sawy NM. Radiation-induced synthesis of tween 80 stabilized silver nanoparticles for antibacterial applications. *J Environ Sci Health A.* 2020;55(10):1210–1217. https://doi.org/10.1080/10934529.2020.1784656.

60. Batista CCS, Albuquerque LJC, Araujo I, Albuquerque BL, Silva FD, Giacomelli FC. Antimicrobial activity of nano-sized silver colloids stabilized by nitrogen-containing polymers: the key influence of the polymer capping. *RSC Adv.* 2018;8:10873–10882. https://doi.org/10.1039/c7ra13597a.

61. Batista CCS, Jäger A, Albuquerque BL, Pavlov E, Stepánek P, Giacomelli FC. Microfluidic-assisted synthesis of uniform polymer-stabilized silver colloids. *Colloids Surf A Physicochem Eng Asp.* 2021;618:126438. https://doi.org/10.1016/j.colsurfa.2021.126438.

62. Liu Y-S, Chang Y-C, Chen H-H. Silver nanoparticle biosynthesis by using phenolic acids in rice husk extract as reducing agents and dispersants. *J Food Drug Anal.* 2018;26(2):649–656. https://doi.org/10.1016/j.jfda.2017.07.005.

63. Mohaghegh S, Osouli-Bostanabad K, Nazemiyeh H, et al. A comparative study of eco-friendly silver nanoparticles synthesis using *Prunus domestica* plum extract and sodium citrate as reducing agents. *Adv Powder Technol.* 2020;31(3):1169–1180. https://doi.org/10.1016/j.apt.2019.12.039.

64. Fernando I, Qian T, Zhou Y. Long term impact of surfactants & polymers on the colloidal stability, aggregation and dissolution of silver nanoparticles. *Environ Res.* 2019;179(Part A):108781. https://doi.org/10.1016/j.envres.2019.108781.

65. Egorova EM, Kaba SI. The effect of surfactant micellization on the cytotoxicity of silver nanoparticles stabilized with aerosol-OT. *Toxicol In Vitro.* 2019;57:244–254. https://doi.org/10.1016/j.tiv.2019.03.006.

66. Rahman A, Kumar S, Bafana A, Dahoumane SA, Jeffryes C. Individual and combined effects of extracellular polymeric substances and whole cell components of *Chlamydomonas reinhardtii* on silver nanoparticle synthesis and stability. *Molecules.* 2019;24(5):956. https://doi.org/10.3390/molecules24050956.

67. Kitherian S. Nano and bio-nanoparticles for insect control. *Res J Nanosci Nanotechnol.* 2017;7:1–9. https://doi.org/10.3923/rjnn.2017.1.9.

68. Heinemann MG, Rosa CH, Rosa GR, Dias D. Biogenic synthesis of gold and silver nanoparticles used in environmental applications: a review. *Trends Environ Anal Chem.* 2021;30. https://doi.org/10.1016/j.teac.2021.e00129, e00129.

69. Bhardwaj SK, Bhardwaj N, Kumar V, et al. Recent progress in nanomaterial-based sensing of airborne viral and bacterial pathogens. *Environ Int.* 2021;146:106183. https://doi.org/10.1016/j.envint.2020.106183.

70. Weng X, Zhang C, Jiang H. Advances in microfluidic nanobiosensors for the detection of foodborne pathogens. *LWT.* 2021;112172. https://doi.org/10.1016/j.lwt.2021.112172.

71. Jiang Y, Zou S, Cao X. Rapid and ultra-sensitive detection of foodborne pathogens by using miniaturized microfluidic devices: a review. *Anal Methods.* 2016;8(37):6668–6681. https://doi.org/10.1039/c6ay01512c.

72. Shahbazi N, Zare-Dorabei R, Naghib SM. Multifunctional nanoparticles as optical biosensing probe for breast cancer detection: a review. *Mater Sci Eng C*. 2021;127:112249. https://doi.org/10.1016/j.msec.2021.112249.

73. Prado M, Espiña B, Fernandez-Argüelles MT, et al. Detection of foodborne pathogens using nanoparticles. Advantages and trends. In: Barros-Velázquez J, ed. *Antimicrobial Food Packaging*. Academic Press; 2016:183–201. https://doi.org/10.1016/B978-0-12-800723-5.00014-0.

74. Wang D, Chen Q, Huo H, et al. Efficient separation and quantitative detection of *Listeria monocytogenes* based on screen-printed interdigitated electrode, urease and magnetic nanoparticles. *Food Control*. 2017;73:555–561. https://doi.org/10.1016/j.foodcont.2016.09.003.

75. Mocan T, Matea CT, Pop T, et al. Development of nanoparticle-based optical sensors for pathogenic bacterial detection. *J Nanobiotechnol*. 2017;15 (1):1–14. https://doi.org/10.1186/s12951-017-0260-y.

76. Kumar V, Chopra A, Bisht B, Bhalla V. Colorimetric and electrochemical detection of pathogens in water using silver ions as a unique probe. *Sci Rep*. 2020;10(1):1–9. https://doi.org/10.1038/s41598-020-68803-8.

77. Wang C, Wang J, Li M, et al. A rapid SERS method for label-free bacteria detection using polyethylenimine-modified au-coated magnetic microspheres and Au@Ag nanoparticles. *Analyst*. 2016;141:6226–6238. https://doi.org/10.1039/C6AN01105E.

78. Popov A, Brasiunas B, Kausaite-Minkstimiene A, Ramanaviciene A. Metal nanoparticle and quantum dot tags for signal amplification in electrochemical Immunosensors for biomarker detection. *Chemosensors*. 2021;9(4):85. https://doi.org/10.3390/chemosensors9040085.

79. Pirzada M, Altintas Z. Nanomaterials for healthcare biosensing applications. *Sensors*. 2019;19(23):5311. https://doi.org/10.3390/s19235311.

80. Jovanska L, Chiu C-H, Yeh I-C, Chiang W-D, Hsieh C-C, Wang R. Development of a PCL-PEO double network colorimetric pH sensor using electrospun fibers containing *Hibiscus rosa sinensis* extract and silver nanoparticles for food monitoring. *Food Chem*. 2022;368:130813. https://doi.org/10.1016/j.foodchem.2021.130813.

81. Ahmed SR, Anwar H, Ahmed SW, Shah MR, Ahmed A, Ali SA. Green synthesis of silver nanoparticles: antimicrobial potential and chemosensing of a mutagenic drug nitrofurazone in real samples. *Measurement*. 2021;180:109489. https://doi.org/10.1016/j.measurement.2021.109489.

82. Luo Q, Lai J, Qiu P, Wang X. An ultrasensitive fluorescent sensor for organophosphorus pesticides detection based on RB-Ag/Au bimetallic nanoparticles. *Sensor Actuat B Chem*. 2018;263:517–523. https://doi.org/10.1016/j.snb.2018.02.101.

83. Chang C, Wang Q, Xue Q, Liu F, Hou L, Pu S. Highly efficient detection of chloramphenicol in water using Ag and TiO_2 nanoparticles modified laser-induced graphene electrode. *Microchem J*. 2022;173:107037. https://doi.org/10.1016/j.microc.2021.107037.

84. Jiao A, Cui Q, Li S, et al. Aligned TiO_2 nanorod arrays decorated with closely interconnected Au/Ag nanoparticles: near-infrared SERS active sensor for monitoring of antibiotic molecules in water. *Sensor Actuat B Chem*. 2022;350:130848. https://doi.org/10.1016/j.snb.2021.130848.

85. Chen R, Sun Y, Huo B, et al. Development of Fe_3O_4@Au nanoparticles coupled to Au@Ag core-shell nanoparticles for the sensitive detection of zearalenone. *Anal Chim Acta*. 2021;1180:338888. https://doi.org/10.1016/j.aca.2021.338888.

86. Wang Z, Zhao H, Gao Q, Chen K, Lan M. Facile synthesis of ultrathin two-dimensional graphene-like CeO_2-TiO_2 mesoporous nanosheet loaded with Ag nanoparticles for non-enzymatic electrochemical detection of superoxide anions in HepG2 cells. *Biosens Bioelectron*. 2021;184:113236. https://doi.org/10.1016/j.bios.2021.113236.

87. Jing M, Zhang H, Li M, Mao Z, Shi X. Silver nanoparticle-decorated TiO_2 nanotube array for solid-phase microextraction and SERS detection of antibiotic residue in milk. *Spectrochim Acta A Mol Biomol Spectrosc*. 2021;255:119652. https://doi.org/10.1016/j.saa.2021.119652.

88. Alex KV, Pavai PT, Rugmini R, Prasad MS, Kamakshi K, Sekhar KC. Green synthesized Ag nanoparticles for bio-sensing and photocatalytic applications. *ACS Omega*. 2020;5(22):13123–13129. https://doi.org/10.1021/acsomega.0c01136.

89. Morais MG, Morais EG, Cardias BB, et al. Microalgae as a source of sustainable biofuels. In: Gupta VK, Treichel H, Kuhad RC, Rodriguez-Cout S, eds. *Recent Developments in Bioenergy Research*. 1st ed. Elsevier; 2020:253–271. https://doi.org/10.1016/B978-0-12-819597-0.00013-1.

Chapter 19

Polyoxometalate: A sustainable material for environmental remediation

Daksha Sharma[a], Varun Rawat[b], Monu Verma[c], and Dipti Vaya[d,*]

[a]Department of Chemistry, Vidhya Bhawan Rural Institute, Udaipur, Rajasthan, India, [b]School of Chemistry, Faculty of Exact Science, Tel Aviv University, Tel Aviv, Israel, [c]Water-Energy Nexus Laboratory, Department of Environmental Engineering, University of Seoul, Seoul, Republic of Korea, [d]Department of Chemistry, Amity School of Applied Sciences, Amity University, Gurugram, Haryana, India

*Corresponding author.

1 Introduction

Urbanization and industrialization cause environmental degradation by releasing toxic substances into the ecosystem.[1,2] Researchers continually strive to search for new materials and methods for environmental remediation. Various semiconductors such as ZnO,[3] TiO_2,[4] CO_3O_4,[5] and carbon-based materials such as carbon nanotube,[6] graphene oxide,[7] C_3N_4,[8] etc., are utilized for remediation.[9] Although these materials are extensively utilized, the efficiency of photochemical processes depends on the bandgap edge and redox potential.

Polyoxometalates (POMs) are among the best materials for redox reactions and possess a suitable bandgap. POMs are special inorganic clusters with various topological structures, usually consisting of early transition metals coordinated with oxygen atoms through co-angle, co-edge, or co-planar mode and forming 3D closed structures.[10] Groups 5 and 6 transition metal ions with higher oxidation states (V and VI) are the usual selection of metal ions. Various POMs have been synthesized that are different from each other concerning the number of atoms per cluster, the existence of a heteroatom, and atomic rearrangements. The general formula of POM anions is $[X_xM_yO_z]^{n-}$, where M is the transition metal (such as Mo, W, Nb, or Ta) and X is the heteroatom (such as P, Si, Ge, Se, or Sb).

POMs include different shapes, sizes, and configurations and classical classification areas as follows[10]: Anderson,[11] Dawson,[12] Keggin,[13] Lindqvist,[14] Silverton,[15] and Waugh[16] types. The first polyoxometalate was discovered in 1826,[17] and it was an ammonium phosphomolybdate ion $[PMo_{12}O_{40}]^{3-}$. In 1934, a similar configuration of phosphotungstate anion was invented, and later it was recognized as the Keggin structure based on his inventor's name.[18] Other analogous structure, Wells-Dawson, was discovered later, along with their characteristics and usability as catalyst. All discovered structures lack symmetry. Later discovered, wheel-shaped molybdenum blue anions and spherical keplerates' POMs were counted as highly symmetric structures. To enhance the properties of POMs, researchers also developed several hybrids that consist of POMs cores along with inorganic/organic materials.[19] These hybrid materials consist of porphyrin ring, graphene, silica, TiO_2, and other metal oxides. These further exhibit potential applications such as photocatalytic degradation, catalysis, CO_2 reduction, water splitting, and medical fields specifically anti-tumor and antiviral therapy. These applications rely on the unique magnetic[20] and optical[21] characteristics of POMs. This chapter begins with the fundamentals of POMs, structures, classification, hybrid compounds, and their recent applications in various fields.

2 Structure of POMs

Structural variability of POMs is associated with a broad range of applications. Following POM classification is given as shown in Fig. 1:

(1) Heteropolymetalates: It consists of a single type of metal, oxygen, and a single type of heteroanion (such as phosphate, silicate, sulfate, etc.). Even literature reported huge deviation from the existing structure.[22,23]
(2) Isopolymetalates: It is comprised of only one kind of metal and oxide.
(3) Molybdenum: It has Mo-based reduced POM nanosized clusters.

Sustainable Materials for Sensing and Remediation of Noxious Pollutants. https://doi.org/10.1016/B978-0-323-99425-5.00021-9

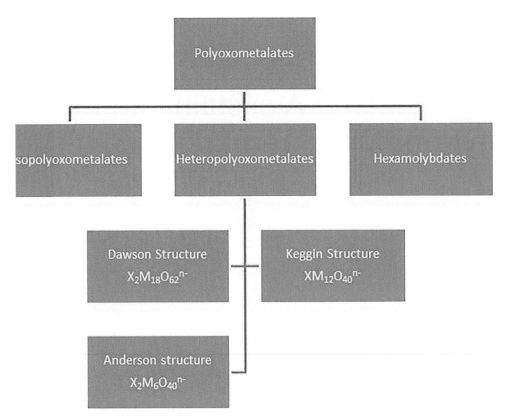

FIG. 1 Classification of polyoxometalates.

The category of POMs includes a variety of anionic multinuclear species with diverse structural characteristics, configurations, and diameters varying from 1 to 5.6 nm. The technique of Miras et al. has been used to assist researchers in making the essential links between the various building block kinds, archetypes, and physical attributes.[24] The following is a broad taxonomy of the POMs family based on the structure and composition:

(a) The first type includes heteropolyanionic species composed of heteroanions like PO_4^{3-}, SO_4^{2-} and SiO_4^{2-} and consists of V, W, or Mo-based metal oxide in the backbone. These clusters are the most popular category. These building blocks are relatively stable, and after rearrangement of heteroanions, they synthesize varieties of archetypes. Therefore, many researchers looked at the modification of catalytic, electronic, and pH-dependent properties of these POMs, with a special concern on the Kegging $[XM_{12}O_{40}]$ and the Wells-Dawson $[X_2M_{18}O_{62}]_n$ (where M = W). Furthermore, it has been investigated that tungsten-based POMs are kinetically inert, and this fact induces the development of derivatives of the Keggin and Dawson type that are kinetically more labile. Mono-, di-, and trilacunary clusters form stable building blocks with bigger aggregates.[25] The synthesis and investigation of lacunary $\{M_{12-n}\}$ and Dawson $\{M_{18-n}\}$, tungsten-based POMs are a vast field of research.

(b) The isopolyanions belong to the second group of the POMs family, and these are composed of a metal oxide backbone with no heteroatoms or heteroanions. As a result, the structure is rather simple compared to the heteropolyanion type. Members of this category are substantially smaller. They have fascinating physical features and can be employed similarly as cluster-based building blocks.[26,27]

(c) Molybdenum (Mo)-based reduced POMs nanosized clusters, also known as Mo-blue and Mo-brown species, make up the third category. Scheele investigated these clusters for the first time in 1783. However, X-ray diffraction tools faced difficulty in detecting their composition and intricate structural properties due to their massive size. This problem was partially solved when Muller et al. revealed the characterization of the pioneer member of the above category in 1995, which exhibited a ring topology, but due to their remarkable size and configuration complexity, even its structural aspects are not identified completely (Mo_{144}).[28] Later, after effortless efforts, scientists investigated the complete characterization of the first member of the Mo-brown species, which exhibited a porous spherical morphology (Mo_{132}) and composed of 132 Mo centers and expressed a higher rate of reduction than their Mo-blue counterparts.

Keggin type was the first to be discovered, and its configuration was found to be shared with distinct center heteroatoms made up of molybdates and tungstates. Tetrahedrally coordinated heteroatoms (P or Si) exist in Keggin and Dawson structures, while octahedral center atom Al exists in Anderson structures.

Lindquist, Keggin, and Wells-Dawson structures, which include 6, 12, and 18 transition metal atoms and have the formulae $M_6O_{192}^{2-}$, $XM_{12}O_{40}^{n-}$, and $X_2M_{18}O_{62}^{n-}$, respectively, are examples of POMs[where M is backbone transition metal atom (Mo, W, and V), X is a heteroatom (P, Al, and Si)]. Preyssler and Wells-Dawson are two forms of POMs that are rarely investigated.

3 Synthesis of polyoxometalates

Numerous POMs complexes are reported in the literature, mostly made up of Mo, W, V, and Nb. The cluster structure of many POMs complexes contains a heteropolyanion, such as phosphorous or sulfur (it is located in tetrahedral SO_4^{2-} position).[29] The Keggin and Dawson POMs types have received the most attention.[30] Furthermore, recent covalent functionalization of classic POMs designs has resulted in novel hybrid species.[31–33]

For the synthesis of POMs, a large range of techniques have been reported. Each form of POMs exhibits a different way of synthesis method. Chemical method, sol-gel, hydrothermal, ultrasonication-assisted, and spin coating are the most efficient procedures for preparing POMs.[13,34,35] These methods result in well-configured POMs nanocomposites with unique properties. Keggin anions are easily synthesized when a precursor of metal alkali salt and heteroatom oxoanions are mixed and then lowered in the pH of the solution. After that, ether extraction was used to separate Keggin anions from excess cations and heteroanions and then re-purified by recrystallization, such as other inorganic materials. The structure of POMs can be determined using the bulk material's XRD pattern or the heteroatoms or metal atom by nuclear magnetic resonance (NMR) spectroscopy.

Lacunary POMs are defective Dawson structures that can be synthesized from pure ones under basic conditions by losing a M=O unit, and pure POMs structures are often uncomplicated. The huge POMs "Mo-brown ball" (a Mo_{132} cluster[36]) and "Mo-blue wheel" (a Mo_{154} cluster[37]) can be segregated using gel electrophoresis, introducing a novel approach to POMs purification.[38]

A different strategy was proposed to synthesize inorganic/organic hybrid POMs. In this strategy, the attachment of the O atom of POMs with organic and inorganic moiety is the superior one. Another way is the attachment through coordination with the nitrogen atom. In a few cases, C-M bond formation is also reported. Few scientists recommended hydrogen bonding which is also helpful in bonding.[39]

4 Role of structure in applications

One of the most valuable POMs is the Keggin ion, $XM_{12}O_{40}^{n-}$. The triangular arrays of oxygen atoms found on POMs surfaces are responsible for the attachment of bulk metal oxides. As a result, POMs are ideal for both homogeneous and heterogeneous catalysis.

A M-O group can be left out of the Keggin ion, which leads to a vacancy in the molecule; otherwise, it is completely symmetric. Keggin-type POMs with mono-lacunary composition have the following chemical formula $XM_{11}O_{39}^{n-}$. It can act as a pentadentate ligand, and it is rigid, hydrolytically stable, thermally durable, and non-oxidizable.

Fig. 2A shows the representation of various POMs families.[40] The comparative ratios of addenda atoms to heteroatoms decide the factors for POMs.[41] Keggin and Dawson configurations have been largely explored among POMs.[42] Other advanced POM structures have been developed in addition to these fundamental ones. To get the desired qualities, POMs would be modulated by constructing the template, addenda atom, hetero-metal atom, and ligands (Fig. 2B).[40,41] The important parameters explored are high Bronsted acidity, feasible redox reactions, better photoelectric properties, thermal stability, and structural variability. POMs are widely used in different applications, including catalysis, sensors, solar cells, magnetism, and medicine.[40,43,44] The thermal decomposition temperature of various commonly used POMs such as $H_3PW_{12}O_{40}$ (PW), $H_4SiW_{12}O_{40}$ (SiW), and $H_3PMo_{12}O_{40}$ (PMo) are 465°C, 445°C, and 375°C, respectively.[45]

5 Applications

Variability in structures, composition, and shapes of POMs are responsible for their broad range of applications. The Keggin ions exhibited reversible reduction, stable at high temperatures, and function as catalysts in various chemical processes. POMs nowadays are used as a green source. The bleaching process of non-chlorine-based wood pulp is carried out with the help of POMs,[46] and similarly, it assists in the removal of pollutants from wastewater.[47] Luminescence is a

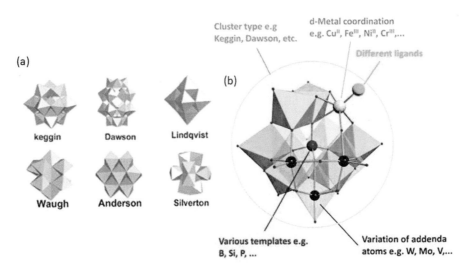

FIG. 2 (A) Classification of POMs. (B) Composition of POMs. *(Reproduced with permission from Herrmann S, Ritchie C, Streb C. Polyoxometalate—conductive polymer composites for energy conversion, energy storage and nanostructured sensors. Dalton Trans. 2015;44:7092–7104. https://doi.org/10.1039/C4DT03763D, John Wiley and Sons under copyright 2020.)*

property of several POMs that help in medical applications such as bioimaging, anti-tumor and antiviral therapy.[11] POMs can initiate and catalyze various processes under exposure to light, including oxidation of alcohols,[48] benzene,[49] and phenol,[50] the oxidative bromination of aromatics or olefins,[51] photoreduction of carbon-dioxide,[52] and photocatalytic H_2 evolution.[53]

5.1 Photocatalysis

POMs, in addition to TiO_2, have been extensively studied as photocatalysts.[54,55] Several new water-resist POMs that consist of hybrid composite photocatalysts were studied, including transition metal-substituted POMs supported with amine-functionalized mesoporous silica/anatase titania.[56] POMs exhibited good photocatalytic or photochemical properties as TiO_2, sometimes even better than this. Lei et al. studied immobilized POMs with H_2O_2 for mineralization of RhB under visible light illumination.[57] An anionic exchange resin was used to immobilize these Keggin-based POMs ($PW_{12}O_{40}^{3-}$) and their lacunary derivatives. As photocatalysts, various POMs, i.e., $PW_{12}O_{40}^{3-}$, $SiW_{12}O_{40}^{4-}$, $P_2W_{18}O_{62}^{6-}$, and $P_2Mo_{18}O_{62}^{6-}$ are being explored.

These different POMs mediate photodegradation of various organic compounds, including rhodamine B(RhB), azo, naphthol blue-black dye, phenol, and nitrophenol[28,57–59] and photoreduction of heavy metal ions.[60–62] Stable gold nanoparticles are prepared when photochemically reduced aqueous solution of Keggin ions ($PW_{12}O_{40}$)[3−] are exposed to auricions. Photocatalytic mercury reduction in aqueous solutions was performed using $PW_{12}O_{40}^{3-}$ or $SiW_{12}O_{40}^{4-}$ as photocatalysts. In the presence or absence of oxygen, resultant activity was investigated with an exposure time of irradiation, the concentration of Hg (II), types of POMs, and the amount of organic substrate.

Gkika et al. investigated POMs, which are clusters of metal oxide mainly containing tungsten, which are effective for the degradation of organic contaminants.[63] The photocatalytic activity of POMs was discovered on a wide range of pesticides, including lindane, bentazone, fenitrothion, and measured decontamination content. Cao et al. invented a POMs-based photocatalyst containing cucurbit [6] uril-via-hydrogen linkages to degrade certain azo dyes.[64] Simultaneously, the catalyst demonstrated good photocatalytic activity in methyl orange (MO) breakdown. Kannan et al. used MnO_2 to support POMs to degrade methylene blue(MB).[65]

Furthermore, POMs have limitations, and they can only perform photoactivity when exposed to UV light, which shows low quantum yield for photochemical reactions. Since the solar spectrum consists only 5% of UV light,[66] POMs were used as catalysts in heterogeneous systems to eliminate different types of pollutants through photo-oxidation. POMs are also useful for homogenous photocatalysis to remove color, phenolic compounds, chloroacetic acids, insecticides, and wood pulp bleaching. Many POMs-based hybrids have exhibited good photocatalytic activity to remove different organic dyes such as MB, MO, RhB, etc., under the solar and UV light illumination.[67]

Yang et al.[68] synthesized octamolybdate metal-organic complex by flexible thioether ligand, copper ion, and Anderson-type POMs. Molybdenum shows six copper metals that exhibit two different oxidation states (I and II). This complex shows a 3D supramolecular structure with hydrogen bond interaction. This shows an electrocatalytic reduction of H_2O_2, confirmed by the CV spectrum and exhibited photocatalytic degradation of MB in 180 min around 80%.[68] On a similar line,

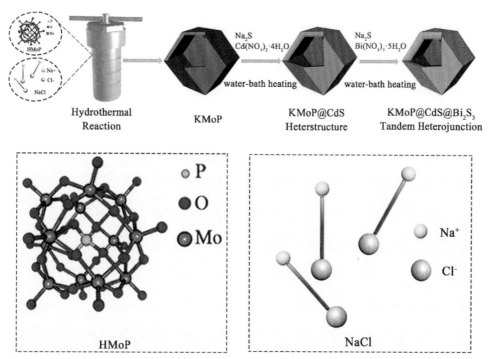

FIG. 3 Formation of KMoP@CdS@Bi₂S₃ heterojunction. *(Reproduced with permission from Cui Y, Xing Z, Guo M, et al. Hollow core-shell potassium phosphomolybdate@cadmium sulfide@bismuth sulfide Z-scheme tandem heterojunctions toward optimized photothermal-photocatalytic performance.* J Colloid Interface Sci. *2022;607:942–953. https://doi.org/10.1016/j.jcis.2021.09.075, Elsevier under copyright 2022.)*

cobalt-based POMs were investigated. This catalyst shows good photocatalytic activity toward MB and RhB dyes. Both dyes show complete degradation within 60 min. This catalyst also was tested against cancer cells and exhibited effective results.[69]

Four different inorganic-organic Wells-Dawson POMs were synthesized using rigid N-heterocyclic ligand (bibb).[70] Each structure's configuration would be determined by the type of polyacid and the coordination mode of ligands. All structures showed 3D supramolecular structure through hydrogen bonding and π-π interaction between layers. These materials are used for photocatalytic degradation of RhB dye.

Another efficient photocatalyst, KMoP@CdS@Bi₂S₃, was reported in the literature.[71] This hollow core-shell structure provides a high surface area, and two sulfides act as a photosensitizer and promote the separation of electrons and holes. The synthesis method of this material is given in Fig. 3. This material expressed efficient photocatalytic reduction of Cr (VI), photocatalytic degradation of toxic tetracycline, and H₂ production.

Photocatalytic degradation of diethyl phthalate (DEP), diallyl phthalate (DAP), and di (2-Ethylhexyl) phthalate (DEHP) was reported by $H_5PMo_{10}V_2O_{40}$@surfactant(n)/graphene hybrid.[34] Resultant, phthalate ester degraded into formic acid, CO_2, and H_2O and mechanistic pathway reported by GC-MS. Degradation efficiency retains 88.1% after 10 cycles. More than 70% of TOC and COD were measured.

α-Keggin POMs ($H_5PV_2Mo_{10}O_{40}$) were combined with porphyrin by ion-exchange method and utilized with photocatalytic degradation of 2-chloroethyl ethyl sulfide.[35] Degradation efficiency was 99%. The sol-gel method was used to synthesis of Na [α-SiW₉O₃₄]/Cu/TiO₂. This compound has 76.53% removal of nitrate with 82.09% N₂ selectivity.[72] Summary of synthesis and applications is presented in Table 1.

5.2 Catalysis

$Na_{11}[CuFeW_{18}O_{62}]\cdot 23H_2O$ is Wells-Dawson-type POMs of the 3D series. $Na_{12}[CuNiW_{18}O_{62}]\cdot 31H_2O$ was synthesized by Deya and Sharma, which was employed in industries as an environmentally friendly and green catalyst in organic processes.[106] Copper substituted and unsubstituted POMs were utilized as catalysts for the oxidation of alkenes by Sadasivan and Patel.[107] Activation energy greatly declines with the help of a catalyst.

Different experimental setups were performed to study the effects of changes observed due to the inclusion of copper atoms and the insertion of organic ligands with POMs. $PW_{10}Cu_2$ was used as a catalyst in similar reactions, and the various

TABLE 1 Summary of synthesis and applications of various POMs.

Polyoxometalates	Synthesis	Applications	Degradation % in time; rate constant
Octamolybdate MOF by Anderson-type POMs[68]	Synthesized through bis (tetrazole)-functionalized thioether ligand	Photocatalysis of MB	80% in 10 min
Cobalt-based POMs $Na_{10}[Co_4(H_2O)_2(VW_9O_{34})_2]\cdot xH_2O$[69]	Self-assembly method	Photocatalysis of MB, RhB, phenol red and MO and anticancer activity	Around 93%
Wells-Dawson POMs + Organic ligand bibb[70]	Hydrothermal method	Photocatalytic degradation of RhB	67.14%
KMoP@CdS@Bi$_2$S$_3$[71]	Hydrothermal method	Photocatalytic degradation of tetracycline, reduction of Cr (VI) and H$_2$ production	95.5% 97.5% 831 ($\mu mol\,h^{-1}$)
$H_5PMo_{10}V_2O_{40}$@surfactant(n)/graphene[34]	Hydrothermal method	Photocatalytic degradation of phthalate acid ester	Max 93%
α-Keggin ($H_5PV_2Mo_{10}O_{40}$) with porphyrin[35]	Ion-exchange method	Photocatalytic degradation of 2-chloroethyl ethyl sulfide	99%
Na [α-SiW$_9$O$_{34}$]/Cu/TiO$_2$[72]	Sol-gel method	Photocatalytic reduction of nitrate	76.53%
$(NH_4)_4[PMo_{11}VO_{40}]$/g-C$_3N_4$[73,74]	Dipping method with wet impregnation	Photocatalysis of MB, X-3B	94.7% in 120 min (0.79 min^{-1})
Carbon-doped phosphotungstate $H_3PW_{12}O_{40}$[75]	Coprecipitation and calcinated at 390°C	Photocatalysis of imidacloprid	76.29% (0.34 h^{-1})
$[H_5O_2]_2[Hpip]_{2.5}[BW_{12}O_{40}]\cdot4H_2O$ or protonated piperazine POMs[76]	Coprecipitation along refluxing	Photocatalysis of M band reduction of nitrite	97.08% after 24 min (4.05 min^{-1})
Pt/H$_3$PMo$_{12}$O$_{40}$/TiO$_2$ nanofibers[77] PMo$_{12}$=H$_3$PMo$_{12}$O$_{40}$	Electrospinning, calcination at 450°C and Pt associated photochemical reduction or deposition	Photocatalysis of MO, tetracycline, bisphenol A	Rate constant 0.011 min^{-1}; 0.043 min^{-1}; 0.00615 min^{-1}
H$_3$PMo$_{12}$O$_{40}$/polymer SiMo$_{12}$O$_{40}$(IPh$_2$)$_4$/polymer[78] Polymer: =poly(trimethylolpropane triacrylate)	Photopolymerization	Photocatalysis of eosin-Y	93% of eosin-Y with H$_3$PMo$_{12}$O$_{40}$/polymer in 87 min (1.069 min^{-1}); and SiMo$_{12}$O$_{40}$(IPh$_2$)$_4$/in 200 min (0.465 min^{-1})
TiO$_2$/POM/Fe$_3$O$_4$@SiO$_2$ microspheres[79]	The thin film was made up by electrostatic layer-by-layer over Fe$_3$O$_4$@SiO$_2$ microspheres	Photocatalysis of MO	83.91% of MO in 100 min (0.84) min^{-1}
Ag$_3$PO$_4$/PMo$_{12}$/GO[80]	Chemical synthesis through precipitation	Photocatalysis of RhB	Ag$_3$PO$_4$/POM/GO rate of 0.18 min^{-1}. Without GO 0.18 min^{-1}; without POMs and GO 0.1 min^{-1}
TiO$_2$/H$_3$PW$_{12}$O$_{40}$[81] ZnO/H$_3$PW$_{12}$O$_{40}$	Sol-gel and Teflon-lined hydrothermal treatment	Photocatalytic degradation of aniline	Rate 0.28×10^{-2} min^{-1}; 0.26×10^{-2} min^{-1}

TABLE 1 Summary of synthesis and applications of various POMs—cont'd

Polyoxometalates	Synthesis	Applications	Degradation % in time; rate constant
$H_3PW_{12}O_{40}/SiO_2$[82] Sensitizer H_2O_2	Sol-gel and calcination at 200°C for 4 h	Photodegradation of MO, methyl red, methyl violet, RhB, MG, MB	% Degradation from 84.6 to 98.3 in 120 min (0.705–0.819 min^{-1})
Ag-TiO$_2$/H$_3$PW$_{12}$O$_{40}$ composite[83]	Sol-gel along with spin coating	Photodegradation of o-chlorophenol (OCP)	82.40% in 240 min^{-1} (0.34 min^{-1}) 80% photochemical stability of composite film
La-H$_3$PW$_{12}$O$_{40}$/TiO$_2$ Ce-H$_3$PMo$_{12}$O$_{40}$/TiO$_2$[84]	Sol-gel and calcination at 400–900°C for 3 h	Photocatalysis of MB	96% MB was degraded by Ce-based POMs (0.96 min^{-1}) and 98% and La-based POMs (0.98 min^{-1}) in 100 min
POMs-phosphonate[85]	Hydrothermal method	Photocatalytic degradation of RhB and anti-tumor activity	86.8%
Co$_3$O$_4$ QDs+glycine [PMo$_{12}$O$_{40}$]$^{3-}$[86]	Chemical method	Photocatalytic oxidation	>99% selectivity
Keggin-based [HPMo$_{12}$O$_{40}$] POMs immobilized on choline hydroxide-based mesoporous silica[87]	Post synthesis method	Catalytic desulfurization of oil	99.7% in 2 h
[Ru^{III}(H$_2$O)SiW$_{11}$O$_{39}$]$^{6-}$[88]	Chemical method	Reduction of CO$_2$	–
Au nanoparticles, Ti-substituted Keggin-type polyoxometalate [PTi$_2$W$_{10}$O$_{40}$]$^{7-}$[89]	One-pot method	Reduction of CO$_2$	CO (2.1 µmol g^{-1} h^{-1}), and CH$_4$ (0.35 µmol g^{-1} h^{-1})
Eosin-Y sensitized aluminum-replaced lacunary Keggin ion[90] α-[AlSiW$_{11}$(H$_2$O)O$_{39}$]$^{5-}$	–	H$_2$ production	Quantum efficiency of more than 10%
Ni-Mo$_2$C/N, P codoped carbon derived by POMs[91]	Supramolecular-confinement pyrolysis strategy.	H$_2$ evolution	Current densities of 500 and 1000 mA cm^{-2}
Ir(III) photosensitized polyoxotunstate[92]	–	H$_2$ evolution	–
POMs[93] [Mn$_4$(H$_2$O)$_2$(VW$_9$O$_{34}$)$_2$]$^{10-}$ sensitizer [Ru(bpy)$_3$]$^{2+}$	Precipitation method	H$_2$ evolution	Rate 28.2 × 10^8 min^{-1}
Polyoxotantalate and metal sulfide[94]		H$_2$ evolution	43.05 mmol h^{-1} g^{-1}
Na$_{12}$[(α-SbW$_9$O$_{33}$)$_2$ Cu$_3$ (H$_2$O)$_3$]·46H$_2$O/ mesoporous TiO$_2$[95]	–	H$_2$ evolution	1284.8 Lmol g^{-1}
K$_8$Na$_8$H$_4$[P$_8$W$_{60}$Ta$_{12}$(H$_2$O)$_4$(OH)$_8$O$_{236}$]· 42H$_2$O, and Cs$_{10}$·5K$_4$H$_5$·5 [Ta$_4$O$_6$(SiW$_9$Ta$_3$O$_{40}$)$_4$]·30H$_2$O-substituted heteropolytungstates[96] Cocatalyst H$_2$PtCl$_6$	–	H$_2$ production	1250 (µmol h^{-1} g^{-1}) 803 (µmol h^{-1} g^{-3})
Cs$_{16}$K$_{16}$Na$_4$[Ta$_{18}$P$_{12}$W$_{90}$(OH)$_6$(H$_2$O)$_2$O$_{360}$]· 24H$_2$O **{Ta$_{18}$P$_{12}$W$_{90}$}**[97] Cs$_{26}$K$_2$H$_2$[Yb$_2$Ta$_{18}$P$_{12}$W$_{90}$(OH)$_6$(H$_2$O)$_{16}$O$_{360}$]· 52H$_2$O **{Yb$_2$Ta$_{18}$P$_{12}$W$_{90}$}**	Hydrothermal synthesis	H$_2$ production	8301.32 (µmol h^{-1} g^{-1}) 6494.4 (µmol h^{-1} g^{-1})
Cs$_5$K$_2$[(Si$_2$W$_{18}$Ta$_6$O$_{78}$)Cr(H$_2$O)$_4$]·8H$_2$O **{CrTa$_6$Si$_2$W$_{18}$}**[98] Cs$_3$K$_4$H$_2$[(Si$_2$W$_{18}$Ta$_6$O$_{78}$)FeCl$_2$(H$_2$O)$_2$]· 15H$_2$O **{FeTa$_6$Si$_2$W$_{18}$}**	–	H$_2$ production	1001.7 (µmol h^{-1} g^{-1}) 743.8 (µmol h^{-1} g^{-1})

Continued

TABLE 1 Summary of synthesis and applications of various POMs—cont'd

Polyoxometalates	Synthesis	Applications	Degradation % in time; rate constant
{$W_{17}Ni$}CNT[99]	Self-assembly and polymerization	H_2 production	94.14 ($\mu mol\ cm^{-2}$)
CdSe QDs+Na_6K_4-Ni_4P_2[100]		H_2 production	138 ($mmol g^{-1}\ h^{-1}$)
Graphene oxide supported Keggin-type [$CoW_{12}O_{40}$]$^{6-}$ polyanion+imidazolium[101]	Hummer method	Water oxidation reaction	OER potential 1.3v
Hybrid of silica aerogel with a composite of Preyssler-type POM and Cr-based MOF[102]	Hydrothermal method	Adsorbent for extraction of natural component	73%
Keggin-based POMs $H_3K_2[Ag_5(DTB)_5][SiW_{12}O_{40}]_2 \cdot Cl_2 \cdot 8H_2O$[103]	Hydrothermal method	Separation of cationic and anionic dyes	–
Ferrocene-rGO-PMo_{12}[104]	–	Detection of acetaminophen	Detection limit 13.27 nM (S/N = 3), sensitivity 36.81 $\mu A\ \mu M$
[$Eu(GeW_{11}O_{39})_2(H_2O)_2$]$^{10-}$ [105]	–	Optical and voltammetric sensor	Detection of IO_3^-, BrO_3^- and NO_3

reaction parameters were improved.[108,109] The di-copper inclusion complex of phosphotungstate was utilized as a heterogeneous catalyst for the selective epoxidation of *cis*-cyclooctene. Here, tert-butyl hydroperoxide (TBHP) acts as an oxidant. Modification of the above complex was performed by introducing cesium salt to increase the efficiency further. Then, catalytic activity was investigated against oxidation of styrene through TBHP. Several operational parameters were tuned, including time, catalyst quantity, substrate mole ratios to TBHP, and temperature.

Similarly, $PW_{11}Cu$ was also investigated against oxidation of styrene and *cis*-cyclooctene (aliphatic and cyclic alkene) and conversion into their respective epoxides[110,111] with the help of oxidant TBHP. To preserve percent selection toward epoxide, the mono-copper-substituted phosphotungstate complex is found to be most sustainable and recyclable.

Co_3O_4 QDs with glycine [$PMo_{12}O_{40}$]$^{3-}$-based nanocomposite are used for catalytic, photocatalytic, and electrocatalytic oxidation of benzyl alcohol and comparing their results indicate that electrocatalytic oxidation product has the highest yield and the maximum yield was 0.55 mmol.[86] The above composite Co_3O_4 QDs help in light absorption and [$PMo_{12}O_{40}$]^3 is responsible for effective charge separation. Keggin-based [$HPMo_{12}O_{40}$] POMs immobilized mesoporous silica using choline hydroxide-based ionic liquid.[87] This material is used in catalytic desulfurization of *n*-octane as model oil contains DBT and H_2O_2.

5.3 Reduction of CO_2

CO_2 reduction is a complicated process whose entire reaction mechanism has yet to be discovered. One of the mechanisms proposed is a CO_2 electron reduction process. In this mechanism, after the adsorption of CO_2 on a catalyst surface, its linear form changes, and the LUMO energy level decreases. This allows electrons from adsorbed molecules to easily migrate to CO_2^- intermediate radical anion.[112,113] This necessitates a high overpotential of 1.9 V compared to the normal hydrogen electrode (NHE).[114] A sequence of multi-electron reactions promoted with protons is also a well-known mechanism. Formaldehyde, carbene, glyoxal, and enol are typical reaction intermediates, and these names are used for other mechanistic routes.[115] Due to their structures, solubility, and stability, POMs are thought to be good homogenous molecular catalysts for electrocatalytic CO_2 reduction.

POMs can transmit photo-induced carriers that contribute to incremental CO_2 reduction and photocatalytic efficiency. Qiao et al. initially created a high-nuclear cluster, $K_4Na_{28}[Co_4(OeH)_3(VO_4)_4(SiW_9O_{34})_4] \cdot 66H_2O$, by extracting pure POMs inorganic units.[116] The catalyst has a higher selectivity of 99.6% for conversion of CO_2 to CO under visible light irradiation through photocatalysis with the help of photosensitizer $Ru[(phen)_3]_2$. According to the proposed mechanism, electrons are transported from excited photosensitizer to POMs, which reduce CO_2 under exposure to visible light illumination.

In a reaction, polyoxoanions induce redox reactions by transferring one or more electrons/protons, and their redox activity can be controlled using appropriate transition metal (TM) cations or tiny organic compounds. The first TM-substituted polyoxoanion $[SiW_{11}O_{39}Co]^{6-}$ of CO_2 reduction was reported into non-polar solvents. There was a lot of research published in the 1990s about the activation of CO_2 molecules by TM-substituted polyoxoanions.[117] The reaction between CO_2 and TM-substituted POMs in non-polar solvents was described by Szczepankiewicz et al.[117] POMs are a novel type of electrolytes with electronic control properties. In non-polar liquids, the performance of TM-substituted POMs is comparable to that of unsaturated modified metal atoms. POMs electrocatalytic CO_2 reduction activity is inhibited when water molecules orient toward the active sites of TM-substituted POMs under an aqueous electrolyte. POMs in the saturated form are analogous to Mo or W oxides, which show no CO_2 reduction activity. CO_2 reduction activity of electrochemical cells containing SiW_9V_3 and $SiW_{11}Mn$ POMs as electrolytes were examined in an aqueous solution.[118–121] These cells stimulate the electrochemical reduction of CO_2 to CO and formic acid. Substituted and hybrid POMs are now employed in various applications instead of pure POMs.

The reaction between POMs with CO_2 in acetonitrile was explored by Girardi et al.[122] The electrolysis of a CO_2 saturated $(TOA)_6[SiW_{11}O_{39}Co (-)]$ di-chloromethane solution using a calomel electrode at 1.5 V (TOA = tetraoctylammonium) was conducted. Hg_2Cl_2/Hg (1 M LiCl) on a functional electrode express that CO_2 may be transformed into CO with faradic efficiency of 18% only and no hydrogen formation as a byproduct.

Metal-organic compounds, such as TM-substituted POMs, showed a CO_2 reduction pathway. Girardi et al. observed CO_2 reduction activity with $[Cp*Rh^{III}(bpy)Cl]^+$ ($Cp*$ = pentamethylcyclopentadienyl anion, bpy = 2,2'-bipyridine) and used $H_2PW_{11}O_{39}$ polyoxoanions to produce a $(TBA)_3[H_2PW_{11}O_{39}Rh^{III}Cp*(OH_2)]$ catalyst for CO_2 reduction in an acetonitrile solution.[120] $[H_2PW_{11}O_{39}\{Rh^{III}Cp*(OH_2)\}]^{3-}$ had lower CO_2 reduction activity than the parent $[Cp*Rh^{III}(bpy)Cl]^+$ catalyst, with a FE of just 4.5% for the reduced product, HCOOH. The FE for the byproduct H_2, on the other hand, was around 68%. The FE for CO_2 reduction to HCOOH for $[Cp*Rh^{III}(bpy)Cl]^+$ was 60% under similar conditions, whereas by product H_2 was only 30%. According to this finding, the substitution of the bpy group with polyoxoanions has a negative impact on CO_2 reduction.

On similar lines, Wang et al.[121] synthesized a range of stable polyoxometalates-metalloporphyrin organic framework (PMOF) catalysts containing Zn-ε-Keggin POMs$\{ε$-$PMo_8VMo_4VIO_{40}Zn_4\}$ and M-tetrakis[4-carboxyphenyl] porphyrin (M-TCPP, M = Co, Fe, Ni) synthesized by a hydrothermal route and further used for CO_2 reduction in aqueous electrolyte solutions.

Manganese carbonyl compounds, in addition to metalloporphyrin compounds, are effective CO_2 reduction catalysts. A carbonyl manganese bipyridine complex $[Mn^I$ (bipyridyl)$(CO)_3Br]$ (MnL) was recently bridged with POMs, resulting in three distinct POMs-MnL (SiW_{12}-MnL, PW_{12}-MnL, and $SiW_{12}MnL$) composites are formed that operate as electrocatalysts.[123] In electrocatalytic CO_2 reduction, POM-based organic-inorganic hybrid materials were also employed. The electrostatic, covalent, or coordination interaction between POMs and organic molecular catalysts can remarkably decline the solubility of POMs and these catalysts in the electrolyte, resulting in the formation of heterogeneous POMs organic molecular catalysts.

Due to the high capacity of electron/proton carriers, POMs operate as electrons/protons buffer solutions, enhancing their solubility and modulating redox activity in various solvent systems. Zhang's coworker employed $PMo_{12}O_{40}^{3-}$ as a stabilizer and electrochemically deposited as Ag $[PMo_{12}O_{40}]^{n-}$ (Ag-PMo) nanocomposite.[124] PMo_{12} polyoxoanions can be added to the nano-Ag surface to lower adsorption energy and help to stabilize the nano-Ag structure. Ag-porous PMo's 3D structure speeds up the catalytic active center and improves catalytic efficiency during CO_2 reduction.

Lai et al.[125] used Keggin-based POMs materials to perform photocatalytic organic waste removal. After being substituted with Ru (III) (photosensitizer), POMs of the Keggin structure catalyzed the photoreduction of CO_2 to CO using tertiary amines as reducing agents, preferring Et_3N.[88] Li et al.[126] created a stable POMs-based metal-organic framework that paired POMs with hydrophobic ligands and demonstrated effective CO_2 photocatalysis.

5.4 H$_2$ production or water splitting

Three processes are usually involved in photocatalytic water splitting with semiconductor photocatalysts: (1) Absorption of sufficient amounts of sunlight by a semiconductor, resulting in the creation of photogenerated charge carriers (2) Then these charge carriers move toward the catalyst surface (3) To evolve hydrogen and oxygen gases, these charge carriers go through redox reactions with $H^+/OH^-/H_2O$ species. POMs are a special type of semiconductor with unique properties.[30,127] As a

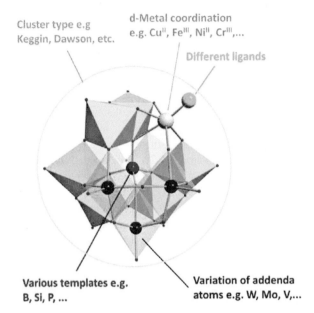

FIG. 4 Catalytic reaction of {Ta₆/CZS} composite. *(Reproduced with permission from Zhou XJ, Yu H, Zhao D, Wang XC, Zheng ST. Combination of polyoxotantalate and metal sulfide: a new-type noble-metal-free binary photocatalyst Na8Ta6O19/Cd0.7Zn0.3S for highly efficient visible-light-driven H2 evolution. Appl Catal Environ. 2019;248:423–429. https://doi.org/10.1016/j.apcatb.2019.02.052, Copyright 2019 Elsevier.)*

result, photocatalytic hydrogen production systems used POMs for absorption of light and transfer it to catalysts. Charge transfer from oxygen to metal occurred after light absorption. This charge transfer creates a hole in the oxygen molecule, acting as a sacrificial agent and lowering the energy level of POMs.

Ioannidis et al.[128] first to discover 1:12 heteropolytungstates of Keggin structures $[XW_{12}O_{40}]^{n-}$ (X = P, Si, Fe, H) for production of hydrogen, in which isopropanol act as sacrificial electron donor, Pt is cocatalyst, under UV light. The H_2 evolution rates of $SiW_{11}Cu$, $SiW_{11}Ni$, $SiW_{11}Co$, and $SiW_{11}Zn$ were 150, 98, 65, and 48 mol g^{-1} h^{-1}, respectively.[129]

Then, Zhou et al.[94] synthesized binary composite $Na_8Ta_6O_{19}/Cd_{0.7}Zn_{0.3}S$ as catalysts with the combination of polyoxotantalates and metal sulfides. These catalyze effective H_2 generation under visible light illumination, and catalytic hydrogen evolution in the absence of a cocatalyst was also observed. The maximum H_2 evolution rate was 43,050 mol - h^{-1} g^{-1} measured (Fig. 4).

Tungsten and nickel-based POMs immobilized at the surface of CNT could act as electrocatalysts due to higher stability and current densities, which are further utilized in H_2 evolution.[99] Four different nickel-substituted POMs coupled with water compatible CdSe QDs and ascorbic acid were reported in the literature recently. These exhibited an efficient H_2 evolution rate due to good light absorber CdSe QDs, charged separation by modified POMs, and ascorbic acid remove holes by capturing these.[100]

Similarly, another investigation of the photocatalytic evolution of hydrogen with TM-substituted titano-tungstates of the Keggin-type $Na_5[MTiW_{11}O_{39}]\cdot xH_2O\{MTiW_{11}\}$ (M = Fe, Co, Zn) as photocatalysts and with PVA as sacrificial electron donor.[130] Li et al. discovered the use of the $K_8Na_8H_4[P_8W_{60}Ta_{12}(H_2O)_4(OH)_8O_{236}]\cdot 42H_2O$, and $Cs_{10.5}K_4H_{5.5}[Ta_4O_6(SiW_9Ta_3O_{40})_4]\cdot 30H_2O$ substituted hetero-polytungstate and the cocatalyst H_2PtCl_6 to catalyze the efficient hydrogen evolution under mercury lamp illumination.[96] Following that, Ta/W mixed-addendum POMs such as $Cs_{16}K_{16}Na_4[Ta_{18}P_{12}W_{90}(OH)_6(H_2O)_2O_{360}]\cdot 24H_2O$, $Cs_{26}K_2H_2[Yb_2Ta_{18}P_{12}W_{90}(OH)_6 (H_2O)_{16}O_{60}]\cdot 52H_2O$, $Cs_5K_2[(Si_2W_{18}Ta_6O_{78})Cr(H_2O)_4]\cdot 8H_2O$ and $Cs_3K_4H_2[(Si_2W_{18}Ta_6O_{78}) FeCl_2(H_2O)_2]\cdot 15H_2O^{97,98}$ were synthesized. The above complexes' average hydrogen production rates were 8301.32, 6494.4, 1001.7, and 743.8 mol g^{-1} h^{-1}, respectively. Polytantalotungstates and poly-oxoniobate were made in the same way.[131,132] The evolution rate of H_2 was measured with Co(III) $(dmgH)_2pyCl$ (dmgH = dimethylglyoxime, py = pyridine). Free of noble metals as cocatalyst and triethylamine (TEA) as a sacrificial electron donor present in three new poly-oxoniobates complexes, namely $KNa_2[Nb_{24}O_{72}H_{21}]\cdot 38H_2O$, $K_2Na_2[Nb_{32}O_{96}H_{28}]\cdot 80H_2O$ and $K_{12} [Nb_{24}O_{72}H_{21}]\cdot 107H_2O$, and the H_2 development rates 5198.87, 5056.43, and 4714.4 μmol h^{-1} g^{-1}, respectively, were measured[133] The synthesis of Keggin-type POMs catalyst $K_7[Co(III)Co (II)(H_2O)W_{11}O_{39}]$ was investigated, and the photocatalytic activity of this catalyst was analyzed under visible light illumination using sacrificial donor TEOA, photosensitizer EY, and cocatalyst Pt.[134] This shows hydrogen evolution rate of 13,395 μmol h^{-1} g^{-1}. Similarly, polyoxotungstate containing tetra-manganese and tetra-nickel were synthesized to further

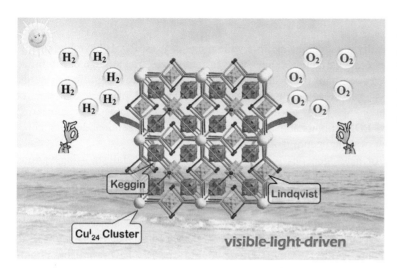

FIG. 5 Dual-functionalized mixed Keggin and Lindqvist-type POM@MOFs as photocatalysts. *(Reproduced with permission from Shi D, Zheng R, Liu C Sen, Chen DM, Zhao J, Du M. Dual-functionalized mixed Keggin- and Lindqvist-type Cu24-based POM@MOF for visible-light-driven H2 and O2 evolution. Inorg Chem. 2019;58:7229–7235. https://doi.org/10.1021/acs.inorgchem.9b00206, Copyright 2019 American Chemical Society.)*

increase efficiency. In the first case, $[Ru(bpy)_3]^{2+}$ and later $[Ir (ppy)_2(dtbbpy)]^+$(4,4′ditertbutyl2,2′dipyridyl)-bis(2phenyl pyridine(1H))-iridium (III)) act as a photosensitizer.[53,93]

POM@MOF composites are made up of specific multifunctional materials that are more efficient in converting solar energy. The physicochemical features of POMs with MOFs' structural variability and modularity make up such a composite.[135] POMs encapsulated in the pores of MOFs will aid in the recyclability and dispersibility of POMs catalysts and the charge separation and migration of photogenerated electrons and holes after excitation with the help of light.[136]

Cao et al.[137] investigated Pt-derivatives POMs, $[(CH_3)_4N]_3[PW_{11}O_{39}\{cis\text{-}Pt(NH_3)_2\}_2]\cdot10H_2O\{\textbf{PW}_{\textbf{11}}\textbf{Pt}_{\textbf{2}}\}$. This catalyst is combined with TiO_2 by modifying the surface of commercially available TiO_2. This further synthesizes the unique heterogeneous catalyst $TiO\text{-}SiNH_2\text{-}PW_{11}Pt_2$. The development of visible light heterogeneous photosystems for water splitting is a new area of research. The first example of a noble metal-free photocatalyst for H_2 evolution is a high-nuclear $\{CuI_{24}(\mu_3\text{-}Cl)_8(\mu_4\text{-}Cl)_6\}$-based POM@MOF that acts as dual-functionalized photocatalyst (Fig. 5). This exhibits highly efficient photocatalytic H_2 evolution $6614\,\mu mol\,g^{-1}\,h^{-1}$ and O_2 evolution $1032\,\mu mol\,g^{-1}$. This material also efficiently conducted electron transport reaction and utilized in solar to fuel conversion.[138]

5.5 Other applications

POMs are inorganic cluster compounds with anti-diabetic, antibacterial, anti-protozoal, antiviral, and anticancer properties. Compared to isopoly substances, hetero-POMs have been examined and explored more extensively in medicine. This is because of their more easily manipulated structural and electrical characteristics, and they are often used in enzyme inhibition.[139]

Some researchers are chosen POMs as an alternative anti-tumor agent, with promising findings in tumor growth suppression.[140–144] POMs are promising new anti-tumor drug, especially for treating difficult-to-treat malignancies. Antimicrobial resistance and viral infections are also worldwide health issues that need to be addressed. This quest for novel medications, such as POMs, is gaining attraction.[145] POM-523 and PM-504 are two POMs compounds that have been synthesized and shown to have antiviral properties. The anti-RNA virus activity of POMs was examined, focusing on anti-respiratory virus activity. Anti-RNA viral agents by using POMs are non-toxic. MXene was combined with Gd- and W-based POMs and utilized in cancer and therapeutic applications.[146]

An efficient water oxidation reaction was observed by graphene oxide supported imidazolium cation with Keggin type $[CoW_{12}O_{40}]^{6-}$ POMs.[101] This catalyst showed a low overpotential and low Tafel plot compared to other existing materials. Due to the strong interaction of the ionic liquid with graphene oxide and POMs expressed extraordinary stability and retained 14 h and 900 catalytic cycles.

Hybrid silica gel with Preyssler-type POMs and Cr-based MOF were synthesized by hydrothermal method.[102] This material acts as a good adsorbent and is used for solid-phase microextraction of volatile organic components of the herb *Ferulago angulata*.

6   Conclusion and future prospects

POMs have been proven to be a green and eco-friendly photocatalyst with a wide range of uses. POMs have high availability, a tendency to perform redox reactions to generate variable valency states, effective electron storage capabilities, which contribute positively to photocatalysis. POMs continue to be hampered by issues such as a significant energy bandgap (prompt UV light absorber), poor specific surface area, a high affinity of dissociation with pH change, low recyclability, and very limited applications under visible or solar light. Various ways are suggested to address these concerns. The photocatalytic efficiency of pure Keggin-based POMs photocatalysts is improved by doping semiconductors, complexing with non-metal elements, and incorporating organic or inorganic backbone.

In fact, scientists have created a large variety of heterogeneous Keggin-based POMs photocatalysts using various matrices, generating self-assembled organic-POMs hybrids and POMs supported over zeolite, SiO_2, porous carbon, and g-C_3N_4. However, due to a lack of desired properties, none of these Keggin-based POMs photocatalysts fulfilled the real usability of industrial applications. A perfect POMs photocatalyst should have a large surface area, efficient electron storage, and feasible transfer of electrons that promote photocatalytic activity and be structurally stable and easily separable for subsequent reactive cycles. Considering the above considerations, hybridization of POMs-based materials over MXene might be a viable way to resolve these troubles. Encapsulating POMs in MXene could result in a new promising photocatalyst for pollutant photodegradation, CO_2 reduction, and H_2 evolution without producing environmental setbacks, as both materials are environmentally friendly. Aside from different modifications, the potential of POMs as a photocatalyst remains untapped. The typical Mo- and W-based POMs photocatalyst is currently the focus of most published reports. The presence of dopants, supported materials, or cocatalyst are frequently used to avoid pH-induced dissociation, but this is hardly ever supported by the fundamental discussion. The stability of the modified POMs-based photocatalyst is usually confirmed by the characterizations and recyclability experiments conducted after catalytic activity, which show the same results as the fresh ones. As a result, in-depth research on surface responses and their importance in preserving unmodified structures and performance is still needed.

7   Challenges

A remarkable challenge with POMs to apply in photocatalysis is its reoxidation. As Khenkin et al.[88] have reported, reoxidation of photoreduced $[PV_2Mo_{10}O_{40}]^{7-}$ in the presence of oxygen converts into $[PV_2Mo_{10}O_{40}]^{5-}$ in acidic media. Even reoxidation of POMs by oxygen is usually kinetically very slow or thermodynamically unfeasible for unsubstituted POMs.

Synthesis in the form of a high crystalline state, single crystal, and proper characterization of material would also be an important milestone to achieve. Modifying POMs with transition metal ions, the organic molecular framework would increase the system's efficiency to a good extent.

Acknowledgment

DV acknowledge thanks to Amity University, Haryana, for encouragement and continuous support.

References

1. Karri RR, Ravindran G, Dehghani MH. Wastewater—sources, toxicity, and their consequences to human health. In: *Soft Computing Techniques in Solid Waste and Wastewater Management*. Elsevier; 2021:3–33.

2. Dehghani MH, Omrani GA, Karri RR. Solid waste—sources, toxicity, and their consequences to human health. In: *Soft Computing Techniques in Solid Waste and Wastewater Management*. Elsevier; 2021:205–213.

3. Meena S, Vaya D, Das BK. Photocatalytic degradation of Malachite Green dye by modified ZnO nanomaterial. *Bull Mater Sci*. 2016;39:1735–1743. https://doi.org/10.1007/s12034-016-1318-4.

4. Alsheheri SZ. Nanocomposites containing titanium dioxide for environmental remediation. *Des Monomers Polym*. 2021;24:22–45. https://doi.org/10.1080/15685551.2021.1876322.

5. Singh M, Vaya D, Kumar R, Das B. Role of EDTA capped cobalt oxide nanomaterial in photocatalytic degradation of dyes. *J Serb Chem Soc*. 2021;86:327–340. https://doi.org/10.2298/JSC200711074S.

6. Wang Y, Pan C, Chu W, Vipin A, Sun L. Environmental remediation applications of carbon nanotubes and graphene oxide: adsorption and catalysis. *Nanomaterials*. 2019;9:439. https://doi.org/10.3390/nano9030439.

7. Verma N, Manju CTS, Vaya D. Role of N-ZnO/GO and Fe2O3-ZnO in photocatalytic activity. In: *AIP*; 2021:020083. https://doi.org/10.1063/5.0061215.

8. Vaya D, Kaushik B, Surolia PK. Recent advances in graphitic carbon nitride semiconductor: structure, synthesis and applications. *Mater Sci Semicond Process*. 2022;137:106181. https://doi.org/10.1016/j.mssp.2021.106181.

9. Khan FSA, Mubarak NM, Tan YH, et al. A comprehensive review on magnetic carbon nanotubes and carbon nanotube-based buckypaper for removal of heavy metals and dyes. *J Hazard Mater*. 2021;413:125375.

10. Huang B, Yang DH, Han BH. Application of polyoxometalate derivatives in rechargeable batteries. *J Mater Chem A*. 2020;8:4593–4628. https://doi.org/10.1039/c9ta12679a.

11. Ito T, Yashiro H, Yamase T. Regular two-dimensional molecular array of photoluminescent Anderson-type polyoxometalate constructed by Langmuir–Blodgett technique. *Langmuir*. 2006;22:2806–2810. https://doi.org/10.1021/la052972w.

12. Suzuki K, Minato T, Tominaga N, et al. Hexavacant γ-Dawson-type phosphotungstates supporting an edge-sharing bis(square-pyramidal) {O2M(μ3-O)2(μ-OAc)MO2} core (M = Mn2+, Co2+, Ni2+, Cu2+, or Zn2+). *Dalton Trans*. 2019;48:7281–7289. https://doi.org/10.1039/c8dt04850a.

13. Lu K, Liebman Peláez A, Wu LC, Cao Y, Zhu CH, Fu H. Ionothermal synthesis of five Keggin-type polyoxometalate-based metal-organic frameworks. *Inorg Chem*. 2019;58:1794–1805. https://doi.org/10.1021/acs.inorgchem.8b02277.

14. Coyle L, Middleton PS, Murphy CJ, Clegg W, Harrington RW, Errington RJ. Protonolysis of [(iPrO)TiMo5O18]3–: access to a family of TiMo5 Lindqvist type polyoxometalates. *Dalton Trans*. 2012;41:971–981. https://doi.org/10.1039/b000000x.

15. Rafiee E, Eavani S. Heterogenization of heteropoly compounds: a review of the structure and synthesis. *RSC Adv*. 2016;6:46433–46466. https://doi.org/10.1039/C6RA04891A.

16. Gong P, Li Y, Zhai C, Luo J, Tian X. Syntheses, structural characterization and photophysical properties of two series of rare-earth-isonicotinic-acid containing Waugh-type manganomolybdates. *CrstEngComm*. 2016;19:834–852. https://doi.org/10.1039/C6CE02428A.

17. Gouzerh P, Che M. From Scheele and Berzelius to Müller: polyoxometalates (POMs) revisited and the "missing link" between the bottom up and top down approaches. *Actual Chim*. 2006;298:9–21.

18. Keggin JF. The structure and formula of 12-phosphotungstic acid. *Proc R Soc London Ser A*. 1934;144:75–100. https://doi.org/10.1098/rspa.1934.0035.

19. Guo H-X, Liu S-X. A novel 3D organic–inorganic hybrid based on sandwich-type cadmium hetereopolymolybdate: {[Cd4(H2O)2(2,2′-bpy)2] cd [Mo6O12(OH)3(PO4)2(HPO4)2]2}[Mo2O4(2,2′-bpy)2]2·3H2O. *Inorg Chem Commun*. 2004;7:1217–1220. https://doi.org/10.1016/j.inoche.2004.09.010.

20. Müller A, Luban M, Schröder C, et al. Classical and quantum magnetism in Giant Keplerate magnetic molecules. *ChemPhysChem*. 2001;2:517–521. https://doi.org/10.1002/1439-7641(20010917)2:8/9<517::AID-CPHC517>3.0.CO;2-1.

21. Paul A, Kentzinger E, Rücker U, Bürgler DE, Brückel T. Field-dependent magnetic domain structure in antiferromagnetically coupled multilayers by polarized neutron scattering. *Phys Rev B*. 2006;73:094441. https://doi.org/10.1103/PhysRevB.73.094441.

22. Greenwood NN, Earnshaw A. *Chemistry of the Elements*. Elsevier; 1997. https://doi.org/10.1016/C2009-0-30414-6.

23. Mattes R. In: Pope VMT, ed. *Heteropoly and Isopoly Oxometalates*. vol. 96. Berlin: Springer-Verlag; 1983:1984. https://doi.org/10.1002/ange.19840960939. XIII, 180 S., Geb. DM 124.00.

24. McAllister J, Miras HN. Building block libraries and structural considerations in the self-assembly of polyoxometalate and polyoxothiometalate systems. In: *Structure and Bonding*; 2017:1–29. https://doi.org/10.1007/430_2017_5.

25. Yaqub M, Walsh JJ, Keyes TE, et al. Electron transfer to a phosphomolybdate monolayer on glassy carbon: ambivalent effect of protonation. *Langmuir*. 2016;55:6929–6937. https://doi.org/10.1021/acs.inorgchem.6b00485.

26. Yaqub M, Walsh JJ, Keyes TE, et al. Electron transfer to covalently immobilized Keggin polyoxotungstates on gold. *Langmuir*. 2014;30:4509–4516. https://doi.org/10.1021/la4048648.

27. Rinfray C, Brasiliense V, Izzet G, et al. Electron transfer to a phosphomolybdate monolayer on glassy carbon: ambivalent effect of protonation. *Inorg Chem*. 2016;55:6929–6937. https://doi.org/10.1021/acs.inorgchem.6b00485.

28. Vinu R, Madras G. Kinetics of simultaneous photocatalytic degradation of phenolic compounds and reduction of metal ions with nano-TiO2. *Environ Sci Technol*. 2008;42:913–919. https://doi.org/10.1021/es0720457.

29. Richardt PJS, Gable RW, Bond AM, Wedd AG. Synthesis and redox characterization of the polyoxo anion, γ*-[S 2 W 18 O 62] 4 - : a unique fast oxidation pathway determines the characteristic reversible electrochemical behavior of polyoxometalate anions in acidic media. *Inorg Chem*. 2001;40:703–709. https://doi.org/10.1021/ic000793q.

30. Long D-L, Burkholder E, Cronin L. Polyoxometalate clusters, nanostructures and materials: from self assembly to designer materials and devices. *Chem Soc Rev*. 2007;36:105–121. https://doi.org/10.1039/B502666K.

31. Izarova NV, Vankova N, Heine T, et al. Polyoxometalates made of gold: the polyoxoaurate [AuIII4AsV4O20]8. *Angew Chem Int Ed*. 2010;49:1886–1889. https://doi.org/10.1002/anie.200905566.

32. Izzet G, Ishow E, Delaire J, Afonso C, Tabet J-C, Proust A. Photochemical activation of an azido manganese-monosubstituted Keggin polyoxometalate: on the road to a Mn(V)-nitrido derivative. *Inorg Chem*. 2009;48:11865–11870. https://doi.org/10.1021/ic902046t.

33. Proust A, Matt B, Villanneau R, Guillemot G, Gouzerh P, Izzet G. Functionalization and post-functionalization: a step towards polyoxometalate-based materials. *Chem Soc Rev*. 2012;41:7605. https://doi.org/10.1039/c2cs35119f.

34. Huo Y, Zhang D, Wu J, et al. Oxidation of phthalate acid esters using hydrogen peroxide and polyoxometalate/graphene hybrids. *J Hazard Mater*. 2022;422:126867. https://doi.org/10.1016/j.jhazmat.2021.126867.

35. Yang Y, Tao F, Zhang L, et al. Preparation of a porphyrin-polyoxometalate hybrid and its photocatalytic degradation performance for mustard gas simulant 2-chloroethyl ethyl sulfide. *Chin Chem Lett*. 2021. https://doi.org/10.1016/j.cclet.2021.09.093. Published online September.

36. Ostroushko AA, Danilova IG, Gette IF, et al. Study of safety of molybdenum and iron-molybdenum nanocluster polyoxometalates intended for targeted delivery of drugs. *J Biomater Nanobiotechnol*. 2011;02:557–560. https://doi.org/10.4236/jbnb.2011.225066.

37. Liu T, Diemann E, Li H, Dress AWM, Mueller A. Self-assembly in aqueous solution of wheel-shaped Mo154 oxide clusters into vesicles. *ChemInform*. 2004;35. https://doi.org/10.1002/chin.200405007.

38. Tsunashima R, Richmond C, Cronin L. Exploring the mobility of nanoscale polyoxometalates using gel electrophoresis. *Chem Sci*. 2012;3:343–348. https://doi.org/10.1039/C1SC00542A.

39. Cao Z, Yang W, Min X, Liu J, Cao X. Recent advances in synthesis and anti-tumor effect of organism-modified polyoxometalates inorganic organic hybrids. *Inorg Chem Commun*. 2021;134(August):108904. https://doi.org/10.1016/j.inoche.2021.108904.

40. Herrmann S, Ritchie C, Streb C. Polyoxometalate—conductive polymer composites for energy conversion, energy storage and nanostructured sensors. *Dalton Trans*. 2015;44:7092–7104. https://doi.org/10.1039/C4DT03763D.

41. Zhang L, Chen Z. Polyoxometalates: tailoring metal oxides in molecular dimension toward energy applications. *Int J Energy Res*. 2020;44:3316–3346. https://doi.org/10.1002/er.5124.

42. Walsh JJ, Bond AM, Forster RJ, Keyes TE. Hybrid polyoxometalate materials for photo(electro-) chemical applications. *Coord Chem Rev*. 2016;306:217–234. https://doi.org/10.1016/j.ccr.2015.06.016.

43. Lv H, Geletii YV, Zhao C, et al. Polyoxometalate water oxidation catalysts and the production of green fuel. *Chem Soc Rev*. 2012;41:7572. https://doi.org/10.1039/c2cs35292c.

44. Zhang H, Wang T, Chen W. Polyoxometalate modified all-weather solar cells for energy harvesting. *Electrochim Acta*. 2020;330:135215. https://doi.org/10.1016/j.electacta.2019.135215.

45. Kozhevnikov IV. Catalysis by heteropoly acids and multicomponent polyoxometalates in liquid-phase reactions. *Chem Rev*. 1998;98:171–198. https://doi.org/10.1021/cr960400y.

46. Gaspar AR, Gamelas JAF, Evtuguin DV, Pascoal NC. Alternatives for lignocellulosic pulp delignification using polyoxometalates and oxygen: a review. *Green Chem*. 2007;9:717. https://doi.org/10.1039/b607824a.

47. Hiskia A, Troupis A, Antonaraki S, Gkika E, Papaconstantinou PK. Polyoxometallate photocatalysis for decontaminating the aquatic environment from organic and inorganic pollutants. *Int J Environ Anal Chem*. 2006;86:233–242. https://doi.org/10.1080/03067310500247520.

48. Troupis A, Hiskia A, Papaconstantinou E. Synthesis of metal nanoparticles by using polyoxometalates as photocatalysts and stabilizers. *Angew Chem Int Ed*. 2002;41:1911. https://doi.org/10.1002/1521-3773(20020603)41:11<1911::AID-ANIE1911>3.0.CO;2-0.

49. Schulz M, Paulik C, Knör G. Studies on the selective two-electron photo-oxidation of benzene to phenol using polyoxometalates, water and simulated solar radiation. *J Mol Catal A Chem*. 2011;347:60–64. https://doi.org/10.1016/j.molcata.2011.07.011.

50. Bonchio M, Carraro M, Scorrano G, Bagno A. Photooxidation in water by new hybrid molecular photocatalysts integrating an organic sensitizer with a polyoxometalate core. *Adv Synth Catal*. 2004;346:648–654. https://doi.org/10.1002/adsc.200303189.

51. Molinari A, Varani G, Polo E, Vaccari S, Maldotti A. Photocatalytic and catalytic activity of heterogenized $W_{10}O_{32}^{4-}$ in the bromide-assisted bromination of arenes and alkenes in the presence of oxygen. *J Mol Catal A Chem*. 2007;262:156–163. https://doi.org/10.1016/j.molcata.2006.08.056.

52. Ettedgui J, Diskin-Posner Y, Weiner L, Neumann R. Photoreduction of carbon dioxide to carbon monoxide with hydrogen catalyzed by a rhenium(I) phenanthroline–polyoxometalate hybrid complex. *J Am Chem Soc*. 2011;133:188–190. https://doi.org/10.1021/ja1078199.

53. Lv H, Guo W, Wu K, et al. A noble-metal-free, tetra-nickel polyoxotungstate catalyst for efficient photocatalytic hydrogen evolution. *J Am Chem Soc*. 2014;136:14015–14018. https://doi.org/10.1021/ja5084078.

54. Hiskia A, Mylonas A, Papaconstantinou E. Comparison of the photoredox properties of polyoxometallates and semiconducting particles. *Chem Soc Rev*. 2001;30:62–69. https://doi.org/10.1039/a905675k.

55. Guo Y, Hu C. Heterogeneous photocatalysis by solid polyoxometalates. *J Mol Catal A Chem*. 2007;262:136–148. https://doi.org/10.1016/j.molcata.2006.08.039.

56. Gamelas JAF, Evtuguin DV, Esculcas AP. Transition metal substituted polyoxometalates supported on amine-functionalized silica. *Transit Met Chem*. 2007;32(8):1061–1067. https://doi.org/10.1007/s11243-007-0277-4.

57. Lei P, Chen C, Yang J, Ma W, Zhao J, Zang L. Degradation of dye pollutants by immobilized polyoxometalate with H2O2 under visible-light irradiation. *Environ Sci Technol*. 2005;39:8466–8474. https://doi.org/10.1021/es050321g.

58. Troupis A, Gkika E, Triantis T, Hiskia A, Papaconstantinou E. Photocatalytic reductive destruction of azo dyes by polyoxometallates: naphthol blue black. *J Photochem Photobiol A Chem*. 2007;188:272–278. https://doi.org/10.1016/j.jphotochem.2006.12.022.

59. Polinarski MA, Beal ALB, Silva FEB, et al. New perspectives of using chitosan, silver, and chitosan–silver nanoparticles against multidrug-resistant bacteria. *Part Part Syst Charact*. 2021;38:2100009. https://doi.org/10.1002/ppsc.202100009.

60. Gkika E, Troupis A, Hiskia A, Papaconstantinou E. Photocatalytic reduction and recovery of mercury by polyoxometalates. *Environ Sci Technol*. 2005;39:4242–4248. https://doi.org/10.1021/es0493143.

61. Mandal S, Selvakannan P, Pasricha R, Sastry M. Keggin ions as UV-switchable reducing agents in the synthesis of Au core–Ag shell nanoparticles. *J Am Chem Soc*. 2003;125:8440–8441. https://doi.org/10.1021/ja034972t.

62. Shanmugam S, Viswanathan B, Varadarajan TK. Photochemically reduced polyoxometalate assisted generation of silver and gold nanoparticles in composite films: a single step route. *Nanoscale Res Lett*. 2007;2:175. https://doi.org/10.1007/s11671-007-9050-z.

63. Gkika E, Kormali P, Antonaraki S, Dimoticali D, Papaconstantinou E, Hiskia A. Polyoxometallates as effective photocatalysts in water purification from pesticides. *Int J Photoenergy*. 2004;6:227–231. https://doi.org/10.1155/S1110662X04000297.

64. Cao M, Lin J, Lü J, You Y, Liu T, Cao R. Development of a polyoxometallate-based photocatalyst assembled with cucurbit[6]uril via hydrogen bonds for azo dyes degradation. *J Hazard Mater*. 2011;186:948–951. https://doi.org/10.1016/j.jhazmat.2010.10.119.

65. Kannan R, Gouse Peera S, Obadiah A, Vasanthkumar S. MnO2 supported POM—a novel nanocomposite for dye degradation. *Dig J Nanomater Biostruct*. 2011;6:829–835.

66. Dehghani R, Aber S, Mahdizadeh F. Polyoxometalates and their composites as photocatalysts for organic pollutants degradation in aqueous media— a review. *Clean (Weinh)*. 2018;46:1800413. https://doi.org/10.1002/clen.201800413.

67. Liu B, Yu Z-T, Yang J, Hua W, Liu Y-Y, Ma J-F. First three-dimensional inorganic–organic hybrid material constructed from an "Inverted Keggin" polyoxometalate and a copper(I)-organic complex. *Inorg Chem*. 2011;50:8967–8972. https://doi.org/10.1021/ic201135g.

68. Yang L, Shen Y, Chen Y, Pan X, Wang X, Wang X. A novel octamolybdate-based metal-organic complex constructed from a bis(tetrazole)-functionalized thioether ligand and an Anderson-type polyoxometalate. *Inorg Chem Commun*. 2019;108(July):107493. https://doi.org/10.1016/j.inoche.2019.107493.

69. Ong BC, Lim HK, Tay CY, Lim TT, Dong ZL. Polyoxometalates for bifunctional applications: catalytic dye degradation and anticancer activity. *Chemosphere*. 2022;286(P3):131869. https://doi.org/10.1016/j.chemosphere.2021.131869.

70. Li D, Tan XL, Chen LL, et al. Four Dawson POM-based inorganic-organic supramolecular compounds for proton conduction, electrochemical and photocatalytic activity. *J Solid State Chem*. 2021;2022(305):122694. https://doi.org/10.1016/j.jssc.2021.122694.

71. Cui Y, Xing Z, Guo M, et al. Hollow core-shell potassium phosphomolybdate@cadmium sulfide@bismuth sulfide Z-scheme tandem heterojunctions toward optimized photothermal-photocatalytic performance. *J Colloid Interface Sci*. 2022;607:942–953. https://doi.org/10.1016/j.jcis.2021.09.075.

72. Wang L, Fu W, Zhuge Y, et al. Synthesis of polyoxometalates (POM)/TiO2/Cu and removal of nitrate nitrogen in water by photocatalysis. *Chemosphere*. 2021;278:130298. https://doi.org/10.1016/j.chemosphere.2021.130298.

73. Zhang D, Liu T, An C, Liu H, Wu Q. Preparation of vanadium-substituted polyoxometalate doped carbon nitride hybrid materials POM/g-C3N4 and their photocatalytic. *Mater Lett*. 2019;262:126954–126965. https://doi.org/10.1016/j.matlet.2019.126954.

74. Zhang M, Lai C, Li B, Xu F, Huang D, Liu S. Unravelling the role of dual quantum dots cocatalyst in 0D/2D heterojunction photocatalyst for promoting photocatalytic organic pollutant degradation. *Chem Eng J*. 2020;396:125343. https://doi.org/10.1016/j.cej.2020.125343.

75. Huang X, Liu X. Morphology control of highly efficient visible-light driven carbon-doped POM photocatalysts. *Appl Surf Sci*. 2019;505:144527. https://doi.org/10.1016/j.apsusc.2019.144527.

76. Jamshidi ALI, Zonoz FM, Wei Y. A new Keggin-based organic-inorganic nanohybrid in the role of a dual-purpose catalyst. *J Chem Sci*. 2020;132:1–37. https://doi.org/10.1007/s12039-020-1739-x.

77. Tan H, Shen W, Wang W, et al. Pt/POMs/TiO2 composite nanofibers with enhanced visible-light photocatalytic performance for environmental remediation. *Dalton Trans*. 2019;48:1–7. https://doi.org/10.1039/C9DT02965F.

78. Ghali M, Brahmi C, Benltifa M, et al. New hybrid polyoxometalate/polymer composites for photodegradation of eosin dye. *J Polym Sci*. 2019;57:1538–1549. https://doi.org/10.1002/pola.29416.

79. Niu P, Wang D, Wang A, Liang Y, Wang X. Fabrication of bifunctional TiO 2/POM microspheres using a layer-by-layer method and photocatalytic activity for methyl orange degradation. *J Nanomater*. 2018;2018:1–8. https://doi.org/10.1155/2018/4212187.

80. Liu G, Zhao X, Zhang J, Liu S, Sha J. Z-Scheme Ag3PO4/POMs/GO heterojunction with enhanced photocatalytic performance for degradation and water splitting. *Dalton Trans*. 2018;47:6225–6232. https://doi.org/10.1039/C8DT00431E.

81. Taghavi M, Ehrampoush MH, Ghaneian MT, Tabatabaee M, Fakhri Y. Application of a Keggin-type heteropoly acid on supporting nanoparticles in photocatalytic degradation of organic pollutants in aqueous solutions. *J Clean Prod*. 2018;197:1447–1453. https://doi.org/10.1016/j.jclepro.2018.06.280.

82. Huang Y, Yang Z, Yang S, Xu Y. Photodegradation of dye pollutants catalyzed by H3PW12O40/SiO2 treated with H2O2 under simulated solar light irradiation. *J Adv Nanomater*. 2017;2:146–152. https://doi.org/10.22606/jan.2017.23002.

83. Lu N, Wang Y, Ning S, et al. Design of plasmonic Ag-TiO2/H3PW12O40 composite film with enhanced sunlight photocatalytic activity towards o-chlorophenol degradation. *Sci Rep*. 2017;7:1–17. https://doi.org/10.1038/s41598-017-17221-4.

84. Shi H, Zhang T, An T, Li B, Wang X. Enhancement of photocatalytic activity of nano-scale TiO2 particles co-doped by rare earth elements and heteropolyacids. *J Colloid Interface Sci*. 2012;380:121–127. https://doi.org/10.1016/j.jcis.2012.04.069.

85. Huang XH, Huang XX, Ying SM, et al. Polyoxometalate-phosphonate compounds: synthesis, structure, photocatalytic and antitumor properties. *J Mol Struct*. 2021;1233:2–7. https://doi.org/10.1016/j.molstruc.2021.130104.

86. Mosleh N, Masteri-Farahani M. Co3O4 quantum dots-polyoxometalate nanocomposites as visible light photoelectrocatalysts for selective oxidation of benzyl alcohol. *J Phys Chem Solid*. 2021;2022(162):110527. https://doi.org/10.1016/j.jpcs.2021.110527.

87. Ortiz-Bustos J, Pérez Y, Del HI. Structure, stability, electrochemical and catalytic properties of polyoxometalates immobilized on choline-based hybrid mesoporous silica. *Microporous Mesoporous Mater*. 2021;321(May). https://doi.org/10.1016/j.micromeso.2021.111128.

88. Khenkin AM, Efremenko I, Weiner L, Martin JML, Neumann R. Photochemical reduction of carbon dioxide catalyzed by a ruthenium-substituted polyoxometalate. *Chem Eur J*. 2010;16:1356–1364. https://doi.org/10.1002/chem.200901673.

89. Liu S, Zhang Z, Li X, Jia H, Ren M, Liu S. Ti-substituted Keggin-type polyoxotungstate as proton and electron reservoir encaged into metal—organic framework for carbon dioxide photoreduction. *Adv Mater Interfaces*. 2018;1801062:1–7. https://doi.org/10.1002/admi.201801062.

90. Liu X, Li Y, Peng S, Lu G, Li S. Photocatalytic hydrogen evolution under visible light irradiation by the polyoxometalate α-[AlSiW11(H2O)O39]5−-eosin Y system. *Int J Hydrogen Energy*. 2012;37:12150–12157. https://doi.org/10.1016/j.ijhydene.2012.06.028.

91. Lu Y, Yuc C, Li Y, et al. Atomically dispersed Ni on Mo2C embedded in N, P co-doped carbon derived from polyoxometalate supramolecule for high-efficiency hydrogen evolution electrocatalysis. *Appl Catal Environ*. 2021;296(March):120336. https://doi.org/10.1016/j.apcatb.2021.120336.

92. Matt B, Fize J, Moussa J, et al. Charge photo-accumulation and photocatalytic hydrogen evolution under visible light at an iridium(III)-photosensitized polyoxotungstate. *Energ Environ Sci*. 2013;6:1504–1508. https://doi.org/10.1039/c3ee40352a.

93. Lv H, Song J, Zhu H, et al. Visible-light-driven hydrogen evolution from water using a noble-metal-free polyoxometalate catalyst. *J Catal*. 2013;307:48–54. https://doi.org/10.1016/j.jcat.2013.06.028.

94. Zhou XJ, Yu H, Zhao D, Wang XC, Zheng ST. Combination of polyoxotantalate and metal sulfide: a new-type noble-metal-free binary photocatalyst Na8Ta6O19/Cd0.7Zn0.3S for highly efficient visible-light-driven H2 evolution. *Appl Catal Environ*. 2019;248:423–429. https://doi.org/10.1016/j.apcatb.2019.02.052.

95. Ma K, Dong Y, Zhang M, Xu C, Ding Y. A homogeneous Cu-based polyoxometalate coupled with mesoporous TiO2 for efficient photocatalytic H2 production. *J Colloid Interface Sci*. 2021;587:613–621. https://doi.org/10.1016/j.jcis.2020.11.018.

96. Li S, Liu S, Liu S, et al. {Ta12}/{Ta16} cluster-containing polytantalotungstates with remarkable photocatalytic H2 evolution activity. *J Am Chem Soc*. 2012;134:19716–19721. https://doi.org/10.1021/ja307484a.

97. Huang P, Qin C, Zhou Y, Hong YM, Wang XL, Su ZM. Self-assembly and photocatalytic H2 evolution activity of two unprecedented polytanta-lotungstates based on the largest {Ta18} and {Ta18Yb2} clusters. *Chem Commun*. 2016;52:13787–13790. https://doi.org/10.1039/c6cc07649a.

98. Huang P, Han X-G, Li X-L, Qin C, Wang X-L, Su Z-M. Self-assembly and photocatalytic properties of Ta/W mixed-addendum polyoxometalate and transition-metal cations. *CrstEngComm*. 2016;18:8722–8725. https://doi.org/10.1039/C6CE01953F.

99. Jawale DV, Fossard F, Miserque F, et al. Carbon nanotube-polyoxometalate nanohybrids as efficient electro-catalysts for the hydrogen evolution reaction. *Carbon*. 2021. https://doi.org/10.1016/j.carbon.2021.11.046. NY Published online.

100. Zhang M, Xin X, Feng Y, Zhang J, Lv H, Yang GY. Coupling Ni-substituted polyoxometalate catalysts with water-soluble CdSe quantum dots for ultraefficient photogeneration of hydrogen under visible light. *Appl Catal Environ*. 2022;303:120893. https://doi.org/10.1016/j.apcatb.2021.120893.

101. Shahsavarifar S, Masteri-Farahani M, Ganjali MR. Design and application of a polyoxometalate-ionic liquid-graphene oxide hybrid nanomaterial: new electrocatalyst for water oxidation. *Colloids Surf A Physicochem Eng Asp*. 2022;632:127812. https://doi.org/10.1016/j.colsurfa.2021.127812.

102. Nazari Serenjeh F, Hashemi P, Rasolzadeh F, Farhadi S, Hoseini AA. Magnetic fiber headspace solid-phase microextraction of Ferulago angulata volatile components using Preyssler-type polyoxometalate/metal–organic framework/silica aerogel sorbent. *Food Chem*. 2022;373(October 2021). https://doi.org/10.1016/j.foodchem.2021.131423.

103. Zhang HY, Liu L, Wang HJ, Sun JW. Asymmetrical modification of Keggin polyoxometalates by sextuple Ag–N coordination polymeric chains: synthesis, structure and selective separation of cationic dyes. *J Solid State Chem*. 2020;2021(296):121986. https://doi.org/10.1016/j.jssc.2021.121986.

104. Han H, Liu C, Sha J, et al. Ferrocene-reduced graphene oxide-polyoxometalates based ternary nanocomposites as electrochemical detection for acetaminophen. *Talanta*. 2021;235:122751. https://doi.org/10.1016/j.talanta.2021.122751.

105. Wang B, Meng R-Q, Xu L-X, Wu L-X, Bi L-H. A novel detection of nitrite, iodate and bromate based on a luminescent polyoxometalate. *Anal Methods*. 2013;5:885–890. https://doi.org/10.1039/C2AY26217G.

106. Deya KC, Sharma V. Well-Dawson type polyoxometallate of 3d series and their industrial use as ecofriendly and green catalysts in organic reactions. *J Indian Chem Soc*. 2011;88:1047–1050.

107. Sadasivan R, Patel A. Unmodified and modified copper polyoxometalates as catalysts for oxidation of alkenes: kinetic and mechanistic investigation. *Inorg Chim Acta*. 2020;510:119757. https://doi.org/10.1016/j.ica.2020.119757.

108. Sadasivan R, Patel A, Ballabh A. Investigation of catalytic properties of Cs salt of di-copper substituted phosphotungstate, Cs7[PW10Cu2(H2O)O38] in epoxidation of styrene. *Inorg Chim Acta*. 2019;487:345–353. https://doi.org/10.1016/j.ica.2018.12.034.

109. Patel A, Sadasivan R. Microwave assisted one pot synthesis and characterization of Cesium salt of di-copper substituted phosphotungstate and its application in the selective epoxidation of cis-cyclooctene with tert-butyl hydroperoxide. *Inorg Chim Acta*. 2017;458:101–108. https://doi.org/10.1016/j.ica.2016.12.031.

110. Patel AU, Sadasivan R. Cs salt of undecatungstophospho(aqua) cuprate(II): microwave synthesis, characterization, catalytic and kinetic study for epoxidation of cis-cyclooctene with TBHP. *ChemistrySelect*. 2018;3:11087–11097. https://doi.org/10.1002/slct.201802258.

111. Sadasivan R, Patel A. Flexible oxidation of styrene using TBHP over zirconia supported mono-copper substituted phosphotungstate. *RSC Adv*. 2019;9:27755–27767. https://doi.org/10.1039/C9RA04892H.

112. Cui X, Wang J, Liu B, Ling S, Long R, Xiong Y. Turning au nanoclusters catalytically active for visible-light-driven CO2 reduction through bridging ligands. *J Am Chem Soc*. 2018;140(48):16514–16520. https://doi.org/10.1021/jacs.8b06723.

113. Di J, Zhu C, Ji M, et al. Defect-rich Bi12O17Cl2 nanotubes self-accelerating charge separation for boosting photocatalytic CO2 reduction. *Angew Chem Int Ed*. 2018;57(45):14847–14851. https://doi.org/10.1002/anie.201809492.

114. Zhang Y, Xia B, Ran J, Davey K, Qiao SZ. Atomic-level reactive sites for semiconductor-based photocatalytic CO2 reduction. *Adv Energy Mater*. 2020;10:1–23. https://doi.org/10.1002/aenm.201903879.

115. Ong WJ, Putri LK, Mohamed AR. Rational design of carbon-based 2D nanostructures for enhanced photocatalytic CO2 reduction: a dimensionality perspective. *Chem Eur J*. 2020;26:9710–9748. https://doi.org/10.1002/chem.202000708.

116. Qiao L, Song M, Geng A, Yao S. Polyoxometalate-based high-nuclear cobalt—vanadium—oxo cluster as efficient catalyst for visible light-driven CO2 reduction. *Chin Chem Lett*. 2019;30:1273–1276.

117. Szczepankiewicz SH, Ippolito CM, Santora BP, et al. Interaction of carbon dioxide with transition-metal-substituted heteropolyanions in nonpolar solvents. Spectroscopic evidence for complex formation. *Inorg Chem*. 1998;37:4344–4352. https://doi.org/10.1021/ic980162k.

118. Zha B, Li C, Li J. Efficient electrochemical reduction of CO2 into formate and acetate in polyoxometalate catholyte with indium catalyst. *J Catal*. 2020;382:69–76. https://doi.org/10.1016/j.jcat.2019.12.010.

119. Lang Z, Miao J, Lan Y, Cheng J, Xu X, Cheng C. Polyoxometalates as electron and proton reservoir assist electrochemical CO2 reduction. *APL Mater*. 2020;8:120702. https://doi.org/10.1063/5.0031374.

120. Girardi M, Platzer D, Griveau S, et al. Assessing the electrocatalytic properties of the {Cp*RhIII}2+-polyoxometalate derivative [H2PW11O39{RhIIICp*(OH2)}]3− towards CO2 reduction. *Eur J Inorg Chem*. 2019;2019:387–393. https://doi.org/10.1002/ejic.201800454.

121. Wang YR, Huang Q, He CT, et al. Oriented electron transmission in polyoxometalate-metalloporphyrin organic framework for highly selective electroreduction of CO2. *Nat Commun.* 2018;9:1–8. https://doi.org/10.1038/s41467-018-06938-z.

122. Girardi M, Blanchard S, Griveau S, et al. Electro-assisted reduction of CO2 to CO and formaldehyde by (TOA)6[α-SiW11O39Co(−)] polyoxometalate. *Eur J Inorg Chem.* 2015;2015:3642–3648. https://doi.org/10.1002/ejic.201500389.

123. Du J, Lang ZL, Ma YY, et al. Polyoxometalate-based electron transfer modulation for efficient electrocatalytic carbon dioxide reduction. *Chem Sci.* 2020;11:3007–3015. https://doi.org/10.1039/c9sc05392a.

124. Guo SX, Li F, Chen L, Macfarlane DR, Zhang J. Polyoxometalate-promoted electrocatalytic CO2 reduction at nanostructured silver in dimethylformamide. *ACS Appl Mater Interfaces.* 2018;10:12690–12697. https://doi.org/10.1021/acsami.8b01042.

125. Lai SY, Ng KH, Cheng CK, Nur H, Nurhadi M, Arumugam M. Photocatalytic remediation of organic waste over Keggin-based polyoxometalate materials: a review. *Chemosphere.* 2021;263:128244. https://doi.org/10.1016/j.chemosphere.2020.128244.

126. Li X-X, Liu J, Zhang L, et al. Hydrophobic polyoxometalate-based metal-organic framework for efficient CO$_2$ photoconversion. *ACS Appl Mater Interfaces.* 2019;11:25790–25795. https://doi.org/10.1021/acsami.9b03861.

127. Nyman M, Bonhomme F, Alam TM, et al. A general synthetic procedure for heteropolyniobates. *Science.* 2002;297:996–998. https://doi.org/10.1126/science.1073979.

128. Ioannidis A, Papaconstantinou E. Photocatalytic generation of hydrogen by 1:12 heteropolytungstates with concomitant oxidation of organic compounds. *Inorg Chem.* 1985;24:439–441.

129. Wang ZL, Lu Y, Li YG, Wang SM, Wang EB. Visible-light photocatalytic H2 evolution over a series of transition metal substituted Keggin-structure heteropoly blues. *Chin Sci Bull.* 2012;57:2265–2268. https://doi.org/10.1007/s11434-012-5050-1.

130. Shang X, Liu R, Zhang G, Zhang S, Cao H, Gu Z. Artificial photosynthesis for solar hydrogengeneration over transition-metal substituted Keggin-type titanium tungstate. *New J Chem.* 2014;38:1315. https://doi.org/10.1039/c3nj01184d.

131. Chiang MH, Williams CW, Soderholm L, Antonio MR. Coordination of actinide ions in Wells-Dawson heteropolyoxoanion complexes. *Eur J Inorg Chem.* 2003;1:2663–2669. https://doi.org/10.1002/ejic.200300014.

132. Ohlin CA, Villa EM, Fettinger JC, Casey WH. The [Ti12Nb6O44]10⁻ ion—a new type of polyoxometalate structure. *Angew Chem Int Ed.* 2008;47:5634–5636. https://doi.org/10.1002/anie.200801883.

133. Huang P, Qin C, Su ZM, et al. Self-assembly and photocatalytic properties of polyoxoniobates: {Nb24O72}, {Nb32O96}, and {K12Nb96O288} clusters. *J Am Chem Soc.* 2012;134:14004–14010. https://doi.org/10.1021/ja303723u.

134. Zhao J, Ding Y, Wei J, Du X, Yu Y, Han R. A molecular Keggin polyoxometalate catalyst with high efficiency for visible-light driven hydrogen evolution. *Int J Hydrogen Energy.* 2014;39:18908–18918. https://doi.org/10.1016/j.ijhydene.2014.09.084.

135. Li R, Ren X, Zhao J, et al. Polyoxometallates trapped in a zeolitic imidazolate framework leading to high uptake and selectivity of bioactive molecules. *J Mater Chem A.* 2014;2:2168–2173. https://doi.org/10.1039/c3ta14267a.

136. Buru CT, Farha OK. Strategies for incorporating catalytically active polyoxometalates in metal-organic frameworks for organic transformations. *ACS Appl Mater Interfaces.* 2020;12:5345–5360. https://doi.org/10.1021/acsami.9b19785.

137. Cao YD, Yin D, Wang ML, et al. Pt-substituted polyoxometalate modification on the surface of low-cost TiO2 with highly efficient H2 evolution performance. *Dalton Trans.* 2020;49:2176–2183. https://doi.org/10.1039/c9dt04446a.

138. Shi D, Zheng R, Sen LC, Chen DM, Zhao J, Du M. Dual-functionalized mixed Keggin- and Lindqvist-type Cu24-based POM@MOF for visible-light-driven H2 and O2 evolution. *Inorg Chem.* 2019;58:7229–7235. https://doi.org/10.1021/acs.inorgchem.9b00206.

139. Stephan H, Kubeil M, Emmerling F, Müller CE. Polyoxometalates as versatile enzyme inhibitors. *Eur J Inorg Chem.* 2013;2013:1585–1594. https://doi.org/10.1002/ejic.201201224.

140. Dinčić M, Čolović MB, Sarić Matutinović M, et al. In vivo toxicity evaluation of two polyoxotungstates with potential antidiabetic activity using Wistar rats as a model system. *RSC Adv.* 2020;10:2846–2855. https://doi.org/10.1039/C9RA09790B.

141. León IE, Porro V, Astrada S, et al. Polyoxometalates as antitumor agents: bioactivity of a new polyoxometalate with copper on a human osteosarcoma model. *Chem Biol Interact.* 2014;222:87–96. https://doi.org/10.1016/j.cbi.2014.10.012.

142. Shah HS, Al-Oweini R, Haider A, Kortz U, Iqbal J. Cytotoxicity and enzyme inhibition studies of polyoxometalates and their chitosan nanoassemblies. *Toxicol Rep.* 2014;1:341–352. https://doi.org/10.1016/j.toxrep.2014.06.001.

143. Li X-H, Chen W-L, Wei M, et al. Polyoxometalates nanoparticles improve anti-tumor activity by maximal cellular uptake. *Inorg Chim Acta.* 2019;486:104–112. https://doi.org/10.1016/j.ica.2018.10.046.

144. Yamase T. Anti-tumor, viral, and bacterial activities of polyoxometalates for realizing an inorganic drug. *J Mater Chem.* 2005;15:4773. https://doi.org/10.1039/b504585a.

145. Flütsch A, Schroeder T, Grütter MG, Patzke GR. HIV-1 protease inhibition potential of functionalized polyoxometalates. *Bioorg Med Chem Lett.* 2011;21:1162–1166. https://doi.org/10.1016/j.bmcl.2010.12.103.

146. Zong L, Wu H, Lin H, Chen Y. A polyoxometalate-functionalized two-dimensional titanium carbide composite MXene for effective cancer theranostics. *Nano Res.* 2018;11:4149–4168. https://doi.org/10.1007/s12274-018-2002-3.

Chapter 20

Smart material-based micro/ nanostructures for the detection and removal of water impurities

Ali Fakhri[a,*], Inderjeet Tyagi[b,*], and Rama Rao Karri[c]

[a]Department of Chemistry, Nano Smart Science Institute, Tehran, Iran, [b]Centre for DNA Taxonomy, Molecular Systematics Division, Zoological Survey of India, Ministry of Environment, Forest and Climate Change, Government of India, Kolkata, West Bengal, India, [c]Petroleum and Chemical Engineering, Faculty of Engineering, Universiti Teknologi Brunei, Bandar Seri Begawan, Brunei Darussalam

[*]Corresponding authors.

1 Introduction

Recent advancements in different domains such as industrial, municipal, automotive, etc., have encouraged researchers across the globe to look for new intelligent materials for the detection and remediation of noxious pollutants.[1] The intelligent materials must possess different characteristics and properties and may be single-handedly applied for multiple applications such as detection and remediation as well as for use in other scientific and technological fields.[2] In 1990, intelligent materials were defined as "materials that respond to the environmental variables at the most optimum situation and ascertain their efficiencies conforming to these changes."[2] Although this was not comprehensible at that time, it was expected to open a broad field in science and technology.[2] Recently, the phrase "intelligent material" has become synonymous with "smart material." It has attracted significant interest from researchers due to the expansion of advanced technologies and the enhanced requirement for novel materials. Thus, smart materials are those materials that show great potential in different aspects of science such as thermal, magnetic, or electric. They are presented in research reports as intelligent, functional, and adaptive materials.[3,4] Further, they are specified as morphological, which means they exhibit multiple behaviors, improving their physical attributes as were as external operating conditions.

Literature evidence suggests that in the year 1940–60, the different smart materials such as rheological fluids,[5] piezoelectric materials,[6] alloys,[7,8] shape memory polymers,[9] etc. Based on the definition, piezoelectric materials are materials that can convert mechanical energy to electrical energy while the opposite is also true for these materials. Further, the shape of the piezoelectric material changes with the electrical impulse or mechanical stress. Shape memory materials also have the ability to change their shape and can regain their main shape under an external heat source. In addition, smart materials such as chromoactive materials change color under different conditions such as light, temperature, and pressure. Magnetorheological materials change properties under a magnetic field and are used in shock absorbers to hinder vibrations in bridges from earthquakes. Photoactive materials are those classified from electroluminescent and phosphorescent materials. The electroluminescents can emit light under electrical impulses, and the phosphorescents emit light after the first source has stopped.[5–10]

This chapter covers different aspects of smart materials, including synthesis, characteristics/properties, categorization, characterization, and multiple applications with a special focus on sensors to detect noxious impurities and on adsorbents to remediate toxic pollutants from aqueous solutions and wastewater. Further, the chapter will provide a single point solution to researchers looking after both sensing and wastewater remediation studies. It will certainly reduce the efforts of researchers working on these aspects.

2 Smart materials

2.1 Smart microscale materials

Nowadays, research on organic/inorganic composites with different potential properties has been expanded. The properties of the composites relate to the characteristics of components. Thus, the term "smart micromaterials" refers to materials with a particle size of 1–1000 μm.[11] The larger particles can act as smart micromaterials, and need similar fabrication methods. To prepare a smart micromaterial, it is necessary to form microparticles or a bulk substrate. Ding et al., using two different polymerization methods, investigated the progress of microstructure formation. They used bottom-up as well as top-down approaches to control the microstructural architecture of the smart micromaterial.[11] Li et al. elucidated the electronic and optical properties of magnetic microparticles under the application of magnetic fields.[12] Moreover, the potential of functionalized microparticles was also explored for use as electronic sensors in optical instruments.[13] Yu et al. presented a new technique for creating oil-based coatings for local corrosion. The new coating was prepared from the micro-capsule, which acts as an excellent barrier to protect metals from corroding agents.[14] In the last decade, nanostructure materials have been extensively studied due to different potential applications compared to microstructure materials.[3] Therefore, limited information is available in the literature about intelligent microstructures.

2.2 Smart nanoscale materials

Smart-based nanoscale materials are extraordinary substances due to their unique properties, such as enhanced thermal, mechanical, optical, and electronic attributes. The specifications of these smart nanoscale materials show them to be outstanding materials for multiple performances via different novel technologies. A detailed overview of the various types of smart nanomaterials with different response mechanisms under various stimuli as well as synthesis methods and applications[15] are shown below.

2.2.1 Synthesis of smart nanomaterials

They can be synthesized via two routes: (I) physical processes, such as fabricated particles from presynthesized polymers, have been used including the layer-by-layer protocol, the grafting method, etc. and (II) chemical processes, which include heterogeneous polymerization techniques.[16] A detailed overview for the synthesis methods is shown below.

Self-assembled micelles: In this technique, the block polymers were selected from different kinds of self-convened materials such as micelles and bilayer structures.

Interaction of nanoparticles and polymers: This method is done by the fabrication of polymers on a nanoparticle surface, and interactions in the colloidal suspension due to the electrostatic and depletion behaviors.

Layer-by-layer: The layer-by-layer technique is used for the deposition of charged molecules such as protein molecules and polyelectrolyte materials.[17] This method is used in the preparation of polymer shells around cores. The particle core is formed by spherical particles with a layer-by-layer gel.

Grafting method: This technique leads to the chemical bonding of functional polymers; the bonds are attached to the core by the sorption process or covalent chemical bonds.[18] Grafting presents responsive nanoparticles, that is, changes with factors such as pH, light, or temperature. This surface is suitable for polymerization chemistries such as plasma-induced, condensation, anionic/cationic, photo- or electrochemical, and radical polymerization. As an example, magnetic nanoparticles are the best materials with excellent biocompatibility and responsiveness. New smart polymer/magnetic nanoparticles have been studied using the separation and adsorption of amino acid enantiomer substrates. Magnetic/polymer nanoparticles have been prepared by the grafting method. The poly(NIPAM-*co*-GMA) attached to the β-cyclodextrin functional was supported on magnetic nanoparticles by surface atom transfer as the polymerization method.[18]

Heterogeneous polymerization: This method is used to prepare dispersed smart nanoscale materials. Several approaches are used for synthesis including emulsion, dispersion polymerization, and precipitation.[19] The heterogeneous polymerization used for preparation of hybrid particles (inorganic/organic materials—noble metal, SiO_2, Al_2O_3, TiO_2, and Fe_3O_4).

2.2.2 Types of smart nanoscale materials

2.2.2.1 Carbon nanotubes

Carbon nanotubes (CNTs) are hexagonal carbon materials shaped as tubes. CNTs are a part of the fullerene structure. CNTs exhibit phenomenal and powerful electronic attributes and are used as effective thermal detectors. Due to their excellent attributes, they have many efficient uses in the nano field. Carbon nanotubes can be classified as single-walled nanotubes (SWCNTs) or multi-walled nanotubes (MWCNTs).[20,21]

2.2.2.2 Chitosan and cellulose

Chitosan and cellulose are linear polysaccharides composed of several glucose saccharides. Chitosan is formed by the deacetylation of chitin, which is the main component of the exoskeletons of crustaceans.[22,23] Cellulose is an essential structural part of the plant wall. These materials are hydrophilic substances with insoluble properties in water and solvents. These materials are biodegradable due to the chiral structure.

2.2.2.3 Smart polymers

Smart polymer materials are polymers grafted to a substrate. Smart polymer materials are agglomerated in order to the polymer chains stretched from the substrate with the volume influences.[24] They include cartilage proteoglycans, neurofilaments, extracellular polysaccharides of bacteria, etc.[25,26] Smart polymers exhibit various properties in solution. The stretched smart polymer has various features such as resistant bio-surfaces of protein, chromatographic instruments, and lubricants.[27-29]

2.2.2.4 Smart hydrogels

Smart hydrogels such as colloidal gels are hydrophobic polymer chains in a water medium. Smart hydrogels are synthesized using synthetic or natural polymers with highly absorbent attributes. Smart hydrogels can sense changes in different mechanisms such as pH, metabolite concentration, or temperature. Smart hydrogels are joined with drug compounds and applied as carriers for drug delivery.[30,31]

2.2.3 Characterization

Smart nanoscale materials are characterized using advanced instrumental techniques as mentioned below[32]:

Scanning system: Atomic force microscopy; magnetic force microscopy; atomic probe microscopy.
Thermodynamic system: Thermal gravimetric analysis; nanocalorimetry.
Optical imaging system: Dynamic light scattering; confocal laser scanning microscopy.
Photo spectrum system: UV-visible spectroscopy; surface plasmon resonance.
Ionic system: X-ray diffraction; Raman spectroscopy; energy dispersive X-ray spectroscopy.
Electron microscopy system: Transmission electron microscopy; scanning electron microscopy; low-energy electron diffraction.
Other miscellaneous systema: Zeta potential measurements; field flow fractionation; tracking analysis.

Fig. 1 presents a clear overview of different characterization techniques.

2.2.4 Factors affecting properties of smart nanoscale materials

Nanoscale materials are responsive to several factors such as pH, temperature, light, chemical interaction via enzymes or solvents, structural changes (particle aggregation), etc.[18] Details of some the factors affecting nanoscale materials are as follows:

2.2.4.1 Light changes

Nanoscale materials possessing photoactive groups such as azo-benzene in the polymer substrate are sensitive to light changes. These are reversible structural conversions using ultraviolet (UV) or visible light. The size, shape, and formation of ionic species are changed due to light irradiation.[33] Moreover, CNTs in the structure of smart nanomaterials display a great photoactive effect.

2.2.4.2 Magnetic field changes

Magnetic fields are applied for drug release. An excellent magnetic smart nanomaterial is Fe_3O_4 nanoparticles, widely referred to as ferro gels. The magnetic response is conducted due to the agglomeration of magnetic particles.[33]

2.2.4.3 pH changes

The pH can change the properties of smart nanomaterials containing polymers with weak polyelectrolytes such as —COOH, —PO_4, or —NH_3 functional agents. The chemical equilibrium is changed with the variation of ionization degree with different pH values in polymers containing smart nanomaterials.[33]

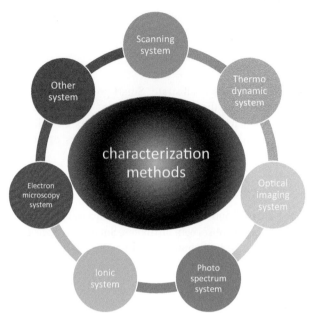

FIG. 1 Different techniques for smart nanomaterials.

2.2.4.4 Temperature changes

Temperature plays a crucial role and its impact varies in different materials. It enhances UV-visible (UV-vis) absorption, the wettability of a surface for smart nanomaterials, and drug release capabilities.[16]

2.2.4.5 Biological and chemical changes

The chemical and biochemical changes have a significant impact on the properties of smart nanomaterials due to the interactions of functional agents between molecules and smart nanomaterial-containing polymers. This selective response is done using polymers containing biological substrates such as antibodies, DNA, proteins, and enzymes. The enzymes are used as a biochemical signalers in polymers. These enzymes can be bio-catalytic converted to bio-products and reason chemical changes in the smart nanomaterials contain polymers with bio-fragments from substrates.[16]

2.2.5 Application of smart nanoscale materials

Due to their unique properties, smart nanoscale materials can be applied in various fields, as shown in Fig. 2. A detailed overview for a better understanding of different applications is provided below.

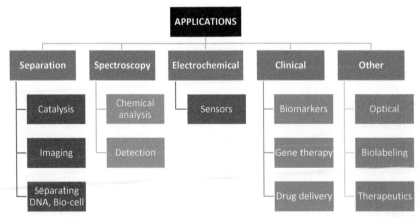

FIG. 2 Usage chart of smart nanomaterials.

2.2.5.1 Separation

Gold nanoparticles (Au NPs) have been applied in novel electrophoresis analysis due to the increased separation imaging.[34] Au NPs are a second stationary phase in the ethylene oxide polymer for DNA separation. The ethylene oxide polymer interacts with DNA in the presence of Au NPs and modifies the sieving ability. The separation of proteins using Au NPs filled with different surfactant compounds has been investigated.[34] The novel metal-organic framework was studied as an efficient smart nano photocatalyst for the photodegradation of dye. The project investigated the characterization, removal of isotherms, and photocatalytic efficiency of the metal-organic framework. Photocatalysis in the metal-organic framework was conducted by e⁻ transfer from the excited organic ligands to the metal cluster of the framework. The metal-organic framework shoes great activity for the recyclability properties.[34]

2.2.5.2 Spectroscopy

Silver nanoparticles (Ag NPs) have great surface plasmon resonance behavior, and they are applied in the highly sensitive technique using surface enhanced Raman scattering.[35] The interference was formed in the direction of the analytes with Ag NPs. Further, the silver was oxidized under an ambient medium with a decrease in plasmonic properties.[35] In addition to these literary evidences about the efficiency of colloidal material from Ag nanoparticles/poly-(N-isopropylacrylamide) for the detection of molecules was also present. Ag NPs have enhanced temperature responsive properties with the poly-(N-isopropylacrylamide). This process traps molecules and locates them near Ag NPs for enhanced plasmon resonance properties and the detection of molecules.[36]

2.2.5.3 Electrochemical

The efficiency of nanoscale materials in the bioreaction process has been studied using the modification of the nanoparticle surface with a protein and amino acids. The functionalized nanoscale material was then applied as an electrosensory for the detection of noxious impurities. The electrode surface was modified with nanoparticles to enhance the electrode surface and the sorption capacity. The electrode modified with CNTs and ZnO NPs have been studied to sense a molecule using electrochemical analysis. The calculations depicted the charge moving from the molecules and zinc oxide was conducted by using electrochemical detecting.[37]

2.2.5.4 Drug delivery/nanomedicine

Smart nanoscale materials have novel efficiencies and play a significant role in the drug delivery process. The literature shows that these synthesized nanoscale materials are used as drug carriers due to biocompatibility, degradability, great stability, and excellent capacity for clinical analysis.[38] For example, silver nanostructures modified on the electrode have been applied to monitor cystic fibrosis. The screen electrode with nanoparticles improved the selectivity and sensitivity of the system.[39]

In another study, Ag NPs functionalized with hydrogel polysaccharides were used for different medical applications. Hydrogel has been applied as a template for the fabrication of silver nanostructures. The glucuronoxylan is a hydrogel smart material with high potential. Glucuronoxylan is used as a drug carrier for different drugs due to its pH-responsive properties.[39] A silver nanostructure was prepared via glucuronoxylan using sunlight.[39] The silver nanostructure/glucuronoxylan was used in wound dressings. The wound mechanism is conducted by using the hydrogel content and potent epithelial tissue cell.

Moreover, sensors with nanoparticles are used to determine Alzheimer's patients. The diagnostic device identifies volatile organic materials in the breath. The sensor has a rapid electrical signal under the characterization of volatile organic materials. Therefore, the biomarker with nanoparticle sensors was the best method to analyze breath due to its simple process and cost-effective properties.[38] This is generally termed rapid disease diagnostic analysis. To increase the therapeutic influence, the drug compounds were transferred to the area of illness in the human body.

3 Nanoscale materials as sensors in water

In recent years, nanotechnology has been rapidly used worldwide with substantial performances in different industries. One of these is sensing, which involves the instruments applied to determine target compounds or to evaluate physicochemical properties such as temperature as well as optical and electronic behaviors, then generate a signal upon detection.

Nanosensors are like macrosensors with nanoscale dimensions, and they can be applied to evaluate signals. Different electrical technologies have expanded, and the smart nanosensor field has taken benefit of this for its own expansion.[40] Nanosensors can operate with signals generated in the nanorange, and analytes are rapidly detected. These activities have led to the efficiencies of different types of nanosensors for various areas, particularly in the environmental field. Further, based on their properties, nanosensors are categorized as optical nanosensors or electrochemical nanosensors. The optical nanosensors are generally applied for the qualitative and quantitative determination of fluorescence sensitivity and intensities while the electrochemical nanosensors detect noxious impurities via electronic or chemical properties as a signal. Thus, nanosensors play a significant role in human life, as they are widely used for the detection of hazardous substances in water. Further, nanomaterials have phenomenal chemical, magnetic, optical, and mechanical properties due to the high surface-to-volume ratio,[40] which is different from the bulk form due to the nanoscale dimension.

3.1 Types of nanosensors

3.1.1 Magnetic

Nanomaterials with an inconvenient arrangement of electrons generally show great magnetic properties. This attribute of nanosensors has been widely studied in several fields such as catalysis, biomedicine, magnetic fluids, and environmental impurity detection. Nano magnetic sensors have been used to detect organic as well as inorganic impurities using different analyses such as the influence of magnetic nanoparticles on the relaxation of water protons by distinguishing the magnetic moment and by determining the existence of the magnetic behaviors of particles by magneto-resistivity.[41]

3.1.2 Electronic

Nanomaterials have exceptional properties due to electronic attributes. The perfect example for an electronic nanomaterial is graphene. It is a two-dimensional single-layer structure that makes the surface available for both interaction and adsorption of the molecule. The functionalization of graphene with CNTs enhances the electronic properties, as the sp^2 hybridization for carbon orbitals in the carbon nanotube leads to free electrons at the tube's surface, which enhances the efficiency.[41] These nanosensors work on the mechanism of current inhibition and enhancement. They are widely applied for the detection of drugs and microorganisms.

3.1.3 Optical

Due to optical attributes such as absorption and light emission along with optical magneto behavior, nanosensors show distinct properties; they are generally termed nano-optical sensors. For example, a fluorescein sensor was able to detect noxious impurities under different environmental variables. It responds to the different stimuli and generates a signal for the detector to monitor the fate or presence of these noxious pollutants. The surface plasmon influence of the metal nanostructures is an active scientific investigation area for preparing nanomaterials as sensors. The surface plasmon influence increases the electric field on the surface of the nanoparticle and the resonant frequency displays the maximum optical properties.[42]

3.2 Materials used as nanosensors

So far, different materials have been used to synthesize nanosensors. Details about different classes of nanomaterials used to prepare nanosensors are shown in Fig. 3, and a detailed overview is provided below.

3.2.1 Carbon nanotubes

CNTs were first developed by Iijima in 1991. They are carbon tubes that are tens of nanometers in diameter and several micrometers in length. It is categorized as Single-walled Carbon nanotube (SWCNTs) and Multi-Walled Carbon nanotube (MWCNTs) that is, covered with single and multiple layers of graphite respectively. CNTs are usually prepared by vapour deposition method or graphite vaporization using a furnace with a gas flow atmosphere. They possess significant properties such as high strength as well as substantial electronic properties due to the hexagonal structure and the free electron of sp^2 hybridization, respectively. The CNT can be easily joined with various organic compounds for different applications.[43]

3.2.2 Nanoparticles

As part of their widespread applications, nanoparticles are generally used for sensing applications. Core-shell nanoparticles contain three layers: (i) the surface layer for functionalization of the particle; (ii) the shell; and (iii) the core for a portion of the particles. Further, they are synthesized using different methods such as spinning, laser pyrolysis, sputtering, and mechanical milling. Nanoparticles can be categorized into several types such as metal nanoparticles, carbon-based nanoparticles, ceramic nanoparticles, polymer nanoparticles, and semiconductor nanoparticles. The important nanoparticle for

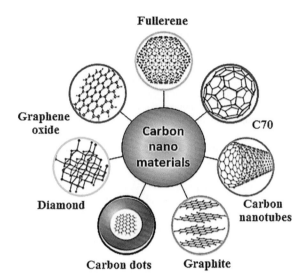

FIG. 3 Various carbon nanomaterials.

sensors is the metal nanoparticle. Metal nanostructures are applied to increase the surface plasmon resonance behavior. The plasmon resonance method is applied in numerous optical sensors.[44]

3.2.3 Nanowires

Nanowires are applied in smart nanosensors. Nanowires are prepared using various methods such as laser ablation, thermal evaporation, and current electrode position.[45] The silicone nanowire is the main nanowire for sensor efficiency. Gold nanostructures are catalyzed via vapor deposition and applied to produce wire-shaped nanostructures as sensors. Nanowires are applied to prepare gas sensors that can determine the quantity of amine.

3.2.4 Fullerenes

The literature shows that due to their inimitable properties, fullerenes are widely used as sensor materials. Fullerenes have a hexagonal and symmetric structure with sp^2 hybridization and are highly stable, even after decomposition.[46] Fullerenes possess properties such as a high surface ratio, electron affinity, and hydrophobic behavior. A considerable number of sensors have been prepared using fullerenes with different nanomaterials as composite materials. A previous study showed the preparation of C_{60}-carbon nanotubes as nanosensors to determine the pyruvic acid substrate. The Cu nanoparticles were supported on the electrode and a composite of Cu/C60-MWCNTs was prepared. The prepared nanocomposites showed great efficiency in paracetamol drug detection.[47] The sensing processes for nanosensors are categorized into four methods based on the detection process: colorimetry, surface plasmon resonance, electrochemical, and surface enhanced Raman scattering. The details of these four analysis methods are shown in Table 1. Previous studies based on sensing noxious pollutants in water are shown in Table 2. The reactivity of target detection and the nanosensor was done using different detection methods. The surface plasmon resonance technique has high detection potential due to the ultrasensitivity and selectivity in pollutant determination from a water medium.

4 Smart materials in wastewater remediation

Several materials have been used in wastewater treatment for the remediation of noxious impurities through different processes such as adsorption,[70–72] membrane filtration,[73] photocatalysis,[74–76] etc. Due to certain limitations such as oxygen diffuse resistance, sludge release, incomplete pollutant removal, and a slow rate,[77] conventional methods have been replaced by smart nanoscale materials in different techniques. A detailed overview (Fig. 4) of smart nanoscale materials in different processes is presented below.

TABLE 1 Analysis of features for sensing water pollution.

Methods	Advantages	Disadvantages
UV-vis analysis	– Low limit of detection and high selectivity – Sensitive to immaterial change – Real-time monitoring	– Low accuracy due to interfering ions
Electrochemical analysis	– Use for multianalyte – Precise for trace amounts	– Temperature sensitive
Colorimetry analysis	– Fast, portable test – Use for cationic and anionic pollutants	– Low sensitivity for biological substrate – Needs several methods
Surface enhanced Raman scattering	– Fingerprint for identifying analyte	– Problem detecting analytes with less affinity – Needs contact for the sensor and analyte

TABLE 2 The main information for different detection analyses for water pollution.

Targets	Sensors	Methods	Detection limit	Refs.
Mn^{2+}	Ag nanoparticles@ sulfoanthranilic acid-dithiocarbamate	Colorimetry	0.01 ppm	48
Melamine	Sulfanilic acid modified Ag nanoparticles	Colorimetry/UV	10.6 nM	49
F^-	Thiobarbituric-capped Au nanoparticles	Colorimetry	10 mM	50
Hg^{2+}	Au nanorods	UV-vis	10 nM	51
Hg^{2+}	Histidine-perylene diimide-bolaamphiphile-Au nanoparticles	SERS	60×10^{-18} M	52
Cr^{4+}	Biphenolbiphenoquinone-Ag NPs	Electrochemical	2.0×10^{-12} M	53
Phenol	Ag@C@Ag modified electrode	Electrochemical	41.5 nM	54
Phorate	Aptamer conjugated Au nanoparticles	UV-vis	0.01 nM	55
NO^{2-}	Au nanoparticles with 4-aminothiophenol	Colorimetry	1 mM	56
As^{5+}	Eu@Au nanoparticles/MMT	Colorimetry	0.01 ppm	57
CV dye	Au nanocone/nickel foam	SERS	0.03 nM	58
Picric acid	Ag nanopillar	SERS	20 ppt	59
Cl^-	Ag nanoparticles	Colorimetry	2 mg/L	60
Bi^{3+}	Dipicolinic acid-Ag nanoparticles	Colorimetry	0.01 mM	61
H_2S	Polyhedral oligomeric-formaldehyde polymer-Ag NPs	Colorimetry	0.2 mM	62
DNB	Glycine-Ag nanoparticles	Electrochemical	1×10^{-7} to 0.1 M	63
Cd^{2+}	Glutathione-Au nanoparticles	Colorimetry	30–70 nM	64
Pb^{2+}	L-Tyrosine-Au nanoparticles	Colorimetry	53 nM	65
S^{2-}	Ag/g-C_3N_4/SiC/polyglutamic acid	Colorimetry	0.15 nM	66
Cr^{+3}	Au-Ag/CuS nanoparticles	Colorimetry	0.5 nM	67
Hg^{2+}	Ag-MnS_2/chitosan	Colorimetry	9.0 nM	68
S^{2-}	Ag/Fe/g-C_3N_4-carrageenan	Colorimetry	0.6 nM	69

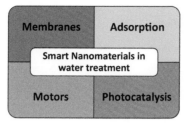

FIG. 4 Schematic of different systems for smart nanomaterials.

4.1 Nanomaterials in membranes

The use of nanomaterials in membrane filters for wastewater treatment is widely accepted and known to be one of the best filtration methods. These nanomaterials provide new functional agents, superior catalytic activity, modified infiltration, and resistance for the membrane process. These filtration systems with membranes are extremely impressive in the elimination of various waste and impurity compounds. The nanomaterials in membranes system are extremely impressive in the removal of metal ions and salts. As a mechanism, the membranes require surface charges due to the interaction between the functional agents with the solute molecules. Therefore, the charged substrate separation with nanomaterials in membrane filtration is related to environmental parameters such as the electrolyte concentration and pH medium.[78,79]

4.2 Nanomaterials in adsorption

Adsorption is the surface phenomenon that takes place due to interaction of adsorbate (noxious impurities) with that of different active sites of adsorbent in the reaction. In other words, the adsorption reaction is extremely reversible as it simplifies the purification of the pollution substrate from the adsorbent compounds. Adsorption is the most widely used method in water purification due to the availability of several adsorbents based on nanoscale structures. Owing to the enhanced active surface area, nanoparticles lead to the increased chemical activity and absorption capability of the adsorbate on the nanoadsorbent surface. The adsorbent-based nanoscale structures can be classified into various adsorbent types based on the essential attributes, such as carbon nanomaterials, graphene, metal oxide nanostructures, and chitosan materials.[80]

Silica was used in the solute separation in the mixture medium using chromatographic analysis. Nano-sized silica is the best candidate for pollutant removal from water due to extremely efficient properties such as a great active surface area, great selectivity, and excellent adsorption ability. Silica nanostructures can be easily prepared with $Si(OR)_4$ (silicon alkoxides).[81]

Chitin is the largest natural polysaccharide, and it can be extracted from crab shells. Chitosan is formed by using the deacetylated chitin polymer. Chitosan is a great adsorbent for the removal of water impurities. Different chitosan adsorbents were applied for the removal of pollutants such as toxic metals, organic compounds, and hydrocarbon substrates.[82,83] Graphene with preferable properties has been used in water purification systems. Graphene oxide can simplify the electrostatic interaction between adsorbents and adsorbate molecules. Graphene oxide is an attractive adsorbent in the adsorption process due to enhanced surface area, high stability, and physically flexible properties.[84,85]

Metal oxide adsorbents have considerable properties such as an active surface area, insolubility, and low produced secondary contaminants. In addition, the high adsorption capacity, facile synthesis, and ecofriendly and cost-effective properties of metal oxide nanoparticles mean these materials can be used to remove impurities from a water aqueous solution.[86–88]

CNTs play a substantial role in water purification systems due to their excellent surface area and numerous surface sites for adsorption. Hydrophobic interactions are done using CNTs. Therefore, CNTs are the most-used adsorbents in the elimination of impurities from water.

4.3 Nanomaterials in photocatalysis

Photocatalysis processes takes place under light irradiation in the presence of photocatalyst. Nanomaterials in photocatalysis were investigated as smart materials for water purification systems due to the excellent surface area and morphological properties. The nanoparticles have discrete quantum influences and morphological characteristics such as electric and optic

TABLE 3 Removal methods for water pollutants.

Pollutants	Nanomaterials	Methods	Refs.
RhB and MO dyes	Au NPs/TiO$_2$/Pt	Motors	90
Different dyes	Chitosan	Membrane	91
Methyl orange	Amine modified multiwalled carbon	Adsorption	92
Cyanide	TiO$_2$/Fe$_2$O$_3$/PAC	Photocatalysis	93
4-Nitrophenol	CoNi@Pt nanorods	Motors	94
Methyl orange	Chitosan/Al$_2$O$_3$/Fe$_3$O$_4$	Adsorption	95
Pb^{2+}	NF with poly-gamma-glutamic acid	Membrane	96
Acid red 88	CuO/ZnO	Photocatalysis	97
Methylene blue	Zero-valent-iron/platinum NPs	Motors	98
Congo red	Ferrous oxide/aluminum oxide	Adsorption	99
Different dyes	TiO$_2$/ZnTiO$_3$/α-Fe$_2$O$_3$	Photocatalysis	100
NH$_3$-N and PO$_4$-P	Polyamide-6/chitosan	Membrane	101
Polydiphenyl ethers	rGO-SiO$_2$-Pt	Motors	102
Rose bengal	Co doped nano TiO$_2$	Photocatalysis	103
Congo red	Silica with ferric oxide	Adsorption	104
Anthraquinone dyes	Polyamide NF	Membrane	105
Reactive red 120	Single-walled carbon nanotubes	Adsorption	106
Hg^{2+}	DNA-Au/Pt tubes	Motors	107
Methyl orange	ZnO/SnO$_2$	Photocatalysis	108
Methylene blue	Magnesium oxide/graphene oxide	Adsorption	109
Methylene blue	Bi$_2$WO$_6$	Photocatalysis	110
Methylene blue	Humic acid/ferric oxide	Adsorption	111
Cr(VI)	SnS$_2$/SnO$_2$	Photocatalysis	112
Cr(VI)	SnS$_2$/polyaniline/N-graphene oxide	Photocatalysis	113
Cr(VI)	SnS$_2$ and polyvinyl chloride	Photocatalysis	114

behaviors. Therefore, nanophotocatalysts can be developed due to the oxidizing ability by generating oxidizing species. They can degrade different compounds under light irradiation as the water treatment method, and water, carbon dioxide, and other inorganic ionic compounds are formed.[89] The removal methods of pollutants from water in previous studies are shown in Table 3.

5 Conclusion

Humans need clean potable water and hence require an efficient water treatment. Smart materials can be used as a material in pollution detection and removal from water sources. With their physicochemical and morphological properties, smart nanomaterials in water treatment were effective at the detection and removal of water pollutants. However, smart micromaterials are less efficient. Surface plasmon resonance with the colorimetric method is an excellent technique for the detection of water pollution due to the ultrasensitivity/selectivity properties in the determination process. Adsorption and photocatalysis systems via smart nanomaterials are the best methods for removal of contaminants with wide application for different pollutants. Based on previous studies, these methods are highly impressive, ecofriendly, and cost-efficient in the detection and removal of water contaminants.

References

1. Armenta S, Esteve-Turrillas FA, Garrigues S, de La Guardia M. Smart materials for sample preparation in bioanalysis: a green overview. *Sustain Chem Pharm*. 2021;21:100411. https://doi.org/10.1016/j.scp.2021.100411.

2. Ting SW, Ye F, Zhang C, Li H. Smart materials for point-of-care testing: from sample extraction to analyte sensing and readout signal generator. *Biosens Bioelectron*. 2020;170:112682. https://doi.org/10.1016/j.bios.2020.112682.

3. Sagdic K, Eş I, Sitti M, Inci F. Smart materials: rational design in biosystems via artificial intelligence. *Trends Biotechnol*. 2022. https://doi.org/10.1016/j.tibtech.2022.01.005.

4. Kumar MK, Rawat A, Jha G. Smart materials and electro-mechanical impedance technique: a review. *Mater Today Proc*. 2020;33:4993–5000. https://doi.org/10.1016/j.matpr.2020.02.831.

5. Olabi AG, Grunwald A. Design and application of magneto-rheological fluid. *Mater Des*. 2007;28:2658–2664. https://doi.org/10.1016/j.matdes.2006.10.009.

6. Gallego-Juarez JA. Piezoelectric ceramics and ultrasonic transducers. *J Phys E Sci Instr*. 1989;22:804. https://doi.org/10.1088/0022-3735/22/10/001.

7. ElFeninat F, Laroche G, Fiset M, Montovani D. Shape memory materials for biomedical applications. *Adv Eng Mater*. 2002;4:91. https://doi.org/10.1002/1527-2648(200203)4:3%3C91::AID-ADEM91%3E3.0.CO;2-B.

8. Buehler WJ, Wang FE. A summary of recent research on the nitinol alloys and their potential application in ocean engineering. *Ocean Eng*. 1968;1:105. https://doi.org/10.1016/0029-8018(68)90019-X.

9. Ratna D, Karger-Kocsis J. Recent advances in shape memory polymers and composites: a review. *J Mater Sci*. 2008;43:254. https://doi.org/10.1007/s10853-007-2176-7.

10. Mustapha KB, Mohamed MK. A review of fused deposition modelling for 3D printing of smart polymeric materials and composites. *Eur Polym J*. 2021;156:110591. https://doi.org/10.1016/j.eurpolymj.2021.110591.

11. Ding H, Zhang Q, Gu H, et al. Controlled microstructural architectures based on smart fabrication strategies. *Adv Funct Mater*. 2020;30(2):1901760. https://doi.org/10.1002/adfm.201901760.

12. Li Z, Yang F, Yin Y. Smart materials by nanoscale magnetic assembly. *Adv Funct Mater*. 2020;30(2):1903467. https://doi.org/10.1002/adfm.201903467.

13. Hwang H, Jeong U. Microparticle-based soft electronic devices: toward one-particle/one-pixel. *Adv Funct Mater*. 2020;30(2):1901810. https://doi.org/10.1002/adfm.201901810.

14. Yu Z, Lim ATO, Kollasch SL, Dong JH, Huang J. Oil-based self-healing barrier coatings: to flow and not to flow. *Adv Funct Mater*. 2020;30(2):1906273. https://doi.org/10.1002/adfm.201906273.

15. Kumar S, Kumari P. Flexible nano smart sensors. In: *Nanosensors for Smart Manufacturing (Micro and Nano Technologies)*; 2021:199–230. https://doi.org/10.1016/B978-0-12-823358-0.00002-2.

16. Chen T, Ferris R, Zhang J, Ducker R, Zauscher S. Stimulus-responsive polymer brushes on surfaces: transduction mechanisms and applications. *Prog Polym Sci*. 2010;35:94–112. https://doi.org/10.1016/j.progpolymsci.2009.11.004.

17. Yong LJ, Shin K, Woong KJ. Tailored layer-by-layer deposition of silica reinforced polyelectrolyte layers on polymer microcapsules for enhanced antioxidant cargo retention. *J Ind Eng Chem*. 2017;58:80–86. https://doi.org/10.1016/j.jiec.2017.09.010.

18. Motornov M, Roiter Y, Tokarev I, Minko S. Stimuli-responsive nanoparticles, nanogels and capsules for integrated multifunctional intelligent systems. *Prog Polym Sci*. 2010;35(1–2):174–211. https://doi.org/10.1016/j.progpolymsci.2009.10.004.

19. Falahati M, Ahmadvand P, Lin Y. Smart polymers and nanocomposites for 3D and 4D printing. *Mater Today*. 2020;40:215–245. https://doi.org/10.1016/j.mattod.2020.06.001.

20. Fakhri A, Rashidi S, Asif M, Tyagi I, Agarwal S, Gupta VK. Dynamic adsorption behavior and mechanism of Cefotaxime, Cefradine and Cefazolin antibiotics on CdS-MWCNT nanocomposites. *J Mol Liq*. 2016;215:269–275. https://doi.org/10.1016/j.molliq.2015.12.033.

21. Fakhri A, Behrouz S, Asif M, Tyagi I, Agarwal S, Gupta VK. Synthesis, structural and morphological characteristics of NiO nanoparticles co-doped with boron and nitrogen. *J Mol Liq*. 2016;213:326–331. https://doi.org/10.1016/j.molliq.2015.09.004.

22. Mahmoodian H, Moradi O, Shariatzadeh B, et al. Enhanced removal of methyl orange from aqueous solutions by poly HEMA–chitosan-MWCNT nano-composite. *J Mol Liq*. 2015;202:189–198. https://doi.org/10.1016/j.molliq.2014.10.040.

23. Gupta VK, Fakhri A, Agarwal S, Sadeghi N. Synthesis of MnO2/cellulose fiber nanocomposites for rapid adsorption of insecticide compound and optimization by response surface methodology. *Int J Biol Macromol*. 2017;102:840–846. https://doi.org/10.1016/j.ijbiomac.2017.04.075.

24. Pathania D, Verm C, Negi P, et al. Novel nanohydrogel based on itaconic acid grafted tragacanth gum for controlled release of ampicillin. *Carbohydr Polym*. 2018;196:262–271. https://doi.org/10.1016/j.carbpol.2018.05.040.

25. Thibbotuwawa N, Singh S, Tong GY. Proteoglycan and collagen contribution to the strain-rate-dependent mechanical behaviour of knee and shoulder cartilage. *J Mech Behav Biomed Mater*. 2021;124:104733. https://doi.org/10.1016/j.jmbbm.2021.104733.

26. Li X, Yu Y, Lian B. Molecular mechanism of increasing extracellular polysaccharide production of *Paenibacillus mucilaginosus* K02 by adding mineral powders. *Int Biodeter Biodegr*. 2022;167:105340. https://doi.org/10.1016/j.ibiod.2021.105340.

27. Zhang C, Xue J, Wang Q. From plant phenols to novel bio-based polymers. *Prog Polym Sci*. 2022;125:101473. https://doi.org/10.1016/j.progpolymsci.2021.101473.

28. Song Z, Song Y, Chen L. Chromatographic performance of zidovudine imprinted polymers coated silica stationary phases. *Talanta*. 2021;239:123115. https://doi.org/10.1016/j.talanta.2021.123115.

29. Song Y, Fukuzawa K, Azuma N. In-situ measurement of temporal changes in thickness of polymer adsorbed films from lubricant oil by vertical-objective-based ellipsometric microscopy. *Tribol Int*. 2021;165:107341. https://doi.org/10.1016/j.triboint.2021.107341.

30. Gupta VK, Tyagi I, Agarwal S, et al. Experimental study of surfaces of hydrogel polymers HEMA, HEMA–EEMA–MA, and PVA as adsorbent for removal of azo dyes from liquid phase. *J Mol Liq*. 2015;206:129–136. https://doi.org/10.1016/j.molliq.2015.02.015.

31. Agarwal S, Sadegh H, Monajjemi M, et al. Efficient removal of toxic bromothymol blue and methylene blue from wastewater by polyvinyl alcohol. *J Mol Liq*. 2016;218:191–197. https://doi.org/10.1016/j.molliq.2016.02.060.

32. Sharma D, Mustansar Hussain C. Smart nanomaterials in pharmaceutical analysis. *Arab J Chem*. 2020;13:3319–3343. https://doi.org/10.1016/j.arabjc.2018.11.007.

33. Mrinalini M, Prasanthkumar S. Recent advances on stimuli-responsive smart materials and their applications. *ChemPlusChem*. 2019;84(8):1103–1121. https://doi.org/10.1002/cplu.201900365.

34. Zhang Z, Yan B, Liao Y, Liu H. Nanoparticle: is it promising in capillary electrophoresis. *Anal Bioanal Chem*. 2008;391:925–927. https://doi.org/10.1007/s00216-008-1930-2.

35. Kumar N, Stephanidis B, Zenobi R, Wain AJ, Roy D. Nanoscale mapping of catalytic activity using tip-enhanced Raman spectroscopy. *Nanoscale*. 2015;7:7133–7137. https://doi.org/10.1039/C4NR07441F.

36. Álvarez-Puebla RA, Contreras-Cáceres R, Pastoriza-Santos I, Pérez-Juste J, Liz-Marzán LM. Au@pNIPAM colloids as molecular traps for surface-enhanced, spectroscopic, ultra-sensitive analysis. *Angew Chem*. 2009;48:138–143. https://doi.org/10.1002/anie.200804059.

37. Vertelov GK, Olenin AY, Lisichkin GV. Use of nanoparticles in the electrochemical analysis of biological samples. *J Anal Chem*. 2007;62:813–824. https://doi.org/10.1134/S106193480709002X.

38. Jotterand F. Nanomedicine: how it could reshape clinical practice. *Nanomedicine*. 2007;2:401–405. https://doi.org/10.2217/17435889.2.4.401.

39. Muhammad G, Ajaz HM, Umer AM, Tahir HM, Zajif HS, Hussain I. Polysaccharide based superabsorbent hydrogel from *Mimosa pudica*: swelling-deswelling and drug release. *RSC Adv*. 2016;6:23310–23317. https://doi.org/10.1039/C5RA23088H.

40. Thakkar S, Dumée LF, Yang W. Nano-enabled sensors for detection of arsenic in water. *Water Res*. 2021;188:116538. https://doi.org/10.1016/j.watres.2020.116538.

41. Paladiya C, Kianoosh Kiani A. Nano structured sensing surface: significance in sensor fabrication. *Sens Actuat B Chem*. 2018;268:494–511. https://doi.org/10.1016/j.snb.2018.04.085.

42. Brahmkhatri V, Pandit P, Kurkuri MD. Recent progress in detection of chemical and biological toxins in water using plasmonic nanosensors. *Trends Environ Anal Chem*. 2021;30:e00117. https://doi.org/10.1016/j.teac.2021.e00117.

43. Yan T, Wu Y, Pan Z. Recent progress on fabrication of carbon nanotube-based flexible conductive networks for resistive-type strain sensors. *Sens Actuator A*. 2021;327:112755. https://doi.org/10.1016/j.sna.2021.112755.

44. Sharma V, Choudhary S, Kumar V. Nanoparticles as fingermark sensors. *TrAC Trends Anal Chem*. 2021;143:116378. https://doi.org/10.1016/j.trac.2021.116378.

45. Comini E. Metal oxide nanowire chemical sensors: innovation and quality of life. *Mater Today*. 2016;19:559–567. https://doi.org/10.1016/j.mattod.2016.05.016.

46. Shetti NP, Mishra A, Aminabhavi TM. Versatile fullerenes as sensor materials. *Mater Today Chem*. 2021;20:100454. https://doi.org/10.1016/j.mtchem.2021.100454.

47. Brahman PK, Pandey N, Topkaya SN, Singhai R. Fullerene–C60–MWCNT composite film based ultrasensitive electrochemical sensing platform for the trace analysis of pyruvic acid in biological fluids. *Talanta*. 2015;134:554–559. https://doi.org/10.1016/j.talanta.2014.10.054.

48. Mehta VN, Rohit JV, Kailasa SK. Functionalization of silver nanoparticles with 5-sulfoanthranilic acid dithiocarbamate for selective colorimetric detection of Mn^{2+} and Cd^{2+} ions. *New J Chem*. 2016;40:4566. https://doi.org/10.1039/C5NJ03454J.

49. Song J, Wu F, Wan Y, Ma L. Colorimetric detection of melamine in pretreated milk using silver nanoparticles functionalized with sulfanilic acid. *Food Control*. 2015;50:356. https://doi.org/10.1016/j.foodcont.2014.08.049.

50. Boken J, Thatai S, Khurana P, Prasad S, Kumar D. Highly selective visual monitoring of hazardous fluoride ion in aqueous media using thiobarbituric-capped gold nanoparticles. *Talanta*. 2015;132:278. https://doi.org/10.1016/j.talanta.2014.08.043.

51. Schopf C, Martin A, Iacopino D. Plasmonic detection of mercury via amalgam formation on surface-immobilized single Au nanorods. *Sci Technol Adv Mater*. 2017;18:60. https://doi.org/10.1080/14686996.2016.1258293.

52. Makam P, Shilpa R, Kandjani AE, et al. SERS and fluorescence-based ultrasensitive detection of mercury in water. *Biosens Bioelectron*. 2018;100:556. https://doi.org/10.1016/j.bios.2017.09.051.

53. Fang W, Jia S, Chao J, et al. Silver nanoparticles/biphenol-biphenoquinone nanoribbons for ultr-trace voltammetric determination of Cr (VI), 5th. In: *International Conference on Researches in Science & Engineering & 2nd International Congress on Civil, Architecture and Urbanism in Asia Kasem Bundit University, Bangkok, 4506*; 2019. https://civilica.com/doc/1115898/.

54. Gan T, Lv Z, Deng Y, Sun J, Shi Z, Liu Y. Facile synthesis of monodisperse Ag@C@Ag core–double shell spheres for application in the simultaneous sensing of thymol and phenol. *New J Chem*. 2015;39:6244. https://doi.org/10.1039/C5NJ00881F.

55. Bala R, Sharma RK, Wangoo N. Development of gold nanoparticles-based aptasensor for the colorimetric detection of organophosphorus pesticide phorate. *Anal Bioanal Chem*. 2016;408:333. https://doi.org/10.1007/s00216-015-9085-4.

56. Pan F, Chen D, Zhuang X, et al. Fabrication of gold nanoparticles/l-cysteine functionalized graphene oxide nanocomposites and application for nitrite detection. *J Alloys Compd*. 2018;744:51. https://doi.org/10.1016/j.jallcom.2018.02.053.

57. Nath P, Priyadarshni N, Chanda N. Europium-coordinated gold nanoparticles on paper for the colorimetric detection of arsenic (III, V) in aqueous solution. *ACS Appl Nano Mater*. 2017;1:73. https://doi.org/10.1021/acsanm.7b00038.

58. Xu F, Lai H, Xu H. Gold nanocone arrays directly grown on nickel foam for improved SERS detection of aromatic dyes. *Anal Methods*. 2018;10(26):3170–3177. https://doi.org/10.1039/C8AY00840J.

59. Hakonen A, Chao WF, Ola AP, et al. Hand-held femtogram detection of hazardous picric acid with hydrophobic Ag nanopillar SERS substrates and mechanism of elasto-capillarity. *ACS Sens.* 2017;2:198–202. https://doi.org/10.1021/acssensors.6b00749.

60. Phoonsawat K, Ratnarathorn N, Henry CS, Dungchai W. A distance-based paper sensor for the determination of chloride ions using silver nanoparticles. *Analyst.* 2018;143:3867. https://doi.org/10.1039/C8AN00670A.

61. Mohammadi S, Khayatian G. Colorimetric detection of Bi (III) in water and drug samples using pyridine-2,6-dicarboxylic acid modified silver nanoparticles. *Spectrochim Acta A Mol Biomol Spectrosc.* 2015;148:405. https://doi.org/10.1016/j.saa.2015.03.127.

62. Zhang Y, Shen HY, Hai X, Chen XW, Wang JH. Polyhedral oligomeric silsesquioxane polymer-caged silver nanoparticle as a smart colorimetric probe for the detection of hydrogen sulfide. *Anal Chem.* 2017;89:1346. https://doi.org/10.1021/acs.analchem.6b04407.

63. Singh S, Meena VK, Mizaikoff B, Singh SP, Suri CR. Electrochemical sensing of nitro-aromatic explosive compounds using silver nanoparticles modified electrochips. *Anal Methods.* 2016;8:7158. https://doi.org/10.1039/C6AY01945E.

64. Manjumeena R, Duraibabu D, Thangavelu Rajamuthuramalingam RV, Kalaichelvan PT. Highly responsive glutathione functionalized green AuNP probe for precise colorimetric detection of Cd^{2+} contamination in the environment. *RSC Adv.* 2015;5:69124. https://doi.org/10.1039/C5RA12427A.

65. Annadhasan M, Muthukumarasamyvel T, Sankar Babu VR, Rajendiran N. Green synthesized silver and gold nanoparticles for colorimetric detection of Hg^{2+}, Pb^{2+}, and Mn2+ in aqueous medium. *ACS Sustain Chem Eng.* 2014;2:887. https://doi.org/10.1021/sc400500z.

66. Zhang N, Qiao S, Gupta VK. Sustainable nano-composites polyglutamic acid functionalized $Ag/g-C_3N_4/SiC$ for the ultrasensitive colorimetric assay, visible light irradiated photocatalysis and antibacterial efficiency. *Opt Mater.* 2021;120:111452. https://doi.org/10.1016/j.optmat.2021.111452.

67. Yang Y, Aqeel AM, Zhang D. Facile synthesis of gold-silver/copper sulfide nanoparticles for the selective/sensitive detection of chromium, photochemical and bactericidal application. *Spectrochim Acta A Mol Biomol Spectrosc.* 2021;249:119324. https://doi.org/10.1016/j.saa.2020.119324.

68. Eskandari L, Andalib F, Gupta VK. Facile colorimetric detection of Hg (II), photocatalytic and antibacterial efficiency based on silver-manganese disulfide/polyvinyl alcohol-chitosan nanocomposites. *Int J Biol Macromol.* 2020;164:4138–4145. https://doi.org/10.1016/j.ijbiomac.2020.09.015.

69. Bahadoran A, Najafizadeh M, Gupta VK. Co-doping silver and iron on graphitic carbon nitride-carrageenan nanocomposite for the photocatalytic process, rapidly colorimetric detection and antibacterial properties. *Surf Interfaces.* 2021;26:101279. https://doi.org/10.1016/j.surfin.2021.101279.

70. Verma M, Tyagi I, Kumar V, Goel S, Vaya D, Kim H. Fabrication of GO–MnO2 nanocomposite using hydrothermal process for cationic and anionic dyes adsorption: kinetics, isotherm, and reusability. *J Environ Chem Eng.* 2021;9:106045. https://doi.org/10.1016/j.jece.2021.106045.

71. Agarwal S, Sadeghi N, Tyagi I, Gupta VK, Fakhri A. Adsorption of toxic carbamate pesticide oxamyl from liquid phase by newly synthesized and characterized graphene quantum dots nanomaterials. *J Colloid Interface Sci.* 2016;478:430–438. https://doi.org/10.1016/j.jcis.2016.06.029.

72. Agarwal S, Tyagi I, Gupta VK, Ghasemi N, Shahivand M, Ghasemi M. Kinetics, equilibrium studies and thermodynamics of methylene blue adsorption on Ephedra strobilacea saw dust and modified using phosphoric acid and zinc chloride. *J Mol Liq.* 2016;218:208–218. https://doi.org/10.1016/j.molliq.2016.02.073.

73. Mashhadi S, Sohrabi R, Javadian H, et al. Rapid removal of Hg (II) from aqueous solution by rice straw activated carbon prepared by microwave-assisted H2SO4 activation: kinetic, isotherm and thermodynamic studies. *J Mol Liq.* 2016;215:144–153. https://doi.org/10.1016/j.molliq.2015.12.040.

74. Agarwal S, Tyagi I, Gupta VK, Fakhri A, Shahidi S. Sonocatalytic, sonophotocatalytic and photocatalytic degradation of morphine using molybdenum trioxide and molybdenum disulfide nanoparticles photocatalyst. *J Mol Liq.* 2017;225:95–100. https://doi.org/10.1016/j.molliq.2016.11.029.

75. Agarwal S, Tyagi I, Gupta VK, et al. Iron doped SnO2/Co3O4 nanocomposites synthesized by sol-gel and precipitation method for metronidazole antibiotic degradation. *Mater Sci Eng C.* 2017;70:178–183. https://doi.org/10.1016/j.msec.2016.08.062.

76. Fakhri A, Behrouz S, Tyagi I, Agarwal S, Gupta VK. Synthesis and characterization of ZrO2 and carbon-doped ZrO2 nanoparticles for photocatalytic application. *J Mol Liq.* 2016;216:342–346. https://doi.org/10.1016/j.molliq.2016.01.046.

77. Lyu Q, Peng L, Zhao J. Smart nano-micro platforms for ophthalmological applications: the state-of-the-art and future perspectives. *Biomaterials.* 2021;270:120682. https://doi.org/10.1016/j.biomaterials.2021.120682.

78. Ahmad KA, Ali MH, Kim J. Metal oxide and carbon nanomaterial based membranes for reverse osmosis and membrane distillation: a comparative review. *Environ Res.* 2021;202:111716. https://doi.org/10.1016/j.envres.2021.111716.

79. Jurado-Sa'nchez B, Wang J. Micromotors for environmental applications: a review. *Environ Sci Nano.* 2018;5:1530–1544. https://doi.org/10.1039/C8EN00299A.

80. Awad AM, Jalab R, Mohammad AW. Adsorption of organic pollutants by nanomaterial-based adsorbents: an overview. *J Mol Liq.* 2020;301:112335. https://doi.org/10.1016/j.molliq.2019.112335.

81. Fakhri A, Rashidi S, Tyagi I, Agarwal S, Gupta VK. Photodegradation of Erythromycin antibiotic by γ-Fe2O3/SiO2 nanocomposite: response surface methodology modeling and optimization. *J Mol Liq.* 2016;214:378–383. https://doi.org/10.1016/j.molliq.2015.11.037.

82. Omer AM, Dey R, Ziora ZM. Insights into recent advances of chitosan-based adsorbents for sustainable removal of heavy metals and anions. *Arab J Chem.* 2022;15:103543. https://doi.org/10.1016/j.arabjc.2021.103543.

83. Eltaweil AS, Omer AM, Abd El-Monaem EM. Chitosan based adsorbents for the removal of phosphate and nitrate: a critical review. *Carbohydr Polym.* 2021;274:118671. https://doi.org/10.1016/j.carbpol.2021.118671.

84. Najafi F, Moradi O, Rajabi M, et al. Thermodynamics of the adsorption of nickel ions from aqueous phase using graphene oxide and glycine functionalized graphene oxide. *J Mol Liq.* 2015;208:106–113. https://doi.org/10.1016/j.molliq.2015.04.033.

85. Fakhri A, Salehpour KD. Synthesis and characterization of MnS2/reduced graphene oxide nanohybrids for with photocatalytic and antibacterial activity. *J Photochem Photobiol B Biol.* 2017;166:259–263. https://doi.org/10.1016/j.jphotobiol.2016.12.017.

86. Ghaedi M, Roosta M, Ghaedi AM, et al. Removal of methylene blue by silver nanoparticles loaded on activated carbon by an ultrasound-assisted device: optimization by experimental design methodology. *Res Chem Intermed.* 2018;44:2929–2950. https://doi.org/10.1007/s11164-015-2285-x.

87. Verma M, Tyagi I, Chandra R, Gupta VK. Adsorptive removal of Pb (II) ions from aqueous solution using CuO nanoparticles synthesized by sputtering method. *J Mol Liq.* 2017;225:936–944. https://doi.org/10.1016/j.molliq.2016.04.045.

88. Dehghani MH, Mahdavi P, Tyagi I, Agarwal S, Gupta VK. Investigating the toxicity of acid dyes from textile effluent under UV/ZnO process using Daphnia magna. *Desalin Water Treat.* 2016;57:24359–24367. https://doi.org/10.1080/19443994.2016.1141327.

89. Leary R, Westwood A. Carbonaceous nanomaterials for the enhancement of TiO_2 photocatalysis. *Carbon.* 2011;49:741–772. https://doi.org/10.1016/j.carbon.2010.10.010.

90. Zhang L, Zhang G, Wang S, Peng J, Cui W. Cation functionalized silica nanoparticle as an adsorbent to selectively adsorb anionic dye from aqueous solutions. *Environ Prog Sustain Energy.* 2016;35:1070–1077. https://doi.org/10.1002/ep.12333.

91. Long Q, Zhang Z, Qi G, Wang Z, Chen Y, Liu ZQ. Fabrication of chitosan nano filtration membranes by the film casting strategy for effective removal of dyes/salts in textile wastewater. *ACS Sustain Chem Eng.* 2020;8:2512–2522. https://doi.org/10.1021/acssuschemeng.9b07026.

92. Sadegh H, Ali GAM, Agarwal S, Gupta VK. Surface modification of MWCNTs with carboxylic-to-amine and their superb adsorption performance. *Int J Environ Res.* 2019;13:523–531. https://doi.org/10.1007/s41742-019-00193-w.

93. Eskandari P, Farhadian M, Nazar ARS, Jeon BH. Adsorption and photodegradation efficiency of TiO_2/Fe_2O_3/PAC and TiO_2/Fe_2O_3/zeolite nano-photocatalysts for the removal of cyanide. *Ind Eng Chem Res.* 2019;58:2099–2112. https://doi.org/10.1021/acs.iecr.8b05073.

94. Garcia-Torres J, Serra A, Tierno P, Alcobe X, Valles E. Magnetic propulsion of recyclable catalytic nanocleaners for pollutant degradation. *ACS Appl Mater Interfaces.* 2017;9:23859–23868. https://doi.org/10.1021/acsami.7b07480.

95. Tanhaei B, Ayati A, Lahtinen M, Sillanpaa M. Preparation and characterization of a novel chitosan/Al2O3/magnetite nanoparticles composite adsorbent for kinetic, thermodynamic and isotherm studies of Methyl Orange adsorption. *Chem Eng J.* 2015;259:1–10. https://doi.org/10.1016/j.cej.2014.07.109.

96. Hajdu I, Bodnar M, Csikos Z, et al. Combined nano-membrane technology for removal of lead ions. *J Membr Sci.* 2012;409:44–53. https://doi.org/10.1016/j.memsci.2012.03.011.

97. Sathishkumar P, Sweena R, Wu JJ, Anandan S. Synthesis of CuO–ZnO nanophotocatalyst for visible light assisted degradation of a textile dye in aqueous solution. *Chem Eng J.* 2011;171:136–140. https://doi.org/10.1016/j.cej.2011.03.074.

98. Lee CS, Gong J, Oh DS, Jeon JR, Chang YS. Zero valent iron/platinum Janus micromotors with spatially separated functionalities for efficient water decontamination. *ACS Appl Nano Mater.* 2018;1:768–776. https://doi.org/10.1021/acsanm.7b00223.

99. Mahapatra A, Mishra BG, Hota G. Adsorptive removal of Congo red dye from wastewater by mixed iron oxide–alumina nanocomposites. *Ceram Int.* 2013;39:5443–5451. https://doi.org/10.1016/j.ceramint.2012.12.052.

100. Mehrabi M, Javanbakht V. Photocatalytic degradation of cationic and anionic dyes by a novel nanophotocatalyst of $TiO_2/ZnTiO_3/\alpha\text{-}Fe_2O_3$ by ultraviolet light irradiation. *J Mater Sci Mater Electron.* 2018;29:9908–9919. https://doi.org/10.1007/s10854-018-9033-0.

101. Ghani M, Gharehaghaji AA, Arami M, Takhtkuse N, Rezaei B. Fabrication of electrospun polyamide-6/chitosan nanofibrous membrane toward anionic dyes removal. *J Nanotechnol.* 2014. https://doi.org/10.1155/2014/278418. 278418.

102. Orozco J, Mercante LA, Pol R, Merkoci A. Graphene based Janus micromotors for the dynamic removal of pollutants. *J Mater Chem A.* 2016;4:3371–3378. https://doi.org/10.1039/C5TA09850E.

103. Malini B, Allen Gnana Raj G. Synthesis, characterization and photocatalytic activity of cobalt doped TiO_2 nano photocatalysts for rose bengal dye degradation under day light illumination. *Chem Sci Trans.* 2018;7:687–695.

104. Wang P, Wang X, Yu S, et al. Silica coated Fe_3O_4 magnetic nanospheres for high removal of organic pollutants from wastewater. *Chem Eng J.* 2016;306:280–288. https://doi.org/10.1016/j.cej.2016.07.068.

105. Askari N, Farhadian M, Razmjou A, Hashtroodi H. Nanofiltration performance in the removal of dye from binary mixtures containing anthraquinone dyes. *Desalin Water Treat.* 2016;57:18194–18201. https://doi.org/10.1080/19443994.2015.1090917.

106. Bazrafshan E, Mostafapour FK, Hosseini AR, Raksh KA, Mahvi AH. Decolorisation of reactive red 120 dye by using single-walled carbon nanotubes in aqueous solutions. *J Chem.* 2013;938374. https://doi.org/10.1155/2013/938374.

107. Wang H, Khezri B, Pumera M. Catalytic DNA-functionalized self-propelled micromachines for environmental remediation. *Chem.* 2016;1:473–481. https://doi.org/10.1016/j.chempr.2016.08.009.

108. Ali W, Ullah H, Zada A, Alamgir MK, Ahmad WMMJ, Nadhman A. Effect of calcination temperature on the photoactivities of ZnO/SnO_2 nanocomposites for the degradation of methyl orange. *Mater Chem Phys.* 2018;213:259–266. https://doi.org/10.1016/j.matchemphys.2018.04.015.

109. Heidarizad M, Sengor SS. Synthesis of graphene oxide/magnesium oxide nanocomposites with high-rate adsorption of methylene blue. *J Mol Liq.* 2016;224:607–617. https://doi.org/10.1016/j.molliq.2016.09.049.

110. Singh VP, Sharma M, Vaish R. Enhanced dye adsorption and rapid photo catalysis in candle soot coated Bi_2WO_6 ceramics. *Eng Res Express.* 2020;1:025056. https://doi.org/10.1088/2631-8695/ab5e93.

111. Zhang X, Zhang P, Wu Z, Zhang L, Zeng G, Zhou C. Adsorption of methylene blue onto humic acid-coated Fe_3O_4 nanoparticles. *Colloids Surf A.* 2013;435:85–90. https://doi.org/10.1016/j.colsurfa.2012.12.056.

112. Zhang YC, Yao L, Zhang G, Dionysiou DD, Li J, Du X. One-step hydrothermal synthesis of high-performance visible-light-driven SnS_2/SnO_2 nanoheterojunction photocatalyst for the reduction of aqueous Cr(VI). *Appl Catal B.* 2014;144:730–738. https://doi.org/10.1016/j.apcath.2013.08.006.

113. Zhang F, Zhang Y, Zhang G, Yang Z, Dionysiou DD, Zhu A. Exceptional synergistic enhancement of the photocatalytic activity of SnS_2 by coupling with polyaniline and N-doped reduced graphene oxide. *Appl Catal B.* 2018;236:53–63. https://doi.org/10.1016/j.apcatb.2018.05.002.

114. Zhang Y, Zhang F, Yang Z, Xue H, Dionysiou DD. Development of a new efficient visible-light-driven photocatalyst from SnS_2 and polyvinyl chloride. *J Catal.* 2016;344:692–700. https://doi.org/10.1016/j.jcat.2016.10.022.

Chapter 21

Metal-organic frameworks for detection and adsorptive removal of pesticides

Partha Dutta[a], Charu Arora[a,*], Sanju Soni[a], Nidhi Rai[a], and Jyoti Mittal[b]

[a]Department of Chemistry, Guru Ghasidas University, Bilaspur, Chhattisgarh, India, [b]Department of Chemistry, Maulana Azad National Institute of Technology, Bhopal, Madhya Pradesh, India

*Corresponding author.

1 Introduction

With the rapid explosion of the population around the globe, there is certainly an increasing demand for food to an unprecedented level. This has sky-rocketed the requirement of agricultural crop production, but in contrast, the fertility of soils of every region, as well as the variety of crop, often does not qualify the standard to meet such demand by depending on classical methodologies of cultivation. There occur situations where the necessity of additives primarily enhances the production in manifold simply by assisting in the removal of barriers in the process.

Among these, two of the barriers are pests and insects which interfere with the growth of the plants by damaging them at various stages starting from seeds to mature. While some plant species are resistant to pests, while most of them fall prey to its devastating outcome destroying the crop yield by disturbing its production. To overcome this effectively, the use of some varieties of chemicals commercially referred to as pesticides and insecticides are applied in regulated quantities. These chemicals function as a growth inhibitor for their target, i.e., the pests, by blocking and destroying their physiological process. These pesticides are synthesized with the intention that they do not affect the plant itself.

Intensive irrigation results in washing away of the residual chemicals as well as their by-products (due to environmental degradation) which are accumulated[1,2] in the soil thereby disrupting the life cycle of the essential microorganisms that assist in maintaining soil fertility. Some of the residues are even non-biodegradable to such an extent that they even get into the food chain thereby causing serious health hazards to higher mammals and human beings.[3,4] So, the prevention of this disaster at present is a major challenge and it needed to be done in such a way that it affects minimally the soil health. One of the promising approaches emerging in the recent development of chemistry and allied branches is the usage of "porous solid materials"[5] specifically employed with the idea of adsorption[6] due to the presence of a very high surface to mass ratio.[7] The designing of microporous and mesoporous[8] materials with pore diameter < 2 nm and 2–50 nm, respectively, for achieving uniform pore size distribution is a field that is receiving interest.[8,9]

2 Structure

Metal-organic frameworks (MOFs) are supra-molecular solid materials consisting of a huge number of organic linkers and inorganic joints, all bounded to metal ions in a very organized fashion. The highly regular framework of these types of compounds via "self-assembly coordination polymers"[10–12] from metal ions as nodes imparts a greater surface area. There also occurs an increase in enthalpy[13] of adsorption for MOFs having an open metal site, whereas adsorption in liquid solution is more complex[14–16] in contrast to gases. The causes for this may be attributed to the factors such as host and guest regarding their polarity, composition, solubility, concentration (adsorptive), and solvent contribution. Porous crystals are useful because of the allowance of molecules through their pore apertures that may be kept for storage, separation, or conversion. The apertures control the size of molecules that may enter the pore.

A glance at the linkers, i.e., the metal nodes give us the choice of the 1st-row transition metal ions[17] especially Cr^{3+}, Fe^{3+}, Cu^{2+}, and Zn^{2+} are very common. Some alternate choices are alkali, alkaline-earth, and rare-earth metal ions. The principal characteristics considered in the choice of metal connectors are the coordination numbers and coordination geometries; the number is ranging from 2 to 7 depending on the nature of the metal (size charge, i.e., oxidation number) resulting from various

Sustainable Materials for Sensing and Remediation of Noxious Pollutants. https://doi.org/10.1016/B978-0-323-99425-5.00019-0

structures whereas the choice of organic linkers containing coordinate groups such as sulphonate, nitriles, etc. The Secondary Building Units[18,19] have intrinsic geometries, having a role in deciding the topology of the metal-organic framework.[20] The ligands having carboxylate groups can form (M-O-C) clusters which are termed as "Secondary Building Units." These Secondary Building Units that are joined together by the linkers may serve as the connecting points resulting in the formation of MOF as a network and are generally quite rigid.[21] The carboxylate ligands lock the metal ions into their positions, giving high stability to the structure as a whole.

3 Synthesis

MOFs are regarded to evolve from the field of coordination chemistry[22] which is primarily composed of central metal ions being surrounded by organic linkers or functionalized organic molecules whose synthetic methodologies[23] that are usually practiced need solvent assistance. This practice was usually a preferred approach in the initial phase of the development of MOF. The principal focus behind different synthetic strategies is to get a single crystal.[24]

With some additional strategies and improvisation in synthetic routes (Fig. 1) for their preparation, the various procedures can be listed under the following categories.

3.1 Conventional (solvothermal) synthesis

This strategy of synthesis is analogous to the conventional[25] methodology of dissolving the raw materials using a common solvent or a mixture of solvents in a vessel and a vigorous stirring followed by solvent removal via drying at room or increased temperatures.[22] The removal of solvent at room temperature requires more time and the overall synthesis becomes slower as compared to synthesis at elevated temperatures which is generally performed[26] at higher than the boiling points of the solvent under autogenous pressure followed by slow cooling. Many reactants undergo changes in morphologies in the nanoscale under these conditions.[27] The primary advantage of these setups is the attainment of desired crystal sizes probably due to slow crystallization. Quite a few MOFs have been prepared under normal conditions by just mixing the starting materials.[22,28]

FIG. 1 Methods of preparation of metal organic frameworks.

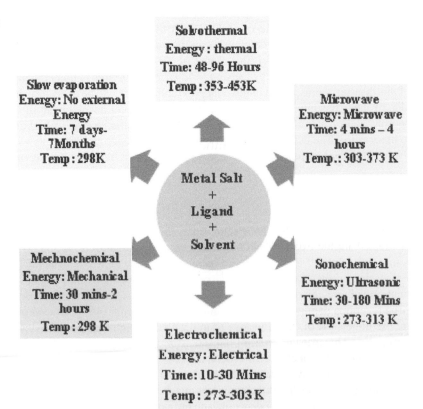

3.2 Microwave-assisted synthesis

The use of microwave region of the electromagnetic spectrum for the introduction of rotational energy in molecules having permanent dipole moment is frequently practiced in synthetic organic chemistry following a well-established protocol; because uniform and homogenous heating are achieved in this case, it is an energy-efficient method, and additionally, this procedure renders faster[29] nucleations of smaller crystals; hence it has better control on the morphology of the MOF synthesis. Microwave-assisted synthesis of metal-organic frameworks has been reported by several researchers.[24,30] Examples include Cr-MIL-100, Cr-MIL-101, Fe-MIL-53, Fe-MIL-101, IRMOF-1, and HKUST-1.[22]

3.3 Sonochemical synthesi

This method utilizes sound energy[31] as the means of excitation of reactants and solvent molecules for the formation of an organized structure of network solid like MOF. The reactant along with the solvent is taken in a Pyrex reactor associated with a sonicating device. The bubbles are formed at very high temperatures and pressure as a result of sonication. This leads to the formation of fine crystals in significantly less time. Examples include HKUST-1, MOF-5,[32] and Mg-MOF-74.[27]

3.4 Electrochemical synthesis

This strategy of synthesis[33] uses metal ions being continuously supplied to the solution. The conduction of electricity causes anode that caused its dissolution, which acts as a metal source, and thus the use of metal salts is avoided. The freshly introduced metal ions and the dissolved linker molecules react among themselves in the reaction medium of conducting salt. The use of protic solvent prevents the deposition of cathode material, but in the process, H_2 is generated. The electrochemical process is favorable for continuity and chances to get a high solid content in comparison to conventional processes. Notable examples are MOF-5,[34] [Zn(1,3-bdc)$_{0.5}$(bzim)], and Al-MIL-53-NH$_2$.[27]

3.5 Mechano-chemical synthesis

The use of mechanical energy to break chemical bonds and subsequent induction of chemical reaction is commonly termed as mechano-chemistry and is a quite interesting branch in modern synthetic chemistry.[35–37] One of the prime advantages of the process is its effectiveness at room temperature and solvent-free conditions, which is why the route is particularly preferred due to environmental issues regarding the regulation of organic solvents. Most of the experiments use metal oxides as reactants and water is produced as a by-product. This route of synthesis has a shorter reaction time with a quantitative yield of smaller crystals. But sometimes, a small amount of solvents is added to assist the grinding of reactants, commonly termed as liquid-assisted grinding (LAG) which helps in the mobility of the reactants down to a microscopic level as well as acts as an agent which decides the structure. However, mechanochemical synthesis is restricted to the synthesis of specific types of MOF and is not favorable for scale-up synthesis. Common examples are being, ZIF-4, ZIF-8, MOF-14, and HKUST-1.[22]

3.6 Dry-gel conversion MOF synthesis

This synthetic strategy is widely practiced in the preparation of zeolite membrane in which a dry aluminosilicate gel crystallizes upon contact with water vapors and volatile amine. The protocol has the advantage of minimization of waste disposal and reduction of reaction volume with an added benefit of a high yield of crystalline zeolites due to the complete conversion of gel. Hence the extension of the idea to achieve MOF synthesis[38,39] is not any surprise. ZIF-8 And Fe-MIL-100 are the two recent examples.[27]

4 Properties

The two chief components responsible for the structural integrity of metal-organic frameworks are: (i) the metal ions that act as nodes and (ii) the organic linkers that make up the mesh network. By varying these two factors, one can achieve interesting properties in a MOF[40] sample. The extensive development in recent times in the field made it possible to cater to fine-tuning in properties associated with these materials, which includes from detection of toxic gases to their adsorption and its chemical conversion to desired species. The principal property supporting the idea of MOF being used as the day-to-day application is its high porosity and organized network. According to Michael Berger Founder of "Nanowerk LLC"—"The physicochemical properties of materials are governed by the synergistic effects of structures and compositions, and

MOFs are fascinating examples of how the unique structure of hollow-structured materials can provide a whole raft of advantageous features. Among them are enhanced surface-to-volume ratio; low density; micro-reactor environment; higher loading capacities; and reduced transmission lengths of mass and charge." The common properties of MOF based on its application can be summarized under the following categories.

4.1 Gas sensors

The sensing and detection of gases in trace or bulk quantities electronically require high-end fabrication as well as delicate functioning of the sensor materials which renders a high cost of operation of the whole setup. The use of metal ion nodes in MOFs as the centers for molecular/ionic species recognition imparts the capability as a gas sensor. It can be very precisely tuned by altering the chemical composition of the MOF, which provides the desired pore size, where specific molecules can enter or leave the surface of a sensor device. NH_2-MIL-101(Al)@ZIF-8 is useful for the detection of Cu (II) very sensitively.[41,42]

4.2 Gas storage

Exceptional porosity for a minimum weight of these materials imparts them a high surface-to-volume ratio—a necessary parameter that highly favors adsorption[43] of molecules. This uniqueness on a bulk scale can reversibly bind gas molecules within them only to be removed on being heated to a slightly higher temperature.[44] The idea is also employed to remove undesired gases from a mixture and is thus considered a potential candidate for gas purification. These include carbon capture via CO_2 sequestration as well as CH_4 capture and removal of greenhouse gases. MFM-300(Al) can selectively store ammonia and filter out nitrogen dioxide gas from their mixture.

4.3 Drug delivery

The selective adsorption property of these materials also finds its use in the pharmaceutical industry where they are used as potential candidates for drug delivery. The release of the pharmaceutically active species, which had been previously adsorbed on to them, delivers to their targets. The entire operation is achieved via temperature, pressure, magnetic field, pH, and ion responses as a trigger.

4.4 Energy storage devices

There is continuous research undergoing in the field of batteries industry for a suitable electrolyte to reduce the issue of flammability in Li-ion batteries. The quest for the solid electrolyte has been fulfilled, whereas the storage capacity is little affected, all by the introduction of MOF in this field. Also, super-capacitor technology requires high conductive electrode materials with enhanced surface area. MOFs made up of metals such as Ni, Co, Cu, Mn, Cr, Zr, Fe, and Zn in combination with organic ligands have been investigated for such applications. The specific capacitance of MOF-based materials ranges from 200 to 1600 F/g.

4.5 Semiconductor and photovoltaics

It is possible to design a thin film of metal-organic framework that is electrically conductive and suitable to be used in electronic material. The substance preferentially used in a photovoltaic cell should have the ability to generate electrons upon irradiation by a suitable wavelength of electromagnetic radiation. The added advantage in MOF in controlling the chemical properties via altering composition helps in monitoring of bandgap to get enhanced performance in semiconductor devices. Due to the presence of a large effective bandgap, metal-organic frameworks exhibit low efficiency of light energy.[45] HKUST-1 was successfully demonstrated by a team of researchers to exhibit high conducting properties comparable to bronze[46] $(Fe_3(THT)_2(NH_4)_3)$.

4.6 Heterogeneous catalysis and electrocatalyst

High porosity, different types of metal ions, and organic linkers are responsible for catalytic properties where surface adsorption is an unavoidable phenomenon. Metal-organic frameworks proved more effective[47–49] in the field of catalysis when compared to traditional zeolites, as the latter have fixed tetrahedral coordination of Si/Al and oxygen linkage resulting in a very small pore diameter that rarely exceeds more than 1 nm as well as harsh reaction condition employed for their synthesis. On the other hand, the mild synthetic strategies followed in MOF synthesis do not disrupt the sensitive functional group being incorporated into the framework. The large surface area consisting of redox-active metal nodes makes these materials promising candidates for electrodes functioning as electrocatalysts.[50]

These catalytic processes can take place via two routes: One is the use of metal-organic frameworks using carbon as support material[51,52] and the second is a direct application of MOFs.[53,54] Some researchers have reported the problem of instability of metal-organic frameworks at reaction conditions.[55] Sometimes metal-organic frameworks are taken as pre-catalyst in the reaction mixture and their conversion to real catalysts takes place during the process of electrocatalysis.[56] So, metal-organic frameworks can be used as electrocatalysts using in situ methods with electrocatalysis. We can improve the conductance of electrodes by incorporating carbon material into metal-organic frameworks.[57] In some cases, metal-organic frameworks can act as self-sacrificing material to produce nanostructured material for electrodes.[58]

4.7 Heavy metal ion and nuclear waste removal

Many MOFs have been synthesized which shows promising ability in the removal of heavy metal ions like lead and mercury directly from contaminated water samples within very shorter times. Likewise, there are reports on the synthesis of MOF, where the metal nodes are deliberately taken as actinides; this strategy can be extended to the removal of nuclear waste.[59] Also, another approach of exchange of metal ions with that of the radio-nuclides has been attempted of course by the use of porosity to trap the nuclear wastes that always remains a viable option. The radioactive organic iodides (a potential hazard) are particularly difficult to capture. Chemical modification with reactive nitrogen at the binding site of metal-organic frameworks makes it capable of binding to organic iodides[60] in an eco-friendly manner.

5 MOF as potential sensor agents

A sensor is a system or assembly of systems that detect changes in the parameter of an environment around it by monitoring the change in the measurable physiochemical property as a stimulus or a signal and conveying that to a more integrated system for processing or even manual analysis. There must be some characteristics necessary for a sensor for its superior function: (a) It should be sensitive only to the property to be detected and (b) it should not affect the property to be measured.[44] Cross-reactivity is a common drawback of all chemical sensors.[61] There is a limited study conducted in this area earlier due to efficient signal transduction due to the insulating behavior of MOFs.[62]

One of the key necessities in designing sensor materials is the concept of molecular recognition, which is best described as interactions between molecules, generally, non-covalent. The shape of the molecules and the type of their surface can be considered as a parameter for information/instruction delivery to them. They can be instructed to assemble themselves into the more complex structure and how to fill a space in a given space. The synthetic strategy is the pivotal key to the coding of the instructions and in each step of the synthesis, this instruction is encoded. Self-assembly indicates the difference between "self" and "non-self," by recognizing and selecting during the process. Different self-assemblies can coexist in self-sorted form or miscegenation. This hybridization takes place if the preferred interaction between the components of different assemblies takes place.[63]

Since the onset of the phenomenon of molecular recognition depends on poor interaction between substrate and receptors, it is very challenging to design such systems that can mimic biological systems which discriminate different substrates.

Metal-organic frameworks containing functional pores are capable of a wide range of molecular interactions[64,65] mentioned in the previous paragraphs. Additionally, these materials are derived from efficient methods which allow rationally tuning the pores to size exclusivity of the desired scale. So, they have all the possible requirements of a sieve of varied molecular dimensions and hence it can sort out different small molecules.[66] The coordination bond energies are also in a moderate range and combined with the flexible nature of coordination geometries provide additional control on their properties. Temperature and pressure are some stimuli that can be used to manipulate externally.

The designing of cubic nets from paddle-wheel structures consisting of linkers such as dicarboxylic[51] acid and pillar bidentate ones are selective of interest. The introduction of $M_2(CO_2)_4$ clusters leads to a 2D sheet-like $M_2(R(COO)_2)_2$ where the metal ion M is predominantly Cu^{2+}, Zn^{2+}, and Co^{2+} and R is the bridging fragment. When these two-dimensional sheets are further pillared using bridging linker L, the resulting structure forms a highly robust microporous pillar with three-dimensional cubic-type structures.

The paddle-wheel cluster derived from Cu as metal ion and fumarate (FMA) as linker form three-dimensional frameworks. The pillaring is furnished using linkers Pyz, 4,4′-Bipy, etc. The resulting composition of the metal-organic frameworks is Cu (FMA)(Pyz)$_{0.5}$, Cu(FMA)(4,4'-Bipy)$_{0.5}$,0.25H$_2$O as well as Cu(FMA)(4,4'-Bpe)$_{0.5}$·0.5H$_2$O. The pore size of the later material is less than that of a CO and N$_2$ molecule but greater than H$_2$ which is why it can differentiate and selectively take up higher amounts of H$_2$.

The petrochemical industry entirely functions on the procedure of purification of crude/raw material to deliver a wide range of high purity and well-furnished substances which are generally hydrocarbon in chemical nature. This process of purification requires the isolation of linear hydrocarbon from branched one to achieve the desired octane rating. The almost absence of dipole moment in these molecules makes it very a difficult candidate for molecular recognition since a substantial amount of van der Waals' interaction is necessary for recognition. Therefore, a successful attempt has been achieved using the cubic net and double framework resulting in a metal-organic framework that has a composition Zn(BDC)(4,4'-Bipy)$_{0.5}$.(DMF).(H$_2$O)$_{0.5}$ where BDC is 1,4-benzene dicarboxylate and DMF is N,N-dimethylformamide. The above-mentioned synthesized material has one-dimensional micropores of about 4.0 Å that is capable of discrimination between linear and branched hydrocarbon. When the gas mixture is passed through a gas chromatographic column packed with this material, the alkanes with a shorter linear chain are eluted first.

Also, the immobilization of the open metal nodes using Lewis acid and Lewis pyridyl bases to sense acetylene, acetone, and even other metal ions has been achieved.

An interesting property of some metal-organic frameworks with Ln^{3+} ions being incorporated within them is luminescence; whose application is thoroughly adopted in sensing applications. The luminescent Eu^{3+} can be incorporated within the porous metal-organic framework selectively in open sites for the interaction of the small solvent molecules by enhancement effect on luminescence. This successfully causes a significant luminescence diminishment and this can detect even small amounts of acetone.

A porous luminescent metal-organic framework Eu(PDC)$_{1.5}$(DMF)(DMF)$_{0.5}$(H$_2$O)$_{0.5}$ has a sensing application of Cu^{2+} ion due to differential binding of the Lewis base pyridyl sites. When Cu^{2+} ion gets adsorbed, the quenching effect of the solvent on the luminescence is detected by the change in intensity. Likewise, methanol molecules that are terminally incorporated in the porous network have a luminescent property.

With the development of nanoscience and technology, the modification of an electrode to determine the trace amount of lead by Pb accumulation on the surface of the electrode has been achieved by the use of the nanocomposites of multi-walled carbon nanotube and metal-organic framework. The measurement is done using differential pulse anodic stripping voltammetry.[67] Cu-MOF supported by Au-SH-SiO$_2$ nanoparticles was used to develop an electrochemical sensor for electrocatalytic oxidation of L-cysteine and the corresponding framework exhibited good electro-catalytic potential for oxidation of hydrazine in its neutral solution.

An amine-functionalized MOF(UiO-66-NH$_2$) was used as an efficient fluorescence sensor to detect DNA. This is a highly selective and sensitive technique to distinguish between complementary and mismatched DNA sequences.[68] A Fe/Zn-MMT film modified glassy carbon electrode-based sensor showed promising results in the detection of tetracycline. It was used for the determination of tetracycline concentration in different animals and animal feeds.[69] The non-alteration of the whole crystal packing but the only local structural change makes Zn-MOF a natural choice for water sensors at the molecular level.[70] Also, polychlorinated dibenzo-p-dioxins based MOFs are potentially useful for sensing applications.[71] MOF derived from Zn(II) and 1,3-bis(benzimidazolyl)benzene ligand could detect acetylacetone with high sensitivity and selectivity.[72]

6 Pesticide removal using MOF

The Food and Agriculture Organization (FAO) has defined pesticide as "a substance or mixture of substances intended for preventing, destroying, or controlling any pest, including vectors of human or animal disease, unwanted species of plants or animals, causing harm during or otherwise interfering with the production, processing, storage, transport, or marketing of food, agricultural commodities, wood and wood products or animal feedstuffs, or substances that may be administered to animals for the control of insects, arachnids, or other pests in or on their bodies." Pesticides include chemicals being used to regulate plant growth, herbicides, chemicals being used for prevention of premature falling of fruit or thinning of fruit, chemicals or substances used to protect crops/products during transportation and storage. Pesticides are very useful to prevent yield loss due to diseases, pests, pathogens, and weeds.[73,74]

Pesticides are generally classified according to the category of pest they reduce/eliminate. They are classified as insecticides and herbicides. Neuroactive insecticides namely neonicotinoids chemically resemble nicotine. Globally, the most widely applied pesticide is Imidacloprid which belongs to the neonicotinoid family. The mode of action of organophosphate and carbamate also attack the nervous system of insecticides. Several pesticides are a synthetic versions of natural pesticidal molecules occurring in plants, e.g., pyrethroid.

Unavoidable use of pesticides in the agricultural sector or the developing countries supporting its need for feedstock requirement due to its exponential population growth has resulted in the accumulation of these chemicals in soil. The effluent contents are toxic and may cause cancer and other diseases on exposure.[75] The higher potency combined with their lesser extent of biodegradability is among the major factors that make them notorious candidates for environmental contaminants. Water-soluble pesticides cause adverse effects on aquatic plants and organisms, while fat-soluble pesticides accumulate and magnify in the food chain and web.[76] Due to their no-polarity as well as a scarcity of functional groups, direct detection of pesticides is challenging utilizing the SERS effect.[77] Exposure to pesticides causes acute and adverse side effects pertaining to the health of human beings, including skin and eye irritation, adverse effects on the nervous system, hormonal disturbance, and problems related to the reproductive system.[78] Sometimes, the residue of antibiotics causes an increase in bacterial resistance.[79]

More than 95% of pesticide chemicals hurt the environment and non-targeted animals. These chemicals reach other areas with the wind as pesticide drift. When pesticides are applied on agricultural land, the pests drift to the areas, where it is still fresh; this poses a serious threat to agricultural activity. The pesticides that are persistent organic pollutants add to soil and flower (pollen, nectar) contamination and finally when washed by water due to irrigation result in water pollution.

Pesticides and their metabolites are hazardous to living beings and have the capability to endanger human health and destroy aquatic animals, and cannot be effectively eliminated from the aqueous systems.[80] Agricultural runoff, direct spray, and leaching are the principal routes via which these chemicals make their way to the hydrosphere.[81] The countermeasures adopted by government agencies primarily rely on flocculation, coagulation, and precipitation separating larger size particles differing in solubility. Ion exchange and membrane separation were also some promising practices in the past. Biodegradation and chemical oxidation are also among other approaches, but due to their slow kinetics, losing their popularity. The most ideal and practical way to deal with the problem of removal is adsorption. Also due to its low cost,[82] simple operation, higher selectivity, environmentally friendly nature, easy recyclability, and availability of several environment-friendly adsorbents materials make this strategy the most accepted one in the recent times. The properties of various adsorbents play a major role in the overall adsorption quality of the desired substance. Activated charcoal is one of the most used adsorbents but the presence of very high porosity with regular arrangement makes the metal-organic frameworks, their composites and numerous materials derived from it the favorite choices considered predominantly in this context. Also, the wide range of thermal stability combined with its chemical inertness adds to the advantages of its competitiveness of robust utility in this field. The allowance of post-synthetic modification especially in functionalization will extend its application in the polar atmosphere and ion exchangers.

Among the numerous varieties of MOFs reported, the Materials Institute Lavoisier (MILs)[81] and bimetallic frameworks[83] are the potential candidates for the removal of pollutants from wastewater. The MIL-101(Cr) prepared by Mirsoleimani-Azizi et al.,[84] using Chromium (III) oxide octahedral trimers and dicarboxylate linker, is a chemically inert and stable molecule. It was found highly efficient (92.5%) to remove pesticide diazinon.[84]

The better thermal and chemical stabilities of Zr-based MOFs are preferred over other ones, and also their stability over a wide range of pH is reported in an aqueous medium.[85] UiO-66 is known for its hydrophilic nature and small cage diameter of 6 Å.[86,87] The interesting property of this material is its magnetic character since it was prepared by a one-pot solvothermal method of immobilization of UiO-66 onto the $Fe_3O_4@SiO_2$ particles. This prepared MOF has been explored for effective removal of triclosan and triclocarban: two extensively used antifungal compounds. Another example of magnetic fabrication of magnetic character is $\gamma Fe_2O_3@C@MoO_3$.[88,89] The UiO-66(Zr) was also found to be efficient in comparison to activated charcoal in the absorptive removal of low concentrations of methyl-chlorophenoxy propionic acid, whereas UiO-67(Zr) was successful in adsorptive removal of organophosphorus pesticides such as glyphosate, glufosinate, dichlorvos, and metrifonate. It was successful in removing ~98% atrazine in 2 min compared to 94% when a comparative study was made by Akpinar and coworkers with ZIF-8(Zn), Uio-66(Zr), UiO-67(Zr), and activated charcoal(F400). The bigger particle size of activated charcoal was suggested to be the reason behind their decreased adsorption rates.[90]

Pristine MOFs were prepared successfully like MIL-53(Cr), UiO-67 having Zr metal ion, ZIF-8 with Co and Cu-BTC. MIL-53(Cr) has successfully removed 2,4-dichlorophenoxyacetic acid.[91] A comparative study on the absorptive performance of activated charcoal, zeolite, and MOF-235(Fe) was performed by De Smedt et al.[92] The test subjects were the pesticides like bentazon, clopyralid, and isoproturon in an aqueous environment; the result was in favor of MOF-235 (Fe) as a beneficial choice. The toxic and carcinogenic pesticide ^{14}C-ethion was shown to be removed successfully by Cu-BTC. Abdelhamid and coworkers investigated adsorptive removal of prothiofos and ethion using ZIF-8(Zn) & ZIF-67(Co). ZIF-8(Zn) exhibited better efficiency for both the pesticide molecules used in the investigation. They also concluded that the difference in the result is due to the strong coordinate bond between prothiofos and ethion and Zn metal ions.

When the pristine MOF was functionalized with some suitable groups, it exhibited a better adsorption capacity. For example, when UiO-66(Zr) was functionalized with cationic sites to transform into UiO-66(Zr)-NMe$_3^+$, it demonstrated a high adsorption capacity of 2,4-diphenoxyacetic acid. An improved adsorption performance to eliminate herbicides such as diuro, alachlor, tebuthiuron, and gramoxone was achieved when MIL-101(Cr) was modified with furan or thiophene to get Cr-MIL-101(Cr)-C, but due to steric hindrance,[93] the urea-functionalized MIL-101(Cr) showed lesser adsorption performance.

The use of MOF composites in the study of adsorptive removal of pesticides added a new dimension. A composite of UiO-67(Zr) and graphene oxide has been reported to remove organophosphorus pesticides efficiently by Zhu et al.[94] Also, magnetic derivatives of composites show better adsorption as the presence of a magnet is helpful for the removal of pesticides. Liu et al.[95] reported the preparation of magnetic composite of Cu-based MOF and iron oxide-graphene oxide-β-cyclodextrin and its removal efficiency against neonicotinoid pesticidal compounds like thiamethoxam, imidacloprid, acetamiprid, nitenpyram, dinotefuran, clothianidin, and thiacloprid from water as a result of a large surface area and hydrophobic inner pore.[95]

7 Future prospects

The initial objective for the preparation of MOFs has gradually broadened to quite a large number of branches. The conventional synthetic strategies were among the widely used schemes, but the ideology of mechano-, sono-, and electrochemical syntheses, as well as microwave-assisted syntheses, is also emerging. The later methodologies are generally applicable to milder reaction conditions for achieving particle sizes with varied properties. The investigation for further development must be focused on a wide variety of compounds available for shaping the MOFs with potential applications. The wide range of composite materials should be considered in this context which majorly emphasizes industrial application.[96] The catalytic studies using metal-organic frameworks are in just the developmental stage where the research primarily covered straightforward and simple applications.[97] It is possible to explore more sophisticated materials having ligands associated with transition metal complexes with multiple metals at the desired spacing between them. Additionally, this could exploit the Lewis acid catalysis of the multi-metallic centers of the MOF. This design may approximate these materials to the catalytic behavior observed with enzymes.[96] The research in the biomedical application of MOF has presently gained much attention and is likely to be considered an alternative in drug delivery applications where still traditional nanoporous materials were being used. The unavailability of sufficient data related to critical issues viz. the stability of MOFs in a moist environment, toxicology, and degree of biocompatibility may be a bit discouraging.[98] However, the achievements in this emerging field have outnumbered the failures. As computationally and practically infinite combinations of the linker and metal ion are possible, there is a wide scope of academic and industrial research in this field.[99]

8 Conclusion

The surface texture and feasibility of diverse post-synthetic modifications have attracted explosive attention.[60] Therefore, the synthesis for scale-up in solvothermal conditions and other synthetic pathways further needs optimization which is necessary for economic production and industrial applications of MOF as an adsorbent.

The process to convert MOF powders to pellets of monolith needs further modifications.[27] The application of MOF@-Plastic mesh for particulate matter removal showed multiple time efficiency when compared to bare plastic mesh.[100] The better knowledge of interactions between the solvent, substrate, and framework is the core area for the achievement of superior adsorption process as well as in catalytic reactions.[48,101] Apart from MOFs for pesticides adsorption, other competitors such as zeolites, mesoporous materials as well as porous organic polymers should be compared on the ground of cost, adsorption capacity, selectivity, reusability, and toxicity. The intrinsic merits of modification of pore structure made them potential candidates as sensing platforms for the development of food contamination. In particular, the versatility of novel MOFs has been magnified by their combination with other functional materials with improved sensing performance.

References

1. Székács A, Mörtl M, Darvas B. Monitoring pesticide residues in surface and ground water in Hungary: surveys in 1990–2015. *J Chem.* 2015;2015:717948. https://doi.org/10.1155/2015/717948.
2. Larramendy M, Soloneski S. *Organic Fertilizers: History, Production and Applications.* London, UK: IntechOpen; 2019.
3. Karri RR, Ravindran G, Dehghani MH. Wastewater—sources, toxicity, and their consequences to human health. In: *Soft Computing Techniques in Solid Waste and Wastewater Management.* Elsevier; 2021:3–33.
4. Dehghani MH, Omrani GA, Karri RR. Solid waste—sources, toxicity, and their consequences to human health. In: *Soft Computing Techniques in Solid Waste and Wastewater Management.* Elsevier; 2021:205–213.

5. Cooper RJ, Hama-Aziz Z, Hiscock KM, et al. Assessing the farm-scale impacts of cover crops and non-inversion tillage regimes on nutrient losses from an arable catchment. *Agric Ecosyst Environ*. 2017;237183–237193. https://doi.org/10.1016/j.agee.2016.12.034.

6. Zhang JP, Zhu AX, Lin RB, Qi XL, Chen XM. Pore surface tailored SOD-type metal-organic zeolites. *Adv Mater*. 2011;23(10):1268–1271. https://doi.org/10.1002/adma.201004028.

7. Dehghani MH, Hassani AH, Karri RR, et al. Process optimization and enhancement of pesticide adsorption by porous adsorbents by regression analysis and parametric modelling. *Sci Rep*. 2021;11(1):1–15.

8. Groen JC, Peffer LAA, Pérez-Ramírez J. Pore size determination in modified micro- and mesoporous materials. Pitfalls and limitations in gas adsorption data analysis. *Microporous Mesoporous Mater*. 2003;60:1–17. https://doi.org/10.1016/S1387-1811(03)00339-1.

9. Mehtab T, Yasin G, Arif M, et al. Metal-organic frameworks for energy storage devices: batteries and supercapacitors. *J Energy Storage*. 2019;21:632–646. https://doi.org/10.1016/j.est.2018.12.025.

10. Batten SR, Champness NR, Chen XM, et al. Terminology of metal–organic frameworks and coordination polymers (IUPAC recommendations 2013). *Pure Appl Chem*. 2013;85(8):1715–1724. https://doi.org/10.1351/pac-rec-12-11-20.

11. Cote AP. Porous, crystalline, covalent organic frameworks. *Science*. 2005;310(5751):1166–1170. https://doi.org/10.1126/science.1120411.

12. Cheetham AK, Rao CNR, Feller RK. Structural diversity and chemical trends in hybrid inorganic organic framework materials. *Chem Commun*. 2006;46:4780–4785. https://doi.org/10.1039/b610264f.

13. Wu D, Navrotsky A. Thermodynamics of metal-organic frameworks. *J Solid State Chem*. 2015;223:53–58. https://doi.org/10.1016/j.jssc.2014.06.015.

14. Piccin JS, Cadaval Jr TRS, De Pinto LAA, Dotto GL. Adsorption isotherm in liquid phase: experimental, modelling and interpretations. In: Bonilla-Petriciolet A, Mendoza-Castillo DI, Reynel-Ávila HE, eds. *Adsorption Processes for Water Treatment and Purification*. Switzerland: Springer; 2017:19–51.

15. Kisliuk P. The sticking probabilities of gases chemisorbed on the surfaces of solids. *J Phys Chem Solids*. 1957;3(1–2):95–101. https://doi.org/10.1016/0022-3697(57)90054-9.

16. Lee JY, Li J, Jagiello J. Gas sorption properties of microporous metal-organic frameworks. *J Solid State Chem*. 2005;178(8):2527–2532. https://doi.org/10.1016/j.jssc.2005.07.002.

17. Ha J, Lee JH, Moon HR. Alterations to secondary building units of metal organic frameworks for the development of new functions. *Inorg Chem Front*. 2020;7(1):12–27. https://doi.org/10.1039/C9QI01119F.

18. Schoedel A. Secondary building units of MOFs. In: Mozafari M, ed. *Metal-Organic Frameworks for Biomedical Applications*. United Kingdom: Elsevier; 2020:11–14.

19. Eddaoudi M, Moler DB, Li H, et al. Modular chemistry: secondary building units as a basis for the design of highly porous and robust metal–organic carboxylate frameworks. *Acc Chem Res*. 2001;34(4):319–330. https://doi.org/10.1021/ar000034b.

20. Kalmutzki MJ, Hanikel N, Yaghi OM. Secondary building nits as the turning point in the development of the reticular chemistry of MOFs. *Sci Adv*. 2018;4(10). https://doi.org/10.1126/sciadv.aat9180.

21. Tranchemontagne DJ, Mendoza-Cortés JL, O'Keeffe M, Yaghi OM. Secondary building units, nets and bonding in the chemistry of metal–organic frameworks. *Chem Soc Rev*. 2009;38(5):1257–1283. https://doi.org/10.1039/b817735j.

22. Stock N, Biswas S. Synthesis of metal-organic frameworks (MOFs): routes to various MOF topologies, morphologies, and composites. *Chem Rev*. 2012;112(2):933–969. https://doi.org/10.1021/cr200304e.

23. Sudik AC, Millward AR, Ockwig NW, Côté AP, Kim J, Yaghi OM. Design, synthesis, structure, and gas (N_2, Ar, CO_2, CH_4, and H_2) sorption properties of porous metal-organic tetrahedral and heterocuboidal polyhedra. *J Am Chem Soc*. 2005;127(19):7110–7118. https://doi.org/10.1021/ja042802q.

24. Chen Z, Hanna SL, Redfern LR, Alezi D, Islamoglu T, Farha OK. Reticular chemistry in the rational synthesis of functional zirconium cluster-based MOFs. *Coord Chem Rev*. 2019;386:32–49. https://doi.org/10.1016/j.ccr.2019.01.017.

25. Zhang B, Luo Y, Kanyuck K, et al. Facile and template-free solvothermal synthesis of mesoporous/macroporous metal–organic framework nanosheets. *RSC Adv*. 2018;8(58):33059–33064. https://doi.org/10.1039/C8RA06576D.

26. Kamal K, Bustan MA, Ismail M, Grekov D, Shariff AM, Pré P. Optimization of washing process in solvothermal synthesis of nickel based MOF-74. *Materials*. 2020;13(12):2741. https://doi.org/10.3390/ma13122741.

27. Lee YR, Kim J, Ahn WS. Synthesis of metal-organic frameworks: a mini review. *Korean J Chem Eng*. 2013;30(9):1667–1680. https://doi.org/10.1007/s11814-013-0140-6.

28. McKinstry C, Cathcart RJ, Cussen EJ, Fletcher AJ, Patwardhan SV, Jan S. Scalable continuous solvothermal synthesis of metal-organic framework (MOF-5) crystals. *Chem Eng J*. 2015;285:718–725. https://doi.org/10.1016/j.cej.2015.10.023.

29. Thomas-Hillman I, Laybourn A, Dodds C, Kingman SW. Realising the environmental benefits of metal–organic frameworks: recent advances in microwave synthesis. *J Mater Chem A*. 2018;6:11564–11581. https://doi.org/10.1039/C8TA02919A.

30. Soni S, Bajpai PK, Arora C. A review on metal-organic framework: synthesis, properties and ap-plication. *Charact Appl Nanomater*. 2020;3(2):87–106. https://doi.org/10.24294/can.v2i1.551.

31. Vaitsis C, Sourkouni G, Argirusis C. Metal-organic frameworks (MOFs) and ultrasound: a review. *Ultrason Sonochem*. 2018;52:106–119. https://doi.org/10.1016/j.ultsonch.2018.11.004.

32. Son WJ, Kim J, Kim J, Ahn WS. Sonochemical synthesis of MOF-5. *Chem Commun*. 2008;47:6336–6338. https://doi.org/10.1039/b814740j.

33. Al-Kutubi H, Gascon J, Sudhölter EJR, Rassaei L. Electrosynthesis of metal-organic frameworks: challenges and opportunities. *Chem Electro Chem*. 2015;2(4):462–474. https://doi.org/10.1002/celc.201402429.

34. Yang H, Liu X, Song X, Yang T, Liang Z, Fan C. In situ electrochemical synthesis of MOF-5 and its application in improving photocatalytic activity of BiOBr. *Trans Nonferrous Metals Soc China*. 2015;25(12):3987–3994. https://doi.org/10.1016/s1003-6326(15)64047-x.

35. Wang Z, Li Z, Ng M, Milner PJ. Rapid mechanochemical synthesis of metal organic frameworks using exogenous organic base. *Dalton Trans*. 2020;49:16238–16244. https://doi.org/10.1039/D0DT01240H.

36. Beamish-Cook J, Shankland K, Murray CA, Vaqueiro P. Insights into the mechanochemical synthesis of MOF-74. *Cryst Growth Des*. 2021;21(5):3047–3055. https://doi.org/10.1021/acs.cgd.1c00213.

37. Klimakow M, Klobes P, Thunemann AF, Rademann K, Emmerling F. Mechanochemical synthesis of metal–organic frameworks: a fast and facile approach toward quantitative yields and high specific surface areas. *Chem Mater*. 2010;22(18):5216–5221. https://doi.org/10.1021/cm1012119.

38. Lu N, Zhou F, Jia H, Wang H, Fan B, Li R. Dry-gel conversion synthesis of Zr-based metal–organic frameworks. *Ind Eng Chem Res*. 2017;56(48):14155–14163. https://doi.org/10.1021/acs.iecr.7b04010.

39. Kim J, Lee YR, Ahn WS. Dry-gel conversion synthesis of Cr-MIL-101 aided by grinding: high surface area and high yield synthesis with minimum purification. *Chem Commun*. 2013;49(69):7647–7649. https://doi.org/10.1039/C3CC44559C.

40. Robatjazi H, Weinberg D, Swearer DF, et al. Metal-organic frameworks tailor the properties of aluminum nanocrystals. *Sci Adv*. 2019;5(2):eaav5340. https://doi.org/10.1126/sciadv.aav5340.

41. Zhang L, Wang J, Ren X, et al. Internally extended growth of core–shell NH-MIL-101(Al)@ZIF-8 nanoflowers for the simultaneous detection and removal of Cu(II). *J Mater Chem A*. 2018;6(42):21029–21038. https://doi.org/10.1039/c8ta07349j.

42. Chen Y, Li S, Pei X, et al. A solvent-free hot-pressing method for preparing metal-organic-framework coatings. *Angew Chem Int Ed*. 2016;55(10):3419–3423. https://doi.org/10.1002/anie.201511063.

43. Murray LJ, Dincă M, Long JR. Hydrogen storage in metal–organic frameworks. *Chem Soc Rev*. 2009;38(5):1294–1314. https://doi.org/10.1039/b802256a.

44. Dincer C, Bruch R, Costa-Rama E, et al. Disposable sensors in diagnostics, food, and environmental monitoring. *Adv Mater*. 2019;31(30):1806739. https://doi.org/10.1002/adma.201806739.

45. Emam HE, Ahmed HB, Gomaa E, Helal MH, Abdelhameed RM. Doping of silver vanadate and silver tungstate nanoparticles for enhancement the photocatalytic activity of MIL-125-NH$_2$ in dye degradation. *J Photochem Photobiol A Chem*. 2019;383:111986. https://doi.org/10.1016/j.jphotochem.2019.111986.

46. Talin AA, Centrone A, Ford AC, et al. Tunable electrical conductivity in metal-organic framework thin-film devices. *Science*. 2014;343(6166):66–69. https://doi.org/10.1126/science.1246738.

47. Jiao L, Wang Y, Jiang HL, Xu Q. Metal-organic frameworks as platforms for catalytic applications. *Adv Mater*. 2017;30(7):1703663. https://doi.org/10.1002/adma.201703663.

48. Henschel A, Gedrich K, Kraehnert R, Kaskel S. Catalytic properties of MIL-101. *Chem Commun*. 2008;35:4192–4194. https://doi.org/10.1039/b718371b.

49. Kitagawa S, Hasegawa S, Horike S, et al. Three-dimensional porous coordination polymer functionalized with amide groups based on tridentate ligand: selective sorption and catalysis. *J Am Chem Soc*. 2007;129(9):2607–2614. https://doi.org/10.1021/ja067374y.

50. Liu S, Tao H, Zeng L, et al. Shape-dependent electrocatalytic reduction of CO$_2$ to CO on triangular silver nanoplates. *J Am Chem Soc*. 2017;139(6):2160–2163. https://doi.org/10.1021//jacs.6b12103.

51. Wang W, Xu X, Zhou W, Shao Z. Recent progress in metal-organic frameworks for applications in electrocatalytic and photocatalytic water splitting. *Adv Sci*. 2017;4(4):1600371. https://doi.org/10.1002/advs.201600371.

52. Zheng W, Tsang C, Lee LY, Wong K. Two-dimensional metal-organic framework and covalent-organic framework: synthesis and their energy-related applications. *Mater Today Chem*. 2019;12:34–60. https://doi.org/10.1016/j.mtchem.2018.12.002.

53. Cheng W, Zhao X, Su H, et al. Lattice-strained metal–organic-framework arrays for bifunctional oxygen electrocatalysis. *Nat Energy*. 2019;4(2):115–122. https://doi.org/10.1038/s41560-018-0308-8.

54. Liu M, Zheng W, Ran S, Boles ST, Lee LYS. Overall water-splitting electrocatalysts based on 2D CoNi-metal-organic frameworks and its derivative. *Adv Mater Interfaces*. 2018;5(21):1800849. https://doi.org/10.1002/admi.201800849.

55. Zheng W, Liu M, Lee LYS. Electrochemical instability of metal–organic frameworks: in situ spectroelectrochemical investigation of the real active sites. *ACS Catal*. 2020;10(1):81–92. https://doi.org/10.1021/acscatal.9b03790.

56. Zheng W, Lee LYS. Metal–organic frameworks for electrocatalysis: catalyst or precatalyst? *ACS Energy Lett*. 2021;6(8):2838–2843. https://doi.org/10.1021/acsenergylett.1c01350.

57. Xu G, Nie P, Dou H, Ding B, Li L, Zhang X. Exploring metal-organic frameworks for energy storage in batteries and supercapacitors. *Mater Today*. 2017;20(4):191–209. https://doi.org/10.1016/j.mattod.2016.10.003.

58. Zhao R, Liang Z, Zou R, Xu Q. Metal-organic frameworks for batteries. *Joule*. 2018;2(11):2235–2259. https://doi.org/10.1016/j.joule.2018.09.019.

59. Li J, Wang X, Zhao G, et al. Metal–organic framework-based materials: superior adsorbents for the capture of toxic and radioactive metal ions. *Chem Soc Rev*. 2018;47(7):2322–2356. https://doi.org/10.1039/C7CS00543A.

60. Li ZX, Zou KY. Controllable syntheses for MOF-derived materials. *Chem Eur J*. 2017;24(25):6506–6518. https://doi.org/10.1002/chem.201705415.

61. Pengpumkiat S, Nammoonnoy J, Wongsakoonkan W, Konthonbut P, Kongtip P. A microfluidic paper-based analytical device for type-II pyrethroid targets in an environmental water sample. *Sensors*. 2020;20(15):4107. https://doi.org/10.3390/s20154107.

62. Campbell M, Dincă M. Metal–organic frameworks as active materials in electronic sensor devices. *Sensors*. 2017;17(5):1108. https://doi.org/10.3390/s17051108.

63. Rebek J. Introduction to the molecular recognition and self-assembly special feature. *Proc Natl Acad Sci USA.* 2009;106(26):10423–10424. https://doi.org/10.1073/pnas.0905341106.

64. Kreno LE, Leong K, Farha OK, Allendorf M, Van Duyne RP, Hupp JT. Metal–organic framework materials as chemical sensors. *Chem Rev.* 2012;112(2):1105–1125. https://doi.org/10.1021/cr200324t.

65. Kumar P, Deep A, Kim KH. Metal-organic frameworks for sensing applications. *TrAC Trends Anal Chem.* 2015;73:39–53. https://doi.org/10.1016/j.trac.2015.04.009.

66. Li HY, Zhao SN, Zang SQ, Li J. Functional metal organic frameworks as effective sensors of gases and volatile compounds. *Chem Soc Rev.* 2020;49(17):6364–6401. https://doi.org/10.1039/c9cs00778d.

67. Lei J, Qian R, Ling P, Cui L, Ju H. Design and sensing applications of metal–organic framework composites. *Trends Anal Chem.* 2014;58:71–78. https://doi.org/10.1016/j.trac.2014.02.012.

68. Yi FY, Chen D, Wu MK, Han L, Jiang HL. Chemical sensors based on metal-organic frameworks. *ChemPlusChem.* 2016;81(8):675–690. https://doi.org/10.1002/cplu.201600137.

69. Gan T, Shi Z, Sun J, Liu Y. Simple and novel electrochemical sensor for the determination of tetracycline based on iron/zinc cations–exchanged montmorillonite catalyst. *Talanta.* 2014;121:187–193. https://doi.org/10.1016/j.talanta.2014.01.002.

70. Chen L, Ye JW, Wang HP, et al. Ultrafast water sensing and thermal imaging by a metal-organic framework with switchable luminescence. *Nat Commun.* 2017;8:15985. https://doi.org/10.1038/ncomms15985.

71. Wang B, Wang P, Xie LH, et al. A stable zirconium based metal-organic framework for specific recognition of representative polychlorinated dibenzo-p-dioxin molecules. *Nat Commun.* 2019;10(1):3861. https://doi.org/10.1038/s41467-019-11912-4.

72. Yao SL, Liu SJ, Tian XM, et al. A Zn (II)-based metal–organic framework with a rare topology as a turn-on fluorescent sensor for acetylacetone. *Inorg Chem.* 2019;58(6):3578–3581. https://doi.org/10.1021/acs.inorgchem.8b03316.

73. Shankar A, Kongot M, Saini VK, Kumar A. Removal of pentachlorophenol pesticide from aqueous solutions using modified chitosan. *Arab J Chem.* 2020;13(1):1821–1830. https://doi.org/10.1016/j.arabjc.2018.01.016.

74. Wang PL, Xie LH, Joseph EA, Li JR, Su XO, Zhou HC. Metal–organic frameworks for food safety. *Chem Rev.* 2019;119(18):10638–10690. https://doi.org/10.1021/acs.chemrev.9b00257.

75. Hasan Z, Jhung SH. Removal of hazardous organics from water using metal-organic frameworks (MOFs): plausible mechanisms for selective adsorptions. *J Hazard Mater.* 2015;283:329–339. https://doi.org/10.1016/j.jhazmat.2014.09.046.

76. Mojiri A, Zhou JL, Robinson B, et al. Pesticides in aquatic environments and their removal by adsorption methods. *Chemosphere.* 2020;253. https://doi.org/10.1016/j.chemosphere.2020.126646, 126646.

77. Wagner M, Andrew Lin KY, Oh WD, Lisak G. Metal-organic frameworks for pesticidal persistent organic pollutants detection and adsorption—a mini review. *J Hazard Mater.* 2021;413:125325. https://doi.org/10.1016/j.jhazmat.2021.125325.

78. Vikrant K, Tsang DCW, Raza N, BalenduShekher G, Kukkar D, Kim KH. The potential utility of metal-organic framework (MOF)-based platform for sensing pesticides. *ACS Appl Mater Interfaces.* 2018;10(10):8797–8817. https://doi.org/10.1021/acsami.8b00664.

79. Zhang QQ, Ying GG, Pan CG, Liu YS, Zhao JL. Comprehensive evaluation of antibiotics emission and fate in the river basins of China: source analysis, multimedia modeling, and linkage to bacterial resistance. *Environ Sci Technol.* 2015;49(11):6772–6782. https://doi.org/10.1021/acs.est.5b00729.

80. Chopra AK, Sharma MK, Chamoli S. Bioaccumulation of organochlorine pesticides in aquatic system—an overview. *Environ Monit Assess.* 2011;173(1–4):905–916. https://doi.org/10.1007/s10661-010-1433-4.

81. Isiyaka HA, Jumbri K, Sambudi NS, Zango ZU, Saad B, Mustapha A. Removal of 4-chloro-2-methylphenoxyacetic acid from water by MIL-101(Cr) metal-organic framework: kinetics, isotherms and statistical models. *R Soc Open Sci.* 2021;8:201553. https://doi.org/10.1098/rsos.201553.

82. Kaur Y, Bhatia Y, Chaudhary S, Chaudhary GR. Comparative performance of bare and functionalize ZnO nanoadsorbents for pesticide removal from aqueous solution. *J Mol Liq.* 2017;234:94–103. https://doi.org/10.1016/j.molliq.2017.03.069.

83. Soni S, Bajpai PK, Mittal J, Arora C. Utilisation of cobalt doped Iron based MOF for enhanced removal and recovery of methylene blue dye from waste water. *J Mol Liq.* 2020;314:113642. https://doi.org/10.1016/j.molliq.2020.113642.

84. Mirsoleimani-Azizi SM, Setoodeh P, Samimi F, Shadmehr J, Hamedi N, Rahimpour MR. Diazinon removal from aqueous media by mesoporous MIL-101(Cr) in a continuous fixed-bed system. *J Environ Chem Eng.* 2018;6:4653–4664. https://doi.org/10.1016/j.jece.2018.06.067.

85. Jiang HL, Feng D, Wang K, et al. An exceptionally stable, porphyrinic Zr metal–organic framework exhibiting pH-dependent fluorescence. *J Am Chem Soc.* 2013;135(37):13934–13938. https://doi.org/10.1021/ja406844r.

86. Ma J, Li S, Wu G, et al. Preparation of magnetic metal-organic frameworks with high binding capacity for removal of two fungicides from aqueous environments. *J Ind Eng Chem.* 2020;90:178–189. https://doi.org/10.1016/j.jiec.2020.07.010.

87. Bagheri AR, Aramesh N, Bilal M. New frontiers and prospects of metal-organic frameworks for removal, determination, and sensing of pesticides. *Environ Res.* 2021;194:110654. https://doi.org/10.1016/j.envres.2020.110654.

88. Shen Y, Jiang P, Wai P, Gu Q, Zhang W. Recent progress in application of molybdenum-based catalysts for epoxidation of alkenes. *Catalysts.* 2019;9(1):31. https://doi.org/10.3390/catal9010031.

89. Zhang F, Hu H, Zhong H, Yan N, Chen Q. Preparation of gamma-Fe$_2$O$_3$@C@MoO$_3$ core/shell nanocomposites as magnetically recyclable catalysts for efficient and selective epoxidation of olefins. *Dalton Trans.* 2014;43:6041–6049. https://doi.org/10.1039/c3dt53105h.

90. Dias EM, Petit C. Towards the use of metal-organic frameworks for water reuse: a review of the recent advances in the field of organic pollutants removal and degradation and the next steps in the field. *J Mater Chem A.* 2015;3:22484–22506. https://doi.org/10.1039/C5TA05440K.

91. Jung BK, Hasan Z, Jhung SH. Adsorptive removal of 2,4-dichlorophenoxyacetic acid (2,4-D) from water with a metal–organic framework. *Chem Eng J.* 2013;234:99–105. https://doi.org/10.1016/j.cej.2013.08.110.

92. De Smedt C, Spanoghe P, Biswas S, Leus K, Van Der Voort P. Comparison of different solid adsorbents for the removal of mobile pesticides from aqueous solutions. *Adsorption.* 2015;21(3):243–254. https://doi.org/10.1007/s10450-015-9666-8.

93. Mondol MMH, Jhung SH. Adsorptive removal of pesticides from water with metal–organic framework-based materials. *Chem Eng J.* 2021;421:129688. https://doi.org/10.1016/j.cej.2021.129688.

94. Zhu X, Li B, Yang J, et al. Effective adsorption and enhanced removal of organophosphorus pesticides from aqueous solution by Zr-based MOFs of UiO-67. *ACS Appl Mater Interfaces.* 2015;7(1):223–231. https://doi.org/10.1021/am5059074.

95. Liu G, Li L, Xu D, et al. Metal-organic framework preparation using magnetic graphene oxide-β-cyclodextrin for neonicotinoid pesticide adsorption and removal. *Carbohydr Polym.* 2017;175:584–591. https://doi.org/10.1016/j.carbpol.2017.06.074.

96. Corma A, Garcia H, Llabresi Xamena FX. Engineering metal-organic frameworks for heterogeneous catalysis. *Chem Rev.* 2010;110(8):4606–4655. https://doi.org/10.1021/cr9003924.

97. Dhakshinamoorthy A, Li Z, Garcia H. Catalysis and photocatalysis by metal-organic frameworks. *Chem Soc Rev.* 2018;47:8134–8172. https://doi.org/10.1039/c8cs00256h.

98. Keskin S, Kızılel S. Biomedical applications of metal-organic frameworks. *Ind Eng Chem Res.* 2011;50(4):1799–1812. https://doi.org/10.1021/ie101312k.

99. Czaja AU, Trukhan N, Müller U. Industrial applications of metal-organic frameworks. *Chem Soc Rev.* 2009;38(5):1284–1293. https://doi.org/10.1039/b804680h.

100. Chen Y, Zhang S, Cao S, et al. Roll-to-roll production of metal-organic framework coatings for particulate matter removal. *Adv Mater.* 2017;29(15):1606221. https://doi.org/10.1002/adma.201606221.

101. Henschel A, Senkovska I, Kaskel S. Liquid-phase adsorption on metal-organic frameworks. *Adsorption.* 2011;17(1):219–226. https://doi.org/10.1007/s10450-010-9317-z.

Index

Note: Page numbers followed by *f* indicate figures and *t* indicate tables.

Printed and bound by CPI Group (UK) Ltd, Croydon, CR0 4YY

08/05/2025

01864940-0001